# Biopharmaceutical Statistics for Drug Development

## STATISTICS: Textbooks and Monographs



Vol. 1: The Generalized Jackknife Statistic, *H. L. Gray and W. R. Schucany*
Vol. 2: Multivariate Analysis, *Anant M. Kshirsagar*
Vol. 3: Statistics and Society, *Walter T. Federer*
Vol. 4: Multivariate Analysis: A Selected and Abstracted Bibliography, 1957-1972, *Kocherlakota Subrahmaniam and Kathleen Subrahmaniam* (out of print)
Vol. 5: Design of Experiments: A Realistic Approach, *Virgil L. Anderson and Robert A. McLean*
Vol. 6: Statistical and Mathematical Aspects of Pollution Problems, *John W. Pratt*
Vol. 7: Introduction to Probability and Statistics (in two parts), Part I: Probability; Part II: Statistics, *Narayan C. Giri*
Vol. 8: Statistical Theory of the Analysis of Experimental Designs, *J. Ogawa*
Vol. 9: Statistical Techniques in Simulation (in two parts), *Jack P. C. Kleijnen*
Vol. 10: Data Quality Control and Editing, *Joseph I. Naus* (out of print)
Vol. 11: Cost of Living Index Numbers: Practice, Precision, and Theory, *Kali S. Banerjee*
Vol. 12: Weighing Designs: For Chemistry, Medicine, Economics, Operations Research, Statistics, *Kali S. Banerjee*
Vol. 13: The Search for Oil: Some Statistical Methods and Techniques, *edited by D. B. Owen*
Vol. 14: Sample Size Choice: Charts for Experiments with Linear Models, *Robert E. Odeh and Martin Fox*
Vol. 15: Statistical Methods for Engineers and Scientists, *Robert M. Bethea, Benjamin S. Duran, and Thomas L. Boullion*
Vol. 16: Statistical Quality Control Methods, *Irving W. Burr*
Vol. 17: On the History of Statistics and Probability, *edited by D. B. Owen*
Vol. 18: Econometrics, *Peter Schmidt*
Vol. 19: Sufficient Statistics: Selected Contributions, *Vasant S. Huzurbazar (edited by Anant M. Kshirsagar)*
Vol. 20: Handbook of Statistical Distributions, *Jagdish K. Patel, C. H. Kapadia, and D. B. Owen*
Vol. 21: Case Studies in Sample Design, *A. C. Rosander*
Vol. 22: Pocket Book of Statistical Tables, *compiled by R. E. Odeh, D. B. Owen, Z. W. Birnbaum, and L. Fisher*
Vol. 23: The Information in Contingency Tables, *D. V. Gokhale and Solomon Kullback*

**Vol. 24:** Statistical Analysis of Reliability and Life-Testing Models: Theory and Methods, *Lee J. Bain*

**Vol. 25:** Elementary Statistical Quality Control, *Irving W. Burr*

**Vol. 26:** An Introduction to Probability and Statistics Using BASIC, *Richard A. Groeneveld*

**Vol. 27:** Basic Applied Statistics, *B. L. Raktoe and J. J. Hubert*

**Vol. 28:** A Primer in Probability, *Kathleen Subrahmaniam*

**Vol. 29:** Random Processes: A First Look, *R. Syski*

**Vol. 30:** Regression Methods: A Tool for Data Analysis, *Rudolf J. Freund and Paul D. Minton*

**Vol. 31:** Randomization Tests, *Eugene S. Edgington*

**Vol. 32:** Tables for Normal Tolerance Limits, Sampling Plans, and Screening, *Robert E. Odeh and D. B. Owen*

**Vol. 33:** Statistical Computing, *William J. Kennedy, Jr. and James E. Gentle*

**Vol. 34:** Regression Analysis and Its Application: A Data-Oriented Approach, *Richard F. Gunst and Robert L. Mason*

**Vol. 35:** Scientific Strategies to Save Your Life, *I. D. J. Bross*

**Vol. 36:** Statistics in the Pharmaceutical Industry, *edited by C. Ralph Buncher and Jia-Yeong Tsay*

**Vol. 37:** Sampling from a Finite Population, *J. Hajek*

**Vol. 38:** Statistical Modeling Techniques, *S. S. Shapiro*

**Vol. 39:** Statistical Theory and Inference in Research, *T. A. Bancroft and C.-P. Han*

**Vol. 40:** Handbook of the Normal Distribution, *Jagdish K. Patel and Campbell B. Read*

**Vol. 41:** Recent Advances in Regression Methods, *Hrishikesh D. Vinod and Aman Ullah*

**Vol. 42:** Acceptance Sampling in Quality Control, *Edward G. Schilling*

**Vol. 43:** The Randomized Clinical Trial and Therapeutic Decisions, *edited by Niels Tygstrup, John M. Lachin, and Erik Juhl*

**Vol. 44:** Regression Analysis of Survival Data in Cancer Chemotherapy, *Walter H. Carter, Jr., Galen L. Wampler, and Donald M. Stablein*

**Vol. 45:** A Course in Linear Models, *Anant M. Kshirsagar*

**Vol. 46:** Clinical Trials: Issues and Approaches, *edited by Stanley H. Shapiro and Thomas H. Louis*

**Vol. 47:** Statistical Analysis of DNA Sequence Data, *edited by B. S. Weir*

**Vol. 48:** Nonlinear Regression Modeling: A Unified Practical Approach, *David A. Ratkowsky*

**Vol. 49:** Attribute Sampling Plans, Tables of Tests and Confidence Limits for Proportions, *Robert E. Odeh and D. B. Owen*

**Vol. 50:** Experimental Design, Statistical Models, and Genetic Statistics, *edited by Klaus Hinkelmann*

**Vol. 51:** Statistical Methods for Cancer Studies, *edited by Richard G. Cornell*

**Vol. 52:** Practical Statistical Sampling for Auditors, *Arthur J. Wilburn*

**Vol. 53:** Statistical Signal Processing, *edited by Edward J. Wegman and James G. Smith*

**Vol. 54:** Self-Organizing Methods in Modeling: GMDH Type Algorithms, *edited by Stanley J. Farlow*

**Vol. 55:** Applied Factorial and Fractional Designs, *Robert A. McLean and Virgil L. Anderson*

**Vol. 56:** Design of Experiments: Ranking and Selection, *edited by Thomas J. Santner and Ajit C. Tamhane*

**Vol. 57:** Statistical Methods for Engineers and Scientists. Second Edition, Revised and Expanded, *Robert M. Bethea, Benjamin S. Duran, and Thomas L. Boullion*

**Vol. 58:** Ensemble Modeling: Inference from Small-Scale Properties to Large-Scale Systems, *Alan E. Gelfand and Crayton C. Walker*

**Vol. 59:** Computer Modeling for Business and Industry, *Bruce L. Bowerman and Richard T. O'Connell*

**Vol. 60:** Bayesian Analysis of Linear Models, *Lyle D. Broemeling*

Vol. 61: Methodological Issues for Health Care Surveys, *Brenda Cox and Steven Cohen*

Vol. 62: Applied Regression Analysis and Experimental Design, *Richard J. Brook and Gregory C. Arnold*

Vol. 63: Statpal: A Statistical Package for Microcomputers – PC-DOS Version for the IBM PC and Compatibles, *Bruce J. Chalmer and David G. Whitmore*

Vol. 64: Statpal: A Statistical Package for Microcomputers – Apple Version for the II, II+, and IIe, *David G. Whitmore and Bruce J. Chalmer*

Vol. 65: Nonparametric Statistical Inference, Second Edition, Revised and Expanded, *Jean Dickinson Gibbons*

Vol. 66: Design and Analysis of Experiments, *Roger G. Petersen*

Vol. 67: Statistical Methods for Pharmaceutical Research Planning, *Sten W. Bergman and John C. Gittins*

Vol. 68: Goodness-of-Fit Techniques, *edited by Ralph B. D'Agostino and Michael A. Stephens*

Vol. 69: Statistical Methods in Discrimination Litigation, *edited by D. H. Kaye and Mikel Aickin*

Vol. 70: Truncated and Censored Samples from Normal Populations, *Helmut Schneider*

Vol. 71: Robust Inference, *M. L. Tiku, W. Y. Tan, and N. Balakrishnan*

Vol. 72: Statistical Image Processing and Graphics, *edited by Edward J. Wegman and Douglas J. DePriest*

Vol. 73: Assignment Methods in Combinatorial Data Analysis, *Lawrence J. Hubert*

Vol. 74: Econometrics and Structural Change, *Lyle D. Broemeling and Hiroki Tsurumi*

Vol. 75: Multivariate Interpretation of Clinical Laboratory Data, *Adelin Albert and Eugene K. Harris*

Vol. 76: Statistical Tools for Simulation Practitioners, *Jack P. C. Kleijnen*

Vol. 77: Randomization Tests, Second Edition, *Eugene S. Edgington*

Vol. 78: A Folio of Distributions: A Collection of Theoretical Quantile-Quantile Plots, *Edward B. Fowlkes*

Vol. 79: Applied Categorical Data Analysis, *Daniel H. Freeman, Jr.*

Vol. 80: Seemingly Unrelated Regression Equations Models : Estimation and Inference, *Virendra K. Srivastava and David E. A. Giles*

Vol. 81: Response Surfaces: Designs and Analyses, *Andre I. Khuri and John A. Cornell*

Vol. 82: Nonlinear Parameter Estimation: An Integrated System in BASIC, *John C. Nash and Mary Walker-Smith*

Vol. 83: Cancer Modeling, *edited by James R. Thompson and Barry W. Brown*

Vol. 84: Mixture Models: Inference and Applications to Clustering, *Geoffrey J. McLachlan and Kaye E. Basford*

Vol. 85: Randomized Response: Theory and Techniques, *Arijit Chaudhuri and Rahul Mukerjee*

Vol. 86: Biopharmaceutical Statistics for Drug Development, *edited by Karl E. Peace*

Vol. 87: Parts per Million Values for Estimating Quality Levels, *Robert E. Odeh and Donald B. Owen*

Vol. 88: Lognormal Distributions: Theory and Applications, *edited by Edwin L. Crow and Kunio Shimizu*

**ADDITIONAL VOLUMES IN PREPARATION**

# Biopharmaceutical Statistics for Drug Development

*edited by*

Karl E. Peace
G. D. Searle & Co.
Skokie, Illinois

MARCEL DEKKER, INC. New York and Basel

ISBN 0-8247-7798-0

MARCEL DEKKER, INC.
270 Madison Avenue, New York, New York 10016

Current printing (last digit):
10 9 8 7 6 5 4 3 2 1

PRINTED IN THE UNITED STATES OF AMERICA

# Preface

This book is intended to reflect statistical aspects of research and development programs in human drug development. It covers biological or basic research applications, chemical research and development applications, preclinical research and development applications, and clinical development applications, all in a single volume.

This book is a resource for scientists who desire or need to know statistical aspects of pharmaceutical drug research and development. It could serve as a desk or library reference, as a textbook for graduate students in biostatistics or other applied statistics programs, as well as a reference for targeted seminars. This book will be useful to scientists engaged in basic biological research—pharmacologists, immunologists, molecular geneticists or cell biologists; those engaged in chemical research and development—particularly pharmaceutical formulation chemists; those engaged in preclinical research and development—pathologists, toxicologists or pharmacokineticists; those engaged in clinical drug development—particularly clinicians; and statisticians providing support to basic research drug discovery programs, clinical development programs and other scientific projects.

Chapters are self-contained, yet connect in a logical way with other chapters. Readers interested in statistical aspects of nonclinical human drug development should read particular chapters of interest from Chapters 2, 3, 4, and 5. Readers interested in statistical aspects of clinical human drug development should first read Chapter 6 and then subsequent chapters.

I wish to express deep and sincere thanks to the authors of the chapters for their remarkable and excellent contributions. Their collective efforts should have a profound impact on the practice of statistics, particularly in the pharmaceutical industry. Special thanks go to the publisher for interest in this project as well as to my previous secretarial staff: Marie Corleto and Jacqueline Swift and to the Word Processing Group at Smith Kline & French Laboratories, particularly to Jerry and Chris, for their invaluable assistance.

Karl E. Peace

# Contents

# Contributors

**Neeti R. Bohidar**, Statistical Services Department, Smith Kline & French Laboratories, Swedeland, Pennsylvania

**J. A. Bolognese**, Clinical Biostatistics and Research Data Systems International, Merck Sharp & Dohme Research Laboratories, Rahway, New Jersey

**Vernon M. Chinchilli**, Department of Biostatistics, The Medical College of Virginia, Virginia Commonwealth University, Richmond, Virginia

**Suzanne Edwards**, Clinical Statistics Department, Burroughs Wellcome Co., Research Triangle Park, North Carolina

**Nancy L. Geller**, Department of Epidemiology and Biostatistics, Memorial Sloan-Kettering Cancer Center, New York, New York

**A. Lawrence Gould**, Biostatistics and Research Data Systems, Merck Sharp & Dohme Laboratories, West Point, Pennsylvania

**John J. Hubert**, Department of Mathematics and Statistics, College of Physical Sciences, University of Guelph, Guelph, Ontario, Canada

**Jerry D. Johnson**, Corporate Regulatory Affairs Department, Syntex Corporation, Palo Alto, California

**Mark A. Johnson**, Computational Chemistry, The Upjohn Company, Kalamazoo, Michigan

**M. K. Kersten**, Biostatistics and Research Data Systems International, Merck Sharp & Dohme Research Laboratories, Rahway, New Jersey

**Gary G. Koch**, Biostatistics Department, School of Public Health, University of North Carolina, Chapel Hill, North Carolina

**Robert C. Kohberger**, Statistics and Data Services, Quality Management, Lederle Laboratories, Pearl River, New York

**Robert T. O'Neill,** Division of Biometrics, Center for Drug and Biologics, Food and Drug Administration, Rockville, Maryland

**Karl E. Peace,*** Statistical Services Department, Worldwide Clinical R&D, Smith Kline & French Laboratories, Swedeland, Pennsylvania

**Stuart J. Pocock,** Department of Clinical Epidemiology and General Practice, The Royal Free Hospital School of Medicine, London, England

**B. E. Rodda,** Clinical Biostatistics and Research Data Systems International, Merck Sharp & Dohme Research Laboratories, Rahway, New Jersey

**Patricia L. Ruppel,** Biostatistics, The Upjohn Company, Kalamazoo, Michigan

**John R. Schultz,** Biostatistics, The Upjohn Company, Kalamazoo, Michigan

**Murray R. Selwyn,** Statistics Unlimited, Inc., Auburndale, Massachusetts

**Richard Simon,** Biometric Research Branch, Division of Cancer Treatment, National Cancer Institute, Bethesda, Maryland

**M.C. Tsianco,** Clinical Biostatistics and Research Data Systems International, Merck Sharp & Dohme Research Laboratories, Rahway, New Jersey

**Wilfred J. Westlake,**† Statistical Services Department, Worldwide Clinical R&D, Smith Kline & French Laboratories, Swedeland, Pennsylvania

---

**Present Affiliation*: G.I. Clinical Studies, Worldwide Biometrics and Medical Writing, Corporate Medical and Scientific Affairs, G. D. Searle & Co., Skokie, Illinois

†*Present Affiliation*: University of Pennsylvania, Philadelphia, Pennsylvania

# Introduction

The chapters of this book and their sequence were selected to reflect the scope of pharmaceutical human drug development. Successful human drug development in the pharmaceutical industry requires aggressive, highly competent and diversified drug discovery, basic research program teams, clinical development teams and the associated research programs and clinical development plans. Such programs and plans must reflect good science and adequately address regulatory requirements of the Food and Drug Administration for approval to market new drugs in the United States.

The first chapter of this book therefore presents historical and current perspectives of new drug development amid a regulatory background. Chapters 2 through 13 reflect, in logical order, the process of human drug development. Compounds must be screened and discovered (Chapter 2) and assessed for pharmacological activity (Chapter 3). A subset of these will be subjected to desirable pharmaceutical formulation (Chapter 4). The safety of a further subset of these will be assessed in animals (Chapter 5) prior to experience in humans (Chapter 6). What happens to a further subset of these compounds once they get into the body is then assessed (Chapter 7). Finally a few of this last subset of compounds merit elevation to clinical development.

Chapters 6 through 13 reflect the clinical development of new drugs for human use. Substantive and statistical aspects of this development in a broad sense are presented in Chapter 6. An aspect of Chapter 6 is to partition clinical trials in the development of new drugs for a variety of indications into those whose primary endpoints of efficacy are quantitative (excluding time-to-event endpoints) and those whose primary endpoints of efficacy are categorical, with oncology trials in a separate category. Chapter 6, therefore, sets the stage for a unified treatment of statistical aspects of design and analysis of trials with quantitative measurements (Chapter 8), with categorical measurements (Chapter 9) and oncology trials (Chapter 10). The assessment of safety from clinical trials sufficient for approval to market new drugs for particular indications is addressed in Chapter 13. Chapter

11 acknowledges the increasing use of sequential procedures in clinical trials and Chapter 12, the increasing use of interval estimation rather than hypothesis testing.

If a new drug is given regulatory approval to be marketed, large quantities of the drug in the desired formulation must be manufactured and stored for future use. Appropriate manufacturing and quality control procedures must therefore be set in place. The last chapter (14) addresses statistical aspects of these important procedures.

# 1 Past and Present Regulatory Aspects of Drug Development

JERRY D. JOHNSON *Syntex Corporation, Palo Alto, California*

## I. INTRODUCTION

In the United States, you have just gone to your doctor for severe stomach pains. After several tests, the doctor concludes that you have a duodenal ulcer. The doctor tells you the changes needed in your diet, tells you to take a week off work, and writes a prescription for a drug. You take this prescription to your local drugstore and purchase a 4-week supply of tablets from the pharmacist. After a few weeks, your ulcer is healed and you are back to your normal way of life. Those little tablets represent a large amount of effort and expense by the pharmaceutical company that manufactured the drug. The tablets were developed over many, many years that began when a chemist first synthesized the chemical molecule that led to the medication that healed your ulcer. After 5 to 10 years of research and development, the pharmaceutical company filed a New Drug Application (NDA) to the U.S. Food and Drug Administration (FDA), who then reviewed the information developed by the company and, after 1 to 3 years, approved the drug to be sold in the United States.

This chapter will describe some of the aspects of the development of a drug in the United States by a pharmaceutical company, focusing on the role played by the FDA. In other countries, there are similar regulatory agencies that play the same role as the FDA but the regulatory process is somewhat different and will not be covered here. Further, this chapter will restrict its discussion to the development of drugs for human use. There are similar processes for the development of drugs for use in animals.

## II. THE DRUG DEVELOPMENT PROCESS

### A. Preclinical Development—Preparation for the IND

In order to ensure that a drug will be safe when given to humans and has certain properties that will benefit humans, the drug goes through many steps of testing before it is given to humans (see Figure 1). After the

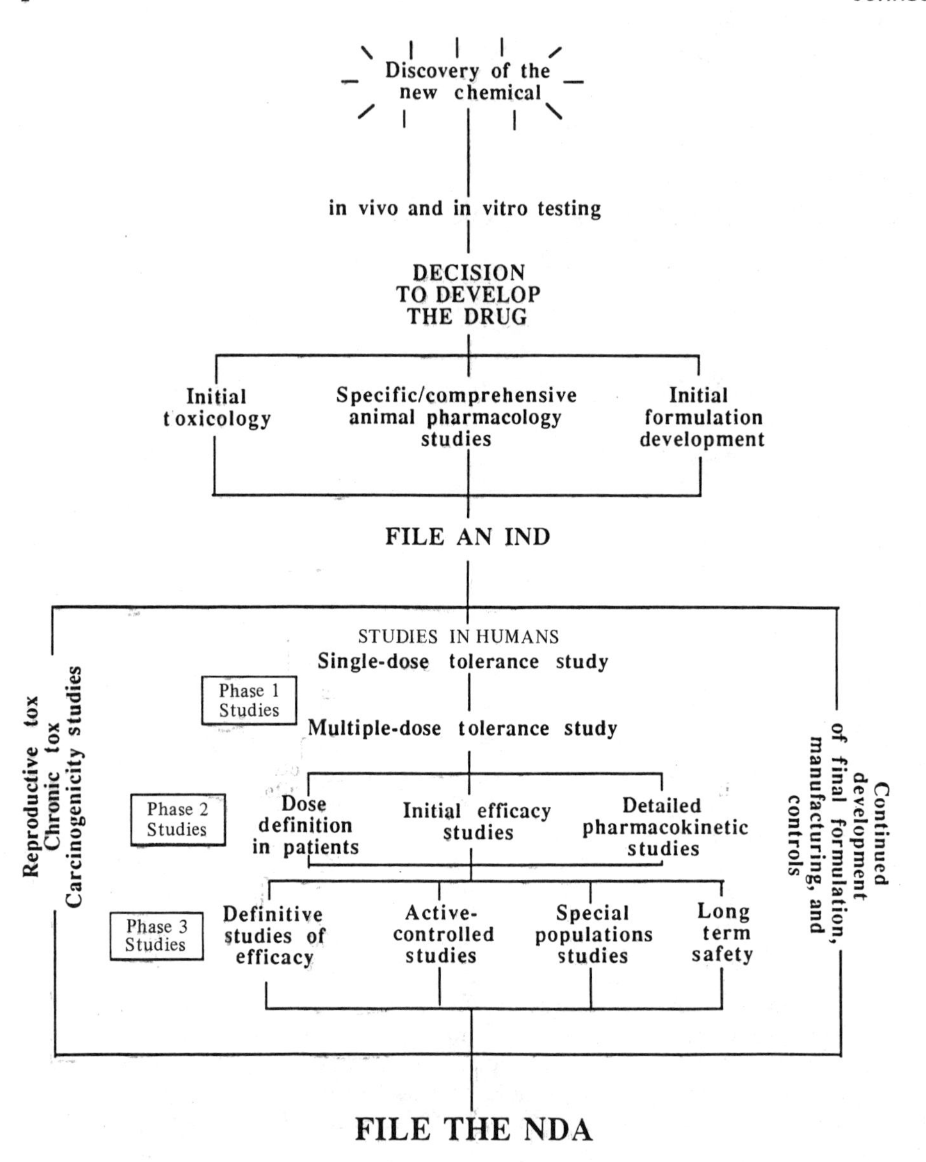


Fig. 1 The drug development process.

drug is first discovered by the chemist, it is tested by basic researchers using both in vivo and in vitro standard assays to determine the drug's pharmacological activities. The specific chemical molecule may have been designed to (or found to) resemble other molecules with known pharmacological activities that will help guide this early research to determine the activities of the drug.

After the basic pharmacological profile of the drug has been determined from animal and in vitro studies, the pharmaceutical company must decide on the feasibility of the drug for further development based on the information developed so far. If a decision is made to continue the development of the drug, various scientific teams in the company's research groups begin initial animal toxicology studies and initial development of a formulation that could be used in humans. More complete animal pharmacology studies are conducted as well.

Once the final decision is made that the drug has adequate potential for being safe and effective in humans, the company makes the decision to develop the drug to a stage where testing may be conducted in humans. To test a drug in humans in the United States, a Notice of Claimed Investigational New Drug Exemption (commonly designated by the initials IND) must be filed to the FDA.* The IND contains all the information known by the company† about the drug; it is organized according to the sections shown in Table 1.

One of the requirements of the FDA is that there be data from animal studies from which the company has "concluded that it is reasonably safe to initiate clinical investigations with the drug" (CFR, 1985). An important part of this preclinical testing is the animal toxicology. The FDA typically has required that the company conduct specific toxicology studies of varying lengths of dosing and in various species to support corresponding studies of varying lengths in humans. The only published guidelines by the FDA are commonly known as the "Goldenthal guidelines" (Goldenthal, 1968). For example, to conduct a clinical study for several days of dosing, these guidelines specify that two animal species each must be tested for 2 weeks in a toxicology study. These guidelines are somewhat out of date, and informal interactions with the FDA have often revealed additions to and modifications of the Goldenthal guidelines.

There are other research efforts that are typically conducted prior to testing the drug in humans. Examples are (1) absorption, distribution, metabolism, and excretion of the drug in various animal species (commonly called ADME studies), and (2) further development of the formulation and of the chemical synthesis. The details of these activities are beyond the scope of this chapter.

---

*Legally the IND is an exemption to the law preventing the shipment of a new drug across state lines (interstate commerce) without prior approval by the FDA. However, most companies file the IND in order to allow the flexibility to test the drug throughout the United States. Also, many U.S. states have their own "FDAs" that would require the company to file a submission similar to an IND in order to test the drug within the state in which the drug is being manufactured.

†The company is often called the "sponsor," which is the term used in many of the regulations and guidelines of the FDA.

Table 1 Outline of the IND[a]

1. Descriptive name of the drug, including chemical name and structure

2. Complete list of the components of the drug

3. Complete statement of the quantitative composition of the drug

4. Description of source and preparation of new drug substance

5. Methods, facilities, and controls used for the manufacturing, processing, and packing of the new drug

6a. Information about the preclinical investigations of the drug on the basis of which the sponsor has concluded that it is reasonably safe to initiate clinical investigations

6b. Complete information about the marketing and/or investigation of the drug outside the United States

6c, 6d. Additional information if the drug is a combination drug or is a radioactive drug

7. Copy of all informational material to be supplied to the clinical investogator(s) including labels and labeling

8. Scientific training and experience of investigators

9. Signed form FD1572 or FD1573 (By signing this form, the investigator agrees to comply with the FDA's regulations)

10. Protocol for the clinical investigation

[a]Based on the IND regulations in CFR (1985). Not complete but intended to provide an overview of the most important elements of an IND. The company is required to include a signed form FD1571, which includes a statement of the regulations under which it is agreed the drug will be studied.

## B. Clinical Testing Under the IND

In addition to information obtained from preclinical testing of the new drug, the original IND must specify the details of the first clinical study to be conducted in humans. The FDA has established fairly specific procedures that must be followed for a clinical investigation to be initiated. Whether the study is the first study to be filed with the original IND or a subsequent investigation, these well-defined procedures must be followed. The information required to be filed to the FDA for subsequent investigations is called an IND Amendment. The information for the original and all subsequent studies includes (1) the clinical study protocol and (2) a form called FD1572 (or FD1573*) signed by the clinical investigator.

---

*FD1572 is used for investigators conducting clinical pharmacology studies of the drug, while FD1573 is used for other clinical studies.

As part of form FD1572 or FD1573, the following information is supplied: (1) the investigator's education, training, and experience, and (2) details regarding the hospital or other medical institution where the investigation will be conducted, special equipment, and other facilities. Usually the information about the investigator's qualifications are provided by enclosing his/her curriculum vitae. The training and experience requirements vary for the different types of studies and drugs. For Phase 1 studies (see below for the definitions of the phases of clinical study), the investigator must be able to evaluate human toxicology and pharmacology. For Phase II, the investigator must be familiar with the conditions to be treated, the drugs used in these conditions, and the methods of their evaluation. For Phase III, in addition to experienced clinical investigators, physicians not regarded as specialists in any particular field of medicine may serve as investigators.

By signing this form, the investigator conducting the study acknowledges that he or she will follow the IND regulations, including obtaining Institutional Review Board (IRB)* approval for the study. Further, the investigator agrees to provide a written statement of the risks and benefits to each subject/patient prior to enrollment into the study, and the subject/patient signs that he or she has read and understood this information, i.e., gives "informed consent."

Unless the company has filed this information to the IND, under U.S. law the company is not allowed to ship the new drug to the clinical investigator to conduct the study. This requirement also pertains to the shipment of an investigational drug to an investigator conducting a study outside the United States.

Once the original IND has been filed to the FDA, the company must wait 30 days to initiate the first study in humans. If the FDA has questions or concerns about the safety of the drug in the initial study, the FDA is obligated to inform the company before this 30-day waiting period has elapsed. The initiation of the study is placed on hold until any questions have been resolved.

The FDA requires that annual progress reports be filed on the IND containing information about the status of clinical studies, including any significant findings in these studies and any significant changes in the informational material supplied to the investigators.

### 1. Phase 1 Studies

Phase 1 is the first stage of clinical testing.† The purpose of Phase 1 is to determine the initial safety information in humans along with the metabolism, absorption, elimination, and other pharmacological action, the preferred

---

*An Institutional Review Board comprises individuals not directly associated with the clinical study. The members are usually physicians, administrators of the institution, and members of the community. The IRB ensures that the patients/subjects are protected from undue harm and are fully instructed as to the risks and benefits of participating in the study.

†Such studies could not have been legally conducted in the United States prior to filing the IND. However, a company may have conducted clinical studies in another country. The information obtained from these studies must be included in the original IND and may be adequate to allow more advanced testing than Phase 1.

route of administration, and safe dosage range (CFR, 1985). The company usually conducts single- and multiple-dose studies in healthy human volunteers rather than in patients for these first clinical studies. This approach is advocated by many, since some of the pharmacological effects of the drug could be confounded by the presence of disease. However, there are instances where it may not be ethical or useful to conduct studies in normal volunteers; in this case, the rationale must be included in the original IND. These studies in normal volunteers may also include sampling of blood to determine the pharmacokinetics of the drug. However, the single- and multiple-dose pharmacokinetic profiles of the drug may be determined from other studies in volunteers during this stage of drug development. The proposed IND regulations (see Section 4 below) state that generally this phase of clinical study involves 20 to 80 subjects. Phase 1 studies are discussed in detail in Chapter 6.

### 2. Phase 2 Studies

Phase 2 studies involve the first experience with the drug in patients with the disease to be treated. Phase 2 begins when the company has established reasonable safety in humans and has defined the range of doses over which the drug is shown to be safe. This phase of clinical testing will typically include dose definition studies, initial efficacy studies, clinical pharmacology studies that require testing in patients, and more extensive pharmacokinetic studies. The proposed IND regulations state that this phase of clinical study involves no more than several hundred patients. Phase 2 studies are discussed in detail in Chapter 6.

### 3. Phase 3 Studies

Phase 3 of clinical development "provides the assessment of the drug's safety and effectiveness and optimum dosage schedules in the diagnosis, treatment, or prophylaxis of groups of (patients) involving a given disease or condition" (CFR, 1985). This phase usually includes the studies designated by the FDA as "adequate and well-controlled" (see Section IV of this chapter). Studies in Phase 3 include

Definitive (or "pivotal") placebo-controlled studies to document the efficacy of the drug.

Active-controlled studies to define better the relative efficacy and safety of the new drug versus other drugs used for the disease under study.

Studies in special populations such as the elderly, renally impaired, and/or hepatically impaired. Specific studies in these populations are usually preferred to subgroup analyses of data from larger, more heterogeneous clinical studies.

For a drug that will be used chronically, long-term studies of a year's duration or longer to define the safety profile of the new drug under extended use. Typically these studies are of uncontrolled open-label design even though there are occasions where controlled double-blind studies are conducted to develop the safety profile of the drug.

Phase 3 usually involves several hundred or even several thousand patients and allows the drug to be fully characterized as to its efficacy and profiled as to its safety in a wide range of controlled and uncontrolled patient situations. Phase 3 studies are discussed in detail in Chapter 6.

### C. The IND Rewrite

The IND process is undergoing some revisions at the FDA. In 1983 the FDA published draft revised IND regulations for comment (Federal Register, 1983) and these are in the final stages of being formally issued. These regulations appear to reduce the amount of control by the FDA over the earlier stages of drug development and even propose that the FDA turn over the responsibility for Phase 1 development to institutional review boards. However, for Phases 2 and 3 the FDA is proposing to increase their involvement in assessing "the scientific quality of the clinical investigations and the likelihood that the investigations will yield data capable of meeting statutory standards for marketing approval." These proposed regulations ask that detailed protocols be submitted for FDA review and, if the company feels that the study design may need to be modified during the study, the protocol should include a statement on how such modifications are to be handled.

The proposed regulations specifically mention the need for end-of-Phase 2 and "pre-NDA" meetings between the FDA and the company. The end-of-Phase 2 meeting is directed at established agreement concerning the overall Phase 3 plan including the specific protocols for the clinical studies. Historically, because of problems of changes in scientific standards and technology, agreements reached at such meetings were on occasion not upheld by the FDA. These proposed regulations address this problem:

> Barring a scientific development that requires otherwise, studies conducted in accordance with the agreement shall be presumed to be sufficient in objective and design for the purpose of obtaining marketing approval for the drug.

The pre-NDA meeting is intended to focus on the format of the NDA but may include discussion of any major unresolved problems, identification of adequate and well-controlled studies to establish efficacy, and discussion of "appropriate methods for statistical analysis of the data."

As this book goes to print, the final revised IND regulations have just been issued and became effective June 17, 1987. Most of the major changes to the present regulations described above remain intact.

## III. HISTORICAL PERSPECTIVES

Before discussing the NDA, which is the culmination of all this research and development of a new drug (see Section V), I will present a capsule view of the major regulatory developments in the United States over the past 100 years and, in so doing, provide a perspective on how the current FDA laws, regulations, and guidelines came to be. Figure 2 shows these major developments chronologically.

### A. Pure Food and Drug Act of 1906

In the late 1800s there was growing concern across the United States about phony remedies available to anyone and adulterated products from the food

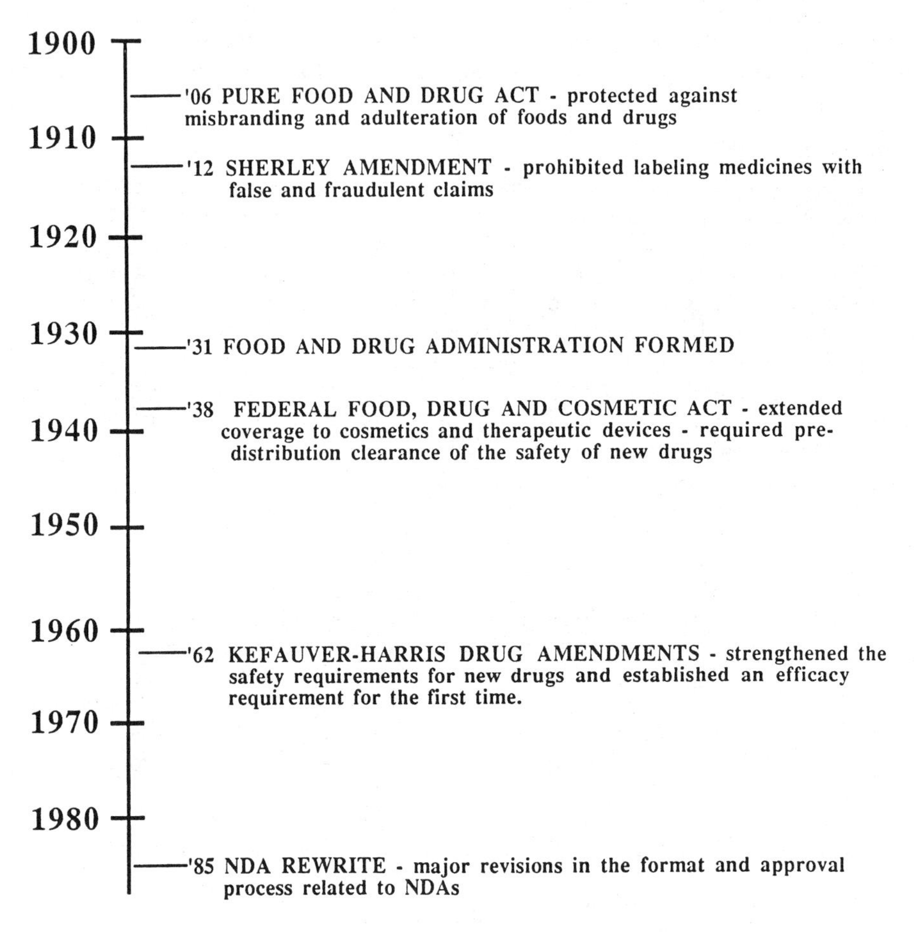


Fig. 2 Significant drug regulatory milestones in the United States.

and meat-packing industries. These products, which had previously been available only on a local basis, were beginning to flow across the nation in interstate commerce. The leader of those who were concerned was the chief chemist of the Department of Agriculture in the U.S. government, Harvey Wiley, who has been called the "father" of the Pure Food and Drug Act of 1906. The "Muckrakers," led by Upton Sinclair, joined with Wiley and other public figures to publicize the problems in the food and drug industries. Finally, in response to the publicity given to the national scandal over impure meat and horrendous conditions in slaughterhouses, Congress passed the 1906 act. The basis of this act, as with all other food and drug acts, is the *regulation of interstate commerce* by the federal government. Compared to today's laws and regulations, this act was very limited in scope. The act protected against misbranding and adulteration of foods and drugs. No preclearance of drugs was required and the government was given no inspection authority. Thus the government had to rely on detective work leading to seizures of illegal goods and criminal prosecution.

This act did not originally regulate the claims made for a product, so the Sherley amendment to the act was passed in 1912, which prohibited labeling medicines with false and fraudulent claims. Some of the reasons for this amendment are evident from the following:

> Campaigns were launched against broad-gauge tonics with panacea claims (one was Humbug Oil), against male-weakness remedies (like Sporty Days Invigorator), and against alleged "cures" for narcotic addiction, cancer and other dread diseases. Most proprieters, when hauled into court, did not contest the misbranding charges, but paid their modest fines, modified their labels slightly, continued their earlier claims in advertising (which the law in no way circumscribed), and went on about their profitable businesses. (Young, 1970).

The Sherley amendment stipulated that false and fraudulent claims for drugs were unlawful. The only problem with this amendment was that a "state of mind" must be proved in order to establish the fraudulence, which turned out to be difficult in many cases.

### B. Major Revisions to the Act in 1938

The 1906 act had many deficiencies that became more apparent as the years passed. The problems were brought to a head with the Elixir of Sulfanilamide disaster in 1937. A liquid formulation of this "sulfa drug" had been developed that contained diethylene glycol, a deadly poison. This formulation had never been tested in humans before its marketing, and it resulted in the deaths of more than 100 people. The incident led to the Federal Food, Drug and Cosmetic Act of 1938, since the FDA was unable to remove the Elixir from the market under the 1906 law. This act contained the "New Drug" concept and vastly overhauled the 1906 act. The new act also covered devices and cosmetics for the first time and provided that *new drugs* could not be marketed in interstate commerce unless they had been established as safe for their intended use. This was the beginning of preclearance of drugs for marketing by requiring pharmaceutical companies to file an NDA, that is, requiring the submission of "full reports of investigations which have been made to show whether or not such drug is safe for use."

The 1938 act also essentially created "prescription drugs." Previously, all drugs had to be labeled with instructions for use by the patient. The federal government felt that certain drugs were too risky to put in the hands of the consumer and hence exempted prescription drugs from this requirement and replaced the wording for patient labeling with the words, "Caution: to be used only by or on the prescription of a physician." This has caused the package insert, which describes the drug to the physician, to play a very important role in the drug regulatory process concerning prescription drugs. Because of the labeling exemption in the 1938 act, the FDA is able to control the advertising of prescription drugs (Young, 1970). The content of an approved package insert represents a baseline from which the FDA can measure and proceed against deviations in a company's promotion of its product.

### C. 1962 Kefauver–Harris Drug Amendments

During World War II certification of insulin and antibiotics was established by law. This meant that the federal government must test a sample from each batch of insulin and antibiotics *before* the batch could be sold. However, other drugs did not, and still do not, fall under such rigid requirements.

During the late 1950s Estes Kefauver, a U.S. senator from Tennessee, chaired the Senate Committee on Antitrust and Monopoly and held investigative hearings into the practices of pharmaceutical companies in the United States. The preclearance requirement for new drugs in the 1938 act had allowed a few companies to monopolize the pharmaceutical market. Senator Kefauver proposed a bill before Congress that would lower prices of drugs and increase competition as well as provide additional protection to the consumer of prescription drugs. As did the Elixir of Sulfanilamide disaster in 1937, the thalidomide disaster in 1961 created enough public support to lead to the so-called 1962 Kefauver-Harris Drug Amendments to the 1938 FFD&C Act. Thalidomide was a sedative that was being marketed in countries other than the United States. Its use by pregnant women, primarily in West Germany, had led to 3500 to 5000 deformed infants. The FDA had not yet approved thalidomide, and credit for this was given to the existing laws and to Dr. Frances Kelsey of the FDA, who was the medical reviewing officer for thalidomide. Nevertheless, the outcry resulting from the thalidomide disaster led to even more tightening of the laws concerning prescription drugs.

Up until 1963, the U.S. drug regulations addressed only the *safety* requirements for obtaining approval for marketing a drug in the United States, and the FDA did not formally receive any information about the new drug until the NDA was filed. Hence the FDA was unaware of how the drug was being tested in humans and hence was unaware of the identity/location of the many clinical investigators across the country who were testing the drug in human clinical studies. This lack of involvement of the FDA was considered a deficiency of the 1938 act. The 1962 amendments removed this deficiency by establishing the IND, which required that the pharmaceutical company provide a statement of the qualifications of investigators and forced these investigators to recognize their obligations when conducting human testing of new drugs (Cavers, 1970).

Other circumstances leading to the Kefauver-Harris Drug Amendments are described by Lasagna (1970) as follows:

1. Initial selection of drugs for clinical trials. Animal toxicity tests varied from company to company and the caliber was sometimes shocking. Often animal testing was postponed until the drug's clinical activity was assured.
2. Poor quality of clinical investigations.
3. FDA legally could not concern itself with efficacy.
4. Information on clinical drug toxicity was of low quality.

Lasagna (1970) went on to summarize the results of the 1962 amendments as follows:

1. Clarification of the safety-efficacy problem [by adding the requirement for proof of efficacy].
2. The requirement that the FDA be apprised of proposed human trials with new drugs has without question almost eliminated the giving of drugs to man on the basis of skimpy animal toxicity tests.

3. Fewer "me-too" drugs, i.e., drugs similar in activity to those already on the market, (have been) introduced due to high costs of preclinical testing and the emphasis on the use of "expert investigators."
4. Written consent by the patient/volunteer has benefited these individuals in regard to their personal rights.
5. Close supervision of drug advertising has led to less flamboyant promotion of drugs.
6. It [became] harder for the smaller firms to develop drugs to market.

## IV. "SUBSTANTIAL EVIDENCE" AND "ADEQUATE AND WELL-CONTROLLED INVESTIGATIONS"

The 1962 amendments included a requirement for "substantial evidence" that, according to statuatory scientific criteria, the drug is effective for its intended use. These amendments state:

> The term "substantial evidence" means evidence consisting of adequate and well-controlled investigations, including clinical investigations, by experts qualified by scientific training and experience to evaluate the effectiveness of the drug involved, on the basis of which it could fairly and responsibly be concluded by such experts that the drug will have the effect it purports or is represented to have under the conditions of use prescribed, recommended or suggested, in the labeling or proposed labeling thereof.

The FDA's regulations (CFR, 1985) go into more detail concerning this topic. The section concerning adequate and well-controlled studies (Section 314.126) states:

> The purpose of conducting clinical investigations of a drug is to distinguish the effect of a drug from other influences, such as spontaneous change in the course of the disease, placebo effect, or biased observation. The FDA considers (these characteristics) in determining whether an investigation is adequate and well-controlled for the purposes of Section 505 of the Act. Reports of adequate and well-controlled investigations provide the primary basis for determining whether there is 'substantial evidence' to support the claims of effectiveness for new drugs and antibiotics. Therefore the study report should provide sufficient details of study design, conduct and analysis to allow critical evaluation and a determination of whether the characteristics of an adequate and well-controlled study are present.

The FDA has for some time interpreted these requirements as meaning that there must be at least *two* adequate and well-controlled studies of patients with the disease being sought for a drug to be approved for that indication. A single, multicenter study is in most cases not sufficient for approval. An exception to this requirement might be made where a major prevention trial involving thousands of patients (e.g., prevention of myocardial infarction) is required to show the drug's efficacy. However, prior agreement with the FDA would be required.

The following excerpt from the regulations presents the characteristics that the FDA has defined to be required for a clinical study to be deemed adequate and well controlled (CFR, 1985):

1. A clear statement of the objectives of the investigation; summary of the proposed or actual methods of analysis in the protocol and the report.
2. The study uses a design that permits a valid conclusion with a control to provide a quantitative assessment of drug effect. The possible controls are:
   a. PLACEBO CONCURRENT CONTROL—Includes randomization and blinding of patients or investigator(s) or both.
   b. DOSE-COMPARISON CONCURRENT CONTROL—Includes randomization and blinding of patients or investigator(s) or both.
   c. NO TREATMENT CONCURRENT CONTROL—Could be used when there are objective measures of the disease available and the placebo effect is negligible.
   d. ACTIVE TREATMENT CONCURRENT CONTROL—A comparison with known effective therapy, e.g., when use of placebo is unethical. Includes randomization and blinding. The report should assess the ability of the study to have detected a difference between treatments. Similarity can mean that either both treatments were effective or that neither were effective. The analysis of the study should explain why the drugs should be considered effective in the study, for example, by reference to results in previous placebo-controlled studies of the active control.
   e. HISTORICAL CONTROLS—comparison with data from other studies.
3. The method of selection of subjects provides adequate assurance that they have the disease being studied.
4. The method of assigning patients to treatment and control groups minimizes bias and is intended to assure comparability of the groups with respect to pertinent variables such as age, sex, . . . Ordinarily in a concurrently controlled study, assignment is by randomization, with or without stratification.
5. Adequate measures are taken to minimize bias . . . the protocol and report of the study should describe the procedures used to accomplish this, such as blinding.
6. The methods of assessment of the subjects' response are well-defined and reliable.
7. The report of the study should describe the results and the analytic methods used to evaluate them, including any appropriate statistical methods. The analysis should assess, among other things, the comparability of the test and control groups with respect to pertinent variables, and the effects of any interim analysis performed.

Much of this discussion is familiar to one trained as a statistician and is the primary link from the laws and regulations to the use of statistics in drug development. As will be discussed in Chapter 7, there are many steps to be taken by the statistician in working with his or her clinical counterparts to ensure that the definitive ("pivotal") studies will be considered adequate and well controlled by the FDA as support for the approval of a new drug for marketing.

## V. THE NDA

### A. What Is an NDA?

A New Drug Application is a formal application to the FDA by a pharmaceutical company to obtain approval to market a new drug in the United States. This application contains (directly or by cross reference) the comprehensive clinical, preclinical, pharmaceutical, and chemical information related to the new drug that has been obtained throughout the many years of research and development. The application is often quite large and may comprise 100 or more volumes. It contains individual reports and data listings from each of the preclinical and clinical studies of the new drug. It contains detailed information concerning the drug substance (the active ingredient intended to furnish the desired pharmacological action) and the drug product (the finished dosage form that contains the drug substance), including a complete description of the chemical synthesis, the process controls used during manufacturing and packaging, and specifications related to, e.g., stability and sterility. The NDA also contains the proposed package insert for the drug, the document that states the claims being proposed for the drug along with any warnings, contraindications, etc. The NDA is filed when the company's research team is confident that an evaluation by independent scientists would result in an agreement that the drug is safe and effective for the uses stated. For example, (1) an oral contraceptive is effective in preventing pregnancy and has an acceptably low level of adverse experiences or (2) an antianginal agent is indeed effective in preventing or at least significantly reducing the number of anginal attacks and does not increase the risk of a myocardial infarction.

### B. NDA Rewrite of 1985—Improved Format of Submissions

In January 1985 the FDA published in the Federal Register what has come to be known as the NDA Rewrite. The objective of the NDA Rewrite was to improve the organization of the NDA and some of the FDA procedures related to drug approval that would accelerate the review process at the FDA. This rewrite was the result of many years of study and interaction between the FDA and the pharmaceutical industry, which was initiated in 1979 when the FDA issued "Concept Papers" for public comment. The FDA stated in its preamble to these new regulations:

> The FDA is revising its regulations . . . to speed up the availability of beneficial drugs to consumers by improving the efficiency of the agency's approval process for new drugs and antibiotic drugs, while improving the already high level of public health protection the drug approval and surveillance processes now provide . . . .

> The improvements will help applicants prepare and submit higher quality applications [NDAs] and permit the FDA to review them more efficiently and with fewer delays . . .. Accordingly the final regulations enable FDA to act as both a public health promoter, by facilitating the approval of important new safe and effective therapies, and as a public health protector, by keeping off or taking off the market drugs not shown to meet safety and efficacy standards (CFR, 1985).

The NDA rewrite revised the previous format of an NDA by requiring individual technical sections for each of the disciplinary reviewers at the FDA and will allow parallel review of an NDA. The prior format for an NDA did not directly allow such parallel reviews to take place since the information for different reviewers was interspersed throughout the application. The NDA also contains an overall summary of the entire NDA that is given to each of the reviewers.

This new organization of an NDA recognizes more than ever before the importance of statistics in the approval process since one of the technical sections is the Statistical Section of the NDA. Previously, the information requiring statistical review was interspersed throughout the NDA. According to these regulations, the Statistical Section is to describe the statistical evaluation of the clinical data in the NDA, including

> 1. A description and analysis of each controlled clinical study, and the documentation and supporting statistical analysis used in evaluating the controlled clinical studies.
> 2. . . . a summary of information about the safety of the drug product, and the documentation and supporting statistical analysis used in evaluating the safety information (CFR, 1985).

Figure 3 shows the organization of an NDA under these new regulations.

Another major change in NDAs resulting from the NDA Rewrite was to significantly reduce the volume of the submission by no longer requiring inclusion of all case report forms collected in clinical studies of the new drug. These new regulations require only that full case report forms be submitted for patients who dropped out of a clinical study or died during a clinical study regardless of drug treatment or apparent cause. All other case report form data are to be presented in the NDA in the form of data listings. However, the FDA reserves the right to request any other case report forms they consider necessary for their review.

In addition to the individual technical sections, the NDA has an overall summary that is given to each of the reviewers. A separate copy of the whole NDA is stored at the FDA and is called the "archival copy." This archival copy contains all the summaries and the technical sections as well as the case report tabulations (data listings). These listings should contain (1) all data from the adequate and well-controlled studies, (2) data from the earliest clinical pharmacology studies, and (3) safety data from all other clinical studies. The archival copy also contains the case report forms for all patients who dropped out of a clinical study or died during a clinical study.

### C. NDA Approval Process

Once the NDA is received by the FDA, the review of the NDA begins. The major reviewers of the NDA are the following:

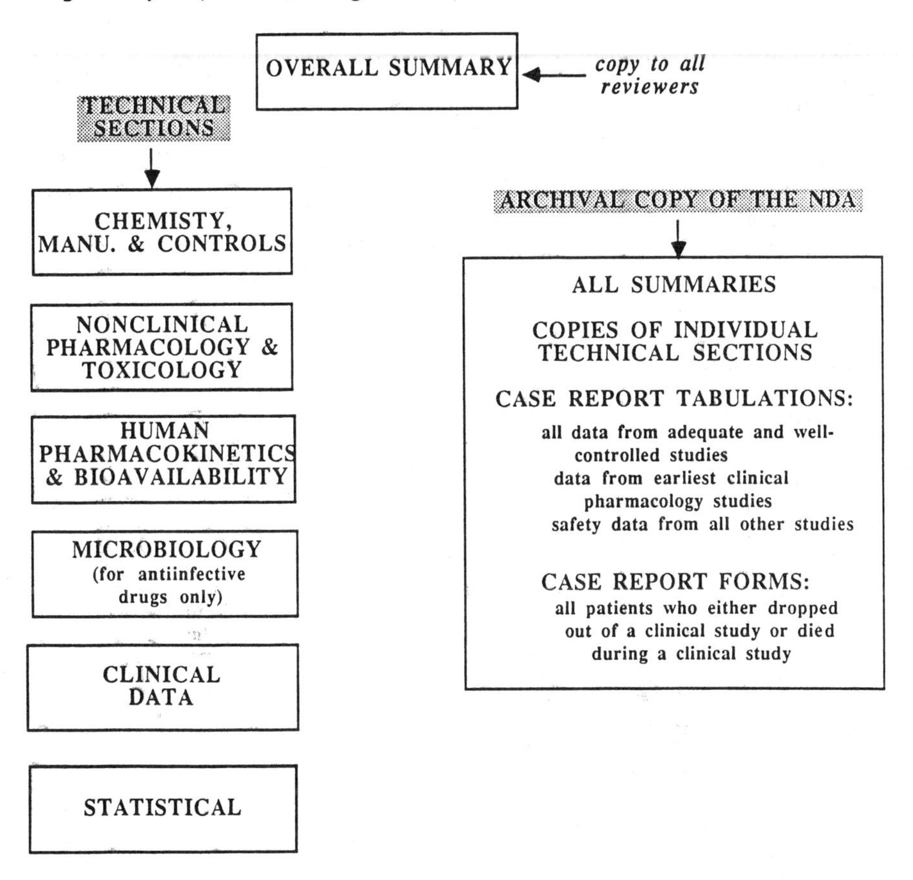


Fig. 3 The "new" NDA.

Medical reviewer—reviews primarily the clinical studies and verifies that the data support the claims in the proposed package insert
Pharmacology reviewer—reviews primarily the preclinical data obtained from animal toxicology, animal pharmacology, and animal ADME studies
Biopharmaceutical reviewer—reviews the human metabolism, bioavailability, pharmacokinetic, and, if present, bioequivalence studies
Chemistry reviewer—reviews the chemistry, manufacturing, and control information
Statistical reviewer—reviews primarily the statistical analyses of the adequate and well-controlled studies as well as the analyses of the human safety data

Often these reviewers will have questions for the company during their review, and they may also request additional information or additional analyses of the data in the NDA.

Once the individual reviewers complete their review, they prepare a written summary of the review and submit it to the FDA division to which the NDA was submitted. With the exception of the statistical and biopharmaceutical reviewers, the reviewers are part of this "reviewing division." The division director and the director's supervisor* review the written reviews and make the decision to either approve the drug for marketing, request even more information from the company, or reject the drug for marketing.

Often the FDA uses advisory committees to perform an additional review of certain important issues related to the approval of the drug. Members of the advisory committees are usually from academic research positions and, in addition to clinical pharmacologists and others performing clinical research, most committees also have a statistician. There are several such committees set up for different therapeutic areas (e.g., the Cardio-Renal Advisory Committee, the Arthritis Advisory Committee, the Gastroenterology Advisory Committee). The meetings of the advisory committees are usually public and provide valuable information to the industry concerning the current trends of thinking in the FDA. Recommendations of the advisory committees are not binding to the FDA, but the FDA usually follows such recommendations.

One of the most important documents in the NDA is the package insert, i.e., the proposed text of the labeling. This document contains all information that will accompany the drug as it is distributed commercially. In the NDA, the package insert must contain annotations to the data in the NDA which support each statement and must be approved by the FDA. Some of the major headings of a package insert are:

1. Description of the drug product
2. Clinical pharmacology
3. Mechanism of action
4. Pharmacokinetics and metabolism
5. Indications and usage
6. Warnings
7. Precautions
8. Carcinogenesis, mutagenesis, impairment of fertility
9. Use in pregnancy
10. Adverse reactions
11. Overdosage
12. Dosage and administration
13. How supplied

Certain headings may be added or deleted depending on the specific drug and/or the specific disease(s) in the indications. Once approved, most subsequent changes in the package insert must be approved by the FDA prior to implementation. Readers who are unfamiliar with package inserts should carefully read some package inserts for marketed drugs in order to appreciate the depth and breadth of information contained therein.

## D. Postmarketing Activities

Promotional activities related to a marketed drug are also highly regulated by the FDA. A company must bear in mind that all claims for a drug require FDA approval before they can be used in any promotional activities. These claims must be supported by studies similar to those in the original NDA,

---

*Either the Director, Office of Drug Research and Review, or the Director, Office of Biologics Research and Review.

often requiring the same attention to study design and analysis that was given to the studies supporting initial approval of the drug.

The specific advertising of a drug is also highly regulated by the FDA. Part 202 of the Code of Federal Regulations (CFR) describes these regulations in detail. Part 314 of the CFR also contains regulations pertaining to advertising by stating:

> The (company) shall submit specimens of mailing pieces and any other labeling or advertising devised for promotion of the drug product at the time of initial dissemination of the labeling and at the time of initial publication of the advertisement for a prescription drug product. Mailing pieces and labeling that are designed to contain samples of a drug product are required to be complete, except the sample of the drug product may be omitted (CFR, 1985).

A detailed discussion of these complicated regulations is beyond the scope of this book.

The reporting of adverse drug reactions is another important postmarketing activity required by the FDA. The FDA's regulations include fairly specific requirements for reporting adverse drug experiences (ADEs) related to a marketed drug. The regulations define an ADE as

> Any adverse event associated with the use of a drug in humans, whether or not considered drug related, including the following: an adverse event occurring in the course of the use of a drug product in professional practice; an adverse event occurring from drug overdose . . .; an adverse event occurring from drug abuse; an adverse event occurring from drug withdrawal; and any significant failure of expected pharmacological action (CFR, 1985).

The company is required to promptly review all ADEs obtained from any source (U.S. or ex-U.S., commercial marketing experience, clinical trials, scientific literature, etc.) and to file to the FDA within 15 working days all reports considered to be serious and unexpected. (See the CFR for the definition of "serious and unexpected.") These reports must be filed on form FDA-1639. The FDA maintains a data base containing the information on these forms and hence is able independently to monitor the safety of a drug while it is being marketed. There are additional FDA reporting requirements for the company to regularly summarize the safety data concerning the drug, including serious and *expected* ADEs. These activities are considered by the FDA to be essential and have resulted in companies developing complex worldwide systems (both procedural and computerized) to obtain reports of ADEs and store and, when required, generate the information on form FDA-1639 for filing to the FDA.

## VI. ROLE OF BIOPHARMACEUTICAL STATISTICS

As is evident in the remaining chapters of this book, it is harder to find areas where statistics (and statisticians) does *not* play an important role among the many application areas of biopharmaceutical research and development than it is to list the areas where it does play a role. Table 2 lists some of the application areas that involve statistics. Statistical design and analysis

Table 2 Statistical Application Areas in Drug Development

| |
|---|
| Screening of new chemical molecules |
| Development of pharmaceutical formulations |
| Determination of biological effects in animal toxicology and pharmacology studies |
| Evaluation of absorption, distribution, metabolism, and excretion of drugs |
| Establishment of bioavailability |
| Evaluation of presence/absence of bioequivalence between two or more formulations of the same drug |
| Design and analysis of clinical studies: |
| Demonstrate tolerance |
| Develop dose-response information |
| Document efficacy in target population |
| Determine adverse experience profile |
| Evaluate potential biological interactions between drugs |
| Validation of assays and manufacturing processes |
| Quality control of manufactured drug products |

are essential tools for the pharmaceutical industry to use in order to properly develop drugs that will be judged approvable by regulatory agencies. Table 2 cannot list the many subtleties of the role played by the statistician in the pharmaceutical industry. This role requires ingenuity and creativity on the part of the statistician, and the rest of this book will provide in-depth information about the many aspects of statistical applications in this industry. As a statistician, I remember the feeling of accomplishment that accompanied the FDA approval of the first drug on which I worked. The close and intense interactions with the clinical and toxicology departments and the many days of burning the night oil were worth it when I realized the important part I had played in ensuring that the drug was safe and effective in humans. To this day, when I talk to someone who is receiving benefit from this drug, I feel good about my contributions. I am sure that most of my colleagues in the industry have had the same experience and share the feelings of accomplishment and contribution. As you read the remaining chapters of this book, remember that the role of the statistician in drug development is regarded as very important and is very rewarding.

## REFERENCES

Cavers, D. F. (1970). The Evolution of the Contemporary Systems of Drug Regulation Under the 1938 Act, *Safeguarding the Public* (John Blake, ed.), Johns Hopkins Press, Baltimore, Md.

*Code of Federal Regulations* (CFR). (1985). Office of the Federal Register, National Archives and Records Administration, Superintendent of Documents, Government Printing Office, Washington, D.C.

Federal Register. (1983). *Federal Register*, 48 (112):26736–26749, June 9.

Goldenthal, E. I. (1968). Current Views on Safety Evaluation of Drugs, *FDA Papers*, 2:13–18, May.

Lasagna, L. (1970). 1938–1968: The FDA, the Drug Industry, the Medical Profession and the Public, *Safeguarding the Public* (John Blake, ed.), Johns Hopkins Press, Baltimore, Md.

O'Reilly, J. T. (1984). *Regulatory Manual Series, Food and Drug Administration*, Shepard's/McGraw-Hill, Colorado Springs, Colo.

Taber, B.-Z. (1969). *Proving New Drugs, a Guide to Clinical Trials*, Geron-X, Inc., Los Altos, Calif.

Young, J. H. (1970). Drugs and the 1906 Law, *Safeguarding the Public* (John Blake, ed.), Johns Hopkins Press, Baltimore, Md.

# 2 Pharmaceutical Lead Discovery and Optimization

JOHN R. SCHULTZ, PATRICIA L. RUPPEL, and MARK A. JOHNSON,
*The Upjohn Company, Kalamazoo, Michigan*

## I. INTRODUCTION

Although the search for new drugs requires intellectual and technological contributions from several scientific disciplines, it remains a highly empirical process. Estimates vary, but the number of new compounds synthesized for each new therapeutic agent is in the range 6000 to 10,000. This high attrition rate is due at least in part to our imperfect knowledge of biological processes and to ever-increasing medical objectives. Oral contraceptives are an example of the latter situation. As more of these agents become available, it is more difficult to find one which will eventually provide meaningful advantages in terms of either safety or efficacy. As a result, there is limited opportunity for economic return.

The drug discovery process can be divided into three parts: (1) lead discovery, (2) lead optimization, and (3) lead development. The objective of this chapter is to highlight areas where statistics can contribute to improved effectiveness of lead discovery and lead optimization. Lead development considerations will be addressed in the following chapters.

### A. Lead Discovery

Leads are chemical compounds which produce biological activities thought to represent therapeutic potential. New leads can be searched for by mass screening or on the basis of theories about disease intervention. Both approaches involve biological testing; the first is often called "random" screening while the second is sometimes called "targeted" screening.

Random or general screens are high-volume routine operations in which large numbers of compounds are initially evaluated for some particular biological action. Either in vivo or in vitro biological test systems may be used. Compounds are drawn, usually without selection, from large inventories and tested in the biological system. The objective is to reliably identify those compounds which possess the desired activity. In general, only a very small proportion of the compounds tested will exhibit useful activity. Several general screens, each designed to detect a specific biological action, may be in operation within the research division of a pharmaceutical company at any given time. It is hoped that the collective output of these screens will provide enough leads to contribute to the discovery process in a meaningful way.

Very few compounds are synthesized specifically for general screening. They are generally prepared for targeted screening on the basis of some distinct rationale. Targeted screens focus on some rather specific biochemical or pharmacological process. Knowledge of these processes will then suggest which class or classes of compounds are most likely to possess the desired activity. These types of compounds will be selected from inventory or synthesized specifically for evaluation in the screen. If the biological theories are correct, the proportion of compounds exhibiting the desired activity should be relatively high.

Since compounds made for one purpose often show other types of activity, the two screening approaches are complementary. Compounds prepared for targeted screens also provide material for the general screens. By the same token, an unexpected novel lead from a general screen may provide new insight on biochemical mechanisms which can be utilized in designing a new targeted screen. The fact remains, however, that the two approaches are fundamentally different and this difference must be considered when the screen is being designed. Further considerations will be given to this issue in the following sections.

### B. Lead Optimization

When a lead compound has been identified, it is then evaluated more thoroughly in other biological test systems. Another process, called lead optimization, also begins with the discovery of a new biologically active chemical. The molecular structure of the lead is varied to increase the level of desirable activity, reduce undesirable activity, or otherwise improve its pharmacologic profile. This approach is based on the concept that the biological properties of a compound are related to its physiochemical properties. Investigation of these relationships has become an exciting area of statistical application.

One approach used to guide modifications of lead compounds is based on quantitative structure-activity relationships (QSAR) analysis. Several statistical methods have been utilized, including regression, cluster analysis, discriminant analysis, and factor analysis (Mager, 1982). The ultimate objective is to predict the biological activity of compounds which are not yet synthesized. In addition to scientific interest, there are obvious economic considerations in achieving this objective. There are many conceptual modifications of the lead compound. Synthesizing only the most promising modifications saves time and reduces both chemical and biological effort.

### C. Biological Test Systems

The success of any screening program hinges on the biological test used to select compounds for further development. Austel and Kutter (1980) give three critical properties:

1. Relevance between the measured biological effect and the required clinical activity
2. Reliability in distinguishing promising compounds from undesirable ones
3. High throughput—particularly for general screens

Other desirable features of the test system were described by Redman (1981). In our experience, the following are particularly important:

1. Generates results within a few days
2. Requires only a few miligrams of compound
3. Remains stable over time
4. Relatively inexpensive in terms of materials and effort
5. Provides automated data management

Selection of the biological model usually involves strategic as well as scientific considerations. Some potential models may be more relevant but too laborious for high-volume operations. Practical issues such as the availability of certain strains of animals or particular types of tissue may also influence model selection. Other considerations may reduce the choice of models further until no further specification can be made without experimental evaluation. With careful planning the results of this evaluation will also provide information needed for the statistical design of the screen.

In intact animal systems, some of the contributing factors which might be evaluated include size or age of animal, sex, strain, dosage level and regimen, route of administration, and pretreatment preparation such as feeding or fasting. The experiments should be designed to compare the responses of both active and inactive agents as these factors are being studied. The level of response and the variability in response are of particular interest. Some of these concepts are illustrated in a following section.

Proper identification of the primary observational or experimental unit is another important issue which must be addressed when the test system is being designed. Depending on the test system, the observational unit may be a cage of animals, a single animal, a muscle strip or organ, a vial of homogenized tissue, or a bacterial culture. This is a general experimental design issue which is particularly important in long-term high-volume screening. Design of the screen and evaluation of the results based on improper identification of the experimental unit will result in misclassification of the compounds. The proper unit of observation will satisfy two conditions: (1) it is the entity to which the treatment is applied and (2) it is independent of the other units.

## II. GENERAL SCREENING

Screening large numbers of compounds for biological activity is a traditional component of pharmaceutical research (Davies, 1958). The National Cancer Institute has also sponsored a large screening program for several decades (Armitage and Schneiderman, 1958; Vendetti et al., 1984). These efforts were a major force in the development of statistical methodology for drug screening. Most of the procedures available today were already developed by the early 1960s and have been reviewed by Federer (1963). A more recent review has been prepared by Bergman and Gittins (1985). They describe most of the important procedures and develop an integrated overview of the subject. Redman (1981) also discusses the literature on screening and addresses many practical issues in implementing a screening program. In this chapter we will discuss only methods with which we have some degree of experience. Although statistical elements of the screening process will be emphasized, it must always be appreciated that they are inextricably tied to the chemical and biological foundations of the screening program.

### A. Decision-Making Considerations

The general principles of acceptance sampling provide the traditional framework for making decisions about the biological activity of compounds tested in a screen (Davies, 1958). In this context the objective is to determine a course of action for each compound rather than estimate its activity. On the basis of screening results the compound will either be dropped from further consideration or passed on for further testing. In the first case the compound will be denoted as "inactive" or "negative" and in the second it will be referred to as "active" or "positive."

The ability of a screen to discriminate between active and inactive compounds is given by its operating characteristic (OC) curve. This gives the probability of declaring a compound positive as a function of its true activity, $\theta$. Following the analogy to acceptance sampling, a test procedure is defined by considering two points on the OC curve. For screening, these points can be defined relative to the observed activity of standard agents in the test system. The first point on the curve is selected on the basis of responses obtained in the test system with an inactive substance. At this level of activity, $\theta_1$, the probability of declaring a compound positive, should be very low. The second point is then chosen according to the response observed with a positive standard representing activity which the screen must detect. This level of activity, $\theta_2$, should be declared positive with high probability. The values $\theta_1$ and $\theta_2$ are obtained during evaluation of the test system described earlier. In cases where no standards exist it may be necessary to define $\theta_1$ and $\theta_2$ by some other means. This could be primarily intuitive, based on knowledge about the system, or it could be defined by some optimality criteria as suggested by Davies (1958), King (1963), or Dunnett (1961).

Alternative designs are examined for the desired target probability levels of declaring compounds positive which have true activity levels $\theta_1$, $\theta_2$. The target probabilities also determine misclassification rates. These errors are of two types and are referred to as "false positives" and "false negatives": a false positive result is obtained when a compound with true activity $\theta_1$ is declared active. A false negative is obtained when a compound with true activity $\theta_2$ is declared inactive. These values are denoted by $\alpha$ and $\beta$, respectively, and are obtained from the OC curve. Further details on the construction and interpretation of OC curves will be given in subsequent examples.

Two other properties of the screen of particular interest are sensitivity and specificity. Given that a compound is active, sensitivity is the conditional probability that the screen will classify it as positive. Specificity is the conditional probability that the screen will call a compound negative given that it is inactive (Redman, 1981). In the ideal case, both of these values should be close to unity.

## B. Operational Issues

A number of fundamental issues must be addressed when a high-volume screening operation is being considered. Several thousand compounds may be screened before even one lead is found; therefore careful consideration must be given to the level of effort applied to each compound. If the effort is too little, a promising lead may be overlooked. On the other hand, fewer compounds will be tested if effort is increased. Promising leads may then be missed simply because they are never tested. Balancing false positive and false negative rates with screen capacity must be addressed at the beginning of the screening program and periodically reviewed during its existence.

It was quickly recognized that many more compounds could be screened with given resources by utilizing a sequential approach. In general, greatest efficiency can be achieved with a full sequential design. This approach is difficult to administer in a high-volume operation, hence extensive use is made of group sequential or multistage designs. In these designs compounds are tested in stages with a fixed number of experimental units in each stage. Typically, a compound may be declared inactive at any stage but must pass through all stages to be declared active. The maximum number of stages for a screen is usually three. This preserves most of the efficiency of the sequential approach without greatly increasing operational effort.

Since a major reason for using multistage procedures is to reduce the number of observations needed to classify a compound, the average sample number (ASN) function is particularly important. This function gives the expected number of experimental units needed to classify compounds according to their true activity level. This function, along with the OC curve, is used to choose the most suitable screening plan. Of alternative plans meeting general screening objectives, the one with the smallest ASN will usually be selected.

A screen must operate on a routine, repetitive basis in order to achieve high output. The stability of the system is therefore another important issue. Some of the factors which reduce stability include seasonal and diurnal variation in the biological material, batch differences in reagents, operator changes, and any number of unknown events. The behavior of the system over time will influence the type of design which can be used. A consistent level of performance is ideal, but a system with some temporal variation may still be useful if it has adequate experimental control. Assessing performance of an established system over time will be discussed further in the section on quality control.

In general screening, compounds are typically tested at a single dose level or concentration. Identifying that dose level is another important process. If the dose is too low, the inherent activity of the compound will not be expressed. If it is too high, the activity may be masked by an undesirable response or even death. When standards are available, they can be tested at various levels in the proposed system to help guide dose selection. Test agents may then be administered at some multiple of a consistently detectable dose of the standard. When a test compound exhibits toxicity at that level, it should be retested at one-half the screening dose.

Of course, if many compounds exhibit toxicity, it may be necessary to reduce the routine screening dose.

Another area which is critical to the success of a mass screening program is related to basic operations. Simplification, standardization, and automation are essential. In this respect, general screening can be viewed as a production system. Any adjustments which can be made to the biological test, screen design, or laboratory procedures which could improve logistics should be considered. In some cases it may be necessary to experimentally evaluate approaches which could contribute to screening efficiency. This is particularly important if there is some risk that the modification could affect the sensitivity and specificity of the screen.

The following example illustrates a number of the points discussed above. Modifications to an existing screen were considered to improve operational procedures and increase screen capacity. Compounds were tested in rats and the objective was to detect oral antidiabetic activity. In order to screen more compounds, it would be necessary to do the work in another laboratory (laboratory B). It was also of interest to evaluate two methods of sample preparation. The old method was rather laborious, while the new method was faster and required less effort. The first experiment was designed to evaluate laboratories and method of sample preparation. This evaluation was done in animals treated with a positive standard as well as animals treated with vehicle only.

The results of the study are given in Table 1. No important differences were observed between laboratories; therefore it was decided to use laboratory B. No differences in sample preparation were apparent within laboratory B, so the new method was accepted.

Nine screening runs were then conducted in laboratory B, using the new method of sample preparation. Each run included eight animals treated with vehicle only (control) and five groups of four rats each treated with standards. Four positive standards and one negative standard were studied. The runs were conducted at approximately weekly intervals. The mean

**Table 1** Comparison of Laboratories and Sample Preparation Procedures on Blood Sugar Levels of Rats

| Laboratory | No. of rats | Preparation | | Treatment | Blood sugar (mg %) | |
|---|---|---|---|---|---|---|
| | | Old | New | | Mean | SD |
| A | 6 | X | — | Vehicle | 68 | 3.7 |
| A | 6 | X | — | Std. A | 53 | 5.4 |
| A | 6 | — | X | Vehicle | 67 | 4.4 |
| A | 6 | — | X | Std. A | 44 | 4.4 |
| B | 6 | X | — | Vehicle | 73 | 2.9 |
| B | 6 | X | — | Std. A | 55 | 6.6 |
| B | 6 | — | X | Vehicle | 70 | 4.4 |
| B | 6 | — | X | Std. A | 51 | 5.4 |

Table 2 Mean Blood Sugar Values and Between-Run and Within-Run Standard Deviations Observed for the Control and Standard Treatments[a]

| No. of rats | Treatment | Dose (mg/kg) | Mean (mg %) | Standard deviation: Between run | Standard deviation: Within run |
|---|---|---|---|---|---|
| 72 | Control | — | 72 | 2.5 | 5.0 |
| 36 | Neg. standard | — | 74 | 5.6 | 9.3 |
| 36 | Standard A | 25 | 54 | 4.0 | 6.4 |
| 36 | Standard A | 50 | 46 | 2.0 | 7.8 |
| 36 | Standard B | 12.5 | 70 | 4.9 | 7.0 |
| 36 | Standard C | 12.5 | 63 | 5.6 | 4.5 |

[a]Nine runs, eight control rats/run, four treated rats/run.

blood glucose values for each of the standard treatments and the controls are given in Table 2 along with the respective between-run and within-run standard deviations. The results show a gradient of activity with the standards. This information provides the basis for designing the screen and will be discussed later.

The test system in the above example generates measurement or continuous data. This is the case with most screening systems, but there are situations in which the responses are dichotomous. When choices can be made, use of measurement data favors efficiency while use of binomial data favors simplicity. Both types of responses will be encountered, and methods have been developed to design screens for each. As a matter of convenience we will discuss them separately.

## C. Binomial Response Variables

Classification of compounds tested in systems with binomial response variables can be accomplished with fraction-defective sampling plans. A single-sample fraction-defective sampling plan is specified by the sample size and the number of defective units that cannot be exceeded. This latter value is called the acceptance number. In drug screening, where responses can be classified as desirable or undesirable, the latter outcome can be thought of as a "defective" outcome. As a further condition, the screening process is viewed as operating in a random manner with sampling of compounds rather than lots. Instead of drawing n items from a lot, n experimental units are randomly assigned to treatment with a given compound. Thus, for a given sample size and critical value, the probability of accepting or declaring a compound active can be given as a function of the binomial parameter p, the probability of an undesirable outcome associated with that compound. This functional relationship is given by the OC curve and it can be computed from the binomial formula or taken from a table of binomial probabilities.

To illustrate this approach, we consider a test system to detect antifertility properties of compounds in hamsters. Pregnant females were injected with the test compound and evaluated for pregnancy status 6 days later. Preliminary results showed that 62 of 68 vehicle-treated or control animals were pregnant at the end of the test period. Two dose levels of a positive standard were also tested. At the low dose 11 of 30 animals were pregnant, and at twice the low dose 1 of 24 animals was pregnant.

The above results were used to guide the design of a screen for routine testing. A single-stage test in which eight animals were treated with each compound was implemented. The compound was declared active if four or fewer of the eight hamsters were pregnant and inactive if five or more of the animals were pregnant. The OC curve for the screen is given in Figure 1. It gives the probability of calling a substance active as a function of the true probability, p, of pregnancy associated with that particular compound. The curve shows that the probability of declaring a compound active which has a true p of 0.2 is 0.99, while a substance with a true p of 0.8 has a probability of 0.05 of being declared active.

Although single-stage plans are more straightforward, multiple-stage procedures are usually more appropriate for drug screening. General methods for obtaining OC curves and ASNs of multiple-stage plans were described by Schultz et al. (1973). These plans are specified by the maximum number of stages, K, the number of units examined at each stage ($n_1, n_2, \ldots, n_k$), the set of acceptance points ($a_1, a_2, \ldots, a_k$), and a corresponding set of rejection points ($r_1, r_2, \ldots, r_k$). At the

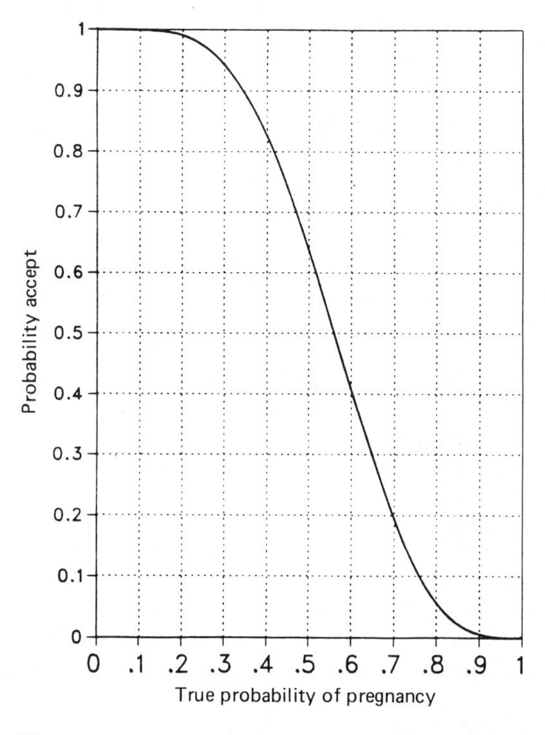


Fig. 1 OC curve for antifertility screen.

terminal stage $a_k = r_k - 1$ and the maximum number of units required is $N_k$. When the gth stage is reached, the test statistic is

$$\sum_{k=1}^{g} S_k$$

where $S_k$ is the number of responses observed in stage k. The sequential procedure is then as follows:

$$\text{if } \sum_{k=1}^{g} S_k \geqslant r_g, \text{ stop sampling and reject } H_0$$

$$\text{if } \sum_{k=1}^{g} S_k \leqslant a_g, \text{ stop sampling and accept } H_0$$

$$\text{if } a_g < \sum_{k=1}^{g} S_k < r_g, \text{ continue to stage } g + 1$$

For any stage, g, with $n_g$ experimental units the sum of the random variables for that stage is $S_g$ and the probability that $S_g = v$ for a specified value of p is given by

$$B_g(v, p) = \binom{n_g}{v} p^v (1 - p)^{n_g - v}$$

Furthermore, the probability that $(S_1 + S_2 + \cdots + S_g) = m$, given that the gth stage is reached, is

$$C_g(m, p) = \sum_{j=E}^{F} C_{g-1}(j, p) B_g(m - j, p)$$

where $E = \max(a_{g-1} + 1, m - n_g)$ and $F = \min(r_{g-1} - 1, m)$. Thus, the probability that $(S_1 + S_2 + \cdots + S_g) = m$ at the gth stage is the probability that $(S_1 + S_2 + \cdots + S_{g-1}) = u$ multiplied by the probability that $S_g = v$ summed over all eligible u, v such that $u + v = m$. This allows $C_g(m, p)$ to be calculated recursively. The eligible values of u and v are determined by the accept and reject points.

The probability of accepting $H_0$ at stage g is then

$$L_g(p) = \sum_{m=a_{g-1}+1}^{a_g} C_g(m, p)$$

and the probability of accepting $H_0$ through the final stage is

$$L(p) = \sum_{g=1}^{K} L_g(p)$$

Similarly, the probability of rejecting $H_0$ at stage g is

$$R_g(p) = \sum_{m=r_g}^{r_{g-1}+n_g-1} C_g(m, p)$$

and the cumulative probability of rejecting $H_0$ is then

$$R(p) = \sum_{g=1}^{K} R_g(p)$$

The average sample number is given by

$$ASN = \sum_{g=1}^{K} N_g[L_g(p) + R_g(p)]$$

where $N_g = (n_1 + n_2 + \cdots + n_g)$.

The above approach was used for an in vitro screen to detect antithrombotic activity (Hearron et al., 1984). In this system active agents protect mice from death following experimentally induced pulmonary thromboembolism. In a series of studies with this model, it was found that 16.7% of 228 animals treated with a positive standard and 79.3% of 227 control animals died. The following two-stage procedure was designed based on this information.

| | Cumulative deaths | | |
|---|---|---|---|
| Stage (g) | $a_g$ | $r_g$ | $n_g$ |
| 1 | ND[a] | 4 | 6 |
| 2 | 6 | 7 | 6 |

[a]No decision.

Compounds must pass through both stages to be declared active. The OC and ASN curves are given in Figures 2 and 3, respectively. The OC curve shows that the probability of declaring a compound active is 0.98 when the agent has a true probability (p) of death equal to 0.2. The probability of declaring compounds active when p = 0.8 is 0.02. The ASN curve shows

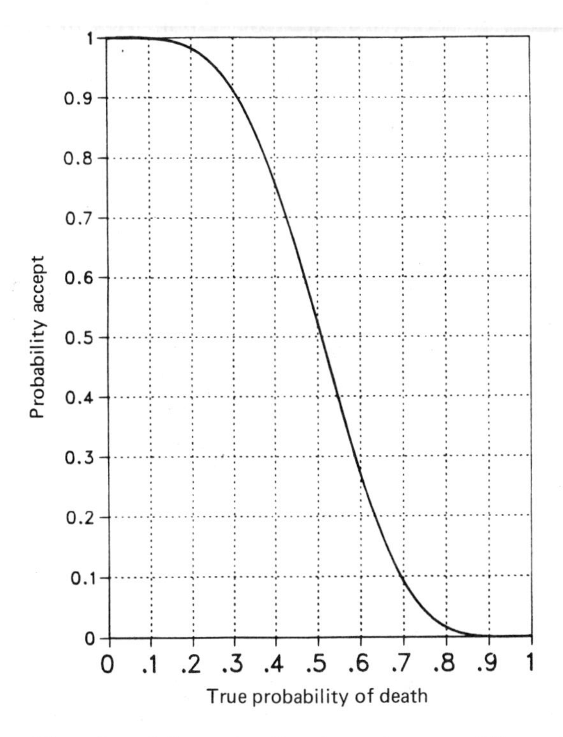


**Fig. 2** OC curve for two-stage antithrombotic screen.

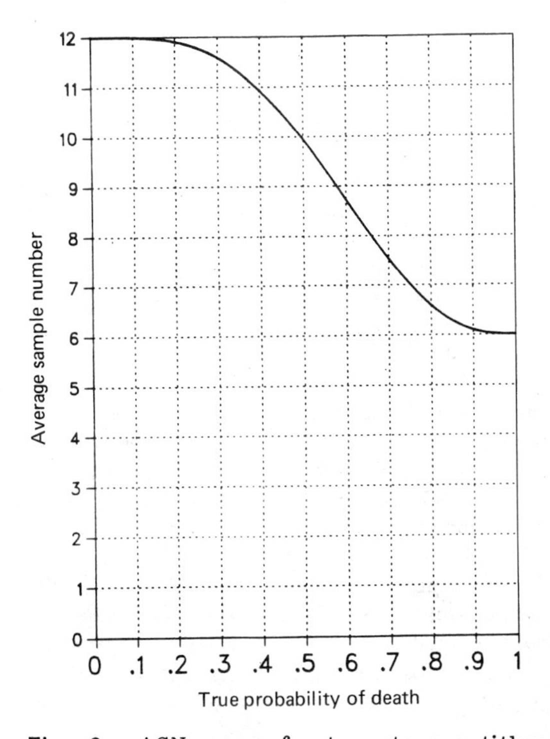


**Fig. 3** ASN curve for two-stage antithrombotic screen.

that compounds with the latter level of activity will be classified with an average of 6.6 mice.

A three-stage test for antiviral agents in experimentally infected mice was described by Schultz et al. (1973). In preliminary studies it was found that 6.0% of 583 animals treated with a positive standard and 84.6% of 625 control animals died during the 7-day observation period. Using these data as a guide, it was decided that when the true probability of death (p) associated with a particular compound under these conditions was 0.25 or less the material would be of interest and when p was 0.60 or greater the compound would be of no interest. It was further determined that the probability of accepting an interesting compound as defined above should be no less than 0.95 and the probability of accepting an uninteresting compound should be no greater than 0.05. Another consideration was that group sizes of 10 animals would result in the most efficient use of facilities and labor.

The decision rules of the selected screening plan are as follows:

| | Accumulated deaths | | |
|---|---|---|---|
| Stage (g) | $a_g$ | $r_g$ | $n_g$ |
| 1 | ND[a] | 6 | 10 |
| 2 | ND | 10 | 10 |
| 3 | 12 | 13 | 10 |

[a]No decision.

With this design, compounds can be declared inactive at any stage but must pass through all three stages to be classified as active. The OC curve given in Figure 4 shows that for compounds with $p = 0.25$ the probability of being declared active is 0.96. On the other hand, the probability of declaring compounds with $p = 0.6$ active is 0.02. The ASN curve in Figure 5 shows that 29.5 and 14.8 mice will be needed to classify compounds with $p = 0.25$ and 0.6, respectively.

Operationally, each screening run consisted of up to 60 groups of 10 mice each. Animals in six of these groups were treated with vehicle only and another six groups were treated with a positive standard. These 12 groups were used to monitor the test system and the remaining groups were used to test new compounds.

The performance of the screen was evaluated on the basis of results obtained in 37 test runs. A total of 1548 unselected compounds were classified. All but 57 of the compounds were declared inactive in the first stage; eight of these were subsequently declared active in the third stage. Excluding the vehicle and standard groups, a total of 16,240 mice were used, giving an average sample size of 10.5. A single-stage plan consisting of 26 animals with accept and reject points of 10 and 11, respectively, gives values which are essentially identical to those in Figure 4. The single-stage plan would have required 24,000 additional animals to classify the 1548 compounds which were examined in the three-stage screen. This clearly illustrates the dramatic saving which can be achieved with multiple-stage designs.

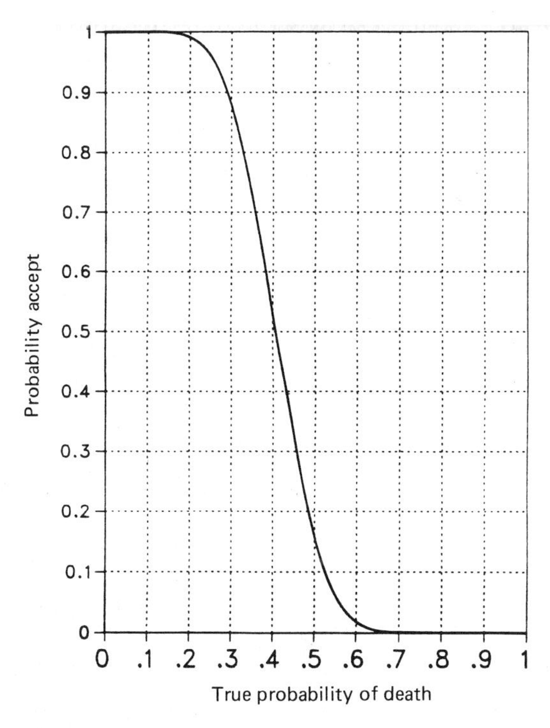


**Fig. 4** OC curve for three-stage antiviral screen.

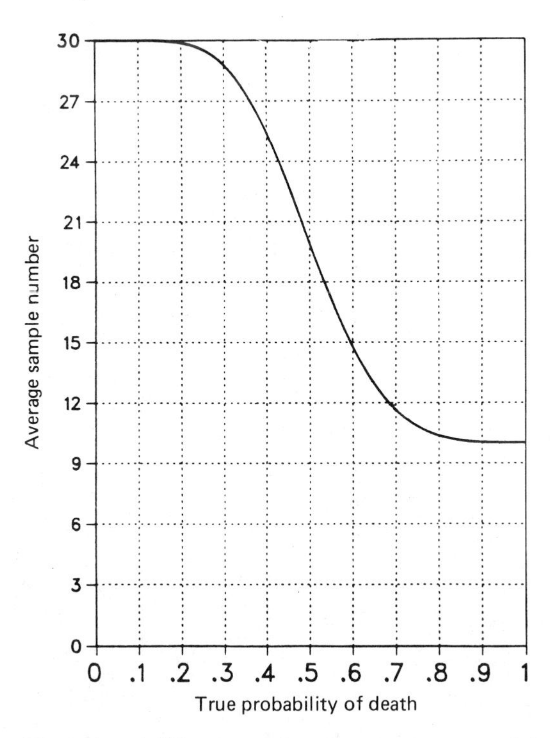


**Fig. 5** ASN curve for three-stage antiviral screen.

The design of these multiple-stage plans can be facilitated by constructing a Wald (1947) test region for given α, β, and location parameters. As a general rule, the number of units tested at each stage depends more on laboratory operations than statistical considerations. Since there are usually operational problems with more than three stages, the Wald region can be closed at that point and the critical values modified until a satisfactory plan is obtained. Most compounds will possess little or no activity; therefore particular attention should be given to the ASN needed to identify uninteresting compounds. Obviously, minimal effort should be given to these agents.

### D. Continuous Response Variables

As in the binomial case, single-stage acceptance sampling procedures for continuous variables could be used for drug screening. Our approach in this situation, however, has been to include experimental controls in each screening run. Responses in experimental units treated with test compounds are compared with those in controls to determine activity. A one-way analysis of variance is carried out on data from each run and test compounds are declared active by the least significant difference (LSD) test.

The general approach can be illustrated with a screen for compounds with hypolipidemic activity in a rat model. After preliminary experiments were conducted to establish test conditions, a series of runs were carried out to establish screen parameters. Data from 19 runs were evaluated. Each run consisted of 30 groups of six rats each. Four of the groups served as vehicle controls, one group was treated with a positive standard, one group with a negative standard, and the remaining 24 groups were used to test unknown compounds.

In 12 of the 19 runs positive standard A was used, while positive standard B was used in the remaining 7 runs. For each of these agents, the data were analyzed as a one-way classification. The objective was to obtain an estimate of the mean response, the within-run standard deviation, and the between-run standard deviation. The same procedure was followed for the vehicle-treated groups and also those treated with the negative standard. All values were transformed to logarithms for analysis. The results of these analyses for one of the screen's endpoints, serum cholesterol, are given in Table 3.

**Table 3** Mean Responses and Within-Run and Between-Run Standard Deviations for Serum Cholesterol

| | Group | | | |
|---|---|---|---|---|
| No. of runs | Vehicle | Neg. control | Std. A | Std. B |
| Mean log | 2.724 | 2.720 | 2.458 | 2.464 |
| Antilog | 530 | 525 | 287 | 291 |
| SD (within) | 0.118 | 0.113 | 0.070 | 0.093 |
| SD (between) | 0.055 | 0.069 | 0.033 | 0.046 |

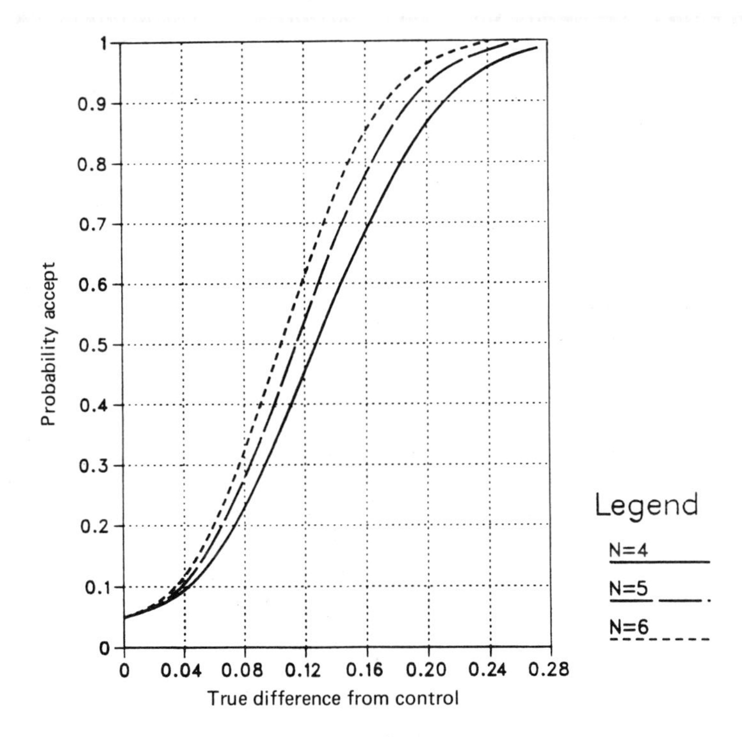


Fig. 6 OC curve for groups of four, five, and six.

The within-run standard deviation was used to construct OC curves for the screen. Figure 6 gives curves for group sizes of four, five, and six animals. In each case, the control group size was set four times larger than that of a test compound. The activity scale for the curves is expressed as the difference between the mean logarithms of control and test populations.

The observed difference between the mean logarithms of cholesterol values for vehicle and standard A was 0.266. Figure 6 shows that the probability of detecting a difference this large is greater than 0.99 with both five and six animals per group and the probability is 0.985 with four rats per group. It was therefore decided to use four rats per group for routine screening. As before, each run consisted of 30 groups. Four groups served as vehicle controls, one group was treated with positive standard, and one group received negative standard. Test drugs were evaluated in the other 24 groups. Thus, including the experimental controls and standards, compounds were classified with five animals each.

Procedures for calculating OC curves and test regions for certain types of two- and three-stage screening procedures were described by Roseberry and Gehan (1964). They considered the case where the variable of interest, x, is normally distributed with unknown mean $\mu$ and known run-to-run standard deviation $\sigma$. The test statistic for these plans, $\Sigma x_i$, is compared with the critical value, $c_i$, to determine if testing terminates or continues. When $\Sigma x_i > c_i$ testing terminates and the compound is rejected. If $\Sigma x_i \leq c_i$, testing continues to the next stage or, if in the last stage, the compound is accepted. As in the methods considered before, a compound may be rejected at any stage but is accepted only at the last stage.

Formulas for calculating $c_i$ for two- and three-stage screens were given in terms of $\gamma_0$, $\mu_0$, and $\sigma$, where $\gamma_0$ is the desired probability of accepting a value of $\mu_0$. Equations for obtaining $c_i$ in a two-stage screen with $\gamma_0 = .95$ are

$$c_1 = \mu_0 + 1.96\sigma$$

$$c_2 = 2\mu_0 + 2.55\sigma$$

In most cases, $\sigma$ will be of the form

$$\sqrt{\sigma_b^2 + \sigma_w^2/n}$$

where $\sigma_b^2$ is the between-stage component of variance and $\sigma_w^2$ is the within-stage component.

Roseberry and Gehan provide tables which can be used to calculate OC curves. Values in the tables are derived from the bivariate and trivariate normal distributions for the two-stage and three-stage plans, respectively. Values from the tables for two-stage plans with $\gamma_0 = .95$ are

| Prob (accept) | 0.99 | 0.95 | 0.90 | 0.70 | 0.50 | 0.30 | 0.10 | 0.05 | 0.01 |
|---|---|---|---|---|---|---|---|---|---|
| $Z_\gamma$ | 2.50 | 1.96 | 1.69 | 1.13 | 0.73 | 0.36 | −0.18 | −0.44 | −0.93 |

The value of $\mu$ corresponding to a given probability of acceptance is obtained with the relation

$$\mu = c_i - Z_\gamma \sigma$$

For a two-stage screen, the ASN can be calculated from

$$\text{ASN}(\mu) = n_1 + n_2 \text{ (prob pass stage 1)}$$

$$= n_1 + n_2 \Phi\left(\frac{c_1 - \mu}{\sigma}\right)$$

where $\Phi$ is the standard univariate normal distribution function and $n_1$ and $n_2$ are the numbers of units examined in stages 1 and 2, respectively.

The case of particular interest occurs when $x_i = \log T/C$, where T and C are average responses of treated and control experimental units, respectively. The log transformation requires corresponding modifications in calculation of the critical values and the test statistic. This will be illustrated in the following example.

Preliminary studies evaluating modifications in the antidiabetic test system were described earlier. For each of the five standards, T/C ratios were obtained by dividing the mean blood sugar value from the four treated animals by the mean value observed in the eight control animals. The averages of these ratios over the nine runs and their standard deviations are given in Table 4. With the exception of the negative standard, the

**Table 4** Mean T/C[a] Ratios and Run-to-Run Standard Deviations Obtained for the Standard Compounds

| Compound | T/C | SD |
|---|---|---|
| Neg. standard | 1.01 | 0.0674 |
| Standard A (25) | 0.75 | 0.0778 |
| Standard A (50) | 0.63 | 0.0550 |
| Standard B | 0.96 | 0.0825 |
| Standard C | 0.86 | 0.0787 |

[a]T/C, treated divided by control.

standard deviations tended to increase with the mean ratio. Analysis of log T/C values resulted in an overall run-to-run standard deviation of 3.7% and the mean ratio for the low dose of standard A was 0.75.

Using standard A at 25 mg/kg as the level of activity the screen should detect, the above data were used to establish parameters for a two-stage screen. For a particular test compound, the ratio of the mean blood sugar values of the eight control and four treated animals for the first run is denoted by $(T/C)_1$ and that for the second run by $(T/C)_2$. Critical values for the screen were obtained as follows:

$$c_1 = \log 0.75 + 1.96(0.037) = -0.0524$$

$$c_2 = 2 \log 0.75 + 2.55(0.037) = -0.1555$$

Taking antilogs of $c_i$, the critical values are

Stage 1—reject if $(T/C)_1 > 0.89$; otherwise continue
Stage 2—reject if $(T/C)_1(T/C)_2 > 0.70$; otherwise accept

Thus, if the T/C ratio for a particular compound is greater than 0.89, it is declared inactive and receives no further testing. This would be the case for the great majority of compounds submitted to the screen. When the ratio is 0.89 or less, the compound is then retested in another run with four more animals. After the second stage, the product $(T/C)_1(T/C)_2$ is calculated and the compound is declared either active or inactive. The OC curve in Figure 7 shows that the probability of a drug with a true T/C of 0.75 being accepted is 0.95, while that of another drug with a true T/C of 0.90 is approximately 0.10. Referring to the ASN curve in Figure 8, an average of 4.3 animals will be needed to classify compounds with a true T/C of 1.0.

A multistage procedure based on ranks to test for the difference between the means of two continuous distributions was described by Hearron et al. (1984). Each stage contains m X and m Y observations. The 2m independent observations are ranked within each stage and the sum of the ranks, S, for the Y values is obtained. At stage g the test statistic

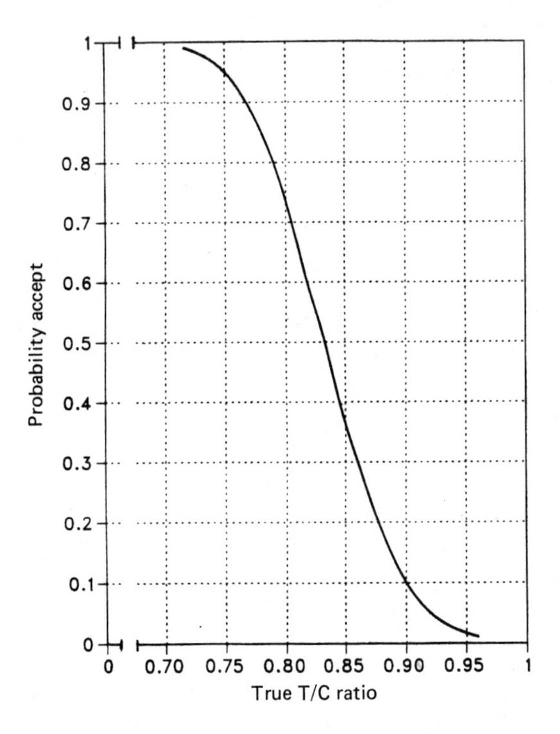


**Fig. 7** OC curve for two-stage antidiabetic screen.

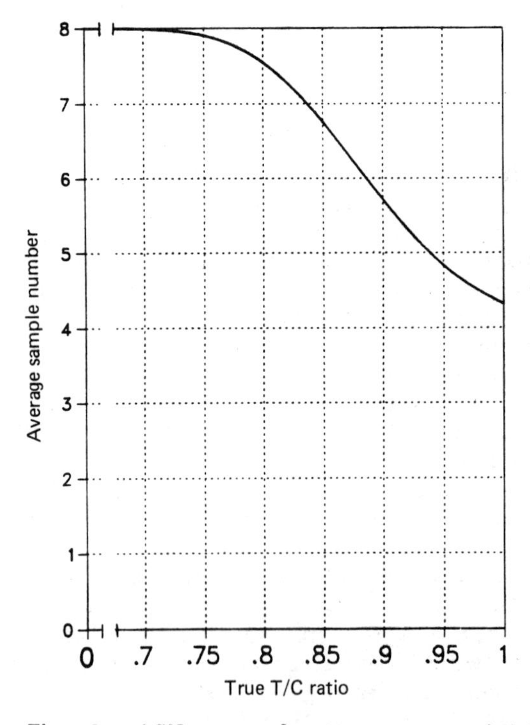


**Fig. 8** ASN curve for two-stage antidiabetic screen.

$$W_g = \sum_{i=1}^{g} S_i$$

is compared with the boundary values of a previously specified test region to determine if testing terminates or continues into the next stage. The case in which the random variables X and Y are normally and independently distributed with respective means $\mu_X$ and $\mu_Y$, common variance $\sigma_2$, and distribution functions F and G was treated by Milton (1970). The hypotheses can then be written

$$H_0: G(\mu) \equiv F(\mu) \qquad \text{and} \qquad H_1: G(\mu) \equiv F(\mu - d)$$

where $d = (\mu_Y - \mu_X)/\sigma \geqq 0$.

For this situation the probability of a given configuration of ranks for the Y sample is obtained by considering the order statistics $(U_1, \ldots, U_{2m})$, $U_1 < \cdots < U_{2m}$, of the random variables $(X_1, \ldots, X_m, Y_1, \ldots, Y_m)$. A random vector of zeros and ones, $\underline{Z} = (Z_1, \ldots, Z_{2m})$, is formed by letting the $Z_i = 0$ (or 1) if $U_i$ is an X (or Y). A fixed vector of zeros and ones, $\underline{z} = (z_1, \ldots, z_{2m})$, then specifies a rank configuration $(s_1, \ldots, s_m)$ of the Y sample. If the normal density with mean $\theta$ and variance 1 is denoted by $f(x - \theta)$, the probability of rank order $\underline{z}$ is

$$P(z; m, d) = P(s_1, \ldots, s_m; m, d) = (m1)^2 \int_R \cdots \int \prod_{i=1}^{2m} f(t_i - Z_i d)\, dt_i$$

where R is the region $-\infty < t_1 \leqq t_2 < \cdots < t_{2m} < \infty$.

A numerical evaluation procedure was described and values of $P(z; m, d)$ were given for $d = 0(0.2)$, 1.0, 1.5, 2.0, and 3.0. The probability of a particular rank sum S for m elements at a specified value of d is then obtained by adding probabilities, $P(z; m, d)$, for all $\underline{z}$ which yield S. This probability is denoted by $P(S; m, d)$ and S extends from $S_L = m(m + 1)/2$ to $S_U = m(3m + 1)/2$.

For a given stage g, the probability that the test statistic $W_g = w$ is computed by the recursive relation

$$C_g(w; m, \theta) = \sum_{i=A}^{B} C_{g-1}(i; m, \theta)\, P(w - i; m, \theta)$$

Thus, the probability that $W_g = w$ at the end of the gth stage is the probability that $W_{g-1} = i$ multiplied by the probability that a rank sum of $w - i$ is observed at the gth stage, summed over all eligible values of i and $w - i$. These eligible values, determined by the accept and reject points, are

$$A = \max(a_{g-1} + 1, w - S_U)$$

$$B = \min(r_{g-1} - 1, w - S_L)$$

The probability of accepting $H_0$ at stage g is

$$L_g(m, \theta) = \sum_{w=a_{g-1}+S_L+1}^{a_g} C_g(w; m, \theta)$$

Similarly, the probability of rejecting $H_0$ for $H_1$ at stage g is

$$R_g(m, \theta) = \sum_{w=r_g}^{r_{g-1}+S_U-1} C_g(w; m, \theta)$$

The total probability of accepting and rejecting is obtained by summing L and R over all N stages. The average number of Y observations is

$$ASN = \sum_{g=1}^{N} gm[L_g(m, \theta) + R_g(m, \theta)]$$

This approach was used in a test for antihypertensive preparations in rats. The test was designed specifically to evaluate a set of preparations which were expected to contain a higher proportion of actives. The screen was therefore targeted toward a very narrow class of agents. Activity in the animal model was judged by lowering of blood pressure. A series of seven indirect pressure readings was taken on day 1 just prior to treatment and a second series was obtained on day 3. For each rat, the median of each series was found and the difference between these two values was obtained. The response of the animal was therefore taken as the difference between the two median values. From previous studies, the standard deviation of this measurement was 23. It was determined that compounds would be of interest if the change in pressure in treated rats was at least one standard deviation less than the change in control rats. A one-sided test was appropriate because only preparations which lower pressure were of interest. A three-stage plan was designed in which groups of six treated rats were each compared with a group of six control rats. The test region is given below.

Decision Points of the Three-Stage Sequential Plan

| Stage | Decision points | |
|---|---|---|
| 1 | 37 | 51 |
| 2 | 81 | 95 |
| 3 | 132 | 133 |

For each rat the blood pressure change is obtained and the values are arranged in rank order from low to high. The sum of the ranks for the treated animals is then compared with the appropriate decision points. At the end of the first stage, the agent is declared active if the sum of the ranks for the treated group is 31 or less and it is declared inactive if the sum is 51 or more. Testing terminates in both cases; the agent is tested in the second stage if the sum is between 31 and 51. The compound may also be classified as active or inactive in the second stage. When the sum of the ranks for the first stage plus the sum of the ranks for the second stage is between 81 and 95, testing continues into the third stage. Note that the plan is closed at the third stage. Agents tested in that stage are declared as either active or inactive.

The OC and ASN curves for this plan are given in Figures 9 and 10, respectively. Figure 9 shows that when the true decrease in blood pressure is zero, the probability of declaring the preparation active is 0.10. When the decrease is 23, the probability of declaring the agent active is 0.91. In this case, a higher $\alpha$ level was selected because there was no particular concern about a few extra false positive results. Figure 10 shows that ASN is 10.9 in each group when the change is 0, reaches 13 when the change is 11.3, and then decreases with increasing activity. Using a single-stage t-test for comparison, 14 animals in each group would be needed to obtain approximately the same levels of $\alpha$ and $\beta$.

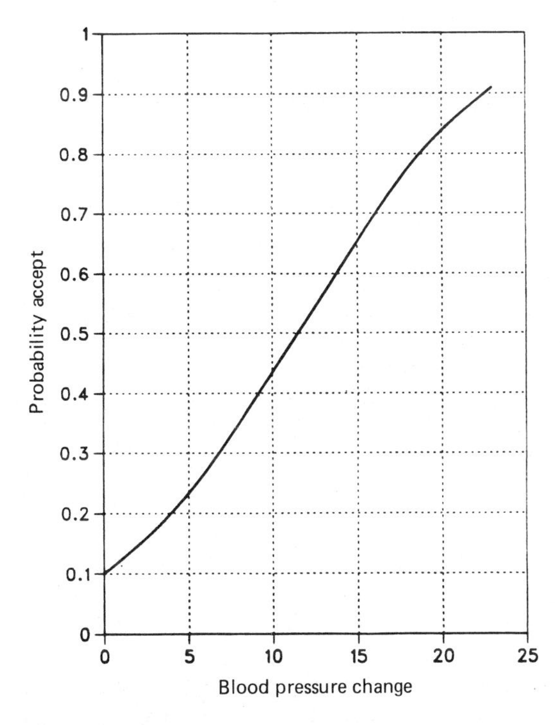


**Fig. 9** OC curve for three-stage blood pressure test.

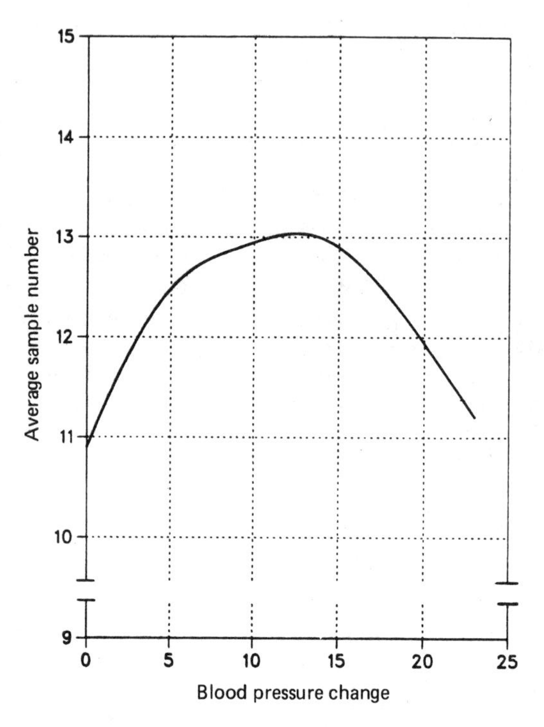


Fig. 10 ASN curve for three-stage blood pressure test.

## E. Quality Control

All of the above screens are designed to operate with certain risks of misclassifying test compounds. These risks are balanced on the basis of location parameters which are specified according to responses obtained with positive and negative standards. It is therefore essential to monitor the stability of the system over time. A location shift in one direction would result in an excess of false positives, while a shift in the other direction would increase the false negative rate. The shift could take place gradually or might occur abruptly with any screening run.

All biological test systems are susceptible to changes in response levels. Pregnancy rates in the antifertility screen are influenced by the male hamsters in the colony. Since some males are replaced on an ongoing basis, the proportion of successful matings could change over time. Any changes in the viral preparation used to infect mice in the antiviral screen would influence baseline death rates. Responses in all the test systems are influenced by the health of the animals and the conditions of their environment. It seems clear that some effort must be devoted to assessing performance of the screens. Depending on the type of screen, up to 20% of the total effort for this purpose is easily justified.

The recommended approach in monitoring stability of the test system is to include both positive and negative standards in each screening run. Gradual changes in the system can be detected through the use of control charts. If it becomes apparent that responses to either the positive or negative standards are shifting in one direction or the other, attempts to determine the cause should be made. In the event that no cause can be identified or the system cannot be restored to the original response levels, it will be necessary to reparameterize the screen. It should be emphasized, however, that this is done only after it is clear that the system can still effectively discriminate between active and inactive compounds.

The standards within each run are also used to judge the performance of the system on that particular occasion. The positive standards should be declared active according to the decision rules for the screen. In the case of multistage systems, first-stage criteria would be used. Failure to detect one or more of the positive standards is immediate cause for concern. Investigation of possible reasons should be initiated and the control charts for the system reviewed. If no reason for a change in the screen's performance is found, consideration should be given to repeating the entire run. The number of standards missed, the difference between the observed response and the critical value, and the false negative level specified for the screen will all affect this decision.

The antiviral screen described earlier included six groups of animals treated with a positive standard and six groups treated with vehicle only. The values of p, the probability of death, obtained from the preliminary tests were used to construct p charts, which were maintained in the laboratory. For each run, the number of deaths in the 60 control animals and 60 animals treated with positive standard was recorded on these charts.

The performance of the screen was also evaluated by considering each of the 12 groups as a "first-stage" test. Recall that a compound is rejected in the first stage when six or more animals die. Of the 222 vehicle-treated groups, 218 had at least six dead animals. None of the 222 groups treated with positive standard had more than five deaths. In the preliminary tests, only 6% of animals treated with the positive standard died. The OC curve for the screen shows that the probability of detecting this level of activity is $>0.99$. With this high probability, observing a positive standard group with six or more deaths would be reason to declare the run invalid unless a cause for the anomaly was apparent.

Four control charts were constructed to monitor performance of the antidiabetic screen. Both the mean and range of the blood sugar values observed in the eight control animals are recorded on a chart for each run. In addition, a chart for the mean response observed in animals treated with 25 mg/kg standard A is maintained. The fourth chart records the T/C ratio for standard A and control. A screening run is considered out of control if the T/C ratio falls outside the $2\sigma$ limits. The upper limit is 0.89, which is also the critical value for stage 1. If an obvious reason for a high T/C ratio for standard A is not found, the entire run is repeated.

When a screen first goes into operation, screening parameters and control charts are usually determined from limited information. In most cases, data from 10 runs which include both positive and negative standards will provide reasonably stable estimates. These values should be updated after the screen has been in routine operation for 3 or 4 months. We recommend that the responses to the standards be analyzed every 6 months thereafter. These data, along with information on the control charts, are then used to determine if changes in screen parameters and/or control chart limits should be made.

## III. TARGETED SCREENING

The test systems discussed in the last section provide results in terms of death, blood pressure, serum cholesterol, or blood glucose. This is typical of general screens in intact animal models. The activity of test compounds is determined on the basis of vital signs, behavior, or blood chemistry measurements. These outcomes are the culmination of complex and interrelated pharmacologic actions within the animal. These events, along with differences in metabolic pathways of the test compounds, can lead to misinterpretation of an agent's potential utility. As a result, test procedures which focus on more specific components of the system are of particular interest. Screens which are directed toward a particular enzyme, receptor, or agonist are defined as targeted systems. Typically, these systems are based on radiolabeling methods; therefore we will focus on statistical aspects of this technology.

Since Berson and Yalow (1954) developed the use of radioimmunoassay for the study of $^{131}$I-labeled insulin in diabetes and Ekins (1960) used the radioligand assay for the measurement of thyroxine in human plasma, radiolabel techniques have become very important in the area of lead discovery. Lefkowitz et al. (1984) discuss the role that radioligand assays played in confirming the existence of two major classes of receptors—the $\alpha$-adrenergic and $\beta$-adrenergic receptors. These receptors are responsible for the actions of important therapeutic agents, the catecholamines. Besides these two classes of receptors, Williams and U'Prichard (1984) review other major receptors which have been established through radioligand techniques. As they point out, many of these receptors are studied in brain tissues because of the abundance of the receptors and the ease with which brain tissue can be prepared for the assay. Using a rational approach, investigators can now focus their attention on a particular physiological or enzymatic pathway which they hope will increase the probability of lead discovery for a particular therapeutic indication. We will approach this topic from the viewpoint of receptor binding studies; however, bear in mind that many of the analysis techniques can be applied to enzyme studies with differences in interpretation of results.

Reactants which combine with a receptor, both antagonists and agonists, are called ligands. Compounds or drugs that react with the receptor and produce some desired biological response, such as muscle contraction, are called agonists. Conversely, drugs that react with the receptor but do not produce the desired response are called antagonists. A common method for studying ligand-receptor binding reactions is by radioactively labeling the ligand and observing the degree of binding that occurs at equilibrium. These systems are often dynamic in nature because the reactants may bind reversibly. That is, they bind and release the receptor at certain association and dissociation rates. The reactions are allowed to reach equilibrium, at which time the ratio of the forward and backward rates becomes constant. These types of studies can be used to assess the degree of activity of a compound at the receptor and, consequently, infer the degree of biological response. In addition, these studies can be used to characterize the receptor itself. For example, the abundance or density of receptor in different tissue sites may be of interest. The receptor may consist of a single class of sites, each capable of reacting identically with the ligand, or the receptor may have multiple subsites. These subsites, as in the opiate receptor, may produce different responses and react differently with the ligand. In addition to identifying compounds which react with the receptor, it would be useful to be able to identify compounds which react with specific subsites.

Radioimmunoassays (RIAs) are used in both drug discovery and drug development. Basically, the system involves maintaining a constant level of radioactively labeled ligand, adding various increasing concentrations of unlabeled ligand, and then measuring the amount of radioactive binding that occurs in a constant amount of protein or tissue homogenate. The resulting dose-response curve is then used for interpolating the concentration of the unlabeled ligand or potency of unknown samples.

Radioligand assays cover a broad range in which radioimmunoassay is a particular type, often referred to as an inhibition, competition, or displacement study. In most cases, the unlabeled ligand is replaced by some other competitive agent. The competition assay is then used to determine the affinity of this agent for the receptor. Another type of radioligand assay is the saturation study. A saturation study is usually performed with the intent of investigating the receptor system itself and the interaction with the control ligand. As in the radioimmunoassay, both types of radioligand assays involve a radioactively labeled ligand and a constant amount of tissue or protein homogenate which contains the receptors of interest.

With the increase in sophisticated computerized software and graphics, radioligand assays are becoming more informative. Autoradiography allows the investigator to perform radioligand assays on whole tissue slices rather than homogenates. Thus, not only is the amount of binding quantified, but also the location of binding is observable. We will not discuss autoradiography here, but the following references may be useful: Altar et al., (1984), Gallistel and Tretiak (1985), and Ramm and Kulick (1985).

## A. Radioligand Assays

All radioligand assays utilize a radioactively labeled ligand and a tissue or protein homogenate. The homogenate contains the receptor of interest and is placed in equal amounts into several tubes or vials. Basically, there are two types of studies from an analysis viewpoint, although they may overlap in interpretation. An inhibition study looks at the effect of another reagent on the radioactively labeled ligand. This type of study is almost always used for estimating the affinity of the reagent for the receptor. An inhibitor may interact with the ligand and receptor in several ways, the most common of which is the classical competitive inhibition. In this situation, the inhibitor actually competes with the ligand for the same receptor site. Thus, it would require greater concentrations of ligand in the presence of the inhibitor to achieve the same amount of binding as in its absence.

The second type of study is a saturation study. In this type of study, the binding is measured for increasing concentrations of radioactively labeled ligand. Ideally, high enough concentrations of ligand are used to saturate the system. From this type of study, we can characterize the receptor system, i.e., find the density of receptor sites in the tissue or protein and also determine the affinity of the ligand for the receptor. The two types of studies may overlap when a constant concentration of another reagent is added to every tube or vial in a saturation study. This may affect the apparent affinity of the ligand or the resulting density estimates. By doing this, the affinity of the reagent for the receptor can be observed.

1. Saturation Studies

As mentioned previously, in a saturation study, increasing concentrations of the radioactively labeled ligand are added to the homogenates. The reactions are allowed to reach equilibrium. The binding of ligand with receptor will result in the formation of a pellet in the tube. This pellet will be separated from the unbound homogenate, washed, and resuspended in another tube. This second tube is then passed through a counter (scintillation, gamma, etc.) which quantifies the amount of receptor bound to radioactively labeled ligand.

The reaction

$$[L] + [R] \underset{k_1}{\overset{k_2}{\rightleftarrows}} [LR]$$

is modeled using the binding isotherm derived by Langmuir (1918)

$$[LR] = \frac{[L]B_{max}}{[L] + K_d}$$

where [LR] and [L] are the concentrations of receptor-bound and free (or unbound ligand, respectively. For ease of presentation, the brackets will be removed from subsequent formulas. $B_{max}$ is the binding parameter, which represents the total receptor density or binding capacity of the tissue or protein. $K_d$ is another binding parameter, which represents the equilibrium dissociation constant of the ligand, the ratio of the backward and forward rates ($k_2/k_1$). This is a measure of the affinity of the ligand for the receptor. The smaller the $K_d$, the more likely the ligand is to bind with the receptor.

In all assays, a certain amount of background or nonspecific binding occurs. This nonspecific binding can be measured by adding a very large concentration of unlabeled ligand to some tubes that contain the radioactively labeled ligand and homogenate. The unlabeled ligand competes with the labeled ligand for the receptor and, because of the very high concentration, essentially overpowers it. The only binding left for the labeled ligand is nonspecific binding. The linear analyses which will be discussed all assume that the data consist of specific binding; that is, the data have been corrected for nonspecific binding. This is done by subtracting the corresponding nonspecific binding from the observed binding. In nonlinear modeling analysis, a nonspecific binding term can actually be added to the model. For example, in the one-site saturation study, the observed total binding would be fit to the following model:

$$LR = \frac{LB_{max}}{L + K_d} + LC$$

and the nonspecific binding tubes would be simultaneously fit to only the nonspecific term

$$LR = LC$$

The nonspecific binding term shows a constant, C, multiplied by the unbound ligand concentration. Thus this term will increase as the concentration of ligand increases.

Prior to the widespread use of the computer, transformations were applied to the Langmuir isotherm to linearize it for analysis. The most common transformations are the Scatchard (1949) and double-reciprocal or Lineweaver-Burke (1934). In a Scatchard analysis, the bound/free ratio is regressed on the bound response. In a double-reciprocal analysis, the 1/bound data are regressed on 1/free. The following linearizations are obtained:

$$\text{Scatchard:} \quad \frac{LR}{L} = \frac{B_{max}}{K_d} - \frac{1}{K_d} LR$$

$$\text{Double-reciprocal:} \quad \frac{1}{LR} = \frac{1}{B_{max}} + \frac{K_d}{B_{max}}\left(\frac{1}{L}\right)$$

The binding parameters, $B_{max}$ and $K_d$, are then estimated from the slope and intercept estimates. A Taylor series approximation can be used to obtain approximate standard errors for these estimates. For example, for Scatchard analysis:

$$\frac{LR}{L} = a + bLR$$

$$B_{max} = \frac{-a}{b}$$

$$se(B_{max}) = \frac{\sqrt{b^2 \, var(a) + a^2 \, var(b) - 2ab \, cov(a, b)}}{b^2}$$

$$K_d = \frac{-1}{b}$$

$$se(K_d) = \frac{se(b)}{b^2}$$

Several papers have been published which compare the various transformations (i.e., Atkins and Nimmo, 1975; Cornish-Bowden and Eisenthal, 1974; Dowd and Riggs, 1965). Figures 11 and 12 show a typical dose-response curve and Scatchard plot for a saturation assay.

There are many problems associated with the linear analysis of the binding isotherm. Some of these, like assuming that the nonspecific binding is known without error, can be circumvented by performing a nonlinear analysis directly on the untransformed data. Most statistical software packages have a nonlinear analysis procedure available. Almost all nonlinear analysis routines require initial parameter estimates. This is not difficult for a saturation study. From the dose-response curve, as shown in Figure 11, $B_{max}$ can be approximated by the upper asymptote and $K_d$ by the concentration of ligand required to achieve 50% of $B_{max}$. A Scatchard analysis could also be performed as a preliminary step.

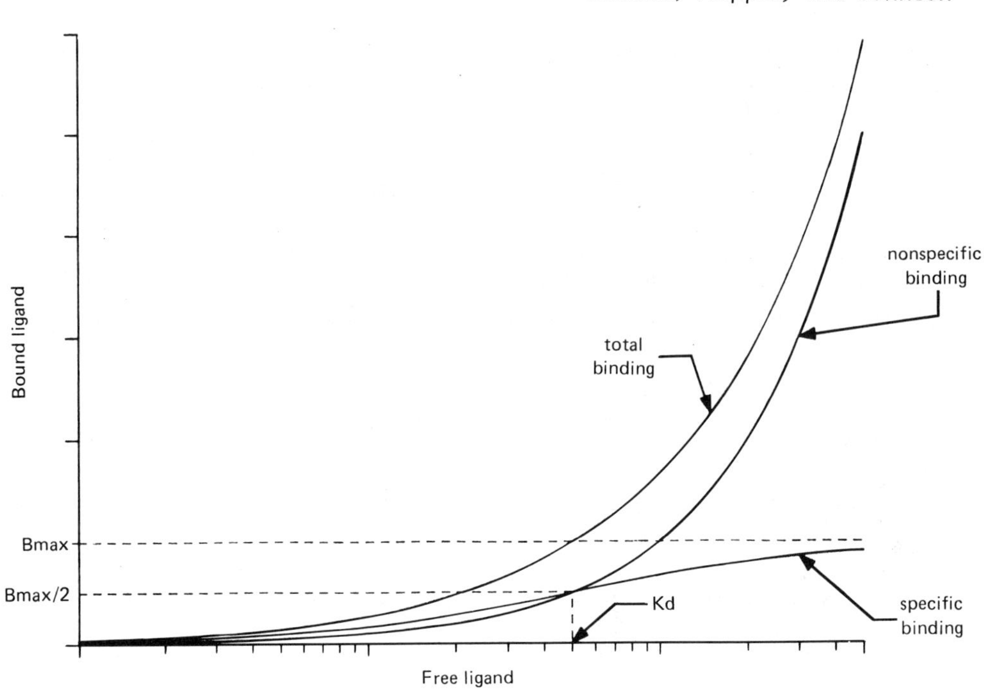


**Fig. 11** Typical dose-response curve from a saturation assay.

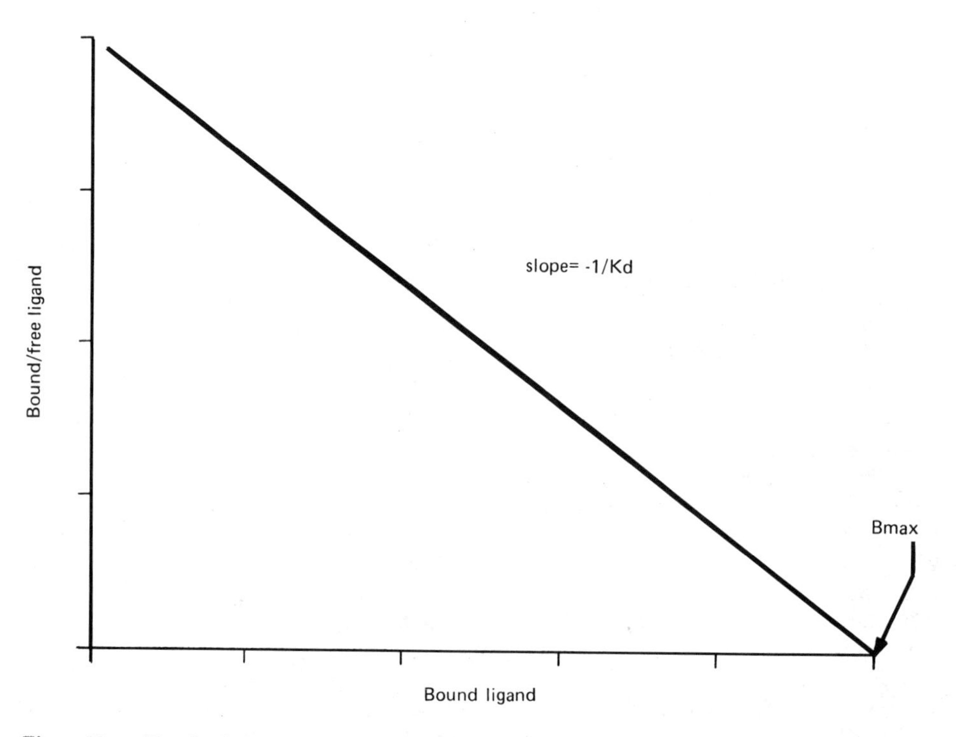


**Fig. 12** Typical Scatchard plot from a saturation assay.

In all radioligand assays, the responses measured are the number of particles emitted by a radioactive substance into a certain locale during a known interval of time. These emissions are Poisson-like in that the variability of the response is proportional to some function of the mean of the response. This property has been observed and discussed many times (e.g., Rodbard et al., 1976; Parker and Waud, 1971). In most radioligand assays, this relationship lies somewhere between the variance being proportional to the mean and being proportional to the square of the mean. This relationship violates the homogeneity assumption made in linear and nonlinear analyses. The recommended approach is to use a weighted analysis. For the linear analyses, Wilkinson (1961) proposed using a Taylor series expansion to define the weights

$$w = \frac{1}{(dy/dLR)^2}$$

For example, in a double-reciprocal analysis

$$y = \frac{1}{LR} \qquad \frac{dy}{dLR} = \frac{-1}{LR^2}, \qquad \text{and} \qquad w = LR^4$$

Initially, the weights are based on the observed values. Improved estimates of the weights can then be obtained in an iterative process by using the predicted values of the observations.

In a nonlinear analysis of data in which the variance is proportional to the mean, weights should be assigned as the reciprocal of the observation. Since most radioligand assays are run with replicate tubes at each dose level, Finney and Phillips (1977) recommended determining the relationship between the mean and variance by fitting the model

$$s^2 = a_0 y^{-a_1}$$

where y and $s^2$ are the mean and variance, respectively, of the replicates. Other variance models have also been proposed and evaluated by Rodbard et al. (1976).

As mentioned earlier, saturation studies are also sometimes used to study inhibitory effects of other reagents. These studies actually result in two dose-response curves, one of the ligand binding alone and one in which the ligand binds in the presence of a constant amount of the other reagent. The displacement in the two curves is an indication of the inhibitory effects of the reagent, i.e., the affinity of the reagent for the receptor. Depending on the type of inhibition involved, the reagent may alter the estimate of $B_{max}$ or $K_d$ or both. Figure 13 shows the effect of a competitive inhibitor on ligand binding both in the raw specific binding and in terms of a Scatchard plot.

### 2. Example

Table 5 shows the data obtained from a saturation assay of an opioid, compound A, binding to the opiate receptor in guinea pig brain. Although the concentration levels were assayed in triplicate, only the replicate means are

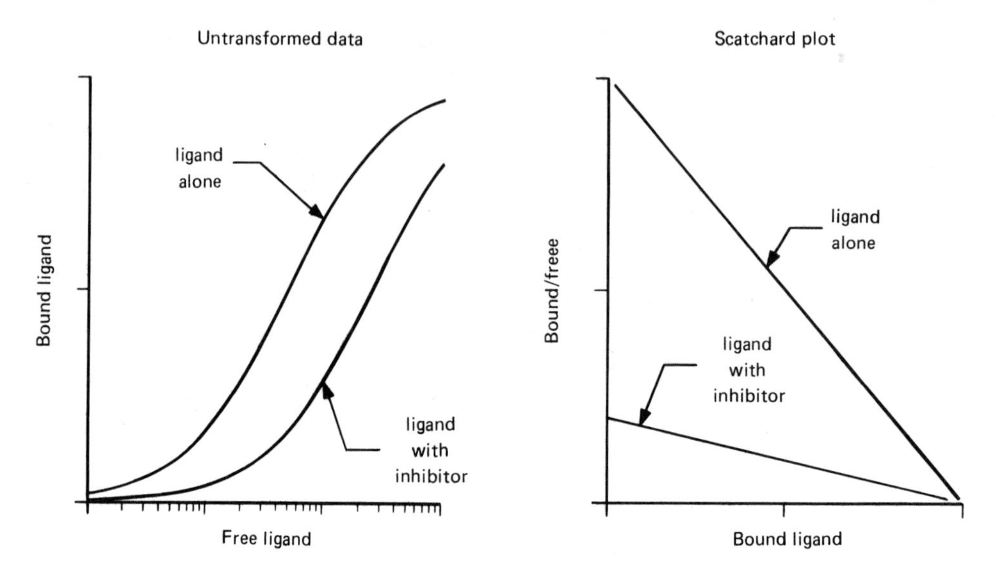


**Fig. 13** Effect of a competitive inhibitor on ligand specific binding.

**Table 5** Saturation Assay of Compound A Binding in Guinea Pig Brain Tissue (Replicate Means Shown)

| Total added (nM) | Free (nM) | Total bound (pmol/g tissue) | Nonspecific (pmol/g tissue) | Specific (pmol/g tissue) | Bound/ free |
|---|---|---|---|---|---|
| 11.9230 | 11.7507 | 18.9519 | 3.8854 | 15.0665 | 1.2822 |
| 9.6621 | 9.5113 | 16.5846 | 3.4215 | 13.1631 | 1.3839 |
| 8.3681 | 8.2331 | 14.8488 | 2.7114 | 12.1374 | 1.4742 |
| 7.3033 | 7.1853 | 12.9787 | 2.4905 | 10.4882 | 1.4597 |
| 6.0007 | 5.8981 | 11.2935 | 1.9771 | 9.3165 | 1.5796 |
| 4.7731 | 4.6834 | 9.8641 | 1.5242 | 8.3398 | 1.7807 |
| 3.6590 | 3.5870 | 7.9248 | 1.2331 | 6.6917 | 1.8656 |
| 2.4174 | 2.3641 | 5.8682 | 0.7820 | 5.0862 | 2.1515 |
| 1.2738 | 1.2417 | 3.5283 | 0.4754 | 3.0529 | 2.4586 |
| 0.9307 | 0.9072 | 2.5809 | 0.3606 | 2.2203 | 2.4474 |
| 0.7301 | 0.7096 | 2.2577 | 0.3251 | 1.9326 | 2.7236 |
| 0.4808 | 0.4670 | 1.5222 | 0.2294 | 1.2928 | 2.7683 |
| 0.2515 | 0.2438 | 0.8519 | 0.1267 | 0.7251 | 2.9746 |

shown. The variance of the replicates was roughly proportional to the mean of the replicates, so the weights in the nonlinear analysis were defined to be the reciprocal of the predicted value. The Scatchard analysis using PROC GLM in SAS resulted in the following estimates:

$$a = 2.2998 \pm 0.0949$$

$$b = -0.0693 \pm 0.0075$$

$$\mathrm{Cov}(a, b) = -0.0007$$

$$B_{max} = 33.19 \pm 2.279 \text{ pmol/g tissue}$$

$$K_d = 14.43 \pm 1.565 \text{ nM}$$

A weighted nonlinear analysis using NONLIN84 yielded these estimates:

$$B_{max} = 37.80 \pm 1.853 \text{ pmol/g tissue}$$

$$K_d = 15.74 \pm 1.299 \text{ nM}$$

$$C = 0.343 \pm 0.008 \text{ ml/g tissue}$$

The results agree fairly well between the two analyses. Figures 14 and 15 show the dose-response curve and Scatchard plot for these data. It is much more apparent in the Scatchard plot how the weighting affects the fit of the data. Note that the lower portion of the curve is strongly emphasized, whereas the upper portion is not.

### 3. Competitive Inhibition

In a typical competitive assay, one set of tubes contains the radioactively labeled ligand binding to the receptor homogenate alone. The reamining tubes have increasing concentrations of another reagent added to them. The other reagent will actively compete with the ligand for the receptor sites. Since the concentration of ligand remains constant, the amount of ligand bound to receptor will decrease as the concentration of the competing reagent increases. The following reactions are taking place:

$$\begin{array}{ccc} R + L & \underset{}{\overset{K_d}{\rightleftharpoons}} & LR \\ + & & \\ I & & \\ K_i \updownarrow & & \\ IR & & \end{array}$$

The concentrations of ligand and inhibitor are denoted by L and I and their equilibrium dissociation constants are $K_d$ and $K_i$, respectively. The modified binding isotherm for modeling this reaction is

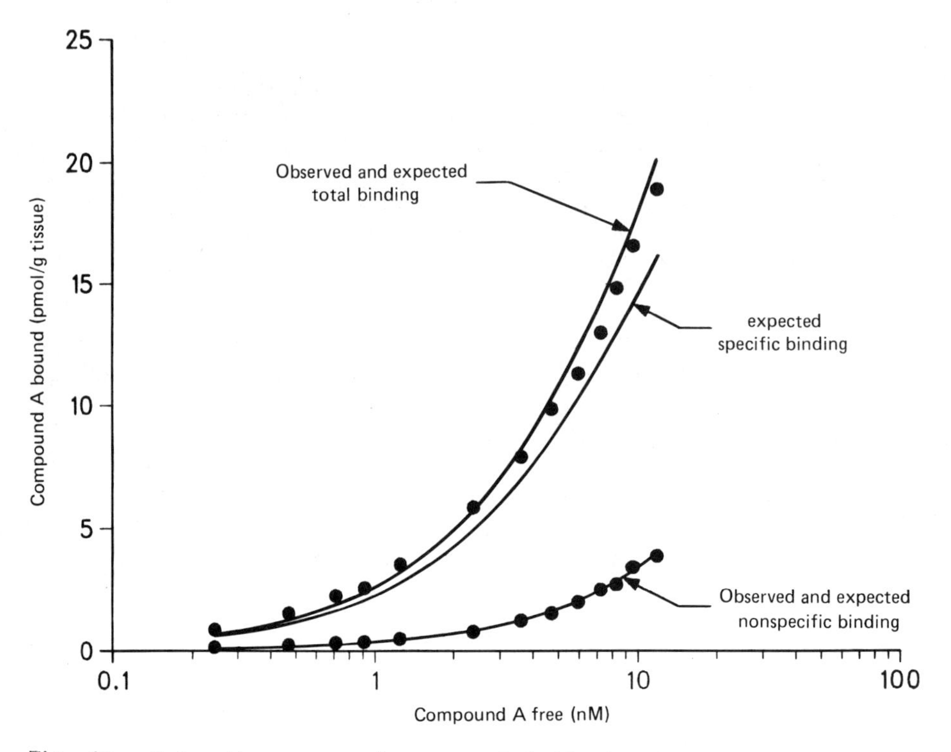


**Fig. 14** Saturation assay of compound A binding in guinea pig brain tissue.

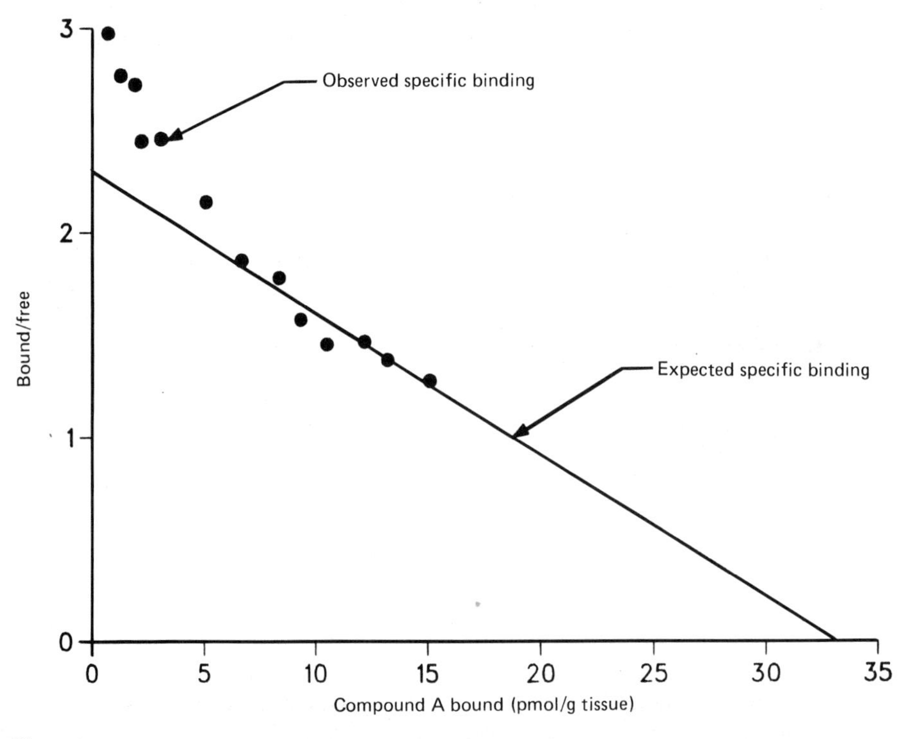


**Fig. 15** Scatchard plot of compound A binding in guinea pig brain tissue.

$$LR = \frac{LB_{max}}{K_d + L + IK_d/K_i} + LC$$

This model can be fit directly to the binding data by using a nonlinear analysis. However, several problems arise due to the design of the assay. Since a constant concentration of ligand is used throughout the study, it is difficult to estimate the ligand dissociation constant with any preciseness. An alternative is to reparameterize the model in terms of the control binding and inhibitor $IC_{50}$. This reparameterization is discussed further in the section on radioimmunassays.

Often the data are transformed to percent of control (ligand binding without reagent) and then a probit or logit analysis is used. The concentration of inhibitor which reduces the ligand binding by 50%, called the $IC_{50}$ or $ED_{50}$, is taken as an indicator of the affinity of the reagent for the receptor. If the investigator has some knowledge of the $K_d$ of the ligand, the Cheng–Prusoff (1973) relationship can be used to convert the $IC_{50}$ to $K_i$, the equilibrium dissociation constant of the inhibitor:

$$K_i = \frac{IC_{50}}{1 + L/K_d}$$

The probit and logit transformations yield essentially the same result for the $IC_{50}$. However, the slope of the logit transformation estimates the Hill coefficient, which is used as an indicator of the existence of multiple sites or cooperativity. Little (1968) has shown that the probit transformation consistently produces larger slope estimates than the logit transformation. Figure 16 shows a typical dose-response curve from an inhibition assay.

4. Example

Table 6 shows the binding of compound A in guinea pig brain in the presence of various concentrations of another opioid, compound B. This drug competes with compound A for the receptor, and so, as the concentration of compound B increases, the amount of compound A bound to the receptor decreases. If the data are fit to the Langmuir binding isotherm directly using a weighted nonlinear analysis, the results are questionable. The estimates obtained from NONLIN84 are

$$B_{max} = 3221 \pm 4{,}172{,}000 \text{ CPM}$$

$$= 6.95 \pm 9002 \text{ pmol/g tissue}$$

$$K_d = 2.177 \pm 6057 \text{ nM}$$

$$K_i = 175.9 \pm 261{,}600 \text{ nM}$$

$$C = 207.1 \pm 15.36 \text{ CPM/nM}$$

$$= 0.447 \pm 0.033 \text{ ml/g tissue}$$

The observed and the expected fit of these data are shown in Figure 17. The variances of these estimates are not only extremely large, but also

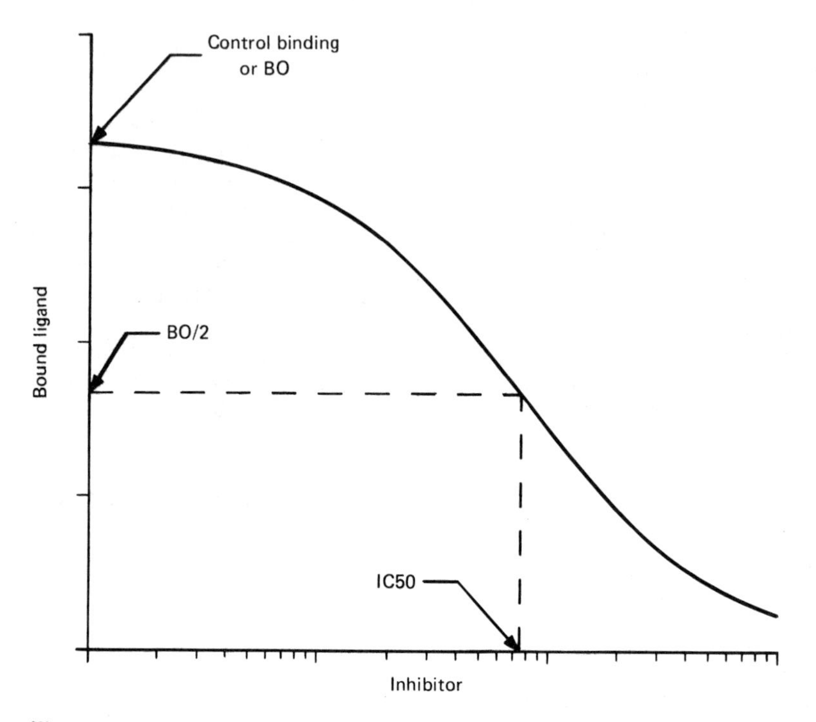


**Fig. 16** Typical dose-response curve from an inhibition assay.

quite different from the ones obtained from the saturation study. More precise estimates are obtained from the reparameterization given in the section on radioimmunoassays.

When the data are analyzed using a probit analysis, only estimates of the $IC_{50}$ and the Hill coefficient are obtained. As mentioned previously, the Hill coefficient is estimated by the slope of the probit-log curve. The following estimates were obtained for this data set:

$$\text{Hill coefficient} = 0.556 \pm 0.035$$

$$IC_{50} = 122.7 \pm 24.29 \text{ nM}$$

Using the Cheng–Prusoff relationship and the $K_d$ estimate obtained from the saturation assay, $K_i$ can be estimated as

$$K_i = 105.6 \pm 20.92 \text{ nM}$$

### 5. Multiple Sites

It has become apparent in many receptor systems that multiple classes of sites actually exist. Molinoff et al. (1981) list several receptors which are believed to consist of multiple sites. Feldman et al. (1972) proposed that these systems be modeled by assuming independence of the sites and modifying the one-site binding models to include additive terms for each site. For example, a two-site saturation assay could be modeled by

**Table 6** Compound A Binding in Guinea Pig Brain Tissue Inhibited by Compound B (Replicate Means)

| Compound B (nM) | CPM bound | Specific | Percent bound |
|---|---|---|---|
| 0.0[a] | 2494.5 | 2069.1 | 100.00 |
| 0.0[b] | 425.3 | — | — |
| 0.1 | 2416.7 | 1991.4 | 96.241 |
| 0.158 | 2705.8 | 2280.5 | 110.217 |
| 0.251 | 2402.8 | 1977.5 | 95.570 |
| 0.398 | 2543.2 | 2117.9 | 102.359 |
| 0.631 | 2228.5 | 1803.2 | 87.146 |
| 1.0 | 2216.4 | 1791.1 | 86.560 |
| 1.58 | 2137.9 | 1712.6 | 82.767 |
| 2.51 | 2125.5 | 1700.2 | 82.169 |
| 3.98 | 1867.6 | 1442.3 | 69.704 |
| 6.31 | 1926.9 | 1501.6 | 72.571 |
| 10.0 | 1732.2 | 1306.9 | 63.160 |
| 15.8 | 1769.8 | 1344.5 | 64.977 |
| 25.1 | 1775.3 | 1349.9 | 65.241 |
| 39.8 | 1781.2 | 1355.9 | 65.528 |
| 63.1 | 1665.6 | 1230.3 | 59.458 |
| 100.0 | 1738.5 | 1313.2 | 63.466 |
| 158.0 | 1528.7 | 1103.3 | 53.323 |
| 251.0 | 1488.3 | 1062.9 | 51.371 |
| 631.0 | 1208.5 | 783.2 | 37.851 |
| 1,000.0 | 1149.4 | 724.1 | 34.994 |
| 1,580.0 | 973.1 | 547.7 | 26.472 |
| 2,510.0 | 832.2 | 406.9 | 19.664 |
| 3,980.0 | 716.1 | 290.8 | 14.054 |
| 6,310.0 | 684.0 | 258.7 | 12.501 |
| 10,000.0 | 613.3 | 188.0 | 9.086 |

[a]2.5 nM compound A.
[b]Nonspecific binding.

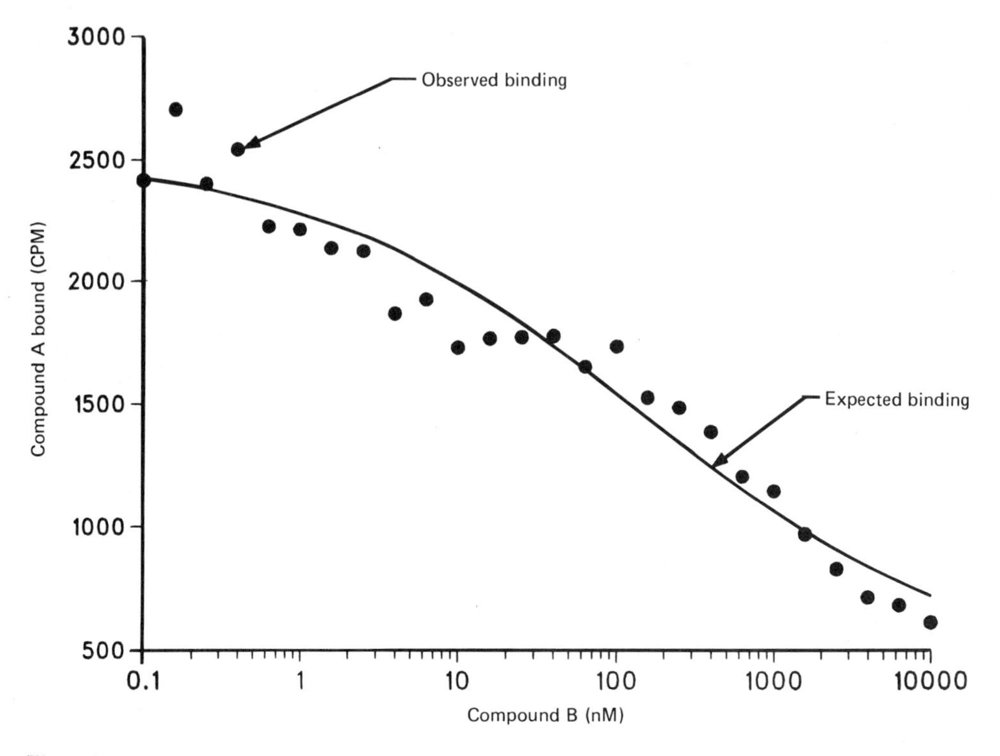


Fig. 17 Compound A binding in guinea pig brain tissue inhibited by compound B.

$$LR = \frac{LB_{max_1}}{K_{d1} + L} + \frac{LB_{max_2}}{K_{d2} + L} + LC$$

The existence of the second site is statistically confirmed by using an F-test on the reduction in the weighted residual error obtained with the full two-site model over the reduced one-site model. As given by Neter and Wasserman (1974), the test statistic which follows an F distribution is calculated by

$$F = \frac{(SSER - SSEF)dfF}{SSEF(dfR - dfF)}, \qquad df1 = (dfR - dfF), \qquad df2 = dfF$$

where SSER, dfR and SSEF, dfF are the weighted error sums of squares and degrees of freedom for the reduced (one-site) and full (two-site) models, respectively.

Although there may be biological evidence for a second site, there are a number of reasons why it may not be validated statistically. First, if the dissociation constants of the reactants are not significantly different for the two sites, they will be considered essentially the same no matter how different the densities are. Second, the range of the observed data must cover enough of the activity at the individual sites to model it. For example, if a small-density site for which the ligand has a high affinity is nearly saturated at the lowest ligand concentration, then there is not enough information to model it. Figure 18 and 19 show the effect of a second site on the dose-response curve and Scatchard plot in a saturation assay. The more different the $K_d$'s, the more pronounced the effect. Finally, caution must be taken when interpreting a curvilinear Scatchard plot as an indication of multiple sites. Weiland and Molinoff (1981) state that, besides the existence of multiple sites, this effect might also be due to violations of the assumptions of no cooperativity and no further ligand-receptor reactions.

## 6. Model Assumptions and Design Problems

These analyses are based on models which make certain assumptions about the underlying reactions taking place. Feldman (1972) pointed out some of the assumptions regarding the reactions and reactants. Instead of independent multiple sites, cooperativity may be taking place in the system. In other words, the affinity of the ligand for the receptor may be dependent on the amount of binding that is taking place. Similarly, the reaction may be multivalent rather than univalent. Relaxation of the assumptions made on the reactants and reactions makes it necessary to use more complex reaction models. Munson (1983) discussed the effect of experimental artifacts on the resulting Scatchard plots. He simulated data which reflected the impact of badly estimated counter background, biological purity of the labeled ligand, purity of the unlabeled ligand, specific activity, recovery efficiency of the bound fraction, and contamination of the bound by free

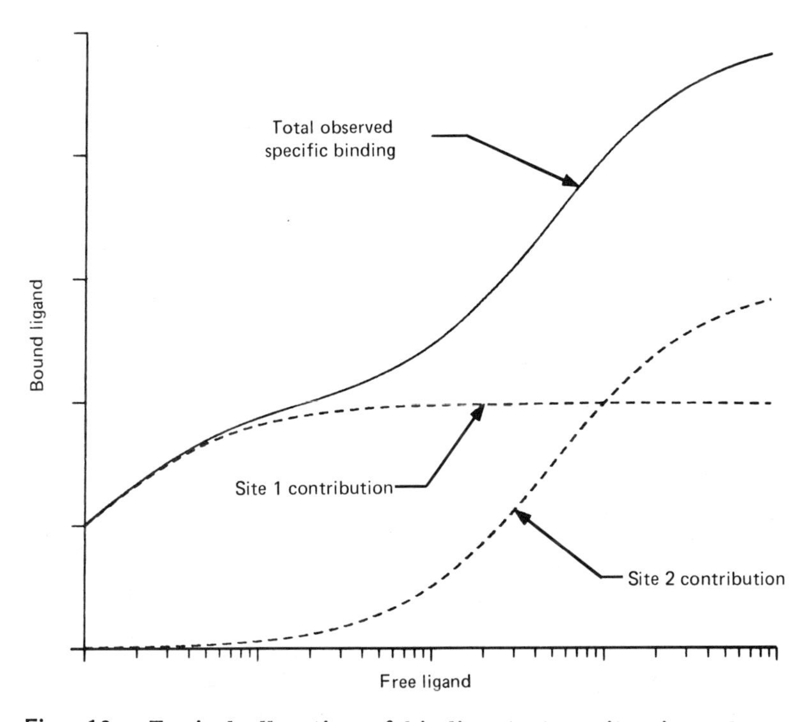


**Fig. 18** Typical allocation of binding to two sites in a dose-response curve from a saturation assay.

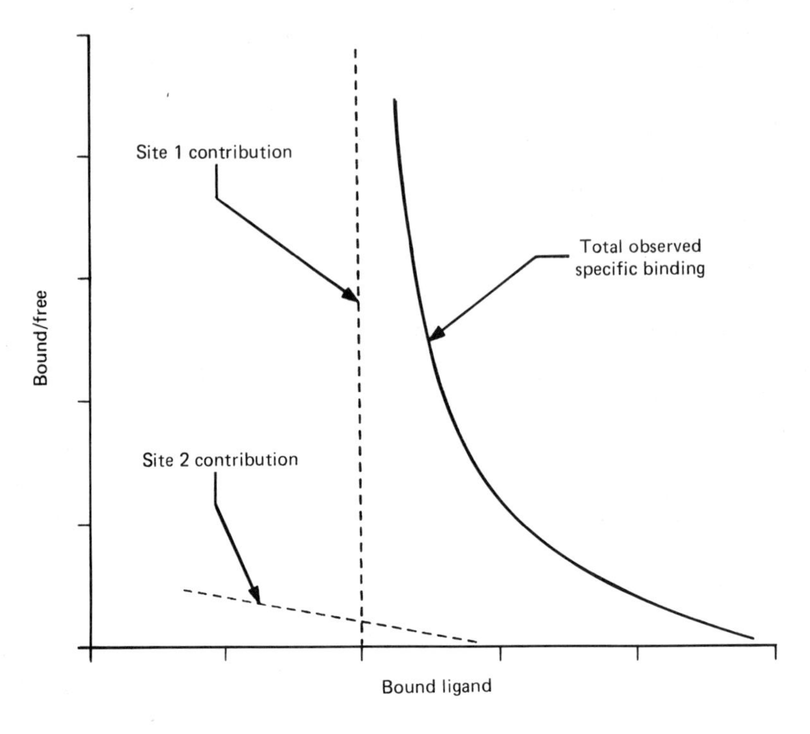


**Fig. 19** Typical allocation of binding to two sites in a Scatchard plot from a saturation assay.

ligand. In some cases, the resulting Scatchard plots could be misinterpreted as indicating multiple sites.

In the assay experiment, usually the investigator measures the amount of radioactively labeled ligand added to the system and the resulting amount bound. Two problems are inherent in this procedure. The first lies in the fact that the model calls for the amount of ligand unbound (free), not the amount added. Munson and Rodbard (1980) actually use the total added ligand to estimate the amount of free ligand using a Newton–Raphson technique. Occasionally, investigators count the amount of unbound ligand in the supernatant, but more often they either just use the total added ligand concentration as the concentration of free ligand or estimate the unbound ligand by subtracting the bound from the total ligand. The levels of bound ligand are usually so small relative to the total ligand that the unbound ligand is essentially the same as the total ligand. The other problem, usually ignored, is that the independent variable also has an error structure associated with it and is not known exactly.

Ferkany and Enna (1982) discuss two properties which they consider essential to the characterization of a particular receptor system. Saturability implies that the concentration of the ligand can be increased to a point where all of the specific binding sites are occupied and only nonspecific binding continues to increase. This saturation point is represented by $B_{max}$, the maximal binding capacity, and is characterized by a leveling off of the specific binding as shown in Figure 11. The dissociation constant ($K_d$) represents the rate at which the dose response reaches the maximal binding capacity. Specificity of the ligand for the receptor of interest should also be confirmed. Specificity in this context means that the ligand is really binding to the physiologically important receptor as assumed. The interactions between compounds known to be biologically active at the site are useful in determining specificity. These two properties, saturability and specificity, are very dependent on the experimental environment, such as incubation time, temperature, and pH. In a single-site system, Currie (1982) shows that the optimal allocation of resources would be to measure ligand binding at concentrations which result in complete saturation and 50% saturation. If the dose-response curve never reaches the saturation point, it is difficult to obtain accurate estimates of $B_{max}$. Klotz (1982) criticizes the estimate of $B_{max}$ that is obtained from the Scatchard plot. He contends that this estimate is a gross underestimate of the true parameter if the data, when plotted on a semilogarithmic scale of LR versue log(L), do not pass the inflection point. Lockwood and Wagner (1983) maintain that this is not necessarily true if methods other than the Scatchard plot are used for analysis.

### B. Radioimmunoassays

As mentioned earlier, radioimmunoassays are very similar to the competitive inhibition assay. Various doses of the unlabeled ligand are used as the inhibitor of the radioactively labeled ligand. The resulting dose-response curve, called the standard curve, is used to interpolate the concentration of unlabeled ligand in samples to which only labeled ligand has been added. The attractiveness of radioimmunoassays stems from the ability to estimate the potency of very small quantities of substances with a fairly high level of precision.

#### 1. Analysis of Radioimmunoassays

Until a few years ago, most radioimmunoassays were analyzed using a linear regression analysis on the logit-log transformation of the data. This transformation first required that the counter data be in terms of percent specifically bound; i.e., the counter data had to have the nonspecific binding subtracted off and then be given as a percent of the control adjusted for nonspecific binding. Several problems arose in this type of analysis. First, we are assuming that the control and nonspecific binding data are known exactly when, in fact, they are also subject to experimental error. Essentially, this meant that the endpoints of the curve were being fixed. Second, dose-response data were sometimes observed to be greater than the control or less than the nonspecific binding. The logit-log analysis could not handle data like this and these data points could only be removed from the analysis.

Rodbard and Hutt (1974) promoted the use of nonlinear analysis of a four-parameter logistic model directly on the raw data. This also allowed the control and nonspecific binding levels to be fit. In addition, data did not have to be discarded for falling outside an appropriate range. As in the radioligand assays, it is necessary to use a weighted analysis. Similarly, the weights can be defined by determining the relationship between the mean and variance of replicate observations.

Since the radioimmunoassay technique is procedurally complicated, it is not unusual for outliers to show up in the data. Several procedures have been suggested for detecting these aberrant observations. An alternative is to use a robust estimation procedure. These methods are not as susceptible as least squres to the effects of outliers, and since they do not rely on a Gaussian error distribution, they may be more efficient than least-squares analysis. Tiede and Pagano (1979) developed a robust nonlinear estimation procedure for the logistic model used in radioimmunoassays. They maintained that when outliers are present, the robust fit is unaffected by their presence, and when there are no outliers, the least-squares and robust fits are indistinguishable.

### 2. Relationship Between the Langmuir Isotherm and the RIA Four-Parameter Logistic

In the competitive assay, the Langmuir binding isotherm can be reparameterized to yield more precise binding parameter estimates. Starting with a slightly more general form of the isotherm, the amount of control binding or binding with only the radioactively labeled ligand can be modeled by

$$LR_0 = \frac{L^n B_{max}}{K_d + L^n} + LC$$

and the amount of binding with the inhibitor, I,

$$LR = \frac{L^n B_{max}}{K_d + L^n + (K_d/K_i)I^n} + LC$$

where $LR_0$ is the concentration of ligand bound to receptor without inhibitor, LR the concentration of ligand bound to receptor with inhibitor, $B_{max}$ the maximal receptor density or binding capacity, $K_d$ the equilibrium dissociation constant of ligand, $K_i$ the equilibrium dissociation constant of inhibitor, C the nonspecific binding constant, and n the Hill coefficient.

The Hill coefficient is often taken as an indicator of the existence of multiple sites or cooperativity. With no cooperativity and a single site, $n = 1$. However, as Weiland and Molinoff (1981) state, a Hill coefficient of 1.0 does not necessarily mean that the binding is a simple bimolecular reaction.

Since a constant amount of radioactively labeled ligand is used throughout the assay, the amount of nonspecific binding, LC, is also constant. Let $B_0$ and B be the amount of specific binding without and with the inhibitor.

$$B_0 = LR_0 - LC$$

$$B = LR - LC$$

Then the $IC_{50}$ is defined as the concentration of inhibitor which reduces specific binding by 50%, that is, $B_0 = 2B$. Substituting back into this equation gives

$$\frac{L^n B_{max}}{K_d + L^n} = \frac{2L^n B_{max}}{K_d + L^n + (K_d/K_i)IC_{50}{}^n}$$

Solving this for $IC_{50}^n$,

$$IC_{50}^{\,n} = (K_d + L^n)\frac{K_i}{K_d}$$

Then the amount of observed binding, LR, can be reparameterized as follows:

$$LR = \frac{L^n B_{max}}{K_d + L^n + (K_d/K_i)I^n} + LC$$

$$LR = \frac{L^n B_{max}}{K_d + L^n}\left[\frac{K_d + L^n}{K_d + L^n + (K_d/K_i)I^n}\right] + LC$$

$$= (B_0 - LC)\left[\frac{1}{1 + I^n[K_d/K_i(K_d + L^n)]}\right] + LC$$

$$= \frac{B_0 - LC}{1 + (I/IC_{50})^n} + LC$$

which is the four-parameter logistic recommended by Healy (1972) and Rodbard and Hutt (1974). Again, since the amount of radioactively labeled ligand is constant throughout the assay, the resulting amount of nonspecific binding is also constant throughout and can be thought of as another parameter. Thus, this model is dependent only on the concentration of inhibitor and the four parameters are $B_0$, the amount of initial or control binding; LC, the amount of nonspecific binding; $IC_{50}$, the concentration of inhibitor which decreases ligand binding by 50%; and n, the Hill coefficient.

3. Example

Although not obtained from a radioimmunoassay, the data from the preceding example will be fit using the radioimmunoassay procedures for the sake of comparison. The following estimates for the $IC_{50}$ and Hill coefficient were obtained from a logit-log analysis:

$$IC_{50} = 144.1 \pm 3.745 \text{ nM}$$

$$\text{Hill coefficient} = 0.421 \pm 0.022$$

Although not very different from the probit analysis, as expected, the estimate of the Hill coefficient is slightly smaller. A weighted nonlinear analysis of the fit of the four-parameter logistic yields

$$B_0 = 2529 \pm 47.38 \text{ CPM}$$

$$= 5.46 \pm 0.10 \text{ pmol/g tissue}$$

Hill coefficient = 0.411 ± 0.027

$IC_{50}$ = 141.1 ± 28.32 nM

LC = 408 ± 28.71 CPM

= 0.88 ± 0.06 pmol/g tissue

These results also agree very well with the probit and logit results.

#### 4. Design Considerations

Chang et al. (1975) have considered the effect of experimental design on the precision of estimates obtained in a logit-log analysis. They recommend that at least eight concentrations of unlabeled ligand or inhibitor be used with at least two replicates at each level. The responses for these levels in terms of percent of control should be between 20 and 80%. Under these conditions, a simple least-squares regression performs as well as a weighted regression. As previously mentioned for the radioligand assays, sensitivity, specificity, accuracy, and precision are all necessary for reliable assay results. Midgely et al. (1969) discuss assay development in terms of these characteristics.

### C. Software

Following is a list of published software or articles about software for analyses discussed. This list is not meant to be complete.

Radioligand assays or enzyme kinetic experiments:

Bianchi, R., Hanozet, G. M., and Simonetta, M. P. (1983). MIR: A Versatile Program for the Statistical Analysis of Enzyme Kinetic Data, *Comput. Programs Biomed.*, 19:189.

Statistical Consultants, Inc. (1986). PCNONLIN and NONLIN 84: Software for the Statistical Analysis of Nonlinear Models, *Am. Stat.*, 40:52.

Munson, P. J. and Rodbard, D. (1980). LIGAND: A Versatile Computerized Approach for Characterization of Ligand-Binding Systems, *Anal. Biochem.*, 107:220.

Radioimmunoassays:

Rodbard, D. (1984). *Cost/Benefit and Predictive Value of Radioimmunoassay* (A. Albertini, R. P. Ekins, and R. S. Galen, eds.), Elsevier Science Publishing, New York, p. 75.

Faden, V. B. and Rodbard, D. (1975). The "Logit-Log" Method and Scatchard Plot, *Radioimmunoassay Data Processing: Listings and Documentation*, 3rd ed., vol. 1, National Technical Information Service, Springfield, Va.

Rodbard, D., Faden, V. B., Hutt, D. M., and Knisley, S. (1975). The Logistic Method and Quality Control, *Radioimmunoassay Data Processing: Listings and Documentation*, 3rd ed., vol. 2, National Technical Information Service, Springfield, Va.

## IV. LEAD OPTIMIZATION

Lead optimization is the process of finding a compound which has some advantage over a related lead. This process could result in a better understanding of the physical-chemical determinants of the newly discovered activity, the reduction of undesirable side effects, experimental verification of the positional requirements of drug-receptor binding, modification of an absorption or metabolic rate, or an increase in the binding coefficient. Underlying this process is the common statistical purpose of characterizing and maximizing a response function.

In general, the response function has the form $g: \Omega \to R$, where $\Omega$ is the set of all compounds and R is the real line. This poses the immediate problem of defining a metric on $\Omega$ (Johnson, 1985). This problem is addressed by defining predictors $p_i: \Omega \to R$, $i = 1, \ldots, n$, of g, so that we can write g(w) as some function $f(p_{w1}, \ldots, p_{wn})$ of these predictors. The function $f: R^n \to R$ now has the form required by response surface methods of statistics and the form that will be used here. Thus, if $Y_w$ is the response variable associated with the activity of compound w in the test system of interest, we will write

$$Y_w = f_\theta(p_{w1}, \ldots, p_{wn}) + e_w$$

where $e_w$ denotes the error in our predictive model and $\theta$ is a k-dimensional vector of parameters.

Obviously, the choice of the predictors is critical and will depend on the response variable Y. The predictors are often termed chemical descriptors and numerous ones have been and continue to be proposed and advocated (Stuper et al., 1979; Kier, 1980; Franke, 1984a). The chemical descriptors proposed by Hansch and Fujita (1964) and be Free and Wilson (1964) have been as broadly accepted as any, and the synthesis of these two approaches as given by Fujita and Ban (1971) will be presented here.

### A. Some General Models

#### 1. Free-Wilson Model

The Free-Wilson model can be utilized when the set of compounds under consideration has a common parent structure. The model further requires specification of a functional group at each position defined by the parent structure. A list of functional groups can be found in Hansch and Leo (1979), and some of the more important classes of functional groups are reviewed in Lemke (1983). The following 2-amino-5-bromo-6-phenyl-4-pyrimidinone parent structure defines the substituent representation of some pyrimidinones in Tables 7 and 8. This representation of structures by means of a parent structure and substituents is called a Markush representation.

If there are q positions on the parent structure, the Free-Wilson model as parameterized by Fujita and Ban (1971) expresses the response variable $Y_w$ by

Table 7 Singly Substituted Pyrimidinone-SFV Inhibition Data[a]

| | | | | | | log (1/C) | |
|---|---|---|---|---|---|---|---|
| Cpd. | X1 | X2 | X3 | X4 | X5 | Obs. | Pred. |
| 1 | H | H | H | H | H | 3.42 | 3.52 |
| 2 | H | H | Cl | H | H | 3.48 | 3.17 |
| 3 | H | H | CH3 | H | H | 3.15 | 3.38 |
| 4 | H | H | C6H5 | H | H | <2.70 | 2.75 |
| 5 | H | Cl | H | H | H | 4.00 | 3.54 |
| 6 | H | NO2 | H | H | H | 3.35 | 3.38 |
| 7 | H | CF3 | H | H | H | 3.22 | 2.97 |
| 8 | H | Br | H | H | H | 3.50 | 3.45 |
| 9 | Cl | H | H | H | H | 3.02 | 3.42 |
| 10 | CH3 | H | H | H | H | 3.67 | 3.45 |
| 11 | H | OCH3 | H | H | H | 4.47 | 4.07 |
| 12 | OCH3 | H | H | H | H | 4.29 | 4.32 |
| 13 | H | H | C(CH3)3 | H | H | <2.70 | 2.68 |
| 14 | H | I | H | H | H | 3.04 | 3.41 |
| 15 | H | F | H | H | H | 3.61 | 3.93 |
| 16 | H | H | F | H | H | 3.25 | 3.44 |
| 17 | F | H | H | H | H | 4.05 | 3.92 |
| 18 | H | OC3H7 | H | H | H | 3.51 | 3.64 |
| 19 | OH | H | Cl | H | H | <2.70 | 4.54[b] |
| 20 | H | OH | H | H | H | <2.70 | 4.54[b] |
| 21 | SCH3 | H | C6H5 | H | H | <2.70 | 3.49[b] |
| 22 | H | H | OCH3 | H | H | <2.70 | 3.51[b] |
| 23 | H | H | COOH | H | H | <2.70 | 3.47[b] |
| 24 | H | H | CN | H | H | <2.70 | 3.74[b] |
| 25 | H | H | OCH2C6H5 | H | H | <2.70 | 2.87[b] |
| 26 | H | H | OH | H | H | <2.70 | 3.78[b] |
| 27 | H | H | N(CH3)2 | H | H | <2.70 | 3.45[b] |

[a]A compound is predicted active (A) if the probability of activity, as predicted by the logistic model, is 0.5 or greater.
[b]Predicted values had the compounds not been predicted inactive by the logistic model.

| log (1/C) | | log P | | Ortho | | | Meta | | Para |
|---|---|---|---|---|---|---|---|---|---|
| Resid. | Act. | Obs. | Pred. | SCH3 | OH | R | OH | R | HA |
| − 0.10 | A | 0.70 | 0.70 | 0 | 0 | 0.00 | 0 | 0.00 | 0 |
| 0.31 | A | 1.59 | 1.41 | 0 | 0 | 0.00 | 0 | 0.00 | 0 |
| − 0.23 | A | 1.05 | 1.26 | 0 | 0 | 0.00 | 0 | 0.00 | 0 |
| − 0.05 | A | — | 2.66 | 0 | 0 | 0.00 | 0 | 0.00 | 0 |
| 0.46 | A | 1.10 | 1.41 | 0 | 0 | 0.00 | 0 | −0.15 | 0 |
| − 0.03 | A | 0.57 | 0.42 | 0 | 0 | 0.00 | 0 | 0.16 | 0 |
| 0.25 | A | 1.53 | 1.58 | 0 | 0 | 0.00 | 0 | 0.19 | 0 |
| 0.05 | A | 1.39 | 1.56 | 0 | 0 | 0.00 | 0 | −0.17 | 0 |
| − 0.40 | A | — | 1.41 | 0 | 0 | −0.15 | 0 | 0.00 | 0 |
| 0.22 | A | — | 1.26 | 0 | 0 | −0.13 | 0 | 0.00 | 0 |
| 0.40 | A | 0.83 | 0.68 | 0 | 0 | 0.00 | 0 | −0.51 | 0 |
| − 0.03 | A | 0.21 | 0.68 | 0 | 0 | −0.51 | 0 | 0.00 | 0 |
| 0.02 | A | — | 2.83 | 0 | 0 | 0.00 | 0 | 0.00 | 0 |
| − 0.37 | A | 1.54 | 1.82 | 0 | 0 | 0.00 | 0 | −0.19 | 0 |
| − 0.32 | A | 0.69 | 0.84 | 0 | 0 | 0.00 | 0 | −0.34 | 0 |
| − 0.19 | A | 0.89 | 0.84 | 0 | 0 | 0.00 | 0 | 0.00 | 0 |
| 0.13 | A | 0.71 | 0.84 | 0 | 0 | −0.34 | 0 | 0.00 | 0 |
| − 0.13 | A | — | 1.75 | 0 | 0 | 0.00 | 0 | 0.45 | 0 |
| − 1.84 | I | — | 0.03 | 0 | 1 | −0.64 | 0 | 0.00 | 0 |
| − 1.84 | I | — | 0.03 | 0 | 0 | 0.00 | 1 | −0.94 | 0 |
| − 0.79 | I | — | 1.31 | 1 | 0 | −0.18 | 0 | 0.00 | 0 |
| − 0.81 | I | 0.72 | 0.68 | 0 | 0 | 0.00 | 0 | 0.00 | 1 |
| − 0.77 | I | — | 0.83 | 0 | 0 | 0.00 | 0 | 0.00 | 1 |
| − 1.04 | I | — | 0.13 | 0 | 0 | 0.00 | 0 | 0.00 | 1 |
| − 0.17 | I | — | 2.36 | 0 | 0 | 0.00 | 0 | 0.00 | 1 |
| − 1.08 | I | — | 0.03 | 0 | 0 | 0.00 | 0 | 0.00 | 1 |
| − 0.75 | I | — | 0.88 | 0 | 0 | 0.00 | 0 | 0.00 | 1 |

Table 8 Di- and Trisubstituted Pyrimidinone-SFV Inhibition Data[a]

| | | | | | | log (1/C) | |
|---|---|---|---|---|---|---|---|
| Cpd. | X1 | X2 | X3 | X4 | X5 | Obs. | Pred. |
| 28 | H | Cl | Cl | H | H | 3.19 | 3.35 |
| 29 | H | Cl | H | Cl | H | 3.14 | 3.14 |
| 30 | Cl | H | H | Cl | H | 3.27 | 3.14 |
| 31 | H | OCH3 | H | OCH3 | H | 4.33 | 4.14 |
| 32 | F | H | F | H | F | < 2.70 | 3.76 |
| 33 | F | H | H | OCH3 | H | < 3.42 | 3.87 |
| 34 | OCH3 | H | H | F | H | < 3.21 | 4.08 |
| 35 | OH | H | H | F | H | < 2.70 | 4.49 |
| 36 | F | H | N | OH | H | 3.31 | 4.2 |

[a]Predictions are based on QSAR models for the data in Table 1.

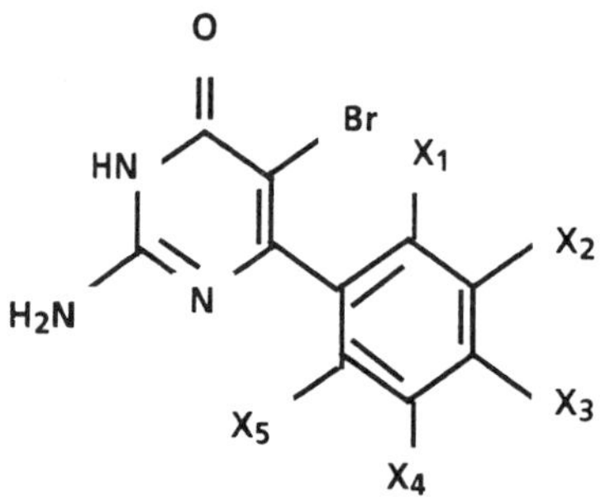

2-amino-5-bromo-6-aryl-4-pyrimidinone parent structure

$$Y_w = \mu_0 + \sum_{ij} \rho_{ij}\delta_{ijw} + \varepsilon_{ijw}$$

where $\mu_0 = E[Y_0]$ is the expected response for a preselected reference compound $w_0$, $\rho_{ij}$ is the effect of the jth substituent at the ith position, and $\delta_{ijw} = 1$ if substituent j occurs at position i on compound w and $\delta_{ijw} = 0$ otherwise. The Free-Wilson model is a main -effects model with no interaction terms under the parameterization $\rho_{ij} = 0$ if substituent j occurs at position i on $w_0$. In QSAR studies this is usually preferred to the classical parameterization $\Sigma_j \rho_{ij} = 0$ for $i = 1, \ldots, q$.

2. Hansch Model

The Hansch model assumes that the response variable Y is a function of particular physical-chemical properties of the compounds. By grouping them into lipophilic (l), electronic (e), and steric (s) categories, we obtain the Hansch equation

| log (1/C) | | log P | | Ortho | | | Meta | | Para |
|---|---|---|---|---|---|---|---|---|---|
| Resid. | Act. | Obs. | Pred. | SCH3 | OH | R | OH | R | HA |
| −0.16 | A | 1.59 | 2.12 | 0 | 0 | 0.00 | 0 | −0.15 | 0 |
| 0.00 | A | — | 2.12 | 0 | 0 | 0.00 | 0 | −0.15 | 0 |
| 0.13 | A | — | 2.12 | 0 | 0 | −0.15 | 0 | 0.00 | 0 |
| 0.19 | A | — | 0.66 | 0 | 0 | 0.00 | 0 | −0.51 | 0 |
| −1.06 | A | — | 1.12 | 0 | 0 | −0.34 | 0 | 0.00 | 0 |
| −0.45 | A | — | 0.82 | 0 | 0 | −0.34 | 0 | 0.00 | 0 |
| −0.87 | A | — | 0.82 | 0 | 0 | −0.51 | 0 | 0.00 | 0 |
| −1.79 | I | — | 0.17 | 0 | 1 | −0.64 | 0 | 0.00 | 0 |
| −0.82 | A | — | 0.17 | 0 | 0 | −0.34 | 0 | 0.00 | 0 |

$$Y_w = \mu + f(l_w) + g(e_w) + h(s_w) + \varepsilon_w$$

An important special case of the Hansch model, in which the lipophilic component is represented by the logarithm of the octanol/water partition coefficient (log P) and f is a quadratic function, is given by

$$Y_w = \mu + a_1(\log P) + a_2(\log P)^2 + \varepsilon_w$$

Physical-chemical properties are usually experimentally determined but are sometimes calculated from the molecular structure. The CLOP program (Pomona Medicinal Chemistry Project, 1984) developed by Hansch and Leo calculates log P values. In a subsequent example, log P values will be estimated using substituent constants.

### 3. Fujita–Ban Model

The Hansch model is generally preferred to the Free-Wilson model because of its wider inferential domain. However, the physical-chemical properties required by the Hansch model are often not available. This problem can be addressed by associating physical-chemical constants with the substituents of the Free-Wilson model. These constants are called substituent constants and are available in data bases (Hansch and Leo, 1979). We will develop the basic substituent constant model and then discuss the definition and interpretation of the more popular constants.

Fujita and Ban (1971) were the first to express the substituent effects from their reparameterization of the Free-Wilson model in terms of substituent constants. In the following equation, the substituent effect $\rho_{ij}$ in the Free-Wilson model is written as a function of lipophilic, electronic, and steric substituent constants:

$$\rho_{ij} = \mu + \alpha_i l_{ij} + \beta_i e_{ij} + \lambda_i s_{ij} + \varepsilon_{ij}$$

Substituting back into the Free-Wilson model and collecting terms, one obtains the Fujita-Ban model

$$Y_w = \mu + \sum_{ij} (\alpha_i l_{ij} + \beta_i e_{ij} + \lambda_i s_{ij}) \delta_{ijw} + e_w$$

$$= \mu + \sum_i \alpha_i l_{iw} + \sum_i \beta_i e_{iw} + \sum_i \lambda_i s_{iw} + e_w$$

where $l_{iw}$, $e_{iw}$, and $s_{iw}$ denote respective lipophilic, electronic, and steric substituent constants associated with the substituent occurring at position i on compound w. The last equation could be generalized to include quadratic expressions of the substituent constants and to allow for more than one lipophilic (electronic/steric) substituent constant for a position. However, these generalizations are seldom needed in practice.

As a final generalization of these models, it should be noted that the Free-Wilson, Hansch, and Fujita-Ban models can be, and frequently are, combined.

## B. Interpretations and Assumptions

### 1. Free-Wilson Model

The models in the preceding section are general enough to support a wide variety of interpretations. The interpretation of the Free-Wilson model is perhaps the simplest. It says that the substituent effects are additive.

It is well known that additivity can be altered by transforming the response variable. In QSAR studies, the response measurement is usually the dose of the compound required to elicit a specified response. For example, the response variable may be the concentration of the compound required to inhibit a specified reaction by 50%. If this inhibitory concentration is denoted by $IC_{50}$, then it is traditional to let $\log(1/IC_{50})$ be the response variable, where the concentration is expressed in moles rather than grams of compound.

### 2. Log P Model

Of the many physical-chemical properties and structural indices that might be used in the Hansch model, we will only discuss the interpretation of the log P parameter as it is related to drug transport. To do so, it is helpful to expand on the events surrounding the action of a drug. Before binding, the drug must reach the receptor site. In doing so, it must pass through one or more biological membranes in transit. The lipophilic parameters, especially log P, correlate well with the rate of passive diffusion through those membranes. The drug might also be trapped in various biological tissues. Very lipophilic drugs tend to concentrate in the fatty tissues.

### 3. Substituent Constants Model

Because of their practical importance, the remainder of this section will be devoted to substituent constants. We will begin by illustrating how the popular $\pi$ substituent constant quantitates a substituent's contribution to the overall lipophilicity of a compound.

Let $\pi_H$ denote the $\pi$ substituent constant for hydrogen, the reference substituent. By convention we set $\pi_H = 0$. Now consider compounds 1 and 2 of Table 7. They differ at only one position, the para position, with the reference substituent at that position on compound 1 and the chloro substituent at that position on compound 2. It is natural to define $\pi_{Cl}$ by

$$\pi_{Cl} = \log_2 P - \log_1 P = 0.89$$

where $\log_i P$ denotes log P for compound i. Other values for $\pi_{Cl}$ are $\pi_{Cl} =$ 0.4 based on the contrast $\log_5 P - \log_1 P$ and $\pi_{Cl} = 0.49$ based on the contrast $\log_{28} P - \log_5 P$. Note that the disubstituted compound, 28, differs from compound 5 in a single position, which again happens to be the para position, and that compound 5 has the reference substituent, H, at that position. A substituent constant in a data base is likely to be the average of such contrasts but is obtained from a variety of compound pairs having possibly different parent structures.

Clearly, the relevance of substituent constants rests on a strong additivity assumption. This additivity assumption often fails unless compounds are broken down into smaller subclasses. Consequently, Hansch and Leo have one set of substituent constants for substituents at aliphatic positions and another set for substituents at aromatic positions. Departures from additivity often diminish the predictability of Fujity-Ban models relative to Hansch models.

There are many substituent constants (Franke, 1984b). We will discuss the interpretation of only the pi ($\pi$) constant, the hydrogen accepting/donating constants, the molar refractivity (MR) constant, the Hammett sigma ($\sigma$) constant, and the Swain-Lupton field (F) and resonance (R) constants given in Tables VI-1 and VI-2 of Hansch and Leo (1979).

The interpretation of substituent constants is usually associated with the interaction of the compound with a receptor site of a macromolecule. This interaction involves the complementary matching of spatial and electronic convexities and concavities of the drug with those of the receptor site. Pockets at the receptor site are often characterized as being either hydrophilic or hydrophobic, depending on whether it is energetically favorable for water to occupy the area or not. The molar refractivity of a substituent is a rough measure of its size, while the partition coefficient of a substituent is a rough measure of its hydrophobicity. Hansch (1982) has shown that if a substituent with a large $\pi$ value is required at a particular position on a parent structure, one expects a hydrophobic pocket at the complementary cavity of the receptor site. On the other hand, if a substituent with a large MR value is required, but one with a large $\pi$ value will not work, one expects a hydrophilic pocket.

The primary determinant of the hydrophilicity of a receptor pocket is the ability of the atoms of the pocket to enter into hydrogen bonds. A hydrogen bond is based on the interaction between a proton donor and a proton acceptor. Thus the hydrogen-donating or -accepting capability of the substituents at a given position often enters into the electronic component required at the complementary position of the receptor site.

Frequently, the reaction center of the drug-macromolecule interaction is located at a position on the parent structure different from the positions of substituent substitution. However, the substituents still can exert electronic effects at the reaction center through resonance and induction of electronic charge. On phenyl substituents, this electronic effect is usually represented by the Hammett sigma constant $\sigma$.

The value of $\sigma$ depends on whether the substituent occupies the meta or para position on the phenyl ring. This dependence of the sigma constant on the position of the substituent is explained by decomposing the electronic effect into a field effect and a resonance effect. Swain and Lupton (1968) have defined a field constant F and a resonance constant R associated with a substituent. The resonance constant contributes to this partially localized electronic effect only if the substituent occurs at specified positions on an aromatic system such as the para position on the phenyl group. Since R is small for groups attached to an aliphatic carbon, the Swain-Lupton field constant F is often used to represent this inductive effect of substituents attached to aliphatic carbons.

#### 4. Caveats

The preceding interpretations help guide one's thought. However, a number of caveats often undermine reliable conclusions. The first caveat results from highly correlated predictive parameters. For example, the pi and MR substituent constants are highly correlated on aliphatic substituents. In this case, if either substituent constant is correlated with activity, the other will be also. Consequently, the conclusions associated with such correlations are not differentiable without further data.

A second caveat results from the frequent inability to deconvolute the absorption, binding, and intrinsic activity components of the response to a drug. (Intrinsic activity is zero if a compound fails to elicit the desired response when it is at a reasonable concentration around the receptor and binds well to that receptor.) Obviously, a compound will not be very active if it is poorly absorbed, does not bind well, or has no intrinsic activity. Without biological test systems that differentiate between these components of the drug response, the preceding types of interpretations are tentative at best.

Without clear interpretations, even the predictive function of these models can become suspect since the random sampling assumption on which statistical prediction is premised is often drastically violated. One cannot circumvent the problem that there is a finite number of "reasonable" substituents and these fall into chemically related subclasses such as alkyl groups, amines, and alkoxy groups.

In spite of these caveats, the Hansch paradigm of QSAR models has proved very useful for a number of reasons. For one, these QSAR models have often proved predictive when the preceding components of the drug response have been deconvoluted. For another, the chemical descriptors in the model have a firm physical-chemical basis.

### C. Selecting a Parent Structure

With few exceptions, the preceding models require the specification of a parent structure and substituents. Three rules facilitate the selection of an appropriate parent structure. First, substituent constants are seldom available on moieties which are substituted ring systems or comprise more than one functional group. Consequently, parent structures should be large enough to exclude such possibilities. Second, a substituent constant is unlikely to be an appropriate quantification of its implied physical-chemical property unless it is attached to either an aliphatic or an aromatic carbon. This rule also leads to larger parent structures.

The larger the parent structure, the smaller the set of compounds representable by that parent structure. Thus the third rule is to minimize the size of the parent structure under the preceding two restrictions. With these rough guides and close collaboration with the chemist, substituent constants can be applied in a meaningful and somewhat routine fahsion.

## D. A Quantitative Pyrimidinone-Antiviral Relationship

The example in this section illustrates the use of the preceding models. The results of the analysis are given in what has become a somewhat standard form. Tables 7 and 8 present the antiviral activities of some 2-amino-5-bromo-6-aryl-4-pyrimidinones as reported in Skulnick et al. (1985) together with the predictive information used in the subsequent equations. The compounds were administered in a single dose intraperitoneally to mice 18 hours prior to a lethal injection of Semliki Forest virus (SFV). The dose ($ED_{50}$) required for 50% survival was estimated by using a probit analysis with 24 mice per group and converted to the corresponding molar concentration by dividing $ED_{50}$ by the molecular weight of the compound.

Octanol/water partition coefficients were measured on some of the compounds and are expressed as logarithms in Table 7. Partition coefficients for the remaining compounds were predicted from the following equation:

$$\text{Pred}(\log P) = \underset{(0.06)}{0.68} + \underset{(0.10)}{0.77} \sum_i \pi_i$$

$n = 15 \quad s = 0.19 \quad r = 0.90$

where $\pi_i$ denotes the $\pi$ constant of the substituent at the ith position. In this and subsequent equations, standard errors of the coefficients are given in parentheses and n, s, and r denote the sample size, standard deviation, and correlation coefficient, respectively. Other predictive variables in Table 7 are the presence/absence of the substituents SCH3 and OH, the Swain-Lupton resonance constant (R), and the indicator of a hydrogen acceptor (HA). The substituent constants were taken from Hansch and Leo (1979).

The di- and trisubstituted compounds in Table 8 were not included in the model fitting for two reasons. The first is the ambiguity in assigning them substituent constants. The rotation of the phenyl ring about the pheny-pyrimidinone bond (refer back to the parent structure) creates a symmetry in the positional assignments in which position 1 corresponds to 5 and position 2 corresponds to 4. The ambiguity in assigning substituents to positions in the case of the multisubstituted compounds is potentially resolvable by using symmetrically substituted compounds. For example, a comparison of the 2,4-dimethoxy compound (31) with the 2-methoxy compound (11) suggests that the methoxy at position 4 has no effect. In this case, the assignment of a methoxy at position 4 is unambiguous. However, there are too few symmetrically substituted compounds to meaningfully resolve such positional effects. Thus, we modeled only the singly substituted compounds. These models will be used to suggest statements one can make concerning the di- and trisubstituted compounds.

Compounds with log(1/C) values of 2.7 or less were declared inactive. The large number of inactive compounds complicates the analysis. One could approach the problem by using models for censored data (Miller, 1981). This entails censoring assumptions, which are difficult to test. Our approach was first to obtain gross predictors of activity versus inactivity by using a logistic model. Then the more traditional linear regression model was used on the compounds *predicted* to be active. Of the singly substituted compounds predicted to be active, only two, compounds 4 and 13, were inactive. They were included in the analysis and assigned a log(1/C) value of 2.7.

### 1. Binary Response Model

The probability that a compound is active is estimated by using the following logistic version of a combination of the Free-Wilson and Fujita-Ban models, where the subscripts o, m, and p denote the ortho, meta, and para positions. Since there were no active compounds when either $SCH3_o$, $OH_o$, $OH_m$, or $HA_p$ equaled one, the maximum likelihood estimates of the coefficients associated with these descriptors diverged to infinity. To circumvent this problem, the quoted coefficients were truncated at −10.3. Standard errors for the coefficients are meaningless to report in this case.

$$\text{Prob(activity)} = \frac{1}{1 + \exp(2.1 - 10.3*SCH3_o - 10.3*OH_o - 10.3*OH_m - 10.3*HA_p)}$$

$$n = 18 \qquad R = 0.66 \qquad X^2 = 23.9 \qquad p < 0.001$$

The development of the model is given by Table 9.

A compound is predicted active by the logistic model if the probability of activity is 0.5 or greater. From Table 7 we see that the correct active/inactive predictions were obtained for all singly substituted compounds except compounds 4 and 13. This could not be accomplished without using the indicator functions $SCH3_o$, $OH_o$, and $OH_m$. Their inclusion in the model has the following justivication. First, $SCH3_o$ has a molar refractivity substituent constant of 13, twice that of the next largest of the other ortho substituents. Second, the OH groups are the only hydrogen-donating groups at the ortho and meta positions. Such justification seems appropriate from an exploratory, as opposed to a confirmatory, perspective.

**Table 9** Development of the Logistic Model

| Model | R | $X^2$ | Sig[a] |
|---|---|---|---|
| $HA_p$ | 0.56 | 13.4 | <0.001 |
| $HA_p$ $SCH3_o$ $OH_o$ $OH_m$ | 0.66 | 23.9 | <0.025 |

[a]Statistical significance of the additional factor adjusted for the presence of the preceding factors.

Looking at the first five di- and trisubstituted compounds, 28 to 32, in which the activity/inactivity is unambiguously assigned and in which the parameters of the model are invariant to the symmetry discussed earlier, we see that all but the trifluoro compound, 32, were correctly predicted to be active. The next two compounds, 33 and 34, were predicted to be active; we know only that these compounds were not very active. The activities of compounds 35 and 36 are either correctly predicted or incorrectly predicted, depending on how one chooses to resolve the symmetry in the substituent assignments.

## 2. Continuous Response Model

The relative activity of the singly substituted compounds predicted to be active by the logistic model can be predicted using a Hansch-Fujita-Ban model. The model is given below where pi equals the measured log P value if available and equals the earlier predicted log P value otherwise. Clearly, the activity decreased with an increase in overall lipophilicity of the compounds and with an increase in the Swain-Lupton resonance constant associated with the ortho and meta substituents. The development of the model is given by Table 10. The correlation matrix is given by Table 11.

$$\log\left(\frac{1}{C}\right) = \underset{(0.18)}{3.79} - \underset{(0.11)}{0.39\text{pi}} - \underset{(0.55)}{1.19R_i} - \underset{(0.37)}{1.19R_2}$$

$n = 18 \qquad r = 0.85 \qquad s = 0.28$

Since the pyrimidinones are known to be poorly soluble, the negative relationship with log P suggests that solubility is a limiting factor among the compounds predicted to be active by the logistic model.

We have no obvious interpretation for the importance of the Swain-Lupton substituent constant in Hansch-Fujita-Ban model. It may serve only as a grouping parameter to "explain" the large effects associated with the alkoxy groups and to some extent the halogen substituents at the ortho and meta positions.

The following picture emerges for the singly substituted compounds. The active compound is expected to have small, non-hydrogen-donating groups with negative R values in the ortho position; similar groups, in the

Table 10 Development of the Hansch-Fujita-Ban Model

| Model | r | $s^2$ | Sig[a] |
|---|---|---|---|
| pi | 0.70 | 0.36 | 0.001 |
| pi $R_2$ | 0.80 | 0.32 | 0.025 |
| pi $R_2$ $R_1$ | 0.85 | 0.28 | 0.05 |

[a]Statistical significance of the additional factor adjusted for the presence of the preceding factors.

Table 11 Correlation Matrix

| | Pi | $R_1$ | $R_2$ |
|---|---|---|---|
| Pi | 1 | — | — |
| $R_1$ | 0.40 | 1 | — |
| $R_2$ | 0.58 | 0.11 | 1 |

meta position, but with fewer restrictions on their size; and non-hydrogen-accepting groups in the para position. Finally, we wish to maximize solubility under these constraints. However, of the non-hydrogen-donating substituents, the methoxy group has about the lowest R value, except for $N(CH_3)_2$, which is roughly the same size as $OC_3H_7$. This substituent would be an excellent test of both models. If both models hold up, these substituents pretty much exhaust the common substituents in the region of optimality. Thus, a termination point for the singly substituted series has about been reached.

Generally speaking, prediction of the activity for the multisubstituted compounds gave mixed results. Good predictions were obtained for the dichloro and dimethoxy compounds, but poor ones for the trifluro compound. Predictions were uniformly bad for compounds 33 to 36, which had mixed functionalities. This strongly suggests the need for symmetrically substituted compounds, since we have already seen how different positions on the phenyl ring require different types of substituents.

### 3. Variable Selection

Clearly, the models discussed above arise from a selection of variables context (Hocking, 1983). This work will not be covered here except to note some aspects of the problem which are characteristic of QSAR analysis. First, the number of predictive variables can approach or exceed the number of compounds, especially if one defines indicator variables to account for changes in the relationship between the response and the initial physical-chemical variables. One can approach this problem by using principal components, clustering the variables and selecting one variable from each cluster, or using principal variables (McCabe, 1984). Second, when dealing with substituent constant models, one often has the opportunity to form subsets of compounds which differ at only one position, say X. By plotting activity versus a substituent constant associated with position X, one can frequently gain a reliable insight into the appropriate model.

## E. Aspects of Experimental Design

### 1. The Modeling Context

As we are viewing it, lead optimization assumes that position X on one compound in a series of analogs corresponds to position X on any other compound in that series. The disubstituted compounds in the preceding example suggest how quickly this assumption can be invalidated. Other mechanisms can invalidate the assumption. Some compounds in the series may be metabolized and others not. Some compounds in the series may form intramolecular bonds which others cannot form. Some compounds may assume conformations which others cannot assume.

Since lead optimization and the interpretations of the preceding models assume that these mechanisms are constant, we must view lead optimization as embedded in a larger context in which such mechanisms are studied. We can divide this larger context into three areas: receptor site modeling, drug transit modeling, and QSAR modeling.

Our interest in receptor site modeling and drug transit modeling only extends to their interface with QSAR modeling, which so characterizes lead optimization. Clearly, any optimum over a series of analogs in which the preceding mechanisms are all constant is not likely to be a global optimum when viewed over a much larger set of compounds. Although it may be important to achieve a local optimum for patent reasons, it is more important to find the global optimum or at least a better local optimum.

Obviously, these larger issues must be kept in mind when deciding on the next set of compounds to test. It may be more useful to test a conformationally constrained analog important to receptor site modeling, or to test a metabolically constrained analog important to drug transit modeling, or to test a structurally new compound important to lead screening than to test a whole series of congeners. This is not to diminish the importance of empirical modeling. As soon as one finds a mechanistically or structurally new lead with good activity, one must empirically explore that region of compound space or some other scientist will.

Four basic approaches to substituent selection at a single position can be distinguished: hypothesis-based experimental designs, sequential designs, computationally based designs, and optimal designs.

## 2. Hypothesis-Based Selection

The varied scientific purposes of hypothesis-based substituent selection can be associated with four phases of substituent selection: selecting substituent constants to study, resolving collinearities in substituent constants, estimating the coefficients of a given QSAR model, and series termination. These phases are not inviolably sequenced as their ordering may suggest; rather each continues to reappear because of the scientific issues in which lead optimization is embedded.

These phases in selection are fairly easy to communicate to the scientist. For example, the $OH_m$ substituent is the only hydrogen-donating group tried at the meta position. The potential for enhancing the model by including other hydrogen-donating groups, such as $NH_3$, at the meta position is straightforward and creates a testable hypothesis. Moreover, this enhancement of the current design offers the scientist considerable flexibility in selecting substituents. By the same token, showing a scientist two collinear substituent constants, both highly correlated with activity, immediately suggests the importance of selecting some substituents that resolve the collinearity. Again, the scientist can be given a whole list of substituents that would resolve the collinearity. This approach to experimental design is transparent to the scientist, easy to implement, and can considerably improve the information content of the data set.

There are some drawbacks to this approach. If the collinear substituent constants in the previous paragraph are not related to activity, the scientist may be reluctant to resolve their collinearity, even if it is explained that the seeming unimportance of these substituent constants may be due to mutual cancellation of their effects. Moreover, the experimental designs resulting from hypothesis-based selection often have little information regarding unforeseen interactions.

### 3. Sequential Designs

Of the sequential search schemes, the topliss tree (Topliss, 1972) has receved the most attention. One descends down the tree with substituents associated with each node by a decision rule based on whether the activity of the last tested compound is less than, equal to, or greater than the activity of the preceding compound. The search ends after three to six steps when the bottom of the tree is reached. The substituents suggested at each branch of the tree are selected to explore the substituent constant space as much as possible given the constraints of the tree.

Darvas (1974) has applied the simplex method of Spendley et al. (1962). A set of substituents are projected on a coordinate space of two substituent constants, usually the $\pi$-$\sigma$ coordinate space. One picks three substituents and tests their associated compounds. Their responses form an inclined triangle in the $\pi$-$\sigma$-z space with z being the response axis. One draws a line segment from the lowest vertex of the triangle that passes through the midpoint of the opposite side. This line segment is projected on the $\pi$-$\sigma$ plane. A new substituent lying close to the extended end of this line segment is chosen so as to form another triangle with the substituents associated with the two most active compounds from the preceding set of three. The compound associated with the newly selected substituent is tested, allowing another inclined triangle to be constructed. This process is iterated until the optimum is achieved.

These sequential schemes will work quite well if the substituent effects are large compared to the error in determining the biological activity of the compounds and if these effects are a function of the selected substituent constants.

### 4. Computationally Based Selection

The computationally based procedures for selecting substituents are largely intuitively defined. The logic of the more popular methods will be covered here. The reader is referred to Martin (1978) for a detailed discussion with excellent tables and examples.

Hansch and Leo (1979) present tables associated with the cluster analysis approach to substituent selection. In this approach, the substituents are clustered with respect to a set of standard substituent constants. One spans the substituent space by selecting a substituent from each cluster. This approach offers the scientist much needed flexibility in selecting the more synthetically accessible substituents. However, it has been shown (Streich et al., 1980) that, based on the determinant of the correlation matrix of the substituent constants, this flexibility can lead to far from optimal designs when the number of substituents being clustered is large.

Wootton (1983) iteratively selects a set of substituents by starting with two which are separated by a prespecified distance, d, in substituent constant space. The next substituent is the one closest to the "center of gravity" of those two under the constraint that it be a least a distance d from both of them. This process is iterated until no substituent can be found that is at least the prespecified distance from the previously chosen ones.

Franke and co-workers (Streich et al., 1980) proposed an iterative method in which a hyperplane is fit to the substituent constant space. The substituents are divided into two sets based on their distance from the hyperplane. This hyperplane and these two sets are iteratively updated after a fixed fraction of the compounds are selected from the two sets by Wooten's approach. In a second approach, the same group (Dove et al., 1980) simply fit a hyperplane to the substituent space and suggest that the scientist visually select a set of well-placed substituents.

### 5. Optimal Experimental Designs

With the notable exception of Austel's (1982) introduction of factorial designs, optimal experimental designs have yet to receive much attention in the area of lead optimization. Consequently, the various optimal designs will not be discussed here. The reader is referred to the review by Steinberg and Hunter (1984). However, a note concerning the introduction fo optimal designs seems appropriate.

The cost of introducing a particular functional group at a particular position varies considerably with the group and the position, and this cost is often poorly estimated. Moreover, the form of the QSAR model is seldom fully known until after the analogs have been made and synthesized. Thus, near-optimal designs with enhanced flexibility regarding the choice of substituents may prove more beneficial than optimal designs. The QSAR field needs a program that accepts a set of substituents that have been tried, develops a near-optimal set of points in substituent constant space to try, and gives the user a choice of additional substituents relevant to each design point.

### 6. Termination of Analog Testing

The last phase of experimental design is the termination of analog synthesis and testing. Termination criteria are explicit if one is using a sequential approach or if one has a reliable QSAR model. However, one should check for predictive variables that have not been varied, for any collinearities among the predictive variables, for outliers whose activities have not been explained, and for interactions which may not be fully understood. One should also consider sampling the substituent space around the optimum a few times to take advantage of the error distribution in the model.

Finally, it should be noted that the goal is not to predict the most active compound. Rather, it is to efficiently find and characterize local optima and to use this information in conjunction with receptor site and drug transit modeling to suggest other local optima.

## REFERENCES

Altar, C. A., Walter, R. J., Jr., Neve, K. A., and Marshall, J. F. (1984). Computer-Assisted Video Analysis of 3H-Spiroperidol Binding Auto-radiogrpahs, *J. Neurosci. Methods*, 10: 173.

Armitage, P. and Schneiderman, M. A. (1958). Statistical Problems in a Mass Screening Program, *Ann. N.Y. Acad. Sci.*, 76: 896.

Atkins, G. L. and Nimmo, I. A. (1975). A Comparison of Seven Methods for Fitting the Michaelis-Menten Equation, *Biochem J.*, 149: 775.

Austel, V. (1982). A Manual Method for Systematic Drug Design, *Eur. J. Med. Chem.*, 17: 9.

Austel, V. and Kutter, E. (1980). Practical Procedures in Drug Design, *Drug Design*, vol. 10 (E. M. Ariëns, ed.), Academic Press, New York, p. 1.

Bergman, S. W. and Gittins, J. C. (1985). *Statistical Methods for Pharmaceutical Research Planning*, Marcel Dekker, New York, p. 105.

Berson, S. A. and Yalow, R. S. (1954). The Distribution of $I^{131}$Labeled Human Serum Albumin Introduced into Ascitic Fluid: Analysis of the Kinetics of a Three Compartment Catenary Transfer System in Man and Speculations of Possible Sites of Degradation, *J. Clin. Invest.*, 33: 377.

Chang, P. C., Rubin, R. T., and Yu, M. (1975). Optimal Statistical Design of Radioimmunoassays and Competitive Protein-Binding Assays, *Endocrinology*, 96: 973.

Cheng, Y. C. and Prusoff, W. H. (1973). Relationship Between the Inhibition Constant (Ki) and the Concentration of Inhibitor Which Causes 50 Percent Inhibition (I50) of an Enzymatic Reaction, *Biochem. Pharmacol.*, 22: 3099.

Cornish-Bowden, A. and Eisenthal, R. (1974). Statistical Considerations in the Estimation of Enzyme Kinetic Parameters by the Direct Linear Plot and Other Methods, *Biochem. J.*, 139: 721.

Currie, D. J. (1982). Estimating Michaelis-Menten Parameters: Bias, Variance and Experimental Design, *Biometrics*, 38: 907.

Darvas, F. (1974). Application of the Sequential Simplex Method in Designing Drug Analogs, *J. Med. Chem.*, 17: 799.

Davies, O. L. (1958). The Design of Screening Tests in the Pharmaceutical Industry, *Bull. Int. Stat. Inst.*, 36: 226.

Dove, S., Streich, W. J., and Franke, R. (1980). On the Rational Selection of Test Series. 2. Two-Dimensional Mapping of Intraclass Correlation Matrices, *J. Med. Chem.*, 23: 1456.

Dowd, J. E. and Riggs, D. S. (1965). A Comparison of Estimates of Michaelis-Menten Kinetic Constants from Various Linear Transformations, *J. Biol. Chem.*, 240: 863.

Dunnett, C. W. (1961). Statistical Theory of Drug Screening, *Quantitative Methods in Pharmacology* (H. DeJonge, ed.), North-Holland, Amsterdam, p. 212.

Ekins, R. P. (1960). The Estimation of Thyroxine in Human Plasma by an Electrophoretic Technique, *Clin. Chim. Acta*, 5: 453.

Federer, W. T. (1963). Procedures and Designs Useful for Screening Material in Selection and Allocation with a Bibliography, *Biometrics*, 19: 553.

Feldman, H. D. (1972). Mathematical Theory of Complex Ligand-Binding Systems at Equilibrium: Some Methods of Parameter Fitting, *Anal. Biochem.*, 48: 317.

Foldman, H., Rodbard, D., and Levine, D. (1972). Mathematical Theory of Cross-Reactive Radioimmunoassay and Ligand-Binding Systems at Equilibrium, *Anal. Biochem.*, 45: 530.

Ferkany, J. W. and Enna, S. J. (1982). Analysis of Biogenic Amines (G. B. Baker and R. T. Coutts, eds.), Elsevier Science Publishing, New York, p. 267.

Finney, D. J. and Phillips, P. (1977). The Form and Estimation of a Variance Function with Particular Reference to Radioimmunoassay, *Appl. Stat.*, 26: 312.

Franke, R. (1984a). QSAR Parameters, *QSAR and Strategies in the Design of Bioactive Compounds* (J. K. Seydel, ed.), VCH, Bad Segeberg, FRG pp. 59–78.

Franke, R. (1984b). *Theoretical Drug Design Methods*, Elsevier, New York, p. 412.

Free, S. M. and Wilson, J. W. (1964). A Mathematical Contribution to Structure-Activity Studies, *J. Med. Chem.*, 17: 395.

Fujita, T. and Ban, T. (1971). Structure-Activity Study of Phenethylamines as Substrates of Biosynthetic Enzymes of Sympathetic Transmitters, *J. Med. Chem.*, 14: 148.

Gallistel, C. R. and Tretiak, O. (1985). *The Microcomputer in Cell and Neurobiology Research* (R. R. Mize, ed.), Elsevier Science Publishing, New York, p. 390.

Hansch, C. (1982). Dihydrofolate Reductase Inhibition. A Study in the Use of X-Ray Crystallography, Molecular Grpahics, and Quantitative Structure-Activity Relations in Drug Design, *Drug Intell. Clin. Pharm.*, 16: 391.

Hansch, C. and Fujita, T. (1964). $\rho$-$\sigma$-$\pi$. A Method for the Correlation of Biological Activity and Chemical Structure, *J. Am. Chem. Soc.*, 86: 1616.

Hansch, C. and Leo, A. (1979). *Substituents Constants for Correlation Analysis in Chemistry and Biology*, Wiley Interscience, New York, p. 339.

Healy, M. J. R. (1972). Statistical Analysis of Radioimmunoassay Data, *Biochem.*, 130: 207.

Hearron, A. E., Elfring, G. L., and Schultz, J. R. (1984). Biopharmaceutical Applications of Group Sequential Deisgns, *Commun. Stat. Theor. Methods*, 13(19): 2419.

Hocking, R. R. (1983). Developments in Linear Regression Methodology: 1959–1982, *Technometrics*, 25: 219.

Johnson, M. A. (1985). A Relating of Metrics, Lines and Variables Defined on Graphs to Problems in Medicinal Chemistry, *Graph Theory with Applications to Algorithms and Computer Science* (Y. Alavi, G. Chartrand, L. Lesniak, D. R. Lick, and C. E. Wall, eds.), Wiley, New York, pp. 457–470.

Kier, L. B. (1980). Molecular Connectivity as a Description of Structure for SAR Analyses, *Physical Chemical Properties of Drugs* (S. H. Yalkowsky, A. A. Sinkula, and S. C. Valvani, eds.), Marcel Dekker, New York, pp. 277–319.

King, E. P. (1963). A Statistical Design for Drug Screening, *Biometrics*, 19: 429.

Klotz, I. M. (1982). Number of Receptor Sites from Scatchard Graphs: Facts and Fantasies, *Science*, 217: 1247.

Langmuir, I. (1918). The Adsorption of Gases on Plane Surfaces of Glass, Mica, and Platinum, *J. Am. Chem. Soc.*, 40: 1361.

Lefkowitz, R. J., Caron, M. G., and Stiles, G. L. (1984). Biochemical, Physiological, and Clinical In0ights Derived from Studies of the Adrenergic Receptors, *N. Engl. J. Med.*, 310: 1570.

Lemke, T. L. (1983). *Review of Organic Functional Groups. Introduction to Medicinal Organic Chemistry*, Lea & Febiger, Philadelphia, p. 131.

Lineweaver, H. and Burke, D. (1934). The Determination of Enzyme Dissociation Constants, *J. Am. Chem. Soc.*, 56: 658.

Little, R. E. (1968). A Note on Estimation for Quantal Response Data, *Biometrika*, 55: 578.

Lockwood, G. F. and Wagner, J. G. (1983). Estimation of Number of Receptor Sites, Binding Capacity, or Vm, *Biopharmaceut. Drug Dispos.*, 4: 397.

Mager, P. P. (1982). Principal Component Regression Analysis Applied to Break Collinearities and Multicollinearities in QSAR, *Zentralbl. Pharm.*, 121: 115.

Martin, Y. C. (1978). *Quantitative Drug Design: A Critical Introduction*, Marcel Dekker, New York, p. 425.

McCabe, G. P. (1984). Principal Variables, *Technometrics*, 26: 137.

Midgley, A. R., Niswender, G. D., and Rebar, R. W. (1969). Principles for the Assessment of the Reliability of Radioimmunoassay Methods (Precision, Accuracy, Sensitivity, Specificity), *Acta Endocrinol. Suppl.*, 142: 163.

Miller, R. G., Jr. (1981). *Survival Analysis*, Wiley, New York, p. 238.

Milton, R. C. (1970). *Rank Order Probabilities*, Wiley, New York,

Molinoff, P. B., Wolfe, B. B., and Weiland, G. A. (1981). Quantitative Analysis of Drug-Receptor Interactions: II. Determination of the Properties of Receptor Subtypes, *Life Sci.*, 29: 427.

Munson, P. J. (1983). Experimental Artifacts and the Analysis of Ligand Binding Data: Results of a Computer Simulation, *J. Receptor Res.*, 3: 249.

Munson, P. J. and Rodbard, D. (1980). LIGAND: A Versatile Computerized Approach for Characterization of Ligand-Binding Systems, *Anal. Biochem.*, 107: 220.

Neter, J. and Wasserman, W. (1974). *Applied Linear Statistical Models*, Richard D. Irwin, Homewood, Ill., p. 87.

Parker, R. B. and Waud, D. R. (1971). Pharmacological Estimation of Drug-Receptor Dissociation Constants. Statistical Evaluation. I. Agonists, *J. Pharmacol. Exp. Ther.*, 177: 1.

Pomona Medicinal Chemistry Project. (1984). Pomona College, Claremont, Calif.

Ramm, P. and Kulick, J. H. (1985). *The Microcomputer in Cell and Neurobiology Research* (R. R. Mize, ed.), Elsevier Science Publishing, New York, p. 390.

Redman, C. E. (1981). Screening compounds for clinically active drugs, *Statistics in the Pharmaceutical Industry* (C. R. Buncher and J. Y. Tsy, eds.), Marcel Dekker, New York.

Rodbard, D. and Hutt, D. M. (1974). Statistical Analysis of Radioimmunoassays and Immunoradiometric (Labeled Antibody) Assays: A Generalized Weighted, Iterative, Least-Squares Method for Logistic Curve Fitting, *Symposium on RIA and Related Procedures in Medicine, International Atomic Energy Agency, Vienna*, Unipub, New York, p. 209.

Rodbard, D., Lenox, R. H., Wray, H. L., and Ramseth, D. (1976). Statistical Characterization of the Random Errors in the Radioimmunoassay Dose-Response Variable, *Clin. Chem.*, 22: 350.

Roseberry, T. D. and Gehan, E. A. (1964). Operating Characteristic Curves and Accept-Reject Rules for Two and Three Stage Screening Procedures, *Biometrics*, 20: 73.

Scatchard, G. (1949). The Attraction of Proteins for Small Molecules and Ions, *Ann. N.Y. Acad. Sci.*, 51: 660.

Schultz, J. R., Nichol, F. R., Elfring, G. L., and Weed, S. D. (1973). Multiple Stage Procedures for Drug Screening, *Biometrics*, 29: 293.

Skulnick, H. I., Weed, S. D., Edison, E. E., Renis, H. E., Wierenga, W., and Stringfellow, D. (1985). Pyrimidinones. 1. 2-Amino-5-halo-6-aryl-4(3H)-pyrimidinones. Interferon-Inducing Antiviral Agents, *J. Med. Chem.*, 28: 1864.

Spendley, W., Hext, G. R., and Himsworth, R. F. (1962). Sequential Application of Simplex Designs in Optimization and Evolutionary Operation, *Technometrics*, 4: 441.

Steinberg, D. M. and Hunter, W. G. (1984). Experimental Design: Review and Comment, *Technometrics*, 26: 71.

Streich, W. J., Dove, S., and Franke, R. (1980). On the Rational Selection of Test Series. 1. Principal Component Method Combined with Multidimensional Mapping, *J. Med. Chem.*, 23: 1452.

Stuper, A. J., Brugger, W. E., and Jurs, P. C. (1979). *Computer Assisted Studies of Chemical Structure and Biological Function*, Wiley, New York, p. 220.

Swain, C. G. and Lupton, E. C., Jr. (1968). Field and Resonance Components of Substituent Effects, *J. Am. Chem. Soc.*, 196: 4328.

Tiede, J. J. and Pagano, M. (1979). The Application of Robust Calibration to Radioimmunoassay, *Biometrics*, 35: 567.

Topliss, J. G. (1972). Utilization of Operational Schemes for Analog Synthesis in Drug Design, *J. Med. Chem.*, 15: 1006.

Venditti, J. M., Westley, R. A., and Plowman, J. (1984). Current NCI Preclinical Antitumor Screening in vivo: Results of Tumor Panel Screening, 1976–1982, and Future Directives, *Adv. Pharmacol. Chemother.*, 20: 1.

Wald, A. (1947). *Sequential Analysis*, Wiley, New York.

Weiland, G. A. and Molinoff, P. B. (1981). Quantitative Analysis of Drug-Receptor Interactions: I. Determination of Kinetic and Equilibrium Properties, *Life Sci.*, 29: 313.

Wilkinson, G. N. (1961). Statistical Estimation in Enzyme Kinetics, *Biochem. J.*, 80: 324.

Williams, M. and U'Prichard, D. C. (1984). Drug Discovery at the Molecular Level: A Decade of Radioligand Binding in Retrospect, *Annu. Rep. Med. Chem.*, 19: 283.

Wootton, R. (1983). Selection of Test Series by a Modified Multidimensional Mapping Method, *J. Med. Chem.*, 26: 275.

# 3 Assessment of Pharmacological Activity

JOHN J. HUBERT *University of Guelph, Guelph, Ontario, Canada*

NEETI R. BOHIDAR *Smith Kline & French Laboratories, Swedeland, Pennsylvania*

KARL E. PEACE *G. D. Searle & Co., Skokie, Illinois*

## I. INTRODUCTION

The need for the assessment of pharmacological activities of candidate compounds, recently screened and synthesized, arises very early in the product research and development process. Pharmacology is the study of the selective biological activity of chemical substances on living matter. A substance has biological activity when, in appropriate doses, it causes a cellular response. It is selective when the response occurs in some cells and not in others. A chemical with selective activity of medicinal value in the treatment of disease is called a *drug*. Its use in the treatment of disease is part of therapeutics. Historically, pharmacology evolved as an essential part of medicine. The need for new compounds with selective activity useful in combating disease is still the strongest and most compelling incentive for continuing investigative new drug research. Most commonly used drugs are classified according to the organ system on which they exert their chief selective action (Bevan, 1976). In the next section, we will describe the actions of drugs based on the organ system classification.

Most drugs are pure chemical compounds of known structure. Many agents commonly used in therapeutics, particularly extracts of plants or animal tissues, are not pure, and the chemical structure of the active principle is not always known. In these cases it is necessary to determine the quantity to be used by observing the biological responses to the drug. The estimation of drug potency by the reactions of living organisms or their components is known as *bioassay*. The principles of bioassay are important not only because the synthesized compounds and valuable drugs must be subjected to bioassay to determine their activities, but also because the same basic principles are applicable to all comparisons of drug effectiveness, including clinical trials of new drugs and their continuing practical evaluation.

Whereas physical and chemical properties such as weight and optical density can be measured with great precision (0.001 to 1%), the errors involved in the measurement of biological properties are much greater (5 to

50%). It is therefore an absolute necessity to design and analyze a bioassay experiment in such a way as to minimize the effects of biological variation by using crossover or blocking strategies.

At one time the biological activity of some drugs was expressed directly in terms of a simple pharmacologic response; for example, the quantity of digitalis extract required to kill a frog was defined as a "frog-unit." Since frogs differ greatly in their species, sensitivity, weight, and other factors, this kind of definition did not provide a sufficiently stable measure for clinical dosage (Bevan, 1976). These units have been replaced by absolute reference standards. A reference standard is a stable sample of the drug maintained in the United States by the Board of Trustees of the United States Pharmacopoeial Convention and distributed to pharmaceutical companies for standardization of their product.

To minimize the effects of biological variation, the responses to the standard and unknown samples should be observed in animals which are as similar as possible in every respect. Factors such as age, sex, genetic strain, body weight, and vendor (farm) can exert a major influence on the sensitivity to drugs. Unless these factors are carefully controlled, the precision of the assay will be reduced and the result may be unreliable. Individual animals should then be assigned to receive standard or unknown preparations on an objectively random basis, such as by use of a table of random numbers. Appropriate statistical experimental designs, such as the randomized block design, Latin square design, and crossover design, should be used. Wherever possible, the biological response should be measured objectively. Instruments for measuring various physiological and pharmacological responses must be accurately calibrated.

Many of our existing drugs are of plant origin, and currently there is a worldwide search for new therapeutically useful plant products. An alternative way to discover new drugs is by empirical screening of chemicals. This time-honored method, routinely used by the pharmaceutical industry, is a relatively easy and inexpensive method for detecting the desired effect. A considerable number of newly marketed drugs are congeners of existing agents (Bevan, 1976). Animal screening, which consists of testing the synthesized compounds primarily in rodents, follows chemical screening. The first test is a simple toxicity measurement, the $LD_{50}$ in mice, which gives some idea of the dose range to be used in subsequent experimentation. The next test is a general or profile screen. The pharmacologist administers a few doses over a wide range to a small number of mice or rats and then observes these animals for a few hours or days. Astute observation at this time can give clues to numerous types of pharmacodynamic effects. If the substance exhibits any interesting pharmacologic activity in the general screen, it is then tested in a system-oriented screen (Bevan, 1976). For example, the compound may be administered to an intact, anesthetized dog to determine whether there is an effect on the cardiovascular system, such as cardiac stimulation, peripheral vasodilation, vasoconstriction, or hypotension. Several in vitro as well as in vivo tests are employed to determine whether the compound has certain commonly recognized pharmacologic qualities, such as adrenergic or cholinergic stimulating properties. Screening of certain types of compounds is disease-oriented, as in the search for anti-inflammatory and anti-infective compounds. Generally, spontaneously hypertensive rats are used for screening antihypertensive agents (Bevan, 1976), and rats with carrageenan-induced foot edema are used for screening anti-inflammatory agents. Biochemical screening involves determination of the effect of a compound on specific enzyme systems. When a compound has shown activity during screening, it must be evaluated. At this stage, the full spectrum of the agent's biological and pharmacological properties and, if possible, its mechanism of action and metabolic fate are defined more explicitly. Once the potential usefulness of an agent has been determined in the laboratory, tests to demonstrate its safety are initiated in laboratory animals.

Since most of the pharmacological parameters, such as potency, dose range, and effective doses, are determined experimentally at various stages of the development, the application of statistics to the design and analysis of bioassay experiments cannot be overemphasized. Biological assay is now a recognized method for the study of drug effects on living matter. Because of this unique usage, bioassay statistics is considered a separate branch of applied statistics, along with chemometrics, econometrics, and psychometrics.

## II. PHARMACOLOGICAL ASSAY METHODS FOR VARIOUS CLASSES OF THERAPEUTIC AGENTS

As mentioned in the Introduction, each therapeutic agent is classified according to the organ system on which it exerts a desirable effect (Bevan, 1976). The various pharmacological assay methods will be presented here according to this classification. The primary purpose of an assay method is to assess the pharmacological activity potential of the candidate compound. If the compound is active, it is retained for further investigation. A pharmacologist, however, develops four or five different assay methods for the same therapeutic class, because the results of these tests provide a broad spectrum of the activity of the compound in question.

Before we proceed to a formal description of the assay methods, it is appropriate here to define the criteria for an efficient assay method—in other words, the properties that a pharmacologist is looking for in an assay method.

### A. List of Criteria for an Efficient Assay Method (Van Arman and Bohidar, 1978)

1. *Validity*. For a human disease, validity requires that drugs proved effective clinically should be effective in the laboratory model, that drugs effective in the model should be effective clinically, that drugs not effective clinically should not be effective in the model, and that drugs inactive in the laboratory should not be active in the clinic. These requirements may not be met for many diseases. However, the model need not ostensibly resemble the clinical diseases; generally, models have some points of similarity with a clinical disease. However, pharmacologists are developing highly sophisticated refined methods that are expected to correlate well with clinical experience.

2. *Reliability*. Here we mean merely how reproducible a given assay method is in the laboratory. If the assay method has high variability, it has low reliability. It is a good laboratory practice to test the assay method periodically by submitting one or two standard drugs at regular intervals.

3. *Sensitivity*. The concept of sensitivity in bioassay is different from that in, for instance, chemical assay. It is the total amount of standard

drug required to achieve a given level of response. For example, the dog's knee joint assay associated with antiarthritic screening requires several dozen milligrams of indomethacin (4 g of aspirin). This assay by our definition is considered to be insensitive. The antipyretic assay (yeast fever) for the same screen is more sensitive since it requires only about 2 mg in rats. In screening methods, it is economical to have good sensitivity.

These are some of the most important attributes to be considered in the selection of an assay method for the development of a drug. The assays considered here are merely examples. They have been selected because they have wide applicability and involve commonly used laboratory species and relatively simple equipment.

## B. Cardiovascular

### 1. Antihypertensive Agents (Van Arman and Bohidar, 1978)

Essential hypertension is associated with an increase in total peripheral resistance. Since peripheral resistance is regulated by the sympathetic nervous system, most antihypertensive agents act to decrease the influence of the sympathetic nervous system on vessel resistance.

Many experimental models of hypertension are available for drug screening and evaluation. Investigators have experimentally elevated blood pressure in rats, mice, rabbits, cats, dogs, and monkeys. These models mimic human hypertension and provide a tool for developing safe and efficacious antihypertensive drugs. The hormonal model consists of renal hypertension, DOCA hypertension, and angiotensin-induced hypertension. Neurogenic hypertension can be produced in animals by surgical procedures which remove or modify neural homeostatic mechanisms. Genetic hypertension has been produced in a strain of rats successfully developed with markedly elevated blood pressure. These rats are called spontaneously hypertensive (SH) rats. They are commercially available and are ideal for screening purposes because of their sensitivity to standard antihypertensive agents. The sensitivity of SH rats to environmental factors such as stress and high sodium diet parallels comparable aspects of human essential hypertension. One drawback of an SH rat screening program is the acknowledged difficulty in reproducing experimental data from one laboratory to another. This may be due to genetic drift accentuated by the large number of separate SH rat breeding sites. Accordingly, a conservative screening approach is suggested within one's own laboratory. The use of SH rats is the most highly recommended screening tool. A compound is considered active if it reduces the mean arterial blood pressure by at least 20 mm Hg.

Some of the drugs successfully tested in the models described above are clonidine, chlorothiazide, L-Dopa, furosemide, hydralazine, methyldopa, propranolol, reserpine, and sotalol.

As far as direct and indirect blood pressure measurements are concerned, the rat is an ideal test model because its tail can be used for indirect blood pressure measurements utilizing a tail cuff. Direct measurement of blood pressure can be achieved by passing a catheter into the caudal artery of the rat and attaching the other end of the catheter to a pressure transducer.

If the primary tests are successful, secondary tests can be accomplished by resorting to the anesthetized dog challenge test.

Statistical analysis will be required for the primary tests as well as the secondary tests. Some background knowledge, such as that given above, will help not only in the development of an appropriate experimental design but also in the appropriate interpretation of the experimental results.

## C. Musculoskeletal (Joint) System

### 1. Antiarthritic Agents (Van Arman and Bohidar, 1978)

Drugs that are used to moderate the pain, swelling, heat, and general discomfort of acute and chronic inflammatory conditions are called analgesic, antipyretic, anti-inflammatory, and antiarthritic agents. The drugs are particularly useful if the pain is of peripheral origin.

Several animal models have been developed for screening, such as the rat, mouse, and dog assays. The following assay methods have been reported in the literature:

1. Carrageenan-induced foot edema. In this assay, male rats are used primarily. The response variable is the degree of swelling of the rat's paw.
2. Granuloma caused by cotton pellet. Again, male rats are used primarily. The weight of the cotton pellet is the response variable.
3. Adjuvant-induced arthritis. Male rats are used primarily. The condition is induced by injecting *Mycobacterium butyricum* into the tail. The response variable is the degree of swelling of the rat's paw. It may be noted here that all drugs known to be active in rheumatoid arthritis that have been tested by this procedure have been found active; in other words, no false negatives have yet been discovered (Van Arman and Bohidar, 1978).
4. Yeats fever. Primarily male rats are used. Brewer's yeast is used as the inducing agent. The response variable is the body temperature of the rat.
5. Inflammation in the dog's knee joint. Obviously, the animal model is the dog. A suspension of sodium urate, injected into one hind knee joint of a dog, develops the inflammatory condition. The response consists of the percent foot pressure of the inflamed foot.
6. Topical inflammation. Primarily female mice are used in this assay. The irritant used topically is a mixture of croton oil, pyridine, and diethyl ether. The response variable is the weight of a circular piece of skin removed from the left ear.

An anti-inflammatory agent must effectively inhibit in animals (models) carrageenan-induced edema, adjuvant-induced polyarthritis, and urate-induced synovitis.

Some of the well-known agents which have been tested by these assay methods and found to be very effective are indomethacin, sulindac, diflunisal, phenylbutazone, thiabendazole, and aspirin.

Because laboratory experiments must be done with animals, some methods must be used that have as many features as possible in common with the human disease. Since the animal model may not completely resemble the human disease, it is best to use a number of different models.

Application of statistics in evaluating the pharmacological effectiveness of a drug is absolutely necessary. This background information will be extremely helpful in the appropriate interpretation of the results. It will also help in the development of appropriate experimental designs.

The primary purpose of presenting the pharmacological assay methods associated with these two classes of therapeutic systems—the cardiovascular system involving antihypertensive agents and the musculoskeletal (joint) system involving antiarthritic agents—is to illustrate in detail the types of activities associated with the pharmacological assessment of candidate chemical compounds. These two illustrations show that the statistician must have a good understanding of the background biological information prior to embarking on bioassay data analysis. This background knowledge will enable a statistician to help the scientists develop more efficient assay methods.

The assay methods associated with other important pharmacological systems, such as the renal system, gastrointestinal system, respiratory/pulmonary system, and central nervous system, have not bee presented because of space and time considerations. However, statisticians should be able to receive this information either from the scientists they are consulting with or from the relevant pharmacological literature.

## III. BIOASSAY, DRUGS, AND DOSES

### A. Introduction

1. *Definition of bioassay.* The experimental determination of the potency or strength of a chemical or biological (organic or inorganic) substance based on the response observed after its administration to living matter (animal or animal tissue) consititues a biological assay.

2. *Objective of bioassay.* The primary purpose of bioassay is to estimate and compare the potencies of the chemical compounds (drugs) under investigation based on an appropriate well-designed experiment. The estimation and comparison of potencies must be accomplished by appropriate statistical methodologies.

3. *Structure of bioassay.* Using the experimental design terminologies, the chemical substance (drug, vaccine, antibiotic, vitamin, hormone) constitutes the "treatment," the biological material (intact animal: rodents, primates; animal tissue: muscle, heart, aorta, embryo) forms the "experimental unit," and the measured observations (heart rate, mean arterial pressure, electrolyte) constitute the "response variable."

4. *Variabilities associated with potency.* The potency of the same drug substance may be altered by the following factors: (1) age (fetal, adult), (2) sex (male, female), (3) species (rodents, primates, humans), (4) route of administration (intravenous, subcutaneous, intraperitoneal), (5) strain (genetic strain, spontaneously hypertensive rats), (6) sham operation (altered anatomy, removal of one kidney, for instance), and (7) the parameter being measured.

### B. Characterization of a Drug and Its Doses

A drug is defined to be a set of doses. Symbolically, we have as follows:

$$A = \{a_1\ a_2\ \cdots\ a_k\}$$

where A denotes the drug and $a_1$, $a_2$, and $a_k$ denote the doses. Note that the symbol "A" is merely a label.

Consider another drug, B, defined as,

$$B = \{b_1\ b_2\ \cdots\ b_K\}$$

We introduce the "null" dose by expressing A and B in the following manner:

$$A = \{a_0\ a_1\ a_2\ \cdots\ a_K\}$$

$$B = \{b_0\ b_1\ b_2\ \cdots\ b_K\}$$

where $a_0$ and $b_0$ represent "saline" or "vehicle". Generally, the doses of a drug may be divided into three major subsets:

$$A = \{a_0\ a_1\ a_2\ \cdots\ a_K \quad a_{K+1}\ \cdots\ a_p \quad a_{p+1}\ \cdots\ a_n\}$$

where $a_0$–$a_K$ represent the very small doses (virtually safe doses, inactive subset of doses), $a_{K+1}$–$a_p$ the pharmacological doses, and $a_{p+1}$–$a_n$ the toxicological (very high) doses.

Usually the doses of a drug are expressed numerically. (Since drugs are merely labels, they are not expressed numerically.) For bioassay studies, the doses of a drug are not selected at random; they are selected on the basis of range of activity in which one is interested. In a typical bioassay, the graded doses of the drug are expected to produce a graded set of responses.

Effectiveness of a dose (drug). A dose of a drug is effective if that dose is significantly different from the control, based on a two-sided Dunnett's test. A drug is effective if it is significantly more potent than the reference standard based on the same range of doses.

## C. Notion of Relative Potency

Comparison of two drugs essentially implies comparison of two sets of doses, such as $\{a_0\ a_1\ a_2\ \cdots\ a_K\}$ vs. $\{b_0\ b_1\ b_2\ \cdots\ b_K\}$. One is tempted to do a multiple-range test (Duncan's test, say) by comparing all possible pairs of doses and interpreting only the relevant comparisons, $a_1$ vs. $b_1$, $a_2$ vs. $b_2$, and so on. This does *not* constitute a bioassay analysis. In fact, this is a severe violation of the principles of bioassay statistics. All doses of A must be compared with all doses of B simultaneously. This is achieved by regression analysis. We compare the parameters of the regression lines: the slopes and the intercepts. Bioassay problems are inverse regression problems. We are not interested in predicting the response ($\hat{Y}$) for a given dose ($X_0$), but we are interested in "predicting" the dose ($\hat{X}$) for a selected value of the response ($Y_0$). We may state that the purpose of a bioassay experiment is to determine that *dose* of a drug which yields a given value of response. Consider an antihypertensive agent. Our interest here will be to estimate that dose of the drug which causes a reduction in mean arterial pressure of 20 mm Hg in spontaneously hypertensive rats.

Now, let $\theta_A$ be the dose of drug A which yields a given value of response ($Y_0$), and let $\theta_B$ be the dose of drug B which yields the same value of response ($Y_0$). If $\theta_A$ is less than $\theta_B$ ($\theta_A < \theta_B$), then, by definition, we say that drug A is more potent than drug B. If $\theta_A$ is equal to $\theta_B$, then, by definition, we say that drug A and drug B are equipotent.

## IV. TESTS OF VALIDITY FOR BIOASSAY ANALYSIS

The most widely used statistical techniques for bioassay analysis are the analysis of variance procedure and the regression analysis. Several assumptions are implicit in these procedures. The role of the tests of validity will be to examine the data for any violation of these assumptions. The point estimation of potency and its confidence intervals are valid if and only if the data meet the implicit assumptions. The following tests of validity are considered.

### A. Equality of Variances of Dose Groups by Levene's Test (Levene, 1971)

Levene's test is a robust test, in that it is not sensitive to the violation of normality assumption.

1. For the completely randomized design (CRD) Levene's test is a one-way analysis of variance on the absolute values of the differences between each observation and the average of its group. Symbolically, we have

$$Z_{ij} = |Y_{ij} - \bar{Y}_i| \qquad i = 1, 2, \ldots, k, \quad j = 1, 2, \ldots, n \text{ (or } n_i\text{)}$$

where $Z_{ij}$ is the absolute value of the deviation from its group mean, $Y_{ij}$ is the jth observation in the ith group, $\bar{Y}_i$ is the mean of the ith group, and n is the number of observation per group ($n_i$ is the number of observations for the ith group). $Z_{ij}$ is called the Levene statistic for the completely randomized design.

2. For the randomized block design (RBD), Levene's test is a two-way analysis of variance (one observation per cell) on the following Levene statistic:

$$Z_{ij} = |Y_{ij} - \bar{Y}_R - \bar{Y}_c + \bar{\bar{Y}}| \qquad i = 1, 2, \ldots, b, \quad j = 1, 2, \ldots, t$$

where $Z_{ij}$ is the calculated Levene statistic for the observation $Y_{ij}$ associated with the ith row and the jth column, $Y_{ij}$ is the raw observation in the ij cell, $\bar{Y}_c$ is the mean of the jth column, and $\bar{\bar{Y}}$ is the overall mean of all the bt data points.

3. For the Latin square (LSQ) design, Levene's test is an analysis of variance associated with a Latin square design based on the following Levene statistic:

$$Z_{ij(k)} = |Y_{ij(k)} - \bar{Y}_R - \bar{Y}_c - \bar{Y}_T + 2\bar{Y}|$$

$$i = 1, 2, \ldots, t, \; j = 1, 2, \ldots, t, \; k = 1, 2, \ldots, t$$

where $Z_{ij(k)}$ is the calculated Levene statistic for the observation $Y_{ij(k)}$ associated with the ith row, jth column, and kth treatment, $Y_{ij(k)}$ is the raw observation in the ij cell, $\bar{Y}_R$ is the mean of tho ith row, $\bar{Y}_c$ is the mean of the jth column, $\bar{Y}_T$ is the mean of the treatment associated with the ij cell, and $\bar{\bar{Y}}$ is the mean of $t^2$ data points.

4. Levene's test for comparing two residual variances associated with the estimation of relative potency consists of the following steps.

I. Let the two drugs be labeled as A and B. Calculate the Y-adjusted values for each observation, based on the following formula:

$$Y_{ADJ(i)} = Y_i - B_c(X_i - \bar{\bar{X}}) \qquad i = 1, 2, \ldots, n_1 + n_2$$

where $n_1$ is the number of pairs in drug group A, $n_2$ the number of pairs in drug group B, $Y_{ADJ(i)}$ the adjusted value for $Y_i$, $B_c$ the common slope for drug groups A and B, assuming parallelism between the two slopes, $X_i$ the dose metameter associated with the $Y_i$ response, and $\bar{\bar{X}}$ the overall mean of the X values associated with both the drugs.

II. Now each drug has a set of Y-adjusted values. Based on the Y-adjusted values just calculated, we conduct a one-way analysis of variance on the $Z_{ij}$, as defined below:

$$Z_{ij} = |Y_{Adj(io)} - Y_{Adj(i)}| \qquad i = 1, 2, j = 1, 2, \ldots, n_i$$

This constitutes the Levene test for comparing two regression residual variances.

## B. Test of Normality (Wilk-Shapiro Test)

The Wilk-Shapiro test provides the following results: (1) test of normality for each group and (2) test of normality for all the groups combined. The procedure is too elaborate to be described here. However, the reference in Wilk and Shapiro (1968) provides the procedure in detail.

## C. Test of Nonadditivity (Tukey's Test)

Since the additivity of rows and columns is a cardinal assumption in the analysis of the randomized block design and Latin square design, we must conduct Tukey's test of nonadditivity prior to the bioassay analysis. The procedures have been described in detail in Tukey (1949).

# V. INVERSE REGRESSION, FIELLER'S THEOREM, AND THE JACKKNIFE METHOD

## A. Inverse Regression

Regression analysis forms the backbone of bioassay analysis. Linear regression of the response measured on the logarithm of the doses selected is generally used. However, in bioassay, the prediction of the response for a given dose is of no interest. The estimation of the dose associated with a given response is of considerable interest. This is what is known as the inverse regression. Let the linear regression equation have the form

$$\hat{Y} = A + BX$$

where $\hat{Y}$ is the estimated value of the response Y, A denotes the intercept, B denotes the slope, and X denotes the log dose.

Then it follows that

$$\hat{X} = \frac{Y_0 - A}{B} = \bar{X} + \frac{Y_0 - \bar{Y}}{B}$$

where $\hat{X}$ is the estimated log dose for a given Y (= $Y_0$).

Our interest will be to find the $(1 - \alpha)$ 100% confidence interval of the true X. There are three ways to approach this problem: (1) the statistical differential (SD) method, (2) the Fieller's theorem (FT) approach, and (3) the jackknife (JK) method.

In the following, we domonstrate for each method the derivation of the $(1 - \alpha)100\%$ confidence interval of a ratio, $R = A/B$, in general terms; where A and B are sample estimates.

## B. Statistical Differential Method

In this method, the variance of a ratio is derived by the Taylor series expansion method as follows:

$$V[f(\theta_1, \theta_2)] = \left(\frac{\delta f}{\delta \theta_1}\right)^2 V(\theta_1) + \left(\frac{\delta f}{\delta \theta_2}\right)^2 V(\theta_2) + 2\left(\frac{\delta f}{\theta_1}\right)\left(\frac{\delta f}{\theta_2}\right) \mathrm{Cov}(\theta_1, \theta_2)$$

where f is a function of two variables, $\theta_1$ and $\theta_2$, V is the symbol for variance, $\delta$ denotes the partial derivative operator, and Cov denotes the covariance between $\theta_1$ and $\theta_2$. Now we have

$$V(R) = \left(\frac{1}{B}\right)^2 V(A) + \left(\frac{A}{B^2}\right)^2 V(B) + 2\left(\frac{A}{B^3}\right) \mathrm{Cov}(A, B)$$

The structure of the $(1 - \alpha)100\%$ confidence interval is

$$R \pm t_{\alpha(d.f.)} \sqrt{V(R)} \qquad \text{(d.f. = degrees of freedom)}$$

It should be noted here that this approach is applicable for large-sample problems only, since V(R) has only a asymptotic properties as an estimator.

## C. Fieller's Theorem Approach (Finney, 1978)

This approach is presented here in general terms. Specific results are presented in subsequent sections.

Assume that A and B are unbiased estimates of the parameters A* and B* and that they are each a linear function of a set of observations with Gaussian distribution. For Fieller's method per se, A and B could be means, differences between means, or regression coefficients calculated from experimental data. For the inverse regression bioassay analysis discussed in Section V,A, A would be the estimate of the unknown but true intercept and B the estimate of the true slope. Let R* be a fixed point in the two-dimensional parameter space. Then

$$R = \frac{A}{B}$$

is an estimator of R*. This relationship is then expressed as a homogeneous equation involving a linear function of two normally distributed variables:

$$A - RB = 0$$

Since the ratio of a squared normal variate and its variance is distributed as a Snedecor F central distribution, with one and $\nu_2$ degrees of freedom, we have

$$[V(A - RB)]F_{1,\nu_2(\alpha)} = (A - RB)^2$$

where $100(1 - \alpha)$ is the desired probability associated with the confidence statement about the parameter R*. Rearranging the expression above as a polynomial in R, we have

$$R^2[B^2 - V(B)F_{1,\nu_2(\alpha)}] - 2R[BA - Cov(A, B)F_{1,\nu_2(\alpha)}]$$
$$+ [A^2 - V(A)F_{1,\nu_2(\alpha)}] = 0$$

The two solutions of R which constitute the upper and lower $100(1 - \alpha)\%$ confidence limits of R are

$$(R_U, R_L) = [AB - Cov(A,B)F_{1,\nu_2(\alpha)}] \pm \{[AB - Cov(A, B)F_{1,\nu_2(\alpha)}]^2$$
$$- [B - F_{1,\nu_2(\alpha)}V(B)][A - F_{1,\nu_2(\alpha)}V(A)]\}^{1/2} \quad /[B^2 - V(B)F_{1,\nu_2(\alpha)}]$$

These limits are valid if and only if the following constraint holds:

$$V(B)F_{1,\nu_2(\alpha)}B^{-2} \ll 1.0$$

The quantity on the left-hand side of the inequality is generally referred to as the "g" value (see Finney, 1978).

### D. Jackknife Method (Miller, 1979; Bohidar, 1987)

It is well known that the variance of a ratio, as derived by the Taylor series expansion method, is not unbiased and has only asymptotic properties. Construction of the confidence limits of a ratio by using Fieller's theorem is appropriate. However, the magnitude of the interval is solely dependent on the magnitude of the g value. This constraint, as well as the fact that the theorem does not specify a variance of the ratio, should lead to a search for an alternative method of approaching the problem. The jackknife method is the best method with a considerable bias-reducing property.

Let $Y_{(j)}$ be the result of making the complex calculation on the portion of the sample that omits the jth subgroup, that is, making it a pool of the (K − 1) subgroup. Let $Y_{all}$, read, Y subscript all, be the corresponding result of the entire sample.

Define pseudovalues as follows:

$$Y_j^* = KY_{all} - (K - 1)Y_{(j)} \qquad j = 1, 2, \ldots, K$$

The jackknifed value Y*, which is our best single result, and an estimate $S_*^2$ of its variance are then given by

$$Y^* = \frac{\Sigma Y_j^*}{K}$$

$$S^{*2} = \frac{\Sigma Y_j^{*2} - [(\Sigma Y_j^*)^2/K]}{K - 1}$$

$$S_*^2 = \frac{S^{*2}}{K}$$

For our situation, we have the following algorithm in general terms. Define

$$L = \log R = \log\left(\frac{A}{B}\right)$$

Now,

$$L_{all} = \log\left(\frac{A_{all}}{B_{all}}\right)$$

Define

$$L_{(1)} = \log\left(\frac{A_{(1)}}{B_{(1)}}\right)$$

$$L_{(2)} = \log\left(\frac{A_{(2)}}{B_{(2)}}\right)$$

$$\vdots$$

$$L_{(K)} = \log\left(\frac{A_{(K)}}{B_{(K)}}\right)$$

where $L_{(i)}$ is computed by excluding the ith pair of $(X_i, Y_i)$, the dose response pair.

Now construct the pseudovalues as follows:

$$L^*_{(1)} = KL_{all} - (K - 1)L_{(1)}$$

$$L^*_{(2)} = KL_{all} - (K - 1)L_{(2)}$$

$$\vdots$$

$$L^*_{(K)} = KL_{all} - (K - 1)L_{(K)}$$

The means of the pseudovalues are calculated as

$$L^{**} = (L^*_{(1)} + {}^*_{(2)} + \cdots + L^*_{(K)})K^{-1}$$

The estimated variance is calculated as

$$S^{*2} = \frac{\Sigma_{i=1}^{K} (L^*_{(i)} - L^{**})^2}{K - 1}$$

The $100(1 - \alpha)\%$ confidence limits are given by

$$(R_U, R_L) = \text{antilog}\left[L^{**} \pm t_{(K-1)(\alpha)} \sqrt{\frac{S^{*2}}{K}}\right]$$

It may be noted here that when the g value is equal to zero, the statistical differential method yields the same result as the Fieller's theorem approach. However, in real-world situations, the g value will not be equal to zero. For this reason, the Fieller's theorem approach is used quite frequently in practice.

## VI. DOSE-RESPONSE RELATIONSHIP: ESTIMATION OF MEDIAN DOSE FOR QUANTITATIVE RESPONSE

### A. Definitions and Numerical Illustration

An important type of relationship encountered in several kinds of biological experiments is that between the dose of a drug and the response elicited. If dose is increased systematically in an isolated tissue or intact animal, a graded response may be obtained. At first there will be a range of doses so low that no response is manifest. Then a higher range of doses elicits responses of increasing magnitude, and finally a maximal response may be attained which cannot be exceeded at any dose. If log dose is plotted on the X axis and response on the Y axis, a symmetrical sigmoid curve is characteristically obtained whose central portion is nearly linear.

For most biological data the logarithmic transformation is most useful. Many measurements yield skewed frequency distributions when X values are plotted directly but fit the symmetrical normal distribution better when logarithms of X values are used. In some cases this can be attributed to a limitation of the possible range of variation in one direction or the other. For example, the mean heart rate in humans is about 70 beats per minute. Deviations to the left are restricted by a lower limit of about 40, whereas possible deviations to the right may be much greater. It is also true that responses to drugs tend to vary in proportion to log dose rather than to dose, so dose-response relationships are routinely plotted with log dose rather than dose on the X axis.

For purposes of statistical analysis, the log dose-response curve may be regarded as a cumulative normal distribution. We consider the responding system to be composed of a population of responsive units whose intrinsic sensitivities to the drug are distributed normally with respect to log dose. As dose is increased, response units of diminishing sensitivity are progressively activated, so that at a given dose all units sensitive to that dose or to any smaller dose respond. The resulting response curve is therefore a plot of the cumulative area of the normal curve. Our primary interest will be to estimate an effective dose ($ED_\alpha$) from the log dose-response curve. The $ED_\alpha$ pertains to that dose of the drug which yields a level of response equivalent to $\alpha$ units. For instance, in antihypertensive studies $ED_{20}$ would be that dose of the agent which would cause a maximal fall of 20 mm Hg in mean arterial pressure (MAP) in spontaneously hypertensive rats. If the response is expressed as a percentage, $ED_{50}$ would be that dose of the drug which would elicit a level of response equivalent to 50%. The procedure for estimating $ED_\alpha$ is illustrated in step-by-step detail in the following.

This is a study of the effect of an antihypertensive agent on maximum fall in MAP in mm HG during a period of 24 hours. The doses of the antihypertensive agents are 20, 40, and 80 mg/kg. The data and the detailed computations are provided in the following.

Let X = log dose and Y = maximum fall in MAP.

| | Group | | | | | | |
|---|---|---|---|---|---|---|---|
| | 20 mg/kg | | 40 mg/kg | | 80 mg/kg | | |
| | (Y) | (X) | (Y) | (X) | (Y) | (X) | Sums |
| | 6 | 1.301 | 12 | 1.602 | 16 | 1.903 | |
| | 6 | 1.301 | 11 | 1.602 | 18 | 1.903 | |
| | 6 | 1.301 | 12 | 1.602 | 19 | 1.903 | |
| | 7 | 1.301 | 12 | 1.602 | 16 | 1.903 | |
| | 4 | 1.301 | 10 | 1.602 | 15 | 1.903 | |
| $\Sigma X_i$ | | 6.505 | | 8.010 | | 9.515 | 24.030 |
| $\Sigma Y_i$ | 29 | | 57 | | 84 | | 170.0 |
| $n_i$ | | 5 | | 5 | | 5 | 15 = N |
| $\bar{Y}$ | 5.8 | | 11.4 | | 16.8 | | 11.33 = Y |
| $\Sigma X^2$ | | 8.463005 | | 12.83202 | 18.107045 | | 39.402070 |

| | Group | | | | | | |
|---|---|---|---|---|---|---|---|
| | 20 mg/kg | | 40 mg/kg | | 80 mg/kg | | |
| | (Y) | (X) | (Y) | (X) | (Y) | (X) | Sums |
| $\Sigma Y^2$ | 173 | | 653 | | 1422 | | 2248 |
| $\Sigma XY$ | 37.729 | | 91.314 | | 159.852 | | 288.895 |

$$[Y^2] = 22.48 - \frac{(170)^2}{15} = 321.333$$

$$[X^2] = 39.402070 - \frac{(24.030)^2}{15} = 0.9060$$

$$[XY] = 288.895 - \frac{(24.030)(170)}{15} = 16.5550$$

$$\frac{[XY]^2}{[X^2]} = \frac{(16.5550)^2}{0.9060} = 302.50$$

where

$$[Y^2] = \Sigma Y^2 - \frac{\Sigma Y^2}{n} = \Sigma(Y - Y)^2$$

$$[XY] = \Sigma(X - \bar{X})(Y - \bar{Y})$$

Note that this "square bracket" notation will be used throughout the chapter

The linear regression model is

$$Y_i = A + BX + E_i \qquad i = 1, 2, \ldots, n$$

where A and B denote the intercept and slope, respectively, X refers to log dose, Y pertains to response, and $E_i$ denotes the experimental error of the ith response.

The estimated regression line is

$$\hat{Y}_i = a + bX_i$$

where a and b are the least-squares estimates of A and B.

Following is the structure of the BANOVA for this study:

| SOV | DF | SSQ | MSQ | F |
|---|---|---|---|---|
| Among doses | 2 | 302.53 | — | |
| Linear regression | 1 | 302.50 | 302.50 | 193*** |

| SOV | DF | SSQ | MSQ | F |
|---|---|---|---|---|
| Curvilinear regression | 1 | 302.53 − 302.50 = 0.03 | 0.03 | <1.0 |
| Within doses | 12 | 18.80 | $1.57 = S^2$ | |
| Total | 14 | 321.33 | | |

BANOVA means bioassay analysis of variance.

$$b = \frac{[XY]}{[X^2]} = \frac{16.5550}{0.9060} = 18.27$$

$$\hat{Y} = \bar{Y} + b(X - \bar{X}) = 11.33 + 18.27(X - 1.602)$$

$$= a + bX = -17.94 + 18.27X$$

We are interested in estimating the dose that will cause a maximal fall in MAP of 10 mm Hg. Now the "inverse prediction" is

$$\hat{X} = \bar{X} + \frac{Y - \bar{Y}}{b}$$

so

$$\hat{X} = 1.602 + \frac{10.0 - 11.33}{18.27} = 1.529$$

$$ED_{10} \text{ (dose)} = \text{antilog } \hat{X} - \text{antilog } 1.529 = 33.8$$

The interpretation is that it would require a dose of 33.8 mg/kg to produce a maximal fall in MAP of 10 mm Hg. It should also be noted here that the data conformed to two tests of validity: (1) that the linear regression is highly significant (F = 193) and (2) that the curvilinear regression is not significant at P = 0.05.

Computation of 95% confidence limits (95% CL):

1. *Statistical differential method* (g = 0 assumed):

$$V(\hat{X}) = \frac{S^2}{b^2}\left[\frac{1}{n} + \frac{(\hat{X} - \bar{X})^2}{[X^2]}\right] = \frac{1.57}{(18.27)^2}\left[\frac{1}{15} + \frac{(1.529 - 1.602)^2}{0.9060}\right]$$

$$= 0.000341$$

$$SE(\hat{X}) = (0.000341)^{1/2} = 0.018 \qquad \text{standard error of } \hat{X}$$

$$t_{0.05,\ 12\ DF} = 2.179 \qquad \text{(tabular t value)}$$

95% CL:

$$\hat{X}_L, \hat{X}_U = 1.529 \pm 2.179(0.018) = 1.529 \pm 0.039$$

$$= 1.490,\ 1.568$$

$$(ED10_L, ED10_U) = (\text{antilog}(1.490), \text{antilog}(1.568))$$

$$= 30.9,\ 37.0$$

2. *Using Fieller's formula*:

$$g = \frac{t^2 s^2}{b^2 [X^2]} = \frac{(2.179)^2 (1.57)}{(18.27)^2 (0.9060)} = 0.025 \qquad (t = \text{tabular } t)$$

$$(\hat{X}_L, \hat{X}_U) = \hat{X} + \frac{g}{1-g}(\hat{X} - \bar{X}) \pm \frac{t}{b(1-g)} \left[ s^2 \left( \frac{1-g}{N} + \frac{(\hat{X} - \bar{X})^2}{[X^2]} \right) \right]^{1/2}$$

$$= 1.529 \frac{0.025}{0.975}(1.529 - 1.602)$$

$$\pm \frac{2.170}{(18.27)(0.975)} \left[ 1.57 \left( \frac{0.975}{15} + \frac{(-0.073)^2}{0.9060} \right) \right]^{1/2}$$

$$= 1.529 - 0.002 \pm 0.041 = 1.527 \pm 0.041$$

$$= 1.486,\ 1.568$$

$ED10_L$: antilog(1.486) = 30.6

$ED10_U$: antilog(1.568) = 37.0

**Table 1** Carrageenan-Induced Foot Edema: Effect of Indomethacin

Dose-response line for indomethacin:

$$y = 38.04 \log x + 33.43$$

in which y is percent inhibition of foot swelling and x is mg/kg dose. Equation is based on 130 groups of six rats each.

Median effective dose for 50% inhibition ($ED_{50}$): 2.73 mg/kg

95% confidence limits of $ED_{50}$: 2.54, 2.92 mg/kg

$g = 0.0058$

Coefficient of variation = 17.9%

Dose-response line for carrageenan, without drugs:

$$y = 0.519 \log x + 0.755$$

in which y is response in ml swelling and x is mg carrageenan.

Table 2 Cotton Pellet Granuloma: Effect of Indomethacin

---

Dose-response line for indomethacin:

$$y = 15.25 \log x + 32.00$$

in which y is percent inhibition of weight gain of pellet and x is mg/kg dose. Equation is based on 101 groups of six rats each.

Median effective dose of indomethacin for 25% inhibition ($ED_{25}$): 0.35 mg/kg

95% confidence limits of $ED_{25}$: 0.23, 0.46 mg/kg

$$g = 0.05$$

Coefficient of variation = 21.2%

Dose-response line for pellets without drugs:

$$y = 159.4 \log x - 188.9$$

in which y is mg gain in weight of pellet and x is mg weight of the pellets at implantation.

---

Table 3 Adjuvant Arthritis: Effect of Indomethacin

---

Dose-response line for indomethacin:

$$y = 57.17 \log x + 84.84$$

in which y is percent inhibition of paw swelling and x is dose in mg/kg. Equation is based on 34 groups of six rats each.

Median effective dose of indomethacin for 50% inhibition ($ED_{50}$): 0.25 mg/kg

95% confidence limits of $ED_{50}$: 0.18, 0.31 mg/kg

$$g = 0.10$$

Coefficient of variation = 27.23%.

Dose-response line for adjuvant without drugs:

$$y = 0.4 \log x + 0.75$$

in which y is ml foot swelling and x is mg of *M. butyricum* in 0.1 ml mineral oil.

---

**Table 4** Yeast Fever: Effect of Indomethacin

---

Dose-response line for indomethacin:

$$y = 1.63 \log x + 0.81$$

in which y is degrees centigrade lowering of fever and x is dose in mg/kg. The equation is based on 48 groups of six rats each.

Median effective dose of indomethacin for 1°C decrease ($ED_1$): 1.31 mg/kg

95% confidence limits of $ED_1$: 1.23, 1.39 mg/kg

$$g = 0.01$$

Coefficient of variation = 11.9%

---

Since g is not assumed to be zero in Fieller's formula a conservative interval is expected. Note that the F statistic ($t^2$) for testing the null hypothesis B = 0 is $b^2/s^2[X^2]^{-1}$. Since $g = F_\alpha/F$, when F is large, g is small. g should be as close to zero as possible (less than 0.15 acceptable).

## B. Comparison of Antiarthritic Assay Methods (Van Arman and Bohidar, 1978)

In Section II we described the various assay methods generally used for testing antiarthritic agents. It would be interesting to examine the $ED_\alpha$

**Table 5** Dog's Knee Joint: Effect of Indomethacin

---

Dose-response line for indomethacin:

$$y = 54.4 \log x + 39.37$$

in which y is percent restoration of the normal foot pressure and x is dose in mg/kg. Equation is based on 104 dogs.

Median effective dose of indomethacin for 50% restoration ($ED_{50}$): 1.57 mg/kg

95% confidence limits of $ED_{50}$: 1.07, 2.08 mg/kg

$$g = 0.15$$

Coefficient of variation = 61%

Dose-response line of sodium urate, without drugs:

$$y = 95.5 - 159.6 \log x$$

in which y is percent of normal foot pressure and x is mg urate crystals injected.

---

**Table 6** Topical Mouse Ear Skin: Effect of Indomethacin

Dose-response line for indomethacin:

$$y = 48.22 \log x + 39.53$$

in which y is percent reduction of swelling from that measured in control mice and x is dose in mg. Equation is based on 54 mice total.

Median effective dose of indomethacin for 50% inhibition ($ED_{50}$): 1.65 mg

95% confidence limits of $ED_{50}$: 1.23, 2.12 mg

$$g = 0.18$$

Coefficient of variation = 36%

Dose-response line for croton oil in vehicle described, without drugs:

$$y = 5.34 \log x + 12.5$$

in which y is mg weight gain of the ear and x is mg of croton oil.

Table 7 Comparison of Six Drugs Tested in Each of Six Different Assays

| Assay | Indomethacin | Sulindac | Diflunisal | Phenylbutazone | Aspirin | Thiabendazole |
|---|---|---|---|---|---|---|
| Carrageenan foot edema | 2.7[a] | 5.5 | 9.8 | 27.7 | 89.2 | 200 |
| Granuloma, cotton pellet | 0.35 | 5.4 | 74 | 40 | 115 | Inactive |
| Adjuvant arthritis | 0.25 | 0.55 | 9.8 | 14 | 67 | 217 |
| Yeast fever | 1.31 | 2.9 | 24 | 24 | 45 | 83 |
| Dog knee joint | 1.57 | 45 | 24 | 12 | 72 | $>150$ |
| Topical, mouse ear | 1.65 | 6.1 | 3 | 4.3 | 5.5 | 12.4 |

[a] All values represent median effective dose (MED).

**Table 8** Comparison of Six Drugs Tested in Each of Six Different Assays

| Assay | Indomethacin | Sulindac | Diflunisal | Phenylbutazone | Aspirin | Thiabendazole |
|---|---|---|---|---|---|---|
| Carrageenan foot edema | 1 | 2 | 3 | 4 | 5 | 6 |
| Granuloma, cotton pellet | 1 | 2 | 4 | 3 | 5 | 6 |
| Adjuvant arthritis | 1 | 2 | 3 | 4 | 5 | 6 |
| Yeast fever | 1 | 2 | 3 1/2 | 3 1/2 | 5 | 6 |
| Dog knee joint | 1 | 5 | 2 | 3 | 4 | 6 |
| Topical, mouse ear | 1 | 5 | 2 | 3 | 4 | 6 |
| Sum of rank orders for drug in the six assays | 6 | 18 | 17 1/2 | 20 1/2 | 28 | 36 |
| Rank order of drug in clinical use as an anti-inflammatory | 1 | 2 | 3 | 4 | 5 | 6 |

and the associated statistics for each assay provided in Tables 1 to 6, which are self-explanatory.

Table 7 provides a comparison of six drugs tested in each of six different assays. Table 8 depicts the rank order of potency of six drugs in six assays. Since each assay method is related to a different aspect of the multifaceted disease, a rank-order approach is appropriate.

## VII. ESTIMATION OF RELATIVE POTENCY

### A. Introduction

In relating the potency of one drug or preparation to that of another, we must satisfy ourselves that the actions of the two are identical. A properly designed test will include a statistical verification of this fact. The verification consists of establishing that the same dose-response curve may be fitted to the two separate drugs. When dealing with a linear log dose-response relationship, this implies only that the two dose-response lines will be parallel. If it is found in practice that they are not, it means quite simply that under the conditions of the test one drug cannot be assayed with any accuracy in terms of the other, since our estimate of relative potency will depend on the particular level of response at which it is made. Figure 1 illustrates the case in which two log dose-response lines are not parallel.

The potency of drug A measured at the 40-unit response level is twice that of drug B, whereas if the measurement is made at the 80-unit response level the potency of drug A is approximately four times that of drug B. When the two lines are parallel (Figure 2), M, the logarithm of the relative potency of the two substances, is the distance between the two lines measured parallel to the abscissa. The value of M, here, remains invariant for any level of the response considered.

When the two dose-response lines do not differ significantly in slope, they are to be regarded as two separate estimates of a common slope relating response to the dose of both preparations. We therefore pool the information from both samples and calculate one value of B in the equation. Here

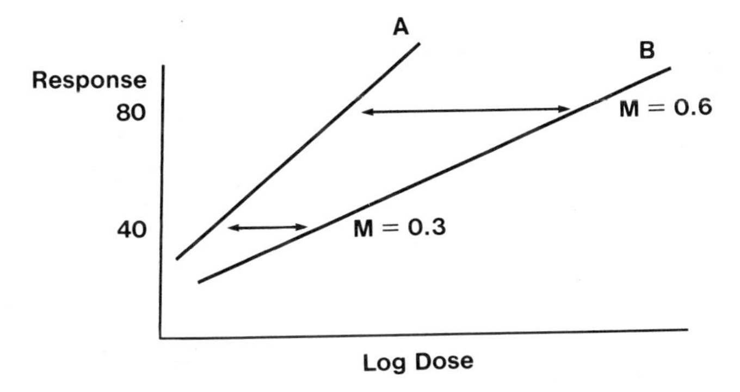


Fig. 1 When log dose-response lines are not parallel, it is impossible to assay one drug in terms of the other.

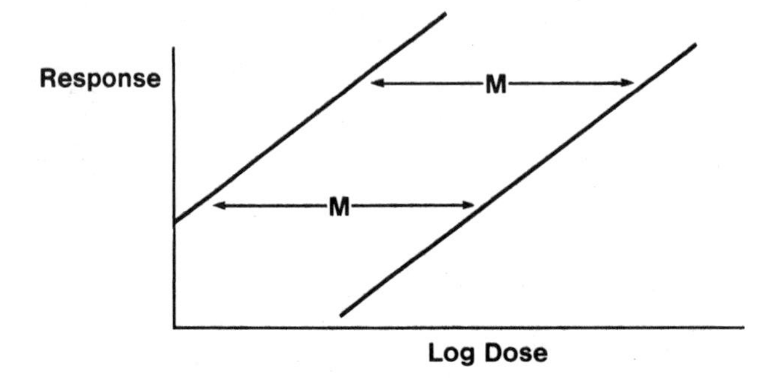


Fig. 2 Parallel log dose-response lines give a constant value for M, the logarithm of the relative potency of the two substances.

$$Y_0 = \bar{Y}_S + B(X_S - \bar{X}_S)$$

and

$$Y_0 = \bar{Y}_T + B(X_T - \bar{X}_T)$$

The relative potency of the two preparations then depends on the difference between the two mean doses and the difference between the two mean responses to all doses. If M is the log ratio of the potency of the unknown (test) to that of this standard, then

$$M = \bar{X}_S - \bar{X}_T + \frac{\bar{Y}_S - \bar{Y}_T}{B}$$

where $\bar{X}_T$ is the mean log dose of the test preparation, $\bar{X}_S$ the mean log dose of the standard, $\bar{Y}_T$ the mean response of the test, and $\bar{Y}_S$ the mean response of the standard. The slope (B) is the common slope associated with the test and the standard preparations.

## B. Algebraic Derivation of M

Relative potency is defined as

$$R = \frac{Z_S}{Z_T}$$

Now,

$$\log R = M = \log Z_S - \log Z_T = X_S - X_T$$

Note that the Z's are the doses in their original units. Our intent will be to substitute the algebraic expressions for $X_S$ and $X_T$ derived from the respective dose-response regression equations (see Figure 3).

Let the two lines be parallel and have a common slope of B. Now we know for the standard preparation that

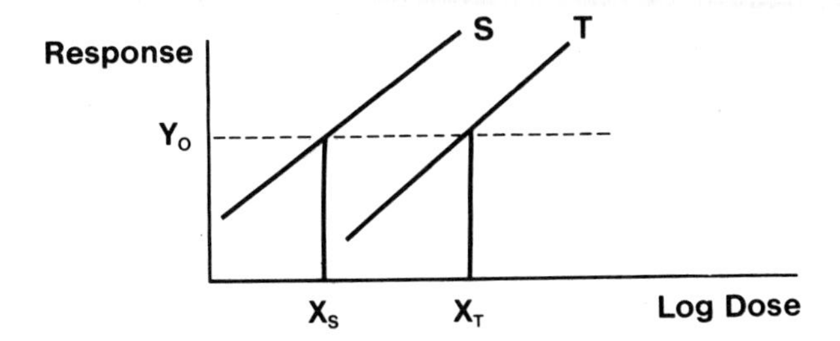


Fig. 3 Algebraic derivation of M.

$$Y_0 = \bar{Y}_S + B(X_S - \bar{X}_S)$$

or

$$X_S = \frac{Y_0 - \bar{Y}_S}{B} + \bar{X}_S$$

Similarly, for the test preparation we have

$$Y_0 = \bar{Y}_T + B(X_T - \bar{X}_T)$$

Solving for $X_T$, we have

$$X_T = \frac{Y_0 - \bar{Y}_T}{B} + \bar{X}_T$$

Now,

$$\log R = M = X_S - X_T = \left[\frac{Y_0 - \bar{Y}_S}{B} + \bar{X}_S\right] - \left[\frac{Y_0 - \bar{Y}_T}{B} + X_T\right]$$

$$= (\bar{X}_S - \bar{X}_T) + \left[\frac{Y_0 - \bar{Y}_S}{B} - \frac{Y_0 - \bar{Y}_T}{B}\right]$$

$$= (\bar{X}_S - \bar{X}_T) + \left[\frac{Y_0 - \bar{Y}_S - Y_0 + \bar{Y}_T}{B}\right]$$

so,

$$M = \bar{X}_S - \bar{X}_T - \frac{\bar{Y}_S - \bar{Y}_T}{B}$$

The antilog of M will yield the relative potency in the original unit. This value is used for the purpose of interpretation.

### C. Example

In antihypertension research in rats one introduces antiotensin I into the femoral vein, and it is converted to angiotensin II by the converting enzymes. The antihypertensive agent causes the system to produce these enzymes and causes the blood pressure to go down. In this study, we have two antihypertensive agents, A and B. The data are the pressure responses obtained by inserting a cannula into the femoral artery and connecting the cannula to a pressure transducer. The rise in blood pressure above the normal rat blood pressure is measured. In this study, as we increase the doses, we find a smaller rise in blood pressure. We are interested in finding the potency of the test preparation (B) relative to the standard preparation (A) and the 95% confidence limits. The detailed calculations are provided in the following.

Bioassay Calculations:

Raw Data

| | A Standard doses (mg/kg) | | | | B Test doses (mg/kg) | | | | |
|---|---|---|---|---|---|---|---|---|---|
| | 10 | 30 | 100 | 300 | 1 | 3 | 10 | 30 | 100 |
| | 48 | 50 | 26 | 20 | 44 | 35 | 23 | 10 | 6 |
| | 49 | 37 | 20 | 14 | 48 | 39 | 32 | 19 | 5 |
| | 52 | 36 | 25 | 12 | 48 | 42 | 33 | 19 | 20 |
| | 53 | 39 | 26 | 16 | 56 | 52 | 48 | 27 | 17 |
| | 34 | 34 | 27 | 15 | 47 | 41 | 33 | 21 | 15 |
| | 50 | 36 | 24 | 11 | 56 | 44 | 28 | 16 | 9 |
| | 58 | 41 | 28 | 18 | | | | | |
| | 48 | 40 | 25 | 16 | | | | | |
| | 46 | 30 | 22 | 14 | | | | | |
| | 56 | 40 | 23 | 13 | | | | | |
| Totals | 494 | 383 | 246 | 149 | 299 | 253 | 197 | 112 | 72 |
| Means | 49.4 | 38.3 | 24.6 | 14.9 | 49.8 | 42.2 | 32.8 | 18.7 | 12.0 |

$$C = \text{correction term} = \frac{(2205)^2}{70} = 69{,}457.5000$$

$$\text{Total SS} = (48)^2 + (49)^2 + \cdots + (9)^2 - C$$

$$\text{Doses SS} = \frac{(494)^2}{10} + \frac{(383)^2}{10} + \cdots + \frac{(72)^2}{6} - C$$

$$\text{Preparations SS} = \frac{(1272)^2}{40} + \frac{(933)^2}{30} - C$$

where SS means sum of squares.

For preparation 1:

$$\Sigma X = 10(1.0000) + 10(1.4771) + 10(2.0000) + 10(2.4771) = 69.5420$$

$$\bar{X} = 1.7386$$

For preparation 2:

$$\Sigma X = 6(0.0000) + 6(0.4771) + 6(1.0000) + 6(1.4771) + 6(2.0000) = 29.7252$$

$$\bar{X} = 0.9908$$

For preparation 1:

$$[X^2] = 10(1.0000)^2 + \cdots 10(2.4771)^2 - \frac{(69.5420)^2}{40} = 12.2762$$

For preparation 2:

$$[X^2] = 6(0.0000)^2 + \cdots + 6(2.0000)^2 - \frac{(29.7252)^2}{30} = 15.0038$$

For preparation 1:

$$[XY] = (1.0000)(494) + \cdots + (2.4771)(149) - \frac{(69.5420)(1272)}{40}$$

$$= -290.6184$$

For preparation 2:

$$[XY] = (0.000)(299) + \cdots + (2.0000)(72) - \frac{(29.7252)(933)}{30} = -297.3122$$

$$\Sigma[X^2] = [X^2]_{PREP1} + [X^2]_{PREP2} = 27.2800$$

$$\Sigma[XY] = [XY]_{PREP1} + [XY]_{PREP2} = -587.9306$$

$$\text{Regression SS} = \frac{(\Sigma[XY])^2}{\Sigma[X^2]} = \frac{(-587.9306)^2}{27.2800} = 12{,}670.9087$$

$$\text{Parallelism SS} = \frac{[XY]^2_{PREP1}}{[X^2]_{PREP1}} + \frac{[XY]^2_{PREP2}}{[X^2]_{PREP2}} - \frac{(\Sigma[XY])^2}{\Sigma[X^2]}$$

$$= \frac{(-290.6184)^2}{12.2762} + \frac{(-297.3122)^2}{15.0038} - \frac{(-587.9306)^2}{27.2800} = 100.4705$$

Curvature SS = Doses SS − PREP SS − Regression SS − Parallelism SS

$$= 98.0875$$

Within doses SS = Total SS − Doses SS = 1747.6333

The BANOVA for this study is presented in Table 9.

*Computation of relative potency estimate and the 95% confidence limits*: Let $M = \log_{10}$ relative potency, $R = 10^M$, and $b_c$ = common slope.

$$b_c = \frac{\Sigma[XY]}{\Sigma[X^2]} = \frac{-587.9306}{27.2800} = -21.5517$$

$$M = \bar{X}_1 - \bar{X}_2 - \frac{\bar{Y}_1 - \bar{Y}_2}{b_c}$$

$$= 1.7386 - 0.9908 - \frac{31.8000 - 31.1000}{-21.5517} = 0.7803$$

$$R = 6.03$$

$$s^2 = \frac{\text{within doses SS}}{\text{within degrees of freedom}} = 28.6497$$

$$t_{0.05,61} = 1.9996$$

$$g = \frac{t^2 \quad s^2}{b_c^2 \quad \Sigma[X^2]} = \frac{(1.9996)^2(28.6497)}{(-21.5517)^2(27.2800)} = 0.0090$$

The 95% confidence limits are

$$M_L, M_U = M + \frac{g}{1-g}(M - \bar{X}_1 + \bar{X}_2)$$

$$\pm \frac{t_{0.05,61}}{b_c(1-g)} \sqrt{\left\{s^2\left[(1-g)\left(\frac{1}{N_1} + \frac{1}{N_2}\right) + \left(\frac{M - \bar{X}_1 + \bar{X}_2}{\Sigma[X^2]}\right)^2\right]\right\}}$$

$$= 0.7806 \pm (-0.1205)$$

$$= 0.6601,\ 0.9011$$

$$R_L, R_U = 4.57,\ 7.96$$

The interpretation is as follows. For R = 6.03, we say that 1 mg/kg of the test preparation, B, is equivalent to 6.03 mg/kg of the standard preparation, A. As an interpretation of the confidence limits, we say that we can assert with a 95% probability that 1 mg/kg of test preparation B is not less potent than 4.57 mg/kg of standard A and not more potent than 7.96 mg/kg of standard A. It should also be mentioned here that since the value of 1.0 is not included in the 95% confidence interval, the two preparations (A and B) are significantly different in their potencies ($P < 0.05$).

**Table 9** Bioassay Analysis of Variance (BANOVA)

| Source of variation | Degrees of freedom | Sums of squares | Mean squares | F Ratio |
|---|---|---|---|---|
| Among doses | 8 | 12,877.8667 | — | — |
| Preparations | 1 | 8.4000 | 8.4000 | 0.293 |
| Common regression | 1 | 12,670.9122 | 12,670.9122 | 442.270 |
| Lack of parallelism | 1 | 100.4533 | 100.4533 | 3.506 |
| Residual (curvature) | 5 | 98.1012 | 19.6202 | 0.685 |
| Within doses | 61 | 1,747.6333 | 28.6497 | — |
| Total | 69 | 14,625.5000 | — | — |

Before performing a bioassay analysis for the data, two tests of validity should be conducted: (1) the Levene test (Levene, 1971), to test for equality of group variabilities as measured by the average absolute mean deviations (AAMDs) and (2) the Wilk-Shapiro test (Wilk and Shapiro, 1968) for normality of each individual group and the combined groups.

Shown below is the ANOVA table for the mean absolute deviations associated with Levene's test. The result shows that there are no significant differences among the groups (overall) with respect to their variabilities. A table of means (AAMD) is also provided below.

ANOVA for Deviations (Levene Test ANOVA)

| Source of variation | DF | SSQ | MSQ | F ratio |
|---|---|---|---|---|
| Among doses | 8 | 93.508190 | 11.600524 | 0.9111 |
| Within doses | 61 | 729.078222 | 11.952102 | — |
| Total | 69 | 822.586412 | | |

Table of Means (AAMD)

| Ascending order of averages | Preparation and doses[a] (mg/kg) | AAMD | Sample size |
|---|---|---|---|
| 1 | S3 A-100 | 1.800000 | 10 |
| 2 | S4 A-300 | 2.100000 | 10 |
| 3 | S2 A-30 | 3.700000 | 10 |
| 4 | T4 B-30 | 3.777770 | 6 |
| 5 | T2 B-3 | 3.000009 | 6 |
| 6 | T1 B-1 | 4.111111 | 6 |
| 7 | S1 A-10 | 4.400000 | 10 |
| 8 | T3 B-10 | 5.166667 | 6 |
| 9 | T5 B-100 | 5.333333 | 6 |

[a]A, standard; B, test.

For completeness only, the results of the Wilk-Shapiro W test (for testing for normality) are provided in Table 10. It may be noted that the individual as well as the combined groups do not show a departure from normality ($P > 0.05$). The computational procedures for this test are described in Wilk and Shapiro (1968).

### D. Tests of Validity

The analysis of variance associated with a bioassay analysis is referred to as BANOVA. The assumptions implicit in BANOVA must be verified and appropriate tests of validity must be conducted to complete the verification process.

Table 10 Test for Normality of Data (Wilk and Shapiro W Statistic)

| Subclass[a] | B value | W statistic | Critical W | Significance | G value[b] | Alpha value | Chi square[c] |
|---|---|---|---|---|---|---|---|
| S1 | 18.664600 | 0.892334 | 0.842000 | N.S. | −0.927663 | 0.176791 | 3.465570 |
| S2 | 15.207500 | 0.924702 | 0.842000 | N.S. | −0.313869 | 0.376810 | 1.952028 |
| S3 | 7.123800 | 0.968483 | 0.842000 | N.S. | 1.078249 | 0.859539 | 0.302719 |
| S4 | 8.066700 | 0.972670 | 0.842000 | N.S. | 1.298122 | 0.902877 | 0.204137 |
| T1 | 10.242600 | 0.840407 | 0.788000 | N.S. | −1.155334 | 0.123977 | 4.175323 |
| T2 | 12.423200 | 0.947815 | 0.788000 | N.S. | 0.603329 | 0.726855 | 0.638056 |
| T3 | 17.568000 | 0.087919 | 0.788000 | N.S. | −0.645670 | 0.259247 | 2.699951 |
| T4 | 12.335700 | 0.967179 | 0.788000 | N.S. | 1.107738 | 0.866012 | 0.287712 |
| T5 | 13.258100 | 0.915506 | 0.788000 | N.S. | −0.072833 | 0.470970 | 1.505923 |
| Sums | | | | | 0.972068 | | 15.231620 |

[a] S, standard (A); T, test (B).

[b] Standard G-test for normality: (G/square root K) where K is the number of groups. Since the observed G (0.324023) is larger than the critical G (−1.645), we would conclude that the data are *normally distributed*

[c] Chi-square test for normality: Since the observed chi-square value (15.231620) is less than the critical chi-square value (28.859338), we would conclude that the data are *normally distributed*.

1. Normality (Wilk-Shapiro test).
2. Homogeneity of variance (Levene test).
3. Independence (randomization).
4. Additivity (tests needed only for randomized block-type designs).
5. Test of linearity. A preliminary graphical investigation will be presumed to have established that, over a range of responses, the regression of response on log dose is practically linear. Violation of this assumption strongly indicates *statistical invalidity* of the assay, that is, inappropriateness of the form of the analysis adopted. A bad choice of doses for either preparation might take most of the observations off the linear portion of the dose-response curve.
6. Parallelism. If the two dose-response regression lines are significantly nonparallel, this clearly indicates *fundamental invalidity* of the assay. The condition of similarity, an essential condition for an analytical dilution assay, demands the parallelism of regression curves when log dose is used as an independent variable. If these curves are linear, nonparallelism of the lines is a violation of the condition of similarity. The initial assumption that the test preparation acted as a dilution of the standard either is inherently false or has been obscured by the presence of an impurity in one preparation.
7. Regression. In any good assay, the mean square for regression should be relatively large. If there was no regression, the dose-response relationship could be of no use for the estimation of potency. No assay should be undertaken without strong prior belief in the existence of a regression. The variance ratio for the regression should be highly significant. In the example, we have $F \simeq 400$. Moreover, unless this ratio is large, the value of g used in this assessment of confidence limits will be large.
8. Preparation. Unlike an ordinary experiment for the comparison of different treatments, no direct interest attaches to the mean difference in response between two preparations. Nevertheless, a large mean square for this component is an indication that the assay is not very satisfactory. A large difference between means will seldom arise unless the response to either the lowest or the highest dose of the test preparation lies far outside the range of responses to the standard. In a well-planned assay this should not happen. If it does, either the range of the doses for the standard preparation should have been wider or that for the test preparation was too wide and extended beyond the region of linearity. A large difference in mean response will cause a decrease in the potency value.
9. g value. The magnitude of the g value should be much less than the value of 1.0. Note that (CI = confidence interval): for values less than or equal to 0.05, the CI tends to be narrow; for values between 0.05 and 0.1, the CI is acceptable; for values between 0.1 and 0.5, the CI is wider; for values between 0.5 and 0.9, the CI is much wider and meaningless; and for values 0.9 or 1 and above, there is *no* estimate of CI.
10. Curvilinearity: must always be nonsignificant. If not, the indications are that the regression curve is in the upper portion of the dose-response curve. The assay will be invalid.

## E. Interpretation of Statistical Results of BANOVA

1. *Preparation*: should not be significant. If significant, it signifies that we have misguessed potency and made a poor choice of doses. If large, it reduces precision of relative potency. It may give rise to apparent lack of parallelism if the linear range of responses has been exceeded.

2. *Common regression*: should be highly significant ($P < 0.001$). Measures the significant of the average slope or regression of response on log dose. Significance of low order indicates a flat slope. This reduces the precision of the assay and gives rise to large confidence limits.

3. *Lack of parallelism*: should not be significant. If significant, it is evidence of fundamental invalidity of the assay. It indicates that the effects of the two drugs are not similar. There will be no unique estimate of potency. Potency will depend on the level of response. Apparent lack of parallelism may be due to a serious misguess of potency, so that the linear range of response may have been exceeded.

4. *Curvature*: should not be significant. If significant, it indicates statistical invalidity. Apparent curvature may be due to a serious misguess of potency, so that the linear range of response may have been exceeded.

## F. Completely Balanced Assay

On assaying the potency of insulin by measuring the fall in blood sugar of rabbits after injection of the drug, groups of animals are taken at random and their blood sugars measured by chemical estimation. The drug is then injected, the fall in blood sugar is measured over a period of several hours by taking successive samples of blood, and the mean percentage fall over ther period of examination is used as the response. Six groups of eight rabbits each are involved, three of them receiving three different doses of the standard and the other three receiving three different doses of the unknown (test). The sources of variation can be expressed as contrasts on the means associated with the standard and test. Therefore, the contrast method of computing sums of squares is used for the analysis of variance.

Assay of the Potency of a Preparation of Insulin Response: % Fall in Blood Sugar

| Standard (S) doses | | | | Test (T) doses | | | |
|---|---|---|---|---|---|---|---|
| 0.25 | 0.50 | 1.0 | | 0.8 | 1.6 | 3.2 | |
| 11.2 | 16.5 | 32.7 | | 19.8 | 37.7 | 45.4 | |
| 21.2 | 23.2 | 14.0 | | 21.7 | 40.7 | 28.6 | |
| 17.7 | 25.6 | 28.9 | | 26.1 | 29.3 | 50.4 | |
| 2.8 | 12.7 | 40.2 | | 32.2 | 48.1 | 47.7 | |
| 27.2 | 29.8 | 35.1 | | 28.5 | 45.6 | 50.0 | |
| 25.1 | 28.4 | 36.2 | | 20.2 | 35.3 | 12.4 | |
| 25.8 | 40.0 | 37.8 | | 35.7 | 14.2 | 39.0 | |
| 2.2 | 2.4 | 38.4 | | 26.1 | 7.9 | 38.1 | |
| 134.2 | 188.6 | 264.3 | 587.1 | 210.3 | 258.8 | 311.6 | 780.7 |

Using the orthogonal contrast coefficients given in Table 11 for the insulin data, we have the results shown in Table 12.

**Table 11** Orthogonal Contrast Coefficients for the Isolation of Individual Effects When Two, Three or Four Doses of the Standard and Test are Considered

| Source of variation | Orthogonal contrast coefficients | | | | | | | | | | | | | | | | | |
|---|---|---|---|---|---|---|---|---|---|---|---|---|---|---|---|---|---|---|
| | $S_1$ | $S_2$ | $T_1$ | $T_2$ | $S_1$ | $S_2$ | $S_3$ | $T_1$ | $T_2$ | $T_3$ | $S_1$ | $S_2$ | $S_3$ | $S_4$ | $T_1$ | $T_2$ | $T_3$ | $T_4$ |
| 1. Preparation | −1 | −1 | +1 | +1 | −1 | −1 | −1 | +1 | +1 | +1 | −1 | −1 | −1 | −1 | +1 | +1 | +1 | +1 |
| 2. Regression | −1 | +1 | −1 | +1 | −1 | 0 | +1 | −1 | 0 | +1 | −3 | −1 | +1 | +3 | −3 | −1 | +1 | +3 |
| 3. Parallelism | +1 | −1 | −1 | +1 | +1 | 0 | −1 | −1 | 0 | +1 | +3 | +1 | −1 | −3 | −3 | −1 | +1 | +3 |
| 4. Curvature of combined curve | | | | | +1 | −2 | +1 | +1 | −2 | +1 | +1 | −1 | −1 | +1 | +1 | −1 | −1 | +1 |
| 5. Opposed curvature of separate curves (parallelism) | | | | | −1 | +2 | −1 | +1 | −2 | +1 | −1 | +1 | +1 | −1 | +1 | −1 | −1 | +1 |
| 6. Double curvature of combined curve (cubic) | | | | | | | | | | | −1 | +3 | −3 | +1 | −1 | +3 | −3 | +1 |
| 7. Opposed double curvature (parallelism) | | | | | | | | | | | +1 | −3 | +1 | −1 | −1 | +3 | −3 | +1 |

Table 12 Application of Orthogonal Contrast Coefficient to Insulin Data: BANOVA[a]

| Source of variation | Orthogonal contrast coefficients | | | | | | Divisor | Sum of products | |
|---|---|---|---|---|---|---|---|---|---|
| | $S_1$ | $S_2$ | $S_3$ | $T_1$ | $T_2$ | $T_3$ | $N_p \Sigma\lambda^2$ | $\Sigma\lambda T$ | SSQ |
| 1. Preparation | −1 | −1 | −1 | +1 | +1 | +1 | 48 | 193.6 | 780.9 |
| 2. Regression | −1 | 0 | +1 | −1 | 0 | +1 | 32 | 231.4 | 1673.3 |
| 3. Parallelism | +1 | 0 | −1 | −1 | 0 | +1 | 32 | −28.8 | 25.9 |
| 4. Curvature of combined curve | +1 | −2 | +1 | +1 | −2 | +1 | 96 | 25.6 | 6.8 |
| 5. Opposed curvature of separate curves | −1 | +2 | −1 | +1 | −2 | +1 | 96 | −17.0 | 3.0 |
| Totals | 134.2 | 188.6 | 264.3 | 210.3 | 258.8 | 311.6 | Error = 125.9 | | |

[a]Here, SSQ = $(\Sigma\lambda T)^2/N_p\Sigma\lambda^2$, $N_p$ is the number of observations per dose (here, eight), $\lambda$ is the contrast coefficient, T stands for totals of the dose group, and SSQ is the sum of squares computed by the contrast method.

## VIII. EXPERIMENTAL DESIGNS IN BIOASSAY

### A. General Considerations

One of the variables in biological assay is always arbitrarily selected. This variable is the dose of the drug which is to be administered to particular groups of test objects. Other variables which may be capable of influencing the course of the test are frequently subject to selection as well. In the description of the majority of biological assays, a considerable amount of space is given to defining conditions under which the test will be made and describing the class of test object suitable for the test. Animals of only one sex and between certain ages and weights may be prescribed. The test must be conducted within certain limits of time and perhaps at certain temperatures, and the response must be measured subject to other conditions, which are more or less rigid. In designing biological assay tests, it is generally true that the more rigid these initial conditions are made, the more accurate is the test.

Once these initial specifications have been fulfilled, it is necessary to see that the allotment of test objects to doses and to any other restricted classes which may be included in the design of the test takes place by the process of randomization. The process of randomization must fulfill the requirement that any particular test object will be as likely to be allotted to any particular class as to any other and, conversely, that any particular class will be as likely to receive any particular test object as any other class. These conditions may, however, be fulfilled within the limits of certain planned restrictions in the design of the test. This important fact is used in designing tests which are balanced so that the effect of a number of variables can be examined simultaneously in subsequent statistical analysis. If, for instance, we are dealing with several litters of rats which are known or believed to exhibit less variation within litters than between litters, we will randomly assign each member of this litter to one of the K dose groups. This procedure will reduce the test variability.

### B. Randomized Designs

Experimental designs which are commonly used in bioassay experiments are the (1) completely randomized design, (2) randomized block design, (3) Latin square design, (4) Greco-Latin square design, (5) balanced incomplete block design, and (6) crossover design. An excellent description and illustration of these designs and an extensive discussion of the principles of planning bioassay experiments are given by Finney (1978, chapters 6, 9, and 10). It may also be instructive to review the first five chapters of Cochran and Cox (1957).

## IX. MULTIPLE BIOASSAY

### A. Introduction

Multiple bioassay refers to the situation in which one is interested in determining the potencies of more than one drug relative to a standard drug. The definition can be broadened to encompass cases involving several test preparations and more than one standard preparation. For this analysis to be valid, the entire experiment must be conducted under one protocol and concurrently. The layout will be as follows:

*Preparations*

| $P_1$ | $P_2$ | $P_3$ | $P_4$ |
|---|---|---|---|
| $D_1, D_2 \cdots D_{K_1}$ | $D_1, D_2 \cdots D_{K_2}$ | $D_1, D_2 \cdots D_{K_3}$ | $D_1, D_2 \cdots D_{K_p}$ |

Here we have p preparations and there are $K_i$ ($i = 1, 2, \ldots, p$) doses for the ith preparation.

The implicit assumptions associated with the BANOVA are (1) homogeneity of variances of the groups (doses), (2) normality, (3) independence, and (4) additivity (RBD, LSQ, etc.).

The implicit assumptions associated with the estimation of relative potency are (1) homogeneity of regression coefficients associated with each preparation (parallelsim), (2) steep common regression, and (3) linearity (curvature not significant).

The $100(1 - \alpha)\%$ confidence limits of the relative potency will be based on the pooled error vairance (if there are no significant differences among the group variances) and common regression (if there are no significant differences among the regression coefficients).

## B. Numerical Illustration

The following numerical illustration is based on a study in which the pharmacologist is interested in determining the potencies of four anti-inflammatory agents relative to a standard. The carrageenan-induced foot edema assay (Section II, C,1) was used in this study. The computational aspects of multiple assay analysis have been presented in detail in Finney (1971). We provide in the following the BANOVA for the multiple assay, the estimates of the four potency values, and their respective 95% confidence limits. It should be noted that, while constructing the 95% confidence interval, one should use the Dunnett t value and not the Student t value. This approach maintains the same prespecified type I error value for each confidence interval.

Bioassay Analysis of Variance (BANOVA)

| Source of variation | Degrees of freedom | Sums of squares | Mean squares | F ratio |
|---|---|---|---|---|
| Among doses | 11 | 86,813.7232 | — | — |
| Preparations | 3 | 21,967.0231 | 7,322.3410 | 15,636 |
| Common regression | 1 | 61,888.1153 | 61,888.1153 | 132.150 |
| Lack of Parallelism | 3 | 225.1877 | 75.0626 | 0.160 |
| Residual (curvature) | 4 | 2,733.3972 | 683.3493 | 1.459 |
| Within doses | 222 | 103,959.9740 | 468.2882 | — |
| Total | 233 | 190,773.6973 | | |

The table shows that the two sources of variation "lack of parallelism" and "curvature" are not significant ($P > 0.05$). This result conforms to the two cardinal assumptions implicit in the estimation of relative potencies. The source of variation "common regression" is highly significant and exceeds the recommended magnitude of 100 for the F value (see Finney, 1978).

The point and interval estimates of each test preparation relative to the standard preparation are as follows:

Case I: Prep. I (standard) vs. Prep. II (test)
The common slope is 68.1482; G = 0.0418
Relative potency = 0.639; 95% confidence limits: (0.4589, 0.9091)
Case II: Prep. I (standard) vs. Prep. III (test)
The common slope is 68.1482; G = 0.0418
Relative potency = 0.2434; 95% confidence limits: (0.1766, 0.9091)
Case III: Prep. I (standard) vs. Prep. IV (test)
The common slope is 68.1482; G = 0.0418
Relative potency = 0.4893; 95% confidence limits: (0.3495, 0.7352)

The analysis clearly shows that the standard preparation is more potent ($P < 0.05$) than the three test preparations.

## X. COVARIANCE ANALYSIS IN BIOASSAY

### A. Introduction

The primary purpose of using the analysis of covariance procedure in bioassay can be demonstrated by considering the thyrotropin assay as an illustration. In this assay guinea pigs are injected for several days with an extract of cattle pituitary glands and are then killed and the weight of their thyroid glands determined. An appropriate experimental design for the test would allow for the elimination of various possible sources of variation such as days of test and strains. However, an experimental design does not necessarily allow for the correction of any concomitant variation exhibited by the animals themselves. The weight of a guinea pig may determine in part the weight of its thyroid gland, whether or not this gland has increased in response to stimulation during the test. Guinea pigs in such a test are normally chosen so as to vary as little as possible in weight, but it is rarely feasible to choose large groups of animals of so nearly the same weight that differences between them in that respect can be neglected. We will now investigate a method by which the influence of the weight of the animal, or any other variable in which we are interested, may be examined. This is the method of *covariance analysis*.

### B. Tests of Validity

Covariance analysis has the following implicit assumptions, which must be examined by resorting to the tests of validity before considering any estimates and/or tests of hypothesis.

1. Normality of adjusted Y's (Wilk-Shapiro test)
2. Homogeneity of variances of groups (Levene test on adjusted Y's)
3. Additivity (only for RBD and LSQ on adjusted Y's)
4. Independence (physical act of randomization)

5. Parallelism (homogeneity of slopes of groups)
6. Linearity
7. X's should not be affected by treatment
8. X's should not be subjected to error and should be fixed
9. Functional relationship between X and Y linear
10. General regression assumptions

A detailed description of the computational aspects of the analysis of covariance is available in Finney (1978, chapter 12).

## XI. BIOASSAY ANALYSIS OF QUANTAL RESPONSE

### A. Quantal Response Distribution

In the class of experiments involving counts, the response variable is discrete and binary or binomial in structure. This response has an interpretation in terms of individual tolerance, which does not apply to other types of response distributions.

If Y describes the two possible outcomes, which could be "alive" (for "success") of "dead" (for "failure") and take on the value 1 (for success) or 0 (for failure) for that dose level, then the dose-response model is

$$P(d) = P[Y = 1 \mid d] = E(Y \mid d)$$

When there are independent replications of the experiment at the same dose level d, then the distribution of the variable $Y = \Sigma_{i=1}^{n} Y_1$ is the total number of successes in the n trials and defines the binomial probability model

$$P[Y = y \mid d] = \binom{n}{y} [P(d)]^y [1 - P(d)]^{n-y}$$

where y = 0, 1, . . ., n and P(d) is defined above. If r denotes an observed value of Y for a specific dose level with n experimental units, then the maximum likelihood estimator of P(d) is r/n for that dose level.

Theoretically, when P(d) is monotone, with P(0) = 0 and $P(\infty) = 1$, it is possible to interpret P(d) as a tolerance distribution function F(d). Suppose that each experiment at dose d is made with a subject which is randomly chosen from a population and that the probability for the response 1 is determined by the value of a nonnegative continuous tolerance variable T which is defined as follows: if $T \leqslant d$ then Y = 1 and if $T > d$ then Y = 0. Then $P(d) = P[T \leqslant d] = F(d) = P[Y = 1 \mid d]$. In this way the model P(d) represents a distribution function F(d) for the tolerance of an individual. Therefore, the purpose of dose-response models is to characterize or assess the tolerance distribution for a certain population.

### B. Dose-Response Models

The following tolerance distributions are quantal dose-response models, and actual applications of these models can be found in Finney (1971, 1978), Tallarida and Jacob (1979), Buncher and Tsay (1981), Armitage (1982), Hubert (1984), Rand (1985), and Bergman and Glittins (1985).

1. The Probit Model

The probit model corresponds to a lognormal tolerance distribution; that is, if $X = \log_{10}[d]$, then

$$P(d) = \int_{-\infty}^{\alpha+\beta X} (2\pi)^{-0.5} \exp\left(\frac{-t^2}{2}\right) dt = \phi\left[\frac{X - (-\alpha/\beta)}{1/\beta}\right]$$

The probit of P(d) is defined as $5 + \phi^{-1}[P(d)]$, where $\phi$ is the cumulative standard normal distribution function. The value 5 is added to the normal percentile in order to avoid negative values. It follows that the probit is a linear function of log[d] = X, the dose metameter.

The following simple example illustrates the weighted least-squares probit method and shows how to estimate $ED_{50}$ with this model. (Most point and interval estimates are determined by the iterative maximum likelihood criteria discussed in the next section.) Let d denote fixed dose levels of a drug such that for each of the k = 7 dose levels r out of n subjects respond. Let p = r/n, the observed mortality rate, say. The transformations are $X = \log_{10}[d]$ and Y = probit[p]. In the following table the weights are $w = f^2/p(1 - p)$, where $f = (2\pi)^{-0.5}\exp[-(Y - 5)^2/2]$ is the ordinate value in a standard normal frequency function for a given value of Y. Also, if $d \neq 0$ and d = 0, replace it by 1/2n; and if a p = 100, replace it by 1 − (1/2n). (This is called the Berkson adjustment.) If d = 0 and p = $c \neq 0$, then replace all other p (which are >c) by (p − c)/1 − c). [This refers to a natural mortality rate and the formula is due to Abbott (1925).] The steps for determining the estimate of $ED_{50}$ are self-evident from the table below and subsequent calculations.

Weighted Probit Calculations

| X | d | n | r | p | Y | w |
|---|---|---|---|---|---|---|
| 1.000 | 10 | 47 | 8 | 0.170 | 4.046 | 0.4552 |
| 1.176 | 15 | 53 | 14 | 0.264 | 4.369 | 0.5503 |
| 1.301 | 20 | 55 | 24 | 0.436 | 4.839 | 0.6307 |
| 1.477 | 30 | 52 | 32 | 0.615 | 5.292 | 0.6174 |
| 1.699 | 50 | 46 | 38 | 0.826 | 5.938 | 0.4585 |
| 1.845 | 70 | 54 | 50 | 0.926 | 6.447 | 0.2854 |
| 1.978 | 95 | 52 | 50 | 0.962 | 6.775 | 0.1853 |

For the weighted least-squares probit line Y = a + bX:

$$b = \frac{\Sigma nwXY - (1/\Sigma nw)(\Sigma nwX)(\Sigma nwY)}{\Sigma nwX^2 - (1/\Sigma nw)(\Sigma nwX)^2}$$

$$= \frac{1232.358 - (1/163.424)(231.765)(841.486)}{342.272 - (1/163.424)(231.765)^2} = 2.868$$

$$a = \frac{\Sigma nwY}{\Sigma nw} - b\,\frac{\Sigma nwX}{\Sigma nw} = \frac{841.486 - (2.868)(231.765)}{163.424} = 1.082$$

$$Y = 1.082 + 2.868X \qquad \log[ED_{50}] = 5 - \frac{a}{b} = 1.366 \qquad ED_{50} = 23.2$$

2. The Logit Model

The logit model corresponds to a logistic distribution of the log tolerance:

$$P(d) = \frac{e^{\alpha+\beta X}}{1 + e^{\alpha+\beta X}}$$

where $X = \log_e[d]$ and the logit, Z, is defined as $Z = \log_e[P(d)/(1 - P(d))]$. Then the logit is a linear function of $\log_e[d]$.

To illustrate the weighted logit method, consider the following calculations of the data set above. The steps in the calculations are detailed in the following table and subsequent calculations. Note that $X = \log_e[d]$, $p = r/n$, $Z = \log_e[p/1 - p)]$, and $w = np(1 - p)$.

Weighted Logit Calculations

| X | d | n | r | p | Z | w |
|---|---|---|---|---|---|---|
| 2.303 | 10 | 47 | 8 | 0.170 | −1.584 | 6.638 |
| 2.708 | 15 | 53 | 14 | 0.264 | −1.025 | 10.302 |
| 2.996 | 20 | 55 | 24 | 0.436 | −0.256 | 13.527 |
| 3.401 | 30 | 52 | 32 | 0.615 | +0.470 | 12.308 |
| 3.912 | 50 | 46 | 38 | 0.826 | +1.588 | 6.609 |
| 4.249 | 70 | 54 | 50 | 0.926 | +2.526 | 3.704 |
| 4.554 | 95 | 52 | 50 | 0.962 | +3.219 | 1.923 |

For the weighted (least-squares) logit line Z = a + bX:

$$b = \frac{\Sigma nXZ - (1/\Sigma w)(\Sigma wX)(\Sigma wZ)}{\Sigma wX^2 - (1/\Sigma w)(\Sigma wX)^2} = \frac{64.729 - (1/55.011)(175.916)(7.096)}{582.399 - (1/55.011)(175.916)^2}$$

$$= 2.118$$

$$a = \frac{\Sigma wZ}{\Sigma w} - b\,\frac{\Sigma wX}{\Sigma w} = \frac{7.096 - (2.118)(175.916)}{55.011} = -6.644$$

$$Z = -6.644 + 2.118X$$

$$\log\,[EC_{50}] = \frac{-a}{b} = \frac{-(-6.644)}{2.118} = 3.137$$

$$EC_{50} = e^{[\log EC_{50}]} = e^{3.137} = 23.0$$

3. The Arcsine Model

If $X = \alpha + \beta \log[d]$, then this model is given by

$$P(d) = \sin^2[X], \qquad 0 \leq X \leq \pi/2$$

with $P(d) = 0$ if $X < 0$ and $P[d] = 1$ if $X > \pi/2$. It follows that $\sin^{-1}[(P(d))^{1/2}]$ is a linear function of $\log[d]$.

The literature is exploding with still other models (including linits, rankits, ridits, and quantits) and approaches (such as sequential and Bayesian) as well as a large class of nonparametric methods.

## XII. MULTIVARIATE APPROACH TO BIOASSAY ANALYSIS

### A. Introduction

The purpose of this section is to introduce another method and a new model that uses the correlations between observations over time to estimate $LD_{50}$. This approach generalizes the approach of Section XI and is based on multivariate growth curve theory. The basic need for this approach is that some experiments are monitored over a sequence of time points, and so the corresponding responses become functions of concentration and time. Applying univariate probit analysis at each time point neglects the dependence on time and the possible effect of a time and concentration interaction on the response. In fact, knowledge of these dependences can be exploited to yield estimates which have statistical validity and greater precision.

We need the following slightly different notation:

$c_j$ = jth dose level, with $j = 1, 2, \ldots, d$
$d$ = number of dose levels
$X_j = \log_{10}[c_j]$ is the jth log dose
$T_i$ = ith time point, with $i = 1, 2, \ldots, t$
$n_{jk}$ = number of sampling units at the jth dose and the kth replication, with $k = 1, 2, \ldots, r$
$z_{ijk}$ = mortality count at the ith time point for the jth dose level and the kth replication

To appreciate this notation consider the data in the following table, in which each of $d = 7$ different dose levels of toxic substance is administered to a group of 10 subjects. The mortality counts per group are recorded after 48 hours (2 days) and again after 3 and 4 days; with these $t = 3$ time points the experiment has $r = 2$ replications.

Observed Mortality Counts

| Block | Time (days) | Dosage units | | | | | | |
|---|---|---|---|---|---|---|---|---|
| | | 0.10 | 0.20 | 0.30 | 0.50 | 1.00 | 2.00 | 2.50 |
| 1 | 2 | 0 | 1 | 1 | 2 | 3 | 3 | 4 |
| | 3 | 1 | 1 | 2 | 3 | 5 | 6 | 7 |
| | 4 | 1 | 1 | 3 | 4 | 8 | 7 | 9 |

Observed Mortality Counts (continued)

| Block | Time (days) | Dosage units | | | | | | |
|---|---|---|---|---|---|---|---|---|
| | | 0.10 | 0.20 | 0.30 | 0.50 | 1.00 | 2.00 | 2.50 |
| 2 | 2 | 0 | 1 | 1 | 2 | 3 | 4 | 4 |
| | 3 | 1 | 1 | 2 | 3 | 5 | 6 | 7 |
| | 4 | 1 | 1 | 2 | 4 | 7 | 8 | 8 |

## B. Transformation

To illustrate one approach to the problem of estimating the $LD_{50}$ for such an experiment, the Freeman-Tukey variance-stabilizing transformation is first made on the mortality rates by letting

$$Y_{ijk} = \frac{1}{2}\left[\text{arcsine}\left(\frac{z_{ijk}}{n_{jk} + 1}\right)^{1/2} + \text{arcsine}\left(\frac{z_{ijk} + 1}{n_{jk} + 1}\right)^{1/2}\right]$$

Then the response variables $Y_{ijk}$ are represented as a polynomial of order $q - 1$ in time and linear in log dose; that is, we assume

$$Y_{ijk} = \sum_{m=1}^{q} (\beta_{m1} + \beta_{m2}X_j + \rho_{mk})T_i^{-(m-1)} + \varepsilon_{ijk}$$

$i = 1, 2, \ldots, t$, $j = 1, 2, \ldots, d$, and $k = 1, 2, \ldots, r$, with $\Sigma_{k=1}^{r}\rho_{ik} = 0$, for each i. (For our example, $i = 1, 2, 3$, $j = 1, 2, \ldots, 7$, and $k = 1, 2$.) The error vector is denoted by $\underline{\varepsilon}_{jk} = (\underline{\varepsilon}_{ijk} \cdots \underline{\varepsilon}_{tjk})$, where the prime denotes the usual transpose of the vector. The $\underline{\varepsilon}_{jk}$ are assumed to be independently and normally distributed with mean zero and covariance matrix $\Sigma$.

## C. Potthoff and Roy Model

This model can be cast in the form of the Potthoff and Roy (1964) growth curve model. Using matrix notation, we can write the model as

$$Y = B\beta A + \varepsilon$$

where the $(t \times dr)$ response matrix (it is $3 \times 14$ in our example) is

$$Y = \begin{bmatrix} Y_{111} & \cdots & Y_{1d1} & & Y_{11r} & \cdots & Y_{1dr} \\ \cdot & \cdots & \cdot & & \cdot & \cdots & \cdot \\ \cdot & \cdots & \cdot & \cdots & \cdot & \cdots & \cdot \\ \cdot & \cdots & \cdot & & \cdot & \cdots & \cdot \\ Y_{t11} & \cdots & Y_{td1} & & Y_{t1r} & \cdots & Y_{tdr} \end{bmatrix}$$

the $(t \times q)$ design matrix for the time variable is

$$B = \begin{bmatrix} 1 & T_1^{-1} & T_1^{-2} & \cdots & T_1^{-(q-1)} \\ 1 & T_2^{-1} & T_2^{-2} & \cdots & T_2^{-(q-1)} \\ \cdot & \cdot & \cdot & \cdots & \cdot \\ 1 & T_t^{-1} & T_t^{-2} & \cdots & T_t^{-(q-1)} \end{bmatrix}$$

The $(q \times r+1)$ parameter matrix is

$$\beta = \begin{bmatrix} \beta_{11} & \beta_{12} & \rho_{11} & \rho_{12} & \rho_{13} & \cdots & \rho_{1(r-1)} \\ \beta_{21} & \beta_{22} & \rho_{21} & \rho_{22} & \rho_{23} & \cdots & \rho_{2(r-1)} \\ \cdot & \cdot & \cdot & \cdot & \cdot & \cdots & \cdot \\ \beta_{q1} & \beta_{q2} & \rho_{q1} & \rho_{q2} & \rho_{q3} & \cdots & \rho_{q(r-1)} \end{bmatrix}$$

The $(r+1 \times dr)$ design matrix for the dose and block effects (it is $3 \times 14$ in our example) is given by

$$A = \begin{bmatrix} 1 & \cdots & 1 & 1 & \cdots & 1 & & 1 & \cdots & 1 & 1 & \cdots & 1 \\ X_{11} & \cdots & X_{d1} & X_{12} & \cdots & X_{d2} & & X_{1(r-1)} & \cdots & X_{d(r-1)} & X_{1r} & \cdots & X_{dr} \\ 1 & \cdots & 1 & 0 & \cdots & 0 & \cdots & 0 & \cdots & 0 & -1 & \cdots & -1 \\ 0 & \cdots & 0 & 1 & \cdots & 0 & & 0 & \cdots & 0 & -1 & \cdots & -1 \\ \cdot & \cdots & \cdot & \cdot & \cdots & \cdot & & \cdot & \cdots & \cdot & \cdot & \cdots & \cdot \\ 0 & \cdots & 0 & 0 & \cdots & 0 & & 1 & \cdots & 1 & -1 & \cdots & -1 \end{bmatrix}$$

and the error matrix is given by $\varepsilon = (\underline{\varepsilon}_{11}, \ldots, \underline{\varepsilon}_{dr})$.

The maximum likelihood (ML) estimator of $\beta$, originally derived by Khatri (1966), can be translated into this matrix setting as follows:

$$\beta = UB' V^{-1} YA' A^*$$

where

$$V = Y (I - A' A^* A) Y'$$

$$U = (B' V^{-1} B)^{-1}$$

$$A^* = (A A')^{-1}$$

The estimator of the dose required to produce 50% mortality for a given time point is the value of the dose that yields a value of $Y = \pi/4$ radians. The value of the log of this estimator is given by

$$X = \frac{\pi/4 - g_1(T)}{g_2(T)}$$

where

$$g_1(T) = \sum_{m=1}^{q} \beta_{m1} T^{-(m-1)}, \qquad g_2(T) = \sum_{m=1}^{q} \beta_{m2} T^{-(m-1)}$$

and $\beta_{mj}$ is the estimator of $\beta_{mj}$, the elements of $\beta$, the ML estimator. Consequently, the estimator of $LD_{50}$ is $10^x$.

### D. Example

For the example we are considering, where $q = 2$, we could define our $3 \times 2$ design matrix by

$$B = \begin{bmatrix} 1 & T_1^{-1} \\ \vdots & \vdots \\ 1 & T_t^{-1} \end{bmatrix} = \begin{bmatrix} 1 & 0.500 \\ 1 & 0.333 \\ 1 & 0.250 \end{bmatrix}$$

With only this degree of representation it can be shown that the estimated model becomes

$$Y_{ijk} = 1.193 - 1.286T_i^{-1} + 0.782X_j - 0.935X_jT_i^{-1} + \hat{\rho}_{1k} + \hat{\rho}_{2k}T_i^{-1}$$

where $\hat{\rho}_{11} = -\hat{\rho}_{12} = 0.0302$ and $\hat{\rho}_{21} = -\hat{\rho}_{22} = -0.0853$.

Confidence limits and confidence bands for $LD_{50}$ can be constructed from a multivariate generalization of Fieller's theorem (see Carter and Hubert, 1984a, for details). The table below gives estimates of $LD_{50}$ for selected values of T. A test of the null hypothesis that the reduced model is adequate against the alternative hypothesis that the quadratic model is necessary can also be done.

Estimates of $LD_{50}$

| Time | Estimate | Confidence interval |
|---|---|---|
| 2.0 | 5.60 | (3.99, 8.76) |
| 2.5 | 1.83 | (1.61, 2.11) |
| 3.0 | 1.02 | (0.93, 1.38) |
| 3.5 | 0.84 | (0.68, 0.91) |
| 4.0 | 0.70 | (0.56, 0.85) |
| 6.0 | 0.49 | (0.37, 0.66) |

Clearly, polynomial growth curve models can be useful in assessing multivariate quantal bioassays over time. Such a method assumes that a smooth curve approximates the relationship between the transformed quantal values and the two explanatory variables, log dose and time. Moreover, this method performs well for interpolation problems and is therefore ideal for estimating $LD_{50}$ values.

### E. General Discussion

The general method of fitting growth curves to experiments with many time points has been thoroughly discussed by Rao (1965); also see Potthoff and Roy (1964) and Grizzle and Allen (1969). They conclude that the response variable can be modeled by orthogonal polyomials with linear, quadratic, and higher-order terms but that it is often the practice to use only a few of the higher-order terms in the analysis.

The choice of a model in inverse time or direct time can be decided by the researcher and the data. In experiments where the subjects become acclimated to the treatments, there is little change in the response variable for large values of T; hence, a choice of inverse time seems appropriate.

In the example the $LD_{50}$ values were calculated with the block effects set to zero. The replicates (blocks) represented two consecutive weeks when the experiment was performed; this is consistant with the observation that the estimates of these effects were small. Testing for block effects can be done by applying standard MANOVA techniques.

An analysis of indirect quantitative assays with correlated data has been given by Box and Hay (1953). Their approach assumes a repeated-measures model; that is, the correlations between observations on the same experimental unit are assumed to be equal. Elashoff (1981) studied the effect of heterogeneity of variances and correlations on this analysis. Vølund (1980) treated the case when these correlations are arbitrary, extending the work of Rao (1954). Multivariate quantal assays have also been studied by Kolakowski and Bock (1981) and Kooijman (1981). The issue of extrabinomial variations has been considered by Segreti and Munson (1981). All these issues are important when analyzing such experiments.

These methods are extendable to nonlinear growth curve models and to the analysis of incomplete data in which measurements are not taken at every time point for each experimental unit. Also, these methods are useful (see, for example, Carter and Hubert, 1984b, 1985; Carter et al., 1985) for generating confidence intervals for the relative potency in multivariate quantitative parallel-line and parabolic bioassays.

## XIII. JOINT DRUG ACTION

### A. Types of Joint Drug Action

There are several types of joint drug action. The most notable and frequently used types are potentiation, inhibition, similar joint action, synergism, and antagonism. A brief description of each type is provided in the following.

1. *Potentiation*. A joint drug action is called a potentiation when a drug with no effect of its own enhances (potentiates) the action of an active drug. This clearly implies that the potentiating drug, when administered alone, does not elicit a response regardless of the size of the

dose employed. Selection of a "no-effect" dose of an otherwise active drug would not qualify it as a potentiator. Further, if the dose(s) of active drug is not sufficient to evoke a response, the potentiating drug, when added, may elicit a response.

2. *Inhibition*. A joint drug action is called an inhibition when a drug with no effect of its own reduces (inhibits) the action of an active drug. The inhibiting drug, when given alone, does not elicit a response regardless of the size of the dose administered. If the dose(s) of an active drug produces only a minimal response, the inhibiting drug, when added, may result in no response.

3. *Similar joint action*. A joint drug action is called a similar joint action if the drugs in question have the same site of action, their log dose-response lines are parallel, one drug behaves as dilution of the other, and one drug can be substituted at a constant proportion for the other. Thus, the response for any mixture is completely predictable given the proportion of the drugs in the mixture and their relative potency.

4. *Synergism* (Finney, 1971). A joint drug action is called a synergism when two or more active drugs produce a joint effect which is greater than that expected on the basis of similar joint action. For synergism to occur, the drugs must have different modes of action. Accordingly, the log dose-response lines may or may not be parallel. In the situation in which a lack of parallelism occurs, a unique estimate of relative potency is not possible.

5. *Antagonism*. A joint drug action is called an antagonism when two or more active drugs produce a joint effect which is less than that expected on the basis of similar joint action. In terms of actions, antagonism is just the opposite of synergism.

Two numerical examples, one for potentiation and the other for synergism, are presented in the following.

## B. Potentiation Example

In this study, the joint drug action of drug $D_2$ (hydralazine) and drug $D_1$ (timolol) on the average maximal recorded fall in the arterial blood pressure (hereafter referred to as response) of spontaneously hypertensive rats within 24 hours (measured at 0, 1/2, 1, 2, 3, 4, 5, 6, 7, 8, 12, 18, and 24 hours) after treatment is studied to evaluate the potentiation of drug $D_2$ by drug $D_1$ (Scriabine et al., 1974). For this, it is necessary to show the following effects:

1. That drug $D_2$ has higher response than the control (saline in these tests)
2. That the response of drug $D_1$ as used in this experiment is similar to that of the control (saline)
3. That drug $D_2$ plus drug $D_1$ has a higher response than drug $D_2$ alone

The design of the experiment is a 6 × 6 Latin square with rows and columns represented by animals and periods of testing, respectively. The animals are treated every 48 hours to remove any possibility of carryover effect. The structure of the Latin square and the numerical values of the response associated with each cell are presented below:

| | | Period of testing | | | | | | | | | | | |
|---|---|---|---|---|---|---|---|---|---|---|---|---|---|
| | | 1 | | 2 | | 3 | | 4 | | 5 | | 6 | |
| | 1 | A | 10 | B | 6 | C | 14 | D | 24 | E | 24 | F | 26 |
| R | 2 | B | 6 | C | 10 | D | 24 | E | 30 | F | 34 | A | 6 |
| A | 3 | C | 18 | D | 26 | E | 24 | F | 40 | A | 10 | B | 6 |
| T | 4 | D | 22 | E | 20 | F | 44 | A | 6 | B | 4 | C | 10 |
| S | 5 | E | 24 | F | 34 | A | 6 | B | 4 | C | 14 | D | 10 |
| | 6 | F | 40 | A | 10 | B | 8 | C | 18 | D | 26 | E | 20 |

The following treatments were considered for comparison:

| Treatment symbol | Description |
|---|---|
| A | Saline p.o. |
| B | Drug $D_1$ 0.5 mg/kg p.o. |
| C | Drug $D_2$ 0.25 mg/kg p.o. |
| D | Drug $D_2$ 1.0 mg/kg p.o. |
| E | Drug $D_1$ 0.5 mg/kg plus drug $D_2$ 0.25 mg/kg p.o. |
| F | Drug $D_1$ 0.5 mg/kg plus drug $D_2$ 1.0 mg/kg p.o. |

The analysis of variance procedure for a Latin square design is used for the statistical analysis. Duncan's new multiple-range test is used to accomplish the pairwise comparisons among the six groups. The statistical results are summarized in the following table.

Average Maximal Recorded Fall in Arterial Blood Pressure Within 24 Hours After Treatment

| | Treatment (p.o.) | Response (mm Hg) |
|---|---|---|
| A | Saline | 8.0 |
| B | Drug $D_1$ 0.5 mg/kg | 5.7[a] |
| C | Drug $D_2$ 0.25 mg/kg | 14.0[b] |
| D | Drug $D_2$ 1.0 mg/kg | 22.0[c] |

Average Maximal Recorded Fall in Arterial Blood Pressure Within 24 Hours After Treatment (continued)

| | Treatment (p.o.) | Response (mm Hg) |
|---|---|---|
| E | Drug $D_1$ 0.5 mg/kg plus drug $D_2$ 0.25 mg/kg | 23.7[d] |
| F | Drug $D_1$ 0.5 mg/kg plus drug $D_2$ 1.0 mg/kg | 36.3[e] |

[a]Not significantly different from saline. Here, it may be noted that a dose of drug $D_1$ 40 times the dose selected for the experiment showed an average response of 7.0, which is not significantly different from that of saline. This demonstrates that the response of drug $D_1$ as used in this experiment is similar to that of the control.
[b]Significantly higher than saline at $P \ll .001$.
[c]Significantly higher than saline at $P \ll .001$.
[d]Significantly higher than drug $D_2$ alone at 0.25 mg/kg at $P \ll .001$.
[e]Significantly higher than drug $D_2$ alone at 1.0 mg/kg at $P \ll .001$.

Analysis of Variance for Latin Square Design on Arterial Blood Pressure Data

| Source of variation | Sum of squares | Degrees of freedom | Mean squares | F ratio |
|---|---|---|---|---|
| Rows | 119.11111 | 5 | 23.84444 | 2.34279 |
| Columns | 231.22222 | 5 | 46.24444 | 4.54367 |
| Treatments | 3911.22222 | 5 | 782.24444 | 76.85808 |
| Error | 203.55556 | 20 | 20.17778 | — |
| Total | 4465.22222 | 35 | | |

The analysis of variance table has been presented above. The analysis conformed to all three tests of validity: Levene's test, the Wilk-Shapiro test, and Tukey's test for nonadditivity.

### C. Synergism Example

The study pertains to the joint drug action of two antihypertensive agents, P and Q, in normotensive rats. There is a reduction in the mean arterial pressure as the dose is increased. The following table provides the basic information about the drugs and their mixture.

| | Drug P alone | | | Drug Q alone | | | $P \cup Q$[a] (1:1 mixture) | | |
|---|---|---|---|---|---|---|---|---|---|
| Dose (mg/kg) | 110 | 121 | 133.1 | 110 | 121 | 133.1 | 83 | 91.3 | 100.4 |
| Log dose = X | 2.0414 | 2.0828 | 2.1242 | 2.0414 | 2.0828 | 2.1242 | 1.9191 | 1.9605 | 2.0017 |

(continued)

| | Drug P alone | | | Drug Q alone | | | PUQ[a] (1:1 mixture) | | |
|---|---|---|---|---|---|---|---|---|---|
| n | 10 | 10 | 10 | 10 | 10 | 10 | 10 | 10 | 10 |
| Means | 11.56 | 9.55 | 6.26 | 11.22 | 8.33 | 6.40 | 7.84 | 5.58 | 4.02 |

[a]P Q denotes the mixture.

The BANOVA for the study is as follows.

| SOV | d.f. | MS | F | Critical $F_\alpha = 0.05$ | Results |
|---|---|---|---|---|---|
| Treatments | 8 | | | | |
| Preparations | 2 | 95.9293 | 6.59 | 3.1 | Sig. |
| Common regression | 1 | 323.4082 | 22.2 | 4.0 | Sig. |
| Lack of parallelism | 2 | 2.8922 | 0.20 | 3.1 | NS |
| Residual (curvature) | 3 | 1.7023 | 0.12 | 2.7 | NS |
| Error | 81 | 14.5574 | | | |
| Total | 89 | | | | |

where SOV denotes source of variation; d.f., degrees of freedom; MS, mean square; Sig., significant at $P < 0.05$; and NS, not significant at $P = 0.05$.

Using the procedures described in Finney (1971):

$b_c = -56.0797$ (common regression slope)

$[X^2] = 0.1028$ $(\Sigma(X - \bar{X})^2)$

$Y_P = 125.9261 - 56.0797X$

$Y_Q = 125.4528 - 56.0797X$

$Y_m = 115.7610 - 56.0797X$

Also, the potency of drug P relative to drug Q is estimated from

$$M_P = \log R_P = \frac{a_P - a_Q}{b_c} = \frac{125.9261 - 125.4528}{-56.0797}$$

$$= -0.0084$$

$$R_P = \text{antilog}(-0.0084) = 0.981$$

The log dose-response line for the mixture expected on the basis of similar joint action will be given by (Finney, 1971)

$$Y^*_m = a_Q + b_c \log(\Pi_Q + R_P \Pi_P) + b_c x$$

where, for a mixture of A:B = 1:1,

$$\Pi_Q = \text{fraction of drug Q in the mixture}$$

$$= \frac{B}{A + B} = \frac{1}{2} = 0.5$$

$$\Pi_P = \text{fraction of drug P in the mixture}$$

$$= \frac{A}{A + B} = \frac{1}{2} = 0.5$$

so that

$$Y^*_m = 125.4528 - 56.0797 \log[0.5 + 0.981(0.5)] - 56.0797X$$

$$= 125.6715 - 56.0797X$$

We then calculate the quantities

$$M^* = \frac{a_m - a^*_m}{b_c} = \frac{115.7610 - 125.6715}{-56.0797}$$

$$R^* = \text{antilog } M^* = \text{antilog } 0.1767 = 1.50$$

Values of $R^*$ are interpreted as follows:

$R^* = 1.0$: similar joint action

$R^* = <1.0$: antagonism

$R^* = >1.0$: synergism

To test for the statistical significance of synergism we first calculate the variance of $M^*$ from

$$V(M^*) = \frac{s^2}{b_c^2(\Pi_Q + R_P\Pi_P)^2}\left\{\frac{\Pi_Q^2}{n_Q} + \frac{R_P^2\Pi_P^2}{n_P} + \frac{(\Pi_Q + R_P\Pi_P)^2}{n_m} + \frac{[\Pi_Q Y_Q + R_P\Pi_P Y_P - (\Pi_Q + R_P\Pi_P)Y_m]^2}{b_c^2[X^2]}\right\}$$

$$= \frac{14.5574}{(-56.0797)^2(0.9905)^2}\left\{\frac{(0.5)^2}{30} + \frac{(0.981)^2(0.5)^2}{30} + \frac{(0.9905)^2}{30}\right.$$

$$\left. + \frac{[(0.5)(8.65) + (0.981)(0.5)(9.12) - (0.9905)(5.82)]^2}{(-56.0797)^2(0.1028)}\right\}$$

$$= 0.00036$$

$$\text{S.E.}(M^*) = \sqrt{V(M^*)} = \sqrt{0.000366} = 0.019$$

$$t = \frac{M^*}{\sqrt{V(M^*)}} = \frac{0.1767}{0.19} = 9.31$$

The critical value of t at P = 0.05 for 81 d.f. is 1.99. Since the observed value of t (9.31) is larger than the critical value of t (1.99), we conclude that there is statistically significant evidence of synergism.

The 95% confidence limits may be obtained as follows:

$$M_L^*, M_U^* = M^* \pm t\sqrt{V(M^*)}$$

$$= 0.1767 \pm 1.99\ (0.019)$$

$$= 0.1389,\ 0.2145$$

$$R_L^*, R_U^* = \text{antilog}(0.1389,\ 0.2145)$$

$$= 1.38,\ 1.64$$

Thus, we conclude that the mixture is 1.50 times more effective than expected on the basis of similar joint action (i.e., is synergistic) with 95% confidence limits of 1.38 and 1.64.

## XIV. DRUG-RECEPTOR INTERACTION

### A. Introduction

This section deals with the development of the statistical methodologies appropriate for the analysis of drug-receptor interaction studies. Drug-receptor interaction belongs to one specific area of general pharmacology known as molecular pharmacology. Molecular pharmacology is concerned with studies of basic mechanisms of drug action on biological systems. It considers molecules as the basic functional units both of the drug and of the system in which the drug acts. Isolated tissues form the primary experimental units in these studies. The statistical methods associated with the analysis of drug-receptor interaction experiments inherently form an integral part of general bioassay statistics. However, the terminologies such as $ED_\alpha$, relative potency, standard drug, test drug, and tolerance distribution are no longer valid. A new set of terminologies, including receptor, agonist, antagonist, competitive, and $pA_2$ will be introduced.

It is well known that most drugs act by combining with a specific cell component in order to produce an effect. This component of the cell is called a *receptor*. In the case of a drug and its receptor, a drug-receptor complex is formed. This complex provides the stimulus for the resulting effect. Such an active drug is said to have intrinsic activity and is termed an *agonist*. Some drugs with no ability to initiate activity of their own act in association with an agonist in such a way that the effect of the agonist is reduced or even eliminated. Such drugs are called *antagonists*. In some cases this inhibition may be overcome completely by increasing the concentration of the agonist. This is the case, for example, with isoproterenol and propranolol, in which (see Figure 4) the antagonism is completely overcome and the curves are parallel. This type of antagonism is called *competitive*. Competitive antagonists have the following three important properties: (1) They give rise to parallel log dose-response curves. (2) The antagonism is competitive so that the curves attain the same maximum effect. (3) Specificity is a prime characteristic of a competitive antagonist.

It is frequently desirable to determine concentrations of the agonist that produce equal effects. For example, consider a situation in which a level of effect, E, is produced by agonist concentration $C_1$ when the agonist alone is used, but a greater agonist concentration $C_2$ is required in the presence of the antagonist. The term *dose ratio* is often used for the ratio $C_2/C_1$. The significance of this ratio will be demonstrated later in the development.

Drugs can be classified into pharmacological families. A family includes all drugs which can interact with the receptors of a certain receptor-effector system. This system of classification is especially applicable for agonists and competitive antagonists. Agonists (receptor activators) and their competitive antagonists do not have the same kind of effect and hence they belong to different families. Yet the members of two such families may be related since they have affinity for the same receptors.

Although one can study the effect produced by a drug and determine a dose-response relation, great care must be exercised in connecting this

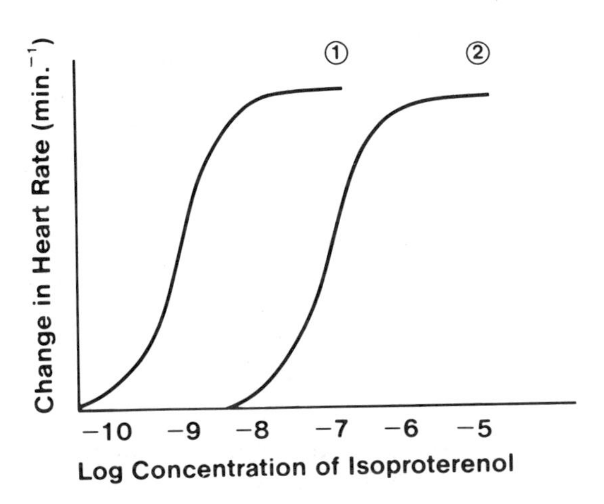


Fig. 4 Dose-response curves. 1 = Isoproterenol alone, 2 = Propranolol plus Isoproterenol. Note the parallel shift to the right with the same maximum effect.

relation with the mechanism at the molecular level. Indeed, much of the controversy that surrounds the several current theories of drug action results from unfounded assumptions connecting effects with bound drug.

The primary objectives of this section are as follows: (1) Clearly define the concepts associated with drug-receptor interaction systems. (2) Derive the fundamental equation for the competitive antagonism. (3) Demonstrate the procedure for estimating $pA_2$ values. (4) List the kinetic assumptions implicit in the estimation of $pA_2$. (5) Describe the various statistical techniques to be used to demonstrate that a drug is indeed a competitive antagonist in the context of the prevalent kinetic theory. (6) Develop an expression for $(1 - \alpha)100\%$ confidence limits for $pA_2$. (7) Develop an ANOVA table to be used routinely in the laboratory to test the various assumptions implicit in the estimation of the $pA_2$ value. (This would be a Finney-type ANOVA used in the analysis of bioassay data.) (8) Develop a test statistic for comparing two $pA_2$ values to be used for assessing the relative strengths of two antagonists or the strengths of the same antagonist in two different receptor environments. (9) Make the structure of statistical techniques (and tests) simple enough to be used *routinely* in the laboratory.

## B. Derivation of the Fundamental Equation for Competitive Antagonism in Drug-Receptor Interaction

There is much similarity between the kinetics of drug-receptor interactions and the enzyme kinetics of substrate-enzyme interactions. The law of mass action is applicable to both situations. According to the law of mass action, the rate of an enzyme-catalyzed reaction per unit volume, denoted by V, has the following expression:

$$V = V_{max}\frac{[S]}{[S] + K_m}$$

where $V_{max}$ is the maximum rate, $K_m$ is the rate constant, and [S] is the substrate concentration. This equation is known as the Michaelis-Menten equation.

Pharmacological studies differ from enzyme-substrate studies in that binding sites are not necessarily receptors, since the term "receptor," when used in pharmacology, represents only binding sites that are capable of initiating a response when stimulated by an agonist drug. Competition between an agonist and an antagonist for a single receptor site must be determined from experiments that produce a pair of dose-response curves (see Figure 4).

When a concentration A of agonist is administered, application of the mass action law gives the fraction of bound receptor $F_A$ in terms of A and $K_A$:

$$F_A = A(A + K_A)^{-1}$$

where $K_A$ is the dissociation constant for the agonist. If the competitive antagonist in concentration B is present, the fractional occupancy by the agonist is diminished to some value $F_B$:

$$F_B = A^*[A^* + K_A(1 + B/K_B)]^{-1}$$

where $K_B$ is the dissociation constant for the antagonist. What one measures in an experiment is the effect and not $F_A$ or $F_B$. Since the relation between effect and F is unknown, it would be difficult to estimate $K_A$ and $K_B$ (Ariens, 1957 and Aranlakshana et al., 1959).

However, assuming only that equal effects imply equal agonist occupancy (occupation theory), we may give the two expressions in the following manner. Now $F_A$ has the following form:

$$F_A = (E_{max}A)(A + K_A)^{-1}$$

and $F_B$, likewise, has the form

$$F_B = (E_{max}A^*)[A^* + K_A(1 + BK_B^{-1})]^{-1}$$

For the same effect, $E_0$, we have

$$E_0 = [E_{max}A][A + K_A]^{-1} = (E_{max}A^*[A^* + K_A(1 + BK_B^{-1})]^{-1}$$

Taking the reciprocal of both sides and rearranging, we have

$$[A^*A^{-1}] = BK_B^{-1}$$

This expression forms the fundamental equation for competitive antagonism in drug-receptor interaction. Schild (1947) denoted the dose ration $A^*A^{-1}$ by X and defined $pA_x$ as the negative logarithm of B, in molar units, that produced the ratio X. To see the relation between $pA_x$ and $K_B$, we take common logarithms of the last equation above, which leads to the following, generally used form:

$$\log_{10}(A^*A^{-1} - 1) = -\log_{10}K_B + \log_{10}B = -\log_{10}X_B - pA_x$$

or

$$pA_x = -\log_{10}(A^*A^{-1}) - \log_{10}K_B$$

When the value of the dose ratio $A^*A^{-1}$ is equal to 2, we have

$$pA_2 = -\log_{10}K_B = \log\left(\frac{1}{K_B}\right)$$

This equation establishes the relationship between $pA_2$ and $K_B$. Since $K_B$ is the affinity, the equation indicates that $pA_2$ is the common logarithm of the affinity constant. The $pA_2$ of a competitive antagonist was defined by Schild (1947) as the negative logarithm of the molar concentration of the antagonist which reduces the effect of a dose of the agonist to that of half the dose. The $pA_2$, the affinity of an antagonist drug, is generally determined experimentally from measurements of the equieffective agonist dose ratio at equilibrium in the absence and presence of a fixed concentration of the antagonist.

### C. Experimental Determination of $K_B$ ($pA_2$)

The dose-response curves associated with the agonist alone and the agonist plus a specific dose of the antagonist are determined experimentally. The agonist is usually added to the tissue bath in a cumulative fashion so that the concentration of the drug in the bath is increased at log intervals. A washout interval is introduced between any two doses. This is achieved by removing the tissue from the bath and allowing the tissue to "recover" (reequilibrate). In practice, several concentrations of the antagonist are used. When a dose of the antagonist is introduced into the tissue bath, sufficient time is allowed for the tissue to equilibrate with the antagonist. This is necessary because it permits the concentrations in the bath and at the receptor site to be equivalent. After the equilibration, the procedure associated with the agonist experiment is repeated.

In practice, $K_B$ is calculated from not just one, but several concentrations of the antagonist. We calculate the dose ratio for each $B_i$ (that is, $A^*A_i^{-1}$ and $B_i$, $i = 1, 2, \ldots, n$). By the least-squares procedure, we obtain the regression of $\log(A^*A^{-1} - 1)$ on $\log B$ as

$$\log(A^*A^{-1} - 1) = \log B - \log K_B$$

which is the logarithmic form of the fundamental equation. Note that $A^*$ and $A$ are estimated by selecting a particular level of response (usually 1/2 the maximum) from the original dose-response curves. When $\log(A^*A^{-1} - 1)$ is plotted against $\log B$, one obtains a straight line of slope unity and intercept $(-\log K_B)$. This type of plot is called a Schild plot. The negative common logarithm of $K_B$ (in molar units) is called the $pA_2$, where $pA_2 = \log K_B$. For example, a value of $K_B$ equal to $10^{-7}$ M with an affinity of $10^{+7}$ gives $pA_2 = 7$. Values of $pA_2$ are frequently used for competitive antagonists instead of their $K_B$ values (Figure 5).

### D. $(1 - \alpha)100\%$ Confidence Limits of $pA_2$

Consider a situation in which we have several tissues assigned to each of the pertinent groups such as agonist, agonist + low dose of antagonist, agonist + middle dose of antagonist, and agonist + high dose of antagonist. Then it would be appropriate to estimate the $(1 - \alpha)100\%$ confidence limits of $pA_2$. Using Fieller's theorem, the limits have the following expression:

$$pA_2(L),\ pA_2(U) = Z + D \pm \left[D^2 - (1 - g)\left\{D^2\left(\frac{g}{(1 - g)^2}\right)\left(\frac{[Z^2]}{n}\right)\right\}\right]^{1/2}$$

where L = lower, U = upper, $Z = -\log_{10}P$, and $W = \log_{10}[C_pC_0^{-1} - 1]$.

Let EC stand for effective concentration; then $C_0$ = EC(1/2 max) of the agonist in the absence of the antagonist and $C_p$ = EC(1/2 max) of the agonist in the presence of the antagonist. In addition, let $pA_2 = R$.

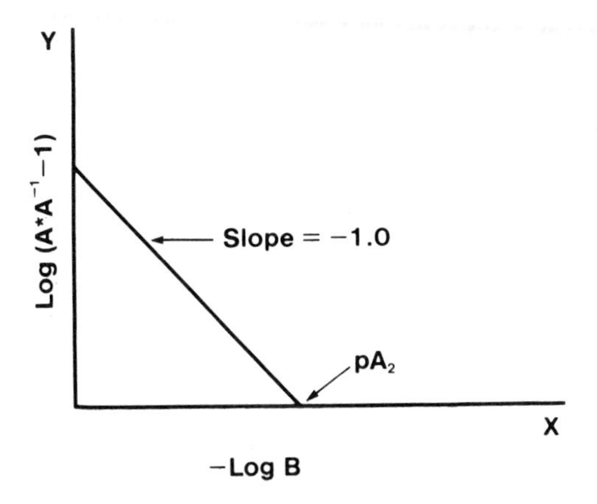


Fig. 5 Values of $pA_2$ used for competitive antagonists.

Schild plot regression is based on the regression of W on Z.

$$D = (R - Z)(1 - g)^{-1}, \qquad g = (t_a s)^2 \theta^{-2} [Z^2]^{-1}$$

where θ is the estimate of the slope, $S^2$ the mean rquare for the error in the ANOVA table, $[Z^2] = \Sigma(Z - Z)^2$, $[W^2] = \Sigma(W - W)^2$, and $t_a$ is the tabular t value for $\Sigma(N_i - 1)$ degrees of freedom at the a level of significance.

### E. Statistical Tests of Validity for Competitive Antagonism

An antagonist is considered to be competitive if it has the following properties: (1) The antagonist must shift the agonist dose-response curve to the right. (2) The agonist dose-response curve and the agonist plus antagonist dose-response curve must attain the same maximum. (3) The linear portions of the dose-response curves are parallel to each other. (4) There are no statistical tests to show the specificity of the antagonist.

The following tests are considered:

1. *Test for equi-maxima.* (a) One may be able to determine the maximum value associated with each tissue response curve graphically. (b) It is general practice to use the following nonlinear regression model:

$$Y = Y_{max}(1 - e^{-RX})$$

where $Y = \log(A^*A^{-1} - 1)$ and X is the concentration in molar units. Nonlinear regression methods should be used to estimate $Y_{max}$.

Now tabulate the $1/2Y_{max}$ value for each tissue in each group and do an analysis of variance after finding the results of the following tests: (a) Levene test for equality of variance, (b) Wilk-Shapiro test for normality, and (c) Tukey test of additivity for randomized block design (if used). (The $1/2Y_{max}$ values may be weighted by their respective variances, if necessary, before the analysis.) At this point, if we have favorable results from the tests in (a), (b), and (c), we proceed to conduct a two-sided Dunnett's multiple comparison test, comparing the control (agonist) with each of the antagonist doses. If there are no significant differences, we conclude that each of the dose-response curves has attained the same maximum value ($P = 0.05$) as that of the agonist.

2. *Equality of slopes of the linear portion of the dose-response curve.* Based on the linear regression analysis, estimate and tabulate the slope values for each tissue in each group. Conduct a two-sided Dunnett's multiple comparison test comparing the control (agonist) with each of the antagonist doses. If there are no significant differences, we conclude that the linear portions of the dose-response curves are parallel ($P = 0.05$).

Now we proceed to estimate an $EC_{50}$ = [1/2 maximum − intercept)/slope] for each tissue in each group. Based on the $EC_{50}$ values, we determine the "dose ratio" values as follows:

$$Y = \log_{10}[C_P C_0^{-1} - 1]$$

The following ANOVA is constructed based on the Y's.

PANOVA-I

| Source | d.f. | Desirable results for validation |
|---|---|---|
| Between doses | $(t - 1)$ | — |
| Linear regression ($\beta = 0$) | 1 | Highly significant |
| Test of slope ($\beta = 1$) | 1 | Not significant |
| Test of curvature | $(t - 2)$ | Not significant |
| Within doses | $\Sigma(K_i - 1)$ | — |
| Total | $(\Sigma K_i - 1)$ | |

3. The slope based on the regression of $\log(A^*A^{-1} - 1)$ on $\log B$ must be equal to −1.0. This test is accomplished by conducting a PANOVA-I (given above).

4. If we conduct a linear regression analysis based on $Y = (A^*A^{-1} - 1)$ and X = concentration of the antagonist in molar units, then we have the graph shown in Figure 6. The ideal curve must pass through the origin with a regression slope of 1.0 (45% angle). Competitive antagonism would require the following.

Simultaneous test of hypothesis:

$$H_0: \begin{bmatrix} B = 1 \\ A = 0 \end{bmatrix} \quad \text{vs.} \quad \begin{bmatrix} B \neq 1 \\ A \neq 1 \end{bmatrix}$$

where A is the intercept and B the slope.

The test statistic is as follows:

$$F = [nA^2 + 2nA\bar{x}(B - 1) + (B - 1)^2 \Sigma x^2](2S^2)^{-1}$$

F = F − test statistic

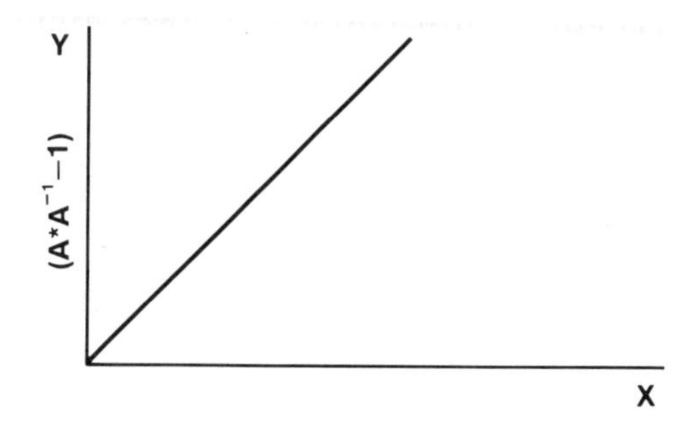


Fig. 6 Linear regression analysis based on $Y = (A^*A^{-1} - 1)$ and X = concentration of antagonist in molar units.

If we choose a value of $\alpha$, based on $F_\alpha$, the $(1 - \alpha)100\%$ confidence ellipse for the simultaneous test can be developed, as shown in Figure 7.

If such a graph of the simultaneous confidence ellipse is maintained in the laboratory, then for a given agonist-antagonist system one can plot the value of the estimated slope and the estimated intercept routinely and see if the point falls in the acceptance region indicating competitive antagonism (Bohidar, 1983).

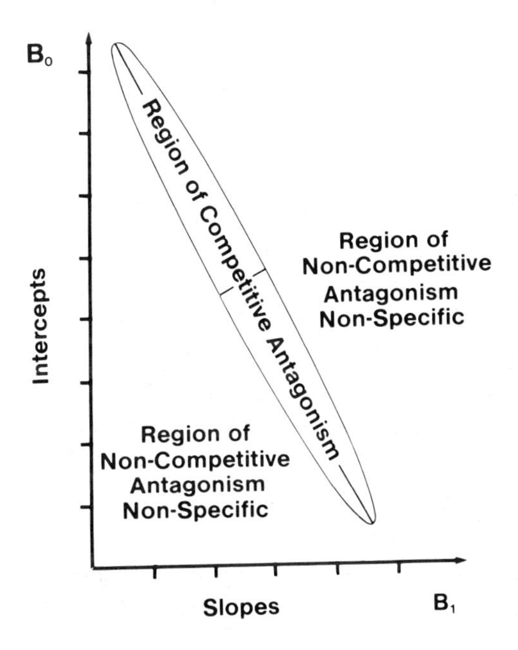


Fig. 7 (95%) Confidence ellipse for $B_0$ and $B_1$ in drug receptor interaction.

### F. Dual-System Validation

The primary motivation for the development of the dual-system validation statistical technique is that the experimenter quite often is interested in comparing two $pA_2$ values arising from an agonist-antagonist system measuring receptor differentiation (say, β-receptors). In a dual-system validation study, however, each system must meet the competitive antagonism validation criteria individually. The individual system which fails to meet the validation criteria cannot be incorporated for further consideration.

In a dual-system validation we have

$$H_0: \begin{bmatrix} A_1 \\ B_1 \end{bmatrix} = \begin{bmatrix} A_{11} \\ B_{11} \end{bmatrix} \quad \text{or} \quad \begin{bmatrix} A_1 - A_{11} \\ B_1 - B_{11} \end{bmatrix} = \begin{bmatrix} 0 \\ 0 \end{bmatrix}$$

Since the two regression parameters are correlated (Anderson, 1958) in each system, a multivariate simultaneous test is appropriate, maintaining the orginal type I error probability intact.

The test statistic for the dual system validation test will be (Bohidar, 1983)

$$\text{Hotelling } T^2_{(\text{REG})} = [t_1^2(1 - R^2)^{-1} - 2R(1 - R^2)^{-1}t_1t_2 + t_2^2(1 - R^2)^{-1}]$$

where $t_1$ is the t-test statistic for testing equality of two slopes, $t_2$ the t-test statistic for testing equality of two intercepts, and R the average correlation between A and B for both systems.

If the Hotelling $T^2_{(\text{REG})}$ test statistic is not significant (say at $\alpha = 0.05$), it is concluded that the two regression lines are not only parallel but also of equi-origin. This simultaneous test essentially achieves the test of concurrence.

### G. Comparison of Two $pA_2$'s Associated with Two Different Systems

The comparison of two $pA_2$ values becomes necessary for the following two important reasons: (1) to establish the relative strengths of two antagonists based on the same agonist and on the same type of tissue and (2) to demonstrate the role of $pA_2$ in receptor differentiation, which can be done only by comparing the two $pA_2$ values associated with the same antagonist. For example, the existence of two kinds of β-receptors (rabbit aorta and duodenum) was demonstrated by comparing two $pA_2$ values associated with the antagonist pronethalol (Furchgott, 1972).

Before we proceed to construct the test statistics for comparing two $pA_2$ values, it is appropriate to demonstrate that the residual variances around the Schild plot are the same. This involves the comparison of variabilities by using Levene's test, which is a robust test in that it does not depend on the cardinal assumption of normality.

Levene's test procedure has been extended to incorporate situations arising in regression analysis. The procedure is as follows. (1) Estimate an adjusted value for each data point by using the formula

$$Y_{ADJ} = Y_i - B_c(X_i - \bar{\bar{x}})$$

where $B_c$ is the common slope for the parallel systems and $\bar{\bar{x}}$ is the mean of all X values belonging to all the systems. (2) Find the absolute mean deviation for each $Y_{ADJ}$ in each system separately. (3) Conduct a one-way analysis of variance on the absolute values (see Section IV).

If we achieve homoscedasticity of the two systems by Levene's test, (F ratio, not significant), then we construct the test statistic for the comparison of the two $pA_2$ values as follows:

$$t = (pA_{2(I)} - pA_{2(II)}[V(pA_{2(I)}) + V(pA_{2(II)})]^{-1/2}$$

where t is the calculated Student t-value and $pA_{2(I)}$ and $pA_{2(II)}$ are the estimated values associated with system I and system II, respectively. Let system I have the regression equation $Y_1 = A_1 + B_1X$ and system II the regression equation $Y_2 = A_2 + B_2X$. If Levene's test is not significant, we can pool their respective residual variances as

$$S_P^2 = \frac{([Y_1^2] - B_1^2[X^2]_1) + ([Y_2^2] - B_2^2[x^2]_2)}{n_1 + n_2 - 4}$$

Now,

$$V(A_1) = S_P^2\left[\frac{1}{n_1} + \frac{\bar{x}_1^2}{[x^2]_1}\right], \qquad V(B_1) = S_P^2\left[\frac{1}{[x^2]_1}\right]$$

$$V(A_2) = S_P^2\left[\frac{1}{n_2} + \frac{\bar{x}_2^2}{[x^2]_2}\right], \qquad V(B_2) = S_P^2\left[\frac{1}{[x^2]_2}\right]$$

The asymptomatic variances of $pA_{2(I)}$ and $pA_{2(II)}$, respectively, are

$$V(pA_{2(I)}) = (B_1^{-2})[V(A_1)] + (A_1^2B_1^{-4})[V(B_1)]$$

$$V(pA_{2(II)} = (B_2^{-2})[V(A_2)] + (A_2^2B_2^{-4})[V(B_2)]$$

The dual-system expressions can be extended to multiple systems involving comparisons of three or more $pA_2$ values.

The numerical example given below pertains to a study involving (1) five doses of an antagonist, (2) histamine receptor in the right atria of guinea pig, (3) chronotropic response, and (4) a single agonist system. The assumptions of equi-maxima, parallel slopes, and a Schild slope of −1.0 are all statistically confirmed. Table 12 provides PANOVA-I for the study. In this study (Table 13) the $pA_2$ value is 7.85 and the lower and upper confidence limits are 7.7 and 8.09, respectively. It may be noted that the value of −1.0 is inclusive in the 95% confidence limits of the Schild slope.

Table 12 PANOVA-I

| Source of variation | d.f. | Sum of squares | Mean squares | F ratio | Stat. results[a] |
|---|---|---|---|---|---|
| Between-antagonist doses | 4 | — | — | — | — |
| Linear REG | 1 | 2.2307 | 2.2307 | 42.062 | *** |
| Test (slope = −1) | 1 | 0.0014 | 0.0014 | 0.0271 | NS |
| Lack of fit | | | | | |
| Quadratic | 1 | | | | |
| Cubic | 1 } = 3 | 0.0155 | 0.0052 | 0.0972 | NS |
| Fourth degree | 1 | | | | |
| Within-antagonist doses | 14 | 0.7425 | 0.0530 | — | — |
| Total | 19 | 2.9887 | | | |

[a]***, highly significant.

Table 13 $pA_2$ Estimation and $(1 - \alpha)100\%$ Confidence Limits

1. Schild plot regression equation:

$$Z^* = 8.0506 - 1.0261(-\log_{10}\text{conc.})$$
$$= 8.0506 - 1.0261W^*$$

2. G value = 0.1093
3. Schild slope: −1.026
4. $pA_2 = \frac{-8.0506}{-1.0261} = 7.85$
5. 95% confidence limits of $pA_2$: lower CL = 7.70
   upper CL = 8.09
6. 95% confidence limits of Schild slope:

   Central value = −1.026

   Lower CL = −1.365

   Upper CL = −0.687

For a dual-system (or multiple-system) validation, the following criteria must be considered: (1) each system must meet its respective validation criteria (single-system validation) and (2) for the estimation of a valid relative $pA_2$ a PANOVA-II, as defined below, must be conducted:

| Source | d.f. | Desirable results for validation[a] |
|---|---|---|
| Antagonist doses | $(t - 1)$ | — |
| Agonist-antagonist system | 1 | NS |
| common Schild slope regression | 1 | ** |
| Parallelism of Schild slopes | 1 | NS |
| curvature (second degree) | 1 | NS |
| Opposed curvature | 1 | NS |
| Within-antagonist doses | $\Sigma(n_i - 1)$ | NS |
| Total | $(\Sigma n_i - 1)$ | |

[a]NS, not significant; **, highly significant.

If the two systems, singly as well as doubly, have met every validation criterion, one would proceed to calculate the relative $pA_2$ for the dual system as follows:

$$\text{Relative } pA_2 = \frac{\text{MCAI/DR}}{\text{MCAII/DR}}$$

where MCAI/DR denotes molar concentration of antagonist I for a selected log dose ratio and MCAII/DR denotes molar concentration of antagonist II for the same selected dose ratio. Here, the log dose ratio, $\log(C_pC_0^{-1} - 1)$, is based on the common agonist molar concentration.

The $(1 - \alpha)100\%$ confidence interval for the relative $pA_2$ can be constructed based on the principles of Fieller's theorem. The explicit expression is not provided here. If the value of 1.0 is included in the interval, it indicates that the two antagonists have "equi-affinity" for the same receptor. It is also possible to combine the two $pA_2$ values ($pA_2$-I and $pA_2$-II) by the method of convex combination (Bohidar, 1984). A clear-cut interpretation will be obtained if the two systems utilize the same agonist system.

## REFERENCES

Abbott, W. S. (1925). Method of computing the Effectiveness of an Insecticide, *J. Econ. Entomol.*, 18: 265–267.

Anderson, T. W. (1958). *An Introduction to Multivariate Statistical Analysis*, Wiley, New York.

Aranlakshana, O. and Schild, H. O. (1959). Some Quantitative Uses of Drug Antagonists, *Br. J. Pharmacol.*, 14: 48–58.

Ariens, E. J. and Van Rossum, J. N. (1957). pDx, pAx and pD Values in the Analysis of Pharmacodynamics, *Arch. Int. Pharmacodyn.*, 110(2–3): 275–299.

Armitage, P. (1982). The Assessment of Low-Dose Carcinogenicity, *Curr. Top. Biostat. Epidemiol.*, Biometrics Suppl., pp. 119–129.

Bergman, S. W. and Gittins, J. C. (1985). *Statistical Methods for Pharmaceutical Research Planning*, Marcel Dekker, New York.

Bevan, J. A., ed. (1976). *Essentials of Pharmacology*, 2nd ed., Harper & Row, New York.

Bohidar, N. R. (1983). "Statistical Aspects of Chemical Assay Validation." Proc. American Statistical Association Annual Meetings. Biopharmaceutical Section, pp. 57–62.

Bohidar, N. R. (1984). Analysis of Randomized Block Design with Inordinate Right Censorship. Experimental Design, Statistical Models, and Genetic Statistics (K. Hinkelmann, ed.) Chapter 9. Marcel Dekker, Inc., New York, pp. 119–140.

Bohidar, N. R. (1985). "Statistical Tests of Validity for Competitive Antagonism in Drug Receptor Interaction," Proc. of American Statistical Association Annual Joint Meetings, Biopharmaceutical Section, pp. 66–71.

Bohidar, N. R. (1987). "Estimation of Apparent Halflife in the Presence of Extensive Enterohepatic Circulation." Proc. American Statistical Association, Biopharmaceutical Section (to appear in June 1987).

Box, G. E. P. and Hay, W. A. (1953). A Statistical Design for the Efficient Removal of Trends Occurring in a Comparative Experiment with an Application in Biological Assay, *Biometrics*, 9: 301–319.

Buncher, C. R. and Tsay, J.-Y. (1981). *Statistics in the Pharmaceutical Industry*, Marcel Dekker, New York.

Carter, E. M. and Hubert, J. J. (1984a). A Growth-Curve Model Approach to Multivariate Quantal Bioassay, *Biometrics*, 40: 699–706.

Carter, E. M. and Hubert, J. J. (1984b). "Covariable Adjustment in Multivariate Bioassay," Proc. of the American Statistical Association (Biopharmaceutical Section), pp. 103–106.

Carter, E. M. and Hubert, J. J. (1985). Analysis of Parallel-Line Assays with Multivariate Responses, *Biometrics*, 41: 703–710.

Carter, E. M., Hubert, J. J., and Walsh, M. N. (1985). "Estimation Methods for Symmetric Parabolic Bioassays," to appear in Symposia on Statistics, University of Western Ontario.

Cochran, W. G. and Cox, G. M. (1957). *Experimental Design*, 2nd ed., Wiley, New York, chaps. 1–5.

Elashoff, J. D. (1981). Repeated-Measures Bioassay with Correlated Errors and Heterogeneous Variances: A Monte Carlo Study, *Biometrics*, 37: 475–482.

Finney, D. J. (1971). *Probit Analysis*, 3rd ed., Griffin, London.

Finney, D. J. (1978). *Statistical Method in Biological Assay*, 3rd ed., Griffin, London.

Furchgott, R. F. (1972). The Classification of Adrenoceptors, an Evaluation from the Standpoint of Receptor Theory, in Catecholamines, *Jeffter's Handbook of Experimental Pharmacology*, vol. 33, Springer, Berlin, p. 283.

Grizzle, J. E. and Allen, D. M. (1969). Analysis of Growth and Dose Response Curves, *Biometrics*, 25: 357–381.

Hubert, J. J. (1984). *Bioassay*, 2nd ed., Kendall-Hunt, Dubuque, Iowa.

Khatri, C. G. (1966). A Note on a MANOVA Model Applied to Problems in Growth Curves, *Ann. Inst. Stat. Math.*, 18: 75–86.

Kolakowski, D. and Bock, R. D. (1981). A Multivariate Generalization of Probit Analysis, *Biometrics*, 37: 541–551.

Kooijman, S. A. L. M. (1981). Parametric Analysis of Mortality Rates in Bioassays, *Water Res.*, 15: 107–120.

Levene, H. (1971). Robust Tests for Equality of Variances, *Contribution to Probability and Statistics. Essays in Honor of Harold Hardling*, Stanford University Press, Stanford, Calif., pp. 278–292.

Miller, R. G. (1979). The Jackknife—a Review, *Biometrika*, 62: 1–17.

Potthoff, R. F. and Roy, S. N. (1964). A Generalized Multivariate Analysis of Variance Model Useful Especially for Growth Curve Problems, *Biometrika*, 51: 313–326.

Rand, G. M. (1985). *Fundamentals of Aquatic Toxicology: Methods and Applications*, McGraw-Hill, Toronto.

Rao, C. R. (1954). Estimation of Relative Potency from Multiple Response Data, *Biometrics*, 10: 208–220.

Rao, C. R. (1965). The Theory of Least Squares When the Parameters Are Stochastic and Its Application to the Analysis of Growth Curves, *Biometrika*, 52: 447–458.

Schild, M. C. (1947). pA, a New Scale for the Measurement of Drug Antagonism, *Br. J. Pharmacol.*, 2: 189–206.

Scriabine, A., Ludden, C. T., and Bohidar, N. R. (1974). Potentiation of Antihypertensive Action of Hydralazine by Timolol in Spontaneously Hypertensive Rats, *Proc. Soc. Exp. Biol. Med.*, 146: 506–512.

Segreti, A. C. and Munson, A. E. (1981). Estimation of the Median Lethal Dose When Responses Within a Litter Are Correlated, *Biometrics*, 37: 153–156.

Srivastava, M. S. and Carter, E. M. (1983). *An Introduction to Applied Multivariate Statistics*, Elsevier/North-Holland, Amsterdam and New York.

Tallarida, R. J. and Jacob, L. S. (1979). *The Dose-Response Relation in Pharmacology*, Springer-Verlag, Berlin.

Tallarida, R. J. and Murray, R. B. (1981). *Manual of Pharmacologic Calculations*, Springer-Verlag, New York.

Tallarida, R. J., Cowan, J., and Adler, N. W. (1980). $pA_2$ and Receptor Differentiation: A Statistical Analysis of Competitive Antagonism, *Life Sci.*, 25: 637–654.

Tukey, J. W. (1949). One Degree of Freedom for Non-Additivity, *Biometrics*, 5: 232–242.

Van Arman, G. C. and Bohidar, N. R. (1978). Antiarthritics, *New Drugs Discovery and Development* (A. A. Rubin, ed.), chap. 1, Marcel Dekker, New York, pp. 1–27.

Wilk, M. B. and Shapiro, S. S. (1968). The Joint Assessment of Normality for Several Independent Samples, *Technometrics*, 10: 825–839.

# 4 Pharmaceutical Formulation Development

NEETI R. BOHIDAR *Research and Development, Smith Kline & French Laboratories, Swedeland, Pennsylvania*

KARL E. PEACE *G. D. Searle & Co., Skokie, Illinois*

## I. INTRODUCTION

A primary aim of the pharmaceutical industry is to provide physicians with newly marketed drugs which are useful in treating patients with various diseases. Toward this end, the industry actively engages in the meticulous efforts of discovery, basic research, pharmaceutical formulation development, clinical development, and delivery of new drugs. This chapter is devoted to a presentation of the statistical design and analysis of experiments associated with the pharmaceutical formulation development of new drugs. Pharmaceutical formulation development forms an integral part of the total drug development process. At the outset, a brief description of the processes leading to pharmaceutical formulation development is given.

Product development begins with the efforts to develop a new drug from a chemical compound which has known biological activity and whose molecular structure is known. By appropriate modification of the molecular structure by a team of chemists, an analog of the compound can be derived which is more active and perhaps more selective in its actions than the parent compound. The structure modification may be as simple as a change of the position of the hydroxyl group on the benzene ring or as complex as replacing the hydroxyl group with a methyl group. Each time a modification is implemented, the biological activity of the analog or new drug is carefully studied in both in vivo and in vitro models. This process of making a new drug starting with a known chemical compound is called the synthesis process. Synthetic chemists, medicinal chemists, and quantitative, analytical, and structural chemists in a pharmaceutical company are actively engaged in the synthesis process of making new drug entities. It should be noted that synthesis is not the only process of obtaining new drug entities. They may be obtained by isolation from natural sources such as plant or animal tissues. They may also be derived by modifying the conditions of certain microbiological fermentation processes with subsequent isolation and characterization of the product (such as a new antibiotic agent).

Initially, the synthetic chemist makes a few milligrams or grams of the new drug for the pharmacologists or microbiologists to test in laboratory animals. Subsequently, large amounts are needed for initial safety and pharmaceutical formulation studies. The primary interest of the chemist who synthesizes the drug the first time is the synthesis rather than the cost or availability of the starting materials. The chemist is only secondarily concerned at this point with whether the process can be scaled up safely and efficiently for the production of equally pure drug. The synthesis may have to be repeated several times before it can be successfully scaled up.

Prior to receiving enough of the new drug for initial pharmaceutical development studies, the pharmaceutical chemist must have worked out preliminary analytical procedures and made a judgment as to the form and composition of the agent that is likely to be suitable for use in humans. For example, a stable succinate may be preferred to a hygroscopic chloride of the agent. The pharmaceutical chemist will use the first few milligrams of the chemical available to explore its solubility and stability characteristics. Such tests are generally referred to as preformulation studies. Accelerated stability tests at elevated temperatures, undertaken by the pharmaceutical chemist, are primarily intended to anticipate drug stability problems that might not show up otherwise until years later at ordinary conditions of shelf life, such as ambient room temperature. The pharmaceutical chemist will also use some of the chemical material to develop specific chemical assay methods for the active ingredient.

It is the primary responsibility of the development pharmacist (USP, 1975) to prepare the first formulation of the new drug to be used for studies in patients. The most common dosage form is the compressed tablet. The second most common dosage form is the capsule. However, the capsule is likely to be the more convenient form for clinical pharmacology and Phase I and Phase II clinical studies. The small amount of drug available early in a development project does not usually permit the formulation of tablet dosage forms. For this reason, hand-filled capsules or a simple solution of the drug in water may be used for initial Phase I clinical studies. For the mass production of dosage units, the tableting of a drug is a much faster and cheaper process. It does, however, require a greater proportion of excipients than a capsule formulation. These materials include lubricants, such as magnesium stearate, that allow the material to flow through the high-speed punches, binding agents that hold the tablet together, and disintegrants that facilitate tablet disintegration in the stomach or beyond. The excipients and the drug itself must be stable under conditions of instantaneous heat and pressure of compression, which must be controlled for product uniformity. Moreover, the physical characteristics of the excipients as well as those of the new drug entity itself can markedly influence the performance of the formulation. Thus, appropriate statistical methodologies for achieving the proper combination of the excipient ingredients and for testing the performance of the formulation should be employed.

In a broad sense, pharmaceutical formulation development involves the research activities of two types of chemical scientists, the pharmaceutical chemist and the development pharmacist. The development pharmacist studies the physical and chemical properties of the excipient ingredients and develops the optimum formulation from one or more appropriately designed experiments. The pharmaceutical chemist develops the specific chemical assay methods for the active and excipient ingredients, conducts chemical assay validation experiments, and establishes the stability characteristics of the product by conducting accelerated stability studies.

This chapter consists of three interrelated but distinct segments. These are pharmaceutical formulation development using statistical optimization, which is addressed in Sections II, III, and IV; chemical assay method validation, addressed in Section V; and statistical aspects of accelerated stability testing, addressed in Section VI.

## II. PHARMACEUTICAL FORMULATION DEVELOPMENT USING STATISTICAL OPTIMIZATION

### A. General Considerations

The pharmaceutical research division of a pharmaceutical company develops a drug delivery system for a selected chemical entity. A drug delivery system (or pharmaceutical dosage form) is composed of two types of chemical ingredients, active and inactive. The drug entity constitutes the active ingredient. The inactive ingredients, the delivery portion of the system, are inert chemicals whose functional properties can be characterized as disintegrant, diluent, binder, and lubricant. For instance, calcium phosphate/lactose, magnesium stearate, granulating gelatin, and starch are incorporated into a dosage form as diluent, lubricant, binder, and disintegrant, respectively. The amount of drug substance to be included in the delivery system is based on the pharmacological and desired therapeutic requirement for the drug. However, the amounts of the excipient ingredients to be incorporated in the system are unknown. The experimental determination of the amounts of excipient ingredients to be incorporated into the drug delivery system, resulting in prespecified product properties, is called the formulation process. This is a most important step in all activities of the pharmaceutical industry. Without a proper formulation, the drug delivery system would fail to provide the expected desirable results. In the not too distant past, the formulation process was considered to be trivial and was accomplished by the subjective judgment of the development pharmacist—the formulator. The performances of drug delivery systems which are developed on the basis of the largely subjective judgment of the formulator have invariably not been satisfactory in the long run. This kind of approach only provides a transient solution to a long-standing complex problem.

The word *optimize* is defined in Webster's dictionary as "to make as perfect, effective, or functional as possible." At present, the term optimization, in pharmaceutical sciences, encompasses three interrelated but distinct processes: (1) formulation, (2) processing, and (3) establishment of monitoring systems. The "formulation process" is defined above. "Processing" implies a sequence of laboratory operations undertaken to achieve the final product based on the product formulation. For instance, the laboratory operations associated with developing a coated tablet comprise the following processes: (1) mixing/blending, (2) granulation, (3) drying, and (4) tablet coating. Establishment of monitoring systems pertains to the activities associated with evaluating the future performance of the optimized formula based on only a small number of tests.

The primary purpose of this section is to outline briefly the design and analysis of pharmacuetical optimization experiments and to present the statistical methodologies associated with the establishment of a monitoring system for evaluating the future performance of the drug delivery system. In the interest of brevity, only the noteworthy features of the statistical techniques will be presented. It is appropriate to note that optimization techniques play the same role in pharmaceutical sciences as "component of variance analysis" in genetics, "multivariate factor analysis" in psychology, and "inverse regression analysis" in pharmacology.

## B. Design of Optimization Experiments

### 1. Formulation Parameters

The development of a pharmaceutical formulation and the associated processes usually involve a number of independent and dependent variables. The independent variables are the formulation and process variables which are directly under the control of the formulator. The dependent variables are the responses or the characteristics of the resulting drug delivery system.

Independent variables may include the level of a given ingredient or the mixing time for a given process step. Excipient ingredients are examples of independent variables. Some are: (1) calcium phosphate/lactose ratio (diluent), (2) starch NF corn (disintegrant), (3) granulating gelatin (binder), (4) magnesium stearate (lubricant), (5) Cab-o-sil (binder), (6) Encompress (diluent), (7) microcrystalline cellulose (binder), and (8) guar gum (disintegrant). Compression pressure is an example of a process variable. Generally, three to five independent variables are considered.

Dependent variables which are generally considered are (1) disintegration time in minutes, (2) hardness (tablet breaking strength) in kilograms, (3) dissolution (%) release in 30 minutes, (4) friability (%) weight loss, (5) thickness uniformity in relative standard deviation (RSD) (%), (6) porosity in micrometers per gram, (7) mean pore diameter in micrometers, (8) weight uniformity in RSD (%), (9) content uniformity in RSD (%), (10) dissolution (%) release in 45 minutes, (11) tablet breakage as the number of chipped tablets, and (12) granulation mean diameter in micrometers. Generally 8 to 10 dependent variables are considered.

There are two general types of optimization procedures—constrained and unconstrained (Bohidar, 1984). Constraints are those restrictions placed on the system due to physical limitations, simple practicality, economic considerations, and/or compendial and regulatory requirements. In unconstrained optimization problems, there are no restrictions. For a given pharmaceutical formulation or drug delivery system one might say, "make the hardest tablet possible." A constrained problem, on the other hand, might be stated as, "make the hardest tablet possible, but it must disintergrate in less than 10 min."

Within the realm of physical reality, and most importantly in the development of drug delivery systems, the unconstrained optimization problem is almost nonexistent. In pharmaceutics, there are many competing restrictions. Hardness and disintegration, as mentioned above, are two examples. It is often necessary to trade off properties, i.e., to sacrifice one characteristic for another. Thus, the primary objective may not be to optimize absolutely but to compromise effectively and thereby produce the best formulation under a given set of restrictions.

An additional complication in pharmacy is that formulations are not usually simple systems (Schwartz et al., 1973). They often contain many ingredients which may interact with one another to produce unexpected, if not unexplainable, results. Therefore not only are the independent variables (X) constrained but the dependent variables (Y) are as well.

### 2. Pharmaceutical Justification for Optimization

As mentioned above, a pharmaceutical dosage form or drug delivery system consists of the active ingredient as well as the excipient ingredients. The effects of the excipient ingredients on the dependent or response variables are of primary interest in formulation experiments (Bohidar, 1984). It is known that compression pressure and starch as a disintegrant influence the disintegration time of a tablet dosage form—compression pressure impedes while starch accelerates. If a disintegration time of 3 minutes is required, because of compendial and/or regulatory requirements, the development pharmacist must know the amounts of the two variables which will yield this level of response. Determination of the amounts constitutes the optimization process. As another example, it is known that magnesium stearate and guar gum influence the dissolution of a dosage form—magnesium stearate impedes and guar gum accelerates dissolution. If one needs to achieve 80% dissolution in 30 minutes because of a certain requirement, the development pharmacist must know the amounts of the two excipients which will bring about the desirable dissolution response. Optimization experiments can be used to answer these inverse regression type questions.

### 3. Stages of Pharmaceutical Activities

Pharmaceutical activities associated with product development are carried out in distinct stages. The pharmaceutical research and development department must maintain their schedule of activities in accordance with the various stages of the overall product development. The following is a brief outline of the stages (showing only the pharmaceutical and clinical activities).

Stage I
- a. Screening and synthesizing
- b. Animal pharmacology and short-term toxicity tests
- c. Preformulation experiments
- d. Analytical technique development

Stage II
- a. Clinical pharmacology studies, Phase I clinical trials
- b. Formulation optimization
- c. Pilot batch for Phase I trials
- d. Synthesis scale-up for Phase I trials
- e. Additional analytical technique development

Stage III
- a. Intermediate-size clinical trials—Phase II
- b. Pharmaceutical scale-up for Phase II trials
- c. Analytical in-process controls
- d. Package testing

Stage IV
- a. Large clinical trials—Phase III
- b. Production-size pharmaceutical batches
- c. Final package selection
- d. Final quality control procedures
- e. Long-term toxicity tests

Stage V. Filing the New Drug Application, which consists of

a. Clinical data
b. Analytical data
c. Label and packaging data
d. Synthesis data
e. Pharmaceutical process data
f. Stability and physiochemical data

4. Design of Optimization Experiments

At the preformulation stage, the development pharmacist conducts any experiments necessary to charactetize the physical and chemical properties of the drug substance and to determine the extent of the interaction between the excipients and the drug substance. He or she selects a set of excipient ingredients which are potentially important in governing the desired response properties of the drug delivery system under consideration. The selection of the excipients is based on the personal knowledge and experience of the pharmacist together with small trial-and-error deterministic experiments. Several range-finding experiments would also be conducted. No formal statistical analyses are performed at the preformulation stage. After selecting a set of excipients and determining their range of activities (experimental region), preformulation operations come to an end.

Before the first formal optimization experiment is conducted, the following steps must be considered:

1. Selection of a Box-Wilson type optimization design (Box et al., 1978; Cochran and Cox, 1957)
2. Determination of the extent of laboratory experimentation needed based on the design (number of central points, repetitions, etc.) (Bohidar, 1984)
3. Specification of the excipient variables and their ranges of activity (Bohidar et al., 1975; Bohidar, 1984)
4. Specification of all response variables and the constraints imposed on them (Bohidar et al., 1975; Bohidar, 1984).
5. Specification of the batch size and the number of tablets per formulation (Bohidar, 1984)
6. Specification of the operational characteristics of each piece of equipment involved, such as AHIBA as the tablet strength tester, Glen Bowl for blending, and USP rotating basket dissolution apparatus (Bavitz et al., 1973, 1974)
7. Selection of the compendial and regulatory requirements for the response parameters (such as 50 rpm, paddle (USP), $H_2$) medium for dissolution) (USP, 1975)
8. Selection of the source of bulk material for active and inactive ingredients (may need blocking if more than one source is involved) (Bavitz et al., 1974)
9. Specification of the number of laboratories participating (Bohidar, 1983)
10. Specification of the number of operators and the number of test days (Bohidar, 1983)
11. Generation of randomization schedules for the order of granulation/compression and analytical determinations of dosage units (Cochran and Cox, 1957)

12. Specification of the appropriate statistical analyses required (Bohidar et al., 1975, 1979; Bohidar, 1984)
13. Identification and procurement of the appropriate computer software and algorithms (Carnahan et al., 1969; Platt et al., 1977)
14. Plan development for postoptimization verification experiments based on candidate optimum formulations (Bohidar, 1984)

Detailed information pertaining to the 14 items listed above is available in the references indicated after each item.

To illustrate some of the detail and procedure of a typical optimization experiment, a production formula (for product P) already in existence is chosen. Briefly, the process involves a starch-paste-gelatin granulation of the drug with diluents. After drying and milling, the disintegrant, the glidant, and the lubricant are added.

The five independent formulation variables selected for this particular study are $X_1$, diluent ratio; $X_2$, compressional force; $X_3$, disintegrant level; $X_4$, binder level; and $X_5$, lubricant level. With the exception of these five variables, everything else in the formulation and processing steps remains constant throughout the study, including (1) the amount of active ingredient, (2) the quantity of starch as paste, (3) the glidant level, (4) the granulating, milling, drying, and blending conditions, and (5) the speed of the compressing machine.

Each formulation consists of a 30,000-tablet batch through the granulating, milling, dry mixing, and lubricating steps. From each batch, 3000 tablets are compressed on a rotary Stokes Model 580 equipped with 35 sets of 0.6 cm (0.25 inch) round, flat beveled-edge punches, without precompression.

The responses measured on the resulting tablets are: $Y_1$, disintegration; $Y_2$, tablet hardness; $Y_3$, dissolution; $Y_4$, friability; $Y_5$, weight uniformity; $Y_6$, thickness uniformity; $Y_7$, tablet porosity; $Y_8$, mean pore diameter; $Y_9$, mean granule diameter; and $Y_{10}$, tablet breakage. Most of these response variables are properties of general interest to tablet formulators, but the list could be varied to fit the particular formulation.

### 5. Fractional Factorial and Augmented Designs

The product P experiment involved five independent variables and represents a five-factor factorial study. For each response variable, we have the following full factorial model:

$$Y_m = \mu + \sum_{i} A_i + \sum_{\substack{i \\ i \neq j}}\sum_{j} A_iA_j + \sum_{\substack{i \\ i \neq j \neq k}}\sum_{j}\sum_{k} A_iA_jA_k + \cdots + E_m$$

where $Y_m$ denotes the mth response variable, m = 1, 2, . . ., n; $\mu$ is the overall mean; $A_i$ denotes the ith factor (main effect), i = 1, 2, . . ., 5; $A_iA_j$ denotes the two-factor interactions, i, j = 1, 2, . . ., 5 ($i \neq j$); $A_iA_jA_k$ denotes the three-factor interactions, i, j, k = 1, 2, . . ., 5 ($i \neq j \neq k$); and $E_m$ denotes the experimental error associated with $Y_m$.

For the purpose of constructing an appropriate and economical design, one might consider each factor as having only two levels, low and high (selected on the basis of the range-finding studies). The number of treatment combinations would be $2^5 = 32$. Due to the large number of treatment combinations which could be tested, a parsimonious fractional factorial experimental design may be chosen. Fractional factorial designs provide practically the same statistical information as the full factorial. Considering each treatment combination as a five-tuple vector point in a five-dimensional Euclidean hyperplane with axes $Z_1, Z_2, \ldots, Z_5$, the following congruential equations in modular algebraic form are obtained:

$$Z_1 + Z_2 + Z_3 + Z_4 + Z_5 \equiv 0 \text{ MOD-2}$$

$$Z_1 + Z_2 + Z_3 + Z_4 + Z_5 \equiv 1 \text{ MOD-2}$$

Each equation yields $2^{5-1} = 16$ solutions when $Z_1$ takes the value of 0 or 1. A one-fourth fractional factorial design, for example, would be based on the solutions of the following simultaneous congruential equations:

$$\begin{array}{l|c|c|c|l} Z_1 + Z_4 + Z_5 \equiv 0 & 1 & 1 & 0 & \text{MOD-2} \\ Z_1 + Z_2 + Z_3 \equiv 1 & 1 & 0 & 0 & \text{MOD-2} \end{array}$$

Each MOD-set (any one of the four sets above) yields $2^{5-2} = 8$ solutions. In practice, fractional factorial designs are often augmented by a composite scheme. Such designs are more efficient than the fractional factorial without augmentation. This point will be discussed in more detail subsequently.

Generally, if all factors are represented by quantitative variables, such as temperature, compression pressure, and amount of an excipient, it is natural to think of the response $Y_i$ as a continuous function ($\Phi$) of the levels of these variables ($X_i$) plus some experimental error (E), expressed as

$$Y_i = \phi(X_{1i}, X_{2i}, \ldots, X_{ki}) + E_i$$

The function $\phi$ is the response surface. Because of the complex nature of the pharmaceutical phenomenon, it is not easy to develop the exact mathematical form of $\phi$. However, experience shows that this function can be approximated satisfactorily, within the experimental region, by a second-order polynomial in the variables $X_i$. One can then fit the polynomial model to the response data and find the levels of the X-variables which would optimize the fitted model (the estimate of the response surface approximated by the polynomial). As an example, the general form of a second-order polynomial model for a two X-variable system is

$$Y = B_0 + B_1X_1 + B_2X_2 + B_{11}X_1^2 + B_{22}X_2^2 + B_{12}X_1X_2 + E$$

To estimate the partial regression coefficients (B) in this model, each X-variable would have to take on at least three different levels—suggesting that a $3^k$ factorial would be required. A major disadvantage of a $3^k$ factorial, however, is that, with more than two X-variables ($K > 2$), the number of runs required in an experiment becomes very large. Confounding in $3^k$ factorials is also usually complex.

To alleviate such situations, Box and Wilson developed new designs (Box and Wilson, 1951) specifically for fitting second-order polynomial response surfaces. These designs, which are called composite designs, are constructed by augmenting the original $2^k$ factorial with 2K + 1 additional treatment combinations (points). For two variables, the points are

$$(0, 0), (-\alpha, 0), (+\alpha, 0), (0, -\alpha), (0, +\alpha)$$

The total number of points is $\theta 2^k + 2K + 1$, where $\theta$ represents the fraction of the full factorial ($\theta = 1$ for full factorial). For 2, 3, and 4 X-variables, the experiment requires 9, 15, and 25 points, respectively, as compared with 9, 27, and 81 in the $3^k$ factorial. The value of $\alpha$ is generally chosen to make the regression coefficients in the augmented system orthogonal to each other. In the pharmaceutical literature, the $\theta 2^k$ and 2K points are called "factorial formulations" and "axial formulations," respectively. The central point is called the "central formulation." The spacing parameter $\alpha$ is calculated as

$$\alpha = \left(\frac{QF}{4}\right)^{1/4}$$

where $Q = [(F + T)^{1/2} - F^{1/2}]^2$, F is the number of factorial formulations, and $T = 2k + 1$.

For the product P optimization study, the design selected is an orthogonal, half-fractional, five-factorial design augmented with 10 (2k) axial points and one central point. The factorial formulations or factorial treatment combinations are generated by using the congruential equation,

$$Z_1 + Z_2 + Z_3 + Z_4 + Z_5 \equiv 0 \text{ MOD } 2$$

yielding 16 treatment combinations. Based on the results of pilot experiments, the coded base level (= 0) and the coded factorial levels (−1;1) are selected. To determine the axial levels, the $\alpha$ value is calculated as

$$F = 16, \qquad T = 11, \qquad Q = (27^{1/2} - 16^{1/2})^2 = 1.4304$$

$$\alpha = \left[\frac{(1.4304)(16)}{4}\right]^{1/4} = (5.7216)^{1/4} = 1.5466$$

Table 1 provides the coded and uncoded values of the levels of the five excipient or independent variables. Table 2 provides the 27 (16 factorial formulations, 10 axial formulations, and 1 central formulation) unique treatment combinations in coded form.

It should be noted that, instead of a full factorial or a fractional factorial, one could also use a highly fractionated design such as a Placket-Burman design, Box-Behnken design, or Rao's "orthogonal array" design. These are useful when there is a large number of X-variables. The rules for augmenting these designs would be similar to those outlined for the full or fractional factorials (Box et al., 1978).

Table 1 Coded and Uncoded Levels of Factors Associated with Product P Experiment[a]

| Factor | $-\alpha$ −1.547 | Factorial level −1 | Baseline 0 | Factorial level +1 | $+\alpha$ +1.547 |
|---|---|---|---|---|---|
| $X_1$ = calcium phosphate-lactose ratio (1 eu = 10 mg) | 24.5/55.5 | 30/50 | 40/40 | 50/30 | 55.5/24.5 |
| $X_2$ = compression pressure (1 eu = 0.5 ton) | 0.25 | 0.5 | 1 | 1.5 | 1.75 |
| $X_3$ = starch disintegrant (1 eu = 1 mg) | 2.5 | 3 | 4 | 5 | 5.5 |
| $X_4$ = granulating gelatin (1 eu = 0.5 mg) | 0.2 | 0.5 | 1 | 1.5 | 1.8 |
| $X_5$ = magnesium stearate (1 eu = 0.5 mg) | 0.2 | 0.5 | 1 | 1.5 | 1.8 |

[a]eu, Factorial spacing; $\alpha$ = 1.547.

### 6. Experimental Procedure

It is essential that the consulting statistician have an understanding of the roles of the independent and dependent variables to be considered in a pharmaceutical formulation optimization study. This not only provides a better understanding of the potential complexities of the study but also permits clearer communication between the statistician and the scientist. These points are illustrated by referring to an existing tablet formulation for product D. This study involved four independent variables ($X_i$) and nine dependent variables ($Y_i$). The experimental design is a four-factor, orthogonal, central, composite second-order design with central points.

Briefly, the independent variables and their functions are:

*Volume of granulating fluid* ($X_1$)—varying the amounts of water used in the granulation operation will help define whether dryer or wetter conditions will improve the tablet properties.

*Screen size* ($X_2$)—varying the screen size will indicate the sensitivity of the tablet properties to the particle size of the granulation.

*Magnesium stearate* ($X_3$)—even though magnesium stearate provides an excellent lubricating effect, tablet properties are usually very sensitive to small changes in magnesium stearate levels. This is an important variable in view of the different screen sizes to be tested.

*Guar gum-starch balance* ($X_4$)—changes in the guar gum level should affect tablet disintegration properties.

Table 2 Experimental Design Points for Product P Experiment

| Formulations | | Factor Levels in Coded Form | | | | |
|---|---|---|---|---|---|---|
| No. | Type | $X_1$ | $X_2$ | $X_3$ | $X_4$ | $X_5$ |
| 1 | Factorial | −1 | −1 | −1 | −1 | 1 |
| 2 | Factorial | 1 | −1 | −1 | −1 | −1 |
| 3 | Factorial | −1 | 1 | −1 | −1 | −1 |
| 4 | Factorial | 1 | 1 | −1 | −1 | 1 |
| 5 | Factorial | −1 | −1 | 1 | −1 | −1 |
| 6 | Factorial | 1 | −1 | 1 | −1 | 1 |
| 7 | Factorial | −1 | 1 | 1 | −1 | 1 |
| 8 | Factorial | 1 | 1 | 1 | −1 | −1 |
| 9 | Factorial | −1 | −1 | −1 | 1 | −1 |
| 10 | Factorial | 1 | −1 | −1 | 1 | 1 |
| 11 | Factorial | −1 | 1 | −1 | 1 | 1 |
| 12 | Factorial | 1 | 1 | −1 | 1 | −1 |
| 13 | Factorial | −1 | −1 | 1 | 1 | 1 |
| 14 | Factorial | 1 | −1 | 1 | 1 | −1 |
| 15 | Factorial | −1 | 1 | 1 | 1 | −1 |
| 16 | Factorial | 1 | 1 | 1 | 1 | 1 |
| 17 | Axial | −1.547 | 0 | 0 | 0 | 0 |
| 18 | Axial | 1.547 | 0 | 0 | 0 | 0 |
| 19 | Axial | 0 | −1.547 | 0 | 0 | 0 |
| 20 | Axial | 0 | 1.547 | 0 | 0 | 0 |
| 21 | Axial | 0 | 0 | −1.547 | 0 | 0 |
| 22 | Axial | 0 | 0 | 1.547 | 0 | 0 |
| 23 | Axial | 0 | 0 | 0 | −1.547 | 0 |
| 24 | Axial | 0 | 0 | 0 | 1.547 | 0 |
| 25 | Axial | 0 | 0 | 0 | 0 | −1.547 |
| 26 | Axial | 0 | 0 | 0 | 0 | 1.547 |
| 27 | Central | 0 | 0 | 0 | 0 | 0 |

Similarly, the dependent variables and the associated measurement procedures are:

*Dissolution* ($Y_1$)—dissolution measurements are accomplished by using a 50 rpm USP Paddle, 900 ml 0.1 *N* HCl procedure. Several tablets (six or more) are tested from each sample bottle.

*Friability* ($Y_2$)—two groups of 20 tablets, collected by withdrawing 4 tablets from each of five sample bottles for each group, are weighed before and after testing in the Roche Tablet Friabilator (25 rpm) at 100 revolutions. If a tablet is chipped or broken, all but the largest portion of the tablet will be counted as weight loss. The data are reported as mean percent weight loss.

*Weight variation* ($Y_3$)—each sample bottle contributes several tablets, whose weights are measured in an analytical balance (Mettler Model H20T) and the relative standard deviation is obtained.

*Disintegration* ($Y_4$)—several tablets are tested from each sample bottle, using the USP disintegration apparatus without disks at 37°C, $H_2O$ medium.

*Thickness variation* ($Y_5$)—the thickness of each tablet used in the weight variation test is measured with an Ames Thickness Comparator. The same gauge is used for all formulations in the study. The relative standard deviation or coefficient of variation is determined.

*Elegance* ($Y_6$)—for this subjective test, a total of 40 tablets are considered. Several tablets are sampled from each bottle. If a tablet shows capping, sticking (rifle marks), picking, lamination, or chipping, it is considered to be flawed. Each tablet without a flaw is granted 1 point.

*Hardness or crushing strength* ($Y_7$)—the hardness of several tablets is determined with the Schlenniger-2E Hardness Tester.

*Loss on drying (LOD)* ($Y_8$)—an attempt is made to maintain the LOD of the dried granulation between 1 and 20% by closely monitoring the drying operation. LOD is determined on the Moisture Computer at 105°C.

*Tonnage* ($Y_9$)—the minimum tonnage pressure (pounds) (applied to the powder mix) required to meet the target tablet thickness is recorded

The total number of tablets to be considered for each of the tests described above would depend on (1) the compendial and regulatory (NDA commitment) requirement, (2) time and cost considerations, (3) laboratory logistics, and (4) statistical sample size determinations.

It is equally important that the consulting statistician have a broad understanding of the entire tableting process. A schematic diagram of the process flowchart appears as Figure 1.

The process consists of the following subprocesses: (1) mixing, (2) granulating, (3) drying, (4) grinding, (5) lubrication, and (6) compression.

*Mixing* consists of (a) placing in a clean, dry, stainless steel bowl of adequate capacity, calcium phosphate dibasic hydrous USP, FD&C Yellow No. 6 Al lake (18% pure dye) and starch NF corn, (b) mixing these powders for 10 minutes at approximately 50 to 60 rpm, (c) passing the mixed powders through a Fitzpatrick Comminutor equipped with a No. 80 stainless steel screen and operated at high speed with impact forward, (d) collecting the milled powders in a clean, dry bowl of adequate capacity and adding this

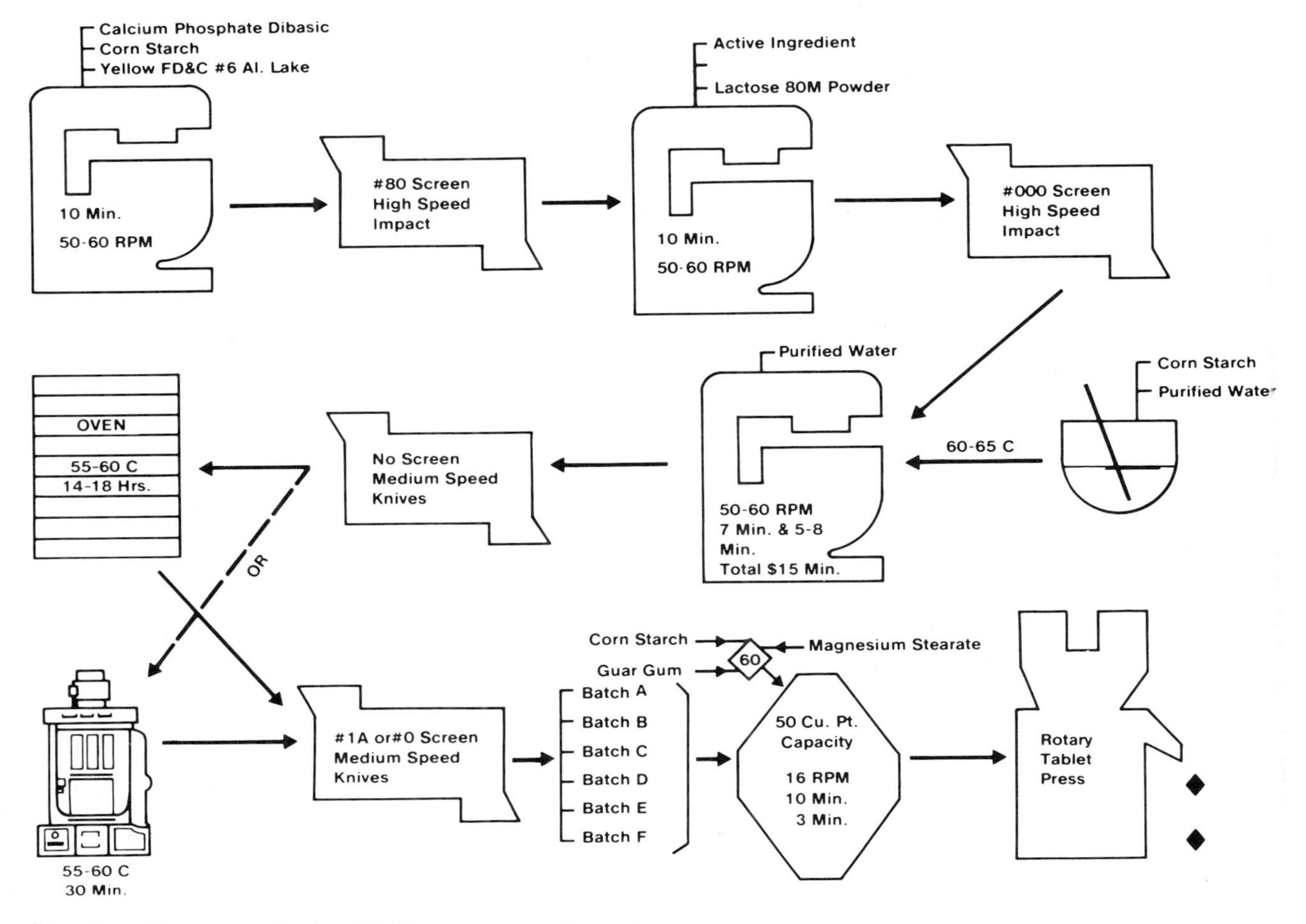


Fig. 1 Pharmaceutical tableting process flow chart.

active ingredient and lactose USP powder, (e) mixing these powders for 10 minutes at approximately 50 to 60 rpm, (f) passing the mixed powders through a Fitzpatrick Comminutor equipped with a No. 000 stainless steel screen and operated at high speed with impact forward, (g) collecting the milled powders in a clean, dry stainless steel bowl of adequate capacity, and (h) remixing the powders for 5 minutes at approximately 60 rpm.

*Granulation* consists of (a) placing purified water (USP) in a suitable jacketed mixing vessel, (b) adding starch NF corn with continuous agitation, (c) heating, with continuous agitation, the suspension until the starch gels, (d) cooling and maintaining temperature of starch paste between 60 and 65°C after gelling, (e) adding all the starch paste to the powder mix at 60 to 65°C and then mixing at 55 rpm for 15 minutes, and (f) passing the wet mass slowly through a Fitzpatrick Comminutor operated at medium speed with knives forward, after the granulation has been completed.

*Drying* consists of (a) spreading the milled granulation thinly on paper-lined trays and placing in a forced-air dryer, (b) drying the granulation at 55 to 60°C (131 to 140°F) for 14 to 18 hours or until LOD at 105°C for 15 minutes is less than 2% but greater than 1%, and collecting the recorder chart of temperature and time for each formulation.

*Grinding* consists of (a) passing the dried granulation through a Fitzpatrick Comminutor equipped with a stainless steel screen and operated with knives forward at medium speed, and (b) collecting the milled granulation in a polyethylene bag.

*Lubrication* consists of (a) weighing a 1069-g portion (equivalent to 5000 tablets) from the appropriate granulation, (b) placing the milled dry granulation into a clean dry V-blender of adequate capacity, (c) passing the following materials through a clean, dry vortisieve fitted with No. 60 bolting cloth screen, and adding the granulation in the V-blender: guar gum NF and starch NF corn, (d) mixing for 10 minutes at 16 rpm, (e) passing through a clean dry vortisieve fitted with No. 60 bolting cloth screen and adding to the granulation in the V-blender, magnesium stearate purified impalpable powder NF, (f) mixing for 3 minutes, and (g) transferring the lubricated granulation into a clean dry polyethylene bag.

*Compressing* involves the following steps. (a) Manesty Beta Press with force feed; compress according to the following specifications. Die: diamond shape, 0.3125 × 0.4218 in. (10/32 × 13.5/32); punch: top—company code; bottom—product code; tablet weight: 10 tablets = 2.40 g (limits 2.30–2.50 g); thickness: 3.70 mm (limits 3.58–3.82 mm); pressure: minimum required to produce the desired thickness; record the pressure setting. (b) Tablet collection: run for 5 minutes to equilibrate settings (weight and thickness) before collection. Collect approximately 1000 tablets successively in each of five polyethylene bags numbered 1 through 5. (c) Remove dust from each sample separately and store the tablets in labeled high-density polyethylene bottles with continuous-thread closures.

### C. Statistical Analysis

The experimental specification associated with product P provides an example for illustrating the statistical methods used in pharmaceutical formulation optimization. Generally, a multiple linear regression model is assumed. This is fitted to the data and an inverse prediction analysis is carried out. In this study, we have five independent variables ($X_1$) and 10 dependent variables ($Y_m$). A second-order polynomial function is fitted to each dependent variable, yielding the following regression equation for *each* variable:

$$Y = B_0 + B_1X_1 + B_2X_2 + B_3X_3 + B_4X_4 + B_5X_5 + B_{11}X_1^2 + B_{22}X_2^2$$
$$+ B_{33}X_3^2 + B_{44}X_4^2 + B_{55}X_5^2 + B_{12}X_1X_2 + B_{13}X_1X_3 + B_{14}X_1X_4$$
$$+ B_{15}X_1X_5 + B_{23}X_2X_3 + B_{24}X_2X_4 + B_{25}X_2X_5 + B_{34}X_3X_4$$
$$+ B_{35}X_3X_5 + B_{45}X_4X_5 \quad (1)$$

where the B's are the least-squares estimates of the partial regression coefficients.

Let W' be a column vector of independent variables and the appropriate polynomial terms,

$$W' = [1\ X_1 \cdots X_5\ X_1^2 \cdots X_5^2\ X_1X_2 \cdots X_4X_5]$$
$$= [W_0\ W_1\ W_2 \cdots W_{21}]$$

Then the estimated partial regression coefficients are obtained by solving the normal equations: W'WB = W'Y. This results in the well-known form $B = (W^1W)^{-1}\ W^1Y$, where $W^1W$ is a positive definite symmetric matrix containing the sum of squares and sum of products of $W_i$'s.

The coefficient of determination ($R^2$) is given in matrix notation as

$$R^2 = [B'W'Y - n\bar{Y}^2][Y'Y - n\bar{Y}^2]^{-1}$$

where $\bar{Y}$ is the mean of the dependent variable.

Once the regression is performed, we proceed to conduct various tests of validity of the regression model.

### 1. Test for Lack of Fit

Generally the center point is repeated several times during the course of the experiment involving different batches, operators, etc. Based on eight repeated points, we have 34 ($1/2\ 2^5 + 2(5) + 8$) total observations. The breakdown of the degrees of freedom (DF) for the analysis of variance (ANOVA) is as follows:

ANOVA

| Source | DF |
|---|---|
| Regression | 20 |
| Linear | 5 |
| Quadratic | 5 |
| Interaction | 10 |

ANOVA (continued)

| Source | DF |
|---|---|
| Residual | 13 |
| Lack of fit (LOF) | 6 |
| Pure error | 7 |
| Total | 33 |

Under on the assumption of a Gaussian error distribution, we have the following lack-of-fit test statistic:

$$F_{6,7} = \frac{\text{Mean SQ(LOF)}}{\text{Mean SQ(ERROR)}}$$

A nonsignificant (say at a nominal $\alpha = 0.05$ level) LOF test statistic is interpreted as the data not contradicting the adequacy of the model. Readers should be advised, however, that the power of the test is almost always low.

### 2. Examination of Residuals and Regression Diagnostic Techniques

At this point, one should examine the residuals and use the techniques associated with "regression diagnostics." A detailed treatment of this topic will be found in most standard regression textbooks such as Draper and Smith (1981). Computer packages such as SAS (PROC REG) could be used to one's advantage, since the program provides almost all the diagnostic techniques and the necessary plots in explicit outputs.

Examining residuals is one of the most important tasks in any regression analysis. There are different types of residuals, for example, unscaled residuals, deleted residuals, standardized residuals, and Studentized residuals. These transformed residuals have different properties, and each can be useful in detecting different patterns in the data. The residual analysis will tell how well the model used describes the data actually observed. One should also study the influence of each data point on the estimation of the regression parameters. One should consider Cook's distance, the Mahalanobis distance, and the Studentized outlier test for detecting the influence points. Partial regression residual plots should also be of great help in this direction.

### 3. Canonical Conversion of Regression Equation

After having examined the adequacy of the model via the test for lack of fit and residual examination, we want to explore the fitted model (surface) in such a way that values of the independent variables are identified which correspond to desired optima on the surface. To facilitate this, it is helpful to convert the regression equation or fitted model to its canonical equivalent. A second-order polynomial equation in n variables has the form of a section of a general n-dimensional conic (ellipsoid). In Cartesian form, the equation of an ellipsoid in n-dimensional space is given by

$$\sum_{i=1}^{n}\sum_{j=1}^{n} a_{ij}Z_iZ_j + \sum_{j=1}^{n} b_jZ_j + C = 0$$

where $(Z_1, Z_2, \ldots, Z_n)$ is a point in n-dimensional space, and the $a_{ij}$, $b_j$, and C are scalar constants.

Using matrix notation (Bohidar, 1984; Wilde and Beightler, 1967), the equation may be written as

$$Z'AZ + b'Z + C = 0$$

where $Z = (Z_1, Z_2, \ldots, Z_n)'$, b is the vector consisting of $b_j$, and A is the symmetric positive definite matrix consisting of the $a_{ij}$. By a suitable translation of axes, $Z = Y - A^{-1}b$, the equation may be simplified to $Y'AY + C = 0$. If the positive vector of a point Y on this hyperellipsoid is the same as the gradient vector at Y, then Y is the principal axes of the hyperellipsoid. The gradient vector is given as $G = AY$. Thus the principal axes are given by the n (nontrivial) vectors Y satisfying $AY = \lambda Y$, which are the eigenvectors of A. The $\lambda$'s are the eigenvalues of the matrix A. These general concepts are applicable to the second-order polynomial regression equation (1) associated with the optimization experiment.

The equation can be expressed in matrix notation (Myers, 1976) as follows:

$$\hat{Y} = B_0 + X'B + X'B^*X$$

where

$$X' = [X_1\ X_2\ X_3\ X_4\ X_5],$$

$$B' = [B_1\ B_2\ B_3\ B_4\ B_5]$$

$$B^* = \begin{bmatrix} B_{11} & & & & \\ 1/2\ B_{21} & B_{22} & & \text{symmetric} & \\ 1/2\ B_{31} & 1/2\ B_{32} & B_{33} & & \\ 1/2\ B_{41} & 1/2\ B_{42} & 1/2\ B_{43} & B_{44} & \\ 1/2\ B_{51} & 1/2\ B_{52} & 1/2\ B_{53} & 1/2\ B_{54} & B_{55} \end{bmatrix}$$

The evaluation of the eigenvalues and the corresponding eigenvectors of matrix $B^*$ is based on the solution of the determinental equation

$$|B^* - \lambda I|$$

The geometry of the multidimensional surface is obtained by considering the direction of the eigenvalues. For five independent variables, there are

32 ($2^5$) different possible combinations of positive and negative signs of the eigenvalues. These are tabulated as follows:

| Possibilities | Eigenvalues $\lambda_1$ | $\lambda_2$ | $\lambda_3$ | $\lambda_4$ | $\lambda_5$ | Geometric interpretation |
|---|---|---|---|---|---|---|
| 1 | − | − | − | − | − | Unique maximum |
| 2 | + | + | + | + | + | Unique minimum |
| 3 | + | − | − | + | − | Saddle point |
| . | . | . | . | . | . | |
| . | . | . | . | . | . | |
| 32 | − | + | − | − | + | |

If either of the first two possibilities obtains, the unique solution (the values of the independent variables which optimize Y) may be obtained by differentiation of Y with respect to the vector X, setting the derivative equal to the vector zero and solving for X. Denoting the solution $X_0$, one obtains:

$$X_0 = B^{*-1}(1/2)B$$

where $X_{01}, X_{02}, \ldots, X_{05}$ are the optimum levels of the process or excipient variables which would yield the unique optimum response. If neither of the first two possibilities obtains, one has to resort to exploratory techniques based on the following canonical function:

$$\hat{Y} = Y_0 + \lambda_1 Z_1^2 + \lambda_2 Z_2^2 + \cdots + \lambda_5 Z_5^2$$

where $Y_0$ is the estimated response at the stationary point, $\lambda_1, \lambda_2, \ldots, \lambda_5$ are the eigenvalues of the B* matrix, and $Z_1, Z_2, \ldots, Z_5$ are the new rigidly rotated coordinates, as discussed subsequently.

The relationship between the rotated coordinates and the original variables is established by the matrix equation

$$Z = V'X$$

where V is the matrix of eigenvectors in columns. The effort here is to find an acceptable local optimum in the absence of the global one. It should be noted that there is a correspondence between the approach given above and a generalization of the scalar calculus applied to multivariate optimization functions (Wilde and Beightler, 1967). The vector B is analogous to the gradient vector—the row vector consisting of the first partial derivatives of the function. The matrix B* is analogous to the Hessian matrix—the matrix consisting of the second-order partial derivatives of the function.

## D. Exploration Considerations

### 1. The Grid Search Procedure and Other Optimization Techniques

Several direct search procedures are available in the literature. A procedure which has been used in optimization of pharmaceutical formulations with much success is the grid search procedure (Davies, 1956; Bohidar, 1984). This method provides, generally, several sets of solutions. Each set is critically examined in the light of practicality, laboratory experience, and regulatory/compendial considerations. Only a very few, say three or four, "optimum formulations" are selected for further study. At this point, it would be instructive to go over the list of the types of optimization techniques available in the literature (Wilde and Beightler, 1967). These are

1. Single-variable optimization (involving only one X and one Y variable)
2. Multivariable optimization (involving several X variables but only one Y variable)
3. Single-variable multiple objective optimization (involving one X variable and several Y variables)
4. Multivariable multiple objective optimization (involving several X variables and several Y variables)

The data structure, in the pharmaceutical formulation optimization, is of the fourth category. However, "multiple objective" implies, in the *general* optimization literature, the construction of one single composite function of all the Y's. This reduces the problem to a multivariable optimization situation (second category) and requires one to develop a weighted function of the Y's. The choice of weights is usually subjective. It is difficult to develop such a composite function in pharmaceutical formulation optimization because the appropriate weights are not known. Also, in pharmaceutical situations, each Y must meet *individually* certain constraints based on compendial, regulatory, and economic considerations. Consequently, constraints imposed on the composite function will not be satisfactory. The type of optimization procedure appropriate for pharmaceutical problems is called "multivariable simultaneous objective optimization" (M-SOOP) (Bohidar, 1984).

### 2. Multivariable Simultaneous Objective Optimization Procedure

The general steps for the M-SOOP procedure are

1. Since the aim is to implement a systematic investigation of the solution space in search of the optimum, which may occur throughout the experimental region, we generate a set of grid search points in the region by using the vector equation

$$X^{(i)} = \dot{\alpha} + \dot{\gamma}_i{}'(\dot{\beta} - \dot{\alpha}) \qquad \text{(dot implies vector)}$$

where $\dot{\gamma}$ is the partition step parameter such that $0 \leqslant \dot{\gamma} \leqslant 1$, $\dot{\alpha}$ represents the coordinates of the starting point, $\dot{\beta}$ represents the coordinates of the end point, and $X^{(i)}$ is the search point (vector).

2. The number of K-dimensional grid points is $(\gamma)^{-K}$.
3. Each five-dimensional (K = 5) grid point $(X_1^{(1)}, X_2^{(1)}, X_3^{(1)}, X_4^{(1)}, X_5^{(1)})$ is substituted into each of the p (number of dependent Variables) $\hat{Y}$-equations (either the regression form of $\hat{Y}$ or the canonical form of $\hat{Y}$). The resultant predicted Y value for each parameter is examined to see whether its magnitude is within the prespecified constraint limits of that parameter. The accept-reject decision is as follows: accept the grid point in question as a solution if *all* the predicted Y values fall within their respective prespecified constraint limits. Reject the grid point as a solution if any one of the predicted Y values does not fall within the constraint limits.

In the case of product P, the ideal formulation would have been the one with a disintegration time ($Y_1$) of less than 1 minute, tablet hardness ($Y_2$) of greater than 12 kg, and dissolution rate ($Y_3$) of greater than 100% in 50 minutes. This set of constraints is expected to be unreasonable, and, indeed, no solution is found until the values are systematically relaxed to a disintegration time of 5 minutes, tablet hardness of 10 kg, and dissolution of 100% at 50 minutes. A grid search utilizing the above constraints produces eight solutions (formulations) that satisfy the conditions. These eight solutions are critically examined in the light of practicality, laboratory experience, economics, and regulatory and/or compendial considerations. Only three or four of these solutions are selected for further study (verification experiment).

## E. Experimental Results

The detailed results of the statistical analysis of an optimization experiment for product T are presented in Tables 3 to 10. Parenthetically, this experiment is identical to that of the product P experiment. The study involves five independent variables and six dependent variables. The experimental design is a five-factor, orthogonal, central, half-fractional factorial composite second-order design. Table 3 provides the coded and uncoded levels of the independent variables. Table 4 depicts the $R^2$ values associated with each response variable. Tables 5 and 6 provide a glimpse of the elements of the B* matrix and B vector for dissolution-15 minutes and disintegration time, respectively. The matrix and the vector are used in determining the five-dimensional stationary point. Table 7 provides the eigenvalues for each of the response variables considered. The magnitude and direction of the $\lambda$'s provide an insight into the form of the stationary point. The table shows that the stationary point for each dependent variable is, in this study, a saddle point, since the eigenvalues do not share the same sign. Consequently, one needs to resort to a systematic exploration of the fitted surface. Table 8 provides the search limits for the independent variables and the constraint limits for the dependent variables. Table 9 shows the elements of the canonical transformation matrix which can be used to convert the orthogonal coordinates to the original coordinates. When the grid search exploration process is conducted entirely on the

Table 3 Compressed (CT) Tablet: Product T

| Independent variable | Coded[a] and uncoded levels | | | | |
|---|---|---|---|---|---|
| | −1.547 | −1 | 0 | +1 | +1.547 |
| $X_1$ = Cab-o-sil (mg) | 0.001 | 0.44 | 1.24 | 2.04 | 2.48 |
| $X_2$ = Encompress/lactose ratio (mg/mg) | 0.98 | 1.26 | 2.15 | 4.19 | 6.68 |
| $X_3$ = Starch disintegrant (mg) | 0.001 | 2.25 | 6.44 | 10.6 | 12.88 |
| $X_4$ = Stearic acid (mg) | 0.001 | 1.22 | 3.44 | 4.66 | 6.80 |
| $X_5$ = Magnesium stearate (mg) | 0.35 | 0.60 | 1.05 | 1.50 | 1.75 |

[a]Coded conversion = C = (A − B)/E, where A = actual, B = base, "0", and E = experimental unit (factorial spacing).

Table 4 Compressed (CT) Tablet: Product T; Coefficient of Determination of the Response Variables

| Response | $R^2$ value (%) |
|---|---|
| 1. Content uniformity | 96.0 |
| 2. Tablet strength | 97.0 |
| 3. Dissolution (15 min) | 98.5 |
| 4. Dissolution (30 min) | 98.3 |
| 5. Dissolution (45 min) | 98.6 |
| 6. Disintegration | 95.0 |

Table 5 Compressed (CT) Tablet: Product T; Elements of B* Matrix and Components of B Vector for Dissolution−15 minutes

$$B^* = \begin{bmatrix} \underset{B_{11}}{-3.75} & \underset{B_{12}}{0.483} & \underset{B_{13}}{0.343} & \underset{B_{14}}{0.621} & \underset{B_{15}}{-1.384} \\ \underset{B_{21}}{0.483} & \underset{B_{22}}{0.829} & \underset{B_{23}}{0.006} & \underset{B_{24}}{0.333} & \underset{B_{25}}{-0.446} \\ \underset{B_{31}}{0.343} & \underset{B_{32}}{0.006} & \underset{B_{33}}{-0.091} & \underset{B_{34}}{0.110} & \underset{B_{35}}{0.032} \\ \underset{B_{41}}{0.621} & \underset{B_{42}}{0.333} & \underset{B_{43}}{0.110} & \underset{B_{44}}{-0.262} & \underset{B_{45}}{0.229} \\ \underset{B_{51}}{-1.384} & \underset{B_{52}}{-0.446} & \underset{B_{53}}{-0.032} & \underset{B_{54}}{0.229} & \underset{B_{55}}{-11.768} \end{bmatrix} \qquad B \text{ vector} = \begin{bmatrix} -2.769 \\ -9.612 \\ 2.061 \\ -3.366 \\ 31.671 \end{bmatrix}$$

**Table 6** Compressed (CT) Tablet: Product T; Elements of B* Matrix and Components of B Vector for Disintegration (in minutes) Time

$$B^* = \begin{bmatrix} 0.511 & 0.084 & -0.143 & -0.174 & 1.498 \\ 0.084 & -0.001 & -0.070 & 0.074 & 0.801 \\ -0.143 & -0.070 & -0.012 & -0.078 & 0.198 \\ -0.174 & 0.074 & -0.078 & 0.135 & 0.438 \\ 1.498 & 0.801 & 0.198 & 0.438 & 0.172 \end{bmatrix} \qquad \text{B vector} = \begin{bmatrix} 0.397 \\ -0.831 \\ 0.704 \\ -0.050 \\ 13.158 \end{bmatrix}$$

**Table 7** Compressed (CT) Tablet: Product T; Direction and Magnitude of Eigenvalues ($\lambda$) Derived from the Determinantal Equation $|B^* - \lambda I| = 0$ for Each Response Variable

| | Eigenvalues[a] | | | | |
|---|---|---|---|---|---|
| Response variable | $\lambda_1$ | $\lambda_2$ | $\lambda_3$ | $\lambda_4$ | $\lambda_5$ |
| 1. Content uniformity | −1.410 | −0.012 | 0.066 | 0.314 | −5.141 |
| 2. Hardness | 0.710 | −0.056 | 0.007 | 0.024 | 0.863 |
| 3. Dissolution (15 min) | −3.694 | 1.040 | 0.002 | −0.364 | −12.020 |
| 4. Dissolution (30 min) | −1.790 | 0.359 | 0.102 | −0.111 | −11.483 |
| 5. Dissolution (45 min) | −1.502 | 0.285 | 0.068 | −0.163 | −8.462 |
| 6. Disintegration | 2.053 | −0.080 | 0.039 | 0.328 | −1.535 |

[a]Note that the five eigenvalues do not share the same sign for any of the six response variables considered in the study.

**Table 8** Compressed (CT) Tablet: Product T; Exploration Ranges for Grid Search

Search limits for independent variables

1. $0.01 \leqslant X_1 \leqslant 2.48$
2. $0.98 \leqslant X_2 \leqslant 6.68$
3. $0.01 \leqslant X_3 \leqslant 12.90$
4. $0.01 \leqslant X_4 \leqslant 6.80$
5. $0.35 \leqslant X_4 \leqslant 1.75$

Table 8 (continued)

| Constraints for dependent variables | |
|---|---|
| 1. Content uniformity | $Y_1 > 100$ |
| 2. Hardness | $Y_2 > 5.0$ |
| 3. Dissolution—15 min | $Y_3 > 80$ |
| 4. Dissolution—30 min | $Y_4 > 90$ |
| 5. Dissolution—45 min | $Y_5 > 95$ |
| 6. Disintegration | $Y_6 < 5$ |

Table 9 Compressed (CT) Tablet: Product T; Elements of Canonical Transformation[a] Matrix; Conversion of X Vector to Z Vector and Vice Versa

$$\begin{bmatrix} Z_1 \\ Z_2 \\ Z_3 \\ Z_4 \\ Z_5 \end{bmatrix} = \begin{bmatrix} 0.965 & -0.108 & 0.089 & -0.150 & -0.163 \\ -0.152 & -0.093 & -0.073 & -0.311 & 0.044 \\ 0.117 & -0.214 & 0.894 & 0.376 & -0.001 \\ -0.068 & -0.262 & -0.434 & 0.859 & 0.020 \\ 0.165 & -0.029 & 0.002 & 0.029 & -0.985 \end{bmatrix} \begin{bmatrix} X_1^* \\ X_2^* \\ X_3^* \\ X_4^* \\ X_5^* \end{bmatrix}$$

[a]Orthogonal matrix $X_i^* = (X_i - X_{oi})$, where $X_{oi}$ is the stationary point of the ith variable.

Table 10 Compressed (CT) Tablet: Product T; Constrained Grid Search Results

| | Optimum formulations[a] | | |
|---|---|---|---|
| | $F_{01}$ | $F_{02}$ | $F_{03}$ |
| $X_1$ | 1.11 | 1.389 | 0.833 |
| $X_2$ | 1.67 | 3.0 | 2.33 |
| $X_3$ | 13.0 | 12.5 | 12.91 |

Table 10 (continued)

| | Optimum formulations[a] | | |
|---|---|---|---|
| | $F_{01}$ | $F_{02}$ | $F_{03}$ |
| $X_4$ | 2.676 | 3.44 | 2.293 |
| $X_5$ | 1.40 | 1.20 | 1.00 |

*Expected response*

$Y_1 = 100$
$Y_2 = 5.2$
$Y_3 = 82.1$
$Y_4 = 91.0$
$Y_5 = 95.1$
$Y_6 = 2.5$

[a]An independent experiment will be conducted based on these selected optimum formulations. It will provide information on the variability associated with the formulations. See Section II,F for details.
[b]$F_{01}$, Optimized formulation No. 1; $F_{02}$, optimized formulation No. 2; $F_{03}$, optimized formulation No. 3.

orthogonal coordinates (rotated coordinates), it is called the canonical grid search process. The results of this process can be transformed back to their original limits by using the canonical transformation matrix. Finally, Table 10 provides the three optimum formulations which yielded responses as close to the established goals as possible. The table also provides the average of the dependent variables as predicted by their respective objective functions.

Figure 2 provides a pictorial representation of the dissolution profiles of the three optimized formulations (Nos. 1, 2, and 3) and of the best unoptimized formulation (subjectively chosen by the pharmacist). It is clearly observed that the performances of two optimized formulations (Nos. 1 and 2) are far superior to that of the unoptimized formulation. It must be mentioned here that the optimized formulation No. 3 is deliberately selected because its results in the optimization experiment are close to the unoptimized formulation and yet one can see that its performance is relatively superior (see 15- and 30-minute responses) to that of the unoptimized formulation.

### F. Postoptimization Validation Experiment

After the termination of the optimization experiment, one additional experiment must be conducted to compare the performances of the three (say)

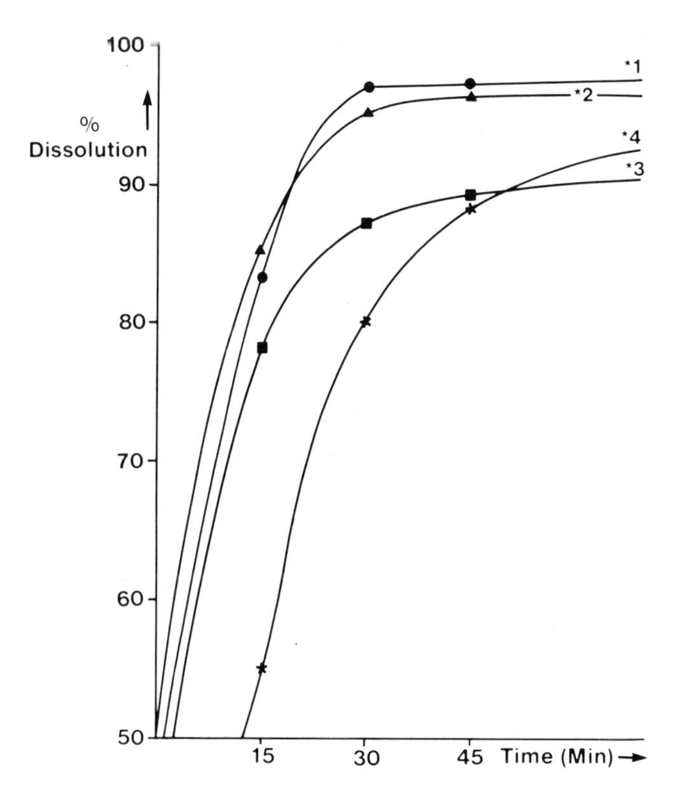


Fig. 2 Mean dissolution profile for product-T (verification experiment). 1 and 2 = optimized formulation, 3 = optimized formulation whose performance is close to the unoptimized (subjective) formulation, and 4 - unoptimized (subjective) formulation.

candidate optimized formulations. A nested classification analysis should be considered for this phase of the experiment. The sources of variation for the analysis would be (1) between formulations, (2) batches within formulations, and (3) tablets/Rx/formulations. Each batch consists of a 4-kg freshly granulated mix batch generating several hundred tablets. It should be noted that not only is the comparison of means undertaken but also the comparison among the variances should be considered. The formulation with the smallest variance should be the candidate of choice. The Levene test (Levene, 1960), the Wilk-Shapiro test (Wilk and Shapiro, 1968), and the general ANOVA test should be performed in this connection. A multiple comparison test should also be performed to accomplish the pairwise comparisons among the formulations.

The experimental results of the verification experiment (Bohidar, 1984) for product P are presented in the following table:

| Response[a] | Results based on validation experiment | Results based on optimization experiment |
|---|---|---|
| 1. Disintegration time (min) | 3.31 | 3.03 |
| 2. Hardness (kg) | 10.08 | 10.38 |
| 3. Dissolution (30 min) | 89.85 | 98.02 |
| 4. Friability (%) | 0.18 | −0.38[b] |
| 5. Thickness (mm) | 2.31 | 2.37 |
| 6. Porosity, pore volume (ml/mg) | 0.0269 | 0.0318 |

[a]The other four responses were not measured in the validation experiment. The results of this experiment met all the compendial limits.
[b]Regression prediction may result in a negative number.

These examples indicate that optimization experiments are necessary in pharmaceutical formulation studies. We have successfully completed several (9 or 10) optimization experiments since we introduced (Bohidar, 1984) the technique as a research tool in pharmaceutical research and development of drug delivery systems.

## III. DEVELOPMENT OF GENERAL MONITORING SYSTEMS

Before describing the statistical procedures associated with the assessment of the scale-up performance of the optimized formulation, we consider the mechanism of developing a monitoring system for the optimized formulation. Monitoring the future performance of an optimized formulation should require only a few tests. Thus, identification of a set of key parameters becomes very important. The information necessary for conducting these tests is available from the optimization experiment data.

### A. Selection of Key Response Parameters

It is time- and cost-effective if only those response parameters which contribute most to the variation among the formulations considered in the optimization experiment (Bohidar et al., 1975) are tested. Since these parameters are interrelated, it is relevant to reduce the dimensionality and to represent the entire system by fewer response measures.

Consider a set of random vectors of responses $(Y_1, Y_2, \ldots, Y_p)$. Construct the associated variance-covariance matrix $\Sigma_y$ which is positive semidefinite, symmetric, and of dimension $p \times p$.

Let $V\ (\nu_{11}, \nu_{12}, \ldots, \nu_{1p})'$ be a p-component column vector such that $V'V = 1$. The variance of $V'Y$ is

$$E(V'Y)^2 = E(V'YY'V) = V'\Sigma V$$

To determine the normalized linear combination $V'Y$ with maximum variance, we must find a $V$ which maximizes $V'\Sigma V$ under the condition $V'V = 1$. Consider

$$\theta = V'\Sigma V - \lambda(V'V - 1)$$

where $\lambda$ is a Lagrange multiplier. The vector of partial derivatives is

$$\frac{\delta\theta}{\delta V} = 2\Sigma V - 2\lambda V$$

Now we set the above expression to zero, that is, $(\Sigma - \lambda I)V = 0$. To get a solution with $V'V = 1$, $\lambda$ must satisfy $|\Sigma - \lambda I| = 0$. The function $|\Sigma - \lambda I|$ is a polynomial in $\lambda$ of degree p with $\lambda_1 \geqslant \lambda_2 \geqslant \cdots \geqslant \lambda_p$. Since

$$V'\Sigma V = \lambda V'V = \lambda$$

the variance of $V'Y$ is $\lambda$. Thus for the maximum variance we should use the largest eigenvalue, $\lambda_1$.

Based on the above theoretical considerations (Anderson, 1958), the reduction in dimensionality is achieved in two steps. (Step 1) We determine the principal component with the largest orthogonal variance ($\lambda_1$). This is achieved by expressing each eigenvalue as a percentage of the trace of the variance-covariance matrix as follows: $(\lambda_1/\Sigma\lambda_i)100$, $(\lambda_2/\Sigma\lambda_i)100$, . . ., $(\lambda_p/\Sigma\lambda_i)100$. An examination of the ratios provides the necessary answer, as shown in the table below. (Step 2) Each of the p principal components is a linear combination of the p original variables. The principal components ($Z_i$) can be expressed as follows:

$$Z_i = \sum_{j=1}^{p} V_{ij} Y \qquad i, j = 1, 2, \ldots, p$$

where $0 \leqslant V_{ij}^2 \leqslant 1$ for each $Z_i$.

Since $\Sigma_{j=1}^{p} V_{ij}^2 = 1$ for each $Z_i$, it is possible to examine the relative magnitude of the components of the eigenvector (Bohidar et al., 1975), some of which may be close to zero. The weight of each original variable in the linear function is measured by the absolute magnitude (or the square) of its coefficient. Thus the variable with a very small weight may not contribute substantially to the overall function.

These two steps of principal component analysis (PCA) are applied to the product P optimization study involving 10 dependent variables. The results associated with the first step, using the variance-covariance matrix, are given in tabular form (Bohidar et al., 1975) in the following.

| Principal component | Eigen-values | Relative information (%) | Cumulative relative information (%) |
|---|---|---|---|
| I | 578.4 | 95.4 | 95.4 |
| II | 23.9 | 3.9 | 99.3 |
| III | 2.2 | 0.4 | 99.7 |
| IV | 2.0 | 0.3 | 100.0 |

The results of the principal component analysis are generally (Bohidar et al., 1975) interpreted by conducting an orthogonal variance analysis (invormation analysis), referred to as step 1 above, and a component structure analysis, referred to as step 2 above. The table below provides the results of the component structure analysis for product P.

The orthogonal variance analysis indicates that the first principal component contributes to 95.4% of the total variation among the formulations. The component structure analysis shows that the dissolution component is weighted 96% (only 4% for the other variables) in the first principal component. The disintegration component is weighted 94% (only 6% for the other variables) in the second principal component, which contributes only 3.9% of the total variation. Additional results of the principal component analysis have been given in Bohidar et al. (1975).

Elements (Components) of Eigenvectors Associated with First Two Principal Components

| Parameter | Principal component I | Principal component II |
|---|---|---|
| Dissolution | 0.98 | 0.22 |
| Porosity | 0.00 | 0.00 |
| Friability | 0.00 | 0.00 |
| Disintegration time | −0.22 | 0.97 |
| Weight uniformity | −0.01 | 0.03 |
| Thickness uniformity | 0.00 | 0.00 |
| Granular mean diameter | 0.00 | 0.00 |
| Tablet breakage | 0.00 | 0.00 |
| Mean pore diameter | 0.00 | 0.00 |
| Tablet breaking strength | 0.01 | 0.12 |
| Relative information (%) $[(\lambda_i/\Sigma\lambda_i) \times 100]$ | 95.4 | 3.9 |

## B. Selection of Key Pharmaceutical Formulation Factors

It is of interest to the research pharmacist to know which excipient and which process variables influence, directly or indirectly, a selected property of a given drug delivery system. This knowledge would provide a basis for controlling key formulation factors to effect a favorable change in the response parameter of interest. For instance, five formulation factors were considered in the development of the product P pharmaceutical tablet formulation. If one is interested in bringing about a desirable change in the dissolution profile of the system considered, then, assuming the existence of a close functional relationship between the five formulation factors and the rate of dissolution, one could assess which of the five factors would play an important role in bringing about the desired change. Likewise, we could ask, "Will a change in the lubrication level significantly improve the dissolution profile of the system?" or "To what extent will a change in the diluent ratio maintain the same dissolution value as we accepted in the initial stage of the development of the process?" Providing answers to these questions is of prime interest in this section (Bohidar et al., 1979).

The interrelation among the process variables and the interrelation between the dependent variable and each of the process variables in the system would have an effect on the selection of the key formulation factors associated with the system. Because of the inseparable closeness of the variables in correlated systems, it would not be easy to identify the key factors without resorting to an objective procedure of elaborate step-by-step numerical examination. One would use "variable selection in regression analysis" (VSR), which attempts to accomplish this goal (Draper and Smith, 1981). Of the several available procedures associated with this analysis, the two most widely accepted procedures are "all possible regression" (APR) and "stepwise regression" (SWR). These procedures can be used independently, and if the results of the two analyses are combined properly, they generally yield a unique set of key formulation factors for a given response parameter. In this section, the individual results of these two procedures are presented, the technique of combining the results is illustrated, and the appropriate interpretation is elucidated. The outcome of the combination essentially eliminates the deficiency associated with either of the procedures when used individually.

In this section, the two selective regression procedures (APR and SWR) are applied to the data associated with the product P optimization experiment (Bohidar et al., 1979). Recall that there were 5 formulation factors and 10 response parameters measured on each of the 27 distinct formulations, VSR analysis was applied to each of the 10 response variables, and a set of key formulation factors was obtained for each parameter. The results of these 10 separate analyses are presented and discussed.

### C. Combining Results of APR and SWR Analyses (CAS)

#### 1. General Procedure

The primary purpose of combining the results of the APR and SWR analyses (Bohidar et al., 1979) is to facilitate a clear interpretation of the findings of the two complementary analyses. The combined results, when used propertly, provide insight into the specific role played by each of the variables selected by the two procedures separately. New quantities are derived as a function of the results of the two analyses (Bohidar et al., 1979). In this section, the procedure for deriving the quantities is described in general terms, primarily symbolically. The appropriate interpretation of the numerical results is discussed in subsections 2 and 3 below.

Consider a multiple regression situation in which there are K independent variables and one dependent variable denoted by $Y^*$. Consider that an APR analysis is conducted. Let $A_1$, $A_2$, $A_3$, and $A_4$ denote, for example, the four sets of X variables whose $R^2$ values are either equal to or close to the maximum $R^2$ value attainable for the dependent variable $Y^*$. Note that the sets may not necessarily be mutually exclusive and the numbers of X variables in the sets may not necessarily be equal.

Now we conduct an SWR analysis on the same K independent variables and the dependent variable $Y^*$. Three separate SWR analyses are conducted.

Let $S(\alpha_1)$, $S(\alpha_2)$, and $S(\alpha_3)$ (where $\alpha_3 < \alpha_2 < \alpha_1$) denote the three sets of X variables selected by each of the three separate SWR analyses based on $\alpha_1$, $\alpha_2$, and $\alpha_3$ levels of significance, respectively. Usually, in practice one considers the values of $\alpha_1$, $\alpha_2$, and $\alpha_3$ to be 0.1, 0.05, and 0.01 levels of significance, respectively. The following relationships generally hold for these sets:

$$S(\alpha_3) \subseteq S(\alpha_2) \subseteq S(\alpha_1)$$

and

$$S(\alpha_3) \cap S(\alpha_2) \cap S(\alpha_1) = S(\alpha_3)$$

where the symbols $\subseteq$ and $\cap$ mean "contained in" and "intersection" of the sets, respectively.

The sets $A_1$, $A_2$, $A_3$, $A_4$, $S(\alpha_1)$, $S(\alpha_2)$ and $S(\alpha_3)$ facilitate the construction of sets ($\theta$) containing a subset of the independent variables, which in turn facilitate combining the results of APR and SWR. It is assumed that none of the $A_i$ nor $S(\alpha_i)$ is null. We now proceed to define the facilitator sets $\theta$.

1. *Minimum central subset* ($\theta_1$): $\theta_1$ is derived by the formula

$$\theta_1 = A_1 \cap A_2 \cap A_3 \cap A_4 \cap S(\alpha_3).$$

2. *Central subset* ($\theta_2$): $\theta_2$ is derived by the formula

$$\theta_2 = A_1 \cap A_2 \cap A_3 \cap A_4 \cap S(\alpha_1) \cap S(\alpha_2).$$

3. Complementary subset ($\theta_3$): the following relationships among $R^2$ values in the APR analysis must hold:

$$R^2(A_1) \geqslant R^2(A_2) \geqslant R^2(A_3) \geqslant R^2(A_4)$$

The following relationship in the SWR analysis must also hold:

$$R^2[S(\alpha_3)] \leqslant R^2[S(\alpha_2)] \leqslant R^2[S(\alpha_1)]$$

In general, we have,

$$R^2[S(\alpha_i)] \leqslant R^2(A_j), \qquad i = 1, 2, 3; \; j = 1, 2, 3, 4$$

Now let $\Omega = S(\alpha_1) \cup A_4$, which is referred to as the whole set, where the symbol U denotes the union of the two sets. Then the complementary subset ($\theta_3$) is derived by the following formula:

$$\theta_3 = [S(\alpha_1) \cap A_4]^c$$

where the symbol c denotes the complement of the set. If $R^2[S(\alpha_1)] = R^2(A_4)$, then set $\theta_3 = \phi$, automatically.

4. *Interchangeable set*: suppose there are two sets associated with APR analysis (say $A_1$ and $A_2$) which have the same number of members and for which $R^2(A_1) \simeq R^2(A_2)$. Now let $\Omega = A_1 \quad A_2$ be the whole set. Then the interchangeable subset ($\theta_4$) is derived by the following formula:

$$\theta_4 = [A_1 \cap A_2]^c$$

When $\theta_4$ is not the null set, $\theta_4$ contains members of both $A_1$ and $A_2$. Let those belonging to $A_1$ be identified as $A_1$* and those belonging to $A_2$ be identified as $A_2$*. The members of $A_1$* are interchangeable with the members of $A_2$*. The advantages of detecting interchangeable variables will be discussed in subsections 2 and 3 below.

Once $\theta_1$, $\theta_2$, $\theta_3$, and $\theta_4$ have been determined, the combined analysis is facilitated through the regression model consisting of the distinct independent variables found in the operations of $\theta_1$, $\theta_2$, $\theta_3$, and $\theta_4$.

### 2. Results

The APR, SWR, and CAS analyses were applied to the data from the product P experiment. They are illustrated for a selected number of parameters in Tables 11 to 13. One table is devoted to each parameter. Each of these tables has four sections. Sections 1, 2, and 3 are devoted to the results of the APR, SWR, and CAS analyses, respectively. Section 4 shows the explicit form of the regression equation involving only the linear terms, its $R^2$ value, and the $R^2$ value for the second-order regression function based on the key process variables selected by APR, SWR, and CAS analyses.

### 3. Discussion

The following general questions arise. (1) How do these results lead to the selection of a set of key process variables? (2) How does one determine the most important variable or variables among the key process variables identified? (3) What is the specific role played by each of the key formulation factors identified by the three procedures employed? To answer the first question, consider the results given in Table 11. One finds from the APR analysis that the maximum attainable $R^2$ value for the dissolution parameter is 66.0%—which involves all the process variables in the system. However, using only the three process variables $X_2$, $X_3$, and $X_5$ (subset $A_4$), one obtains an $R^2$ value of 64.8%, which is only 1.2% less than the maximum attainable. This implies that $X_1$ and $X_4$ are not contributing substantially to the variation in the dependent variable under consideration. The SWR analysis confirms these findings by reproducing some of the variables associated with subset $A_4$ in APR analysis. It also independently demonstrates that $X_5$ and $X_2$ are the two most important process variables in the system with respect to the dissolution parameter. However, using only these two variables in the regression equation, an $R^2$ value of 62.2%, which is 3.8% less than the maximum $R^2$ is obtained. The process variables in subset $A_4$ of the APR analysis indicate that including $X_3$ in the regression equation along with $X_5$ and $X_2$ not only enhances the $R^2$ value, but also brings it close to the maximum attainable. The answer to the second question is implicit in the results of SWR analysis. The results of $S(\alpha_3 = 0.01)$ suggest that $X_5$ is the most important formulation factor of

**Table 11** Results of APR, SWR, and CAS Analyses; Parameter: Dissolution in 30 Minutes ($t_{30}$)

| Section | Analysis | Sets | Formulation factors | $R^2$ (%) |
|---|---|---|---|---|
| 1 | APR | $A_1$ | $X_1, X_2, X_3, X_4, X_5$ | 66.0 |
| | | $A_2$ | $X_2, X_3, X_4, X_5$ | 65.6 |
| | | $A_3$ | $X_1, X_2, X_3, X_5$ | 65.2 |
| | | $A_4$ | $X_2, X_3, X_5$ | 64.8 |
| 2 | SWR | SWR($\alpha_1$ = 0.10) | $X_2, X_5$ | 62.2 |
| | | SWR($\alpha_2$ = 0.05) | $X_2, X_5$ | 62.2 |
| | | SWR($\alpha_3$ = 0.01) | $X_5$ | 51.3 |
| 3 | CAS | Minimum central subset | $X_5$ | |
| | | Central subset | $X_2, X_5$ | |
| | | Complementary subset ($A_4$) | $X_3$ | |
| | | Interchangeable subset | $\phi$ | |
| 4 | Structural regression equation and $R^2$ values | | | |
| | $\hat{Y} = 69.91 - 37.3X_5 - 17.48X_2 + 4.24X_3$ | | | |
| | $R^2$ = 64.8% (with $X_2$, $X_3$, and $X_5$) | | | |
| | $R^2$ = 87.2% (for the second-order regression function) | | | |

Table 12 Results of APR, SWR, and CAS Analyses; Parameter: Disintegration

| Section | Analysis | Sets | Formulation factors | $R^2$ (%) |
|---|---|---|---|---|
| 1 | APR | $A_1$ | $X_1$, $X_2$, $X_3$, $X_4$, $X_5$ | 83.6 |
| | | $A_2$ | $X_2$, $X_3$, $X_4$, $X_5$ | 83.5 |
| 2 | SWR | SWR($\alpha_1$ = 0.10) | $X_2$, $X_3$, $X_4$, $X_5$ | 83.5 |
| | | SWR($\alpha_2$ = 0.05) | $X_2$, $X_3$, $X_4$, $X_5$ | 83.5 |
| | | SWR($\alpha_3$ = 0.01) | $X_2$, $X_3$, $X_4$, $X_5$ | 83.5 |
| 3 | CAS | Minimum central subset | $X_2$, $X_3$, $X_4$, $X_5$ | |
| | | Central subset | $X_2$, $X_3$, $X_4$, $X_5$ | |
| | | Complementary subset | $\phi$ | |
| | | Interchangeable subset | $\phi$ | |
| 4 | Structural regression equation and $R^2$ values | | | |
| | $\hat{Y} = 1.45 + 9.98X_2 - 2.30X_3 + 6.90X_4 + 6.49X_5$ | | | |
| | $R^2$ = 83.5% (with $X_2$, $X_3$, $X_4$, and $X_5$) | | | |
| | $R^2$ = 91.8% (for the second-order regression function) | | | |

**Table 13** Results of APR, SWR, and CAS Analyses; Parameter: Hardness (Tablet Breaking Strength)

| Section | Analysis | Sets | Formulation factors | $R^2$ (%) |
|---|---|---|---|---|
| 1 | APR | $A_1$ | $X_1, X_2, X_3, X_4, X_5$ | 70.9 |
| | | $A_2$ | $X_2, X_3, X_4, X_5$ | 70.9 |
| | | $A_3$ | $X_2, X_3, X_5$ | 68.7 |
| | | $A_4$ | $X_2, X_4, X_5$ | 68.7 |
| 2 | SWR | SWR($\alpha_1 = 0.10$) | $X_2, X_5$ | 66.4 |
| | | SWR($\alpha_2 = 0.05$) | $X_2, X_5$ | 66.4 |
| | | SWR($\alpha_3 = 0.01$) | $X_5$ | 44.8 |
| 3 | CAS | Minimum central subset | $X_5$ | |
| | | Central subset | $X_2, X_5$ | |
| | | Complementary subset ($A_4$) | $X_3$ or $X_4$ | |
| | | Interchangeable subset | $X_3 \Leftrightarrow X_4$ | |
| 4 | Structural regression equation and $R^2$ values | | | |

$$\hat{Y} = 6.09 + 1.65X_2 + 0.53X_4 - 2.35X_5$$

$R^2 = 68.7$ (with $X_2$, $X_4$, and $X_5$)

$R^2 = 83.3\%$ (with the second-order regression function)

the key variables identified. The answer to the third question is found in the results of CAS analysis and requires some elaboration.

In the CAS analysis, "central subset" consists of the variables which are most important among the key formulation factors identified. "Minimum central subset" provides the most indispensable variables among the variables associated with central subset. Note that it is possible for the minimum central subset to contain the same variables associated with central subset. This would imply that the variables in central subset cannot be decomposed further into dispensable and indispensable key variables. In such cases, it may be inferred that all the variables in the central subset are contributing jointly (and inseparably) to the variation in the dependent variable. This phenomenon was observed in 8 of the 10 parameters studied, e.g., in disintegration (Table 12). An exception to this phenomenon is found in the results given in Table 11, in which case the minimum central subset is not identical to the central subset, which allows the identification of the single most indispensable variable.

The primary purpose of the "complementary subset" is to enhance the $R^2$ value of the regression equation in which these variables and the variables associated with central subset appear. High $R^2$ values lead to increased precision in the estimation of regression parameters and consequently enhance the predictive efficiency of the estimated regression equation.

The interpretation of "interchangeable subset" is now considered. To illustrate, the numerical example given in Table 13 will be used. The key factors influencing tablet hardness are $X_2$, $X_4$, and $X_5$. By changing these variables, one can effect a desirable change in the response variable. However, since the CAS analysis shows that $X_4$ is interchangeable with $X_3$ (see sets $A_3$ and $A_4$), one could elect to use $X_2$, $X_3$, and $X_5$ in the regression equation instead of $X_2$, $X_4$, and $X_5$. Controlling the variables in either set would bring about essentially the same improvement in the response variable. The practical utility of the interchangeability property can easily be demonstrated by using the results given in Table 14. It may be observed that one of the variables influencing dissolution is $X_3$. By making appropriate changes in $X_2$, $X_3$, and $X_5$, dissolution requirements could easily be met. Since $X_4$ is not a key formulation factor for dissolution, it would be advisable to impose changes in $X_4$ (rather than $X_3$) to bring about an improvement in hardness if one is interested in bringing about changes in both dissolution and hardness. By following this strategy, one does not disturb the dissolution profile previously attained.

There may also be economic advantages, since one of the interchangeable variables may be more economical to use than the other. Detection of interchangeable variables is therefore extremely important, and the CAS analysis is the appropriate tool for detecting such variables.

In light of these discussions, one should analyze the remaining seven response variables, summarize the analyses as in Tables 11–13, and examine the results in detail. Only the general principles of interpreting the results and the mechanism of arriving at a correct conclusion were presented in the discussion of the analyses of dissolution, disintegration, and hardness.

A summary of the salient results of APR, SWR, and CAS analyses for the 10 parameters is presented in Table 14. It is observed that compression pressure, magnesium stearate, and calcium phosphate/lactose ratio are identified as the key factors in 9, 7, and 6 of the 10 parameters,

**Table 14** Summary of Results of APR, SWR, and CAS Analyses

| Parameter | Calcium phosphate/ lactose ratio $X_1$ | Compression pressure $X_2$ | Starch (corn) disintegration $X_3$ | Granulating gelatin $X_4$ | Magnesium stearate $X_5$ |
|---|---|---|---|---|---|
| 1. Disintegration | | a | a | a | a |
| 2. Dissolution ($t_{30}$) | | a | a | | a |
| 3. Hardness | | a | | a | a |
| 4. Weight | a | a | | | a |
| 5. Friability | a | a | | a | a |
| 6. Thickness | a | a | a | a | |
| 7. Porosity | | a | | | |
| 8. Tablet breakage | a | a | | a | a |
| 9. Granular mean diameter | a | | | a | a |
| 10. Mean pore diameter | a | a | a | | |

[a]Represents key formulation factor for the parameter.

respectively. Further, most of the "interchangeability" is observed between starch disintegrant and granulating gelatin. It may also be noted that the calcium phosphate/lactose ratio is not identified as a key factor for disintegration, dissolution ($t_{30}$), or hardness—the most important tablet properties for the system under consideration (Bohidar et al., 1975). It is clearly observed that compression pressure and magnesium stearate are the key factors for dissolution, disintegration, and hardness. Cornstarch, on the other hand, is the key factor for disintegration and dissolution only. Finally, granulating gelatin is identified as a key factor for disintegration alone.

On the basis of these findings, one can begin to analyze the mechanism of action of the key factors in the system (Bohidar et al., 1979). For example, the hydrophobic nature of magnesium stearate ($X_5$) causes a decrease in dissolution and an increase in disintegration time. Its presence also causes a reduction in tablet hardness, probably by preventing the binding of the granules which apparently are cohesive in its absence. Prevention of adequate binding is important in friability and tablet breakage, as well. Compressional force ($X_2$) obviously affects tablet hardness and works in opposition to the effect of magnesium stearate. For dissolution and disintegration, compressional forces works in the same direction as magnesium stearate; this is perhaps accomplished by its effect on tablet pore volume. It is interesting to note that the only key factor for pore volume response is the compression pressure, $X_2$. As the pressure is increased, the pore volume decreases.

The cornstarch ($X_3$) is indeed performing its function and is shown as a key factor in disintegration. The fact that it is also a key factor in dissolution seems to indicate that disintegration and dissolution are closely related. The gelatin in the granulating solution ($X_4$) is a key factor in granule diameter and increases it, as expected.

The calcium phosphate/lactose ratio ($X_1$), although it has little or no effect on the dissolution, disintegration, or tablet hardness response, is a key factor in friability and tablet breakage. Apparently, the granules formed with a higher level of lactose have better cohesive and binding properties or are more elastic than those with a higher level of calcium phosphate. This may be related to the effect the calcium phosphate/lactose ratio has on granule diameter, since larger and perhaps stronger granules are produced at higher lactose levels.

It should be emphasized that those factors which are demonstrated to be the key factors in one formulation may not necessarily be so in another because of variations in physicochemical interactions in different systems. The results are applicable only to the system studied.

In conclusion, it has been shown that VSR analysis consisting of APR, SWR, and CAS analyses plays an important role in identifying the key formulation factors contributing substantially to the variation in the response parameters (Bohidar et al., 1979). Imposing controls on the subsequently identified key factors enables one to effect a favorable change in the response variable (Bohidar et al., 1975)—the desired characteristics of the drug delivery system.

### D. Process Validation by Calibration Combining PCA and CAS Results

The statistical results provided in the previous section are used in monitoring the future performance of the optimized formulation. These results could also be used in accomplishing optimization process validation by a calibrationlike procedure, in the following manner. Suppose that dissolution ($Y_{PCA}$) has been selected as one of the key response parameters by the PCA analysis (Section III,A) and that magnesium stearate ($X_{CAS}$) has been selected as one of the key process variables by the CAS method (Section III,C). The following regression equations could be developed based on the data associated with the optimization study.

Regression equation (1):

$$Y_{PCA} = B_0 + B_1 X_{CAS}$$

Regression equation (2):

$$Y_{PCA} = B_0 + B_1 X_{CAS} + B_2 X_{CAS}^2$$

Regression equation (3):

$$Y_{PCA} = B_0^* + B_1 X_{CAS} + B_2 X_{CAS}^2 + \Sigma B_i X_i X_{CAS}, \qquad i \neq CAS$$

where $B_0^*$ represents all the other terms in the second-order regression equation given in Section III,C.

Consider that we have a batch manufactured at the development phase or production phase of the process. The dissolution values are measured. Now we proceed to predict inversely the average amount of magnesium stearate available in the batch, using the best fitting regression equation given above. If the optimized amount of magnesium stearate falls within the $(1 - \alpha)\%$ confidence limits associated with the inversely predicted estimate, it will be inferred that the process is operating (adequately) within the experimental variation. If the optimized amount falls outside the limits, the sources of variation which might have caused the situation should be investigated thoroughly. It should be noted that the process variation could have been caused by such sources as new equipment, bulk material, operator, and ambient environmental conditions. The optimized formulation may not necessarily be the contributing factor. The effects of all factors on the final product must be examined carefully and the necessary strategies developed to minimize their influence.

## IV. SCALE-UP CONSIDERATIONS: DEMONSTRATION BATCH

The development pharmacist, at this stage, has an optimized formulation whose performance capability has been adequately proved on a laboratory scale. The research team now proceeds to the production laboratory to conduct the operations necessary to scale up the optimum formulation. The team carefully evaluates the magnitudes of the processing parameters so as to reach the level desired in a normal production batch (Rx). Usually, three or four production batches are compressed for later use in process validation. These production batches are called demonstration batches. A demonstration batch is a scaled-up production batch with a weight of 1200 to 1300 kg. The size of the demonstration batch depends on the capacity of the blender (GEMCO or Double Cone). The tableting process proceeds through an expanded version of blending, granulating, drying (electric or fluid bed), grinding (comminuting), lubricating, and finally compressing. The compression process is generally accomplished by either a Manesty Beta Press, Stokes, Nova, or Fette compressor.

### A. The Paddle Sample

During compression, "paddle" samples consisting of 20 to 25 tablets are collected from each side of the compressor every 20 minutes. A paddle sample is a sample of tablets removed from the top of the collecting pan when the compression process is in progress. It is called a paddle sample because the tablets are withdrawn by the operator with the help of a plastic paddle, which is usually the size and shape of a table tennis paddle.

In an 8-hour production operation, a paddle sample should collect approximately 1000 tablets from both ports of the compressor. Two sample bottles each containing approximately 300 tablets are sent to the analytical testing laboratory for the extended release tests, as defined below.

In production jargon, a paddle sample refers to a collection of uncoated tablets, and a "batch" sample refers to a collection of coated tablets. (In this text, the word paddle sample will be used to denote both kinds of samples.) All the paddle samples collected during the course of a compression process are thoroughly mixed, and a random sample (100 tablets) of the composite paddle sample is subjected to the release tests: (1) dissolution, (2) disintegration, (3) hardness (tablet breaking strength), (4) weight uniformity, (5) thickness uniformity, (6) content uniformity, and (7) composite "assay" for the active ingredient based on 10 packets, each packet containing the grindings of 10 (powder form) tablets, and other exterior assessments such as elegance and defects.

The magnitudes of the observed responses are confirmed against (1) the target values set during the formulation phase and (2) USP/NF compendial requirements such as (a) content or dose uniformity—values to be within 85–115% of the proposed label claim (the amount of active drug in the particular formulation), and (b) assay values to be within 95–105% of the label claim. In addition, the observed responses are used to propose specification limits expressed in terms of Q values (e.g., 80% in 30 minutes for dissolution, 8 kg for hardness) which will be incorporated into the NDA.

To study the effect of time, hardware, operator, etc. on the production process, quality control charts would have been established for weight variation, hardness, thickness, disintegration, and elegance (defects), with two-sigma limits and three-sigma limits properly marked on the charts. Every 30 minutes during the production of a demonstration batch these five tests are conducted. The results are plotted on the quality control chart. If the value falls within the 2-sigma limits, the process is carried on. If the value falls outside the 2-sigma limits, the machine is stopped and an adjustment is made. If the value falls outside the 3-sigma limits, all the tablets produced up to this point (or between this and the previous checking point) are isolated. Such talbets do not become part of the demonstration batch.

To meet compendial requirements, one may develop inspection sampling schemes using either single or double acceptance sampling plans (USP, 1975) and carry out dissolution, disintegration, weight variation, content uniformity, and composite assay tests.

## B. Process Validation of Scale-Up Batch

The amount of each excipient ingredient in a scale-up batch is a constant multiple of the amount of the same ingredient in the optimized formulation. For instance, if the amount of magnesium stearate is 1.0 mg per 100 mg of active drug in the optimized tablet, then 1% of the total weight of the scale-up demonstration batch (say 1000 kg) will be the weight of magnesium stearate.

However, since the hardwares and the amounts of the ingredients are different in scale, the performance characteristics of the optimized formulation could be different from that of the scale-up version. Consider that we have three types of formulations: (1) production formulation (PRD), (2) optimized formulation (OPT), and (3) an old formulation (OLD) developed on the subjective judgment of the formulator. Consider a random production batch, recently manufactured. It is of interest to determine whether the performance of this batch is closer to (1), (2), or (3). This would be an acid test of the performance of the production formulation. Based on (say) three PCA-selected response variables, we have the following configuration of the data:

| Batches, times, or bottles | $F_{opt}$ | | | $F_{old}$ | | | $F_{prod}$ | | |
|---|---|---|---|---|---|---|---|---|---|
| | $P_1$ | $P_2$ | $P_3$ | $P_1$ | $P_2$ | $P_3$ | $P_1$ | $P_2$ | $P_3$ |
| 1 | . | . | . | . | . | . | . | . | . |
| 2 | . | . | . | . | . | . | . | . | . |
| | . | . | . | . | . | . | . | . | . |
| | . | . | . | . | . | . | . | . | . |
| | . | . | . | . | . | . | . | . | . |
| | . | . | . | . | . | . | . | . | . |
| n | . | . | . | . | . | . | . | . | . |

$F_{opt}$, Optimized formulation group; $F_{old}$, subjective formulation group; $F_{prod}$, production formulation group. $P_1$, $P_2$, and $P_3$ are the three PCA-selected parameters.

Based on these three groups and three PCA selected parameters, a boundary function analysis, which is an extended version of a discriminant analysis incorporating more than two groups, is conducted.

Two preliminary tests are required. The first addresses the null hypothesis of homogeneity of group variance-covariance matrices:

$$H_0: \Sigma_{OPT} = \Sigma_{OLD} = \Sigma_{PROD}$$

The test of $H_0$ is accomplished by the multivariate extension of Levene's (1960) test. The second addresses the null hypothesis of equality of the group mean vectors:

$$H_0: M_{OPT} = M_{OLD} = M_{PROD}$$

The test is a MANOVA test for comparing the means and can be accomplished by Wilk's $\Lambda$ test (Anderson, 1958):

$$\Lambda = \frac{|E|}{|E + H|}$$

or Roy's largest $\Lambda$-test (Anderson, 1958):

$$\theta_1 = \frac{\Lambda_{max}}{1 + \Lambda_{max}}$$

where $\Lambda_{max}$ is the largest of all $\Lambda$'s based on the solution of the following determinantal equation:

$$|HE^{-1} - \Lambda I| = 0$$

and where E is the variance-covariance matrix of the within group source of variation and H is the variance-covariance matrix of the between-group source of variation.

If the above tests show significant group mean differences and also homogeneity of the three variance-covariance matrices, the following steps are carried out to accomplish the boundary function analysis.

1. Construct the following two matrices:

$$\Sigma_A = \begin{bmatrix} A_{11} & A_{12} & A_{13} \\ A_{21} & A_{22} & A_{23} \\ A_{31} & A_{32} & A_{33} \end{bmatrix}$$

$$\Sigma_B = \begin{bmatrix} B_{11} & B_{12} & B_{13} \\ B_{21} & B_{22} & B_{23} \\ B_{31} & B_{32} & B_{33} \end{bmatrix}$$

where $A_{11}$, $A_{22}$, and $A_{33}$ are the between-group sum of squares, $A_{12}$, $A_{13}$, and $A_{23}$ are the between-group sum of products, $B_{11}$, $B_{22}$, and $B_{33}$ are the within-group sum of squares, and $B_{12}$, $B_{13}$, and $B_{23}$ are the within-group sum of products associated with the three parameters.

2. Find $\Sigma_B^{-1}$.
3. Compute $\Sigma_B^{-1}\Sigma_A$ (note the order of the product).
4. Solve for the determinantal equation, $|\Sigma_B^{-1}\Sigma_A - \Lambda I| = 0$.
5. Find $\Lambda_{max}$ among the eigenvalues $\Lambda_1$, $\Lambda_2$, and $\Lambda_3$ of $\Sigma_B^{-1}\Sigma_A$.
6. (a). Find the eigenvector of $\Sigma_B^{-1}\Sigma_A$ based on $\Lambda_{max}$, $\upsilon_{11}$, $\upsilon_{12}$, and $\upsilon_{13}$.

(b). Construct the general boundary function for the three groups as follows:

$$\theta = \upsilon_{11}T_1^* + \upsilon_{12}T_2^* + \upsilon_{13}T_3^*$$

where $T_1^*$, $T_2^*$, and $T_3^*$ are the averages of the three response variables, $P_1$, $P_2$, and $P_3$.

7. Construct the individual boundary function for each group:

$$\theta_{OPT} = \upsilon_{11}T^*_{OPT(1)} + \upsilon_{12}T^*_{OPT(2)} + \upsilon_{13}T^*_{OPT(3)}$$

$$\theta_{OLD} = \upsilon_{11}T^*_{OLD(1)} + \upsilon_{12}T^*_{OLD(2)} + \upsilon_{13}T^*_{OLD(3)}$$

$$\theta_{PRD} = \upsilon_{11}T^*_{PRD(1)} + \upsilon_{12}T^*_{PRD(2)} + \upsilon_{13}T^*_{PRD(3)}$$

where $\theta_{OPT}$, $\theta_{OLD}$, and $\theta_{PRD}$ are the respective boundary values.

8. Consider that $\theta_{OPT} > \theta_{OLD} > \theta_{PRD}$.

The construction of the boundary points proceeds as follows: The boundary point for OPT and OLD is computed as

$$\delta_{OPT,OLD} = 1/2(\theta^*_{OPT} + \theta^*_{OLD})$$

The boundary point for OLD and PRD is computed as

$$\delta_{OLD,PRD} = 1/2(\theta^*_{OLD} + \theta^*_{PRD})$$

where $\theta^*_{OPT}$, $\theta^*_{OLD}$, and $\theta^*_{PRD}$ are the respective estimated boundary values of the three groups.

Consider a random batch (R). Calculate $T^*_{r1}$, $T^*_{r2}$, $T^*_{r3}$ and then construct

$$\theta_r = \upsilon_{11}T^*_{r1} + \upsilon_{12}T^*_{r2} + \upsilon_{13}T^*_{r3}$$

If $\theta_r < \delta_{OPT,OLD}$, then the group membership of the random batch is the OPT group. If $\theta_r > \delta_{OPT,OLD}$ but $\theta_r < \delta_{OPT,OLD}$, then the group membership of the random batch is the OLD group. If $\theta_r > \theta_{OLD,PRD}$, then the group membership of the random batch is the PRD group. With this analysis, one will be able to locate the group membership explicitly.

## C. Numerical Considerations

The numerical example involves three parameters, dissolution, disintegration, and hardness, and two groups, the PRD group with the scaled-up optimized formula (n = 25) and the OLD group with the subjective formula (n = 27). The numerical structure of the $\Sigma_B^{-1}\Sigma_A$ matrix is as follows:

$$\Sigma_B^{-1}\Sigma_A = \begin{bmatrix} 88897.4 & 9135.09 & 421.92 \\ 9135.09 & 6090.01 & 148.24 \\ 421.92 & 148.24 & 30.95 \end{bmatrix}^{-1} \begin{bmatrix} 35334.4 & 9829.7 & 980.6 \\ 9829.7 & 2734.5 & 272.8 \\ 980.6 & 272.8 & 27.2 \end{bmatrix}$$

The eigenvalues of $\Sigma_B^{-1}\Sigma_A$ matrix, based on the determinantal equation $|\Sigma_B^{-1}\Sigma_A - \Lambda I| = 0$, are $\Lambda_1 = 1.401898$, $\Lambda_2 = 0.0$, and $\Lambda_3 = 0.0$. The eigenvector associated with $\Lambda_{max} = 1.401898$ is

$$\upsilon^1 = (\upsilon_1 = 0.18342875,\ \upsilon_2 = 0.000743539,\ \text{and}\ \upsilon_3 = 0.001671487$$

The general boundary function has the expression

$$\theta = 18.342875T_1 + 0.0743539T_2 + 0.1671487T_3$$

The boundary value for the PRD group is 148.262 and the boundary value for the OLD group is 115.422. The boundary point between the two groups is 131.84 (note that the $\upsilon_i$'s have been multiplied by 100). Since the components of the eigenvectors are proportional weights, multiplying by a constant does not change the results. The individual $\theta$ values are provided in the following for each group:

*PRD group*:

| Observation: | 1* | 2 | 3 | 4* | 5 | 6 | 7 | 8 | |
|---|---|---|---|---|---|---|---|---|---|
| θ value: | 117.5 | 133.8 | 167.3 | 111.8 | 134.8 | 151.8 | 151.8 | 151.7 | |
| Observation: | 9 | 10* | 11* | 12 | 13 | 14 | 15 | 16* | |
| θ value: | 135.8 | 128.2 | 123.1 | 134.9 | 160.4 | 169.7 | 159.6 | 119.7 | |
| Observation: | 17 | 18* | 19 | 20 | 21 | 22 | 23 | 24 | 25 |
| θ value: | 156.3 | 150.0 | 166.1 | 173.7 | 166.5 | 142.6 | 169.0 | 166.0 | 164.5 |

*OLD group*:

| Observation: | 26 | 27 | 28 | 29 | 30 | 31 | 32 | 33 | 34 |
|---|---|---|---|---|---|---|---|---|---|
| θ value: | 122.4 | 108.9 | 113.2 | 110.0 | 120.5 | 124.0 | 111.5 | 122.9 | 108.0 |
| Observation: | 35 | 36 | 37 | 38 | 39 | 40 | 41 | 42 | 43 |
| θ value: | 104.8 | 110.8 | 109.2 | 110.7 | 141.4 | 116.9 | 120.7 | 129.8 | 114.2 |
| Observation: | 44 | 45 | 46 | 47 | 48 | 49 | 50 | 51 | 52 |
| θ value: | 109.9 | 113.7 | 118.1 | 117.6 | 110.8 | 113.2 | 111.6 | 110.4 | 110.8 |

The θ values derived from the general boundary function analysis show that the batches of the PRD group numbered 1, 4, 10, 11, and 16 do not confirm to the characteristics of the PRD group. The causes of the non-confirmation should be investigated. The problem might have originated from various sources such as scale-up process, hardware, blending uniformity, and other process problems. It should be noted that batch 39 of the OLD group does not conform to its group characteristics. It is observed that batches 1, 4, 10, 11, and 16 are behaving more like the OLD group than the PRD group, and batch 39 is behaving more like the PRD group than the OLD group. This numerical example has been presented here only as an illustration of the things that can be accomplished by using this statistical procedure.

## V. CHEMICAL ASSAY METHOD VALIDATION

### A. General Considerations

One of the preformulation activities of the pharmaceutical chemist is to develop a specific chemical assay method for the active ingredient in the drug delivery system. Occasionally, it is necessary to develop assay methods for degradant products as well. After the development of a specific analytical assay method, appropriate validation experiments must be conducted to demonstrate that the method is, indeed, valid. This section is devoted to the design and analysis of assay method validation experiments (Bohidar, 1983).

### B. Validation Parameters

Parameters reflecting validation aspects of a chemical assay method are accuracy, precision, bias, sensitivity, specificity, interference, repeatability, reproducibility, and reliability. General interpretations and operational definitions are presented (Bohidar, 1983) for these parameters rather than precise "mathematical" definitions.

1. *Accuracy*. Accuracy is the closeness of the test results obtained by the assay method to the true value. Operationally, if there is no significant difference ($\alpha = 0.05$) between the average value of the recovered amounts and the actual amount added, the assay method is considered to be accurate.

2. *Precision*. Precision is a measure of the degree of closeness of the recovered amounts. This is generally estimated by standard deviation (S); standard error ($S/\sqrt{n}$, where n is the number of samples); or coefficient of variation or relative standard deviation ($S/\bar{Y}$, the ratio of the standard deviation and the mean).

3. *Bias*. Bias is precisely given by

$$B = \bar{R} - P$$

where B is the bias, $\bar{R}$ the average of the recovered amounts, and P the actual amount present. To test the hypothesis of no bias (bias = 0) statistically, one constructs the 95% confidence limits on B as follows:

$$B \pm t_{(n-1)(\alpha=0.05)} \sqrt{S^2/n}$$

where B is the observed difference between the average of the recovery values and the known reference value, t is Student's t value for (n−1) degrees of freedom with $\alpha = 0.05$, $S^2$ is the square of the standard deviation (S) [$S = [\Sigma(Y - \bar{Y})^2/(n - 1)]^{1/2}$, $\bar{Y}$ = average value], and $S/\sqrt{n}$ is the standard error of the assay with a sample size of n. Limits that do not include the value of zero indicate that the method is biased. This demonstrates the interconnection between bias and precision in the definition of accuracy.

4. *Sensitivity*. Sensitivity is the accuracy of the method for a wide range of potencies encompassing low as well as high potencies. Figure 3 depicts the characteristic sigmoid sensitivity curve, which clearly defines the three regions of the three degrees of sensitivity: low-sensitivity regions occupying the two extreme ends of the sigmoid curve and a high-sensitivity region between the two extreme regions. Generally, it is the linear portion of the sigmoid curve which is highly sensitive to modest changes in potency. The statistical test of linearity is one of the cardinal tests of validity of the assay method.

5. *Specificity*. The assay method must be specific to the particular chemical entity for which the method has been developed. The method should be absolutely free from interferences by substances that are known to be present in the product as well as the appropriate process-related substances, excipients, and known degradation products. There is no specific statistical methodology to detect specificity. However, the quantitative expressions of precision and accuracy should provide adequate criteria for detecting the nonspecific behavior of the method.

6. *Interference*. As mentioned before, excipient ingredient and degradation products may cause interference. The level of these possible interferences should be tested at the maximum levels that can occur in the product. If the integrity of the analyte peak is in question, the peak should be examined by an alternate analytical procedure.

7. *Repeatability* (Mandel, 1984). This terminology has been well defined and documented in the context of testing material, such as the stress

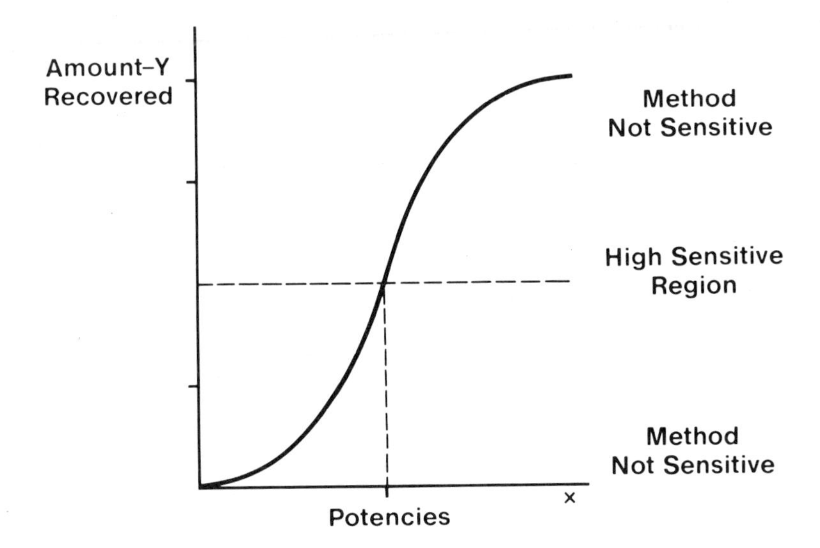


Fig. 3 Characterization of degree of method sensitivity.

test of natural rubber. It pertains to the variation between two repeated observations measured within the same laboratory. The concept of repeatability may be extended to pharmaceutical situations. The variation between two repeated assay values within the same lot or within the same day of test should constitute a good measure of repeatability. An assay method meets the repeatability criterion if 95% of the absolute difference between any two values does not exceed the value of $(1.96\ \sqrt{2}S)$.

8. *Reproducibility* (Mandel, 1984). This terminology has also been well defined and documented in the context of testing material. It pertains to the variation between two test results obtained in different laboratories. The concept of reproducibility may also be extended to pharmaceutical situations. The variation between two assay values representing two different lots or two different days should constitute a good measure of reproducibility. An assay method meets the reproducibility criterion if 95% of the differences in question, taken in absolute value, do not exceed the value of

$$1.96\ \sqrt{2(\sigma_D^2 + \sigma^2/n)}$$

where n is the number of observations in each lot/day, $\sigma_D^2$ the component of variance due to between days or lots, and $\sigma^2$ the component of variance due to within days or lots. These quantities are estimated by the analysis of variance procedure.

9. *Reliability*. Subtle differences in the factors governing the ambient conditions within the laboraotry may bring about significant changes in the assay values. These factors may include the source, age, concentration of the reagents, rate of heating, thermometer errors, and so on. To detect the influential sources of variation, a two-level factorial experiment is conducted and the results are statistically analyzed to obtain some indication of the sources causing variation. If the assay procedure is "rugged" (sometimes this test is called a test of ruggedness), it will be immune to variation of the ambient conditions and the assay results will not be

significantly altered. If the results are altered, it will be necessary to determine the limits of these conditions in which the alterations are at a minimum. Another way to solve the problem is to conduct an optimization study (much as one does in the formulation study) to determine the levels of the factors which minimize significant changes in the response. An experimental design for the test of ruggedness will be provided in Section E below.

## C. General Planning and Experimental Design

The strategy for conducting an assay validation experiment is to consider the following groups (samples) for test (Bohidar, 1983): (1) excipient + 50% active drug, (2) excipient + 80% active drug, (3) excipient + 100% active drug, (4) excipient + 120% active drug, (5) excipient + 150% active drug, (6) excipient (placebo alone), and (7) active drug.

The first five groups provide information on the general validity of the assay method and the sensitivity of the procedure. Groups 6 and 7 provide information on interference and specificity of the method.

### 1. Methods of Sampling (Bohidar, 1983)

There are four general methods for selecting the samples. These are the (1) homogeneous tablet grind method, (2) compressed tablets from potency batch method, (3) spiked-placebo recovery method, and (4) standard addition method. The homogeneous tablet grind method consists of grinding 10 to 20 tablets into a homogeneous powder and then taking sample weights of the homogeneous grind equivalent to 50% of the theory (active drug), 80% of the theory (active drug), 100% of the theory (active drug), 150% of the theory (active drug), placebo alone, and active alone. The compressed tablets from potency batch method consists of compression five batches of tablet conforming to 50, 80, 100, 120, and 150% of the theory, respectively. The batch size should be large enough to yield 200 to 300 tablets. This is the ideal method of sampling. The spiked-placebo recovery method requires pure active ingredients to be added to a mixture of product excipients. This composite mixture is assayed and the results obtained are compared with the expected amount. The standard addition method requires a sample to be assayed (= amount S); then add an amount of active ingredient (= amount a) and assay the resulting sample (amount S + a). The difference between the two assay results [(S + a) − S = a] is compared to the expected amount. In routine assay validation experiments, the spiked-placebo recovery method is generally used.

### 2. Experimental Design

In this section, appropriate and routinely used experimental designs for the validation studies are described. The purpose of the design is to provide an unbiased estimate of the validation parameter and its variance. In the description of the designs, the detailed mechanics of randomization and replication are not presented explicitly. An outline of the structure of the analysis of variance arising from these designs includes the sources of variation and the partition of the total degrees of freedom. The appropriate tests of validity are also incorporated. The explicit expression of the ANOVA model is not provided, since the structure of the model is obvious from the ANOVA table.

For the purpose of description, let the designations A, B, C, D, and E pertain, respectively, to the five potency groups E + 50%, E + 80%, E + 100%, E + 120%, and E + 150% (where E means excipient). The following experimental designs are proposed.

1. Randomized block design (RBD)

(a) Three-day test with three repeats: the following schematic diagram refers to a 3-day test with three repeated assay values. The test sequence has been randomized.

Protocol

| | Test sequence | | | | |
|---|---|---|---|---|---|
| Days[a] | 1 | 2 | 3 | 4 | 5 |
| I | C ≡ | A ≡ | D ≡ | B ≡ | E ≡ |
| II | B ≡ | C ≡ | A ≡ | E ≡ | D ≡ |
| III | D ≡ | E ≡ | B ≡ | C ≡ | A ≡ |

[a]The days may not be consecutive.

The analysis of variance reflecting sources of variation appears below. Days are considered as a fixed effect since many logistical contraints preclude them being selected at random. However, if days were considered as random, then the appropriate error term for assessing the significance of potency would be the days by potency interaction term.

ANOVA

| Source of variation | DF | |
|---|---|---|
| Days (D) | 2 | |
| | | Test of linearity |
| | | Test of curvature |
| Potency (P) | 4 | Test of slope = zero |
| | | Test of slope = one |
| | | Test of intercept = zero |
| D × P | 8 | |
| Within days | 30 | |
| Total | 44 | |

The tests referred to in the upper portion of the ANOVA table above are the tests of validity. The test of linearity should be highly significant. All the other tests given above should not be significant. The day × potency interaction should not be significant.

(b) Three-day test with two repeats: the protocol and the test sequence are similar to that given above but there are only two repeated assays per group. The structure of the ANOVA is as follows:

ANOVA

| Source of variation | DF | |
|---|---|---|
| Days (D) | 2 | |
| Potency (P) | 4 | Test of linearity<br>Test of curvature<br>Test of slope = zero<br>Test of slope = one<br>Test of intercept = zero |
| D × P[a] | 8 | |
| Within days | 15 | |
| Total | 29 | |

[a]The day × potency interaction should not be significant.

2. Latin square design. This design is usually complex and is never used for routine analysis. The design is used only for special studies. Consider a situation in which it is intended to do a 5-day test involving different operators and/or different lots. A 5 × 5 Latin square can be utilized with the following protocol.

(a) 5-Day test with duplicates:

| | Operator/lots/labs | | | | |
|---|---|---|---|---|---|
| Days | $L_1$ | $L_2$ | $L_3$ | $L_4$ | $L_5$ |
| I | B = | D = | A = | E = | C = |
| II | E = | C = | B = | D = | A = |
| III | C = | B = | D = | A = | E = |
| IV | A = | E = | C = | B = | D = |
| V | D = | A = | E = | C = | B = |

ANOVA

| Source of variation | DF |
|---|---|
| Days | 4 |
| Lots | 4 |

ANOVA (continued)

| Source of variation | DF | |
|---|---|---|
| Potency | 4 | Test of linearity<br>Test of curvature<br>Test of slope = zero<br>Test of slope = one |
| Interaction[a] | 12 | |
| Within cells | 25 | |
| Total | 49 | |

[a]The interaction should not be significant.

(b) A Latin square design with no duplicates may be used. It provides only 12 degrees of freedom for the error.

(c) If the protocol calls for only three levels of potency, two 3 × 3 Latin squares could be used and a combined analysis could be conducted. These 3 × 3 Latin squares may be used with or without duplicates. This provides 18 degrees of freedom for the error with duplicates.

## D. General Statistical Analysis of Validation Experiments

The general statistical analysis of validation experiments is presented here in two parts: single-system validation and dual- or multiple-system validation.

### 1. Single-System Validation

The single-system validation primarily pertains to a single assay method. The experimenter is interested in validating this method without reference to any other methods. The statistical procedure will be illustrated with a numerical example. Data from a simple "amount-present, amount-recovered" (P-R) type experiment are used. This will provide an appreciation of the validation analysis without the intricacies of the complex designs considered in Section C above.

Gravimetric Determination of Calcium in the Presence of Magnesium

| No. | CaO present (mg) X | CaO found by new method (mg) Y |
|---|---|---|
| 1 | 20.0 | 19.8 |
| 2 | 22.5 | 22.8 |

Gravimetric Determination of Calcium in the Presence of Magnesium (continued)

| No. | CaO present (mg) X | CaO found by new method (mg) Y |
|---|---|---|
| 3 | 25.0 | 24.5 |
| 4 | 28.5 | 27.3 |
| 5 | 31.0 | 31.0 |
| 6 | 33.5 | 35.0 |
| 7 | 35.5 | 35.1 |
| 8 | 37.0 | 37.1 |
| 9 | 38.0 | 38.5 |
| 10 | 40.0 | 39.0 |

These series of observations are taken by a laboratory on material whose contents are known accurately by design. There are 10 samples. The random variable Y (dependent variable) is considered as the laboratory measurement of the sample; the true amount in the sample is regarded as the X variable (independent variable). The analysis of a P-R experiment for validating a new method falls in the realm of regression analysis. One of the assumptions of the regression analysis is that the X's are fixed and without error. In practice, the value of X is never known exactly, and so it is sufficient to have the error in X small compared to the variability associated with Y.

A brief description of the assay method is as follows. (1) The Ca is precipitated with oxalic acid. (2) The calcium oxalate, $CaC_2O_4$, so formed is filtered and ignited. (3) Since oxalic acid is organic, it burns on ignition, leaving calcium oxide, CaO. This residue is weighed and the result calculated as calcium carbonate, $CaCO_3$. (4) These results are expressed as mg $CaCO_3$ per liter and reported. The following statistical tests will be performed for the validation of the assay method:

1. Least-squares analysis of the data and estimation of the regression equation

   $$\hat{Y} = A + BX$$

2. Based on the variability ($S^2_{Y.X}$) around the regression line, one would be able to conduct the following tests of hypothesis: (a) $H_0$: $B^* = 0$ (slope = 0), (b) $H_0$: $B^* = 1$ (slope = 1), (c) $H_0$: $A^* = 0$ (intercept = 0), where $A^*$ and $B^*$ denote the population intercept and slope, respectively.

The scatter diagram and the least-squares regression line are presented in Figure 4. The logic behind the test in 2(a) is that no regression analysis is valid unless the test in 2(a) is highly significant. The

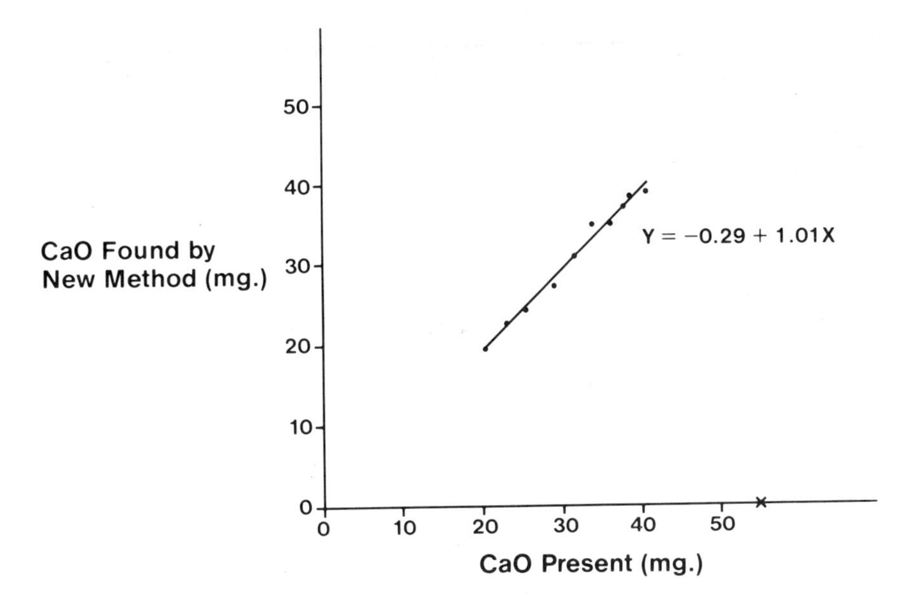


Fig. 4 Least squares regression line and scatter diagram.

logic behind the test in 2(b) is that, if the assay is performing ideally, then any change in "CaO present" should reflect a corresponding amount of change in "CaO found." The logic behind the test in 2(c) is that, if the assay is performing ideally, we should get no CaO in the assay when we have not put any CaO in the sample! These three logical statements are statistically translated to the three tests of hypothesis noted above. In the present study we find the following: (1) Regression equation: $Y = -0.2928 + 1.0065X$. (2) Test of hypothesis in 2(a) shows that the slope is highly significant ($P < 0.001$). This implies that our regression analysis is valid. (3) Test of hypothesis in 2(b) shows that the true slope is not significantly different from one ($\alpha = 0.05$). (4) Test of hypothesis in 2(c) indicates that the true intercept is not significantly different from zero ($\alpha = 0.05$). This implies that the line is passing through the origin. (5) The analysis also shows that the coefficient of determination ($R^2$) is 98.2%. This implies that almost all the variation in the assay response amount found is explained by a linear relationship between response and amount present. The statistical formulas are presented in Section V,D,4.

*Joint test of hypothesis*: The tests given in 2(b) and 2(c) are currently practiced in pharmaceutical laboratories. However, because we make a judgment about the performance of the assay based on the outcome of both the tests and because the two test statistics are correlated (parameters are correlated), according to the statistical principle, we must do a simultaneous test of hypothesis in the following manner:

$$H_0: \begin{bmatrix} B^* = 1 \\ A^* = 0 \end{bmatrix} \qquad H_A: \begin{bmatrix} B^* \neq 1 \\ A^* \neq 0 \end{bmatrix}$$

The probability of type I error is inflated if we do two independent tests and use the results to make a simultaneous inference. The explicit expression of the F-statistics is given in Section V,D,4. The observed F-ratio is 0.074. The result implies that, in fact, the assay is valid ($\alpha$ = 0.05). A confidence ellipse for the simultaneous test is presented in Figure 5, and the regions of "accuracy" and "bias" are delineated.

### 2. Dual-System Validation/Multiple-System Validation

The primary motivation for the development of dual/multiple-system validation statistical techniques is that the experimenter is often interested in comparing (1) two assay methods such as the standard and an alternative

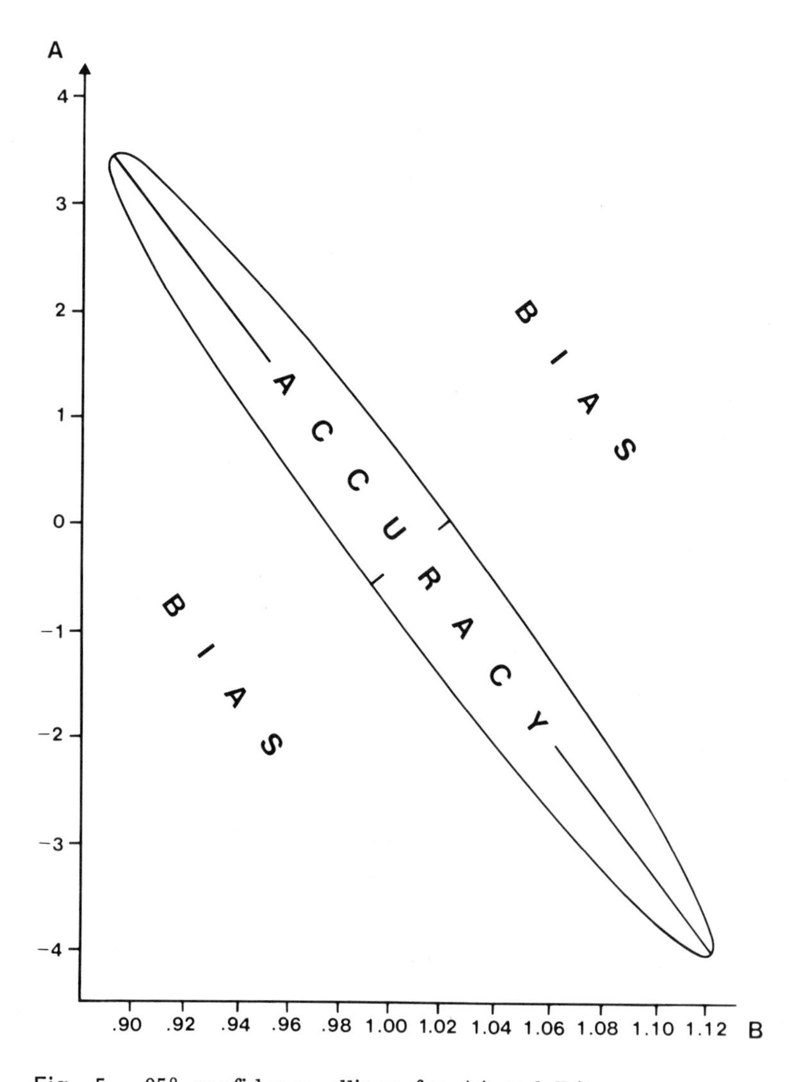


Fig. 5 95% confidence ellipse for A* and B*.

method developed in his or her laboratory, (2) the same method tested in two different laboratories, (3) the original protocol and a modified protocol, such as a change in the mixture proportion of the solvent in high-performance chromatography (HPLC), or (4) the same assay procedure conducted by two different operators.

In a multiple-system validation study, each system must meet the assay validation criteria individually. The individual system which fails to meet the validation criteria cannot be incorporated for further consideration in the multiple-system validation study.

The original CaO study is extended to incorporate two other systems for the purpose of intercomparison among the three systems, $R_1$, $R_2$, and $R_3$. The following table presents the statistical results of the individual assay validation tests,

Individual Assay Validity

| | Test hypothesis | Results | | |
|---|---|---|---|---|
| | | $R_1$ | $R_2$ | $R_3$ |
| Slope: | $B^* = 0$ | NS[a] | NS | NS |
| Intercept: | $A^* = 0$ | NS | NS | NS |
| Slope: | $B^* = 1.0$ | NS | NS | NS |
| Both: | $A^* = 0$, $B^* = 1$ (Joint hypotheses) | NS | NS | NS |

[a]NS, Not significant at $\alpha = 0.05$.

The statistical results given above indicate that, in fact, each of the systems is performing according to the established validation criteria. The next step is to compare the candidate systems, based on the following criteria: (1) test of parallelism among the slopes of the candidate systems (test of equality of slopes), (2) test of equi-origin (test of equality of Y intercepts), (3) simultaneous test of parallelism and equi-origin involving all three systems, and (4) test of intrinsic variability among the systems.

The first test, test of parallelism, involves the following comparisons: (1) BI versus BII, (2) BI versus BIII, and (3) BII versus BIII Duncan's new multiple-range test should be used so that an overall error rate at 5% for all comparisons can be maintained. The second test, test of equality of intercepts, involves the following comparisons: (1) AI versus AII, (2) AI versus AIII, and (3) AII versus AIII. Duncan's new multiple-range test should be used so that an overall error rate at 5% for all comparisons can be maintained.

The third test, simultaneous test of parallelism and equi-origin, (test of concurrence) (Bohidar, 1983) involves a unique extension of the Hotelling $T^2$ to test regression parameters (the usual Hotelling $T^2$ is used for testing two groups based on several correlated variables). For a two-system comparison, the hypothesis to be tested is

$$H_0: \begin{bmatrix} A^*_1 & = & A^*_{111} \\ B^*_1 & & B^*_{111} \end{bmatrix} \quad \text{or} \quad \begin{bmatrix} A^*_1 - A^*_{111} \\ B^*_1 - B^*_{111} \end{bmatrix} = \begin{bmatrix} 0.0 \\ 0.0 \end{bmatrix}$$

Since the two regression parameters are correlated in each system, a multivariate simultaneous test is a better test, maintaining the type I error intact. The reason for considering only $R_1$ and $R_3$ systems is provided in the following.

The following statistical results were obtained.

Statistical Results[a]

| Tests | $R_1$ versus $R_2$ | $R_1$ versus $R_3$ | $R_2$ versus $R_3$ |
|---|---|---|---|
| Intercept | NS | NS | NS |
| Slope | * | NS | * |
| Concurrence Hotelling $T^2_{REG}$ | * | NS | * |

[a]NS, Not significantly different (at P - 0.05). *, Significantly different (P < 0.05).

In this set of tests, it is found that the slope of $R_2$ is not parallel to the slope of $R_1$ or that of $R_3$. However, $R_1$ and $R_3$ passed all the tests of validity considered.

3. Test of Intrinsic Variabilities (Bohidar, 1983)

This involves the comparison of system variabilities by Levene's test. It is essentially a comparison among the average absolute mean deviations (AAMD) of the systems. Levene's (1960) test is a robust test; that is, it does not depend on the cardinal assumption of normality. Levene's test (Bavitz et al., 1973; Bohidar, 1983) procedure has been extended to situations arising in regression analysis. The purpose here is to test for equality of the system variabilities. The necessary steps are as follows.

1. An adjusted-Y value ($Y_{ADJ}$) is calculated for each data point associated with the systems by using the formula

   $$Y_{i,ADJ} = Y_i - B_c(X_i - \bar{\bar{X}})$$

   where $B_c$ is the common slope for all parallel systems, $\bar{\bar{X}}$ is the mean of all the X values associated with all the systems, and $X_i$, $Y_i$ are the original P-R data.
2. The procedure in part 1 reduces each system to a single column of numbers as follows:

| System I | System II | System III |
|---|---|---|
| $Y_{1,ADJ}$ | $Y_{1,ADJ}$ | $Y_{1,ADJ}$ |
| $Y_{2,ADJ}$ | $Y_{2,ADJ}$ | $Y_{2,ADJ}$ |
| $Y_{n,ADJ}$ | $Y_{n,ADJ}$ | $Y_{n,ADJ}$ |

Now, conduct an ordinary Levene's test based on the above set of data. The multiple comparisons among the system AAMDs can be accomplished by utilizing Duncan's new multiple-range test procedure. The system with the smallest AAMD will be considered the most desirable system. This information will be an adjunct to the other results given in Section V,D,2.

### 4. Statistical Addendum

This section contains all the statistical formulas used in the development of the validation tests.

1. Let $\Sigma(X - \bar{X})^2 = [X^2]$
   Let $\Sigma(Y - \bar{Y})^2 - [Y^2]$
   Let $\Sigma(X - \bar{X})(Y - Y) = [XY]$
2. $B = [XY][X^2]^{-1}$
3. $A = \bar{Y} - B\bar{X}$, $\bar{Y} = (\Sigma Y)(n^{-1})$, $\bar{X} = (\Sigma X)(n^{-1})$
4. $Y = A + BX$
5. t-test: $H_0$: $B - 0$; $t = \{B^2[X^2]\}^{1/2} S^{-1}_{Y \cdot X}$
6. $S^2_{Y \cdot X} = \{[Y^2] - B^2[X^2]\}(n - 2)^{-1}$
7. t-test: $H_0$: $B = 1$; $t = \{(B - 1)^2[X^2]\}^{1/2} S^{-1}_{Y \cdot X}$
8. t-test: $H_0$: $A = 0$; $t = (A)[S^2_{Y \cdot X}(n^{-1} + \bar{X}^2[X^2]^{-1}]^{-1/2}$
9. $R^2 = 100 \times B^2[X^2][Y^2]^{-1}$
10. Simultaneous test: $F = [nA^2 + 2n\bar{X}A(B - 1) + (B - 1)^2 \Sigma X^2](2S^2_{Y \cdot X})^{-1}$
11. Hotelling $T^2_{REG} = [t_1^2(1 - r^2)^{-1} - 2r(1 - r^2)^{-1} t_1 t_2 + t_2^2(1 - r^2)^{-1}]$

where

$t_1$ = t-test statistic for testing equality of two slopes

$t_2$ = t-test statistic for testing equality of two intercepts

$r$ = average correlation between A and B involving both systems

$$t_1 = (B_1 - B_2)[S^2_{P \cdot Y \cdot X}([X^2]_1^{-1} + [X^2]_2^{-1})]^{-1/2}$$

$$t_2 = (A_1 - A_2)[S^2_{P \cdot Y \cdot X}(n_1^{-1} + n_2^{-1} + \bar{X}_1^2[X^2]_1^{-1} + \bar{X}_2[X^2]_2^{-1}]^{-1/2}$$

$$r = -\bar{X}S^2_{P \cdot Y \cdot X}[X^2]^{-1}$$

$$S^2_{P \cdot Y \cdot X} = \text{pooled estimated residual variances}$$

Note: Since this Hotelling $T^2$ statistic is associated with regression coefficients, it would be appropriate to denote this as Hotelling $T^2_{REG}$.

### E. Validation of Compendial Assays

Here, consideration is extended to the specifics of submissions to the compendia for new or revised analytical methods, in general terms. Validation of an assay method is the process by which it is established, by laboratory studies, that the performance characteristics of the method meet the requirements of the intended analytical goal. Based on their respective validation requirements, assay methods are divided into three categories, which are defined as follows:

Category I: analytical methods for quantitation of major components of bulk drug substances or active ingredients in finished pharmaceutical products
Category II: analytical methods for determination of impurities in bulk drug substances or degradates in finished pharmaceutical products, involving quantitative assays as well as limit tests
Category III: analytical methods for determination of performance characteristics such as dissolution and content uniformity

For each assay category, different analytical information is needed (Jensen et al., 1986). The usual requirements are listed in the following table.

Experimental Data Needed for Assay Validation

| Analytical performance parameter | Assay category I | Assay category II | | Assay category III |
|---|---|---|---|---|
| | | Quantitative | Limit tests | |
| Precision | Yes | Yes | No | Yes |
| Accuracy | Yes | Yes | [a] | [a] |
| Limit of Detection[b] | No | No | Yes | [a] |
| Limit of Quantitation[b] | No | Yes | No | [a] |
| Selectivity | Yes | Yes | Yes | [a] |
| Range | Yes | Yes | [a] | [a] |
| Linearity | Yes | Yes | No | [a] |
| Ruggedness[b] | Yes | Yes | Yes | Yes |

[a]May be required, depending on the nature of the specific test.
[b]Terminologies to be defined below.

Limit of detection (Jensen et al., 1986) is the lowest concentration of analyte in a sample which can be detected but not necessarily quantitated under the stated experimental conditions. Thus, limit tests merely substantiate that the analyte concentration is above or below a certain level.

Limit of quantitation (Jensen et al., 1986) is a parameter of quantitative assays for low levels of compounds in sample matrices, such as impurities in bulk drug substances and degradation products in finished pharmaceuticals. It is the lowest concentration of the analyte in a sample which can be determined with acceptable precision and accuracy under the stated experimental conditions.

The ruggedness of an analytical method is the degree of reproducibility of test results obtained by analysis of the same samples under a variety of normal test conditions, such as different laboratories, different analysts, different instruments, and different days. Ruggedness is normally expressed as the lack of influence on test results of operational and environmental variables of the analytical method.

It is appropriate here to propose an experimental design for a test of ruggedness. Consider as many as seven factors each at two levels. These factors could be represented by volume of solution, time of waiting, different lots of reagent, different concentration, and so on.

Let A, B, C, D, E, F, and G denote the nominal levels of the seven factors. Let a, b, c, d, e, f, and g denote the alternate levels of the same seven factors. There are $2^7$ or 128 different treatment combinations. However, using a fractional factional scheme, it would be possible to test seven factors in only eight determinations (Box et al., 1978) as follows:

Design of Experiment for Test of Ruggedness in Assay Validation Testing
7 Subtle Conditions in 8 Determinations

| Experimental condition | Treatment combinations | | | | | | | |
|---|---|---|---|---|---|---|---|---|
| | 1 | 2 | 3 | 4 | 5 | 6 | 7 | 8 |
| 1. Temperature | A | A | A | A | a | a | a | a |
| 2. Relative humidity | B | B | b | b | B | B | b | b |
| 3. Time to reaction | C | c | C | c | C | c | C | c |
| 4. Age of solution | D | D | d | d | d | d | D | D |
| 5. Day | E | e | E | e | e | E | e | E |
| 6. Concentration | F | f | f | F | F | f | f | F |
| 7. Solvent | G | g | g | G | g | G | G | g |
| Assay value | $Y_1$ | $Y_2$ | $Y_3$ | $Y_4$ | $Y_5$ | $Y_6$ | $Y_7$ | $Y_8$ |

Computationally, the effect of A is estimated by

$$E(A) = [1/4(Y_1 + Y_2 + Y_3 + Y_4)] - [1/4(Y_5 + Y_6 + Y_7 + Y_8)] = (\bar{A} - \bar{a})$$

$$E(B) = 1/4[(Y_1 + Y_2 + Y_5 + Y_6) - (Y_3 + Y_4 + Y_7 + Y_8)] = (\bar{B} - \bar{b})$$

and so on.

The size of these differences ($\bar{A} - \bar{a}$, $\bar{B} - \bar{b}$, $\bar{C} - \bar{c}$, . . .) determines the degree of effectiveness of the factors—the larger the magnitude of the difference, the greater is the influence of the factor on the assay value. Obviously, a rugged method should not be affected by subtle changes that will almost certainly be encountered within and between laboratories. An estimate of the standard deviation is obtained by the following formula:

$$\text{Standard deviation} = \left(\frac{D_a^2 + D_b^2 + \cdots + D_g^2}{8 - 1}\right)^{1/2}$$

where the $D_i$'s ($i = a, b, \ldots, g$) are the differences such as $(\bar{A} - \bar{a})$, $(\bar{B} - \bar{b})$, and $(\bar{C} - \bar{c})$.

### F. Standardized Procedures: Design and Analysis of Validation Experiments

#### 1. The Greenbrier Procedure*

In this procedure the design of experiment has been standardized in three respects (Platt, 1977). (1) Only three specific levels of the active ingredient must be selected, 100%, 100 + 1.5D%, and 100 − 1.5D%, where D = 0.5 (upper specification limit − lower specification limit of USP). For example, for an ingredient such as calcium pantothenate in multivitamin (USP), the limits may be 90 to 200% of the labeled amount. In this case, then, the "100% theory" would be .5(90 + 200) = 145%, D would be .5(200 − 90) = 55, and the lower and upper limits would be 62.5 and 227.5%, respectively. (2) The results are in terms of the relative amount of the drug recovered and not the instrumental readings. (3) The procedure requires 3 days of test with duplicates, involving a total of 18 determinations. The structure of the design is as follows:

| | Levels | | |
|---|---|---|---|
| Days | 100 − 1.5D% | 100% | 100 + 1.5D% |
| I | = | = | = |
| II | = | = | = |
| III | = | = | = |

The analysis of the experiment has also been standardized. The procedure defines a quantity denoted by U, which is a measure of the overall uncertainty

$$U = (B_{max} + S^2)^{1/2}$$

*The procedure is called the Greenbrier procedure because the paper was first presented at Greenbrier, N.C. The work was undertaken by the Quality Control section of Pharmaceutical Manufacturer Association in 1973.

where $B_{max} = Max(B + B', B - B')$, $B = (\bar{Y} - 100)$, $B' = D(b - 1)$, where b is the estimated slope of the line, $b = (\bar{Y}_1 - \bar{Y}_3)/3D$, and $S^2$ is the residual variance from regression plus within "level" variation.

To compare with an alternative assay, one calculates

$$R = \frac{U_1}{U_2}$$

where $U_1^2 = B_{max(1)} + S_1^2$ and $U_2^2 = B_{max(2)} + S_2^2$.

If the confidence limits of R include the value of 1.0, then there would be sufficient indication to believe that the two assay methods are equivalent.

It may be noted here that the quantity R is difficult to deal with. The distribution of log R is examined by a Monte Carlo method. The asymptotic variance of log R is derived by the Taylor series expansion (sometimes called the statistical differential method). The antilog of the limits log R ± $2\sqrt{V(\log R)}$ provides an estimate of the confidence interval for R. Since the procedure depends largely on the asymptotic variance, the method is a large-sample approximation at best. The design aspects of the Greenbrier procedure, however, are optimal and practical. An alternative approach would be the Rosemont method, which is described next.

### 2. The Rosemont Procedure*

The salient features of the Rosemont procedure (Bohidar, 1983) are as follows.

1. The experimental design aspects of the procedure are similar to those of the Greenbrier procedure. The decomposition of the total degrees of freedom for the ANOVA is:

| Source of variation | DF | |
|---|---|---|
| Day (D) | 2 | Test of linearity |
| Potency (P) | 2 | Test of curvature |
| D × P[a] | 4 | Test of slope = 0 |
| Within days | 9 | Test of slope = 1 |
| Total | 17 | Test of intercept = 0 |

[a]The day × potency interaction should be tested first. If it is not significant, it signifies that the potency slopes are parallel from day to day.

2. In this procedure, one is not restricted to three levels of potency; more than three levels may be included. The Rosemont design and analysis

*The procedure is called the Rosemont procedure because the paper was first presented at Rosemont, Pa. (1982). The work was sponsored by the American Chemical Society, American Pharmaceutical Association, and USP Committee on Assay Validation.

can provide information on sensitivity, specificity, and interference in addition to the information on the other routinely used validation parameters (see Sections V,C and V,D).

3. For a single-system validation, the procedure requires the joint test of hypothesis ($A = 0$ and $B = 1$) (see Section V,E).

4. For a dual-system validation, the procedure requires the test of concurrence based on the Hotelling $T^2_{REG}$ statistic (see Section V,F,1).

5. For a dual-system validation, the procedure recommends a test of intrinsic variability, which can be accomplished by the Levene test as (Bavitz et al., 1973; Bohidar, 1983) extended to regression situations.

6. This procedure requires the test of parallelism and the test of equi-origin (intercept) prior to the test of concurrence. It may be noted that most of the discussions in the early part of this section (V) pertain primarily to the Rosemont procedure.

7. *Test of concurrence*: The Rosemont procedure provides the test of concurrence for dual- or multiple-system validation in a two-part test system. The two parts are called the simultaneous test of concurrence and the conditional test of concurrence. The simultaneous test of concurrence (Hotelling $T^2_{REG}$) has been given in Section V,D,4. The conditional test of concurrence will be described below in detail.

*a. Dual System*

Let the two estimated regression lines be denoted by

$$Y_{1i} = \bar{Y}_1 + B_1(X_i - \bar{X}_1) \qquad i = 1, 2, \ldots, N_1$$

$$Y_{2i} = \bar{Y}_2 + B_2(X_i - \bar{X}_2) \qquad i = 1, 2, \ldots, N_2$$

The following tests are performed. First, test $H_0$: $\sigma^2_{Y \cdot X \cdot 1} = \sigma^2_{Y \cdot X \cdot 2}$ by using the Levene test (Bavitz et al., 1973; Bohidar, 1983) extension to regression. Assuming that there is no significant difference between the two AAMDs, estimate the pooled residual variance as

$$S^2 = \frac{(N_1 - 2)S^2_{Y \cdot X \cdot 1} + (N_2 - 2)S^2_{Y \cdot X \cdot 2}}{(N_1 - 2) + (N_2 - 2)}$$

Then, based on the pooled estimate of the residual variance, test $H_0$: $\beta_1 = \beta_2$ by using the following t-test:

$$t = (B_1 - B_2)\{S^2([X^2]_1^{-1} + [X^2]_2^{-1})\}^{-1/2}$$

where $[X^2] = \Sigma(X - \bar{X})^2$.

Assuming that there is no significant difference between the slopes, that is, the lines are parallel, estimate the common slope as

$$B^* = \frac{[X^2]_1 B_1 + [X^2]_2 B_2}{[X^2]_1 + [X^2]_2}$$

The variance of $B^*$ is equal to $(S^2/[X^2]_1 + [X^2]_2)$.

Now let the expressions for the two population regression lines with a common slope be

$$Y_{1i} = \mu_1 - \beta\bar{X}_1 + \beta X_i \qquad i = 1, 2, \ldots, N_1$$

$$Y_{2i} = \mu_2 - \beta\bar{X}_2 + \beta X_i \qquad i = 1, 2, \ldots, N_2$$

Assuming that the population intercepts are equal, we have

$$\mu_1 - \beta\bar{X}_1 = \mu_2 - \beta\bar{X}_2$$

or, solving for $\beta$, we have

$$\beta = \frac{\mu_1 - \mu_2}{\bar{X}_1 - \bar{X}_2}$$

If the population intercepts are equal and if the ratio $(\mu_1 - \mu_2)/(\bar{X}_1 - \bar{X}_2)$ is equal to $\beta$, a common slope of $\beta_1$ and $\beta_2$, we would say that, in fact, the two population regression lines are not different.

Let $B^{**} = (\bar{Y}_1 - \bar{Y}_2)/(\bar{X}_1 - \bar{X}_2)$, an estimate of $\beta$.

Now we have two estimates of $\beta$, one an estimate of a common slope when the two slopes are parallel and the other an estimate of a slope when the intercepts are the same. These two estimates are independent since one pertains to "within-slope" variation and the other to "between-slope" variation. The test strategy will be that if the two regression lines are identical or coincidental, $\beta^*$ must be equal to $\beta^{**}$ ($H_0$: $\beta^* = \beta^{**} = \beta$).

The variance of $B^{**}$ is

$$V(B^{**}) = \frac{\sigma^2}{(\bar{X}_1 - \bar{X}_2)^2}\left(\frac{1}{N_1} + \frac{1}{N_2}\right)$$

The variance of the difference $(B^{**} - B^*)$ is

$$V(B^{**} - B^*) = \sigma^2\left[\frac{1}{(X_1 - X_2)^2}\left(\frac{1}{N_1} + \frac{1}{N_2}\right) + \frac{1}{[X^2]_1 + [X^2]_2}\right]$$

The test statistic t is equal to

$$t_{(N_1 + N_2 - 4)} = \frac{(B^{**} - B^*)}{\sqrt{V(B^{**} - B^*)}}$$

If the test is not significant, it indicates that there is sufficient evidence that, indeed, the two regression lines are coincidental. This test is called the conditional test of concurrence.

*b. Multiple System*

The dual-system strategy can be extended to the test of concurrence for a multiple system (more than two regression lines).

Let there be p regression lines. The form of the estimated regression line is

$$Y_{ji} = \bar{Y} + B_j(X_i - \bar{X}_j) \qquad j = 1, 2, \ldots, p, \; i = 1, 2, \ldots, N_j$$

The necessary steps are as follows.

1. Conduct Levent's test. If the variances are homogeneous, compute the pooled variance as

$$S^2 = \frac{(N_1 - 2)S_1^2 + \cdots + (N_p - 2)S_p^2}{(N_1 - 2) + \cdots + (N_p - 2)}$$

2. Based on the pooled variance, test the parallelism of the p number of slopes.

3. If the regression lines are parallel as indicated by the test above, then estimate a common slope as

$$B^* = \frac{\Sigma_{i=1}^{p} [X^2]_i B_i}{\Sigma_{i=1}^{p} [X^2]_i}$$

4. If the regression lines are coincidental, then

$$\mu_1 - B\bar{X}_1 = \mu_2 - B\bar{X}_2 = \cdots = \mu_p - B\bar{X}_p$$

must be true. The concurrence of p parallel regression lines is tested by establishing the linearity of the regression line based on the p points, $(\bar{X}_1, \bar{Y}_1)$, $(\bar{X}_2, \bar{Y}_2)$, . . ., $(\bar{X}_p, \bar{Y}_p)$. If the hypothesis of linearity is not rejected, the slope of the regression line, B**, is compared with the slope B*.

It may be noted that in the conditional case, since one of the hypotheses tested depends on the other, one may not be comfortable with the status of the type I error. However, this situation does not arise for the simultaneous test, since only one hypothesis is tested. The statistical techniques associated with the Rosemont method are directly or indirectly available in the SAS computer package.

## G. Design and Analysis of Experiments for Estimation of Assay Precision Involving Multilocation and Multilevel Test

A multinational pharmaceutical company markets products worldwide. Manufacture and control of these products occur at many locations in the world.

Analytical methods to determine the potency of the active ingredient(s) in the formulation are usually developed at a single location. Such development involves the normal analytical concerns of specificity, accuracy, and other validation parameters. Specificity and accuracy can be determined in the laboratory developing the method. Precision is best determined in more than one laboratory in order to reflect the situation that will occur when the formulation is manufactured at several locations.

Frequently, precision is estimated by one analyst on two or three potency levels on several days. This approach is not sufficient since it involves only one analyst (usually the most competent in the method), one laboratory, and known samples. The approach outlined in this section should not be confused with the Greenbrier and Rosemont procedures, which are used primarily to validate an alternative assay method against an acceptable method for the same ingredient.

This section outlines a scheme for determining the assay precision in a manner that represents the day-to-day variations in more than one laboratory. The statistical methods used to analyze the data and to estimate the precision of the method are presented. Actual laboratory experiences and data are reflected in this presentation.

## 1. Experimental

The experimental design should have the following features. (1) Three or four laboratories should participate. (2) The volume of analysis should not be burdensome. (3) Samples should contain more than one level of active ingredient(s). (4) Samples should be close enough in potency not to require adjustments in the assay procedure, thus allowing submission on a "blind" basis.

Small lots of tablets (400 to 500) are prepared by a development pharmacist. Where the formulation is other than tablets, a batch of similar size can be used. Where more than one potency of the formulation is planned for marketing, the midpotency product is chosen. Three lots of tablets are prepared at approximately 90, 100, and 110% of the label claim for single-ingredient products. In the instances where only the pure compound is involved, vials are filled at 90, 100, and 110% of the target fill by using an an analytical balance. For tablets containing two active ingredients, five lots are prepared as shown on Table 15. For tablets containing three active ingredients, three lots are prepared by using a Latin square design configuration as shown in Table 16.

The experimental design involves three or four laboratories analyzing two or four samples daily until each laboratory has performed 16 analyses. Thus a total of either 48 (for three laboratories) or 64 (for four laboratories) samples is required. Most experience has been with four laboratories, and in all cases four analyses are done daily. The two-analyses-daily scheme is designed for lengthy analytical methods. The three levels

**Table 15** Percent of Claim

| Entity | Lot 1 | Lot 2 | Lot 3 | Lot 4 | Lot 5 |
|---|---|---|---|---|---|
| A | 90 | 110 | 100 | 100 | 100 |
| B | 100 | 100 | 100 | 90 | 110 |

Table 16 Percent of Claim

| Entity | Lot 1 | Lot 2 | Lot 3 |
|---|---|---|---|
| A(1) | 90 | 100 | 110 |
| A(2) | 100 | 110 | 90 |
| A(3) | 110 | 90 | 100 |

of concentration are each assigned a set of random numbers from 1 to 64 (or 1 to 48) such as those illustrated in Table 17. Samples from each level are subdivided and identified with a random number. At least 10 tablets are included in each sample. The samples are then assigned to the four (or three) laboratories on the basis of a balanced randomization scheme as shown in Table 18. Such a scheme guarantees "blindness" of the samples. Samples are sent to each analyst with the following instructions and information: (1) the analytical procedure, (2) a report form showing sample numbers for each day of analysis, (3) the reference standard for the drug, (4) instructions to report all values on a milligrams per gram basis to avoid any problems of weight difference, and (5) instructions to perform the analysis no more frequently than 1 day per week and to make up fresh reagents and standards for each day of analysis.

Of the four laboratories involved, two or three should be familiar with the specific method and one or two should have used the method. In all

Table 17 Random Number Assignment

| | | | | | |
|---|---|---|---|---|---|
| | | 90% level | | | |
| 21 | 13 | 33 | 6 | 10 | 63 |
| 31 | 46 | 4 | 30 | 34 | 18 |
| 42 | 19 | 23 | 2 | 5 | 12 |
| 24 | 39 | 9 | | | |
| | | 100% level | | | |
| 58 | 47 | 22 | 53 | 43 | 51 |
| 36 | 50 | 44 | 37 | 57 | 61 |
| 7 | 11 | 17 | 27 | 8 | 29 |
| 45 | 32 | 48 | 54 | | |
| | | 110% level | | | |
| 52 | 60 | 38 | 28 | 64 | 40 |
| 62 | 49 | 3 | 16 | 26 | 14 |
| 56 | 41 | 1 | 59 | 35 | 25 |
| 20 | 15 | 55 | | | |

**Table 10** Serial Numbers of Bottles Randomly Assigned to Each Analyst on Each Day of Test

| Day of test | Serial numbers | |
|---|---|---|
| | *Laboratory A* | *Laboratory B* |
| $D_1$ | 38-50-17-45 | 63-14-49-60 |
| $D_2$ | 24-37-13-10 | 12-21-58-22 |
| $D_3$ | 30-59-34-3 | 29-43-36-28 |
| $D_4$ | 41-15-18-47 | 16-53-6-23 |
| | *Laboratory C* | *Laboratory D* |
| $D_1$ | 9-27-7-35 | 19-5-2-48 |
| $D_2$ | 31-33-44-51 | 11-4-8-54 |
| $D_3$ | 61-62-42-52 | 57-64-32-1 |
| $D_4$ | 55-20-40-39 | 25-56-26-46 |

cases at least one and sometimes two laboratories are outside the United States. A single experienced analyst is used in each laboratory, however.

2. Numerical Example

The numerical results associated with the estimation of the precision of the chemical assay of product Q tablets will be used to illustrate the statistical methodology involved. The data received from the four laboratories, A, B, C, and D, are properly tabulated. The entire data set is arranged according to the potency levels, 1, 2, and 3, irrespective of laboratories and days. In this case, we have 21, 21, and 22 assay values, respectively, for levels 1, 2, and 3. Since we are interested in estimating the assay precision, we proceed to compare the variabilities associated with the potency levels by using Levene's test for homogeneity of average absolute mean deviations associated with the potency levels. It should be noted that a desirable feature of Levene's (1960) test is that it is applicable to sets of data whose underlying population is nonnormal. Duncan's new multiple-range test is used to accomplish the pairwise comparisons among the potency levels. Other multiple-range tests could also be used as long as per comparison error rates are highlighted rather than preservation of an overall "experimentwise" error rate. In this example, we find no differences ($P = 0.05$) among the potency levels as far as their variabilities are concerned.

The formula given in Table 20 is used to estimate the pooled within-potency-level variance ($S^2$). Based on this variance, we calculate the quantity $\{(2\sqrt{S^2}/\bar{X}) \times 100\}$, called the precision of the assay (sometimes it is called the relative precision of the assay), where $\bar{X}$ represents the arithmetic average of the potency level under consideration.

Table 19 Statistical Comparison of Variabilities

| | Comparison among levels over all laboratories | | |
|---|---|---|---|
| | Level I | Level II | Level III |
| AAMD | 4.9 | 6.6 | 6.0 |
| N | 21 | 21 | 22 |

| | Comparison among laboratories over all levels | | | |
|---|---|---|---|---|
| | Lab A | Lab B | Lab C | Lab D |
| AAMD | 40.7 | 56.2 | 53.6 | 37.6 |
| N | 16 | 16 | 16 | 16 |

For the example shown, the averages and precision values for potency levels 1, 2, and 3 are found to be (552.1, 3.3%), (622.0, 2.9%), and (685.5, 2.7%), respectively. This essentially completes the calculation.

At this point, one may be interested in comparing the variabilities associated with the laboratories, A, B, C, and D. In this case there are no differences among the laboratories ($\alpha = 0.05$) with respect to their variabilities. If the analysis showed significant differences among the potency levels and/or the laboratories, it would be advisable to investigate the reasons for such findings.

Note that a relative precision value of 5% or less is considered an acceptable value in most experiments.

This experimental protocol seems to be a reasonable and valid approach for practical determination of assay precision in pharmaceutical formulations. The statistical calculations are not complex, and the analysis time in any laboratory is minimal. The preparation, subdivision, and distribution of samples require care and attention but are not overly time-consuming.

Table 20 Formula for the Pooled Variance

$$S^2 = \frac{(n_1 - 1)S_1^2 + (n_2 - 1)S_2^2 + (n_3 - 1)S_3^2}{n_1 + n_2 + n_3 - 3}$$

$S_1^2$ = sample variance of potency level I with $n_1$ assay values

$S_2^2$ = sample variance of potency level II with $n_2$ assay values

$S_3^2$ = sample variance of potency level III with $n_3$ assay values

## H. Assay Validation of Multicomponent Mixture

### 1. Beer-Lambert Law

Assay validation of a multicomponent mixture arises when, during stability determinations on drug formulations, one discovers the presence of not only intact drug but also degradation products. Chromatographic procedures, although selective, are very time-consuming and are subject to problems of column variability. Recent advances in spectrophotometers and laboratory computers suggest that it might be possible to perform stability determinations by simultaneous multicomponent analysis on the unseparated mixture of intact drug and low-level degradation products. Multicomponent assay validation essentially pertains to testing the validity of the Beer-Lambert law (Connors, 1975) for multicomponent mixtures. When a compound is irradiated with electromagnetic radiation of an appropriate wavelength, it absorbs energy. If the intensity of the incident light is $I_0$ and that of the transmitted light is T, the absorbance (A) of the sample is given by $A = \log(I_0/T)$. It is known that the absorbance of a sample is proportional to the product of the molar concentration (C) and the path length of the cuvette (P), leading to the relationship

$$A = C \cdot P \cdot E$$

where E is the molar absorptivity (Connors, 1975). It is also known that if the Beer-Lambert law is applicable to a mixture of k solutes, the measured absorbance of the mixture can be expressed as the weighted sum of the absorbances of the individual solutes. The statistical model for the Beer-Lambert law can be expressed as

$$A = C_1B_1P_1 + C_2B_2P_2 + \cdots C_KB_KP_K + E$$

where A is the absorbance of the mixture under test, $B_i$ the absorptivity of solute i, i = 1, 2, . . ., K, $P_i$ the length of the ith cell containing the mixture (if the length of the cuvette is kept constant, then $P_1 = P_2 = \cdots = P_K$), E the experimental error, and $C_i$ the concentration of the ith solute in the mixture. For $P_1 = P_2 = \cdots P_K$, we have $A = \Sigma_{i=1}^{K} C_iB_i + E$, i = 1, 2, . . ., K.

Now if the mixture under test and each of the K solutes (standards) are measured for absorbance at the same n wavelengths, we can write the statistical model in the following way:

$$A_i = C_0 + C_{1i}B_{1i} + C_{2i}B_{2i} + \cdots C_{Ki}B_{Ki} + E_i$$

where i = 1, 2, . . ., n and n > K, $A_i$ is the dependent variable, $B_1$, $B_2$, . . ., $B_K$ are the K independent variables, $C_0$ is the intercept, $C_1$, $C_2$, . . ., $C_K$ are the respective concentrations of the solutes, which are the unknowns to be estimated, and $E_i$ is the Gaussian error associated with $A_i$ with the usual statistical assumptions. If the $B_i$'s are fixed and measured without error (or, as in practice, the variations of the B's are much smaller than those of the A's), the standard least-squares approach is considered for the estimation of the coefficients $C_1$, $C_2$, . . ., $C_K$.

The basic formulation in matrix notation is

$$A = BC + E$$

The simultaneous set of normal equations is

$$(B'B)C^* = B'A$$

where $C^*$ is an estimate of C.

Since $B^1B$ is a positive-definite symmetric matrix with nonsingularity properties, there is a unique solution for the C's:

$$C^* = (B'B)^{-1}B'A$$

It may happen that the absorbances are less reliable in some parts of the spectrum than in others. This would mean that the variances of the observations are not all equal. In the model, $EX(E) = o$ and $V(E) = W\sigma^2$ (and *not* $I\sigma^2$), where EX is the expected value and V the variance. Now the normal equations become

$$(B'W^{-1}B)C^* = B'W^{-1}A$$

The solutions then become

$$C^* = (B'W^{-1}B)^{-1}B'W^{-1}A$$

2. Test of Validity of Beer-Lambert Law Based on P-R Experiment

Consider that we have three components, I, E, and M, where I is the intact drug and E and M are the two thermal degradation products. Known amounts of I, E, and M are mixed with a solvent, such as acetonitrile. This constitutes the mixture solution. The solutions for the individual entities are prepared in the same way. The individual concentrations are known and maintained constant throughout the experiment. The mixture and the individual component solutions are subjected to spectrophotometric analysis. The spectra are analyzed in derivative mode in the range of suitable wavelengths (280 to 310 nm, say). Here the concentrations of I, E, and M are fixed (2.5, 0.1, and 0 mg/ml, respectively, say). The absorbance values are tabulated as follows:

| Wave-length | Mixture, Y | I, $X_1$ | E, $X_2$ | M, $X_3$ |
|---|---|---|---|---|
| 280 | — | — | — | — |
| 282 | — | — | — | — |
| 284 | — | — | — | — |
| 285 | — | — | — | — |
| . | . | . | . | . |
| 310 | — | — | — | — |

*a. Design of Experiment*

Historically, the set of mixtures to be tested in a validation experiment was chosen subjectively. For this reason, we are providing an objective, coherent method of selection. This involves generating a set of mixtures with K components by using an orthogonal central composite K-factor fractional combination scheme. The total number of mixtures would be $\theta 2^K + 2K + 1$, where $\theta$ is the selected fraction of the full factorial combinations ($\theta = 1$ for a full factorial).

To demonstrate the validity of the Beer-Lambert law, we must undertake the following tests of validity.

1. To test the adequacy of the regression model, we must examine several test statistics and several kinds of plots, such as $R^2$, raw residuals $(Y - \hat{Y})$, deleted residuals, standardized residuals, Studentized residuals, leverage value, Cook's distance, Mallows Cp statistics, Mahalanobis distance, various types of residual plots, and various partial regression residual plots.

2. If the model is shown to be reasonably adequate, then we conduct the following test,

$$H_0: C_0 = 0$$

This implies that there is no background noise.

3. Individual concentration test:

$$H_0: C_i = C_{i0}$$

$$H_0: C_e = C_{e0}$$

$$H_0: C_m = C_{m0}$$

where $C_{i0}$, $C_{e0}$, and $C_{m0}$ are the actual amounts of the three solutes that are present in the mixture and $C_i$, $C_e$, and $C_m$ are the true partial regression coefficients. None of the above tests should be significant if the method is valid. The following t-tests are used:

$$t = \frac{(C_i^* - C_{i0})}{[v(C_i^*)]^{1/2}} \quad \text{with } n - 4 \text{ DF}$$

$$t = \frac{(C_e^* - C_{e0})}{[v(C_e^*)]^{1/2}} \quad \text{with } n - 4 \text{ DF}$$

$$t = \frac{(C_m^* - C_{m0})}{[v(C_m^*)]^{1/2}} \quad \text{with } n - 4 \text{ DF}$$

4. Simultaneous concentration test. Test statistics:

$$\text{Test } H_0: \begin{bmatrix} C_i - C_{i0} \\ C_e - C_{e0} \\ C_m - C_{m0} \end{bmatrix} = \begin{bmatrix} 0 \\ 0 \\ 0 \end{bmatrix}$$

$$F = \frac{\begin{bmatrix} C_i - C_{i0} \\ C_e - C_{e0} \\ C_m - C_{m0} \end{bmatrix}' \begin{bmatrix} V(C_i) & COV(C_i, C_e) & COV(C_i, C_m) \\ COV(C_e, C_i) & V(C_e) & COV(C_e, C_m) \\ COV(C_m, C_i) & COV(C_m, C_e) & V(C_m) \end{bmatrix}^{-1} \begin{bmatrix} C_i - C_{i0} \\ C_e - C_{e0} \\ C_m - C_{m0} \end{bmatrix}}{2S^2}$$

where $S^2$ is the residual variance about the regression hyperplane. This test should not be significant.

### *b. Situation in Which the $B_i$'s are Multicollinear and/or Subject to Experimental Error*

It is reasonable to assume that multicomponent spectrophotometric absorbance data may demonstrate multicollinearity and experimental error in the B's.

The appropriate methods for the analysis of this type of data are principal component regression, latent root regression, and ridge regression. One or more methods should be undertaken if such a situation arises.

These procedures are presented in several well-known textbooks (such as Draper and Smith, 1981). They are essentially multivariate procedures applied to univariate regression analysis.

For situations in which the concentrations of the components are unknown and several dependent variables have been measured, the validation methodologies described above are not directly applicable. One must have recourse to a circular pathway to accomplish the validation test requirements.

The multivariate methodology called factor analysis (Malinowski, et al., 1980) must be applied in this situation. The analytical chemistry situation in which factor analysis is applicable is illustrated in the following. Consider a data matrix (A)

$$A = \begin{bmatrix} a_{11} & a_{12} & a_{13} & a_{14} \\ a_{21} & a_{22} & a_{23} & a_{24} \\ a_{31} & a_{32} & a_{33} & a_{34} \\ a_{41} & a_{42} & a_{43} & a_{44} \\ a_{51} & a_{52} & a_{53} & a_{54} \\ a_{61} & a_{62} & a_{63} & a_{64} \end{bmatrix}$$

representing the absorbances of four different mixtures containing the same number of components measured at six wavelengths. The columns represent the four mixtures and the rows the six wavelengths. Such spectral measurements are obtained from a chemical kinetics study if the samples of the reaction mixture are collected at four different times during the experiment. The problem here is to determine the number of components present and to ascertain their concentrations. Factor analysis provides a solution to this problem. Recognizing that absorbance data for multicomponent mixtures can be approximated by the Beer-Lambert law, it would be possible to interpret the "factors" chemically. Once the number of components has been determined and their respective concentrations have been ascertained, a P-R multicomponent mixture experiment should be conducted with these components and concentrations. All the tests of validation can now be applied to these data. In addition to these tests, a CAS analysis (Bohidar et al., 1979) should be conducted to determine which of the components are contributing substantially to the variation in the multicomponent mixtures.

## VI. THERMAL STABILITY ACCELERATED TESTING

### A. Introduction

Drug molecules are subject to decomposition. Pharmaceutical preparations of drug molecules can lose their drug content by this decomposition and also by volatilization of their active ingredient. Loss of potency inevitably follows these losses. Therefore, chemical stability, storage conditions, and shelf life become very important parameters in the preparation of drug formulations.

### B. Chemical Kinetics

Decomposition reactions follow the law of mass action and can display zero-order, first-order, or multiple-order kinetics, depending on the nature and concentrations of the reactants. The law of mass action states that the rate at which a reaction proceeds is proportional to the active masses of the reacting substances (Hajratwala and Dawson, 1977). Basically, the rate of reaction is a function of the probability of collision of the reacting molecules. In solutions, the active masses are represented by the molar concentrations of the reactants raised to a power equal to the number of molecules taking part in the reaction. If we consider the reaction,

$$aC \longrightarrow \text{products}$$

where a is the number of reacting molecules and C is their concentration, the rate of reaction, dC/dt, is given by

$$\frac{-dC}{dt} = K$$

or

$$-\int dC = \int K\,dt$$

or

$$C = -Kt + \text{constant}$$

If the initial concentration is $C_0$ (when t = 0), then

$$C = -Kt + C_0$$

The reaction proceeds at a constant rate, independent of the concentration of the reactant.

In first-order reactions, a = 1, so

$$\frac{-dC}{dt} = KC \qquad \text{and} \qquad \int \frac{dC}{C} = -\int K\,dt$$

or

$$\log_e C = -Kt + \text{constant}$$

When t = 0, $C = C_0$, we have

$$\log_e C = -Kt + \log_e C_0$$

The reaction proceeds at a rate proportional to the (log) concentration of the reactant.

In second-order reactions, a = 2, so

$$\frac{-dC}{dt} = KC^2 \qquad \text{or} \qquad \int C^{-2}\,dC = -\int K\,dt$$

or

$$\frac{-1}{C} = -Kt + \text{constant}$$

1. Effect of Temperature

Elevated temperature is a fundamental problem in storage of pharmaceuticals because of instability at elevated temperatures. Arrhenius showed that the effect of temperature on the reaction rate is given by

$$\frac{d \log_e K}{dT} = \frac{E}{RT^2}$$

where E is the energy of activation per gram mole. From this,

$$aK = \frac{E}{R}\int \frac{1}{T^2}\,dt = -\frac{E}{R}\frac{1}{T} + \text{constant}$$

so

$$\log K = \frac{E}{2.303R}\ \frac{1}{T} + \frac{\text{constant}}{2.303}$$

If K is determined at a number of different values of T, a graph of log K against 1/T will be a straight line of slope −E/2.303R. This is called an Arrhenius plot.

The Arrhenius plot concept will be used in developing the prediction model to estimate the thermal stability of products.

## C. Prediction of Thermal Stability via the Arrhenius Equation

In developing new pharmaceutical products it is often necessary to predict degradation rates at marketing temperatures from data collected on accelerated degradation at elevated temperatures. The prediction of degradation rate is usually based on the Arrhenius equation. Here we are considering a weighted least-squares method for predicting stability. In addition, a statistical test is proposed to test the validity of the Arrhenius assumption.

The functional relationship between time and concentration of a drug stored under constant conditions is dependent on the order of reaction and a rate constant which determines the speed of reaction. An example of a typical situation is the first-order reaction, given below, where the logarithm of concentration (C) at time t is linearly related to time by

$$\log_e C_t = \log_e C_0 + K_T t$$

where T is the temperature at which storage took place and $K_T$ is the rate constant.

The Arrhenius relationship (McBride, et al., 1952)

$$\log_e K_T = \gamma + \delta T^{-1}$$

states that the logarithm of the reaction rate is a linear function of the reciprocal of the absolute temperature. If the assumption is valid, this relationship allows data taken at elevated temperatures to be used to predict the degradation rate at room temperature and hence estimate the shelf life of the drug. The rate constants obtained at the elevated temperatures can be used to estimate the parameters in the Arrhenius equation, which in turn can be used to estimate the reaction rate at room temperature (or any other desired temperature).

## D. Statistical Development of Weighted Regression Analysis for Predicting Thermal Stability

We must consider two functional relationships for this regression analysis: (1) the functional relationship between time and concentration of a drug or other substance stored at a given temperature and (2) the functional relationship between temperature and time-dependent rate of degradation per unit time.

### 1. Characterization of the Functional Relationship Between Time and Concentration of a Drug

(A) Usually we would expect the functional relationship between time and concentration of a drug to be exponential. The mathematical model can be expressed as

$$C_t = C_0 e^{-b_1 t} \tag{2}$$

where $C_t$ is the concentration of the drug at time t, $C_0$ the initial concentration of the drug, $b_1$ the rate of degradation, and t the time.

The regression model for the statistical analysis would require a linearization of the model in (2). For a first-order reaction rate, this is achieved by expressing the final concentration in the natural logarithmic scale ($\log_e C_t$). Thus we have

$$\log_e C_t = \log_e C_0 - b_1 t \tag{3}$$

This, in turn, can be expressed in terms of our familiar regression notation as $Y = b_0 + b_1 t$, where Y is the $\log_e C_t$, given above, $b_0$ is the intercept ($\log_e C_0$), $b_1$ is the linear regression coefficient, and t is the time.

The final data layout at the conclusion of an experiment is:

| Time (days) | Final concentration | $\log_e$ of final concentration |
|---|---|---|
| $t_1$ | $C_{t_1}$ | $Y_1$ |
| $t_2$ | $C_{t_2}$ | $Y_2$ |
| . | . | . |
| . | . | . |
| . | . | . |
| $t_N$ | $C_{t_N}$ | $Y_N$ |

The least-squares estimates of the regression parameters are

$$b_0 = \bar{Y} - b_1 \bar{t}$$

$$b_1 = \frac{\Sigma ty - (\Sigma t)(\Sigma Y)/N}{\Sigma t^2 - (\Sigma t)^2/N}$$

$$\bar{t} = \frac{\Sigma t}{N}$$

where N is the number of time points considered in the experiment. The residual variance is estimated by

$$S^2_{y \cdot t} = \frac{[Y^2] - b_1^2[t^2]}{N - 2}$$

where

$$[Y^2] = \sum_{i=1}^{N} (Y - \bar{Y})^2 \qquad [t^2] = \sum_{i=1}^{N} (t - \bar{t})^2$$

The estimate of the regression line is

$$\hat{Y} = b_0 + b_1 t$$

Now by substituting a given value of t, say $t_0$, we get an estimate of the log concentration of the drug associated with the given time ($t_0$). (By taking the antilogarithm of $Y_0$ we would be able to express the concentration in the original units.)

$$Y_0 = b_0 + b_1 t_0$$

The variances of the estimates then would be

$$V(b_0) = S^2_{y \cdot t}\left[\frac{1}{n} + \frac{(\bar{t})^2}{[t^2]},\right] \qquad V(b_1) = \frac{S^2_{y \cdot t}}{[t^2]}$$

$$V(Y_0) = S^2_{y \cdot t}\left[\frac{1}{n} + \frac{(t_0 - \bar{t})^2}{[t]^2}\right]$$

(B) Suppose we repeat part (A), for several temperatures, then we would have a rate of degradation ($b_1$) for each temperature, Consider K temperatures, then, we have the following, (let $S^2_{y \cdot t} = S^2$):

| Temperatures | Rate of degradation | Number of time points | Variance of rate of degradation |
|---|---|---|---|
| $T_1$ | $b_1$ | $N_1$ | $S_1^2/[t^2]_1$ |
| $T_2$ | $b_2$ | $N_2$ | $S_2^2/[t^2]_2$ |
| . | . | . | . |
| . | . | . | . |
| . | . | . | . |
| $T_K$ | $b_K$ | $N_K$ | $S_K^2/[t^2]_K$ |

Suppose we show statistically [by the Levene (1960) test or some other relevant test] that the residual variances associated with each temperature are homogeneous; then we may pool these variances in the following way:

$$\text{Pooled residual variance} = S_P^2 = \frac{\Sigma_{i=1}^{K} [Y^2]_i - b_i^2 [t^2]_i}{\Sigma_{i=1}^{K} (N_i - 2)}$$

### 2. Characterization of the Functional Relationship Between Rate of Degradation per Unit of Time and Temperature

Here the mathematical model can be expressed as the Arrhenius relationship,

$$b_T = e^{\gamma+(\delta/T)} \tag{4}$$

where $b_T$ is the rate of degradation associated with temperature (T) obtained from the part A analysis, $\gamma$ and $\delta$ are the parameters of equation (4), and T is the absolute temperature.

The regression model for the statistical analysis is

$$\log_e b_T = \gamma + \frac{\delta}{T} = \gamma + \delta T^{-1}$$

This, in turn, can be expressed in terms of our familiar regression notation as follows:

$$R = B_0 + B_1 \frac{1}{T} = A + BX$$

where R is the natural logarithm of $b_T$, A the intercept (= $B_0$), B the change in the rate of degradation for a unit change in the reciprocal absolute temperature, and X the reciprocal absolute temperature (1/T).

The precision with which the K rates of degradation ($b_t$) are estimated would not necessarily be the same. The regression analysis which takes into account the heterogeneity of variances in the dependent variable is called weighted regression analysis and is used routinely for analyzing stability data. Nonlinear regression analysis may be used with proper weighting of the dependent variable.

## E. Accelerated Testing: An Example

The objective here is to determine the reaction rate at a lower temperature by using the Arrhenius equation and reaction rates determined at higher temperatures. As far as the planning of the experiment is concerned, one should store the drug preparation at least at four different elevated temperatures and determine the concentration of drug initially and at least at four suitable time periods for each temperature. Avoid excessively high temperatures that may cause precipitation or coagulation.

### 1. Arrhenius Equation

The Arrhenius equation states that the log reaction rate is linearly related to the reciprocal of the absolute temperature. A weighted least-squares procedure is used to obtain this equation.

The calculations are illustrated using the following data:

| °C | $T^a$ | $b_1{}^b$ | $[t^2]$ | n | n − 2 | $S^2$ |
|---|---|---|---|---|---|---|
| 90.3 | 363.46 | 1.105 | 549 | 4 | 2 | 0.085 |
| 79.8 | 352.96 | 0.3351 | 1,349 | 4 | 2 | 0.145 |

| °C | $T^a$ | $b_1^b$ | $[t^2]$ | n | n − 2 | $S^2$ |
|---|---|---|---|---|---|---|
| 69.5 | 342.66 | 0.08352 | 19,373 | 5 | 3 | 0.253 |
| 59.1 | 332.26 | 0.02973 | 28,221 | 5 | 3 | 0.137 |

[a] T, Absolute temperature = °C + 273.16
[b] $b_1$, Reaction rate = %/hour.

The various quantities required for the weighted regression analysis are as follows:

1. Absolute temperature: $T_1$, $T_2$, $T_K$
2. Reciprocal of absolute temperature: $T_1^{-1}$, $T_2^{-1}$, $T_K^{-1}$
3. Rate of degradation from different temperature study: $b_1$, $b_2$, $b_K$
4. $\text{Log}_e(b_i)$: $\log_e b_1 = R_1$, $\log_e b_2 = R_2$, $\log_e b_K = R_K$
5. Variance $(R_i)$: $Sp^2/b_1^2[t_1^2]$, $Sp^2/b_2^2[t_2^2]$, $Sp^2/b_K^2[t_K^2]$
6. $W_i$: $b_1^2\Sigma(t - \bar{t})^2$, $b_2^2\Sigma(t - \bar{t})^2$, $b_K^2\Sigma(t - \bar{t})^2$

The following numerical example explicitly demonstrates the use of the statistical procedure outlined above.

The combined variance is $S_c^2 = \Sigma(n_i - 2)S_c^2$. Numerically, we have

$$S_c^2 = \frac{(2)(0.085) + (2)(0.145) + (3)(0.253) + (3)(0.137)}{2 + 2 + 3 + 3} \quad 0.163$$

Let

$$X = \frac{1}{T} \times 10^3, \qquad Y = \log_{10} b_1, \qquad W = b_1^2[t^2] = \text{weight}$$

Then

| X | Y | W | WX | WY |
|---|---|---|---|---|
| 2.751 | 0.0434 | 669.95557 | 1843.04777 | 29.07607 |
| 2.833 | −0.4748 | 151.48440 | 429.15531 | −71.92479 |
| 2.918 | −1.0782 | 135.13598 | 394.32679 | −145.70361 |
| 3.009 | −1.5268 | 24.94400 | 75.05650 | −38.08450 |
| Sums | | 981.51995 | 2741.58637 | 226.63683 |

$$[X^2] = \Sigma WX^2 - \frac{(\Sigma WX)^2}{\Sigma W} = 7662.51199 - \frac{(2741.58637)^2}{981.61995} = 4.69941$$

$$[XY] = \Sigma WXY - \frac{(\Sigma WX)(\Sigma WY)}{\Sigma W} = -663.53408 - \frac{(2741.58637)(-226.63683)}{981.51995}$$

$$= -30.49097$$

$$[Y^2] = \Sigma WY^2 - \frac{(\Sigma WY)^2}{\Sigma W} = 250.65684 - \frac{(-226.63683)^2}{981.51995} = 198.32551$$

$$\bar{X} = \frac{\Sigma WX}{\Sigma W} = \frac{2741.58637}{981.51995} = 2.79320$$

$$\bar{Y} = \frac{\Sigma WY}{\Sigma W} = \frac{-226.63683}{981.51995} = -0.23090$$

$$B_1 = \frac{[XY]}{[X^2]} = \frac{-30.49097}{4.69941} = -6.48824$$

$$S_a^2 = \frac{[Y^2] - [XY]^2/[X^2]}{N - 2} = \frac{198.32551 - (-30.49097)^2/4.69942}{2} = 0.24616$$

K = number of temperatures = 4

One test for adequacy of the Arrhenius fit (McBride et al., 1952) may be carried out as follows:

$$F = \frac{S_a^2}{S_c^2} = \frac{0.24616}{0.163} = 1.51$$

An F-test is carried out with $V_1 = K - 2 = 2$ degrees of freedom and $V_2 = \Sigma(N_i - 2) = 10$ degrees of freedom.

Since the observed F (1.51) is less than the critical F (4.10), we conclude that the data fit the Arrhenius equation within the limits of experimental error. Then,

$$\hat{Y} = \bar{Y} + B_1(X - \bar{X}) = -0.23090 - 6.48825(X - 2.79320)$$

$$= B_0 + B_1 X = 17.89208 - 6.48825X$$

$$V(\hat{Y}) = S_a^2\left[\frac{1}{\Sigma W} + \frac{(X - \bar{X})^2}{[X^2]}\right] = 0.24616\left[\frac{1}{981.51995} + \frac{(X - 2.79320)^2}{4.69941}\right]$$

To estimate the log reaction rate at 23°C and 95% confidence limits, we proceed as follows:

$$X = \frac{1}{23 + 273.16} \times 10^3 = 3.38$$

$$\hat{Y} = 17.89208 - 6.48825(3.38) = -4.0382$$

$$V(\hat{Y}) = 0.24616\left[\frac{1}{981.51995} + \frac{(3.38 - 2.79)^2}{4.69941}\right] = 0.07508$$

$t = 4.303 =$ Student's t at $\alpha = 0.05$ for $N - 2 = 2$ degrees of freedom

$$\hat{Y}_L, \hat{Y}_U = \hat{Y} \pm tV(\hat{Y}) = -4.0382 \pm 4.303\ (0.274)$$

$$= -5.2172, -2.8592$$

Then the reaction rate will be given by

$$B_1(B_{1L}, B_{1U}) = \text{antilog}\ [\hat{Y}(\hat{Y}_L, \hat{Y}_U)$$

$$= \text{antilog}[-4.0382\ (-5.2172, -2.8592)]$$

$$= 0.0154\ (0.0010, 0.2335)\ \%\ \text{per week}$$

## REFERENCES

Anderson, T. W. (1958). *An Introduction to Multivariate Statistical Analysis*, Wiley, New York, pp. 272-287.

Bavitz, J. F., Bohidar, N. R., Karr, J. I., and Restaino, F. A. (1973). Performance of Tablet Breaking Strength Testers: I. Intralaboratory Comparison and Prediction, *J. Pharm. Sci.*, 62(9): 1520-1524, 1973.

Bavitz, J. F., Bohidar, N. R., and Restaino, F. A. (1974). Disintegrability Characteristics of Three Selected Tablet Excipients, *Drug Dev. Commun.*, 1(4): 331-347.

Bohidar, N. R. (1967). "Conditional Effects and Interactions in Symmetrical Factorial Confounding with Application to Biology," 12th Conf. on Design of Experiment in Army Research Development and Testing, vol. 12, pp. 207-220.

Bohidar, N. R. (1983). "Statistical Aspects of Chemical Assay Validation," Proc. of American Statistical Association Annual Joint Meetings, Biopharmaceutical Section, pp. 57-62.

Bohidar, N. R. (1984). "Application of Optimization Techniques in Pharmaceutical Formulation—An Overview," Proc. of American Statistical Association Annual Joint Meetings, Biopharmaceutical Section, pp. 6-13.

Bohidar, N. R., Restaino, F. A., and Schwartz, J. B. (1975). Selecting Key Parameters in Pharmaceutical Formulations by Principal Component Analysis, *J. Pharm. Sci.*, 64(6): 966-969.

Bohidar, N. R., Restaino, F. A., and Schwartz, J. B. (1979). Selecting Key Pharmaceutical Formulation Factors by Regression Analysis, *Drug Dev. Ind. Pharm.*, 5(2): 175-216.

Bohidar, N. R., Bavitz, J. R., and Shiromani, P. K. (1986). Formulation Optimization for a Multiple Potency System with Uniform Tablet Weight, *Drug. Dev. Ind. Pharm.*, 12(10): pp. 1503-1510.

Box, G. E. P. (1954). The Exploration and Exploitation of Response Surfaces: Some General Considerations and Examples, *Biometrics*, 10: 16-60.

Box, G. E. P. and Hunter, J. S. (1957). Multifactor Experimental Designs for Exploring Response Surfaces, *Ann. Math. Stat.*, 28(1): 195.

Box, G. E. P. and Wilson, K. B. (1951). On the Experimental Attainment of Optimum Conditions. *J. R. Stat. Soc. Ser. B*, 13: 1–45.

Box, G. E. P., Hunter, W. G., and Hunter, J. S. (1979). *Statistics for Experimenters*, Wiley, New York.

Carnahan, B., Luther, H. A., and Wilkes, J. O. (1969). *Applied Numerical Methods*, Wiley, New York, pp. 255–259.

Cochran, W. G. and Cox, G. M. (1957). *Experimental Design*, 2nd ed., Wiley, New York.

Connors, K. A. (1975). *A Textbook of Pharmaceutical Analysis*, 2nd ed., Wiley, New York.

Davies, O. L. (1956). *Design and Analysis of Industrial Experiments*, 2nd ed., Hafner, New York.

Draper, N. R. and Smith, H. (1981). *Applies Regression Analysis*, 2nd ed., Wiley, New York, pp. 313–352.

Fisher, R. A. (1947). *The Design of Experiment*, 4th ed., Oliver & Boyd, Edinburgh and London.

Hajratwala, B. R. and Dawson, J. E. (1977). Kinetics of Indomethacin Degradation I: Presence of Alkali, *J. Pharm. Sci.*, 66: 27–29.

Jensen, E. H. and Members of the PMA Compendial Assays Validation Committee. (1986). Current Concept for the Validation of Compendial Assays, *Pharmacopeial Forum*, March–April, pp. 1241–1245.

Kempthorne, O. (1952). *The Design and Analysis of Experiments*, Wiley, New York.

Kendall, E. C. (1975). *Cortisone*, Scribner's, New York.

Levene, H. (1960). Robust Tests for Equality of Variances, *Contribution to Probability and Statistics, Essays in Honor of Harold Hotelling* (I. Olkin, S. Ghuaye, W. Hoeffding, W. Madow, and H. Mann, eds.), Stanford University Press, Stanford, Calif., pp. 278–292.

Malinowski, E. R. and Howery, D. G. (1980). *Factor Analysis in Chemistry*, Wiley, New York, pp. 24–35.

Mandel, J. (1972). Repeatability and Reproducibility, *J. Qual. Technol.*, 4(2): 74–85.

Mandel, J. (1984). *The Statistical Analysis of Experimental Data*, Wiley, New York.

McBride, W. R. and Villars, D. C. (1952). An Application of Statistics to Reaction Kinetics, *Anal. Chem.*, 26: 901.

Myers, R. H. (1976). *Response Surface Methodology*, Virginia Polytechnic Institute and State University, pp. 127–175.

Platt, R. and Members of the PMA Standing Committee on Statistics. (1977). "The Greenbrier Procedure," Final Report to the PMA/QE Section.

Schwartz, J. B., Flamholz, J. R., and Press, R. H. (1973). Computer Optimization of Pharmaceutical Formulation, I. General Procedure, *J. Pharm. Sci.*, 62(7): 1165–1170.

USP. (1975). *United States Pharmacopeia*, 19th revision, USP Convention, Inc., Rockville, Maryland.

WHO Expert Committee on Specification for Pharmaceutical Preparation. (1975). World Health Organization Technical Report, Series No. 567, WHO, Geneva.

Wilde, D. J. and Beightler, C. S. (1967). *Foundations of Optimization*, Prentice Hall, Englewood Cliff, N.J.

Wilk, M. B. and Shapiro, S. S. (1968). The Joint Assessment of Normality of Several Independent Samples, *Technometrics*, 10: 825–839.

# 5 Preclinical Safety Development

MURRAY R. SELWYN *Statistics Unlimited, Inc., Auburndale, Massachusetts*

## I. INTRODUCTION

Before a drug can be tested in a clinical program in humans, it must first undergo a substantial amount of safety testing in laboratory animals in order to reduce the potential risk to human subjects. Short-term studies in animals precede the initial studies in humans, with longer studies conducted to permit further testing.

The proper design, conduct, analysis, and interpretation of animal safety studies are critical to the successful introduction of a new drug entity. First and foremost is the predictive power of animal testing for toxicity in humans. When the results of animal studies indicate potentially serious side effects, drug development is either terminated or suspended pending further laboratory investigations of the problem. At this point, development of backup compounds (e.g., analogs) may be undertaken to replace the initial candidate. When results are "clean," drug development proceeds in the normal manner. Second, to conduct a comprehensive safety assessment program, companies must invest a substantial amount of resources in personnel, facilities, time, and money. In particular, each long-term study takes at least 3 years before final results can be obtained and costs as much as $1,000,000.

Current standards for toxicity testing have evolved only modestly since the 1960s and most of today's standard protocols follow along the lines of guidelines published at that time (D'Aguanno, 1973; World Health Organization, 1966). Thus, the program for safety evaluation is well structured and the range of experimental design issues is perhaps more limited than in the clinical area.

Although the basic set of studies and their designs have not changed much in the past 20 years, the actual conduct of studies has been greatly influenced by the adoption of U.S. Food and Drug Administration (FDA) Good Laboratory Practice (GLP) regulations in 1978 (U.S. FDA, 1978). The GLPs require extensive record-keeping for nonclinical safety studies.

In addition, they emphasize the need for statistical concepts such as randomization and proper statistical analysis of data (p. 60018). The GLPs also provide for a distinct quality assurance unit (QAU) to independently monitor studies. Thus, while the interest of the GLPs is to ensure the reliability of study data, they also increase the overhead associated with the conduct of these studies.

The initial studies on drug toxicity are single-dose or *acute* studies. Following these initial studies, multiple-dose studies of limited duration are conducted. Range-finding studies are typically of about 2 weeks duration; they provide information on the type of effects to be expected with multiple dosing and data useful in dose selection for longer-term studies. Repeat dose studies of 30 to 90 days duration are called *subacute* (sometimes also *subchronic*) studies. For the purpose of exposition, range-finding and subacute studies will be treated together, since their design and analysis aspects are very similar. Studies in which drugs are administered over prolonged periods of time are called *chronic* studies. These studies are often conducted for the better part of an animal's lifetime, especially in rodents. When one of the major objectives of a chronic study is to determine whether the drug causes an increase in tumors in laboratory animals, and in particular in malignant tumors, the study is also termed a *carcinogenicity* study. The timing of these studies in relation to the Phase I–III clinical program is depicted in Table 1.

In addition to the above studies, reproductive studies which measure the effects of drug administration on general reproductive performance and

**Table 1** Relation of Safety Assessment to Clinical Program[a]

| Duration of human study | Toxicology requirement |
|---|---|
| | Phase I and Phase II |
| Up to 2 weeks | Acute studies in three or four species (one nonrodent) plus 2-week subacute studies in two species (one nonrodent) |
| Up to 3 months | Acute studies plus 4-week subacute studies in two species (one nonrodent) |
| 6 months or more | Acute studies plus 3-month subacute studies |
| | Phase III |
| Single dose to several days | Same as Phases I and II |
| Up to 3 months | 3-month subacute studies in two species (one nonrodent) |
| 6 months or more | 6-month subacute studies in two species (one nonrodent) |

[a]*Source*: Traina (1983).

on the development of the embryo, fetus, and newborn and conducted before Phase III.

Since the mid-1970s, a battery of short-term in vitro (e.g., test tube) tests have also been introduced which complement the usual program of in vivo (whole live animal) experiments described above.

In the remainder of this chapter, we provide further details about the animal safety studies mentioned above, with particular emphasis on the statistical aspects of these studies. For an overview of these studies from a toxicologist's perspective, the reader is directed to the review article by Traina (1983). In Section II we discuss the design and analysis of acute studies. Section III presents corresponding material for subacute studies, and Section IV discusses chronic/carcinogenicity studies. Reproductive studies are covered in Section V. In Section VI, aspects of in vitro testing are presented.

## II. ACUTE STUDIES

Acute studies provide the first indications of the safety/toxicity of a new drug. Initial studies are conducted in mice and rats.

The major objective of these studies is to establish what dose levels of the drug will be lethal to the test animals. Lethality can be characterized in several ways, e.g., the greatest dose which is not lethal, the smallest dose which is 100% lethal, or the dose which kills a specified percentage of the animals. The first two of these depend on sample sizes in the experiment, whereas the latter is independent of sample size. The most common measure of acute toxicity is the median lethal dose or $LD_{50}$, i.e., the dose which kills 50% of the animals.

In addition to information on the $LD_{50}$, acute studies also provide information useful in predicting toxic reactions following acute overdose in humans. When drugs are administered both intravenously and orally, acute studies provide information about drug absorption. Results from acute studies are also helpful in dose selection for subsequent studies.

The typical design of an acute study consists of administering the drug to the same number of animals at several dose levels. Other approaches have been proposed, such as sequential testing with different doses given to individual animals, and doses adjusted up or down depending on the outcome (Dixon, 1965), but such designs are not in common use in the pharmaceutical industry.

The purpose of a statistical analysis is to provide point and interval estimates for the $LD_{50}$ and/or other percentages. The analysis is accomplished by assuming a mathematical model for the death probability. Perhaps the most common such model is the probit model:

$$P(d) = \Phi(A \log_{10}(d) + B) \tag{1}$$

where P(d) is the probability of death given dose d, $\Phi$ is the normal cumulative distribution function, and A and B are parameters of the model.

Instead of the normal distribution function, other probability models such as the logistic or Weibull could be used. For moderate percentages, and especially for the median, similar results are obtained with most such models. Moreover, with small experiments, such as those typical of most acute studies, one cannot distinguish between these various models.

Parameters of the model in equation (1) are estimated via maximum likelihood. Given maximum likelihood estimates A and B, the $LD_{50}$ is calculated in the following manner.

Let l be $\log_{10}(LD_{50})$. Since $\Phi(Al + B) = .5$, $Al + B = 0$, or $l = -B/A$. Therefore the maximum likelihood estimate of l is given by

$$\hat{l} = -\hat{B}/\hat{A}$$

Confidence interval estimates for l are obtained by a fiducial argument. For any l, Al + B is just a linear combination of maximum likelihood estimates. Its asymptotic variance is therefore

$$V_{AA}l^2 + 2lV_{AB} + V_{BB} \tag{2}$$

where $V_{AA}$, $V_{BB}$, and $V_{AB}$ are the asymptotic variances of A, B, and the asymptotic covariance of A and B. Therefore with appropriate probability,

$$\frac{(\hat{A}l + \hat{B})^2}{V_{AA}l^2 + 2lV_{AB} + V_{BB}} \leqslant Z^2 \tag{3}$$

where Z is a suitable percentile from the standard normal distribution. Solving the quadratic inequality (3) for l gives

$$\frac{\hat{l} + (\hat{l} + V_{AB}/V_{AA})g}{1 - g}$$

$$\pm \frac{[V_{BB} + 2\hat{l}V_{AB} + \hat{l}^2V_{AA} - g(V_{BB} - V_{AB}^2/V_{AA})]^{1/2}Z}{\hat{A}(1 - g)} \tag{4}$$

where $g = Z^2V_{AA}/\hat{A}^2$.

The final step in the process is to convert back to the dose scale by taking antilogs (base 10).

Suppose, for example, the following proportions are observed at doses 1, 3, 10, 30, and 100:

2/10, 3/10, 5/10, 7/10, 8/10

The maximum likelihood estimates of A and B are $\hat{A} = .889$ and $\hat{B} = -.881$ with covariance matrix

$$\begin{bmatrix} V_{AA} & V_{AB} \\ V_{AB} & V_{BB} \end{bmatrix} = \begin{bmatrix} .0798 & -.0789 \\ -.0789 & .1142 \end{bmatrix}$$

The maximum likelihood estimate of the $\log_{10}(LD_{50})$ is therefore .881/.889 = .99 and the estimated $LD_{50}$ is $10^{.99} = 9.8$.

Substituting in (4) gives .991 ± .535 for 95% confidence limits. On the original dose scale, these limits become 2.9 and 33.6.

For points other than the median, the procedure works similarly. For any percentage p, set $\Phi(Al + B) = p$. Thus,

$$Al + B = Z_p$$

and

$$\hat{l} = \frac{Z_p - \hat{B}}{\hat{A}} \qquad (5)$$

Since p is fixed, $Var(\hat{A}l + \hat{B} - Z_p)$ is again given by (2), and confidence limits are given by (4), with the difference that now $\hat{l}$ is determined by the more general equation (5).

Continuing with the above numerical example, suppose that we are also interested in a point estimate and confidence limits for the $LD_{10}$. Using (5),

$$\hat{l} = \frac{-1.28 + .881}{.889} = -.45$$

Substituting this value into (4) yields confidence limits on the log scale of $-1.36 \pm 1.56$. The estimated $LD_{10}$ is therefore $10^{-.45} = .4$ with 95% confidence limits of .001 to 1.6.

To illustrate the effects of dose selection and sample size on $LD_{50}$ estimates and their confidence limits, consider the sets of data presented in Table 2. Comparison of results from data set 2 with those from data set 1 shows the increase in precision associated with larger sample sizes. Data sets 3 and 4 compare two three-point designs and indicate that doses tightly clustered around the $LD_{50}$ do not provide sufficient information about A and B to provide real solutions to the quadratic equation implied by (3). Comparison of data set 5 with data set 1 shows that although estimates and confidence limits can sometimes be obtained without intermediate dose points, the resulting confidence interval will generally be wider than when those doses are included. Finally, data set 6 shows the effect of the slope on the confidence limits. With a steep slope as in data set 6, the confidence interval is considerably narrower than with a shallower slope, as in data set 1.

In generalizing inferences from these six artificial examples, it must be kept in mind that they are not intended to be representative of results typical of acute studies. Notice that each observed incidence at dose 10 is 50% and that incidences strictly increase with dose for each data set. Therefore, these results clearly underrepresent the amount of variation one would expect in practice. As a consequence, $LD_{50}$ estimates obtained from real experiments would be more variable than indicated in Table 2 and confidence intervals would also be wider. In fact, the reason that the confidence limits in data set 3 are not farther apart is that, for these data, the probit model fits the data exactly.

Despite the caveats, the general recommendations arising from these examples do provide guidance on experimental designs for acute studies.

**Table 2** Estimated $LD_{50}$ and Confidence Limits for Several Data Sets

| Data set | Response at dose | | | | | Estimated $LD_{50}$ | Lower 95% confidence limit | Upper 95% confidence limit |
|---|---|---|---|---|---|---|---|---|
| | 1 | 3 | 10 | 30 | 100 | | | |
| 1 | 2/10 | 3/10 | 5/10 | 7/10 | 8/10 | 9.8 | 2.9 | 33.6 |
| 2 | 1/6 | 2/6 | 3/6 | 4/6 | 5/6 | 9.8 | 1.5 | 63.8 |
| 3 | 1/6 | — | 3/6 | — | 5/6 | 10.0 | .4 | 232.0 |
| 4 | — | 2/6 | 3/6 | 4/6 | — | 9.7 | —[a] | —[a] |
| 5 | 2/10 | — | 5/10 | — | 8/10 | 10.0 | 1.4 | 73.6 |
| 6 | 0/10 | 1/10 | 5/10 | 9/10 | 10/10 | 9.8 | 5.8 | 16.4 |

[a]Confidence limits (at 95% level) cannot be obtained for this set of data.

Whenever possible, at least four or five doses should be studied, with doses selected to bracket the likely value of the $LD_{50}$. When there is considerable uncertainty about the $LD_{50}$, it is better to employ wider spacing between doses (see also Finney, 1971, pp. 496–499).

## III. SUBACUTE STUDIES

Repeat-dose studies are performed to determine whether repeated drug administration adversely affects animals in these studies and to characterize the nature of the effects. Subacute studies are conducted in at least one rodent species, often the rat, and in one nonrodent species, usually the dog or monkey. The duration of these studies is typically 30 or 90 days. Subacute stidies have several objectives: to examine toxicity in specific organs and organ systems (e.g., liver, heart, lung), to determine differences in species and sex sensitivities, to characterize dose-response, and to observe for drug tolerance and/or drug accumulation.

### A. Design Aspects

The most common design for subacute studies includes an untreated control group and three dose groups: at low, intermediate, and high levels of the drug. The highest dose is selected to produce overt toxicity. Traina (1983) recommends using multiples of the anticipated human clinical doses (10, 30, and 100× in rats; 5, 15, and 50× in nonrodents). The route of administration should be the same as the intended route in humans.

Sample sizes in subacute studies are typically small, especially for nonrodent studies. In keeping with standard scientific practices, animals should be randomized to treatment groups according to a well-defined stochastic (chance) procedure.

As stated earlier, the major objectives of subacute studies are to determine whether animals are adversely affected by repeated drug administration and, if this is the case, to characterize the nature of the dose-response. In Section III,C, we describe statistical methods for hypothesis testing which are useful in analyzing data from subacute studies. Such data analyses help the toxicologist to characterize compound toxicity in these studies.

Very often, repeat-dose studies will include pretreatment or baseline measurements on the animals. It is important to examine the baseline data to ensure group comparability. Section III,B discusses statistical methods for testing for equality at baseline.

### B. Tests for Equality at Baseline

Since one of the major objectives of subacute studies is to determine whether the drug causes adverse effects in animals on repeated dosing, it is therefore important, when differences are found, that they can be attributed to the drug and not to extraneous factors. This is one of the reasons behind randomization. When differences are found at a certain time point, are these drug effects or were similar differences present before treatment, i.e., at baseline?

Statistical tests for equality at baseline test the null hypothesis that measurements in all treatment groups come from the same population. The most common alternative hypothesis is that not all the group means are the same. Two statistical tests will be described in this section. The analysis of variance (ANOVA) F-test assumes that individual measurements are normally distributed, that the groups have equal variances, and that, if there are differences between groups, these are reflected in the group means. The second test is nonparametric; i.e., the normality assumption is not necessary.

The assumption of normality may appear at first quite restrictive. In practice, variables are often analyzed by taking appropriate transformations (e.g., log, square root) to induce approximate normality on the transformed variables. Thus, the choice between parametric and nonparametric methods is not always clear. In very small studies* it is especially difficult to verify the basic assumptions of homogeneity of variance and normality, thus making the choice of statistical method one of personal preference.

We first describe the ANOVA F-test. This procedure should be available in almost every computerized statistical package, but we will provide the details for its implementation anyway. We use the same notation in Section III,C in our exposition of trend tests.

Assume I groups with $n_i$ observations in the ith group, i = 1, 2, . . ., I. Let $y_{ij}$ be the jth observation in the ith group. The F statistic for treatment group equality is given by the ratio of the between-group mean square to the within-group mean square,

$$F = \frac{SS_B/(I - 1)}{SS_W/(N - 1)} \tag{6}$$

where $N = \Sigma_{i=1}^{i=I} N_i$ and the between- and within-group sums of squares are

$$SS_B = \sum_{i=1}^{i=I} n_i(\bar{y}_i - \bar{y})^2 \tag{7}$$

and

$$SS_W = \sum_{i=1}^{i=I} \sum_{j=1}^{j=n_i} (y_{ij} - \bar{y}_i)^2 \tag{8}$$

where the mean for the ith group is $\bar{y}_i = \Sigma_{j=1}^{j=n_i} y_{ij}/n_i$ and the grand mean is $\bar{y} = \Sigma_{i=1}^{i=I} n_i \bar{y}_i/N$.

---

*One approach to the question of transformation of data in small studies is to accumulate data from a number of such studies into a single data base. By examining the distributional properties of variables in the data base, a general plan can be formulated regarding transformations or the use of nonparametric techniques.

The quantities in (6), (7), and (8) are displayed in an ANOVA table. For example, consider the data given in Table 3. The ANOVA table for these data are presented in Table 4. If these were baseline data, we would conclude that the groups were not significantly different at baseline at the .05 level of significance.

The Kruskal-Wallis test (e.g., see Hollander and Wolfe, 1973) is a nonparametric test similar to the F-test, but based on ranks. Consider the data structure just described. Let $r_{ij}$ be the rank of $y_{ij}$ with observations ranked from smallest to largest (i.e., the smallest observation gets a rank of 1 and the largest observation gets a rank of N). Let $\bar{r}_i$ be the mean rank in the ith group. Then the Kruskal-Wallis statistic is given by

$$KW = \frac{12}{N(N+1)} \sum_{i=1}^{i=I} n_i \left( \bar{r}_i - \frac{N+1}{2} \right)^2 \tag{9}$$

assuming that there are no ties in the data. When ties occur, the most common approach (see Hollander and Wolfe, 1973) is to use the statistic

$$KW' = \frac{KW}{1 - \Sigma_{j=1}^{j=g} (t_j^3 - t_j)/(N^3 - N)} \tag{10}$$

where g is the number of sets of tied observations and $t_j$ is the number of observations in the jth set, j = 1, 2, . . ., g. Tied observations receive the average rank among the set of ties. When there are no ties in the data, (10) reduces to (9). Either form of the Kruskal-Wallis statistic is asymptotically distributed as a $\chi^2$ with I − 1 degrees of freedom. For the sample data in Table 3, the $\chi^2$ statistic (10) is $\chi_3{}^2 = 7.18$ (p = .07).

Another approach to correcting for ties in the data was proposed by Quade (1966), who suggested obtaining the ranks of the data and then performing an analysis of variance on the ranks. Since almost all statistical packages contain ANOVA, this approach is easily implemented. For the sample data in Table 3, the F statistic on ranks is $F_{3,16} = 3.24$ (p = .05).

Suppose that a test for baseline equality has been conducted and it indicates statistically significant differences between groups. How should one proceed in subsequent analyses of the data?

First, it must be kept in mind that when many parameters are being analyzed statistically, it is likely that some significant differences will arise

**Table 3** Sample Data Set

| Group number | Observations |
|---|---|
| 1 | 1.5, 2.3, 2.8, 2.7, 3.9 |
| 2 | 3.5, 2.8, 3.2, 3.5, 2.3 |
| 3 | 3.3, 3.6, 2.6, 3.1, 3.2 |
| 4 | 3.7, 3.8, 3.7, 3.4, 3.8 |

Table 4 Analysis of Variance (ANOVA) for Sample Data of Table 3

| Source | Degrees of freedom | Mean square | F | P |
|---|---|---|---|---|
| Between groups | 3 | .914 | 3.10 | .057 |
| Within groups | 16 | .295 | — | — |
| Corrected total | 19 | — | — | — |

due to chance. Are these differences biologically meaningful? Do all observations fall within the normal range expected for the parameter? Are the differences so large as to compromise the study?

Second, one should examine the individual animal data to see whether observed differences are being generated as result of one or two unusual observations. If so, are there data errors? Can the unusual observations be explained in some way?

Finally, if there is nothing suspect about the data and the differences are not so large as to compromise the study, one must plan a strategy for subsequent analyses. Perhaps the most common approaches to adjusting for baseline differences are based on utilizing baseline measurements in further statistical analyses, e.g., by analyzing differences from baseline and/or treating the baseline measurement as a covariate.

In closing our discussion of tests for baseline equality, we return to the assumptions required for our analyses. Both the parametric approach and the nonparametric approach assume equality of variance among the groups. The F-test assumes normality as well.

Both tests are impaired when groups have different variances.* Apparently the F-test is more sensitive to unequal variances than to nonnormality (see Scheffe, 1959).

We present two procedures for checking the assumption of equal group variances. The first is Bartlett's test (Snedecor and Cochran, 1980). Assume I groups each of size n. Bartlett's test compares the statistic

$$B = \frac{(n - 1)\left[I \log \bar{s}^2 - \Sigma_{i=1}^{i=I} \ln s_i^2\right]}{1 + (I + 1)/(3I(n - 1))} \tag{11}$$

with a $\chi^2_{I-1}$. Here $s_i^2$ is the sample variance in the ith group, $\Sigma_{j=1}^{j=n}(y_{ij} - \bar{y}_i)/(n - 1)$, and $\bar{s}^2$ is the average of the $s_i^2$. The rationale for the statistic B can be understood by noticing that the numerator of B is $(n - 1)I$ times (log of the arithmetic mean of the $s_i^2$ – log of the geometric mean of the $s_i^2$). When the $s_i^2$ are all close together, the arithmetic mean and geometric mean will be close. When the $s_i^2$ are discrepant, the

*If in fact groups have different variances, then one may argue that they are not equivalent at baseline.

arithmetic mean will be much larger than the geometric mean and hence lead to a large value of B.

The B statistic can be compared to a $\chi_{I-1}^2$ as a test of significance.

For the sample data in Table 3, the within-group sample variances are .758, .263, .133, and .027. Bartlett's statistic (11) has the value 8.56 ($p = .036$), giving some indication that the variances are unequal.

Levene (1960) proposed another ingenious test for the equality of group variances. Consider each observation's absolute deviation from its group mean, $|y_{ij} - \bar{y}_i|$. In groups with small variances, observations will be clustered about group means and therefore the $|y_{ij} - \bar{y}_i|$ will be small. In contrast, in groups with large variances, the $|y_{ij} - \bar{y}_i|$ will be large. Therefore, to test for equality of group variances, Levene suggested performing an ANOVA on the $|y_{ij} - \bar{y}_i|$!

For the sample data set of Table 3, the F statistic on the $|y_{ij} - \bar{y}_i|$ is $F_{3,16} = 1.94$ ($p = .16$). In this example, Levene's method does not give as strong an indication of heterogeneity of variance as does Bartlett's test.

If there is an indication of unequal variances among the groups, how should the data analysis proceed?

Often the cause of the heterogeneity of variances can be observed by plotting the data. Sometimes one or two unusual observations lead to unusually large group variances. Again, suspect observations should be investigated. Sometimes variances are large because the observations themselves are large. If one plots sample variances or sample standard deviations versus group means and a clear-cut relationship is observed, then a transformation of the original data, say using logs or square roots, will often stabilize the variances among the groups.

Another approach is to use "robust" statistical methods, i.e., methods which are insensitive to the unequal variances (e.g., see Welch, 1951; Brown and Forsythe, 1974).

### C. Tests for Trend

After baseline measurements are made, drug treatment begins. At periodic intervals or at study termination, measurements are made on the animals (see Table VII of Traina, 1983). In this section, we discuss statistical methods for the analysis of such subsequent measurements, taking into account the study design and likely responses.

In particular, we define a collection of "trend tests." These tests (1) take into account the ordering of the treatment groups, (2) provide an "overall" test at a predefined level of significance (e.g., .01 or .05), (3) allow for subsequent statistical tests comparing subsets of groups (e.g., control, low, and intermediate), and (4) have high power when the group means are approximately ordered (i.e., increase or decrease) with dose.

The rationale for employing trend tests rather than procedures with more general alternatives is that we believe that certain alternatives are more likely than others. Specifically, it can be argued that, in many cases, responses will be more severe with increasing dose levels. Several such dose-response functions are depicted in Figure 1. Figure 1a shows a dose-response which is essentially linear. In Figure 1b, the response decreases with dose but flattens out at the highest level. Figure 1c presents a situation where only the high dose exhibits a response different from the control.

The various dose-response functions depicted in Figure 1 share the property of monotonicity; i.e., there are no changes in direction. The

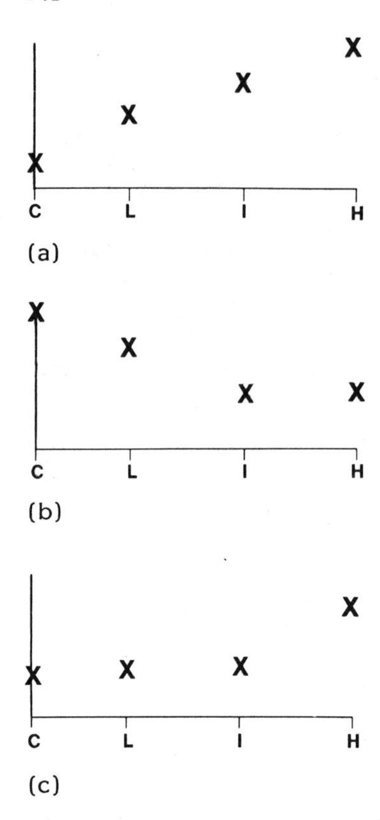


Fig. 1 Likely dose-response functions.

collection of trend tests discussed below is specifically designed to have high power against monotone alternatives.

In a study with three nonzero dose groups, trend tests can be implemented according to the strategy depicted in Figure 2. First, an overall trend test is performed. If the result is not significant, no further statistical tests are conducted, and the null hypothesis of no difference among treatment means is accepted. If a significant result is obtained in the overall trend test, then one concludes that there is an effect at the high dose and statistical testing continues. Testing is stopped whenever a nonsignificant result is obtained.

Whenever this procedure is followed, evidence of an effect at any level will imply effects at all higher levels tested. Conversely, if there is no evidence of an effect at a given level, then it is also concluded that there are no effects at all lower levels tested. Thus, interpretation of results from trend tests implemented in this manner will be clear-cut.

By analogy with our discussion of tests for equality at baseline, we present both parametric and nonparametric tests for trend. These make the same assumptions as did the baseline tests: equal group variances with potential group differences (i.e., a trend in dose) reflected by a shift in the group means. The parametric test also assumes normality.

For the purpose of exposition, we assume that we are interested in a test of equal group means versus the alternative that the means are nondecreasing with increasing dose, with at least one inequality. Assuming four treatment groups, the null and alternative hypotheses are

$$H_0: \mu_1 = \mu_2 = \mu_3 = \mu_4$$

versus

$$H_A: \mu_1 \leqslant \mu_2 \leqslant \mu_3 \leqslant \mu_4$$

with at least one inequality. Corresponding tests for decreasing means and for a two-sided alternative are similarly represented. The details of these situations follow from the test for increasing trend.

Perhaps the simplest trend test is one based on linear contrasts of the sample means. Let such a linear contrast be denoted by

$$\sum_{i=1}^{i=4} c_i \bar{y}_i \qquad \text{where} \qquad \sum_{i=1}^{i=4} c_i = 0$$

Such linear contrasts provide natural representations for comparisons of interest. With four treatment groups, the following set of contrast coefficients might be used in trend tests:

| $c_1$ | $c_2$ | $c_3$ | $c_4$ | Interpretation of contrast |
|---|---|---|---|---|
| −1 | 0 | 0 | 1 | High versus control |
| −1 | 1/3 | 1/3 | 1/3 | Average of all treated groups versus control |
| −3 | −1 | 1 | 3 | Equal spacing between groups |

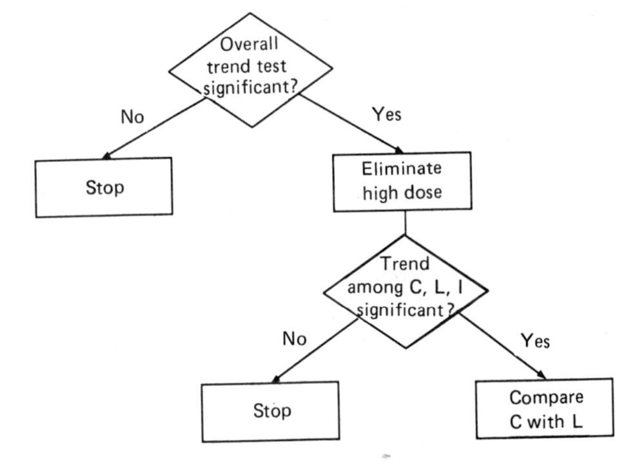


**Fig. 2** Strategy for trend tests (four groups).

For any such contrast, $\Sigma_{i=1}^{i=4} c_i \bar{y}_i$, its variance is $\Sigma_{i=1}^{i=4} c_i^2 \text{Var}(\bar{y}_i)$ which can be estimated by $\Sigma_{i=1}^{i=4} c_i^2 s^2/n_i$, where $s^2$ is the pooled estimate of the population variance $\sigma^2$. A test for trend is then defined as

$$T = \frac{\Sigma_{i=1}^{i=4} c_i \bar{y}_i}{\left(\Sigma_{i=1}^{i=4} c_i^2 s^2/n_i\right)^{1/2}} \tag{12}$$

which, under the null hypothesis, is distributed as a Student-t random variable with $\nu$ degrees of freedom, where

$$\nu = \sum_{i=1}^{i=4} n_i - 4$$

For the data in Table 3, assume that group 1 is the control, group 2 is the low dose, and so on. Then the linear contrasts defined above result in

| $c_1$ | $c_2$ | $c_3$ | $c_4$ | T | P (one-sided) |
|---|---|---|---|---|---|
| −1 | 0 | 0 | 1 | 3.03 | .004 |
| −1 | 1/3 | 1/3 | 1/3 | 2.35 | .016 |
| −3 | −1 | 1 | 3 | 2.96 | .005 |

It is easily seen that the above comparisons are considerably more significant than the overall analysis of variance F-test (p = .06). Since the four treatment means (2.64, 3.06, 3.16, 3.68) are ordered from smallest to largest, any statistical test which takes this ordering into account will be much more powerful for detecting treatment effects.

Another parametric test for trend is the "trend version" of the F-test. Recall that the F statistic (6) with I groups and N − I degrees of freedom for errors is

$$F = \frac{SS_B/(I - 1)}{SS_W/(N - 1)}$$

and that the total sum of squares is $SS_T = SS_B + SS_W$. The idea in the $\bar{E}^2$ test (Barlow et al., 1972), if one is testing for an increasing trend, is to "amalgamate" means so that they are nondecreasing and then calculate the between-group sum of squares. The amalgamated means are the maximum likelihood estimates under $H_A$. The test statistic is

$$\bar{E}^2 = \frac{SS_B \text{ (after amalgamation)}}{SS_T} \tag{13}$$

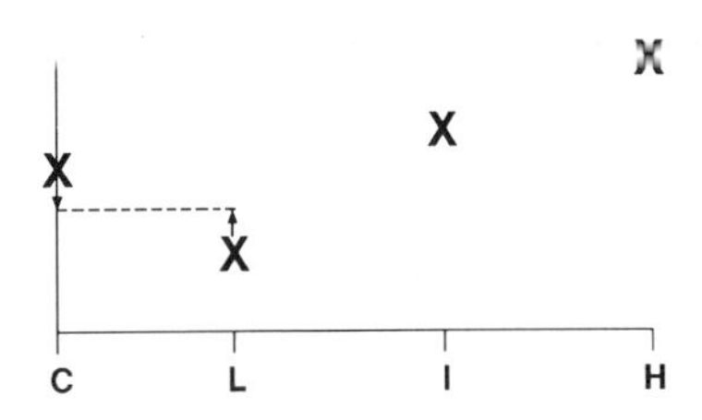


Fig. 3 Illustration of amalgamation of means.

What do we mean by amalgamation? Suppose we observed the set of means depicted in Figure 3. They would be nondecreasing except for the slight decrease between the control and low-dose means. Since our goal is to amalgamate whenever means decrease, we therefore amalgamate these two groups so that they have the same means (see graphical illustration in Figure 3).

In practice, such amalgamations must be performed whenever the monotonicity rule is violated. One algorithm for accomplishing this is the "pool-the-adjacent-violators" algorithm (Barlow et al., 1972). Table 5 displays several examples of the amalgamation process to produce nondecreasing means. In the first case sample means are initially increasing so that no amalgamation is necessary. In the second case the means are decreasing, and the end result is that all groups must be amalgamated into a single group. In cases three and four, only a single amalgamation each is necessary.

The $\bar{E}^2$ statistic takes the value .367 ($p = .010$) for the sample data of Table 3.

The final test for trend is a nonparametric version of $\bar{E}^2$. The test is like $\bar{E}^2$ since one calculates amalgamated means. It is also like the Kruskal-Wallis test since it is based on ranks. The test statistic (Barlow et al., 1972) is

$$\bar{\chi}^2_{\text{rank}} = \frac{12}{N(N+1)} \sum_{i=1}^{i=4} n_i \left( R_i^* - \frac{N+1}{2} \right)^2 \tag{14}$$

Table 5 Some Examples of Amalgamation

| Before amalgamation | | | | | | | | After amalgamation | | | | | | | |
|---|---|---|---|---|---|---|---|---|---|---|---|---|---|---|---|
| Means | | | | Sample sizes | | | | Means | | | | Sample sizes | | | |
| 1 | 2 | 3 | 4 | 5 | 5 | 5 | 5 | 1 | 2 | 3 | 4 | 5 | 5 | 5 | 5 |
| 4 | 3 | 2 | 1 | 5 | 5 | 5 | 5 | | 2.5 | | | | 20 | | |
| 1 | 3 | 2 | 4 | 6 | 6 | 4 | 4 | 1 | 2.6 | | 4 | 6 | 10 | | 4 |
| 1 | 1 | 3 | 1 | 4 | 6 | 2 | 2 | 1 | 1 | | 2 | 4 | 6 | | 4 |

where $R_1^*$, $R_2^*$, $R_3^*$, and $R_4^*$ are the amalgamated mean ranks! The $\bar{\chi}^2_{rank}$ statistic for the sample data is 7.15 (p = .013).

Other tests for trend in addition to those discussed above are available, both parametric (e.g., see Williams, 1971, 1972) and nonparametric (e.g., Jonckheere, 1954; Shirley, 1977).

All these trend tests assume homogeneity of variance. For a discussion of trend tests which allow unequal group variances, see Roth (1984).

## IV. CHRONIC/CARCINOGENICITY STUDIES

Studies which exceed 90 days in duration are generally considered to be chronic studies (Traina, 1983). These studies are similar in design to subactue studies except that the doses employed are lower and hence less stressful to the test animals.

Chronic studies are conducted in rodents and in at least one nonrodent species. A study in dogs or monkeys will be a 1-year study. Studies in rats or mice will typically last 2 years. Since the rodent studies consider tumor incidence as an important endpoint, these also serve as carcinogenicity studies.

Laboratory tests are performed at 3-month intervals during the course of the study. Most 2-year studies also include a 1-year sacrifice for a small proportion of animals.

Statistical methods useful for the analysis of data from the in-life portion of carcinogenicity studies are the same as those for subacute studies, as discussed in Sections III,B and III,C. In Sections IV,A through IV,E, therefore, we discuss design and analysis aspects unique to carcinogenicity studies.

### A. Design Aspects

The typical design for a carcinogenicity study is the same as that used for a subacute study: an untreated or vehicle control group and three dosed groups at low, intermediate, and high levels of the drug. In selecting the highest dose, the aim is to select a dose which will produce only slight or moderate side affects and limited lethality. The dose used in the National Toxicology Program (NTP) is the "maximum tolerated dose" (MTD), which is defined by Haseman (1984) as "the highest dose of the test agent during the chronic study that can be predicted not to alter the animals' normal longevity from effects other than carcinogenicity." Often this definition is supplemented with (Sontag et al., 1976)

> MTD is the highest dose that causes no more than a 10% weight decrement, as compared to appropriate control groups and does not produce mortality, clinical signs of toxicity or pathological lesions (other than those that may be related to a neoplastic response) that would be predicted to shorten the animals' lifespan.

Although this definition is acceptable in principle, several authors (e.g., Ciminera, 1985) argue that it is often difficult to predict toxicity in chronic studies from shorter-term results. Excessive mortality in the high-dose group generally complicates the statistical analysis of tumor data (see

Section IV,D). They therefore recommend using 1/2 – 2/3 MTD in the high-dose group.

Once the highest dose is determined, the low and intermediates doses are generally specified as fractions of this dose. Haseman (1984) recommends 20 to 30% for the low dose and 50% for the intermediate dose.

Calculation of adequate sample sizes for carcinogenicity studies is difficult because of the potential multiplicity of endpoints and variety of available statistical procedures. Despite this, properties of the "standard design" have been investigated using the simplest statistical procedure: Fisher's exact test for a 2 × 2 table of lifetime tumor incidence. Examination of Table 1 of Haseman (1984) shows that studies with 50 aminals per sex per dose group have very low power to detect differences in tumor rates. This situation is only slightly improved with 100 animals per sex per dose group. Because of this, careful consideration should be given to sample size selection based on likely spontaneous rates and differences to be detected for the tumor types of greatest interest.

Given the magnitude, importance, and cost of a chronic/carcinogenicity study, careful consideration should also be given to other design aspects. One potentially important factor may be the correspondence between treatment assignment and cage location. Ciminera (1985) suggests possible effects of cage location (top, middle, bottom) on mortality and/or leukemias and proposes designs which block accordingly.

In any case, the statistician and study director should have sufficient knowledge of the properties of their testing facility to identify any potential sources of bias, i.e., potentially confounding factors, and eliminate them or minimize their effects.

## B. Survival Estimates and Tests

One of the primary questions in long-term studies is how the drug affects the longevity of the animals. In addition to the interest in survival in its own right, differences in survival are important with regard to analyses related to carcinogenicity (see Section IV,D).

### 1. Survival Estimates

A useful summary of the survival experiences in each treatment group is provided by the sample survival function. If there are no intermediate sacrifices, then in a group of size n, the sample survival function would be decremented by 1/n each time a death occurs. Even with censoring of death times for some animals (e.g., because of intermediate sacrifices) the survival can easily be estimated.

Consider just a single treatment group with death times $t_1 < t_2 < \cdots < t_l$. At the ith such time, let $R_i$ be the number of animals at risk, i.e., the number alive just prior to $t_i$, and let $m_i$ denote the number of deaths at time $t_i$. Then the survival function is estimated as the product-limit estimate (Kaplan and Meier, 1958):

$$\hat{S}(t) = \prod_{i:t_i \leqslant t} \left(1 - \frac{m_i}{R_i}\right) \tag{15}$$

An indication of the uncertainty in S is provided by Greenwood's formula (Greenwood, 1926):

$$\hat{\mathrm{Var}}[\hat{S}(t)] = \hat{S}^2(t) \sum_{i:t_i \leqslant t} \frac{m_i}{R_i(R_i - m_i)} \tag{16}$$

Notice that (15) is a nonparametric estimate of the survival function. If one assumes a particular parametric form for the survival function (or equivalently the death distribution), then techniques for parametric estimation of parameters such as maximum likelihood can be used (e.g., see Kalbfleisch and Prentice, 1980). Of course, intermediate and final censoring must be taken into account.

## 2. Tests for Difference in Survival

In testing for difference in survival, consider the ith death time ($t_i$). At $t_i$, the data in an experiment with control and three dose groups may be summarized in a 2 × 4 table:

| | Control | Low | Intermediate | High | |
|---|---|---|---|---|---|
| Number of deaths at $t_i$ | $m_{01}$ | $m_{1i}$ | $m_{2i}$ | $m_{3i}$ | $m_i$ |
| | | | | | |
| Number at risk | $R_{0i}$ | $R_{1i}$ | $R_{2i}$ | $R_{3i}$ | $R_i$ |

Given the total number of deaths $m_i$ at time $t_i$ and the number at risk in each group ($R_{0i}$, $R_{1i}$, $R_{2i}$, $R_{3i}$), the distribution of the ($m_{0i}$, $m_{1i}$, $m_{2i}$, $m_{3i}$) is hypergeometric under the null hypothesis. By comparing the set of observed deaths to its expectation, assuming no group differences, a statistical test is constructed as follows.

Let $\underline{O}_i$ be a row vector of observed death at time $t_i$ and $\underline{e}_i$ be the corresponding expected value under the null hypothesis. Further, let $\underline{V}_i$ be the variance-covariance matrix under the hypergeometric model. These are defined as

$$e_{ji} = \frac{m_i R_{ji}}{R_i}$$

and

$$V_{jk}(i) = \begin{cases} \dfrac{R_{ji}(R_i - R_{ji})m_i(R_i - m_i)}{R_i^2(R_i - 1)} & \text{for } j = k \\[2ex] \dfrac{-R_{ji}R_{ki}m_i(R_i - m_i)}{R_i^2(R_i - 1)} & \text{for } j \neq k \end{cases}$$

Since all marginal totals are fixed, one can sum over the death times to obtain the overall logrank statistic

$$LR = (\underline{O} - \underline{e})\underline{V}^{-1}(\underline{O} - \underline{e})^T \qquad (17)$$

where $\underline{O} = \Sigma_{i=1}^{i=1} \underline{O}_i$, $\underline{e} = \Sigma_{i=1}^{i=1} \underline{e}_i$, and $\underline{V} = \Sigma_{i=1}^{i=1} \underline{V}_i$. In practice, since each $(\underline{O}_i - \underline{e}_i)$ sums to zero, it is convenient to drop one of the components, e.g., the one corresponding to the control group.

Under the null hypothesis LR is distributed as a chi-square with 3 degrees of freedom.

The form of the logrank statistic (17) provides a test with the global alternative hypothesis that the survival curves differ among the groups. A test of interest in chronic/carcinogenicity studies is concerned with a potential dose-related trend in survival. Accordingly, a trend version of the logrank test (Tarone, 1975) is given by

$$Z = \frac{[\underline{d}\ (\underline{O} - \underline{e})]}{(\underline{d}\ \underline{V}\ \underline{d}^T)} \qquad (18)$$

where $\underline{d} = (d_1, d_2, d_3)$ are the doses in the experiment.

## C. Tests for Lifetime Tumor Incidence

The simplest analysis of tumor data from carcinogenicity studies is based on the lifetime incidences of tumor in each treatment group. Lifetime incidences are appealing because they are simple to estimate, have an easy interpretation, and have a long history of use. Some of their drawbacks are discussed in Section IV,D on time-adjusted methods.

### 1. Fisher's Exact Test

Among statistical tests of lifetime tumor incidence, perhaps the one which is most widely used is Fisher's exact test, which compares tumor proportions with data presented in the form of a 2 × 2 table. For example, consider the comparison between a single treatment group and the control group:

| | Treated | Control | |
|---|---|---|---|
| Animals with tumor | a | b | a + b |
| Animals without tumor | c | d | c + d |
| Total examined | a + c | b + d | n |

Since we are interested in testing whether the tumor rate in the treated group is equal to the tumor rate in the controls (null hypothesis) against the (alternative) hypothesis that the tumor rate is higher in the treated group, we want to use a one-sided statistic.

When the set of marginal totals a + b, a + c, b + d, c + d is regarded as fixed, then under the null hypothesis of no treatment effect, the distribution of the cell entries in the 2 × 2 table is hypergeometric. This allows us to determine how "extreme" the observed table is when the null hypothesis holds.

What do we mean by extreme? Certainly "more extreme" tables would have more animals with tumor in the treated group (relative to controls), i.e., cell entries a + 1, a + 2, . . . .

Therefore, the p value for Fisher's exact test of the above 2 × 2 table is given by

$$p = \sum_{x=a}^{x=\min(a+b,\ a+c)} \frac{(a+b)!(a+c)!(b+d)!(c+d)!}{n!x!(a+b-x)!(a+c-x)!(d-a+x)!} \tag{19}$$

Consider for example the 2 × 2 table

| | Treated | Control |
|---|---|---|
| Tumor | 6 | 0 |
| No tumor | 44 | 50 |

The p value from Fisher's exact test for this table is

$$p = \frac{6!\,50!\,50!\,94!}{100!\,6!\,0!\,44!\,50!} = \frac{50!\,94!}{100!\,44!}$$

$$= \frac{50 \cdot 49 \cdot 48 \cdot 47 \cdot 46 \cdot 45}{100 \cdot 99 \cdot 98 \cdot 97 \cdot 96 \cdot 95} = .013$$

## 2. Trend Tests

While Fisher's exact test is useful in comparing a single treated group to controls, we continue to argue that tests of trend are more appropriate than pairwise comparisons (see discussion in Section III,C). Analogously to the 2 × 2 table, data for three dosed groups and control can be summarized in a 2 × 4 table:

| | Control ($d_0 = 0$) | Low ($d_1$) | Intermediate ($d_2$) | High ($d_3$) | |
|---|---|---|---|---|---|
| Animals with tumor | $x_0$ | $x_1$ | $x_2$ | $x_3$ | $x$ |
| Animals without tumor | | | | | |
| | $n_0$ | $n_1$ | $n_2$ | $n_3$ | $n$ |

To assess whether tumor rates increase with increasing dose, one could plot the observed incidences (the $x_i/n_i$) versus dose (the $d_i$). One measure of the strength of the relationship between dose and tumor incidence is the correlation between them. In fact, the Cochran-Armitage test for trend (Cochran, 1954; Armitage, 1955) is based on this correlation. The test statistic is of the form

$$CA = \frac{\Sigma_{i=1}^{i=3} x_i d_i - \hat{p}\Sigma_{i=1}^{i=3} n_i d_i}{[\hat{p}(1 - p)(\Sigma_{i=1}^{i=3} n_i d_i^2 - (\Sigma_{i=1}^{i=3} n_i d_i)^2/n)]^{1/2}} \qquad (20)$$

where $\hat{p} = x/n$, the overall proportion of animals with tumor in the experiment.

The statistic CA is compared to a standard normal to assess significance. This test can also be derived as a score test assuming a logistic dose-response model (Tarone and Gart, 1980).

As an illustration of the CA test, consider the 2 × 4 table:

| | Dose 0 | 1 | 2 | 3 |
|---|---|---|---|---|
| Animals with tumor | 0 | 1 | 3 | 6 |
| Animals examined | 50 | 50 | 50 | 50 |

The CA statistic has the value 2.90 ($p = .002$). Notice that this statistic is considerably more significant than Fisher's exact test comparing the high dose to controls. By including intermediate results from the two middle groups, the overall dose-response and test of significance are strengthened. On the other hand, if the two middle groups exhibit incidences both higher or both lower than the two extreme groups, the value of CA would not be as large as it is with a strictly increasing dose-response.

While the distribution theory for the Cochran-Armitage test depends on large-sample properties of the test statistic, alternative trend tests based on permutations can also be employed (e.g., see Bickis and Krewski, 1985). When the total number of tumors (x) is small, the calculations for permutation tests are manageable. In such case, they provide a closer analog to Fisher's exact test.

### 3. Historical Controls

When carcinogenicity studies are conducted as a routine part of the safety testing of new drug entities, these studies are often conducted in the same testing facility and under identical or very similar protocols. One would therefore expect that the control groups would respond similarly from experiment to experiment. Is it possible to make use of information outside the realm of the current experiment in drawing inferences about the effects of the drug under study?

In practice, this is routinely done by all scientists in a somewhat informal way since they use their judgment and experience in interpreting experimental results. In particular, such judgments lead to the inequivalence between "biological" and "statistical" significance. The use of formal statistical methods incorporating historical incidences can be thought of as a replacement for and hopefully an improvement on their use in informal and subjective ways.

The goal of doing this is to sharpen the inferences that can be made through the use of historical control incidences. The historical control data will be especially useful with rare tumors or when the effects of treatment are uncertain based on the current experiment alone. For historical experiments to be relevant and useful, several conditions should be met:

1. The historical experiments should be conducted in the same species and strain, preferably with animals from the same breeder.
2. Protocols for all experiments should be virtually identical. Any discrepancies should not be expected to affect the outcome. In particular, lengths of the studies should be the same.
3. The response variable of interest (tumor) should have a consistent definition throughout the studies.
4. The experiments should be conducted in the same laboratory with primarily the same investigators [pathologist(s)].
5. There should be no other conditions which would cause one to eliminate any of the historical experiments (e.g., differences in survival).

These conditions are quite restrictive. While some of the factors mentioned above are less important than others, some are critical in constructing a valid set of historical studies. In reviewing tumor incidences obtained from studies in the National Cancer Institute/National Toxicology Program, Haseman et al. (1984) cite species, strain, sex, duration of study, protocol, laboratory, nomenclature, quality assurance procedures, and type of control group (e.g., untreated or corn oil gavage) as important. In these studies, supplier and differences in pathologists within laboratories were less important.

Before any formal statistical analysis incorporating historical controls is conducted, it is advisable to examine both the incidence data themselves and data on survival as a check on condition 5 above. Plots of tumor incidence versus study date will provide information on time trends in tumor incidences. Plots of survival curves and/or other measures of survival (e.g., median survival times or proportion surviving at 2 years) for all studies can also be informative. Formal significance tests can be performed as discussed in Section IV,B.

Several statistical methods for incorporating historical control incidences are currently available. Tarone (1982) assumes a logistic model in dose for the tumor incidence rate and assumes that the probability of tumor in controls is distributed according to a beta distribution from experiment to experiment. He then derives a modified form of the Cochran-Armitage statistic (20) as his test statistic. The modified statistic depends on the parameters of the beta, which are estimated via maximum likelihood from the historical data. Hoel (1983) similarly assumes an underlying beta model for the probability of tumor and derives an exact conditional test for an elevated incidence rate in a treated group, where the conditioning is on the observed tumor count in the current controls. Dempster et al. (1983) also assume a dose-response model of the logistic form. Their approach to experiment-to-experiment variability in control tumor rates assumes that the logit of the tumor rate (or log-odds of tumor) is a normally distributed variable. Dempster et al. implement a fully Bayesian procedure and their test for a treatment effect is based on comparing the estimated parameter in the logistic model to its approximate posterior standard deviation.

With any of these approaches, the degree to which the historical controls influence the analysis depends on the experiment-to-experiment variability. With little experiment variability, results are repeatable and hence the historical data can be relied on. On the other hand, with large variability from experiment to experiment, the results are not as repeatable, and hence the historical controls do not receive as much weight in the analysis.

Hoel (1983) also proposes that historical data can be used in a quality control sense; this idea is currently being pursued by Krewski et al. (1985).

### D. Time-Adjusted Tests for Tumors

The use of lifetime tumor incidences is clearly an attractive way to summarize tumor data. These overall proportions are easy to interpret. They can be compared with previous or future results (as in Section IV,C), and they lead to simple statistical tests. Overall proportions can be misleading, however, when survival distributions are different across treatment groups in an experiment.

To see this, consider two simple examples. Suppose we can classify deaths as "early deaths" or "late deaths" as in the following 2 × 2 tables:

"Early deaths"

| | Control | Treated |
|---|---|---|
| Tumor | 0 | 0 |
| No tumor | 30 | 10 |

\+

"Late deaths"

| | Control | Treated |
|---|---|---|
| T | 10 | 20 |
| NT | 10 | 20 |

=

Total

| | Control | Treated |
|---|---|---|
| T | 10 | 20 |
| NT | 40 | 30 |

Note that for both treated and control, there are no tumors observed for early deaths and the tumor rate is 50% among animals classified as dying late in the study. When summarized in a single 2 × 2 table, tumor rates are 20% in controls and 40% in the treated group, but really there is only a difference in survival, not in tumor rates.

As a second example, consider

"Early deaths"

| | Control | Treated |
|---|---|---|
| Tumor | 0 | 0 |
| No tumor | 10 | 40 |

+

"Late deaths"

| | Control | Treated |
|---|---|---|
| T | 10 | 10 |
| NT | 30 | 0 |

=

Total

| | Control | Treated |
|---|---|---|
| T | 10 | 10 |
| NT | 40 | 40 |

Here tumor rates are considerably different, but this is masked when using overall proportions because of differences in survival. Although these examples are admittedly extreme, they do point out the difficulties in using overall proportions when there are differences in survival.

Even when the survival experiences of the groups are very similar, routine use of time-adjusted methods may be preferable to the analysis of simple proportions (Ryan, 1985).

When the time at risk, i.e., time to death or sacrifice, is utilized, then the proper statistical analysis also depends on the "context" of tumor observations. In Sections IV,D,1 and IV,D,2 we discuss time-adjusted statistical methods appropriate when the type of tumor can be globally classified as either rapidly lethan (Section IV,D,1) or nonlethal (Section IV,D,2) for all animals in an experiment. In Section IV,D,3 we deal with situations in which no such global assessment can be made.

### 1. Rapidly Lethal Tumors

When a type of tumor is considered to be "rapidly lethal," i.e., the tumor quickly leads to death of the animal, then the appropriate basis for judging tumor incidence at any point in time t at which there are deaths with tumor is the number of animals "at risk" at that time. The risk set is the set of animals alive just prior to t. For example, suppose there were 40 animals alive at day 649, two of which die on day 650, one from tumor. Then the risk set at day 650 would consist of the 40 animals alive just prior to that time.

Analysis of rapidly lethal tumors can be performed in the same way as survival analysis. Instead of looking at deaths from all causes, however, one considers only deaths from the tumor type of interest. For each time $t_i$ at which a tumor death occurs, a 2 × 4 table is constructed as in

Section IV,B,2 with events of interest being the number of tumor deaths at $t_i$. Both the logrank and Tarone's test for trend can be applied to these data.

An alternative to the logrank test is the generalized Wilcoxon (Gehan, 1965), which weights each $(O_i - E_i)$ by the number at risk $R_i$. Thus tumors that occur early in a study are given more importance than tumors that occur toward the end of the study.

## 2. Nonlethal Tumors

By nonlethal or incidental tumors, we mean tumors that do not lead to death of the animal, but are undetectable until death. Skin tumors or tumors detectable by palpation would not be in this category since they could be detected at time of occurrence. In fact, these types of tumors can be statistically analyzed in the same way as rapidly lethal tumors.

For nonlethal tumors, an appropriate basis for judging tumor incidence cannot be from live animals, since the tumor is detectable only at death. Therefore the most common approach to analyzing nonlethal tumors has been to consider animals dying at about the same time and to group these observations together for the purpose of estimating tumor prevalence. Hoel and Walburg (1972) suggested the idea of using time intervals and considering the number of animals in each group dying in an interval and, of those, the number with tumor. Again assuming four treatment groups, the data could be summarized in a set of 2 × 4 tables:

| | Control | Low | Intermediate | High |
|---|---|---|---|---|
| Number of animals with tumor | | | | |
| Number dying in ith interval | | | | |

By again considering all marginal totals as fixed, a Mantel-Haenszel statistic (Mantel and Haenszel, 1959; Mantel, 1963) is constructed and used for an overall test. To illustrate this concept, Hoel and Walburg employed 100-day intervals (0–99, 100–199, etc.) to define their sets of tables.

An alternative approach to defining intervals was provided by Peto et al. (1980). The motivation for their approach is as follows. If the tumor is truly nonlethal, then animals that die should be representative of living animals. The probability of having a tumor should increase the longer an animal lives. Therefore, the tumor prevalence rate should be increasing as a function of time.

Peto's algorithm for defining intervals can be characterized as follows:

1. Without regard to dose group membership, order all death times and indicate tumors by T and nontumor deaths by N. If tumor deaths and nontumor deaths occur at the same time, put T's before N's.

2. For the initial set of intervals, each except the first starts with T's and ends with N's. The next interval starts with the next set of T's.
3. For each initial interval, calculate the observed tumor prevalence. If the prevalence of two adjacent intervals does not increase, amalgamate these intervals.
4. Repeat step 3 until the prevalence for the set of remaining intervals is increasing.
5. Study termination is treated separately.

To illustrate the algorithm, consider the following sequence of tumor (T) and nontumor deaths (N):

N N T N N N T N N T T N N T N N N

The initial intervals are obtained by starting each interval except the first with T's and ending with N's:

N / T N N N / T N N / T T N N / T N N N

0 1/4 1/3 2/4 1/4

Since the prevalence decreases from the fourth (2/4) to the fifth (1/4) interval, these are amalgamated:

N N / T N N N / T N N / T T N N T N N N

0 1/4 1/3 3/8

Since tumor prevalence is now increasing, no further amalgamation takes place.

While Peto's algorithm produces intervals with an overall increasing prevalence, the prevalence is not necessarily increasing in each group. Further, when all deaths in an interval occur in the same group, the data from that interval do not contribute to the Mantel-Haenszel statistic (see Selwyn, 1986).

We now briefly describe two alternative approaches for analyzing nonlethal tumors. Dinse and Lagakos (1983) propose a logistic model:

$$\log\left[\frac{p(t, d)}{1 - p(t, d)}\right] = \alpha + \beta d + \gamma t \tag{21}$$

where p(t, d) is the probability of tumor at time t and dose d. Tests for a dose effect are based on the score test for $\beta$. Dinse (1985a) presents simulation results comparing the logistic score test to the Peto et al. method.

Another approach to the analysis of nonlethal tumors was proposed by Selwyn et al. (1985). The major difficulties in applying Peto's algorithm arise when the survival experiences are different across treatment groups. Selwyn et al. suggested using separate sets of time intervals for the different groups to estimate prevalences. They define "weighted prevalence" estimates for each treatment group, which can then be used to test for group differences.

### 3. Incorporating Cause of Death

When tumors are not uniformly rapidly lethal or uniformly nonlethal, then the methods described in Sections IV,D,1 and IV,D,2 cannot be applied directly. If, however, one can classify deaths for each animal into one of three categories—(1) animal dies with tumor as cause of death, (2) animal dies with tumor but from another cause, or (3) animal dies without tumor—then the intermediate results obtained via the lethal and nonlethal analyses can be combined into an overall test incorporating cause of death.

Peto et al. (1980, p. 371) suggest combining the O − E scores from the logrank (LR) or Tarone analysis with the O − E scores from the nonlethal or prevalence (P) analysis and also combining their variances. Thus an overall test can be based on

$$\frac{O_{LR} - E_{LR} + O_P - E_P}{(V_{LR} + V_P)^{1/2}} \tag{22}$$

Here the O's and E's are formed by summing over times and doses, weighting the total for each dose by $d_j$. Similarly, the V's are calculated as in the denominator of (18).

Lagakos and Louis (1985) further segment the O, E, and V for prevalences into natural or sacrifice deaths. This distinction is important only if there is one or more planned interim sacrifice time (e.g., at 1 year) in the study and tumors are detected in the sacrificed animals.

Several authors have investigated the joint nonparametric estimation of tumor mortality and tumor prevalence when cause of death can be determined for each animal. Kodell et al. (1982) derive maximum likelihood estimates of these functions when the ratio of disease resistance to survival is monotone decreasing. Dinse and Lagakos (1982) and Turnbull and Mitchell (1984) discuss methods of obtaining nonparametric maximum likelihood estimates without this restriction. In addition, parametric approaches to this problem have been studied.

## E. Unresolved Issues

Throughout Section IV we have concentrated on the statistical analysis of a single type of tumor. In most carcinogenicity studies, however, histopathological evaluations are conducted on 30 to 40 organs and tissues. Since as many as 40 statistical tests may be conducted simultaneously, the overall type I error rate for a study may be considerably higher than the nominal level employed in an individual statistical test. This potential problem was first recognized by Salsburg (1977), but its importance has been debated (e.g., see Fears et al., 1977).

Haseman (1983) undertook an investigation of the experimentwise error rate, using data from 25 studies conducted under the National Toxicology Program. Haseman found that overall type I error rates are acceptable if statistical tests are conducted at the .05 level for rare tumors ($\leq$1–2% historical incidence) and at the .01 level for common tumors. Such a recommendation seems to provide a straightforward approach to controlling the experimentwise error rate. A more sophisticated approach is that of Meng and Dempster (1985), who propose a simultaneous analysis of multiple tumor types based on Bayesian methods.

In Section IV,D,3, we showed how cause of death determined for each animal can be incorporated into the statistical analysis. Pathologists differ considerably in their willingness to assign causes of death and in their belief in the certainty of these assignments. Several attempts have been made to take this uncertainty into account, e.g., by using a four-point scale: tumor cause of death, tumor probably cause of death, tumor probably not cause of death, tumor not cause of death (see Peto et al., 1980; Dinse, 1985b; Racine-Poon and Hoel, 1984).

Scientists also disagree on the relevance of historical information in analyzing a randomized experiment. Some scientists feel that the best and perhaps only appropriate control group is the one in the current experiment. Others feel that historical data are relevant. A list of restrictions which the historical experiments must meet has been provided in Section IV,C,3, and we feel that the focus is on data which should be very comparable to the current experiment.

The direction for statistical analysis in the future will be to routinely incorporate time-to-rumor. At present, the alternative strategies for doing this include:

1. Make assumptions about the global lethality of the tumor; if the tumor is rapidly lethal or exclusively nonlethal, methods are available.
2. Assign cause of death for each animal; either parametric or nonparametric models can then be used.
3. Redesign experiments to incorporate periodic serial sacrifices.

These options and their implications are discussed in the review article by McKnight and Crowley (1984). In this chapter, we have concentrated on options 1 and 2, since most carcinogenicity studies currently conducted in the pharmaceutical industry have no provisions for frequent interim sacrifices. Current statistical research in this area is being undertaken (Dewanji and Kalbfleisch, 1985; Louis, 1985). If the potential information derived from such designs is great enough to outweigh the additional costs involved in implementing them, then study designs should be amended in this way.

## V. REPRODUCTIVE STUDIES

Reproductive studies are carried out to assess the drug's effect on fertility and conception and on the fetus and developing offspring. Among reproductive studies most commonly performed in pharmaceutical research, three types or "segments" are standard: a segment I study of fertility and general reproductive performance, a segment II teratological study, and a segment III study of perinatal and postnatal drug effects. Each segment corresponds to a separate time segment before and during the gestation period. The studies are typically performed in rats and have four dose groups (untreated controls and low, intermediate, and high levels of the drug).

In a segment I study, a minimum of 12 male rats per group are administered the drug for at least 60 days. At least 24 virgin female rats per group also receive the drug for at least 14 days. For both males and females, body weights and feed intakes are recorded weekly. The males and females are mated to obtain at least 20 pregnant females per does group. One-half of the females per group are allowed to litter, and for these a record is made of the gestation length, number of dead fetuses in the uterus, number of stillbirths, number of viable newborns, and weight of each. For these dams (mothers), treatment is continued for 35 days. The dams' body weights and feed intakes are measured weekly. The survival and individual weights of the pups are recorded on days 7, 14, 21, 28, and 35.

For the remaining one-half per dose group, the females are killed on day 13 of gestation. The parameters measured are the weight of the uterus and its contents, number of corpora lutea, number of implants, number of resorptions, and number of viable embryos.

In a segment II study, untreated male and untreated virgin female rats are mated to obtain at least 24 pregnant females per dose group. The females are administered the drug from day 6 through day 15 of gestation. Body weights are measured on days 0, 6, 13, 16, and 20; feed intake is measured daily.

The dams are killed and the fetuses delivered by cesarean section 1 or 2 days prior to term. At this time, measurements are taken on the weight of the uterus and its contents, number of corpora lutea, number of implants, number of resorptions, number of live and dead fetuses, and weight of each live fetus and placenta. One-third of the fetuses per litter are examined for visceral abnormalities and two-thirds of the fetuses per litter are examined for skeletal abnormalities.

In a segment III study, untreated male and untreated virgin female rats are also mated to obtain at least 24 pregnant females per dose group. The females are administered the drug from day 15 of gestation through parturition. Body weights are measured on days 0, 6, 13, 15, and 20; feed intake is measured daily.

The females deliver their litters. The length of gestation, number of dead fetuses in the uterus, number of stillbirths, number of viable newborns, and their weights are recorded. Treatment of the dams is continued for 5 weeks. The body weights of dams are recorded weekly and their feed intakes recorded daily. Survival and pup weights are assessed on days 4, 7, 21, 28, and 35.

The parameters measured in segment I, segment II, and segment III studies can be classified into four types for the purpose of statistical analyses:

1. Measurements made on dams, e.g., body weights, feed intake
2. "Count data," e.g., number of corpora lutea, number of implants, number of resorptions
3. "Proportion or percentage data," e.g.,

$$\text{Percent resorptions} = \frac{\text{number of resorptions}}{\text{resorptions + viable implants}}$$

4. "Pup parameters," e.g., pup weights, placenta weights

Parameters of type 1 or 2 can be analyzed by parametric or nonparametric trend tests as discussed in Section III. In some cases, transformations will be necessary with the parametric approach. In Section V,A, we present statistical methods for pup data and in V,B we discuss statistical methods for percentages/proportions.

## A. Analysis of Pup Parameters

A unique feature of reproductive studies is the fact that treatment is administered to the dam, whereas some measurements are made on their offspring, the pups. It is well known that pups from the same litter are more likely to be similar than pups from different litters. Statistical methods for analyzing pup data, therefore, must account for the correlation between pups from the same litter.

In analyzing continuous data, the statistical model which corresponds to this biological phanomenon postulates two sources of variation: the dam (variance $\sigma_1^2$) and the pup (variance $\sigma_2^2$). Thus the total variation for each pup measurement is $\sigma_1^2 + \sigma_2^2$ and the correlation between any two pups in the same litter is $\rho = \sigma_1^2/(\sigma_1^2 + \sigma_2^2)$. In a litter of size N, the variance of the litter mean is $\sigma_1^2 + \sigma_2^2/N$.

"Extreme cases" occur when the dam variance dominates the pup variance ($\sigma_1^2 \gg \sigma_2^2$) and vice versa ($\sigma_2^2 \gg \sigma_1^2$). In the former case $\rho$ is close to 1, so that all measurements within a litter are very similar. Also,

$$\sigma_1^2 + \sigma_2^2/N \approx \sigma_1^2$$

so that each litter contributing to a treatment group mean receives about the same weight.

In the latter case, $\rho$ is close to 0 and

$$\sigma_1^2 + \sigma_2^2/N \approx \sigma_2^2/N$$

so that weights are roughly proportional to litter sizes (since weighting is by the inverse of the litter variance).

The idea of estimating $\sigma_1^2$ and $\sigma_2^2$ and then using them to get a weighted average for each treatment group in an experiment (with weights inversely proportional to $\sigma_1^2 + \sigma_2^2/N$) was first proposed by Healy (1972). Healy then used weighted t-tests to test for treatment differences. The variances $\sigma_1^2$ and $\sigma_2^2$ were estimated by equating observed and expected mean squares in an ANOVA table.

Dempster et al. (1984) suggested several refinements to Healy's approach. Noting that pup weights are larger in small litters and that male pups are heavier than females, these authors introduced litter size and sex as covariates in the analysis. In addition, they employed a restricted maximum likelihood method for estimating $\sigma_1^2$ and $\sigma_2^2$, rather than the ANOVA approach. The following example, taken from the Dempster et al. paper, illustrates the important asepcts of the method.

Pup weights from an experiment with a control group and two dose groups are summarized as follows:

| Treatment group | Number of litters | Number of pups | Mean birth weight[a] (g) |
|---|---|---|---|
| Control | 10 | 131 | 6.32 |
| Low dose | 10 | 126 | 5.93 |
| High dose | 7 | 65 | 5.89 |

[a]Weighted by number of pups per litter.

Inspection of the group means indicates that both treated groups have lower birth weights than the controls. Differences from the control group means are −.39 for the low dose and −.43 for the high dose.

Figure 4 displays litter means as a function of litter size. Note that average birth weights are lower in larger litters and also that the high-dose litters have fewer pups. Table 6 presents the results for the statistical analysis of these data by the method of Dempster et al. (1984).

Both covariates, litter size and sex, are highly significant, as are the two treatment contrasts. The estimated difference between the low dose and control is −.43 (similar to the original estimate of −.39). The estimated difference for the high dose is −.86, considerably larger than the naive estimate of −.43. Estimates of $\sigma_1^2$ and $\sigma_2^2$ are .163 and .097 for this set of data.

Experience with the Dempster et al. method for pup parameters has shown that estimates of treatment effects and covariates are not very

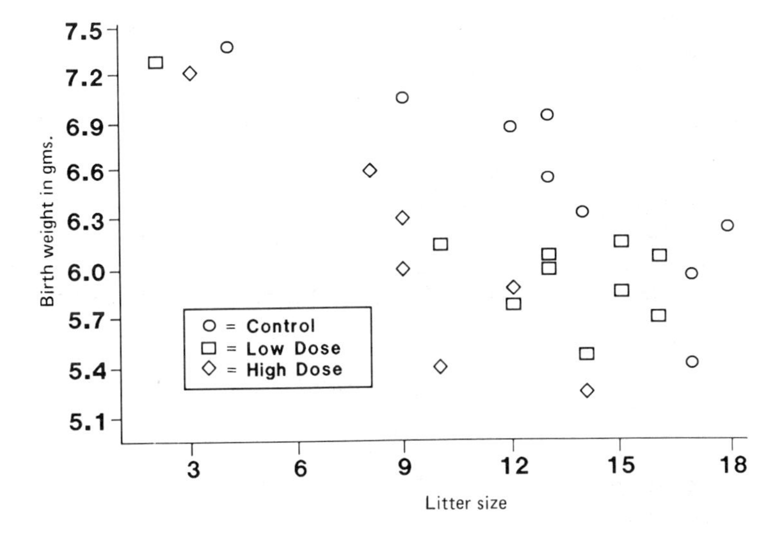


Fig. 4 Plot of litter means versus litter size. (Data from Dempster et al., 1984).

Table 6 Analysis of Pup Weights[a]

| Parameter | Estimate | Standard deviation | Normal value |
|---|---|---|---|
| Litter size | −.129 | .019 | −6.86 |
| Sex (F-M) | −.359 | .047 | −7.56 |
| Control | 8.310 | .274 | 30.36 |
| Low dose-control | −.429 | .150 | −2.85 |
| High dose-control | −.859 | .182 | −4.72 |

[a]*Source*: Dempster et al. (1984).

sensitive to different $\sigma_1^2$ and $\sigma_2^2$ values. However, incorporation of covariates for litter size and sex is very important. By including these effects in the model, the statistical analysis becomes much more powerful for detecting treatment differences, as demonstrated by the above example.

### B. Analysis of Litter Proportions

In analyzing the percentage or proportion in each litter with or without a certain abnormality, litter effects may also be present. For this reason, the standard binomial model may not hold. In fact, empirical investigations have demonstrated such extrabinomial variation (Haseman and Soares, 1976).

There are three basic approaches to dealing with extrabinomial variation. The first is to employ a parametric model which allows for it. The second is to use transformations to induce approximate normality. The last is to use nonparametric techniques.

A common model used in the reproductive study setting is the beta-binomial (Williams, 1975; Kupper et al., 1986). The idea is that for each litter the observed proportion, say x/n, estimates a true rate p. The p's may not be common across all litters in a group, but vary because of a litter effect. The approach assumes that (1) given p, x is binomial, and (2) the p's vary according to a beta distribution.

Hence, the name beta-binomial. The beta distribution provides a rich set of distributions to characterize the distribution of p: distributions with most of the probability close to 0 (e.g., for rare abnormalities), distributions with most of the probability close to 1 (e.g., for survival), and unimodal densities.

One transformation which is applied to binomial data is the arcsine square root transformation. This method has been used with proportions in reproductive studies.

Lastly, one can employ nonparametric approaches such as analyzing ranks of proportions (e.g., $\bar{\chi}^2_{\text{rank}}$ test on proportions). Gladen (1979) compares several of these methods.

As a numerical example, consider the set of day 4 survival data presented by Williams (1975):

| Group | Survival on day 4 |
|---|---|
| Control | 13/13, 12/12, 9/9, 9/9, 8/8, 8/8, 12/13, 11/12, 9/10, 9/10, 8/9, 11/13, 4/5, 5/7, 7/10, 7/10 |
| Treated | 12/12, 11/11, 10/10, 9/9, 10/11, 9/10, 9/10, 8/9, 8/9, 4/5, 7/9, 4/7, 5/10, 3/6, 3/10, 0/7 |

Application of the beta-binomial model to these data results in a $\chi^2$ statistic (1 degree of freedom of 5.77 ($p < .02$ for a two-sided test).

For this particular example, application of the beta-binomial model seems to give a more appropriate result than two of the methods recommended by Gladen (1979). Analyzing the ranks of the x/n data yields a $\chi_1^2$ statistic of 1.96 ($p = .16$). Use of the arcsine square root transformation on the observed proportions followed by a t-test gives a t value of 1.78 ($p = .09$). With the arcsine square root method, there is also strong indication of unequal group variance, but even when this is taken into account the p value is unaffected.

While the beta-binomial approach is superior for the above example, the selection of the best method of analysis for general use still needs further study.

Diagnostic plotting techniques can be used to visually assess the adequacy of models such as the beta-binomial. Dempster et al. (1983) propose modified P-P plots for this purpose. Figures 5 and 6 present modified P-P plots for the above numerical example. Figure 5 shows that both the binomial and the beta-binomial provide an adequate fit to the control group data. For the treated group, Figure 6 portrays a much better fit for the beta-binomial as compared to the standard binomial model.

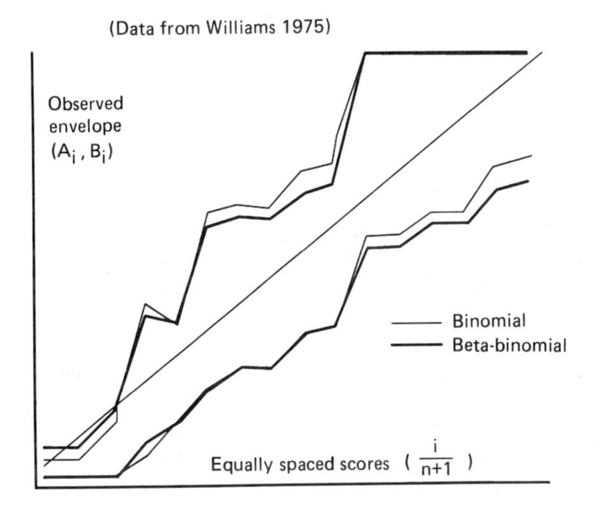


**Fig. 5** Modified P-P plot for controls. (Data from Williams, 1975.)

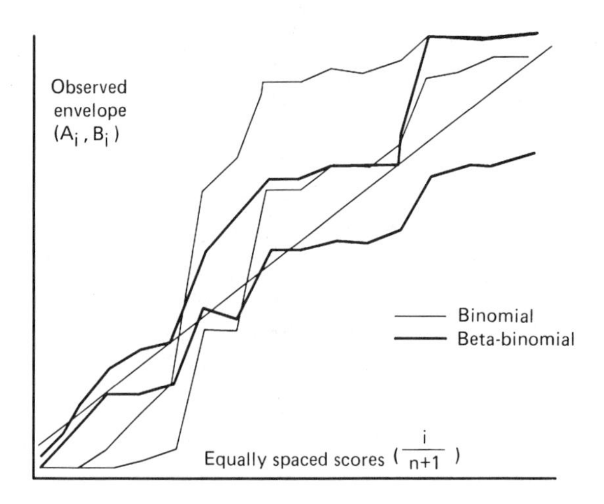


Fig. 6 Modified P-P plot for treated group. (Data from Williams, 1975.)

## VI. IN VITRO STUDIES

It is now common practice to include one or more in vitro tests for mutagenicity as part of the safety evaluation program for a new drug. One rationale for doing this is the association between the outcomes of such tests and carcinogenicity as determined by bioassays in animals (McCann et al., 1975). Although the strength of the association apparently depends on the class of compounds studied (Krewski et al., 1982), it is likely that such short-term tests have some predictive power.

Perhaps the most common test in use today is the Ames *Salmonella* microsome test (Ames et al., 1975). In Section VI,A we discuss statistical methods for the Ames test. In VI,B we present methods of analysis for other tests whose endpoints are plate counts and in VI,C methods where the endpoint is dichotomous.

### A. The Ames Test

The Ames test is based on the existence of auxotrophic and prototrophic strains of *Salmonella typhimurium*. These differ in that the prototrophic strains are able to synthesize histidine, an amino acid essential to growth, whereas the auxotrophic strains do not have this ability. The auxotrophic strains arise typically through mutation, but are capable of reverting back to prototrophy.

The experiment consists of treating replicate plates containing the auxotrophic strains of *S. typhimurium* and a very limited amount of histidine with several levels of the test compound. Untreated and positive controls are also included. The response variable is the number of revertant colonies (i.e., prototrophic strains) per plate at the various dose levels of the drug. (See Brusick, 1980, for sample protocol.)

A standard model for analyzing count data in microbiology is the Poisson model. Margolin et al. (1981) specify the assumptions under which the Poisson model holds for the Ames test. However, the same authors indicate that extra-Poisson variation is common among replicate plates within laboratories. To account for this extra variation, they allow the parameter of the Poisson model to vary according to a gamma density. The resulting distribution of plate counts becomes negative binomial.

To take into account the potential mutagenic and toxic effects as a function of dose administered, they further model the mean of the negative binomial $\mu$ as $N_0P_D$, where $N_0$ is the number of microbes placed on the plate and $P_D$, a function of the administered dose D, is the probability that a plated microbe yields a revertant colony. Several parametric forms of $P_D$ are studied, each including terms for toxicity as well as for mutagenicity. The parameters of each model are $\alpha$ (spontaneous mutation rate), $\beta$ (mutation rate per unit dose), and $\gamma$ (a measure of drug toxicity). Tests for mutagenicity are based on comparing the estimate of $\beta$ to its asymptotic standard deviation. In addition, Margolin et al. propose checks on suspect observations through the negative binomial model.

## B. Analysis of Count Data

The response variable in many in vitro test systems is a count. As mentioned earlier, a common model for count data is Poisson. Since the variance and the mean of a Poisson are equal, the ratio of the sample variance to the mean sample can be used to check the Poisson assumption. It is often found that extra-Poisson variation exists, as noted above for the Ames test. In addition to the specific models mentioned, several other approaches have been proposed.

In analyzing count data from the Chinese hamster ovary/hypoxanthine-guanine phosphoribosyltransferese (CHO/HGPRT) mutagenesis assay, Snee and Irr (1981) exhibited evidence of extra-Poisson variation in studies conducted over a period of 18 months with 29 chemicals. Their approach was to find a transformation (Box and Cox, 1964) which would induce approximate normality and equal variances on the transformed counts when analyzed via a two-way ANOVA model. The transformation of the original counts (x's)

$$y = (x + 1)^{.15} \tag{23}$$

was judged satisfactory for this purpose when extensive diagnostics were examined on the transformed data.

In analyzing dose-response experiments Snee and Irr utilized t-tests, comparing the results at each dose level with control, and linear contrasts for linear, quadratic, and higher-order effects in dose. Although linear contrasts are probably satisfactory, trend tests such as the $\bar{E}^2$ test (see Section III,C) should probably be applied to the transformed data rather than pairwise t-tests for the reasons given in Section III,C. If there is potential toxicity at high doses, alternatives other than those with strictly monotonic means should be considered.

Margolin et al. (1986) also demonstrate extra-Poisson variability in sister chromatid exchange counts from CHO cells in an in vitro cytogenic assay. Through Monte Carlo simulation, Margolin et al. investigated the properties of several methods of analysis for data from this assay. They concluded that among the methods studied, the Cochran-Armitage trend test was superior. Similar to the CA statistic for binomial data (20), the trend statistic for count data is of the form

$$\frac{\Sigma_{i=0}^{i=k} d_i(y_i - n_i\bar{y})}{\left[\bar{y}\Sigma_{i=0}^{i=k} n_i(d_i - \bar{d})^2\right]^{1/2}} \tag{24}$$

where $y_i$ is the sum of the observed counts at dose $d_i$, $n_i$ is the number of replicates at dose $d_i$, and $\bar{y}$ and $\bar{d}$ are the means of the y's and the d's.

## C. Analysis of Binomial Data

Methods for the analysis of binomial data arising from mutagenicity experiments are studied by Collings et al. (1981), Margolin et al. (1983), and also Margolin et al. (1986).

In comparing the results at a single dose to a control, Collings et al. (1981) recommend the use of a large-sample normal theory test based on the observed difference in proportions when the number of replicates (flasks, test tubes) is large and the event probability is moderate. If the observed proportion of mutants is $X_T/N_T$ for the treated group and $X_C/N_C$ for the control group, then the test statistic is of the form

$$\frac{X_T/N_T - X_C/N_C}{[\hat{P}(1 - \hat{P})(1/N_C + 1/N_T)]^{1/2}} \tag{25}$$

where $\hat{P} = (X_T + X_C)/(N_T + N_C)$.

In dose-response situations, the Cochran-Armitage trend test (20) is recommended, as is an isotonic trend test (Collings et al., 1981).

## REFERENCES


Ames, B. N., McCann, J., and Yamasaki, E. (1975). Methods for Detecting Carcinogens and Mutagens with the *Salmonella*/Mammalian-Microsome Mutagenicity Test, *Mutat. Res.*, 31: 347–364.

Armitage, P. (1955). Tests for Linear Trends in Proportions and Frequencies, *Biometrics*, 11: 375–386.

Barlow, R. E., Bartholomew, D. J., Bremner, J. M., and Brunk, H. D. (1972). *Statistical Inference Under Order Restrictions*, Wiley, New York.

Bickis, M. and Krewski, D. (1985). Statistical Design and Analysis of the Long-Term Carcinogenicity Bioassay, *Toxicological Risk Assessment*, vol. 1 (D. B. Clayson, D. Krewski, and I. Munro, eds.), CRC Press, Boca Raton, Fla.

Box, G. E. P. and Cox, D. R. (1964). An Analysis of Transformations (with Discussion), *J. R. Stat. Soc. Ser. B*, 26: 211–252.

Brown, M. B. and Forsythe, A. B. (1974). The Small Sample Behavior of Some Statistics Which Test the Equality of Several Means, *Technometrics*, 16: 385–389.

Brusick, D. (1980). *Principles of Genetic Toxicology*, Plenum, New York.

Ciminera, J. (1985). Some Issues in the Design, Evaluation, and Interpretation of Tumorigenicity Studies in Animals, *Proceedings of the Symposium on Long-Term Animal Carcinogenicity Studies: A Statistical Perspective*, American Statistical Association, Washington, D.C., pp. 26–35.

Cochran, W. G. (1954). Some Methods for Strengthening the Common $\chi^2$ Test, *Biometrics*, 10: 417–451.

Collings, B. J., Margolin, B. H., and Oehlert, G. W. (1981). Analyses for Binomial Data, with Application to the Fluctuation Test for Mutagenicity, *Biometrics*, 37: 775–794.

D'Aguanno, W. D. (1973). Drug Toxicity Evaluation. Pre-Clinical Aspects, *Introduction to Total Drug Quality*, No. 017-012-00220-0, Government Printing Office, Washington, D.C.

Dempster, A. P., Selwyn, M. R., and Weeks, B. J. (1983). Combining Historical and Randomized Controls for Assessing Trends in Proportions, *J. Am. Stat. Assoc.*, 78: 221–227.

Dempster, A. P., Selwyn, M. R., Patel, C. M., and Roth, A. J. (1984). Statistical and Computational Aspects of Mixed Model Analysis, *Appl. Stat.*, 33: 203–214.

Dewanji, A. and Kalbfleisch, J. D. (1985). Non-Parametric Methods for Survival/Sacrifice Experiments, *Proceedings of the Symposium on Long-Term Animal Carcinogenicity Studies: A Statistical Perspective*, American Statistical Association, Washington, D.C., pp. 100–106.

Dinse, G. E. (1985a). Testing for a Trend in Tumor Prevalence Rates: I. Nonlethal Tumors, *Biometrics*, 41: 751–770.

Dinse, G. E. (1985b). Evaluating Tumor Prevalence, Lethality, and Mortality, *Proceedings of the Symposium on Long-Term Animal Carcinogenicity Studies: A Statistical Perspective*, American Statistical Association, Washington, D.C., pp. 91–99.

Dinse, G. E. and Lagakos, S. W. (1982). Nonparametric Estimation of Lifetime and Disease Onset Distributions from Incomplete Observations, *Biometrics*, 38: 921–932.

Dinse, G. E. and Lagakos, S. W. (1983). Regression Analysis of Tumor Prevalence Data, *Appl. Stat.*, 32: 236–248.

Dixon, W. J. (1965). The Up-and-Down Method for Small Samples, *J. Am. Stat. Assoc.*, 60: 967–978.

Fears, T. R., Tarone, R. E., and Chu, K. C. (1977). False-Positive and False-Negative Rates for Carcinogenicity Screens, *Cancer Res.*, 37: 1941–1945.

Finney, D. J. (1971). *Statistical Methods in Biological Assays*, 2nd ed., Griffin, London.

Gehan, E. A. (1965). A Generalized Wilcoxon Test for Comparing Arbitrarily Singly-Censored Samples, *Biometrika*, 52: 203–224.

Gladen, B. (1979). The Use of the Jackknife to Estimate Proportions from Toxicological Data in the Presence of Litter Effects, *J. Am. Stat. Assoc.*, 74: 278–283.

Greenwood, M. (1926). A Report on the Natural Duration of Cancer: The Errors of Sampling of the Survivorship Tables, *Reports on Public Health and Medical Subjects*, 33: 23–25.

Haseman, J. K. (1983). A Re-Examination of False-Positive Rates for Carcinogenesis Studies, *Fundam. Appl. Toxicol.*, 3: 334–339.

Haseman, J. K. (1984). Statistical Issues in the Design, Analysis and Interpretation of Animal Carcinogenicity Studies, *Environ. Health Perspect.*, 58: 385–392.

Haseman, J. K. and Soares, E. R. (1976). The Distribution of Fetal Death in Control Mice and Its Implications on Statistical Tests for Dominant Lethal Effects, *Mutat. Res.*, 41: 277–288.

Haseman, J. K., Huff, J., and Boorman, G. A. (1984). Use of Historical Control Data in Carcinogenicity Studies in Rodents, *Toxicol. Pathol.*, 12: 126–135.

Healy, M. J. R. (1972). Animal Litters as Experimental Units, *Appl. Stat.*, 21: 155–159.

Hoel, D. G. (1983). Conditional Two-Sample Tests with Historical Controls, *Contributions to Statistics: Essays in Honor of Normal L. Johnson* (P. K. Sen, ed.), North Holland, Amsterdam, pp. 229–236.

Hoel, D. G. and Walburg, H. E. (1972). Statistical Analysis of Sruvival Experiments, *J. Nat. Cancer Inst.*, 49: 361–372.

Hollander, M. and Wolfe, D. (1973). *Nonparametric Statistical Methods*, Wiley, New York.

Jonckheere, A. R. (1954). A Distribution-Free k-Sample Test Against Ordered Alternatives, *Biometrika*, 41: 133.

Kalbfleisch, J. D. and Prentice, R. L. (1980). *The Statistical Analysis of Failure Time Data*, Wiley, New York.

Kaplan, E. L. and Meier, P. (1958). Nonparametric Estimation from Incomplete Observations, *J. Am. Stat. Assoc.*, 53: 457–481.

Kodell, R. L., Shaw, G. W., and Johnson, A. M. (1982). Nonparametric Joint Estimates for Disease Resistance and Survival Functions in Survival/Sacrifice Experiments, *Biometrics*, 38: 43–58.

Krewski, D., Clayson, D., Collins, B., and Munro, I. C. (1982). Toxicological Procedures for Assessing the Carcinogenic Potential of Agricultural Chemicals, *Genetic Toxicology* (R. A. Fleck and A. Hollaender, eds.), Plenum, New York.

Krewski, D., Smythe, R. T., and Colin, D. (1985). Tests for Trend in Binomial Proportions with Historical Controls: A Proposed Two-Stage Procedure, in press.

Kupper, L. L., Hogan, M. D., Portier, C., and Yamamoto, E. (1985). The Impact of Litter Effects on Dose-Response Modeling in Teratology, *Biometrics*, 42: 85–98.

Lagakos, S. W. and Louis, T. A. (1985). Statistical Analysis of Rodent Tumorigenicity Experiments, *Toxicological Risk Assessment*, vol. 1 (D. B. Clayson, D. Krewski, and I. Munro, eds.), CRC Press, Boca Raton, Fla.

Levene, H. (1960). Robuts Tests for Equality of Variances, *Contributions to Probability and Statistics*, (Ingram Olkin et al., eds.) Stanford University Press, Stanford, Calif., pp. 278-292.

Louis, T. A. (1985). Adaptive Sacrifice Plans for the Carcinogen Bioassay, *Proceedings of the Symposium on Long-Term Animal Carcinogenicity Studies: A Statistical Perspective*, American Statistical Association, Washington, D.C., pp. 36-41.

Mantel, N. (1963). Chi-Square Tests with One Degree of Freedom, Extensions of the Mantel-Haenszel Procedure, *J. Am. Stat. Assoc.*, 58: 690-700.

Mantel, N. and Haenszel, W. (1959). Statistical Aspects of the Analysis of Data from Retrospective Studies of Disease, *J. Natl. Cancer Inst.*, 22: 719-748.

Margolin, B. H., Kaplan, N., and Zeiger, E. (1981). Statistical Analysis of the Ames *Salmonella*/Microsome Test, *Proc. Natl. Acad. Sci. U.S.A.*, 78: 3779-3783.

Margolin, B. H., Collings, B. J., and Mason, J. M. (1983). Statistical Analysis and Sample-Size Determinations for Mutagenicity Experiments with Binomial Response, *Environ. Mutagen.*, 5: 705-716.

Margolin, B. H., Resnick, M. A., Rimpo, J. Y., Archer, P., Galloway, S. M., Bloom, A. D., and Zeiger, E. (1986). Statistical Analyses for In Vitro Cytogenic Assays Using Chinese Hamster Ovary Cells, *Environ. Mutagen.*, 8: 108-204.

McCann, J., Choi, E., Yamasaki, E., and Ames, B. N. (1975). Detection of Carcinogens as Mutagens in the *Salmonella*/Microsome Test: Assay of 300 Chemicals, *Proc. Natl. Acad. Sci. U.S.A.*, 72: 5135-5139.

McKnight, B. and Crowley, J. (1984). Tests for Differences in Tumor Incidence Based on Animal Carcinogenesis Experiments, *J. Am. Stat. Assoc.*, 79: 639-648.

Meng, C. Y. K. and Dempster, A. P. (1985). A Bayesian Approach to the Multiplicity Problem for Significance Testing with Binomial Data, *Proceedings of the Symposium on Long-Term Animal Carcinogenicity Studies: A Statistical Perspective*, American Statistical Association, Washington, D.C., pp. 66-72.

Peto, R., Pike, M., Day, N., Gray, R., Lee, P., Parish, S., Peto, J., Richards, S., and Wahrendorf, J. (1980). Guidelines for Simple Sensitive, Significance Tests for Carcinogenic Effects in Long-Term Animal Experiments, annex to *Long-Term and Short-Term Screening Assays for Carcinogens: A Critical Appraisal*, IARC Monogr. Suppl. 2, pp. 311-426.

Quade, D. (1966). On Analysis of Variance for the k-Sample Problem, *Ann. Math. Stat.*, 37: 1747-1758.

Racine-Poon, A. H. and Hoel, D. G. (1984). Nonparametric Estimation of the Survival Function When Cause of Death is Uncertain, *Biometrics*, 40: 1151-1158.

Roth, A. J. (1984). Robust Trend Tests Derived and Simulated: Analogs of the Welch and Brown-Forsythe Tests, *J. Am. Stat. Assoc.*, 78: 972-980.

Ryan, L. M. (1985). Efficiency of Age-Adjusted Tests in Animal Carcinogenicity Experiments, *Biometrics*, 41: 525–532.

Salsburg, D. (1977). Use of Statistics When Examining Lifetime Studies in Rodents to Detect Carcinogenicity, *J. Toxicol. Environ. Health*, 3: 611–628.

Scheffe, H. (1959). *The Analysis of Variance*, Wiley, New York.

Selwyn, M. R. (1986). The Hoel-Walburg Data: Revisited, *Proceedings of the Biopharmaceutical Section, ASA*, American Statistical Association, Washington, D.C., in press.

Selwyn, M. R., Roth, A. J., and Weeks, B. J. (1985). The Weighted Prevalence Method for Analyzing Nonlethal Tumor Data, *Proceedings of the Symposium on Long-Term Animal Carcinogenicity Studies: A Statistical Perspective*, American Statistical Association, Washington, D.C., pp. 85–90.

Shirley, E. (1977). A Non-Parametric Equivalent of William's Test for Contrasting Increasing Dose Levels of a Treatment, *Biometrics*, 33: 386–389.

Snedecor, G. W. and Cochran, W. G. (1980). *Statistical Methods*, 7th ed., Iowa State University Press, Ames.

Snee, R. D. and Irr, J. D. (1981). Design of a Statistical Method for the Analysis of Mutagenesis at the Hypoxanthine-Guanine Phosphoribosyl Transferase Locus of Cultured Chinese Hamster Ovary Cells, *Mutat. Res.*, 85: 77–93.

Sontag, J. M., Page, N. P., and Safiotti, U. (1976). *Guidelines for Carcinogen Bioassay in Small Rodents*, DHHS Publ. (NIH) 76-801, National Cancer Institute, Bethesda, Md.

Tarone, R. E. (1975). Tests for Trend in Life Table Analysis, *Biometrika*, 62: 679–682.

Tarone, R. E. (1982). The Use of Historical Control Information in Testing for a Trend in Proportions, *Biometrics*, 38: 215–220.

Tarone, R. E. and Gart, J. J. (1980). On the Robustness of Combined Tests for Trends in Proportions, *J. Am. Stat. Assoc.*, 75: 110–116.

Traina, V. M. (1983). The Role of Toxicology in Drug Research and Development, *Med. Res. Rev.*, 3: 43–72.

Turnbull, B. W. and Mitchell, T. J. (1984). Nonparametric Estimation of the Distribution of Time to Onset for Specific Diseases in Survival/Sacrifice Experiments, *Biometrics*, 40: 41–50.

U.S. Food and Drug Administration (U.S. FDA) (1978). Nonclinical Laboratory Studies. Good Laboratory Practice Regulations, *Federal Register*, 43: 59986–60025.

Welch, B. L. (1951). On the Comparison of Several Mean Values: An Alternative Approach, *Biometrika*, 38: 330–336.

Williams, D. A. (1971). A Test for Differences Between Treatment Means When Several Dose Levels Are Compared with a Zero Dose Control, *Biometrics*, 27: 103–117.

Williams, D. A. (1972). The Comparison of Several Dose Levels with a Zero Dose Control, *Biometrics*, 28: 519–531.

Williams, D. A. (1975). The Analysis of Binary Responses from Toxicological Experiments Involving Reproduction and Teratogenicity, *Biometrics*, 31: 949–952.

World Health Organization (WHO). (1966). Principles for Pre-Clinical Testing of Drug Safety, *Tech. Rep. Ser.*, 341: 3–22.

# 6 Clinical Development

B. E. RODDA, M. C. TSIANCO, J. A. BOLOGNESE, and M. K. KERSTEN
*Merck Sharp & Dohme Research Laboratories, Rahway, New Jersey*

## I. CLINICAL DEVELOPMENT OVERVIEW

### A. Introduction

This chapter describes the clinical development program as a whole, with emphasis on the scientific, statistical, medical, regulatory, and management issues which can affect the scientific validity of a program.

The objective of clinical research in the pharmaceutical industry is to determine the value of a drug or biologic in the prevention, treatment, or diagnosis of disease. This assessment is made by evaluating benefits of the agent relative to its risks and undesirable effects. The effort will be meaningful only if the program has a sound scientific foundation.

Because of their training, statisticians are uniquely concerned with the scientific validity of the individual clinical trials which comprise the development program and of the program as a whole. Schor (1965) has described the role of biostatistics in clinical research in terms of "a series of steps an investigator must go through in order to produce valid results." Individual trials must be designed to provide scientifically valid answers to specific, medically important questions. The overall program must be planned to provide answers to a broader range of questions which cover the entire spectrum of pharmacology, safety and efficacy, various populations, and perhaps several indications.

Statistics plays an essential role in planning these programs, in addition to designing individual studies. The constructive interaction between statisticians and physicians is critical in the development of well-defined study objectives and in the design of clinical programs which can meet those objectives. Various sources of bias haunt clinical trials, and effective interaction between statistician and physician is important in reducing or eliminating specific sources of bias. This cooperative effort is also necessary in defining the appropriate target population and ensuring that

---

*With the technical assistance of D. Vattelana.

it is the population actually sampled, determining the most appropriate experimental design, and selecting sound methods of analysis which will permit definitive inferences. Because of their intimacy with study design, the data collected, and the methods of analysis, statisticians are the primary architects in determining the precise conclusions that a study will support.

The clinical development program is influenced by both medical theory and medical practice. Medical theory provides the biologic rationale for clinical development. Medical practice—the ways in which physicians deal with their patients—influences clinical trial procedures. When designing clinical trials, statisticians and physicians occasionally disagree over clinical trial procedures. Statisticians tend to favor procedures intended to increase scientific validity, and physicians tend to favor procedures on the basis of medical ethics and generally accepted medical practices. When such conflict arises, statisticians and medical researchers must communicate and understand each other's concerns. Only by doing so can they hope to develop a study plan which is both medically and scientifically acceptable.

The introduction of new drugs and biologics into humans and the subsequent clinical development of such compounds through clinical trials are closely regulated by government agencies in the United States and abroad. The U.S. regulatory environment has been discussed in detail in Chapter 1. During the clinical development of a compound, regulatory agencies and sponsors seek to avoid undue risks to the patients/subjects who participate in clinical trials. At the end of the development process, agencies must review the documentation provided by the sponsor and judge the safety and efficacy of new compounds with respect to their approvability for use in specific indications. Some foreign agencies also determine the price which can be charged and/or the amount which government health plans will reimburse patients for medications. The intent of government regulations and guidelines is to promote clinical development programs which will provide an adequate body of scientifically sound data on which an agency can base its decisions and will provide clear safety and efficacy data related to the agency's unique national population.

No regulatory agency in the world offers complete guidance for new drug development, and this is appropriate. The U.S. Food and Drug Administration (FDA) provides written guidelines for the design of clinical trials for many classes of drugs. Generally, these guidelines promote a scientifically and medically sound program, but they are guidelines, not regulations, and as such are not binding on the sponsor. Nevertheless, the guidelines provide a suggested basis for formulating a development program of a new therapeutic agent. These guidelines are not necessarily consistent with foreign regulatory, scientific, or medical perspectives, and this must be considered when multinational registration is anticipated.

Agencies ultimately judge the safety and efficacy of new agents after scientific and clinical evaluation of the completed program. In most cases, a favorable judgment will require a sound statistical basis for the planning, evaluation, and interpretation of the program, although most agencies do not have formal departments of statistical reviewers at this time.

The remainder of this introductory section describes the phases of clinical development and some regulatory considerations in more detail. The concepts of an overall clinical development plan and a clinical research protocol (the plan for one study in the program) are also considered.

## B. Phases of Clinical Development

Clinical development of new therapeutic agents is loosely classified into four sequential phases. Phases I, II, and III have been described in Chapter 1. These phases may overlap temporally, and some studies may not clearly belong to just one phase.

Phase I ordinarily involves the initial use of the drug in humans. These are clinical pharmacology studies designed to characterize the drug's pharmacologic effect. They involve "normal" volunteer subjects in order to reduce the risk of serious toxicity problems and avoid confounding pharmacologic and disease effects. Because this is the introduction of the drug in humans, participants in the earliest Phase I studies are very closely monitored. Once comfort against acute toxicity is established, early dose-ranging studies in subjects typically follow. These are intended to determine tolerable doses.

Other Phase I studies will be necessary to determine bioavailability, to compare the bioavailability of a variety of different formulations, and to identify the dynamic and metabolic actions of the drug. Some of these Phase I studies may be carried out at different times, sometimes concurrently with Phase II or Phase III trials. The number of subjects in a Phase I study typically varies from 4 to 20 and may amount to 100 for the entire phase.

Phase II consists of clinical pharmacology studies in patients (Phase IIA) and the earliest controlled clinical trials in patients (Phase IIB). The former studies are intended to estimate the effective dose range, including characterization of the dose-response curve and determination of the minimal effective dose. They may also include pharmacokinetic studies in patients. The latter studies are intended to provide an initial demonstration of the safety and efficacy of the drug at the dose or doses developed in the clinical pharmacology studies. In many respects Phase IIB and Phase III studies are indistinguishable.

Patients selected for early Phase II studies ordinarily come from a relatively restricted population. They should have the disease under investigation, but be relatively free of other confounding conditions, and the consequent need for concomitant therapies, because the latter often have the potential for interference in the assessment of the investigational drug. The patient population may be gradually broadened in later Phase II and Phase III studies, but usually not to the extent of the population that will ultimately use the marketed product. As in Phase I, patients in Phase II studies are closely monitored for potential toxicity and unanticipated pharmacologic effects. The number of patients in Phase II trials is greater than in Phase I, but infrequently exceeds a few hundred patients for the entire phase.

Phase III is an extension of Phase II into expanded controlled comparative and noncomparative clinical trials. These are intended to provide more precise information on adverse effects, provide confirmation of efficacy, including whether or not tolerance or tachyphylaxis exists, and develop a meaningful clinical experience with a variety of control (placebo or competitive) agents. The Phase III patient population may be less restricted than the early Phase II population, and the numbers of patients involved may range from several hundred to more than a thousand. Patients may be less intensely observed than during Phase I and II studies. These studies, in addition to including more patients than the earlier phases, are usually much longer in duration. Typically, Phase III studies of a drug for a chronic condition may include hundreds of patients for more than a year. The pivotal studies needed to gain approval of a drug by the FDA typically will be Phase III trials.

Phase IV consists of postapproval trials that are conducted for one of several distinct purposes. First, they may be voluntarily undertaken to clarify the incidence of rare adverse effects or to explore a specific pharmacologic effect. Second, they may be long-term studies to determine the effects of the drug on morbidity and mortality. A third purpose, and one which is growing quite rapidly, is as a condition of approval to supplment premarketing data in cases where, in the public interest, the drug has been released by the FDA, or another regulatory agency, for widespread usage prior to acquisition of all data which might ordinarily be obtained before marketing (see Wardell et al., 1979a, 1979b). Fourth, they may be clinical trials in patient populations which are not ordinarily studied in detail during Phases II and III, for example, children, the elderly, and patients with commonly associated conditions who were excluded from earlier clinical trials. And finally, they may be clinical trials in indications for which the drug will be used, but has not been approved. For example, an analgesic which has been approved for use in mild to moderate pain and which has prostaglandin synthetase-inhibiting effects may be studied for a potential indication in dysmenorrhea. (Studies of new claims for approved drugs are also often referred to as Phase V studies.)

### C. Regulatory Considerations Prior to Initiating Clinical Trials

By the time a drug candidate is introduced into humans for the first time it may be several years old and will have undergone intensive pharmacologic and toxicologic evaluations. These have been described in detail in other chapters in this volume. The principal objective of these early evaluations is to ensure a safety margin great enough that the anticipated effectiveness of the compound far outweighs the potential harm that may be anticipated. It must be remembered that at the earliest stages we are in complete ignorance concerning the effect the drug will have on humans. The positive and negative responses to a drug by human beings may be very different from the pharmacologic and metabolic profiles observed in other species.

To ensure adequate safety and to maintain consistency of control at this very critical stage in the developmental process, each country has a set of regulations which define what needs to be done before the first dose is administered to humans. Some countries have rather broad and flexible guidelines, and others have very stringent ones.

In the United States, the Food, Drug and Cosmetic Act states that a new drug may not be introduced into interstate commerce unless a New Drug Application (NDA) is filed and approved by the FDA. Since this would not permit the investigation of new drugs to obtain the data necessary to file an NDA, the act called for regulations which would permit exemptions from this requirement in the case of new drugs intended solely for investigational use by qualified experts. The New Drug Regulations require a drug developer to request such exemption by filing a "Notice of Claimed Investigational Exemption for a New Drug," which is commonly called an IND. It is primarily intended to assure the FDA that the developer has sufficient information about the drug from laboratory and animal studies to conclude that it is reasonably safe to initiate human studies in the United States.

The IND contains information on the chemical structure; mode of administration; manufacturing, processing, and packing controls to maintain the identity, strength, quality, and purity of the compound; preclinical safety studies in animals; and any previous clinical experience (e.g., a summary of foreign experience if the drug has been studied or marketed abroad, or, if the drug is a combination, a summary of experience with the ingredients). In addition, the IND contains an outline of planned investigations, including initial protocols, identification of investigators and their qualifications, and informational material investigators will receive, and assurances of institutional review board review and approval of the proposed studies. If the drug developer has not heard to the contrary within 30 days of filing the IND, the clinical program may begin. The IND and NDA are discussed more extensively in Chapter 1.

Although the preclinical statistician will often have a very significant involvement in the pharmacology and toxicology studies which comprise the IND, the clinical biostatistician's primary role with an IND is the development of protocols for individual trials. He or she is also intimately involved in the planning of the overall clinical program.

Clinical Study Authorization (CSA) is a general term used for the foreign counterpart of the IND. Keeping abreast of changes in foreign regulatory requirements (with more than 20 major agencies) is difficult, and no attempt will be made here to discuss the specific requirements for a CSA in any one foreign country. Some countries, like the United States, require formal approval to study a drug within their borders and/or approval to import clinical supplies. A few countries do not require a CSA, but some of these require a notice of intent to conduct studies. Approvals differ from one country to another and may be on an overall basis or on a study-by-study basis. Depending on the individual country's requirements, the information necessary to obtain a CSA may vary from a small subset of that required for an IND to an even greater amount of information in some countries.

The time required to prepare a CSA and the time required to obtain approval also vary substantially from one country to another. They typically range from a few days to several months. Variation in the requirements for beginning the first clinical studies complicates the timing of the initiation of an international clinical development program.

## D. Regulatory Guidelines and Their Influence on Clinical Development

Clinical development must be founded on good scientific practice. However, for many issues there is not universal agreement on what constitutes good science. Differences of opinion exist among statisticians, among medical researchers, among pharmaceutical companies, and among regulatory agencies. In addition to the differences within these classes, there are differences among them as well.

Regulatory guidelines provide a useful framework for consistency. The FDA publishes guidelines for the development of specific classes of drugs (e.g., "Guidelines for the Clinical Evaluation of General Anesthetics," FDA 82-3053, 1982), and these may be ordered from the Superintendent of Documents, U.S. Government Printing Office, Washington, D.C. 20402.

"General Considerations for the Clinical Evaluation of Drugs" (FDA 77-3040, 1977) is an important companion piece to the class-specific guidelines and contains suggestions which are applicable to investigational drug studies for most classes of drugs. Many of these published sets of guidelines make fairly specific recommendations about such things as the major questions which the development program should address, the types of study designs which should be employed, key efficacy and safety variables, patient populations, and medical procedures. Because of their detail, these guidelines might be misconstrued as a set of instructions which must be rigidly obeyed. However, the FDA makes its position clear in the following paragraphs, which appear in the foreword of each set of class-specific guidelines.

> The purpose of these guidelines is to present acceptable current approaches to the study of investigational drugs in man. These guidelines contain both generalities and specifics and were developed from experience with available drugs . . . .
>
> These guidelines are not to be interpreted as mandatory requirements by the FDA to allow continuation of clinical trials with investigational drugs or to obtain approval of a new drug for marketing. These guidelines, in part, contain recommendations for clinical studies which are recognized as desirable approaches to be used in arriving at conclusions concerning safety and effectiveness of new drugs; and in the other part they consist of the views of outstanding experts in the field as to what constitutes appropriate methods of study of specific classes of drugs. In some cases other methods may be equally applicable or new methods may be preferable, and for certain entirely new entities it is possible that the guidelines may be only minimally applicable.

In spite of this disclaimer, it is prudent to consider these guidelines carefully prior to initiating a clinical development program and to consider the consequences of any major departures from them. The spirit of these guidelines is to provide a solid basis for a program which will be well founded from both a scientific and clinical perspective.

In addition to class-specific guidelines which shape the structure of clinical development programs, the FDA has published new regulations and new draft guidelines which reshape the presentation of results in an NDA:

New Drug and Antibiotic Regulations; Final Rule (FDA, 1985a)
Draft Guideline for the Format and Content of the Statistical Section of an Application (FDA, 1985b)
Draft Guideline for the Format and Content of an Application Summary (FDA, 1985c).
Draft Guideline for the Format and Content of the Clinical Data Section of an Application (FDA, 1986)

These guidelines, which are also discussed in Chapter 1, describe "an acceptable" format for an NDA and make suggestions with regard to informational content, organization, data display, and analysis. They represent major changes in NDA format, submission of safety update reports, and the use of foreign data (among other major changes). These changes will be of particular importance in the planning and timing of future clinical development programs. Revisions to these most recent draft guidelines are anticipated in 1987.

In general, foreign regulatory agencies provide less detailed written regulations and guidelines than does the FDA. There is a tendency for foreign regulations to be relatively specific about preclinical information required and often less specific about clinical information. In addition, most foreign agencies do not employ their own statisticians, although many do consult with academic statisticians. Consequently, foreign agencies are influenced by statisticians to a lesser degree than is the FDA. The prominence of statisticians within the FDA has enhanced the influence of statisticians within the U.S. pharmaceutical industry.

A clinical development program intended to support registration in the United States alone should be planned in accordance with FDA regulations and guidelines, but may depart from those guidelines when warranted by good scientific practice. Of course, it is wise to discuss such departures with the FDA before adopting them, and the new regulations encourage discussion of such issues.

However, a development program for the United States is not usually sufficient to obtain worldwide registration. A worldwide development program needs to include investigators from key countries; it may need to include comparative agents not in use in the United States; and it almost certainly needs to include compromises which overcome international differences of opinion about medical ethics, good medical practice, and good scientific practice.

## E. Concept of a Clinical Development Plan

A clinical development plan (CDP) is an outline of the clinical development program. The objective in writing a CDP is to define the totality of studies needed to register the drug either in a particular country or worldwide. This description of the information that belongs in a CDP is written from the latter, broader perspective.

A CDP is necessary for early, macroscopic management decisions: Does the planned program make sense—can it meet our objectives? To permit this type of decision making, the CDP should include the following types of information:

Medical background and rationale for the program
Program objectives
A description of all studies necessary to meet those objectives
Estimated personnel requirements
Estimated drug requirements
Estimated target dates
Estimated costs

The medical background and rationale should discuss the prevalence and importance of the indication(s) targeted for drug usage and the role of the new drug in treatment or prevention. The advantages of the new therapy, possible limitations, and the likelihood of achieving objectives should be considered.

Examples of typical program objectives are:

*Phase I/II*

To evaluate the safety, tolerability, pharmacodynamics, and pharmacokinetics in healthy volunteers and/or patients
To establish the dosage range to be used in Phase III
To provide preliminary evaluation of efficacy in selected populations
To evaluate particular pharmacodynamic interactions with other specific drugs

*Phase III*

To demonstrate efficacy for each indication
To establish safety of long-term and/or intermittent use
To evaluate additional drug interactions

The description of studies should include the number and kinds of subjects/patients required, a summary of power considerations, expected locations and estimated numbers of clinics involved, dosage and duration of therapy, identification of control agents, study design, and key efficacy and/or safety measures.

The estimation of timing should include identification of critical points in the developmental path and their target dates. Dates for filings of CSAs and/or INDs, study-start dates, last-patient-entered dates, last-case-report-in-house dates, study-summary-written dates, and submission dates should be estimated.

Creation of such an outline requires input from all of the many groups which will support the program and communication among them. To be successful, the worldwide development program requires the right studies with the right designs, the right numbers of patients within and across studies, and the right geographic distribution of studies with the right control agents. The time, resources, and costs required for development must strike the right balance with the potential gain from, and likelihood of, a positive outcome. The CDP should provide enough information to permit an early decision on whether the program is well conceived and should be pursued.

## F. Concept and Content of a Clinical Research Protocol

A protocol is a clear, detailed plan for the conduct of a study. Good science mandates a good protocol. A poorly planned, poorly written, insufficiently detailed protocol will lead to inconclusive results. A well-planned, well-written, sufficiently detailed protocol maximizes the chances for a conclusive study.

The following is an outline for a study protocol.

I. Title
II. Site
III. Investigators
IV. Institutional review board or review committee
V. Summary
VI. Background and rationale
VII. Objectives
VIII. Patient definition
IX. Study design and treatment definition

X. Concurrent treatment
XI. Clinical and laboratory measurements for safety and efficacy
XII. Planned data analysis (and other statistical considerations)
XIII. Administrative aspects
  A. Labeling of study material
  B. Preparation of case report forms
  C. Duration of study
  D. Instructions to the recorder or study secretary
  E. Record retention
XIV. Informed consent
XV. Confidentiality
  A. Confidentiality of data
  B. Confidentiality of patient records
XVI. Adverse experiences
XVII. Signature of the investigators

Protocols are used to define the critical details of a study. Investigators will use the protocol as a set of instructions for carrying out the study. (Some studies may use an even more detailed operations manual for this purpose.) Reviewers will use the protocol in assessing the scientific validity of the study and regulatory compliance. For these purposes protocols need to contain basic identifying information; a clear statement of objectives; a clear identification of the patient population sampled; a description of study design (including methods of blinding and randomization), course of treatment, permissible concurrent therapies, and the measurements to be collected; statistical information regarding planned analyses and power with respect to primary efficacy and/or safety variables; descriptions of any planned interim analyses; the questions to be answered in the statistical analysis; instructions regarding study administration; and information required in order to be in compliance on patient protection issues such as informed consent and institutional review board approval.

The FDA has provided direction concerning the essential features of study protocols in several documents:

Adequate and Well-Controlled Clinical Investigations (FDA (1970): 21 CFR 314.111)
Protection of Human Subjects (FDA (1975): 45 CFR Part 46)
General Considerations for the Clinical Evaluation of Drugs (FDA 77-3040)
Draft Guideline for the Format and Content of the Statistical Section of an Application (FDA, 1985b)

## II. CLINICAL PHARMACOLOGY STUDIES

### A. Introduction

Prior to drafting the first protocol for human study, a large body of preclinical information is accumulated from animal studies. The introduction of a new drug into humans is never realized until there is evidence of therapeutic potential and a wide margin of safety. The details of this development are presented elsewhere in this volume.

The selection of doses to be investigated in initial human studies is based on available toxicology and pharmacology studies in animals. The maximum dose permitted for humans is usually about 1/20 of the lowest toxic dose in animal tests. If particular species are known to be sensitive to a given class of drugs, then the maximum dose allowed in humans may be greater than the toxic dose in animals.

For doses intended to elicit a pharmacologic response, the lowest dose level studied is usually expected to show no therapeutic effect and is generally smaller (1/20 to 1/100) than the no-effect dose in animals; the highest dose is set by extrapolation of the maximum tolerated dose (dose units per kilogram) from animal studies to humans of 70 kg, or by clinical assessment. If the window between intended pharmacologic dose and maximum permitted dose is narrow, again higher doses may be considered, depending on the risk/benefit ratio.

Clinical pharmacology trials comprise the initial safety and tolerability studies of new drug entities or new dosage forms and evaluate the drug's pharmacology prior to large-scale trials in patients with the target disease. Section II,B distinguishes between five types of clinical pharmacology studies: safety and tolerance studies, bioavailability/bioequivalence studies, pharmacodynamic studies, pharmacokinetic studies, and dose-ranging/dose-response studies. Objectives, definitions, designs, and analytical approaches specific to each type of study are discussed. General considerations in the design and analysis of these studies are described in Section II,C.

## B. Objectives and Designs of Clinical Pharmacology Studies

### 1. Safety and Tolerance Studies

#### *a. Objectives*

Once a new drug has been tested and characterized in laboratory animals and shows potential to be efficacious and safe in humans, it must be tested in humans. Given the uncertainty of an untried drug, initial studies in humans are necessarily small ($n \sim 8-16$). Typically, these studies are single-dose studies in young, healthy male subjects and have as their objective the gathering of early information related to safety and tolerability. Later, multiple-dose safety studies are conducted. Information on clinical adverse experiences (AEs) and vital signs is often collected several times during a 24- or 48-hour postdose observation period. Physical examinations, electrocardiograms, and collection of blood and urine for laboratory safety determinations are generally done at least once during the postdose observational period as well as prestudy. These data are used to make clinical judgments on the safety of the drug. Because of the limited objectives and observational nature of these studies, statistical analyses are often not done.

An additional objective of these studies is often the collection of pharmacokinetic or pharmacodynamic information to obtain a gross picture of the drug's action and to aid in the design of future studies. Here, the nature of analyses is best seen as hypothesis generating, rather than inferential. For efficacy purposes in these studies, lowering the type II error at the expense of increasing the type I error is important, since one wishes to avoid stopping the drug program if the drug really works.

#### *b. Single-Dose Safety Studies—Designs*

Single-dose safety and tolerability studies may use several different designs. In the *rising single-dose study*, each subject receives a single dose of drug and then, if no critical safety problems occurred at the lower dose, is titrated to the next higher single dose. Typically, three or four incremental doses are scheduled to be given, with a placebo treatment randomized into sequence.

Example of rising-single dose study allocation schedule.

| Subject | Dose level (mg) 100 | 200 | 300 | 400 |
|---|---|---|---|---|
| 1 | A | A | P | A |
| 2 | A | A | P | A |
| 3 | A | P | A | A |
| 4 | P | A | A | A |
| 5 | A | A | A | P |
| 6 | P | A | A | A |
| 7 | A | P | A | A |
| 8 | A | A | A | P |

A, Active; P, placebo.

A variation on this design is the *alternating-panel rising single-dose (or dose-ranging) study*. In this design, subjects are assigned to one of two or three panels. Rising incremental doses are determined, and subjects in panel I receive the lowest dose, subjects in panel II receive the next higher dose, etc. When all panels have received their "low" dose, then panel I is given the next dose. Panels are assigned doses until all predetermined doses have been given, or until a dose is given which appears to cause important safety problems. Usually each panel will be assigned two to four doses with placebo randomized into the sequence.

Example of an alternating-panel rising single-dose (dose-ranging) study.

| | Subject numbers | | | | | | | | | | | | | | | |
|---|---|---|---|---|---|---|---|---|---|---|---|---|---|---|---|---|
| Dose level (mg) | Panel I | | | | | | | | Panel II | | | | | | | |
| | 14 | 2 | 7 | 10 | 16 | 3 | 6 | 12 | 1 | 5 | 15 | 13 | 4 | 11 | 9 | 8 |
| 100 | A | A | A | A | A | P | P | P | | | | | | | | |
| 200 | | | | | | | | | A | A | A | A | A | P | P | P |
| 300 | P | P | P | A | A | A | A | A | | | | | | | | |
| 400 | | | | | | | | | P | P | P | A | A | A | A | A |
| 500 | A | P | A | P | A | P | A | A | | | | | | | | |
| 600 | | | | | | | | | A | P | A | P | A | P | A | A |

| Dose level (mg) | Subject numbers | | | | | | | | | | | | | | | |
|---|---|---|---|---|---|---|---|---|---|---|---|---|---|---|---|---|
| | Panel I | | | | | | | | Panel II | | | | | | | |
| | 14 | 2 | 7 | 10 | 16 | 3 | 6 | 12 | 1 | 5 | 15 | 13 | 4 | 11 | 9 | 8 |
| 700 | P | A | P | A | P | A | A | A | | | | | | | | |
| 800 | | | | | | | | | P | A | P | A | P | A | A | A |

A, Active drug; P, placebo. Panel I will complete a dose level at least 24 hours before panel II begins. Within panels, there will usually be an interval of at least 48 hours between treatment levels per subject.

Generally, these two designs are "placebo-controlled" in that at each dose level one to three subjects are randomly assigned to receive placebo and three to five subjects receive drug. Thus, the studies can be done "double-blind" with respect to actual treatment, but not with respect to "dose level." In addition, comparison of the data for actively treated subjects to the data for those receiving placebo may help to detect gross irregularities, extraordinary or gross changes due to drug, or changes in the experimental situation. Extensions to this type of placebo control include the *rising single-dose crossover study or alternating-panel, rising single-dose (dose-ranging) crossover study*, where each subject receives drug *and* placebo at each dose level.

Example of an alternating-panel rising single-dose crossover study sample allocation schedule.

| Subject | Panel | | Period: day of study | | | | | | | | | | | |
|---|---|---|---|---|---|---|---|---|---|---|---|---|---|---|
| | | Dose level: | A | A | B | B | C | C | D | D | E | E | F | F |
| | | Day: | 1 | 2 | 1 | 2 | 1 | 2 | 1 | 2 | 1 | 2 | 1 | 2 |
| 1 | 1 | | 10 | P | | | P | 50 | | | P | 250 | | |
| 2 | 1 | | P | 10 | | | P | 50 | | | 250 | P | | |
| 3 | 1 | | P | 10 | | | 50 | P | | | P | 250 | | |
| 4 | 1 | | 10 | P | | | 50 | P | | | 250 | P | | |
| 5 | 2 | | | | 20 | P | | | P | 100 | | | 500 | P |
| 6 | 2 | | | | 20 | P | | | 100 | P | | | P | 500 |
| 7 | 2 | | | | P | 20 | | | P | 100 | | | 500 | P |
| 8 | 2 | | | | P | 20 | | | 100 | P | | | P | 500 |

Number values correspond to milligrams of active drug given; P, placebo.

Parallel designs might also be used to examine the safety of various doses. However, one must be cautious about giving a relatively high dose of an untested drug to a group of subjects who have not been tested previously for tolerance to the drug at lower doses.

Other designs are possible but are not used as frequently as the ones described above. A rising dose design where all subjects receive all doses as well as placebo is shown below. This design could be double-blind (drug vs. placebo), but need not be if the objective is based on laboratory assays or measurements which will be carried out by a blinded observer.

Example of a rising single-dose crossover study.

| | Period | | | | | |
|---|---|---|---|---|---|---|
| Subject | 1 | 2 | 3 | 4 | 5 | 6 |
| 1 | 2.5 | 5 | 10 | 20 | P | 40 |
| 2 | 2.5 | 5 | P | 10 | 20 | 40 |
| 3 | 2.5 | 5 | 10 | 20 | 40 | P |
| 4 | P | 2.5 | 5 | 10 | 20 | 40 |
| 5 | 2.5 | 5 | 10 | P | 20 | 40 |
| 6 | 2.5 | P | 5 | 10 | 20 | 40 |
| 7 | 2.5 | 5 | 10 | 20 | P | 40 |
| 8 | 2.5 | 5 | 10 | P | 20 | 40 |
| 9 | 2.5 | P | 5 | 10 | 20 | 40 |
| 10 | 2.5 | 5 | P | 10 | 20 | 40 |
| 11 | 2.5 | 5 | 10 | 20 | 40 | P |
| 12 | P | 2.5 | 5 | 10 | 20 | 40 |

Numbers indicate amount (mg) of active drug; P, placebo.

Another possibility is the *parallel-panel rising dose design*. This design is generally not used because reliance on one to three subjects for controls may yield misleading results if these subjects are not representative for any reason.

Example of parallel-panel rising single-dose design.

| | | Dose level (mg) | | | |
|---|---|---|---|---|---|
| Panel | Subject | 10 | 20 | 40 | 80 |
| I | 1 | A | A | A | A |
| | 2 | A | A | A | A |
| | 4 | A | A | A | A |
| | 6 | A | A | A | A |
| | 7 | A | A | A | A |
| | 9 | A | A | A | A |
| | 10 | A | A | A | A |

| | | Dose level (mg) | | | |
|---|---|---|---|---|---|
| Panel | Subject | 10 | 20 | 40 | 80 |
| II | 3 | P | P | P | P |
| | 5 | P | P | P | P |
| | 8 | P | P | P | P |

A, Active drug; P, placebo.

*c. Multiple-Dose Safety Studies—Designs*

After a single-dose study has been completed, generally a multiple-dose safety study is carried out to examine the safety of repeated dosing with the drug. The data collected are usually similar to data collected for the single-dose safety study, but observations are collected on the last dosing day and on at least one intermediate day, as well as on the first day. Special objectives to be examined in these studies are gross changes in the drug's effect (safety, pharmacokinetic, pharmacodynamic) with multiple dosing. Designs used in these studies include *parallel time-lagged* [placebo versus drug (dose x) versus drug (dose y), etc.] and *panel-type*. In a parallel time-lagged study, each group of subjects receives a different dose and the groups are treated in increasing dose sequence. In a panel-type design, subjects in each panel receive a different dose of drug for several days, and one subject within the panel receives placebo. Typical study size is 8 to 16 subjects.

Example of parallel-panel multiple-dose time-lagged study.

| | | Dose level (mg) | | |
|---|---|---|---|---|
| Panel | Subject | 250 (period 1) | 500 (period 2) | 1000 (period 3) |
| I | 1 | P | | |
| I | 2 | A | | |
| I | 3 | A | | |
| I | 4 | A | | |
| II | 5 | | P | |
| II | 6 | | A | |
| II | 7 | | A | |
| II | 8 | | A | |
| III | 9 | | | A |
| III | 10 | | | A |
| III | 11 | | | A |
| III | 12 | | | P |

P, Placebo for 7 days; A, active drug for 7 days. Each panel completes treatment before next panel begins.

Example of parallel-panel rising multiple-dose design.

| Treatment | Duration (days) | Study days |
|---|---|---|
| *Group A* | | |
| 100 mg | 7 | 1–7 |
| 250 mg | 7 | 8–14 |
| 500 mg | 7 | 15–21 |
| Placebo | 7 | 22–28 |
| *Group B* | | |
| Placebo | 28 | 1–28 |

Example of alternating-panel multiple-dose rising dosage study.

| | | Study day | | | | | | |
|---|---|---|---|---|---|---|---|---|
| Panel | AN | 1 | 2 | 3 | 4 | 5 | 6 | 7 |
| I | 1 | A | A | A | C | C | C | |
| | 2 | A | A | A | C | C | C | |
| | 3 | A | A | A | C | C | C | |
| | 4 | P | P | P | P | P | P | |
| II | 5 | | B | B | B | D | D | D |
| | 6 | | P | P | P | P | P | P |
| | 7 | | B | B | B | D | D | D |
| | 8 | | B | B | B | D | D | D |

P, Placebo; A, 100 mg; B, 250 mg; C, 500 mg; D, 1000 mg.
Note: a washout period could be inserted between dose increases.

*Parallel multiple-dose* designs are generally not used, because safety and tolerability must be shown at lower multiple dosings before higher multiple dosings are given.

Example of parallel multiple-dose study.

| | Treatment | | |
|---|---|---|---|
| | 250 mg b.i.d. | 500 mg b.i.d. | Placebo b.i.d. |
| Subject: | 1, 5, 8, 9 | 3, 6, 7, 12 | 2, 4, 10, 11 |

Treatment given for 7 days.

Additional safety studies are usually done for each modification of the drug itself or its route of administration. For example, safety studies may be done for the intravenous (i.v.), transdermal, intramuscular (i.m.), and oral (p.o.) administration of single or multiple doses of a given drug.

#### d. Analysis of Safety and Tolerability Studies

Most of the rising dose safety studies and their variations have a limited number of subjects. Data can be plotted and the minimum, median, and maximum for each variable can be tabulated by dose/panel etc. to facilitate the search for gross trends over time which might be attributable to drug. Usually, hypothesis testing should be viewed as exploratory; the resulting p values should not be regarded as firm support for conclusions, but rather as suggestive of areas for examination in future studies. Six subjects per group (panel) is the minimum generally needed (unless specifically planned in the protocol) to consider doing formal hypothesis testing. With 6 to 10 subjects nonparametric (rank) tests may be more appropriate than parametric ones, because a test for normality (or normal probability plot) may be unreliable. If data on 10 or more subjects are available, parametric analysis may be considered, provided the results of tests for normality (tested by Shapiro-Wilk, etc.) are acceptable. This could maximize power in these small studies. Within-group changes are usually examined, and often both parametric and nonparametric analyses are done (if sample size warrants analysis). If differing results are found, the reason should be explored and explanation attempted. Qualitative comparisons between panels may also be done.

To evaluate safety, adverse experiences should be tabulated by dose and also by subject. Clinicians will use the AE data as well as other safety data to draw "soft" conclusions about the safety of the drug. Power statements in the protocol may be based on the probability of missing an AE (i.e., the probability of observing none or, equivalently, one minus the probability of observing one or more occurrences of an AE) given a particular underlying occurrence rate. For example, if an AE occurs with a 20% incidence, then the sample size of $n = 10$ subjects has probability $(1 - .2)^{10} = .11$ of yielding no occurrences. Equivalently, it has probability $1 - (1 - .2)^{10} = .89$ of yielding at least one occurrence. Typically, these probabilities should be included in the protocol.

### 2. Bioavailability/Bioequivalence Studies

#### a. Definitions and Objectives

The study of bioavailability of a formulation attempts to characterize both the amount of active drug which gets from the administered dose to the site of pharmacologic action and the rate at which it gets there. This topic is covered extensively elsewhere in this volume. Bioavailability is usually estimated in the following manner: a formulation is administered to subjects and then drug concentrations in blood and urine samples are measured at various prespecified times. The blood-time profile is then characterized relative to a standard (i.v. or solution). Three parameters important in describing bioavailability are area under the plasma (or serum) concentration-over-time curve (AUC), observed maximum concentration ($C_{max}$), and observed time to maximum ($T_{max}$). Urinary recovery may also be important if the drug is excreted primarily through the kidney.

A bioequivalence study is a comparative bioavailability trial designed to demonstrate the equivalence of two formulations, e.g., a new to a current formulation, or a standard formulation given under different physical or environmental conditions. Although a bioequivalence study attempts to demonstrate the equivalent bioavailability of two or more formulations, the key underlying issue is whether the formulations result in equivalent therapeutic or pharmacodynamic effects.

To demonstrate equivalent therapeutic effect of a formulation when given as a tablet and as an oral solution would require a fairly extensive clinical trial. If we can assume that two formulations of a drug resulting in essentially equivalent blood level profiles over time will elicit equivalent therapeutic effects, a comparative bioavailability trial of a smaller size can be conducted in normal volunteers or patients to indirectly evaluate therapeutic equivalence.

Objectives often addressed in bioavailability/bioequivalence trials are listed below:

- Do differences in formulations (tablet vs. capsule), coating processes, or inert ingredients alter the bioavailability of the active ingredient?
- Is a generic drug formulation equivalent to the brand name drug formulation?
- What is the effect of food ingested near the time of dosing?
- Do age and/or specific disease state affect bioavailability?
- Is bioavailability affected by different routes of administration (for instance, i.m. versus i.v.)?
- Do interactions exist with other drugs which might be administered concurrently?
- How does bioavailability differ between steady-state (of multiple doses) and single-dose conditions?

The objective will determine how close the blood level-time curves should be in order to satisfy our definition of equivalence. For example, when comparing two formulations that differ only in the coating process, with the current formulation being the one used for all Phase III trials submitted for the NDA and the new process being proposed as the final market image, it would be important for bioavailabilities to be very similar, with nearly superimposable blood concentration-time profiles.

If, on the other hand, a drug has a high incidence of gastrointestinal (GI) adverse effects which are reduced when the compound is administered in a fed rather than a fasting condition, it may not be as important for the serum concentration-time curves to be nearly superimposable following administration under the two conditions.

Thus, although the FDA generally defines bioequivalence as no more than a 20% difference in AUC and $C_{max}$ between standard and test formulations, in some situations it is desirable for bioavailability estimates to be even closer. In others, slightly larger differences in serum profiles may be tolerated if serum concentrations remain at levels sufficient to produce adequate therapeutic effects.

#### *b. Design of Bioavailability/Bioequivalence Trials*

In a typical bioavailability trial, single doses of a standard and a test formulation(s) are administered to healthy volunteer subjects. Serum or plasma (and urine) are then assayed for drug content at a sequence of times following administration.

Usually all subjects receive all formulations in some type of crossover design. Crossover trials are advantageous in general because they remove intersubject variation from the error term, allowing a more sensitive test. This is particularly important in bioequivalence studies because of the considerable variation among subjects with respect to factors influencing drug concentration in the blood (e.g., volume of distribution and metabolism). Furthermore, the major drawback of crossover designs—undetected carryover effects of the treatments—can be assessed directly at the zero-time blood sample beginning the next treatment period,* or by estimation in a design balanced for carryover effects. Carryover effects can be eliminated or minimized by allowing an adequate washout between periods.

Although rarely used, parallel designs in which each group of patients receives just one of the formulations may be used, especially when intrasubject variability is not much smaller than between-subject variation. A parallel design is obviously necessary when comparing the bioavailability of a drug in normal subjects versus the elderly or patients with a particular disease state.

Bioavailability studies in general, and bioequivalence studies in particular, should be designed with sample sizes large enough to detect differences of 20% in mean AUC. Precision of detection should be 80% power (or greater in some instances) at the 5% significance level.

*c. Analysis*

Instead of a point-by-point or repeated-measures analysis of the series of drug concentrations over time, characteristics of the blood level sequence such as $T_{max}$, $C_{max}$, AUC, half-life, and urinary recovery are usually analyzed.

Differences between treatments in pharmacokinetic variables are usually analyzed by parametric or nonparametric methods appropriate for crossover or parallel designs. AUCs may be log-transformed, while half-life of the drug is often analyzed by using the inverse (harmonic) transformation. The log-transformation is used for analyzing AUCs when comparisons of one treatment with another are desired (e.g., when testing $H_0$: $AUC_a/AUC_b = 1$). The inverse transformation is usually used to analyze half-life because half-life is often estimated as a constant times the reciprocal of an estimated pharmacokinetic rate constant.

The null hypothesis being tested in a classical parametric or nonparametric model is that of no treatment difference, whereas in bioavailability/bioequivalence studies the hypothesis of interest is usually whether the test treatment is within 20% (or any clinically meaningful difference) of the current formulation. Thus, a null hypothesis of two formulations being "equivalent" may be rejected because the test is sensitive enough to detect a difference ($p < 0.05$) of 5 to 10%, which may not be clinically important.

Estimation of the difference in bioavailability of the test preparation relative to the standard is an alternative or additional approach often used. The bioavailability ratio of the test relative to a standard formulation is

---

*Although it can be determined that blood concentrations have returned to zero at the end of washout, the effects of previously receiving drug on how the drug may be handled by the body cannot be measured directly unless the design permits adjustment for carryover effects.

estimated for each individual as the ratio of his or her AUCs or urinary recoveries, or it may be determined by the pharmacokineticists taking into account AUCs as well as clearances of the drug. A 95% confidence interval is then calculated on the ratio of means (using log-transformed data and exponentiating the endpoints of the interval). If differences of 20% or less are considered bioequivalent, the 95% confidence interval should be contained within the interval (0.8, 1.25).

Claims of equivalence based on confidence interval estimation are often supported by the Bayesian method proposed by Rodda and Davis (1980). This method calculates the posterior probability that the true difference in treatments is, for example, 20% or less given the observed results of the study. The posterior probability can be calculated on the bioavailability ratio as well as other pharmacokinetic variables. In a bioequivalence study, a 90% posterior probability that the true difference between two treatments in bioavailability is less than 20% supports a claim of equivalence.

A distribution-free method for calculating confidence intervals or posterior probabilities was described by Steinijans and Diletti (1985). Hauck and Anderson (1984) described a hypothesis test for the null hypothesis of nonequivalence versus the equivalence alternative.

A list of methods for documenting bioequivalence (in order of preference) is given below:

1. Nonrejection of $H_0$: $\mu_A = \mu_B$ versus $H_a$: $\mu_A \neq \mu_B$ when the study had 80% power to reject $H_0$ if $\mu_A/\mu_B$ is outside (.8, 1.25)
2. A 95% confidence interval for $\mu_A/\mu_B$ contained in (.8, 1.25)
3. Posterior probability of at least .9 that $\mu_A/\mu_B$ is contained in (.8, 1.25)

Some international regulatory agencies may request symmetric confidence intervals (Westlake, 1976).

### 3. Pharmacodynamic Studies

#### a. *Objectives*

Pharmacodynamics may be defined as the study of the biochemical and physiological effects of drugs and their mechanisms of action. For most purposes, however, when conducting a pharmacodynamic study, the primary function is to observe and/or quantify one or more effects of the drug or the actions thought to be the basis of these effects. In an antihypertensive study this may mean monitoring diastolic and systolic blood pressure, in a gastric antisecretory study measuring gastric acid output and stomach pH, or in an antipyretic study taking body temperature. The ultimate objective of a pharmacodynamic study is to determine whether the drug is capable of eliciting a clinically meaningful response. Other objectives may be to characterize the time course of response and/or to identify the variables, e.g., dosage or dosing intervals, which influence the response.

#### b. *Design*

Pharmacodynamic studies may be conducted either in healthy volunteers or in patients from a clearly defined target population. Since a high degree of intersubject variability is associated with a pharmacodynamic study, a crossover design with adequate washout intervals separating treatment periods may be preferable. A parallel design rather than a crossover design should be employed if the treatment creates a "permanent" or sustained change in a subject's condition, as in an antipyretic or analgesic study.

The variability associated with an observed pharmacodynamic effect has numerous sources. Factors influencing the response which should be taken into accout or controlled include weight, age, diet, concomitant medication, circadian rhythm effects, and the presence, severity, and time course of existing pathological conditions. However, even when all known sources of variation are considered, pharmacodynamic effects are usually not identical in all subjects or in a given subject on different occasions. Whether parallel or crossover in design, pharmacodynamic studies should be placebo-controlled, may be either single or repeated dose, and may include a range of doses.

To demonstrate a pharmacodynamic effect, the sample size should be large enough that a specified difference from placebo, either in the proportion of responders or in the magnitude of response, can be detected with adequate power at an appropriate significance level. Contingent on the magnitude of the specified difference and the anticipated variation in response, 10 to 18 subjects are often sufficient to meet this requirement in a crossover study with quantifiable response. If prior information is available, the mean square error (from models similar to those intended for use in the study being designed) is used in the power computation. However, as is usually the case, when data from a placebo group alone are available, the power computation for a one-sample test of the difference from placebo may be performed with an arbitrary within-subject correlation, often 0.5. This is accomplished by using $\rho = .5$ in the following formula: $V(x - y) = V(x) + V(y) - 2\rho\sigma_x\sigma_y$, where $V(x) = \sigma_x^2 = V(y) = \sigma_y^2$ = between-subject variance, x = posttreatment value, and y = pretreatment value. To detect a specified difference in the pharmacodynamic effect from placebo in a parallel study, the standard deviation from the placebo group can be used in the power computation for a two-sample test. A parallel study with quantifiable or dichotomous response typically requires 20 to 40 subjects per treatment group. If, however, the primary objective is to detect differences in response between doses, prior relevant pharmacodynamic data are required, and increases in sample size may be necessary for both parallel and crossover design studies.

Measurements of the pharmacodynamic effect should be taken not only following treatment administration but also prior to treatment administration. For each individual, a series of baseline measurements is needed to demonstrate clearly that a patient has the condition and degree of severity required for admittance to the study and for continued participation in each treatment period. The number of response measurements and times at which they are to be taken will vary for each drug tested; however, as many response measurements should be taken as needed to completely characterize the drug's effect.

Since the fate of the drug may be dependent on the accuracy and reliability of these measurements, the best techniques available should be employed. In addition to monitoring the pharmacodynamic effect or response, determinations of drug concentrations in appropriate biological samples (plasma, urine, saliva, stool, etc.) may be performed. Sampling should be scheduled in at least a subset of the same times at which response measurements are scheduled to facilitate the analysis and interpretation of the blood level-response relationship.

### *c. Analysis of Pharmacodynamic Studies*

Differences in pharmacodynamic effect or response between treatments may be analyzed at each time point or collectively (repeated measures) via standard parametric/nonparametric techniques appropriate for crossover or parallel designs. The average of the baseline measurements may be used as a covariate or to determine the change in response from baseline. A confidence interval and/or posterior probability may be calculated. The posterior probability is the probability that the true difference in mean response between two active treatments is, for example, greater than a clinically meaningful difference given the observed results. Posterior probability and confidence intervals are computed in addition to tests of significance, since sample sizes employed may not be sufficient to detect differences of small magnitudes between active treatments. The relationships between response, time, and blood drug concentration may be investigated by constructing plots of response versus time, response versus concentration, and/or concentration versus time. Correlation and regression analysis and between-dosage-group comparisons may be performed where appropriate.

Pharmacodynamic effects may also be assessed by examining the proportion of patients at each response measurement time for which a clinically meaningful response (CMR) is observed. For an antisecretory study, a CMR may be defined as an increase in stomach pH above 3.5, whereas for an antihypertensive drug, a CMR may be defined as a decrease in diastolic blood pressure of at least 10 mmHg. Clinical efficacy should further be characterized by (1) the time required to reach a CMR, i.e., onset of action, (2) the magnitude and time of peak response, and (3) the period of time for which a CMR is continuously sustained, i.e., duration of action. Blood drug concentrations at the times during which a CMR is demonstrated may also be examined.

If, with reasonable assurance, the minimum blood drug concentration required for effective therapy can be reliably estimated, then the time at which this concentration is first reached and the time above it may also be estimated. These two blood level parameters may be of greater practical importance than the more commonly estimated blood level parameters such as AUC, $T_{max}$, and $C_{max}$, since they are directly associated with clinical efficacy. Once again, correlation and regression analysis and between-dosage-group comparisons may be performed where appropriate.

## 4. Pharmacokinetic Studies

Pharmacokinetics deals with the absorption, distribution, metabolism, and elimination of a drug. Studies encountered include dose-dependent studies, steady-state studies, drug interaction studies, radiolabeling studies, and studies in special populations. Usually drug levels at each time point are summarized by pharmacokinetic parameters such as AUC, $C_{max}$, $T_{max}$, clearance, and half-life.

### *a. Dose-Dependent Kinetics*

With many drugs, the pharmacokinetic profile is linearly related to dose. If this is the case, the area under the plasma concentration curve divided by dose should be constant for all doses. Goodness of fit tests may be used to evaluate the hypotheses of dose-independent kinetics.

Dose-dependent pharmacokinetic studies are, almost invariably, crossover studies with the number of periods equal to the number of doses being tested. Sometimes, an additional i.v. dose is added. The sequences of doses should be balanced for carryover effects. If this is not feasible because safety is unknown, then a study should be done to determine safe doses prior to the dose-dependent kinetic study. A dose-dependent kinetic study should not be done in a rising dose design, because the purpose of the study is precisely to compare doses, and every effort should be made to keep those factors which are confounded with dose to a minimum. Suggested steps in the analysis are:

A graph or table of unadjusted mean or median plasma levels over time.
Summary statistics for pairwise ratios or ratios relative to a standard.
Dose-adjusted plasma levels, AUC, $C_{max}$, and urinary recovery. (Note: clearances should not be adjusted.)
An analysis of the between-dose differences in log-transformed dose-adjusted plasma levels, AUC, and urinary recovery (testing the hypothesis that differences on the log scale are zero, which is equivalent to ratios in the untransformed scale equaling one).
Appropriate analyses for a crossover design. This might include a multiple comparison procedure.

Other techniques which might be considered include:

Analysis of log-transformed unadjusted plasma levels, AUC, etc., if the spacing between doses is multiplicative; for example, dose 2 = 2 × dose 1 and dose 3 = 2 × dose 2 (testing the hypotheses that the ratios dose 2/dose 1 = 2 and dose 3/dose 1 = 4.*
Use of Fieller's theorem to analyze raw ratios rather than log-transformed ratios, but this may be difficult computationally.

*b. Steady-State Kinetics*

Chronic multiple dosing is often scheduled so that (1) a certain level of drug in the blood is achieved and (2) once this level is achieved, the crude rate of drug appearance in the blood equals that of drug elimination. Such a situation is called *steady state*. Sometimes, steady state cannot easily be obtained because the drug induces or, more rarely, suppresses its own metabolism. Hence, clearance of the drug may change with chronic dosing, causing plasma concentrations to change over time.

Normally, steady-state studies are open label (i.e., not single- or double-blind). Blood and urine samples are taken at specified times for up to 2 to 4 days following a single dose, after which chronic dosing is administered. Sampling is repeated at regular trough intervals (before drug intake) during multiple dosing and at the same times after the last dose as after the single dose.

The primary comparisons of interest are between first dose and last dose values for each pharmacokinetic variable measured. Ratios of last

---

*This is done because "linear pharmacokinetics" is demonstrated by observed increases in AUC of the same proportion as increases in dosage.

dose values to first dose values are usually estimated with point and interval estimates. The last-to-first dose AUC ratio is usually referred to as the "accumulation" ratio. A summary of the trough values over time is also required.

Suggested steps for analyses:

Tabulate data in such a way that each relevant comparison can easily be made (e.g., multiple-dose AUC versus single-dose AUC).
Use transformations when appropriate, e.g., logs for the AUC, reciprocals for the half-life.
Do tests appropriate for paired data.
Optionally, support the results with power or posterior probabilities.

### c. *Pharmacokinetic Drug Interaction Studies*

The purpose of pharmacokinetic drug interaction studies is to determine whether coadministration of drug A with a second drug, B, alters the absorption profile of either drug A or drug B. This section deals specifically with pharmacokinetic interaction; pharmacodynamic interaction is a special case of pharmacodynamic studies.

The usual design is a three-period crossover, with treatments A, B, and AB given in sequences balanced for carryover effects. However, such a study obviously cannot be *analyzed* with the usual models for a three-period crossover design because the plasma level data for drug B in subjects given drug A equal zero and vice versa. The easiest way to proceed is to do two sets of paired bioequivalence analyses, one for A vs. AB and the other for B vs. AB.

Suggested steps for analyses:

Do one-sample tests for A vs. AB and B vs. AB using appropriately transformed data.
Calculate power for the one-sample tests.
Calculate posterior probability for the one-sample tests.
Alternatively, use the following analysis of variance model as described by J. L. Ciminera et al. (Biometrics, in press). The paired t-test (model I) could be expanded to account for factors as indicated in model II.

Subdesign for treatment B (similar for treatment A).

| | | Drug levels | | | |
|---|---|---|---|---|---|
| SEQ | Subjects | PER 1 | PER 2 | PER 3 | Period pair |
| 3 | 3, 7 | B | | AB | 1 |
| 5 | 2, 12 | AB | | B | 1 |
| 4 | 4, 8 | B | AB | | 2 |
| 6 | 5, 9 | AB | B | | 2 |
| 1 | 6, 10 | | B | AB | 3 |
| 2 | 1, 11 | | AB | B | 3 |

| Model I (paired t-test) | | Model II | | |
|---|---|---|---|---|
| Source | DF | | | DF |
| | | *Whole plot* (between subjects) | | |
| SUBJ | 11 | PP | | 2 |
| TRMT | 1 | SEQ(PP) | | 3 |
| Error | 11 | SEQ | 1 | |
| Total | 23 | SEQ*PP | 2 | |
| | | SUBJ(SEQ(PP)) | | 6 |
| | | *Subplot* (within-subject) | | |
| | | TRMT | | 1 |
| | | PER (initial, last)[a] | | 1 |
| | | Residual | | 10 |
| | | PER*PP[a] | 2 | |
| | | TRMT*PP[a] | 2 | |
| | | TRMT*SUBJ(SEQ(PP))[a] | 6 | |
| | | Total | | 23 |

[a]These may be pooled to form the Error term, resulting in a test equivalent to the paired t-test.
Note: SUBJ, subjects; TRMT, treatments; PP, period pair; SEQ, sequence; PER, period.

### 5. Dose-Ranging/Dose-Response Studies

#### *a. Definitions and Objectives*

The dose-response relationship is studied to seek the therapeutically useful dosage range, that is, the range in which the drug is both effective and safe. The lower endpoint of the range is the minimum dose *above* which clinically useful efficacy is demonstrated by one or more efficacy response variables. The upper endpoint is the maximum tolerated dose, i.e., that dose up to which tolerance is clinically acceptable and above which it is unacceptable. If the lowest clinically useful dose is much lower than the maximum tolerated dose, then the drug is said to have a wide therapeutic window (range). If the two are close, then the therapeutic window is narrow. If the maximum tolerated dose is lower than the minimum effective dose, then the drug will probably not be therapeutically useful for that specific effect.

Usually the dose-response function, f(d), is nondecreasing, and this is nearly always assumed unless specific reasons exist to indicate otherwise.

Dose-ranging studies seek to determine the set of doses d such that f(d) is strictly increasing. Dosages from this range are then selected for inclusion in a dose-response study to identify the nature of the dose-response relationship. The usual dose increment is a doubling from one dose to the next higher dose.

The minimum effective dose can be thought of as the minimum dose which produces a clinically meaningful response. Similarly, the maximum effective dose is the dose beyond which no clinically meaningful increase in response is observed.

The aim of early dose-ranging studies is to determine the minimum and maximum effective doses, i.e., the "effective dose range." The range between minimum effective dose and maximum effective dose is the target for characterization of the dose-response curve in later dose-response studies.

Once the effective dose range is found, the time course of effect may be of interest.

Useful parameters related to the time-response relationship are onset of action, peak effect, time to peak effect, duration of action, and area under the time-response curve. Each of these parameters may be assessed as a response in relation to dose, and the dose-response relationships for these parameters may or may not be similar.

The clinical investigation of the dose-response curve can be broken down into three stages:

| Stage | Objective |
|---|---|
| 1. Dose-ranging study | Estimate the minimum and maximum effective doses, i.e., the range of doses for which the dose-response is strictly increasing. |
| 2. Dose-response existence | Decide whether there is a clinically meaningful increase in response between the minimum and maximum effective doses. |
| 3. Dose-response characterization | Describe the shape of the dose-response relationship. |

The distinctions among these objectives must be considered in the design of dose-ranging/dose-response studies. Often, due to inability to obtain a large enough sample, dose-response studies only answer the objective of dose-response existence and cannot adequately describe the shape of the dose-response relationship. Furthermore, the clinical scenario from which the "response" variable is obtained must be well defined in relation to the clinical objective(s) in Phase III. For example, a single-dose study measuring forced expiratory volume in a single second ($FEV_1$) after Leukotriene $D_4$ ($LTD_4$) challenge following various doses of a leukotriene antagonist may or may not yield information about the usefulness of the antagonist in the clinical treatment of asthma. The minimum effective dose from a single-dose $LTD_4$ challenge study may be a useful lowest dose in a multiple-dose dose-response study in asthmatic patients, but the single-dose maximum effective dose may be irrelevant to that for multiple dosing. In general, the response (or dose-response relationship) of a pharmacodynamic variable may not correlate well enough with the ultimate target clinical response.

*b. Dose-Ranging/Dose-Response Designs*

The up-and-down design and the rising-dose design are the two major designs for dose-ranging. The major objective of these designs is to identify the range of doses which yield a clinically meaningful response.

*Up-and-down design.* The up-and-down design (Bolognese, 1983) is a sequential-type design which involves assigning each subject's dose on the basis of the previous observation. For each patient's first administration, dose is determined from the dose and response of the previous patient. Subsequent administrations, if designed, are determined from the patient's own previous dose and response. If the previous observation was a nonresponse, then the next dose is increased to the next higher level. If the previous observation was a response, then the next dose is decreased to the next lower level. Thus, the doses given tend to migrate toward the effective dose range.

Selection of the actual doses used must be individualized, depending on prior knowledge of the drug or drug class. The usual increment is a doubling or halving from one dose to the next.

*Rising-dose design.* In a rising-dose design, there are two major variations. *Fixed sequence* of doses requires that each subject receive all doses in the sequence. Placebo can be inserted in the sequences and should be balanced with period. Patients are then randomized to sequences. The washout between periods must be long enough for comparisons among doses and placebo to be valid. In the *titration study*, patients are started with a fixed low dose and then titrated upward until a response is obtained. In these designs the following must be considered:

- Dosage is confounded with time period; thus, any dose-response interpretation must be of a hypothesis-generating nature, not of a conclusive nature.
- In the titration-type study, if enough information is available after the first n doses, then the dose-response relationship may be examined.
- The time course of effect after a single dose can be examined after the maximum dose in the titration study and after each dose in the fixed-sequence study if sufficient washout is included.
- Dosage levels are selected on the basis of prior information about the drug or drug class; the usual increment is a doubling from one dose to the next.
- The analysis of the latter should show the distribution of doses achieved.
- These designs usually have an escape clause based on safety.

The rising-dose titration study has the advantage of mimicking what is often done in practice. This type of design is also able to cover a broad range of doses. One must guard against overinterpretation of the data derived from such a study, because of the confounding with time. Another problem with this approach is that these studies are usually not blinded. As a consequence, rather than following the protocol, some investigators may alter the dose titration rules.

The rising-dose fixed-sequence study overcomes many of the problems that we have with the titration approach. In particular, with placebo given randomly in the rising dose sequence, these studies can be carried out under double-blind conditions. Moreover, with this design we can obtain complete data for all subjects. The cost, however, is that if we have not selected the appropriate doses, the study will have to be repeated. There is also a potential problem with carryover effects. These problems often limit this type of design to rising single-dose studies.

*Designs for dose-response* characterizations can be either parallel or crossover. The latter is usually either a complete Latin square or a balanced incomplete block (BIB) design, which must be balanced for carryover effects.

The parallel (completely randomized) design is usually the design of choice. Patient availability and other practical considerations often make crossover studies attractive. Nonetheless, we should not ignore the consequences of carrying out a crossover study in a situation where carryover effects may be present. Thus, designs balanced for carryover effects should always be employed.

The low-dose filter design is a special case of the parallel design. In this design all the patients are "filtered" through the lowest dose. Then the patients who respond to this low dose are continued on this dose. The remaining patients are randomized to higher dose groups. The rationale for this type of approach is that, while there is a subgroup of patients who will respond to almost any dose that is given, the patient population we are concerned with comprises those patients who do not respond easily but who, in practice, will have to be titrated up to gain control. Thus, this design identifies dose-response characteristics in this patient population of interest.

### c. *Sample Size and Power*

For the up-and-down designs, unfortunately, there is no standard methodology for assessing sample size. We are left to Monte Carlo simulation to check the adequacy of the planned study. For the more conventional designs, examine the power to detect dose-to-dose differences and include enough subjects to detect the anticipated difference between the low and high doses. This procedure is conservative and is preferable to making assumptions about linearity of dose response and approaching the power analysis from a regression viewpoint.

### d. *Analyses*

For all but the up-and-down design, parametric regression for analysis of variance models will maximize sensitivity. When the normality assumption is questionable, the rank transformation is useful. When the homogeneity of variance assumption is questionable, the log, square, square root, etc. transformations may be useful rather than resorting to the rank transformation.

Analyses for parameters representing the time-response relationship can be built into any of the above designs within the context of the designs. Note that for balanced designs, comparisons of time-response parameters among doses can be made. However, for the up-and-down design, which yields severely unbalanced (with respect to doses) data, comparisons among doses are difficult. Furthermore, these comparisons are invalid because the observations are not obtained independently; administered doses depend on previous response(s), thus, the responses are not necessarily independent from dose to dose. In this design, however, time-response parameters after single doses can be estimated for the combined distribution of observed doses by selecting the maximum dose given to each subject and performing usual one-sample analyses.

## C. General Considerations in Design and Analysis of Data from Clinical Pharmacology Studies

### 1. Data Reduction

When data are collected over time, an analysis may often be done at each time point. Summary measures may also be employed such as area under the curve, magnitude and time to peak response, or duration of predefined response.

### 2. Considerations in Carrying out ANOVA

Analysis of variance is usually the first choice for analysis of continuous or nearly continuous data from clinical pharmacology studies; when the assumptions are satisfied, it yields the most information. In these studies, sample size is usually small and type I errors committed will be uncovered in later confirmatory clinical trials. Type II errors are therefore most crucial, so the analysis should seek to maximize power.

Assumptions of equal variance and normality of error residuals are regularly analyzed, and often plots of residuals (and/or squared residuals) may augment or replace these tests. If departures are severe, data transformations may be sought, e.g., log, square root, inverse, and ranks. Nonparametric methods may also be employed if the design permits.

Multiple comparisons controlling for experimentwise error are generally not done because of the concern for maximizing power when the nature of the study is exploratory. Pairwise tests are used to maximize sensitivity to differences. When firm conclusions are necessary, multiple comparisons to control for experimentwise error may be desirable.

In the analysis of clinical pharmacology studies, estimation is usually more important than classical hypothesis testing. Therefore, the $\alpha = .05$ level of significance for ANOVA, contrasts, and pairwise comparisons may be softened. Reporting of actual p values which are less than .2, .15, or .1 is often done to give reviewers/readers a "feel" for the significance of the observed differences.

Preliminary testing of higher-order interactions and other nuisance effects may be done to support pooling them with error, which results in increased sensitivity.

### 3. Crossover Studies

The two-treatment, two-period design is often analyzed by Grizzle's method (Grizzle, 1965, 1974) and other designs are available for two treatments if carryover effects are suspected (e.g., BAA, ABB design). The Grizzle analysis can be carried out with any standard program, such as SAS GLM, and the ANOVA is presented below. If the sequence effect is not large, this treatment effect test is valid.

| Source | DF |
|---|---|
| Sequences[a] | 1 |
| Subjects (sequences)[b] | S − 2[c] |

| Source | DF |
|---|---|
| Treatments | 1 |
| Periods | 1 |
| Error | S − 2 |
| Total | 2S − 1 |

[a]A/B and B/A.
[b]Error term for sequences.
[c]S is the number of subjects.

In designs with more than two periods and/or two treatments balanced for carryover effects, these effects may be evaluated as illustrated in the following example. The usual model (e.g., three treatments, three periods, 12 subjects) is

| Source | DF |
|---|---|
| Subjects | 11 |
| Periods | 2 |
| Treatments | 2 |
| Carryover | 2 |
| Error | 18 |
| Total | 35 |

If carryover effects are not found to be important, then this effect can be pooled with the residual variability, and the following model may be used (e.g., three periods, three treatments, 12 subjects):

| Source | DF |
|---|---|
| Subjects | 11 |
| Treatments | 2 |
| Periods | 2 |
| Error | 20 |
| Total | 35 |

Data from crossovers may also be analyzed nonparametrically by Koch's method (1972), Gart's test (1969), McNemar's test, or ordered categorical methods.

### 4. Posterior Probability

Posterior probability is the probability that the true difference between means is at least (or at most) a specified magnitude, given the observed results of the study (that is, $\overline{x_1}$, $\overline{x_2}$, $n_1$, $n_2$, and SD). For further reading see Rodda and Davis (1980) and Steinijans and Diletti (1985). This method is used as an alternative to calculating power when no significant difference is found between means for variables of major objectives, or when a claim of "equivalence/similarity" is sought. This calculation may be an effective approach for the evaluation of bioequivalence and drug interaction studies.

## III. PHASE IIB/III: GENERAL CONSIDERATIONS

Some clinical pharmacology studies may be regarded as Phase II studies. These will be referred to as Phase IIA studies. Phase II also includes the first mid-size to large studies intended to demonstrate safety and efficacy. These will be referred to as Phase IIB studies. This section covers topics of general interest in Phase IIB and Phase III of clinical development. In many respects, Phase IIB and Phase III trials are indistinguishable. Section IV discusses examples of challenging questions that arise in Phase III clinical trials for specific indications.

### A. Phase III Objectives

By the time Phase III trials are initiated, Phase II trials will have provided some evidence of efficacy and safety. However, the Phase II trials will not have established efficacy for all indications hoped for in the drug's eventual claim structure. The patient population in Phase II trials may have been highly restricted to minimize patient risk until adequate clinical experience was available to safely study a broader patient population. The numbers of patients in Phase II trials permit only relatively imprecise estimates of the incidence of adverse experiences, and the Phase II trials may not provide any evidence of long-term safety. The Phase II trials almost certainly lack sufficient information on special subpopulations such as the elderly and the renally impaired.

Typical objectives in Phase III are to provide conclusive evidence of efficacy for specific indications, to gather evidence of safety and efficacy in the broader target population(s) that will be referenced in the claim structure and package circular, to obtain more precise estimates of the incidences of different types of adverse experiences, to obtain evidence of long-term safety, and to obtain evidence of safety and efficacy in special subpopulations.

### B. Definitions of Populations

The design of Phase III clinical trials must take into consideration all populations for which claims are desired. However, the populations defined by the inclusion/exclusion criteria of Phase III protocols do not, in general, include representation of all portions of the populations targeted for claims.

The exclusion criteria of Phase III trials are generally less restrictive than those of Phase II trials. Often, some Phase II exclusions which were intended to avoid safety problems can be relaxed in Phase III trials. Phase III studies, unlike Phase II studies, may need to ensure representation of several distinct populations in order to establish claims in each population (e.g., in patients with uncomplicated versus complicated urinary tract infections). Protocols should clearly identify the trial's target population. Inclusion criteria should include diagnostic criteria for determining the presence of the disease under investigation. Any other specific requirements for entry (e.g., disease of particular severity or nonresponse to prior therapy) and specific exclusion criteria should be listed in the protocol.

Sackett (1983) and Armitage (1983) both point out the need to strike a balance between efficiency and generalizability when defining target populations for clinical trials. A narrowly defined trial target population offers advantages in terms of lower variance, greater precision, greater power, smaller sample size, and hence smaller costs. For example, it may be possible to use studies that are restricted with respect to disease severity to substantiate an unrestricted claim, while using fewer patients than would have been necessary in unrestricted studies (e.g., it may be possible to convince the medical/regulatory community of general efficacy in the treatment of angina by studying patients who are severely enough afflicted to have five or more attacks per week). Too narrow a trial target population will not permit generalization to the populations targetted for claims, and may make the time necessary to recruit patients prohibitively long.

Even if study designers give far greater weight to generalizability than to efficiency, the trial's target population will exclude portions of the population targeted for claims. For example, Phase III trials will obviously not include patients who are unwilling to enter clinical trials and may not include patients with characteristics that are very likely to interfere with the assessment of safety or efficacy (e.g., alcoholism, drug abuse, and severe concomitant illnesses).

Availability of patients is an important practical consideration in defining target populations for trials. According to "Lasagna's law" (Calimlim and Weintraub, 1981) the incidence of the disease under study will drastically decrease once the study begins and will not return to its previous level until completion of the study. One major determinant of recruitment rates is patient willingness to enter clinical trials.

It is important to recognize that the patients in a clinical trial are never a random sample of the trial's target population. The investigators who participate are not a random sample of investigators, and the patients who are treated by an individual investigator are not a random sample of patients. It is hoped that the patients in the trial will be representative of the target population. Unwillingness of patients to enter a trial is a major cause of nonrepresentative samples. In some contexts, an unduly large share of the patients willing to enter a trial are nonresponders to prior therapies. This can bring about what Lasagna termed the "Lazarus phenomenon." An unduly large share of nonresponders to prior therapies can bias results against a new therapy.

The fact that the patients in clinical trials are not randomly sampled has implications pertinent to the dialogue between proponents of estimation and proponents of hypothesis testing. Because samples are not random, estimation, although of great practical value, has a weak theoretical foundation. As a result, the questions "What population has actually been sampled?" and "How is it related to the target population?" must be carefully considered when interpreting confidence intervals. However, the probability theory underlying hypothesis testing is unaffected by the nonrandomness of samples—the nonrandomly selected patients are randomly allocated to treatment, thereby preserving the theoretical foundation of hypothesis testing (between treatments). When a statistical test leads to rejection of the hypothesis of equality of treatments, it can be concluded that the treatments differ in *some* population, that is, in the population actually sampled.

In the end, the relationships between the population actually sampled and the trial target population and between the trial target population and the claims target populations must be considered when interpreting results of both estimation and hypothesis testing. The inferences or generalizations from Phase III data to the claim's target population will be, in any event, nonstatistical. They must be medically reasonable.

### C. Registration Considerations

Approval of a new drug in the United States requires substantial evidence of effectiveness. Substantial evidence has been defined by the FDA as consisting of adequate and well-controlled investigations by qualified experts. Consistent with the general scientific demand of reproducibility of results, this definition has been taken to mean that the effectiveness of a drug should be supported by more than one well-controlled trial. The FDA's draft clinical guidelines (FDA, 1986) state that "Ordinarily, therefore, the clinical trials submitted in an application will not be regarded as adequate support of a claim unless they include studies by more than one independent investigator." The guidelines go on to explain that rare exceptions to this standard are possible.

Many clinical development programs contain some studies which cannot be regarded as adequate and/or well controlled. The studies which are capable of satisfying this definition are referred to as "pivotal" studies. Reasonably conducted, but nonpivotal, studies may be regarded by the FDA as corroborative of the results of the pivotal studies. Planners of a clinical development program for the United States must plan for an adequate number of potentially pivotal studies. These will be Phase IIB or Phase III studies.

Potentially pivotal studies should be designed to have substantial power (perhaps 80 to 95%) against realistic alternatives when using the 5% significance level. Temple (1983) noted that the scientific community has generally agreed on the standard of a 5% significance level and that the FDA and most of regulated industry have accepted this standard. Furthermore, very careful consideration must be given to the nature of the controls in potentially pivotal studies. The FDA's new regulations recognize five types of control:

Placebo concurrent control
Dose-comparison concurrent control
No treatment concurrent control
Active treatment concurrent control
Historical control

Published in the Federal Register along with the new regulations (FDA, 1985a) is a lengthy discussion by the FDA of comments it received on draft versions of the new regulations. The above list is a reordering of past lists, and the FDA received several comments expressing concern that this represented a preferential order. The FDA responded that it did not consider one type of control to be necessarily preferable to another and that it had listed the types of control in roughly descending order of ease of interpretation. The FDA stated that it recognizes that "ethical and practical considerations will play a central role in the type of study selected, a decision that will ordinarily depend upon the type and seriousness of the disease being treated, availability of alternative therapies and the nature of the drug and the patient population."

In spite of these remarks, serious difficulties of interpretation can arise from positive-controlled studies (see Temple, 1983; Hsu, 1983). Planners of a U.S. development program must carefully consider the possibility of placebo-controlled pivotal studies. The choice of control is discussed in Section III,H. Because of international differences in attitudes toward the ethics of placebo control in some contexts, worldwide program planners must consider the relative ease of involving U.S. or foreign investigators in placebo-controlled trials when deciding where to conduct these trials.

Other countries have regulatory requirements which must be considered in planning the Phase III portion of a worldwide program. Some countries require a minimum of 1 year to establish long-term safety and efficacy. Some countries require special analyses of important subgroups of the patient population (e.g., the elderly or the renally impaired). The patients who provide the data necessary to satisfy such requirements generally come from the Phase III program, which must continue long enough to provide enough data to satisfy these types of constraints. They may be studied under separate protocols, or a summary of experience pooled across protocols may be adequate. Of course, the exact nature of the constraints depends on the type of drug and disease involved.

## D. Designs

Typically, pivotal Phase IIB/III trials employ multicenter, parallel, double-blind, controlled, randomized complete block designs with centers as blocks. If the drug is intended for chronic usage, the duration of therapy often ranges from 3 to 12 or more months. In chronic studies, dosage is often titrated to the "optimal" dose for each patient. Chronic studies may have study time periods in addition to a randomized double-blind period (e.g., periods in which specified concomitant therapies are added, periods in which nonresponders are crossed over to competing therapies, and open-label periods on the new drug). The pivotal portions of such studies are the randomized double-blind periods. The remaining periods are information-gathering exercises which may be useful in responding to ad hoc questions and in providing noncomparative long-term data.

Multicenter designs are common in Phase IIB/III trials because of the need to enter adequate numbers of patients within reasonably short time periods. Time is perhaps more precious within the pharmaceutical industry than in academic and government research. It has been estimated that only 1 in 10,000 compounds makes it all the way from the laboratory bench to approved use in humans. The protected patent life of a compound is limited, and the profits from today's research on successful products will pay for tommorrow's research. The possibility of generic competition makes it vital to maximize the length of the patent-protected marketing period.

Consequences of the use of multicenter designs are discussed in detail in Sections III,E and III,F.

There are severe risks inherent in planning that relies on the acceptance of a crossover rather than a parallel trial as a pivotal study. The pros and cons of crossover trials have been discussed extensively elsewhere (BEMAC, 1976; O'Neill, 1977, 1978; Hills and Armitage, 1979; Brown, 1980). Because of problems caused by dropouts, it is practical that crossovers be of relatively limited duration. Crossovers assume a constant underlying disease state. They offer sample size/power advantages over parallel studies only if it can be safely assumed that there are no differential pharmacologic or psychologic carryover effects. Today, the constant disease state and carryover assumptions receive extensive scrutiny. The degree to which the scientific/regulatory community will accept crossover results depends on how convincingly it can be argued that these assumptions are met. The statistical test of carryover effects may not be adequately convincing. Brown (1980) showed that in the two-period crossover, the typical sample size does not permit an adequately powerful test of carryover effects. This implies the need for a strong clinical rationale for these two assumptions.

The clinical development program cannot possibly answer every question with parallel, randomized, controlled, double-blind studies. It may be reasonable to address lower-priority, tangential questions with crossover designs and also to use the crossover for exploration and corroboration.

Filter designs are sometimes used in Phase III trials. These are simply parallel designs in which the patient population has been determined by employing an active therapy baseline to identify nonresponders to that therapy. Only the nonresponders go on to a randomized comparative period involving the new drug and control (either in addition to or instead of the active baseline therapy).

Double-blind pivotal trials are preferable to open or single-blind trials. Under rare circumstances, blinding is not possible or is extremely impractical. If this is the case, extreme care must be taken to ensure that treatment allocation is, in fact, randomized. The issue of blinding is discussed further in Section III,G.

Group sequential designs are infrequently employed in Phase III trials. When they have been used, it has most often been in studies where the endpoint is death. These designs are discussed in detail in a later chapter. On the surface they appear to have great theoretical and practical value. However, there are subtle practical and theoretical problems with these designs in an industrial setting which have prevented their widespread use. On the practical side, data flow in a sequential trial requires procedures very different from those used for the vast majority of trials. This is not an insurmountable problem, but it contributes to inertia. A second practical problem is recruitment of investigators. It is difficult to get investigators to agree to be involved with a study when the length of time, number of patients, and amount of money involved are indeterminate. Finally, the potential for correlation between results and the time at which they are received from investigators (early or late in the study) poses a theoretical problem for sequential methods. The example which will be discussed in Section III,F, in which investigators with favorable results sent in data first and those with unfavorable results followed, illustrates one aspect of this problem. A second aspect of the problem can arise when each investigator has a learning curve and late data from an individual investigator are of higher quality than early data.

Phase III programs also include many nonpivotal studies. Examples include studies intended to show similarity with a variety of active controls for marketing or pricing purposes; open, uncontrolled studies intended to provide a greater volume of safety data in a greater variety of patients; small exploratory studies intended to provide insight but not definitive answers to specific, scientifically interesting questions; and small studies by national experts.

Dropouts are a greater problem in long-term trials than in short-term trials. If a drug is intended for chronic use and is to be studied in a long-term trial, sample size estimates should be adjusted for dropout rates so that adequate numbers of patients can be expected to complete the trial. Therapy-related dropouts have great potential for biasing analyses. This is discussed in Section III,L.

Randomization at each center may or may not be stratified in Phase IIB/III trials. Assuming that the allocation numbers at each center are generated in blocks of one or two times the number of treatments, it is trivial to stratify within centers over a single two-level factor (the low level gets low allocation numbers; the high level starts with the highest numbers and proceeds down). However, because of the complexities of conducting multicenter studies with many strata, stratification in these large studies is usually limited.

## E. Multicenter Studies

Phase IIB/III trials often require one or more hundreds of patients to have a reasonable likelihood of arriving at a definitive conclusion. It is usually not feasible to accumulate such large numbers of patients at a single center within a reasonable time frame. Thus, multicenter trials are usually necessary in Phase IIB/III trials.

Compared with single-center trials, multicenter trials have many advantages and many disadvantages, and some of the advantages can also be viewed as disadvantages. For example, a multiclinic study will allow estimation of the consistency of response from one investigator to another. But if the interinvestigator variability is large, or if there is a treatment-by-investigator interaction, pooling the results analytically may become a complex issue.

We know that treatment-by-investigator interaction exists in multicenter studies, and we do not want to get rid of it. We want to characterize this interaction and use it to develop comprehensive conclusions which describe what a patient can anticipate if he or she takes this medication for a specific indication. The objective is to consolidate all the information presented by the various components of a study and reach global conclusions. Only a multiclinic study will allow this breadth of inference. Such broad conclusions cannot usually be gleaned from a study conducted at a single center.

Unfortunately, the success of a multiclinic study is dependent on the least effective investigator. These investigators often provide poor quality data which will compromise the interpretation and are often the latest to submit their data to the data coordinating center, thus creating even more administrative difficulties.

Another issue which can compromise the effectiveness of a multicenter study is investigators deviating from the protocol at their convenience. In a single-center study, deviations from the protocol are not desired, but because the study will stand on its own and will not be combined with those from other centers, conclusions can sometimes still be gleaned for a modified objective. In multicenter trials, on the other hand, the essence of poolability is the core protocol which is followed by all investigators. Any deviation from this common protocol can wreak havoc with the ultimate conclusions. To minimize this effect and also provide investigators with an opportunity to do some unique research, a core protocol can be followed by all investigators with each providing identical information and having the option of following some type of unique subprotocol in addition.

Another advantage of the multicenter study is that it can more clearly reflect the population in which the product will ultimately be used. Different centers will have different patient demographics. They will differ with respect to standard therapies, concomitant conditions, hospital or clinic procedures, and a variety of other characteristics.

The objectives of any clinical trial, be it an individual study or a multiclinic trial, are to estimate the treatment effects and to draw conclusions *relative to the study objectives*. Clinical trials are designed to estimate effects with a specified precision or to reach a decision with a certain power.

The importance of the study objectives cannot be understated. These objectives might be worded, for example, as "to determine whether the antihypertensive effect of drug A is at least 10 mm Hg greater than the antihypertensive effect of placebo in patients with mild to moderate hypertension." Any analysis which does not produce very specific conclusions concerning this objective is inadequate. (In fact, the answer must be either "yes" or "no" to truly satisfy this study objective.) In a multiclinic study, effects must be estimated, and conclusions drawn which address the objectives—*regardless* of the existence of investigator effects or treatment-by-investigator interactions. The existence of these effects may complicate the analysis, but they do not permit statisticians to abrogate their responsibility to draw conclusions.

### F. Treatment-by-Investigator Interactions

Many statisticians feel that if there is an important treatment-by-investigator interaction, pooling is not appropriate. In general, statisticians are taught that interactions are bad and that transformations or other techniques should be used to minimize interactions.

In the presence of interaction, pooling is inappropriate from a purely analytical standpoint. However, from a pragmatic point of view, the need to draw a general conclusion is independent of the results. Note that the *need* for conclusions is not dependent on the results. The conclusions themselves are certainly dependent on the results and will be critically dependent on whether there are investigator effects or treatment-by-investigator interactions. For example, let the objective of a study be to determine whether there is a clinically important difference between the active therapy and, say, a placebo and to estimate what that effect is. The most convenient, and perhaps least expected, outcome would be for each of the investigators to produce the desired results. This means that the treatment effect would be comparable for all investigators and that its estimate would also be similar for each of the investigators. However, it is the unusual study that has this simple result.

A more reasonable result, and certainly one which is encountered more frequently, is that there is investigator-by-treatment interaction to a degree. Suppose the desired result is observed by six of nine investigators in a multiclinic study and there does not seem to be a treatment effect in the results of the remaining three investigators. For simplicity, assume that there is consistency within each of these two sets. The analysis will certainly partition the positive investigators from the other investigators and yield a *result* such as "the results of six of the nine investigators exhibited a significant treatment effect when pooled; however, the results of the remaining three investigators did not exhibit a comparable effect. Because of this discrepancy these results could not be pooled in the analysis." Unfortunately, this type of statement does not address the objective of the study.

This example, although perhaps provoking *some* controversy among statisticians, is not as controversial as the following example. In the previous data set the desired result was observed by a majority of the investigators. A more difficult problem develops when the desired result is observed by a minority of investigators, but for those investigators the result is unequivocal.

In the nine-investigator study, suppose two investigators found the desired result and the remaining seven did not observe any treatment effect.

Actual data from a clinical study are represented below. These data represent the overall therapeutic response stated by the investigator and are categorized into good or excellent and none, poor, or fair. The top of the following table presents the results from two of the nine investigators who participated in this study. Based on these data, there is a convincing drug effect. However, for seven of the investigators there was no clear drug effect, as shown by the data at the bottom of the table. Because of the influence of the two positive investigators, the pooled drug effect is extremely clear, and any statistical test would produce a very small p value.

| | Number of patients with specified therapeutic response | |
|---|---|---|
| | G + E | N + P + F |
| Investigators 1 and 2 | | |
| Drug | 14 | 2 |
| Placebo | 1 | 16 |
| Investigators 3 to 9 | | |
| Drug | 33 | 28 |
| Placebo | 37 | 27 |

When the study was designed, it was anticipated that the results would be observed by a great majority of the investigators, rather than what actually occurred. Obviously, when the data are analyzed, there is a clear investigator-by-treatment interaction. In this case, the results of two investigators show a statistically significant effect and those of the rest of the investigators show no significant difference between the two treatments.

Again, this begs the conclusion, and in this example drawing that conclusion is much more difficult. If there were no good pharmacologic reason for the treatment effect, one might say that no clinically important effect was observed and effectively pool all the results mentally, with the overall conclusion among the investigators watered down.

On the other hand, patients of selected investigators might anticipate a very positive drug effect, and this must be reflected in the conclusions. However, to pool mentally all of these data and say that there *is* a drug effect would probably *not* be appropriate here, because of the small proportion of investigators with whom this was observed. Perhaps a more appropriate conclusion in this example would be, "The results of this study are consistent with the premise that drug A has a therapeutic effect in the appropriate indication; and that although the degree of response may vary substantially among practicing physicians, clinically important effects do occur in many patients."

The results of this particular study presented an additional statistical problem which is often encountered. As in most studies, data from these investigators arrived sequentially—and in the order presented. As the incoming results were reviewed, the positive results were observed and there was a temptation to terminate the study. In this study, the program was not interrupted, but it should be noted that data from the better investigators usually arrive earlier than data from poorer investigators.

Estimating the treatment effects in these circumstances can also be a difficult issue. Should the estimate of the treatment effect be the pooled estimate? Since the average patient cannot expect to see any effect, perhaps the effect should be estimated only in the patients who respond. A choice must be made.

There are two schools of thought concerning the issue of whether the investigator factor should be considered as a fixed or random effect in the model. Some feel that investigators should always be considered as a random effect in the analysis, since it is the population of potential prescribers to which we usually want to infer. If investigators were considered a fixed effect, the inference would be limited to the investigators who participated in the study. However, even proponents of the random effects philosophy often include investigators as a fixed effect, because their specification as a random effect would reduce the power of statistical tests used in the analysis.

In contrast, many statisticians feel that investigators are never truly a random sample of the ultimate population of interest and, because of this, should be treated as a fixed effect in the analysis. The assumption necessary for the ultimate inference is that even though the investigators are not a random sample from the target population, they are representative enough that conclusions reached from the sample of investigators extend to the population of physicians who might use the drug.

In any case, the conclusion from the study requires a statement which will permit patients to know whether or not the drug works and to have an estimate of its effect when it is prescribed by their physicians—and their physicians will almost certainly not be part of the study sample. Therefore, conclusions will almost always have to be based on the pooled effect from all investigators.

If there is a treatment-by-investigator interaction, clearly there is a question of pooling when performing hypothesis tests. The existence of a pure investigator effect is not usually considered important for this issue. In fact, an investigator effect in the absence of an interaction will have no theoretical impact on the testing of hypotheses. However, it can be very problematic in the estimation of drug effects.

The data below represent the percentage of patients with a central nervous system (CNS) adverse experience observed during a parallel study of three different drugs. There are approximately 120 patients in each of these groups, and it is clear that for one set of investigators the frequency of adverse experiences was well under 10%, while for the other investigators the frequency approached 25% of the sample. Because there is no treatment-by-investigator interaction, it is fair to compare the three treatments across investigators and to conclude that there is no important difference in the adverse experience profiles of these three drugs.

| | Percentage of patients with CNS AE[a] | | |
|---|---|---|---|
| | Drug A | Drug B | Drug C |
| Investigators 1-4 | 2 | 2 | 7 |
| Investigators 5-8 | 26 | 25 | 24 |

[a]Approximately 15 patients per investigator per treatment.

The difficulty comes in estimating the incidence of CNS adverse experiences. When analyzing these data, a great deal of attention was paid to identifying any characteristics which might explain this dichotomy. However, the only difference noted was that investigators 1-4 were U.S. investigators and investigators 5-8 were foreign investigators. This suggests a potentially easy resolution of the problem—for domestic labeling say that in the United States there is a 2% incidence of CNS adverse experiences and outside the United States there is a 25% incidence of adverse experiences. However, this still begs the issue. Uniform labeling should be the rule, unless there is a clear and logical rationale for the contrary. And, unless these differences can be explained, there is no reason to draw such a definitive conclusion. In this case, the conclusion may have to be that the proportion of patients experiencing CNS adverse experiences with this drug can be as high as 25%—even though no intermediate incidences were observed.

This example demonstrates again that, even though analytical pooling would not be appropriate to gain an overall estimate, inferential pooling must be done to reach a definitive conclusion.

Multicenter studies are designed with the assumption that results will be pooled to reach a conclusion across the spectrum of patients and physicians to whom the new drug is directed. When performing the analysis of a multicenter study, pooling should be assumed unless there is clear and interpretable evidence that pooling would be inappropriate. The examples presented demonstrate that although simple averaging of data may not be an appropriate pooling technique in some studies, the totality of patient experience must be included in developing conclusions.

## G. Blinding

For most drug classes, comparative Phase IIB/III trials are usually double-blind. Double-blinding is widely recognized as an important method of reducing investigator/observer and patient bias. Double-blinding may be accomplished by providing all treatments in identical images or by requiring each patient to receive a product image of each therapy, only one of which is an active agent.

In single-blind studies, the investigator is not blinded. This risks the possibility that investigators will selectively rather than randomly allocate patients to treatment. Furthermore, it risks the possibility of biased data collection. The latter risk may be avoided by employing a blinded observer other than the investigator. In open studies, neither the investigator nor the patient is blinded. The unblinded patient can introduce patient biases (e.g., related to personal expectations about the drug's effects).

An accusation that unblinded investigators might wantonly violate an allocation schedule or bias data collection is an insult to clinicians. When explaining to clinicians the importance of blinding investigators, it is helpful for the statistician to be equipped with illustrations of how investigators can innocently fall into this bias trap. Consider the following examples.

In a randomized, open, multiclinic comparison of an angiotensin converting enzyme (ACE) inhibitor and a β-blocker, the percentages of patients who had used β-blockers at entry into the study and the percentage who had never before received antihypertensive therapies were:

| | ACE inhibitor group[a] | β-Blocker group[a] |
|---|---|---|
| Prior β-blocker | 54% | 33% |
| No prior antihypertensives | 23% | 48% |

[a]Approximately 90 patients/group.

These differences were not the work of one or two investigators. Nearly every investigator had an extra one or two ACE inhibitor patients with prior β-blockers and one or two extra virgin β-blocker patients. When study monitors were questioned about possible explanations for the differences, their immediate reply was that the investigators were probably reluctant to allocate prior nonresponders to β-blockers to the β-blocker group. This explanation seems plausible.

Here, the open study design appears to have confronted many of the investigators with an ethical dilemma: "This patient hasn't responded to β-blockers in the past. Should I allocate him to a β-blocker now? What harm can it cause if I just assign this one patient to the ACE inhibitor group?"

A very similar situation arose in a randomized, open multiclinic comparison of two antibiotics. Prior to randomization, etiologic pathogens were tested for susceptibility to both antibiotics. There were 21 patients with pathogens susceptible to only one of the two drugs (not always the same drug). All 21 patients received the correct drug—the one to which their pathogen was susceptible. Here, the protocol did not instruct investigators how to proceed in a situation which could easily have been anticipated and which confronted investigators with an ethical dilemma: "How can I randomize this patient when I know he has a high probability of being cured on drug A but not on drug B?"

In addition to the risk of nonrandomized allocation, unblinded investigators increase the risk of biased data collection. For example, an investigator who expects a particular type of adverse experience with drug A and not drug B might *explicitly* ask members of group A, but not group B, whether this type of adverse experience had occurred. Such uneven methodology can bias the data that are collected.

The scientific validity of studies is best preserved by using double-blinding whenever possible. In the absence of double-blinding, one should make every effort to preserve the randomization (e.g., explain its importance to investigators and/or hide the identity of the next treatment to be allocated until after it has been allocated). Protocols should attempt to anticipate ethical dilemmas (and other sources of bias) which might arise and provide approptiate instruction aimed at preserving randomization and minimizing bias. Once data have been collected, comparability of the treatment groups should be thoroughly investigated.

A third level of blinding is possible which goes beyond double-blinding in avoiding potential sources of bias. In-house or triple-blinding refers to situations in which the sponsor is also blinded as data are received in-house from investigators, until key decisions with a potential for bias (e.g., exclusions from per protocol analyses or identification of questionable data values which need to be discussed with investigators) have been made. If in-house blinding is attempted at all, it should not merely involve replacement of treatment group names with abstract labels (e.g., A and B). Variables which permit separation of patients into groups corresponding to the (unnamed) treatment groups should be totally inaccessible. Otherwise, it is often too easy to guess the treatment group name from group characteristics (e.g., group A had the low heart rate so it must be the β-blocker group).

## H. Selection of Control Agents

The specific objective of a Phase IIB or Phase III clinical trial will usually dictate the control regimen very specifically. Five types of control are used in clinical studies: concurrent placebo, lower active dose, concurrent no therapy group, active control, and historical control. In Phase IIB studies, where the primary objectives are usually to develop preliminary efficacy and safety data on a new formulation, the primary control regimens are placebo, lower doses, and historical.

Historical controls may appear to be the easiest because they involve no concurrent therapy. However, the assumptions necessary for their use, and the complexities of comparing the sample under study with that from which the historical information was obtained, can result in outcomes which are not clearly interpretable and will usually be subject to criticism by the scientific community. Historical controls can be useful for exploratory purposes early in a program. Knowledge of the patient's outcome under a historical regimen with *very high* probability enhances the usefulness of historical controls.

For example, it is well known that untreated patients with onchocerciasis have no remission in the number of microfilaria which may be biopsied from their skin. Therefore, any treatment which would reduce the concentration of microfilaria would easily be identified with a small number of patients and without the need for a concurrent placebo control group. Unfortunately, it is the exceptional condition which has a sufficiently predictable course to allow the use of historical information as the control basis.

In the absence of historical controls, there are only two control regimens which permit a definitive conclusion that the proposed new product is more effective than no treatment. A concurrent placebo control clearly provides this comparative information, and in the majority of clinical research programs Phase IIB will include placebo-controlled studies. The other option is clear characterization of a dose-response curve. If it can be demonstrated that increasing the dose from one level to another in a sequential manner results in a greater response at each succeeding level, then there is at least one dose beyond which a clinically important effect occurs and thus a comparison with a negative control would be unnecessary. This type of study is often a rather large study and one which cannot be begun prior to the assurance that a particular dose is, in fact, different from a placebo control. Thus, in most clinical programs, both the dose-response program and studies with placebo controls are conducted.

Placebos may be chemically inert (although there are documented cases of patients having anaphylactic reactions to components of placebos), but they are not without effects on the patient. Placebos relieve pain, heal ulcers, reduce inflammation, provide euphoria, and cause adverse experiences. It is very difficult to characterize the true efficacy and safety profile of a new therapeutic agent without some concurrent comparison against a placebo control agent.

The use of active controls in Phase II trials is somewhat limited compared to their use in Phase III. In Phase II clinical studies, active control agents may be used as a second control agent, in addition to a negative control, to provide an initial estimate of the utility of the new agent compared to a standard therapeutic modality. This active agent ordinarily would be selected to represent the standard therapy for the indication under study. In Phase II it is usually inappropriate to compare the new agent directly with an active control without also comparing the two with a placebo.

In many clinical conditions, the variability of the measurements or the condition itself and the resultant variation in response to active control will not permit conclusions that the new formulation is effective, even though similarity with an approved product can be demonstrated. Analgesics are prime examples. From 30 to 70% of patients with mild to moderate pain will experience substantial relief from a placebo tablet. Therefore, if a new analgesic were compared to aspirin, for example, and both provided relief of pain in 60% of the patients, there would be no assurance that the new agent was effective, even though it is well known that aspirin is valuable for reducing mild to moderate pain. The exclusive use of active controls without a placebo regimen in a study is usually reserved for later phases.

When selecting control agents for Phase III trials, potentially pivotal studies must first be identified. A development program for the United States requires at least two pivotal studies and many foreign agencies would also expect two pivotal studies, although their definition of pivotal might differ slightly from the U.S. definition. Sometimes the Phase IIB program will provide two pivotal studies. If this is the case, there is greater latitude in the Phase III program to focus on marketing and pricing issues in the selection of control agents. This will be discussed later in this section. More often than not, the Phase III program must provide one or more pivotal studies (e.g., the placebo-controlled Phase II trials may have been too small to be considered pivotal). If it is possible to demonstrate superiority over an active control or a lower dose of the new drug and if it can also be argued that the active or lower-dose control group fared no worse than a placebo group would have, then it might be reasonable to plan a program that relies on active or lower-dose controlled studies for pivotal support. However, the sample sizes required to demonstrate superiority over active agents may be prohibitively large. For the disease and drug under investigation, there may be no ethical difficulty in conducting placebo-controlled trials. If this is the case, placebo control is probably optimal in terms of permitting results with a high degree of scientific credibility while using practically attainable sample sizes.

If there are ethical problems with placebo control, possibilities for creative design compromises which permit at least a limited placebo comparison should be considered. One example of this is a study in which the new drug and placebo are compared as add-ons to existing therapies (e.g., an ACE inhibitor versus placebo as add-on therapy in congestive heart failure patients treated with digitalis and diuretics). This approach may be impractical if the additional effects are small or if the new and existing therapies have similar mechanisms of action (Temple, 1983). If an add-on therapy comparison with placebo is under consideration, one should consider whether or not it would be advantageous to first filter out responders to existing therapies.

Another approach, suggested by Pledger (1986), is to make effective use of escape clauses. In some contexts it might be possible to persuade investigators and patients to participate in placebo-controlled trials that include escape clauses, such as:

> If, in the opinion of the investigator, treatment has failed and the patient requires replacement therapy, the patient will be withdrawn from the study and receive appropriate therapy. Study therapy will be regarded to have failed.

Analyses of data from such a study could include comparisons of withdrawal rates due to therapy failure as a major efficacy analysis. However, this method of analysis may be less powerful than a method which attempts to score both study completers and dropouts on a common scale and compare treatments with respect to these scores. The method described by Gould (1980) can be used for this purpose. This method is discussed in Section III,L.

If all else fails and positive controls must be utilized for pivotal support, yet superiority of the new drug cannot be demonstrated, there are serious interpretation problems. These are discussed in detail by Temple (1983). Basically, in order to use equivalence to a positive control as pivotal evidence of efficacy, it must be possible to argue *convincingly* that the positive control was indeed effective in the study in question and that poor study methodology did not introduce excess variability into the data and blur the distinction between an inferior new drug and the active control.

In Phase III, apart from pivotal studies which may use placebo, there is much greater use of active control agents than in Phase II. This is true for at least two reasons. First, Phase III studies are often of longer duration than Phase I or II studies, and in many cases it is unethical to maintain a patient on an inactive medication for an extended period of time.

The second reason is the marketing need to determine the position of the new product among competitive agents.

The actual selection of active control agents will rest on several factors. If the sponsor believes it has a superior product, a head-to-head comparison with the product which the sponsor feels will be the major competition will be useful to determine whether the new therapy is in any way superior to the current standard. These studies have the built-in weakness that, in many cases, the advantage of the new therapy over the existing therapy is minimal, and consequently a study with 100 or 200 patients will not be able to distinguish between the two. This will produce a weak result of "no significant difference." Since it was not designed to detect realistic differences, the study does not identify any differences and, on the other hand, does not convincingly confirm equivalence.

Other considerations in the selection of active control agents for Phase III clinical studies are the requirements of individual departments of health. In many countries, socialized medicine dictates that the price paid for a therapy is set by the government, and thus a sponsor who wishes a specific price for a new drug must compare it with the drug of choice in that individual country. To gain a higher price, the sponsor must demonstrate some superiority over that drug of choice. Because the chosen therapy varies from one country to another, this will require a number of clinical trials with a number of different control agents. Each must provide enough information to convince the individual country that the new product is appropriately positioned against the specific drug of choice in that country.

Although several control agents may be used in a Phase III program, and the results may not be pooled for comparative analyses, there is an advantage in conducting studies with similar protocols. This provides an opportunity to combine the safety data and give a useful description of the safety characteristics of the new product. At the completion of the program, the sponsor has a substantial reservoir of information on the new product and a fraction of that information on the competitors. However, the competitors are usually better established and there should be some historical information available on these products. On completion of a Phase III program of this type, the result should be adequate safety information on the new drug and, in addition, a reasonable amount of information with which to compare it against a number of control agents.

Although analytical pooling may be inappropriate for this type of program, the ability to intellectually pool the data and develop an accurate safety profile for the drug is important. In addition, the positioning of the safety profile with respect to the variety of control agents used in the program should provide an understanding of the relative safety profile of the new agent.

### I. Estimation Versus Hypothesis Testing

The use of control agents focuses objectives, analyses, and interpretations on comparisons. This creates a substantial emphasis on the value of hypothesis testing in the evaluation of studies. Although it is critically important to be able to distinguish between test formulations and the appropriate controls with a clear knowledge of type I and type II errors, there is often not enough emphasis on estimation of the treatment effects. It cannot be understated that the objective of clinical research programs is not only to compare a new agent with an existing standard, but also to clearly describe the effect that can be anticipated when that agent is administered in the patient population.

Statisticians often perform a variety of transformations on data to provide data with desirable distributional characteristics for their analysis. Relating the results of the analysis on these transformed data to estimates of the therapeutic response can sometimes be very difficult. It is important to design each clinical study not only with comparative objectives, but also with the objective of providing an unbiased and accurate estimation of both the positive and negative effects of the clinical regimen.

### J. Unequal Allocation to Treatments

The principal objective of most Phase I and Phase II clinical trials is the demonstration of efficacy, or mechanism of action, because the test agent is usually compared against a placebo or other control and a definitive distinction between the control and the test agent is required. Sample sizes are usually determined to provide maximum power. This means that equal numbers of patients would be allocated to both the test and control agents.

However, in Phase III clinical programs, an additional objective is to gain as much experience as possible with the new agent. This objective is somewhat inconsistent with distinguishing between the two agents, since maximal experience would be obtained if no patients received the control and maximal power is obtained when the control is assigned to the same number of patients as the treatment. To accommodate this, many Phase III programs are designed with unequal allocation of treatment. The table below provides an approximation to the efficiency factors for varying ratios of test to standard allocations.

Efficiency Factors for Unequal Allocation

| Allocation ratio of test to standard | Efficiency |
|---|---|
| 3:2 | 1.04 |
| 2:1 | 1.12 |
| 5:2 | 1.22 |
| 3:1 | 1.33 |
| 4:1 | 1.57 |
| 5:1 | 1.80 |
| 6:1 | 2.04 |

If an adequate amount of information about the efficacy and safety compared with the appropriate control agents has been acquired in earlier trials, or if power in the present trial will be adequate in spite of modest inefficiencies, then an unequal allocation may be an optimal schedule for providing additional patient experience in Phase III clinical programs. Since many agencies around the world have requirements of certain numbers of patients for certain periods of time, this is often the only way to acquire an adequate amount of information in a reasonable period of time and with a minimum exposure of patients to experimental therapies.

### K. Studies in Special Populations

Specific subpopulations are often excluded from earlier phase studies for several reasons. For example, lack of negative teratogenicity in animal studies would exclude women in whom pregnancy might be a possibility. Patients with renal impairment would be excluded from a study of a drug known to have a significant pharmacologic effect on the kidney. The elderly and children are also routinely excluded from clinical trials. The question is, will these patients be denied the use of the new therapy simply because they have not been included in earlier clinical studies?

In some cases, the answer is yes. A tragic example of ignoring these situations was the phocomelia observed in the offspring of women exposed to thalidomide during pregnancy. Clearly, recommending a new therapy for use in pregnant women without a clean safety profile in pregnant animals of other species would not be reasonable. In other cases, the therapeutic orphans are created not because of a safety concern but because of a trialistic convenience. An example of this would be the exclusion of elderly patients from Phase III clinical programs.

It may or may not be necessary to plan separate Phase III studies in special subpopulations. If there is sufficient commonality among (several) Phase III protocols, it may be adequate to pool data across protocols to create a descriptive summary of experience in a subpopulation (e.g., elderly persons who may have entered studies as protocol violators or in a restricted age range such as up to 70 years of age). Such a pooled summary would not be used for the purpose of doing comparative analyses or drawing definitive conclusions. Given sufficient commonality and an adequate number of patients, a pooled summary may help the sponsor and/or regulatory agencies heuristically gauge the potential for problems within the subpopulations and decide whether additional studies are needed (perhaps as postmarketing surveillance).

### L. The Dropout Problem

The proportion of patients who withdraw from long-term clinical trials can be substantial. If the reasons for withdrawal are related to therapy (i.e., ineffective therapy or adverse experience) and treatment groups differ with respect to either proportion of therapy-related withdrawals or time until therapy-related withdrawal, then dropouts can be a serious source of bias in comparisons of treatment efficacy. This situation is especially likely to occur in moderately long-term (e.g., 3 to 6 months) placebo-controlled trials in which, due to the nature of the disease under study (e.g., congestive heart failure, CHF), substantial percentages of placebo patients (e.g., 10 to 30%) can be expected to experience obvious deterioration during the course of the trial. Any effective drug is likely to result in a smaller percentage of patients with obvious deterioration and create an imbalance in withdrawals due to ineffective therapy. Possibly offsetting this imbalance, the active drug group could have a higher percentage withdrawing due to adverse experiences, but it is also possible that the active group could have a similar or lower percentage withdrawing due to adverse experiences (in situations when many adverse experiences are further manifestations of ineffective therapy, e.g., in CHF).

In the situations described above, comparative analyses of only those patients who completed the study can be grossly misleading. This type of analysis risks selecting patients who were less serious at baseline and/or whose natural disease course was deteriorating relatively slowly, and selecting this type of patient relatively more frequently in the placebo group than in the active therapy group. Under the conditions described, this type of analysis would tend to (1) bias results upward in both groups, but (2) more so in the placebo group, thereby biasing comparative results against active drug.

Endpoint analyses have been commonly used to deal with withdrawals. These use the last recorded values of withdrawn patients as data for observation points after the patients withdrew. This implies an assumption that beneficial and adverse effects are stationary in time. In contexts where it is clear that withdrawals due to ineffective therapy and adverse experiences represent negative outcomes (e.g., in CHF; see Pledger and Hall, 1982, for an example in which the negativity of such withdrawals might be questionable), the validity of the stationarity assumption can be assessed by examining the data values which have been carried forward in time past withdrawal dates. In a 12-week study measuring changes in exercise duration in CHF, it is absurd to analyze data that impute increases or modest decreases in exercise duration at week 12 for patients who died of CHF weeks earlier. It is especially absurd to do this twice as often for placebo patients as for actively treated patients.

Gould (1980) describes a method for accounting for dropouts in efficacy analyses. Basically, study completers and dropouts are assigned scores on a single ordinal scale. These ordinal scores are then analyzed by any appropriate method. Starting with the range of possible measured values for study completers, dropouts due to ineffective therapy and adverse experiences are assigned a very bad score, whose value is chosen to be more extreme than any measured value. If patients withdrew because they were cured, they are assigned an extremely good score. The assigned and measured scores are then ranked to produce ordinal scores. The method can be modified to carry forward data for patients who withdrew for reasons unrelated to therapy. As noted earlier, this method might be very useful in situations in which early use of escape clauses is proposed as a compromise to permit ethical placebo control.

## M. Interim Evaluations

Ethical considerations, i.e., the desire to minimize patient exposure to an inferior therapy, can motivate interim analyses of ongoing trials. In addition, a sponsor has a responsibility to monitor safety and be alert to any need to stop an ongoing trial in the interest of patient safety. However, to a great extent the motivation for interim awareness of trial results comes from corporate management. Millions of dollars may be at risk, and if a product is going to fail, it is best to know as early as possible so that losses can be minimized. Preparation of a submission may require the hiring and training of new staff. Management may wish to base hiring decisions on the progress of ongoing trials. Preparation for many postsubmission activities may have to begin while Phase III trials are ongoing (e.g., development of scaled-up manufacturing processes). Key decision points entailing increased expenditures may be reached before completion of Phase III trials, and management may wish to assess current results and the prospects that ongoing trials will culminate in a successful result before committing to greater expenditures.

It is well known that, as one monitors an ongoing trial, repeated significance testing with a fixed sample size procedure can inflate the probability of type I error. However, this is not an issue in every interim evaluation, and care must be taken to discern when, in fact, it is an issue. Whether control of type I error is of concern or not depends on the purpose of the interim evaluation and the decisions that are possible as a result of the evaluation.

Suppose, for example, that a difference in favor of the sponsor's new drug A over a comparative agent B represents a successful result and that equivalence or the superiority of B represents an unfavorable result. If an interim evaluation is to be performed and the sole purpose is to provide an early opportunity to stop the trial and abandon a failing product (i.e., under no circumstances would the study be terminated early because A appears successful), then repeated testing does not inflate the probability of type I error, because a type I error could not be made at the interim evaluation(s). In this context interim evaluations only affect the probability of type II error, which is increased. The power of the decision rule used to stop early and abandon a failing product should be assessed, but here control of power is more an issue to the sponsor than to the scientific/regulatory community. This example illustrates the fact that repeated significance testing does not necessarily imply inflation of the probability of type I error.

Next, consider a slightly different example, in which only inferiority of the new drug A to the comparative agent B represents an unsuccessful outcome; the sponsor hopes to show the superiority of A to B, but can accept equivalence as a successful outcome as well. Suppose, for simplicity, that the null hypothesis of equality is tested using a statistic which follows a symmetric distribution with one tail representing inferiority and the other superiority of A over B. Here, type I error can occur in two ways, either with the false declaration of success or with the false declaration of failure. If, as in the preceding example, interim evaluations will only result in early termination in the event of the inferiority of A, then the probability of a false declaration of success is not inflated by repeated testing. Only the probability of a false declaration of failure is inflated. It is perhaps useful for the sponsor to be aware of this, but adjustments to control this portion of the probability of type I error are not crucial. In this context the magnitude of the difference between A and B has a greater impact on the outcome. If the response rate to A is 60% and the response rate to B is 75%, marketing A in competition with B may become difficult, regardless of whether this difference is statistically significant.

An interim evaluation may be conducted for the purpose of making management decisions which are completely external to the study (e.g., hiring decisions), without any intention of terminating the study. If there truly was no possibility of early termination, again no adjustments to the final analysis are necessary. However, an understanding of sequential concepts can be helpful in presenting interim results to management and in reconciling differences between interim and final results.

Used properly, the interim evaluations described thus far are all relatively benign in that they have no (important) impact on the control of type I error. On the other hand, it is possible to misuse interim evaluations and completely distort results. For example, 10 trials could be started simultaneously and early interim evaluations could be used to decide to terminate the trials with unfavorable results while continuing those with favorable results. In the end, the large trials in the clinical development program would tend to be more favorable than the small trials and the selection process being employed could lead to falsely positive results.

If a sponsor conducts interim evaluations that permit early termination due to a successful outcome, then inflation of the probability of type I error is an important consideration in interpreting results. In a later chapter, group sequential methods (as well as pure sequential methods) and stochastic curtailing are discussed. These techniques can be useful in dealing with this problem. However, there are practical difficulties in conducting (group) sequential trials which have limited their use. These difficulties may include:

Data flow problems (perhaps more severe in industry than elsewhere)
Difficulties in enrolling investigators in studies which involve variable commitments with respect to time, number of patients, and amount of money
Difficulty in defining a single response variable on which to base a stopping rule (mortality studies excluded)
Treatment-by-time interactions (learning curves create differences between early and late treatment differences)
Investigator-by-treatment-by-time interactions in multicenter trials (data come in early from the best investigators)

Although these difficulties limit the planned use of group sequential designs, group sequential methods can be useful in interpreting the result of unplanned interim evaluations. Assuming that there would have been only one or two interim evaluations, we can make an approximate assessment of significance by using these methods. Stochastic curtailing could also be useful in this situation. In the end, either technique is likely to represent only one part of a complex and subjective decision process.

Finally, it is interesting to speculate on whether advances in computer technology will enhance the usefulness of either pure or group sequential designs. Perhaps personal computers in investigators' offices, clever (error-detecting) data entry screens to promote high-quality data throughout the study, and automatic modems which periodically telecommunicate new data to a central office will make these designs more attractive in the future.

## N. Equivalence Trials

The Phase III development program often includes actively controlled trials. It is important to know whether the objective of such a trial is to demonstrate therapeutic equivalence (rather than superiority). If demonstration of equivalence is the objective, the study must be designed accordingly.

The word "equivalence" must be given clear meaning (e.g., equivalent *to within 5 mm Hg*). Equivalence may be defined in terms of differences or ratios (e.g., $|\mu_A - \mu_B| < 5$ or $.95 < \mu_A/\mu_B < 1.05$). Distinction should be made between clinically important differences and a reasonable definition of equivalence. A difference in response rates of 20% (e.g., 65%

versus 85%) may be clinically important, but it may be unreasonable to regard all smaller differences as indicative of equivalence. If differences of 20% or more are clinically important, perhaps equivalence should be defined as differences of no more than 10%. The nature of the underlying disease and the risks and benefits associated with therapy must be considered before defining equivalence. Mortality rates of 3 and 13% would almost certainly not be considered equivalent; mortality rates of 50 and 60% would probably not be considered equivalent; response rates of 50 and 60% to analgesic therapy would probably be considered equivalent.

The methods used to analyze bioequivalence studies can be applied to Phase III studies. The major differences between bioequivalence and Phase III equivalence studies are that the latter often are parallel rather than crossover studies, often involve much larger sample sizes, and often are confused with studies intended to show a difference (i.e., there is more confusion about the objective of demonstrating equivalence in Phase III than in clinical pharmacology studies). Possible methods include use of

- A test of the null hypothesis of no difference with adequate power against all alternatives that should not be considered equivalent
- A confidence interval for $\mu_A/\mu_B$ contained in an appropriate interval around 1 such as (.8, 1.2); this might be obtained through ANOVA of log-transformed data or an application of Fieller's theorem
- A suitably large posterior probability that $\mu_A/\mu_B$ is contained in an appropriate interval around 1 (Rodda and Davis, 1980)
- A suitably large posterior probability that $P_A/P_B$ is contained in an appropriate interval around 1 (Aitchison and Bacon-Shone, 1981)

Temple (1983) discusses difficulties in interpreting results of positive controlled trials. One of the issues to be considered in these trials is whether or not the active control was actually effective in this particular trial (e.g., were homeopathic doses of the control agent used?). Another issue is that poor study methodology can inflate variances and make detection of differences difficult. This type of flaw is indirectly detectable with the above methodologies (because low power, wide confidence intervals, and low posterior probabilities are a likely result).

## IV. PHASE III: SPECIFIC INDICATIONS

### A. Challenging Questions

At the completion of a clinical trial, the statistician's principal responsibility shifts from one of design and monitoring to one of analysis and interpretation. The mathematical tools necessary to perform the science of statistics are learned in most comprehensive graduate programs. However, the art of statistics in clinical trials is a skill which can only partially be taught. Specific concerns in the analysis of a variety of types of clinical trials are presented elsewhere in this text. However, listed below are various drug classes and some of the questions that provide the statistician with challenges in the analytical phase.

#### 1. Anti-Infectives

Although anti-infectives would seem to be a rather simple analytical issue, since the patients are either cured or not cured, a principal issue is one of evaluability. Patients will often not have infections documented by appropriate bacteriology or, later in the study, not have the presence or absence of the infection documented properly. The principal measure of efficacy is bacteriologic outcome. Clinical outcome, which is correlated with bacteriologic outcome, is a secondary measure. Nonevaluable patients make intention-to-treat analyses of bacteriologic outcomes impossible. Clinical outcomes of nonevaluable patients are examined in place of intention-to-treat analyses of bacteriologic outcomes.

In addition, many antibiotic studies must be conducted on an open basis because of inability to completely blind the investigator. The consequence of this may be selective allocation on the part of the investigator and willful (though perhaps ethically motivated) violation of the allocation schedule. This provides a conundrum to the statistician in interpretation, since both efficacy and safety will be biased by this selection.

Baseline susceptibility is often available prior to randomization. Ethics may dictate that patients only be randomized if their pathogens are susceptible to both study drugs. However, if the agents being compared have vastly different spectra of antimicrobial activity, the logic of restricting their comparison to subpopulations where susceptibility predicts they will both work is highly questionable.

The statistical considerations in evaluating prophylactic studies are many and complex and cannot be considered here. They can arise, for example, when trying to show that a new anti-infective agent is effective in preventing infections associated with surgery.

## 2. Cardiovascular Agents

This class of drugs is very broad and includes drugs for hypertension, congestive heart failure, angina, and, recently, for the prevention of mortality and heart attacks. Evaluation of blood pressure is relatively straightforward; however, dose determination and the definition of an acceptable therapeutic response can be issues which confront the statistician. This is due to recent downward shifts in the levels of blood pressure at which clinicians would begin to treat patients and to doubts about what risk/benefit ratio is appropriate in a silent, chronic disease with serious long-term effects.

Angina is a uniquely variable condition which can appear as an occasional attack in some patients and more than two attacks per day in others. A classical measure of efficacy in angina patients is reduction in the number of attacks by 50%. Since this reduction in a patient who has 15 attacks per week compared with a patient who has 3 attacks per week may reflect a very different response to medication, the pooling of this information into a common conclusion may be very difficult. In addition, the placebo response rate to angina is very high, and although this indicates that prolonged placebo periods would be necessary for a clear evaluation, such periods are considered unethical.

Evaluation of congestive heart failure is also a difficult statistical challenge. Although definitions may be clearly written, the interpretation of these definitions may vary from country to country, thus leading to difficulty in a clean pooling of the results. Also, because of the nature of congestive heart failure, the ability of many investigations to measure improvement precisely will limit the ability of a clinical program to distinguish an active therapy from even a negative control. Medical ethics limits the possibilities for placebo control in the disease. Comparisons with placebo as add-on therapy to digitalis and diuretics can be severely muddled if the doses of digitalis and diuretics are permitted to change. Furthermore, unless conducted by experienced investigators, results of exercise testing can be highly variable. Investigators must be experienced enough to encourage patients to exercise to *objective* endpoints.

### 3. Analgesics

The class of analgesic drugs provides a classic case where a placebo is necessary in literally all phases of clinical research, because placebo response to analgesics can range from 30 to as much as 70%. A study without a negative control will always be open to criticism. A principal difficulty in evaluating analgesic studies is the measurement obtained. Many books have been written on this subject, and different analgesiologists favor a variety of measurements, such as visual analog scales and discrete ordinal scales (e.g., none, a little, some, a lot, terrible). The preferred analytic procedures for evaluating these measures also differ among statisticians. The various indices used, and some of the composite scores which can be derived, have presumed and misunderstood statistical properties. Additional questions concern patients who drop out. Since pain is a condition which patients will not tolerate, many will drop out of studies at the time they feel the analgesic has not worked. Other patients, becoming aware that their pain is gone, will no longer record the result. A rule concerning how these patients are handled is a principal responsibility of the statistician. An additional problem with measuring response to analgesics is that patients confound the onset of, and degree of, analgesia. For this reason, mild analgesics such as aspirin, which have a rapid onset of action, are often evaluated as being better than stronger analgesics such as codeine, which have slower onsets of action. Comparing these two is often difficult.

### 4. Hypolipidemic Agents

The primary dependent variables of interest are

Total cholesterol (TCHOL) = low density lipoproteins (LDL)
+ high density lipoproteins (HDL)
+ very low density lipoproteins (VLDL)

Usually TCHOL and HDL are measured. LDL is obtained by computing TCHOL − HDL − VLDL; however, VLDL is difficult to assay, requiring ultracentrifugation. It can be estimated as (see Friedewald et al., 1972) 0.2 times plasma triglycerides (TRIGL), which are easier to measure. The relationship VLDL = 0.2 × TRIGL may be used for TRIGL values up to about 400 to 500 mg/dl. Beyond this range the relationship may not hold, so VLDL should not be estimated for large TRIGL values.

Hypolipidemic agents may alter this relationship because their effects on VLDL and TRIGL may be different. The VLDL-TRIGL relationship and the agent's effect on it should be examined at some point in the agent's development.

Since baseline levels of the cholesterol variables may differ over a two- to threefold range, analysis of percent change and/or analysis of covariance (with baseline level as covariate) should be considered. Therapeutically useful hypolipidemic agents will typically reduce TCHOL, LDL, VLDL, and TRIGL and increase HDL. All of these effects are therapeutically desirable, but most important among them is LDL reduction. Definition of categories of LDL response should include some relationship to percentiles adjusted for age and sex. For example, three categories of response might be:

LDL $<$ 50th percentile
50th percentile $\leqslant$ LDL $\leqslant$ 75th percentile
LDL $>$ 75th percentile

Categories of total cholesterol response should be related to TCHOL $<$ 200 or 235 mg/dl (Lipid Research Committee guidelines).

Some hypolipidemic agents have a history of poor patient compliance. If a new agent is likely to have less noncompliance, differences between the new and the old agents could be a mixture of pharmacologic and compliance effects. To reach precise conclusions, it might be necessary to distinguish between these two types of effects.

#### 5. Others

Oncolytic agents are discussed in another chapter in this book. The objective of many studies of these agents is very different from that in most clinical trials of other drugs, since a positive outcome is a delay in the endpoint rather than a change in it. Studies designed to show a delay in the time to rejection of an organ transplant have such an objective. These studies require use of appropriate methods of survival analysis.

Studies designed to show the safety and efficacy of fixed ratio combinations present a whole array of special problems which cannot be addressed here. Some of these problems are discussed in Abdelbasit and Plackett (1982), Beaver (1986), Voundra (1986), and Leung and O'Neill (1986).

### B. Summary

In summary, the role of statistics in pharmaceutical clinical trials is extremely varied, and the statistician's contributions to the design, implementation, evaluation, and interpretation of these studies in the pharmaceutical industry is essential.

Studies of new drugs cannot be effective without the considered input of statistical experts at each stage. Human variability, variations in new drug regulations among countries, different medical practices in one nation compared to another, and the variety of scientific and marketing objectives provide a challenge which can only be met with the extensive and considered use of both the art and science of statistics.

## ACKNOWLEDGMENTS

The authors acknowledge the helpful information provided by Ms. Donna Stock, Mr. Joe Antonello, Ms. Kathy Lipschutz, Dr. Joe Ciminera, and other colleagues at MSDRL.

## REFERENCES

Abdelbasit, K. M. and Plackett, R. L. (1982). Experimental Design for Joint Action, *Biometrics*, 38: 171.

Aitchison, J. and Bacon-Shone, J. (1981). Bayesian Risk Ratio Analysis, *Am. Stat.*, 35: 254.

Armitage, P. (1983). Exclusions, Losses to Follow-Up and Withdrawals in Clinical Trials, in *Clinical Trials* (S. H. Shapiro and T. A. Louis, eds.), Marcel Dekker, New York, p. 99.

Beaver, W. T. (1986). "Clinical Trials for Combination Treatments: A Clinician's View," presented at ASA 146th Annual Meeting, Chicago, Ill.

BEMAC, (1976). Minutes of the FDA Biometric and Epidemiologic Methodology Advisory Committee.

Bolognese, J. A. (1983). A Monte Carlo Comparison of Three Up-and-Down Designs for Dose Ranging, *Controlled Clin Trials*, 4: 187.

Brown, B. W., Jr. (1980). The Crossover Experiment for Clinical Trials, *Biometrics*, 36: 69.

Calimlim, J. F. and Weintraub, M. (1981). Selection of Patients Participating in a Clinical Trial, in *Statistics in the Pharmaceutical Industry* (C. R. Buncher and J. Y. Tsay, eds.), Marcel Dekker, New York, p. 107.

Ciminera, J. L., Bolognese, J. A., and Gregg, M. H. (1987). The Statistical Evaluation of a Three-Period Two-Treatment Crossover Pharmacokinetic Drug Interaction Study, *Biometrics*, in press.

FDA (1970): 21 CFR 314.111. *Code of Federal Regulations*, Office of the Federal Register, National Archives and Records Administration, Superintendent of Documents, Government Printing Office, Washington, D.C.

FDA (1975): 45 CFR Part 46. *Code of Federal Regulations*, Office of the Federal Register, National Archives and Records Administration, Superintendent of Documents, Government Printing Office, Washington, D.C.

FDA 77-3040 (1977). Superintendent of Documents, Government Printing Office, Washington, D.C.

FDA 82-3053 (1982). Superintendent of Documents, Government Printing Office, Washington, D.C.

FDA (1985a). "New Drug and Antibiotic Regulations; Final Rule," Docket No. 82N-0293, *Code of Federal Regulations*, Office of the Federal Register, National Archives and Records Administration, Superintendent of Documents, Government Printing Office, Washington, D.C.

FDA (1985b). "Draft Guideline for the Format and Content of the Statistical Section of an Application," Docket No. 85D-0246, Center for Drugs and Biologics, Office of Drug Research and Review (HFN-100), FDA, Rockville, Md.

FDA (1985c). "Draft Guideline for the Format and Content of an Application Summary," Docket No. 85D-0247, Center for Drugs and Biologics, Office of Drug Research and Review (HFN-100), FDA, Rockville, Md.

FDA (1986). "Draft Guideline for the Format and Content of the Clinical Data Section of an Application," Docket No. 85D-0467, Center for Drugs and Biologics, Office of Drug Research and Review (HFN-100), FDA, Rockville, Md.

Friedewald, W. T., Levy, R. I., and Frederickson, D. S. (1972). Estimation of the Concentration of Low-Density Lipoprotein Cholesterol in Plasma, Without Use of the Preparative Ultracentrifuge, *Clin. Chem.*, 18: 499.

Gart, J. S. (1969). An Exact Test for Comparing Matched Proportions in Crossover Designs, *Biometrika*, 56: 75.

Gould, A. L. (1980). A New Approach to the Analysis of Clinical Trials with Withdrawals, *Biometrics*, 36: 721.

Grizzle, J. E. (1965). The Two-Period Change-Over Deisgn and Its Use in Clinical Trials, *Biometrics*, 21: 467.

Grizzle, J. E. (1974). *Biometrics*, 30: 727. Correction to the above 1965 article.

Hauck, W. W. and Anderson, S. (1984). A New Statistical Procedure for Testing Equivalence in Two-Group Comparative Bioavailability Trials, *J. Pharmacokinet. Biopharmaceut.*, 12: 83.

Hills, M. and Armitage, P. (1979). The Two-Period Crossover Clinical Trial, *Br. J. Clin. Pharmacol.* 8: 7.

Hsu, J. P. (1983). "The Assessment of Statistical Evidence From Active Control Clinical Trials," Proceedings of Biopharmaceutical Section, ASA 143rd Annual Meeting, Toronto, Canada, p. 12.

Koch, G. G. (1972). The Use of Nonparametric Methods in the Statistical Analysis of the Two-Period Change-Over Design, *Biometrics*, 28: 577.

Leung, H. M. and O'Neill, R. T. (1986). "Statistical Assessment of Combination Drugs: A Regulatory View," presented at ASA 146th Annual Meeting, Chicago, Ill.

O'Neill, R. T. (1977). "Current Status of Crossover Designs," presentation to Pharmaceutical Manufacturer's Association members in Statisticians Meeting, General Session, Arlington, Va.

O'Neill, R. T. (1978). "Subject-Own-Control Designs in Clinical Drug Trials: Overview of the Issues with Emphasis on the Two Treatment Problem," presentation at ECDEU meeting, Key Biscayne, Fl.

Pledger, G. W. (1986). "Active Control Trials: Do They Address the Efficacy Issue?," presented at ASA 146th Annual Meeting, Chicago, Ill.

Pledger, G. W. and Hall, D. (1982). *Biometrics*, 38: 276. Correspondence: "Withdrawals from Drug Trials."

Rodda, B. E. and Davis, R. L. (1980). Determining the Probability of an Important Difference in Bioavailability, *Clin. Pharmacol. and Ther.*, 28: 247.

Sackett, D. L. (1983). On Some Prerequisites for a Successful Clinical Trial, in *Clinical Trials* (S. H. Shapiro and T. A. Louis, eds.), Marcel Dekker, New York, p. 65.

Schor, S. S. (1965). Statistics: An Introduction, *J. Trauma*, 5: 515.

Steinijans, V. W. and Diletti, E. (1985). Generalization of Distribution-Free Confidence Intervals for Bioavailability Ratios, *Eur. J. Clin. Pharmacol.*, 28: 85.

Temple, R. (1983). "Medical Issues in Active Control Trials," Proc. of Biopharmaceutical Section, American Statistical Association (ASA) 143rd Annual Meeting, Toronto, Canada, p. 1.

Voundra, W. (1986). "Combination Drugs: A Legal View," presented at ASA 146th Annual Meeting, Chicago, Ill.

Wardell, W. M., Tsianco, M. C., Anaveker, S. N., and Davis, H. T. (1979a). Postmarketing Surveillance of New Drugs: I. Review of Objectives and Methodology, *J. Clin. Pharmacol.*, 19: 85.

Wardell, W. M., Tsianco, M. C., Anaveker, S. N., and Davis, H. T. (1979b). Postmarketing Surveillance of New Drugs: II. Case Studies, *J. Clin. Pharmacol.*, 19: 169.

Westlake, W. J. (1976). Symmetrical Confidence Intervals for Bioequivalence Trials, *Biometrics*, 32: 741.

# 7 Bioavailability and Bioequivalence of Pharmaceutical Formulations

WILFRED J. WESTLAKE* *Smith Kline & French Laboratories, Swedeland, Pennsylvania*

## I. INTRODUCTION

When a medication is administered to a human subject the active drug goes through various stages. The first is absorption, the second is distribution, the third is metabolism, and the fourth is elimination.

### A. Absorption

A drug may be administered in a number of different ways. It may be injected (either intravenously or intramuscularly); it may be taken orally as a tablet, capsule, solution, or liquid suspension; or it may be given by one of the less frequently encountered routes such as sublingually, as a suppository, or by transdermal patch. The route of administration is a key element in the extent to which the drug is absorbed and consequently in the successful administration of the drug. For example, some drugs are very poorly absorbed when taken orally, so injection may be necessary. For a drug given orally, the formulation of the drug (the composition of the tablet, the excipients used in its manufacture, the degree of compression of the tablet, and so on) may vitally affect the degree to which the drug is absorbed into the bloodstream. The overriding concern is that of getting the active drug to the site of action in the body. The latter is often not known and, even when it is, the particular organ or tissue is not directly amenable to sampling, so one cannot determine by direct means how much of the drug is reaching the site of action.

### B. Distribution

In general, absorption of a drug implies absorption into the circulatory system so that the drug is distributed throughout the body by the bloodstream and presumably thereby to the site of action. Consequently, our knowledge of the absorption of a drug is mainly established by taking blood samples and assaying them for drug content.

*Present affiliation:* University of Pennsylvania, Philadelphia, Pennsylvania.

### C. Metabolism

In its passage through the body—and particularly through the liver—the drug is acted on by enzymes which metabolize it and convert the original active drug to various metabolites which are, in general, inactive, although there are some drugs whose metabolites are thought to have activity similar to that of the original "parent" drug. A particular problem with orally administered drugs is the so-called first-pass effect, in which the absorption into the bloodstream via the portal vein to the liver and thence to the bloodstream results in extensive metabolism. In this case, the active drug is rendered inactive at a fairly early stage.

### D. Elimination

By elimination, we mean the removal of the active drug from the body. If metabolites are inactive, metabolism itself may be seen as elimination. The other major routes of elimination are, of course, from the bloodstream via the kidney into the urine and by fecal excretion. Urine samples can be assayed for parent drug and the various metabolites and thus provide evidence relative to the amount of drug that was absorbed into the bloodstream. When no reliable assay method is available for blood analysis, analysis of urine is sometimes used as an indicator of how much drug was absorbed into the circulatory system. However, it does not provide a direct estimate of the amount of drug that reached the bloodstream unchanged (i.e., before metabolism).

## II. BLOOD-LEVEL TRIALS

An examination of the fate of a given formulation of a drug when administered to a human subject is generally accomplished by carrying out a "blood-level trial." In its simplest form, one dose of the formulation is administered to each of a number of human subjects, and for each subject blood samples are taken sequentially over time and assayed for drug content. The result, for any given subject, is a sequence of blood drug concentrations over time which delineates the progress of the drug through the circulatory system. A typical blood-level profile might look like that shown in Figure 1.

The choice of sampling times is critical if an accurate characterization of the time course of the drug is required. Typically, sampling will be at fairly frequent time intervals in the ascent to the peak blood level and in the neighborhood of the peak and will then be at longer time intervals during the final decaying portion of the curve. Sampling will be continued until the drug concentration is negligible or at least until an adequate characterization of the decay has been obtained. The final decay of the concentration curve is often approximately exponential and can thus be characterized by the half-life. The rate at which drug is eliminated from the body determines the extent to which drug levels can build up with repeated dosing; the half-life is, consequently, an important parameter which plays a key role in determining the appropriate dosing schedule for a drug.

The time course of a drug in a biological organism comes under the all-embracing title of "pharmacokinetics." Once blood-level profiles of the drug in a number of subjects have been obtained, the data can be used in

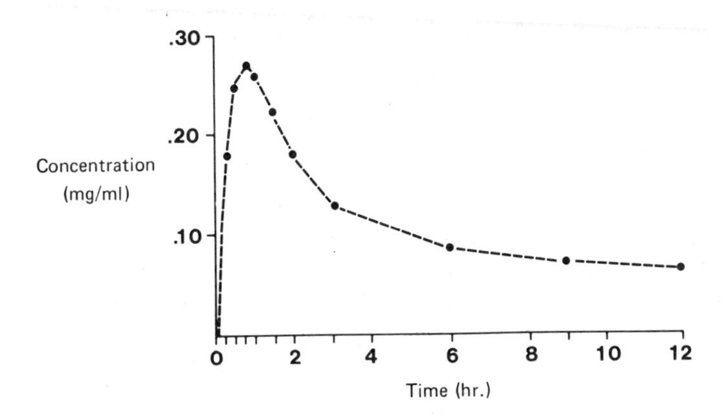


Fig. 1 Typical blood-level profile.

a number of ways. One course is to examine the bioavailability of the formulation or, if different formulations are to be examined, whether they are bioequivalent. These terms will be defined shortly, and a discussion of bioequivalence will be the principal object of this chapter. Another course of action is to use the blood-level data to fit pharmacokinetic models. This subject will now be briefly discussed.

## III. PHARMACOKINETIC MODELS

The idea behind pharmacokinetic modeling is that the biological organism in question (in this case, the human body) can be represented by a simplified mathematical model with various undetermined parameters. The model can then be fitted to the blood-level data, thus providing estimates of these parameters. The most widely used models are compartmental models—so called because the organism is assumed to be represented by a series of compartments with linear transfers between them. One of the simplest and most frequently encountered in the literature is the two-compartment open model. This is shown in Figure 2, and it is assumed that the rate of transfer from a compartment is proportional to the amount of drug in that compartment with the proportionality (or rate) constants shown. With the example shown, the description of the system reduces to a simple set of ordinary linear differential equations, and the solution for the concentration of drug in the blood compartment is

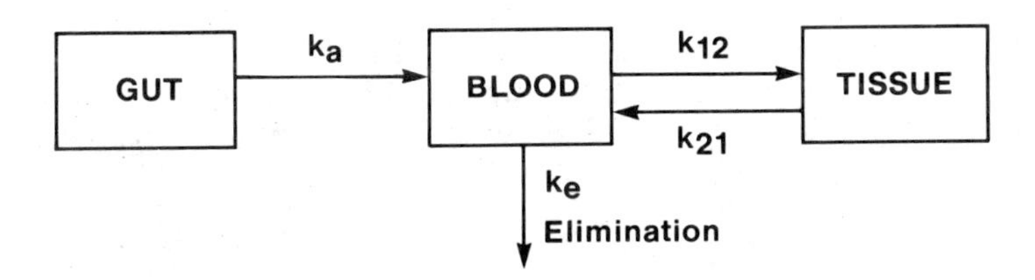


Fig. 2 Two-compartment open model.

$$C(t) = \frac{k_a D}{V}\left[\frac{(\alpha - k_{21})}{(\alpha - \beta)(k_a - \alpha)} e^{-\alpha t} + \frac{(k_{21} - \beta)}{(\alpha - \beta)(k_a - \beta)} e^{-\beta t} - \frac{(k_a - k_{21})}{(k_a - \alpha)(k_a - \beta)} e^{-k_a t}\right]$$

where D represents the amount of drug absorbed, V the volume of distribution of the central (or blood) compartment, and $\alpha$ and $\beta$ are determined by the relations $\alpha\beta = k_e k_{21}$ and $\alpha + \beta = k_e + k_{12} + k_{21}$. Given a series of values of C(t) over time, the above model can be fitted to the data to obtain estimates of $k_a$, $k_{12}$, $k_{21}$, $k_e$, and D/V. The fitting is an exercise in nonlinear regression and, as is customary with the fitting of sums of exponentials, the estimated values of the parameters may be very different from the true values—even assuming, of course, that the human body obligingly conforms to this very simplified model. For a particularly striking demonstration of this fact the reader may refer to Lanczos (1956).

The important question is: if a compartmental model is fitted to a sequence of blood-level data, to what practical use will the model be put? This was the question posed by Westlake (1971), and although the fitting of compartmental models is still a popular pursuit, there appears to be no completely satisfactory answer to this question. The obvious answer is that the model can be used to predict drug concentrations over time in any compartment of the system and, thanks to the linearity of the system, such predictions can easily be made not only for single doses of the drug but also for multiple-dose regimens with doses of varying magnitude. The two major points made in the reference, however, were:

1. If one wants to predict concentrations from various dosing regimens in the blood compartment, then simple superposition techniques which depend only on the linearity of the system and not on the particular form of the model are superior, being simpler, more general, and more accurate.
2. If prediction of the concentration in any other compartment which is not sampled directly is required (e.g., the tissue compartment in the two-compartment model described earlier), the considerable possible error in the estimation of the rate constants makes such predictions extremely unreliable.

Consequently, the practical utility of such models is still questionable and they will not be considered further in this chapter. On the other hand, the reader should be aware that there are pharmacokineticists who think otherwise and they might characterize my point of view as idiosyncratic. However, I have yet to be convinced by their arguments. For readers interested in pursuing the subject further, the elementary text by Gibaldi and Perrier (1975) gives a good introduction; the *Journal of Pharmacokinetics and Biopharmaceutics* (published bimonthly) has many examples of pharmacokinetic modeling.

## IV. BIOAVAILABILITY

We turn now to the subject of bioavailability. The simplest definition of the bioavailability of a formulation of a drug is the amount of drug that it makes available to the circulatory system in its unchanged form. Other definitions refer not only to the amount of drug but also to the rate at which it is made available. In general, it might be agreed that the definition is the amount of drug made available together with other, less precisely characterized factors among which rate of absorption is perhaps the most relevant. The latter might be of importance, for example, with analgesics or antibiotics, where a rapid onset of therapeutic effect is desirable. For a discussion of bioavailability from a statistical viewpoint, reference may be made to the papers of Metzler (1974) and Westlake (1979).

If a blood-level trial with an orally administered drug is performed in a number of subjects, it is clear that a measure of speed of absorption might be obtained by examining the time at which the peak blood level is attained. However, how does one estimate the amount of drug absorbed into the circulatory system? If the drug is eliminated (i.e., metabolized and/or excreted) solely from the so-called blood compartment and if the elimination is linear with rate constant $k_e$, then it can be shown that the amount of drug absorbed is equal to the time integral of the drug concentration from zero time to infinity divided by the product of $k_e$ and V, the volume of distribution of the blood. For a simple proof, see Westlake (1970). The product $k_eV$ is known as the clearance and the time integral of concentration is usually referred to as the "area under the blood-level curve." From this result it is immediately apparent that, in general, we do not know the absolute bioavailability of a drug formulation but only its relative bioavailability, since the equation

$$\text{Amount of drug absorbed} = \frac{\text{area under the blood level curve}}{k_e V}$$

contains an unknown term, $k_eV$, which is unique to the given subject. The exception, of course, is where the drug is administered intravenously; in this case the drug is, by definition, 100% available to the circulatory system. Consequently, administration of the drug intravenously and by a second route (e.g., orally) to the same subject on separate occasions enables the absolute bioavailability by the second route to be estimated by comparison of the areas under the blood-level curves. The only assumption necessary is that the clearance $k_eV$ of any given subject remains the same for the two separate tests. The absolute bioavailability of a formulation is of interest in the development of a new drug. The pharmaceutical company responsible for the drug may have formulated it for oral administration and will be interested in determining its absolute bioavailability; if it is low, the finding may prompt attempts to adjust the formulation or to reformulate entirely to improve bioavailability.

It is important, at this point, to reemphasize that the bioavailability of a formulation of a drug is dependent on two main factors, namely the drug substance itself and the formulation. Some drugs given orally, for example, are much better absorbed than others. Drugs subject to extensive first-pass metabolism, for example, may show poor bioavailability with any formulation administered orally. In fact, some drugs can only be given intravenously because of their poor absorption by other routes. The second factor, formulation, is also of critical importance. As mentioned earlier, the excipients, the degree of compression used in tablet formulations, etc. can greatly affect the absorption of the drug, and there are numerous cases of radical differences in bioavailability of different formulations of a given drug.

## V. BIOEQUIVALENCE

The concept of bioequivalence of different formulations of a given drug is of considerable importance. If two formulations are deemed bioequivalent they may be pharmaceutical equivalents (i.e., similar dosage forms made, for example, by different manufacturers) or pharmaceutical alternatives, which is to say different dosage forms (e.g., tablet or capsule) whose rate and extent of absorption are essentially the same when administered under similar conditions to human subjects. The key concept is that bioequivalent formulations should lead to the same therapeutic effect. A direct demonstration of the same therapeutic effect of different formulations of a drug would require a full-scale clinical trial in which their therapeutic effects could be compared. In bioequivalence trials, one attempts to circumvent this direct approach and approach the problem by conducting a blood-level trial in which it is demonstrated that the formulations give rise to essentially equivalent blood-level profiles in human subjects. The principle underlying this concept is important, namely that two formulations of a drug that result in essentially equivalent blood-level profiles over time should elicit equivalent therapeutic effects.

There are at least two reasons why the subject of bioequivalence currently arouses so much interest. Without doubt, that of most immediate concern is the question of so-called generic equivalents. In the United States when the patent expires on a drug product, other manufacturers can plan to market the drug under its generic or chemical name. They must first seek approval from the Food and Drug Administration (FDA) and demonstrate the efficacy of their respective formulations. However, rather than repeat the long and expensive procedure of demonstrating efficacy through clinical trials, the manufacturers may achieve approval by showing the bioequivalence of their formulation to that of the original patent-holder. In this way, a blood-level trial demonstrating bioequivalence provides a bridge to the clinical trials performed by the original patent-holder and, consequently, to the efficacy of the original formulation. The generic manufacturer can submit a so-called Abbreviated New Drug Application in which a bioequivalence trial is the key element. By doing this the generic manufacturer is able to bypass the considerable expense and time required to demonstrate efficacy de novo. A second reason for the interest in the subject of bioequivalence as demonstrated by blood-level trials is that of the originator of the drug who may have decided to formulate the compound in a particularly convenient form, e.g., a tablet. It will be of considerable interest to ascertain how efficient this formulation is by conducting a bioequivalence trial comparing it with the most available form conceivable, in general, an intravenous injection or, for orally administered formulations, a solution or suspension.

## VI. DESIGN OF BIOEQUIVALENCE TRIALS

The task of the statistician is to assist in the design of a bioequivalence trial in which, typically, the blood-level profile of a drug arising from a single administration of a single dose of formulation A is to be compared with that from a similar administration of formulation B. Given a number of human subjects, two basic designs are possible: a parallel-groups design or a crossover (changeover) design.

### A. Parallel-Groups Design

In a parallel-groups trial, an even number of subjects is divided randomly into two equal groups, one group receiving one formulation of the drug and the other group the second formulation. In most bioequivalence trials, one formulation will be the "standard" and the other the "test" formulation; and the object of the trial is to determine whether the test formulation is bioequivalent to the standard. The parallel-groups concept can be readily generalized to more than two groups, and in this case one formulation will generally be the standard and several test formulations will be compared with it. Typically, each subject is administered a single dose of the appropriate formulation and blood samples are taken periodically and assayed for drug content to give a blood-level profile of drug concentration over time, as described earlier. Exactly how this sequence of blood levels should be analyzed statistically will be discussed later in the chapter. However, in order to examine the design characteristics of the trial, let us suppose that just one characteristic of the sequence is to be analyzed; as an example, the area under the curve has already been suggested as a key element in the comparison of the formulation with respect to absorption. The basic analysis of variance (ANOVA) for a parallel-groups design with n subjects in each of two groups—essentially, a simple t-test—would then be as shown in Table 1.

Note that the error term for testing the difference between formulations is the variation among subjects within a formulation. One of the most striking features of bioavailability data is the enormous differences that can occur among human subjects—not only in their size, weight, and presumably blood volume but also in the way they metabolize a drug. Consequently, in a parallel-groups design, the error sum of squares is likely to be large and the test for a difference between formulations will be insensitive. This fact has led to a strong preference for the crossover or changeover design in bioequivalence trials.

### B. Two-Period Crossover Design

Consider again the comparison of two formulations in an even number of subjects randomly divided into two equal groups. As before, the members of one group each receive a single dose of the standard formulation and the members of the other group a single dose of the test formulation. After a suitable "washout" period, during which all traces of the drug should have disappeared from the body, each subject receives a dose of the alternate formulation (the second period) and the experiment is repeated.

**Table 1** ANOVA for Parallel-Groups Design

| Source of variation | Degrees of freedom (d.f.) |
|---|---|
| Formulations | 1 |
| Error | 2(n − 1) |
| Total | 2n − 1 |

Table 2 Crossover Design with 10 Subjects

| Subject No. | Period 1 | Period 2 |
|---|---|---|
| 1 | S | T |
| 2 | S | T |
| 3 | T | S |
| 4 | S | T |
| 5 | T | S |
| 6 | T | S |
| 7 | T | S |
| 8 | S | T |
| 9 | T | S |
| 10 | S | T |

An example of a crossover design in 10 subjects comparing a standard and test formulation is shown in Table 2. With 2n subjects and two formulations, the basic ANOVA takes the form shown in Table 3.

The subjects sum of squares can be further partitioned into 1 degree of freedom for "sequence" (either standard followed by test or vice versa) and 2(n − 1) degrees of freedom for subjects within sequence. The critical feature is that the variation among subjects has now been removed from the error term, a maneuver which should result in a more sensitive test of the difference between formulations. For a detailed exposition of the ANOVA and the underlying model one should consult Grizzle (1965).

The underlying model for the two-period crossover trial as proposed by Grizzle:

$$Y_{ijk} = \mu + S_{ij} + \pi_k + F_l + \lambda_l' + e_{ijk}$$

Table 3 ANOVA for Crossover Design with 2n Subjects

| Source of variation | Degrees of freedom (d.f.) |
|---|---|
| Subjects | 2n − 1 |
| Periods | 1 |
| Formulations | 1 |
| Error | 2(n − 1) |
| Total | 4n − 1 |

where $S_{ij}$ is the effect of the jth patient in the ith sequence, $\pi_k$ the effect of the kth period, $F_l$ the effect of the lth formulation, and $\lambda_l'$ the residual (or carryover) effect of the l'th formulation. Note that the carryover effect is, essentially, the sequence effect, which can be tested against the sum of squares within sequence. If this carryover effect exists, then it confounds the test on formulations. Grizzle proposed a strategy based on first testing for carryover effect; if this turned out to be significant at a certain level (e.g., 0.10 or 0.15), he recommended basing the analysis on first-period data only. For a further discussion of this strategy reference may be made to Brown (1980). My own experience with a large number of comparative bioavailability trials has led me to believe that significant carryover effects (at the 0.05 level) tend to occur in about 5% of the trials; in other words, I believe that carryover effects do not normally exist. There are several reasons why this assumption is a reasonable one. First, subjects in bioavailability trials are usually normal human volunteers whose physical condition is unlikely to change radically from one period of a crossover trial to another, unlike patients whose disease state may progress during the course of even a short trial. Second, the trial usually involves only single doses of the drug formulations and is thus unlikely to result in a marked change in a subject's physical condition. Third, a washout period is scheduled between the periods of the crossover trial which allows for elimination of all drug from the system; and a direct check of the presence of drug is, of course, obtained from the zero-hour blood level. Wagner (1974) has, however, raised the possibility that an unmeasured metabolite might still be present in the body and affect the biotransformation at the succeeding administration.

### C. Extensions of the Two-Period Crossover Design

The basic crossover principle has a natural extension to the comparison of more than two formulations, the most likely application being a bioavailability trial in which several test formulations are compared with a standard. For example, in a trial comparing test formulations $T_1$, $T_2$, and $T_3$ with a standard formulation S, one would use a sequence of Latin squares and each subject would receive all four formulations with suitable washout intervals intervening between their administration. Thus, a group of four subjects might receive the formulations in the sequence shown in Table 4.

**Table 4** Crossover Design Comparing Four Formulations

| Subject | Period 1 | Period 2 | Period 3 | Period 4 |
|---|---|---|---|---|
| 1 | $T_2$ | S | $T_3$ | $T_1$ |
| 2 | $T_3$ | $T_1$ | S | $T_2$ |
| 3 | S | $T_2$ | $T_1$ | $T_3$ |
| 4 | $T_1$ | $T_3$ | $T_2$ | S |

### D. Balanced Incomplete-Block Designs

A further extension of the crossover concept is possible in the case in which subjects do not receive all formulations but at least two of them. Westlake (1974) suggested several reasons why it might not be convenient to administer all test formulations to each subject. Chief among these are

1. If a large number of formulations are to be compared (and, in practice, "large" may be as few as three) testing of all of them in each subject may require the drawing of too many blood samples, exceeding the amount of blood that is medically advisable.
2. If the washout time is substantial (e.g., a week or more) the administration of all formulations to each subject may make the trial too time-consuming.
3. The more times a subject must return for testing, the greater the likelihood of dropping out of the trial with the concomitant problem of missing data.

These considerations suggest that, in certain circumstances, the idea of conducting a crossover trial in a balanced-incomplete-block design (BIBD) might present an attractive alternative to the more familiar Latin square design. The crossover principle remains intact and the variation among subjects is still separated out from the error term that is used to test the differences among formulations. An example of a BIBD comparing four formulations in 12 subjects with each subject receiving only two of the formulations is shown in Table 5. Note that the first six subjects receive the six possible pairs of formulations that can be chosen from the four formulations to be tested while the second six subjects receive the six pairs in reverse order so that there is a balance of formulations over periods.

Table 5 BIBD Design with 12 Subjects

| Subject No. | Period 1 | Period 2 |
|---|---|---|
| 1 | S | $T_3$ |
| 2 | $T_1$ | $T_2$ |
| 3 | $T_2$ | S |
| 4 | $T_3$ | $T_1$ |
| 5 | $T_2$ | $T_3$ |
| 6 | S | $T_1$ |
| 7 | $T_1$ | $T_3$ |
| 8 | $T_2$ | $T_1$ |
| 9 | $T_3$ | S |
| 10 | $T_1$ | S |
| 11 | $T_3$ | $T_2$ |
| 12 | S | $T_2$ |

### E. Steady-State Trials

This concludes our discussion of basic designs for bioavailability trials. It has been assumed throughout that each subject receives a single dose of a formulation and that the formulations under examination are compared via the blood-level sequences generated by the single doses. A variation on this type of trial is the "steady-state" bioavailability trial. Here, one compares the blood-level sequences after repeated dosing when a steady state in the blood level has been achieved. As an example, suppose that we wish to compare two formulations of a drug that would normally, for proper therapeutic effect, be dosed twice a day, say at 8:00 a.m. and 8:00 p.m. If dosing at this 12-hour interval is carried out for a few days, the concentration of drug in the bloodstream will eventually reach a steady state in which the blood-level profile over each 12-hour cycle will simply repeat itself. From a purely theoretical point of view, of course, true steady state will not be reached in a finite time; in practice, steady state will be reached in a certain number of days dependent mainly on the half-life of the drug. The comparison of formulations will then be carried out relative to the blood-level profile over one cycle of the steady-state condition (in the example, over a 12-hour cycle).

## VII. ANALYSIS OF BIOEQUIVALENCE TRIALS

### A. Formulation of the Problem

In the preceding section dealing with the experimental design of bioequivalence trials, the basic ANOVA was discussed and treated as though one were analyzing a single univariate measurement corresponding to the formulation. However, the data set for each administration of a formulation to a subject is not univariate but, rather, multivariate, comprising as it does a sequence of blood levels over time. The trial can be viewed as a repeated-measures experiment with repeated measurements of blood drug concentration being made over a series of points in time. A number of papers have treated bioequivalence trials as repeated-measures experiments or as multivariate problems; Westlake (1974), Wallenstein and Fisher (1977), and Redman (1968), who proposed the use of principal components, are examples. However, after a number of years of dealing with this subject, I now no longer find any of these approaches convincing. We must return, I think, to the underlying reason for the bioequivalence trial, namely that one is trying to demonstrate the therapeutic equivalence of two (or more) formulations by the indirect route of comparing blood-level profiles. Consequently, I am led to the principle, espoused most recently in a review paper (Westlake, 1979), that comparison of the formulations should be made with respect to only those characteristics of the blood-level sequence that possess some meaningful relation to the therapeutic effect of the drug. As an example, consider area under the blood-level curve. It was noted earlier that under quite general conditions this area is proportional to the amount of drug absorbed. Since the amount of drug that is made available to the circulatory system is a key characteristic of a drug formulation, it follows that the area will almost always be a prime candidate for analysis.

Two other characteristics of the blood-level sequence are usually picked out for analysis: the peak drug concentration attained ($C_{max}$) and the time at which this peak occurs ($T_{max}$). For many drugs these are important features of the blood-level profile that are clearly related to the therapeutic use of the drug and thus, in my opinion, qualify for statistical analysis. $C_{max}$ may be important by virtue of its connection with both therapeutic and toxic effects. In some cases, it may be important that it reaches a certain threshold level to induce the therapeutic effect; in others, it may be important to ensure that it remains below another threshold to avoid toxic effects. Drugs for which these two thresholds are relatively close together are said to possess a narrow therapeutic window and clearly present particular problems. The other parameter, $T_{max}$, may be of importance with drugs for which it is important to reach the peak concentration—and presumably the peak therapeutic effect—as rapidly as possible. Analgesics and antibiotics are two such classes of drugs that spring readily to mind. However, with drugs for which the onset of therapeutic action is not apparent until repeated dosing over a period of time has occurred, $T_{max}$ would seem to be of scant interest. A number of psychotropic drugs fall in this category. Where the relationship of the drug concentration in the blood to therapeutic effect is poorly understood one can certainly make a reasonable case that $C_{max}$ and $T_{max}$ together with the area provide a basic description of the blood-level profile and can thus reduce the analysis of a repeated-measures trial to that of three univariate characteristics. Naturally, when other characteristics of the profile are known to have a meaningful relation to therapeutic effect they should also be regarded as candidates for analysis.

## B. Methods of Analysis

### 1. Introduction

In what follows it will be assumed that a number of characteristics are to be analyzed separately (i.e., as univariate quantities), that we are dealing with the comparison of just two formulations (standard and test), and that for the purpose of the discussion we can confine the development to a simple two-way crossover trial in which it is desired to analyze the area under the curve. Little generality will be lost by the assumptions and they will enable us to concentrate more clearly on the essential problem: how should one analyze a bioequivalence trial?

We have already examined the basic ANOVA for a two-way crossover trial, and it is immediately apparent that the ratio of the formulations mean sum of squares to the error mean sum of squares gives an F-statistic to test the null hypothesis that $\mu_t = \mu_s$. For many years this was the standard statistical test applied to bioequivalence trials, and it still continues to be performed by many who are responsible for their analysis. In the present case it provides a test of whether the mean amount of drug absorbed from the test formulation is identical to the mean amount absorbed from the standart. My own view is that in the real world, as opposed to the realm of mathematical abstraction, the test of the simple null hypothesis of identity is of no interest since the answer is known in advance of the trial: the null hypothesis is always false! One could hardly expect the mean amounts of drug absorbed from two batches of the same formulation manufactured under closely controlled conditions to be identical—very nearly equal, perhaps, and differing by only a fraction of a percent, but not identical. Detection of the difference then becomes simply a function of sample size. Consideration of this point leads one to realize that the probable magnitude of the difference is the critical factor and this will be the basis of all the methods discussed later. For the present, however, consider two anomalies that occur when the test of the simple null hypothesis is performed. They are

1. A large difference between the formulations which is, nevertheless, not statistically significantly at some level, say 0.05
2. A small difference, probably of no therapeutic importance whatsoever, that is shown to be statistically significant

The first case suggests lack of sensitivity in the analysis, the second, an excess of it. Naturally, the trial will have been planned to have just the required sensitivity; but as every statistician knows, the actual data may have characteristics quite different from those of the data on which the sample size was based.

It is of some interest to trace the FDA's response to these two anomalous situations. Early on, it appears to have been recognized that a finding of no statistical significance in the first case was not necessarily evidence of bioequivalence and the agency began to ask for a retrospective examination of the power of the test of the null hypothesis. Specifically, it was mandated that the test of equivalence have at least an 80% power of detecting a 20% difference between $\mu_t$ and $\mu_s$, where 20% was, apparently, arbitrarily chosen to represent the minimum difference that could be regarded as of therapeutic importance. No such numerical criterion appears to have been adopted for the second case, and my understanding is that when a very small difference is shown to be statistically significant at the 0.05 level, the question is often decided on clinical grounds by determining that while the actual difference between the formulations is statistically significant it is so small that one can assume that the formulations are nevertheless bioequivalent. In other words, the null hypothesis is true even though the statistical analysis has decided otherwise. This is equivalent to saying that one will believe the results of the statistical analysis if they accord with preconceived ideas of bioequivalence and common sense but will ignore them if they do not. A more enlightened response would be to realize that if one is forced into this irrational stance then perhaps the statistical methods employed were not appropriate. That the testing of the simple null hypothesis is not appropriate should be abundantly clear; fortunately, a number of statisticians have realized this and have proposed methods which *are* relevant to the problem of determining bioequivalence. These will now be discussed in detail.

## 2. Confidence Intervals

Westlake (1972) was the first to suggest the use of confidence intervals as a means of testing for bioequivalence. However, the rather unorthodox method of modifying the confidence interval described in this paper contains an inaccuracy which was corrected in a later paper (Westlake, 1976). Briefly, the method proposed is as follows.

Using the error variance from the ANOVA and the difference in the sample means for the standard and test formulations, form a $(1 - \alpha)$ confidence interval, where $\alpha$ is typically 0.05, for the difference in population means, $\mu_t - \mu_s$

$$k_1 < (\mu_t - \mu_s) < k_2$$

Clinically, it will have been decided that the two formulations can be considered bioequivalent provided that

$$K_1 < (\mu_t - \mu_s) < K_2$$

Then the decision rule is to accept bioequivalence if $k_1 > K_1$ and $k_2 < K_2$, that is, if the confidence limits lie wholly within the acceptable range for bioequivalence. In practice, the FDA has tended to make its pronouncements on the acceptable range for bioequivalence in a proportional form (for example, the ±20% rule) such that bioequivalence is accepted if

$$.80 < \frac{\mu_t}{\mu_s} < 1.20$$

If the parameter undergoing analysis is logarithmically transformed initially, the confidence interval will appear as an interval for $\mu_t/\mu_s$. Later in this chapter it will be argued that there are compelling reasons, a priori, why several of the more important parameters should undergo logarithmic transformation, so that in these cases the confidence intervals arise quite naturally in the desired form.

A refinement of the proposed decision procedure based on a $(1 - \alpha)$ confidence interval was described in Westlake (1976). It was proposed that, where the equivalence range is given symmetrically about the null value (e.g., ±20%), the confidence interval could likewise be adjusted to be symmetrical about the null value. This rather unorthodox procedure has the effect of (1) increasing the manufacturer's chance of demonstrating equivalence of the test formulation to the standard formulation and (2) increasing the confidence coefficient, which will actually always be greater than the nominal value of $(1 - \alpha)$. Finally, Westlake (1981) suggested a further modification in this approach which has the merit of attempting to put the proposed rule for acceptance of bioequivalence on the same logical basis as the FDA's well-established rule for acceptance of efficacy in the clinical trial of a new drug entity. The suggestion was to base the decision rule on a conventional confidence interval, as outlined at the beginning of this section, but to use a confidence coefficient of $(1 - 2\alpha)$ rather than $(1 - \alpha)$, with $\alpha$, again, being typically chosen to be equal to 0.05. I quote from the reference cited:

> What protection should the regulatory agency seek against approving a new formulation that is not bioequivalent to the standard? Current practice in approving new drugs for efficacy presents a helpful analogy. In this case, one is usually attempting to demonstrate the efficacy of a new drug by testing against a placebo, and it is customary in the U.S.A. to insist that in this test of the null hypothesis (identity of drug and placebo) a statistically significant result at the $\alpha$ level (traditionally 0.05) be obtained as proof of efficacy. My interpretation of this is that the regulatory agency is attempting to ensure that if the drug is really the same as placebo there is only a low probability, 0.05, that it will be approved. Note, however, that since the drug would never be approved for being less efficacious than placebo, the test and its associated critical region should be

> one-sided. It seems to me that a similar policy should prevail in the decision criterion for bioequivalence. A suitable rule might be: if the difference in means of the two formulations is actually $\Delta$, where $\pm\Delta$ is the allowable range for bioequivalence, then the probability that the $(1 - \alpha)$ confidence interval falls within $\pm\Delta$ should be acceptably small (say 0.05). That is, in the borderline case the probability of accepting the new formulation as bioequivalent to the standard should be small. (Westlake, 1981)

It is easily seen that with $(1 - \alpha)$ confidence limits this probability is less than $\alpha/2$. Consequently, in order to maintain the parallelism with the FDA's requirements for efficacy testing, one should use a $(1 - 2\alpha)$ or 90% confidence interval.

It is instructive at this point to consider how one would determine the sample size for a crossover trial in which the results will be given in the form of a $(1 - 2\alpha)$ confidence interval. The development sheds light on several aspects of the problem, including the interests of the manufacturer and the regulatory agency. Suppose that a crossover trial comparing the test and standard formulations is to be carried out in n subjects (n even). Assume an error variance of $\sigma^2$ with the estimate $s^2$ from the ANOVA based on $(n - 2)$ degrees of freedom. With a standard notation,

$$(\bar{x}_t - \bar{x}_s) \sim N\left((\mu_t - \mu_s), \frac{2\sigma^2}{n}\right)$$

and the result of the analysis is the $(1 - 2\alpha)$ confidence interval

$$(\bar{x}_t - \bar{x}_s) - \sqrt{\left(\frac{2}{n}\right)}\, st_{1-\alpha} < (\mu_t - \mu_s) < (\bar{x}_t - \bar{x}_s) + \sqrt{\left(\frac{2}{n}\right)}\, st_{1-\alpha}$$

A desirable criterion, from the manufacturer's point of view, might be to pick n such that there is a high probability $(1 - \beta)$ that the limits of the confidence interval are within the acceptable range, $\pm\Delta$, when $\mu_t = \mu_s$. That is, we require that the two inequalities

$$(\bar{x}_t - \bar{x}_s) + \sqrt{\left(\frac{2}{n}\right)}\, st_{1-\alpha} < \Delta$$

$$(\bar{x}_t - \bar{x}_s) - \sqrt{\left(\frac{2}{n}\right)}\, st_{1-\alpha} > -\Delta$$

are true. These inequalities may be expressed in the form

$$\left| (\bar{x}_t - \bar{x}_s) / \sqrt{\left(\frac{2}{n}\right)}\, s \right| < \Delta / \sqrt{\left(\frac{2}{n}\right)}\, s - t_{1-\alpha}$$

If $\mu_t = \mu_s$, the term enclosed in the absolute value sign has the t-distribution with $(n - 2)$ degrees of freedom. To ensure a greater than $(1 - \beta)$ probability that the above inequality holds, one requires

$$\Delta / \sqrt{\left(\frac{2}{n}\right)}\, s - t_{1-\alpha} > t_{1-\beta/2}$$

or

$$n > \frac{2(t_{1-\alpha} + t_{1-\beta/2})^2 s^2}{\Delta^2}$$

We might note here that when one plans a trial to detect a difference of $\Delta$ at the $\alpha$ level with power $(1 - \beta)$ the familiar formula for sample size estimation is

$$n > \frac{2(t_{1-\alpha/2} + t_{1-\beta})^2 s^2}{\Delta^2}$$

which shows how the roles of $\alpha$ and $\beta$ are switched when the issue is equivalence. A desirable criterion from the point of view of the regulatory agency might be that if $(\mu_t - \mu_s) = \Delta$—the borderline case—the probability of accepting the new formulation will be suitably low. The inequality on which the sample size calculation was based can be written

$$-2\Delta / \sqrt{\left(\frac{2}{n}\right)}\, s + t_{1-\alpha} < \frac{(\bar{x}_t - \bar{x}_s - \Delta)}{\sqrt{(2/n)}s} < -t_{1-\alpha}$$

If $(\mu_t - \mu_s) = \Delta$, the expression in the middle has the t-distribution with $(n - 2)$ degrees of freedom and the probability that the inequality holds is clearly less than $\alpha$. This result is independent of the sample size n and, consequently, the regulatory agency is assured that the probability of accepting a formulation just on the borderline of equivalence is always less than $\alpha$, as stated earlier. An interesting by-product is that determination of the sample size becomes the interest of the manufacturer, not of the regulatory agency.

### 3. Nonequivalence as the Null Hypothesis

Anderson and Hauck (1983) formulated the null hypothesis as

$$H_0: \mu_t - \mu_s \leqslant A \quad \text{or} \quad \mu_t - \mu_s \geqslant B$$

where A and B are the lower and upper limits of the allowable range for bioequivalence. In most applications, A and B will be symmetrical about the null value as in the $\pm\Delta$ used earlier. The alternative hypothesis, $H_A$, is then

$$H_A: A < \mu_t - \mu_s < B$$

or the hypothesis of bioequivalence.

As the authors point out, their procedure has the advantage of bringing the problem in line with traditional practice, in which the event of interest is that of finding a statistically significant difference. In the present case, rejection of the null hypothesis at some significance level (e.g., 0.05) is evidence in favor of rejecting bioinequivalence and thus of accepting bioequivalence. Their test statistic is

$$T = \frac{(\bar{x}_t - \bar{x}_s) - (A + B)/2}{s\sqrt{(1/n_t + 1/n_s)}}$$

where $x_t$ and $x_s$ are assumed to be normally distributed with means $\mu_t$ and $\mu_s$, respectively, and common variance $\sigma^2$. The symbol s denotes the sample standard deviation based on the appropriate number of degrees of freedom and $n_t$ and $n_s$ the sample sizes for the test and standard formulations. For a crossover trial $n_t$ and $n_s$ are, of course, equal, and for a symmetrical bioequivalence region (e.g., $\pm\Delta$) the term $(A + B)/2$ in the above statistic is equal to zero.

The quantity T has a noncentral t-distribution with noncentrality parameter

$$\delta = \frac{(\mu_t - \mu_s) - (A + B)/2}{\sigma\sqrt{(1/n_t + 1/n_s)}}$$

The test will be to reject $H_0$ if T is small enough. It is required to choose $C_1$ and $C_2$ such that

$$P[C_1 < T < C_2 \mid (\mu_t - \mu_s) = B, \sigma] = P[C_1 < T < C_2 \mid (\mu_t - \mu_s) = A, \sigma] = \alpha$$

and it is clear from the symmetry that a single C $(= C_2 = -C_1)$ can be chosen. Thus the region for rejection of $H_0$ is chosen so that at the borderline of inequivalence, the probability of rejection of $H_0$ is some suitably small value, $\alpha$. This is, therefore, the type 1 error at the boundaries: if $(\mu_t - \mu_s)$ is either less than A or greater than B the type 1 error is clearly less than $\alpha$.

The exact distribution of T when $(\mu_t - \mu_s)$ is equal to either A or B is not known because of the dependence of the statistic on an unknown $\sigma$. If $\sigma$ were known one would compute

$$P[|T| < |\text{Observed value of } T| \mid (\mu_t - \mu_s) = B, \sigma] = \rho$$

and would reject the null hypothesis whenever $\rho \leqslant \alpha$.

Anderson and Hauck considered several approximations of which the following appears to be the most satisfactory. At the boundary point, $(\mu_t - \mu_s) = B$, for example,

$$\delta = \frac{(B - A)}{2\sigma\sqrt{(1/n_t + 1/n_s)}}$$

One approximates the unknown $\sigma$ by the known s and can thus treat $\delta$ as a known constant, $\hat{\delta}$. Then, with this approximation,

$$T - \hat{\delta} = \frac{(\bar{x}_t - \bar{x}_s) - B}{s\sqrt{(1/n_t + 1/n_s)}}$$

which has a central t-distribution when $(\mu_t - \mu_s) = B$.

The required probability that T is less than the observed value of T, say $T_0$, is given by evaluating the probability that

$$-|T_0| - \hat{\delta} < T - \hat{\delta} < |T_0| - \hat{\delta}$$

which can be obtained from tables of the cumulative t-distribution. Anderson and Hauck examined the adequacy of the approximation by simulation methods with values of $n_t$, $n_s$ ranging over 10, 20, and 50. Over this range, performance characteristics of the test appeared to be quite satisfactory, and since 10 might represent a reasonable lower limit on sample size, their method would appear to give a perfectly satisfactory approach to the analysis of bioequivalence trials. Their paper also discussed the determination of the appropriate sample size. A second publication, Hauck and Anderson (1984), gave an application of their proposed method. It should also be noted that Schuirmann (1981) has formulated the problem of analyzing bioequivalence trials in exactly the same manner, although his test of the null hypothesis proceeds in a somewhat different manner. It is unfortunate that his development has not, to date, appeared in print; one interesting feature is that his decision rule results in exactly the same decision concerning bioequivalence as the rule based on a conventional $(1 - 2\alpha)$ confidence interval discussed earlier.

## 4. Bayesian Methods

The papers by Selwyn et al. (1981) and Selwyn and Hall (1984) developed a Bayesian approach to the problem of testing for bioequivalence. They discussed the application of the methodology to a two-way crossover trial in which the model is assumed to be

$$Y_{ijk} = \mu + \pi_k + F_l + S_{ij} + e_{ijk}$$

with the notation that was described earlier in this chapter. No carryover effect was included in the model, although they indicated how their methodology could be adapted to handle this case. The Bayesian approach requires the assumption of a prior probability distribution for $\mu$, P, F, and two variances related to $S_{ij}$ and $e_{ijk}$. The posterior probability density is then, according to Bayes' theorem, proportional to the product of the prior density and the likelihood function, the latter being a function of the data collected in the trial. Choice of the prior probability density is, of course, a familiar problem, and the authors took the usual route of assuming a noninformative prior. In fact, they compared four noninformative priors of the form

$$p(\mu, P, F, \sigma_e^2, \sigma_A^2) \propto \sigma_e^{-r}\sigma_A^{-r}, \qquad r = 1, 2$$

and

$$p(\mu, P, F, \sigma_e^2, \sigma_A^2) \propto \sigma_e^{-r} \sigma_A^{-r} \qquad \text{for } \sigma_A^2 > \sigma_e^2, \ r = 1, 2$$

After the posterior probability density has been calculated, one can compute the posterior probability that the two formulations are bioequivalent (e.g., $.80 < \mu_t/\mu_s < 1.20$) by integrating the posterior probability density over the corresponding region. If this probability exceeds some appropriate value, typically 0.95, one would conclude bioequivalence.

Selwyn et al. also gave a brief discussion of the experimental design implications of the Bayesian approach. An attractive feature is that one need not attempt to estimate the appropriate sample size before undertaking the bioequivalence trial but can proceed sequentially. The posterior probability density after a number of stages of the trial is proportional to the product of the prior density and the likelihood functions from each of the stages of the trial. Two less attractive features of the method are its inherent dependence on an arbitrary prior distribution and the computational difficulties involved in the numerical integration required in evaluating the posterior probability density over the region of bioequivalence. With respect to the first point, the authors examined the four different noninformative priors given above. Although the results obtained were quite close, they *were* different, and this would pose problems if one prior gave a posterior probability of bioequivalence $>0.95$ while another prior gave $<0.95$. In this case, what would be the regulatory agency's conclusion concerning bioequivalence? My own view is that where one party (manufacturer) is in the position of having to prove a point to the satisfaction of a second party (regulatory agency), arbitrary assumptions, on which the parties could disagree, should be kept to a minimum. The Bayesian approach introduces an arbitrary assumption that the other methods described earlier do not require.

The Bayesian approach has also been discussed by Rodda and Davis (1980). While they took essentially the same position as Selwyn et al., their mathematical development was much simpler because they introduced an approximation, replacing $\mu_s$ by $\bar{x}_s$. Reference can also be made to the papers by Mandallaz and Mau (1981) and Fluehler, Hirtz, and Moser (1981), which discussed the use of confidence intervals and provided some Bayesian interpretation.

## 5. Logarithmic Transformations of Parameters

Earlier it was stated that a strong case could be made for the logarithmic transformation of the area under the blood-level curve and of other parameters usually subjected to analysis. A useful consequence of this transformation is that confidence intervals constructed following ANOVA will be on ratios of means rather than differences, which fits well with the proportional form in which the FDA gives its rules (e.g., the ±20% rule).

It will be recalled that earlier in the chapter it was shown that, under fairly general conditions, the area under the curve could be shown to satisfy the equation

$$\text{Area under curve} = \frac{\text{amount of drug absorbed}}{k_e V}$$

where $k_eV$ is the product of the elimination rate constant and the volume of distribution for a given subject. One is interested in analyzing the area under the curve as a measure of the amount of drug absorbed. However, there is a multiplicative term, $k_eV$, which is a function of the subject. The standard ANOVA treats subject as an additive effect; and since a logarithmic transformation of the area brings the quantity $k_eV$ additively into the above equation, it seems appropriate to make the logarithmic transformation before performing the ANOVA. A similar although slightly weaker case can be made for a logarithmic transformation of any drug concentration data such as $C_{max}$. In this case one observes that, although a logarithmic transformation of the expression for drug concentration as given earlier in the discussion of the two-compartment open model will not yield any immediately recognizable form, the volume of distribution (a subject-dependent quantity) is always multiplicative. Thus a logarithmic transformation of concentration data has a certain justification when the drug concentration data are to be analyzed using an ANOVA model with additive effects. If one wished to be meticulous, analysis of the data twice—once with and once without the transformation—followed by plotting of residuals could be instructive. No corresponding justification for logarithmic transformation of the other well-known parameter, $T_{max}$, seems possible, and a confidence interval on the difference of $T_{max}$ for standard and test formulation following an ANOVA on the untransformed data seems appropriate. In this case, however, it is difficult to accept that the FDA proportional rule (e.g., within ±20%) is really meaningful for a parameter like $T_{max}$. For a drug that peaks early following administration (e.g., at 30 minutes) the rule means that the test formulation must peak within ±6 minutes of the standard formulation to be accepted as bioequivalent, whereas for a drug that peaks late (e.g., at 5 hours) the test formulation must peak within 1 hour of the standard formulation. It seems to me that for a parameter like $T_{max}$ the ±20% rule has little meaning; a rational decision rule would probably depend heavily on the nature of the drug and, if such a rule could be devised, would be expressed in terms of an absolute difference rather than proportionally.

## 6. Comparison of Methods

Three methods of analyzing bioequivalence trials have been presented, each of which attacks the problem of testing for bioequivalence in a meaningful way: the use of confidence intervals, the setting up of the null hypothesis as nonbioequivalence, and Bayesian methods. Several of the papers referenced have given comparisons of the methods in varying degrees of detail, and mention should also be made of the paper by Rocke (1984) and the correspondence occasioned by his paper (Anderson and Hauck, 1985; Selwyn and Hall, 1985; Rocke, 1985). Metzler (1985) in an as yet unpublished paper also compares the power characteristics of these methods by simulation methods. The requirements of the manufacturer and the regulatory agency are, of course, best met by a method that maximizes the probability of declaring bioequivalence while giving the regulatory agency a guaranteed small probability that bioinequivalent formulations (e.g., $|\mu_t/\mu_s - 1| \geqslant .20$) are accepted as bioequivalent. Not surprisingly, most of these examinations of the operating characteristics referred to found very little difference of real importance among the three methods discussed. I have already expressed my own concern at the arbitrary element that is necessarily introduced into the Bayesian approach, and my preference is for the use of a $(1 - 2\alpha)$ confidence interval, not least because of the demonstrable parallelism of this approach with the FDA's well-established practice in adjudicating efficacy trials.

### 7. Multiple Comparisons

So far, the discussion of the analysis of bioequivalence trials has proceeded as though there were just one characteristic of the blood-level sequence, for example, area under the curve, that required analysis. In fact, there may be several parameters, including $C_{max}$ and $T_{max}$ as the most familiar. In this case the question arises as to whether the $(1 - 2\alpha)$ confidence coefficient should apply to three or four confidence intervals jointly or whether each of the confidence intervals should have its individual $(1 - 2\alpha)$ confidence interval. Similarly, in the Anderson-Hauck approach, should the $\alpha$ level apply to each parameter individually or to all parameters jointly? A similar concern applies with Bayesian methods. Presumably an answer to these questions depends on the requirements of the regulatory agency which is responsible for setting up the rules concerning bioequivalence. If statements that are valid jointly are required, Bonferroni inequalities offer a useful and very simple way of handling the problems for all the methods discussed. For example, with the confidence interval approach and with three parameters (e.g., area under the curve, $C_{max}$, and $T_{max}$) one would construct a $(1 - 2\alpha/3)$ confidence interval for each to obtain a joint statement with confidence coefficient greater than $(1 - 2\alpha)$.

A second type of situation in which the issue of multiple comparisons is sometimes raised is where several test formulations are compared with a standard formulation in a Latin square or BIBD design. The multiple comparison of interest are then those of each test formulation against the standard. In this case I think the issue is quite clear, notwithstanding the fact that in numerous instances in the literature adjustments for multiple comparisons have been made. One is not normally interested in making the statement that either all of the test formulations are equivalent to the standard or they are not all equivalent—which is the question addressed by a joint confidence interval statement, for example. Rather, one is interested in assessing each test formulation relative to the standard individually and the situation is similar to that which obtains when a separate bioequivalence trial is carried out for each test formulation. The difference is that economy has been achieved by utilization of the same data from the standard formulation for each of the comparisons. Quite clearly, no adjustment to the confidence coefficient or significance level for multiple comparisons is appropriate.

### 8. Tolerance Intervals

To this point, the discussion of statistical analysis of bioequivalence trials has centered on the testing of the difference (or ratio) of means for various parameters derived from the blood-level sequence. An additional rule has been promulgated in various pronouncements by the FDA which goes something as follows. In a crossover trial comparing a test and a standard formulation one forms the ratio, for each subject, of the areas under the curve for the test to the standard formulation. Then at least 75% of these ratios must lie between .75 and 1.25 in order to demonstrate bioequivalence. The rule is a curious one: it specifies a requirement on the data from the very limited number of subjects in the particular bioequivalence trial and completely ignores the statistical aspects of inference to the population in general. It should also be noted that the underlying model is a considerably simplified version of that discussed earlier; in particular, there is an inherent assumption that there are no carryover or period effects.

Understandably, this rule has occasioned much criticism, and for a representative example of such one may refer to the paper by Haynes (1981) in which some of the rule's numerous shortcomings are detailed. It is unfortunate that those responsible for the rule did not, apparently, seek appropriate statistical advice since it is clear that, flawed though it may be, the rule is an attempt to come to grips with an extremely important issue. Suppose, for example, that a physician wanted to switch the medication for a patient from the standard formulation to a newly marketed generic formulation. This might be done to minimize the expense to the patient or to satisfy a requirement of the patient's health insurance policy to prescribe generic versions of drugs. The physician might formulate a requirement of the new formulation along the lines that in a large percentage of patients the amount of drug absorbed from the new formulation should be within 10 or 20%, for example, of the amount absorbed from the standard formulation on which the patient had been stabilized. The underlying principle is clear, namely that the switch to a new formulation should not result in any important change in the therapeutic effect on the patient; and it is equally clear that the FDA "75/75–125" rule, as it is sometimes known, aims at this specific problem. Statements of the type implied by this FDA rule are, of course, well known to statisticians and are available through the construction of tolerance intervals, as suggested in Westlake (1979). A typical tolerance interval statement would be of the form: the probability is $\beta$ that a proportion of at least $\gamma$ of the subjects in the population will have an area under the curve for the test formulation that lies between $p_1$ and $p_2$ of the area obtained with the standard formulation. Note that in order to satisfy the FDA rule, $\gamma$ would be equal to 0.75 while $p_1$ would have to be greater than 0.75 and $p_2$ less than 1.25. Note also, however, the introduction of a probability $\beta$ (which might be 0.95, for example) and also the fact that the statement is now an inference to the population at large. Distribution-free tolerance intervals (see, for example, Kendall and Stuart, 1961) can readily be constructed and have the advantage of not requiring that the ratios of the areas conform to a normal distribution. A disadvantage, on the other hand, is that for a given sample size one cannot specify in advance both the probability, $\beta$, and the proportion of subjects, $\gamma$. In the previous example they were 0.95 and 0.75, respectively. As an example, suppose there were 12 subjects in the bioequivalence trial and that the 12 ratios of areas when ordered in increasing magnitude extended from 0.78 to 1.29. Then if we choose to make a tolerance interval statement based on the highest and lowest ratios, it turns out to be as follows: there is a 0.95 probability that at least 66% of the subjects in the population will have a mean area under the curve for the test formulation that lies between 0.78 and 1.29 of the mean area obtained with the standard formulation. If one is prepared to assume normality of the ratios, $\beta$ and $\gamma$ can both be postulated in advance and there are numerous tables available; again one may refer to Kendall and Stuart (1961) for details.

Finally, it should be repeated that this direct treatment of ratios for individual subjects presupposes a simplified statistical model and may not be appropriate in all cases.

## VIII. CONCLUDING COMMENTS

The concepts of absorption, distribution, metabolism, and elimination of a drug administered to a human subject have been briefly described. The subject of pharmacokinetic modeling has also been briefly touched on and my reasons for skepticism as to its practical value have been given. Design and analysis of comparative bioavailability, or bioequivalence, trials has been discussed with particular emphasis on the statistical analysis—the latter with good reason. For too many years the analysis of bioequivalence trials has been marked by the uncritical accaptance of standard methods of hypothesis testing which have no relevance to the particular problem posed by these trials. More recently, methods have been proposed which do attack the problem in a meaningful way, and it is to these that much of the chapter has been devoted. The temptation to moralize is hard to resist: perhaps the greatest contribution that a statistician can make to the solution of a problem in the real world is to ensure that the problem has been correctly formulated and that the statistical methodology employed is appropriate to the formulation.

## REFERENCES

Anderson, S. and Hauck, W. W. (1983). A New Procedure for Testing Equivalence in Comparative Bioavailability and Other Clinical Trials, *Commun. Stat.—Theory Methods*, 12: 2663–2692.

Anderson, S. and Hauck, W. W. (1985). On Testing for Bioequivalence: Letter to the Editor, *Biometrics*, 41: 563.

Brown, B. W. (1980). The Crossover Experiment for Clinical Trials, *Biometrics*, 36: 69–79.

Fluehler, H., Hirtz, J., and Moser, H. A. (1981). An Aid to Decision-Making in Bioequivalence Assessment, *J. Pharmacokinet. Biopharmaceut.*, 9: 235–243.

Gibaldi, M. and Perrier, D. (1975). *Pharmacokinetics*, Marcel Dekker, New York.

Grizzle, J. E. (1965). The Two-Period Changeover Design and Its Use in Clinical Trials, *Biometrics*, 21: 467–480.

Hauck, W. W. and Anderson, S. (1984). A New Statistical Procedure for Testing Equivalence in Two-Group Comparative Bioavailability Trials, *J. Pharmacokinet. Biopharmaceut.*, 12: 83–91.

Haynes, J. D. (1981). Statistical Simulation Study of New Proposed Uniformity Requirement for Bioequivalency Studies, *J. Pharm. Sci.*, 70: 673–675.

Kendall, M. G. and Stuart, A. (1961). *The Advanced Theory of Statistics*, vol. 2, Hafner, New York, pp. 518–521 and 128–130.

Lanczos, C. (1956). *Applied Analysis*, Prentice-Hall, Englewood Cliffs, N.J., chap. 4.

Mandallaz, D. and Mau, J. (1981). Comparison of Different Methods for Decision Making in Bioequivalence Assessment, *Biometrics*, 37: 213–222.

Metzler, C. M. (1974). Bioavailability: A Problem in Equivalence, *Biometrics*, 30: 309–317.

Metzler, C. M. (1985). "Statistical Methods for Deciding Bioequivalence of Formulations," paper presented at the 39th National Meeting of the American Pharmaceutical Association Academy of Pharmaceutical Sciences, Minneapolis, Minn., October 23, 1985.

Redman, C. E. (1968). Biological Applications of Principal Components, *Biometrics*, 24: 235 (abstr.).

Rocke, D. M. (1984). On Testing for Bioequivalence, *Biometrics*, 40: 225–230.

Rocke, D. M. (1985). On Testing for Bioequivalence: Response to Correspondence, *Biometrics*, 41: 563.

Rodda, B. E. and Davis, R. L. (1980). Determining the Probability of an Important Difference in Bioavailability, *Clin. Pharmacol. Ther.*, 28: 247–252.

Schuirmann, D. L. (1981). On Hypothesis Testing to Determine if the Mean of a Normal Distribution Is Continued in a Known Interval, *Biometrics*, 37: 617 (abstr.).

Selwyn, M. R. and Hall, N. R. (1984). On Bayesian Methods for Bioequivalence, *Biometrics*, 40: 1103–1108.

Selwyn, M. R. and Hall, N. R. (1985). On Testing for Bioequivalence: Letter to the Editor, *Biometrics*, 41: 561–563.

Selwyn, M. R., Dempster, A. P., and Hall, N. R. (1981). A Bayesian Approach to Bioequivalence for the 2 × 2 Changeover Design, *Biometrics*, 37: 11–21.

Wagner, J. G. (1974). Use of a Clinical Pharmacokinetics Laboratory for Bioavailability Assessment, *Clinical Pharmacokinetics: A Symposium* (G. Levy, ed.), American Pharmaceutical Association, Academy of Pharmaceutical Sciences, Washington, D.C., p. 154.

Wallenstein, S. and Fisher, A. C. (1977). The Analysis of the Two-Period Repeated Measurements Crossover Design with Application to Clinical Trials, *Biometrics*, 33: 261–269.

Westlake, W. J. (1970). Time Integral of Drug Concentration in the Central (Plasma) Compartment, *J. Pharm. Sci.*, 59: 722.

Westlake, W. J. (1971). Problems Associated with Analysis of Pharmacokinetic Models, *J. Pharm. Sci.*, 60: 882–885.

Westlake, W. J. (1972). Use of Confidence Intervals in Analysis of Comparative Bioavailability Trials, *J. Pharm. Sci.*, 61: 1340–1341.

Westlake, W. J. (1974). The Use of Balanced Incomplete Block Designs in Comparative Bioavailability Trials, *Biometrics*, 30: 319–327.

Westlake, W. J. (1976). Symmetrical Confidence Intervals for Bioequivalence Trials, *Biometrics*, 32: 741–744.

Westlake, W. J. (1979). Statistical Aspects of Comparative Bioavailability Trials, *Biometrics*, 35: 273–280.

Westlake, W. J. (1981). Bioequivalence Testing—A Need to Rethink (Reader Reaction Response), *Biometrics*, 37: 591–593.

# 8 Clinical Efficacy Trials with Quantitative Data

VERNON M. CHINCHILLI *Medical College of Virginia, Virginia Commonwealth University, Richmond, Virginia*

## I. INTRODUCTION

This chapter focuses on the statistical methodology available for analyzing clinical efficacy trials with quantitative data. The experimental designs considered are fixed and nonsequential, and it is assumed that the continuous response variables of interest cannot be expressed as survival/ failure times. The latter topics are discussed in Chapters 10 and 11, respectively. The proper design of clinical trials is discussed in Chapter 6, although some aspects of experimental design are presented here which directly affect the validity of the statistical analyses, such as in crossover experiments.

Usually, the proofs of statistical results in this chapter are referenced rather than derived. Numerical examples or references to accessible examples are provided throughout the chapter. When available, references to computer programs in the Statistical Analysis System (SAS) 1985 and the Biomedical Programs (BMDP) 1983 for conducting the appropriate statistical analyses are given. Some programs written by the author in the PROC MATRIX language of SAS are provided in an appendix.

The material presented in this chapter is not new, but it reflects the author's approaches to the analysis of quantitative response variables in clinical trials. Obviously, some knowledge of statistics is necessary to understand this chapter. In particular, it is assumed that the reader has some familiarity with univariate and multivariate linear models, which are reviewed briefly in Sections II and III.

## II. UNIVARIATE LINEAR MODELS

Most of the statistical analyses applied to continuous response variables in clinical trials fall into the class of univariate linear models. A univariate linear model serves as the general framework for regression, analysis of variance (ANOVA), and analysis of covariance (ANCOVA). Typically, in clinical trials comparisons are needed among a fixed number of treatment groups with respect to some specified response variable in the presence of other prognostic and/or confounding variables. So although estimation of effects is important in clinical trials, the emphasis is placed on hypothesis testing.

### A. Parametric Analysis

For convenience, we express the univariate linear model in terms of a full-rank design matrix $\underline{X}$, whose components represent main effects, interactions, block effects, regressors, and/or covariates:

$$\underline{Y} = \underline{X}\underline{\beta} + \underline{\varepsilon} \tag{1}$$

In this model, $\underline{Y}$ is the $n \times 1$ vector of observed responses, $\underline{X}$ is the $n \times r$ design matrix of known values, $\underline{\beta}$ is the $r \times 1$ vector of fixed but unknown parameters, and $\underline{\varepsilon}$ is the $n \times 1$ vector of unobservable random errors. In the standard set of parametric assumptions, the rows of $\underline{\varepsilon}$ are independently and identically distributed as $N(0, \sigma^2)$, where $\sigma^2 > 0$ is unknown. For the present we assume that all of the components of $\underline{\beta}$ are fixed and not random.

As can be found in any standard text on regression (e.g., Draper and Smith, 1981) or linear models (e.g., Searle, 1971), under these assumptions the maximum likelihood (ML) estimate is the same as the least-squares (LS) estimate and is given by

$$\hat{\underline{\beta}} = (\underline{X}'\underline{X})^{-1}\underline{X}'\underline{Y} \tag{2}$$

Any testable hypothesis about the treatment groups can be generalized to the form $H_0$: $\underline{C}\,\underline{\beta} = \underline{0}$, where $\underline{C}$ is a known, full-rank, $c \times r$ matrix ($c \leqslant r$). Under the same set of assumptions, the likelihood ratio (LR) test of this null hypothesis rejects $H_0$ for large values of

$$F = \frac{(\underline{C}\,\hat{\underline{\beta}})'\{\underline{C}(\underline{X}'\underline{X})^{-1}\underline{C}'\}^{-1}(\underline{C}\,\hat{\underline{\beta}})}{cs^2} \tag{3}$$

where

$$s^2 = (n - r)^{-1}\underline{Y}'\{\underline{I}_n - \underline{X}(\underline{X}'\underline{X})^{-1}\underline{X}'\}\underline{Y} \tag{4}$$

is an unbiased estimate of $\sigma^2$, $\underline{I}_n$ is the $n \times n$ identity matrix, and F follows an F-distribution with c and $n - r$ degrees of freedom, respectively.

Usually it is of interest to test a number of these hypotheses, especially if the levels of one factor represent treatment groups. In order to control for the overall significance level (false-positive rate), simultaneous inference or multiple comparison procedures are employed. It is not possible to review in this chapter all such procedures that have been proposed within the context of the univariate linear model. Miller (1981) has provided a thorough review of simultaneous inference. Briefly, there are four general types of multiple comparisons that are often used in the univariate linear model. Scheffé's technique for multiple comparisons controls the overall significance level for all possible linear combinations $\underline{C}'\underline{\beta}$, where $\underline{C} \in C$, a d-dimensional subspace of $R_p$. Then, for a significance level $\alpha$, $0 < \alpha < 1$,

$$P[\underline{C}'\underline{\beta} \in \underline{C}'\hat{\underline{\beta}} \pm (ds^2F_{1-\alpha,d,n-p})^{1/2}\{\underline{C}'(\underline{X}'\underline{X})^{-1}\underline{C}\}^{1/2}, \text{ for all } \underline{C} \in C] = 1 - \alpha \quad (5)$$

For example, if there are r groups in a one-way ANOVA model and interest lies in testing contrasts about the groups, then $d = r - 1$. The Scheffé approach is not quite as powerful as the other approaches if only a few comparisons are desired. For the Bonferroni method (and analogously, the Sidák method), the user identifies the number of desired tests, say it is k, and conducts an F-test using (3) above at the $\alpha/k$ significance level for each test to achieve an overall significance level of $\alpha$. This works relatively well for small values of k ($\leqslant 10$).

There are a variety of procedures for conducting all possible pairwise comparisons of the levels of a main effect, such as Fisher's least significant difference test, Tukey's Studentized maximum modulus, and the Newman-Keuls and Duncan multiple-range tests. There is no general agreement among statisticians as to the best approach for this class of multiple comparisons (GLM of SAS 1985 provides 12 different multiple comparison procedures). Finally, if the levels of a main effect contain a control or standard level, Dunnett's test (and others like it) controls for the overall significance level when comparing each level to the control or standard level.

In a random effects linear model, some or all of the components of $\underline{\beta}$ are assumed to be normally distributed random variables, independent of $\underline{\varepsilon}$. Then the construction of F-tests for various effects depends on the expected values of mean squares, which are functions of the variances of the random elements. In balanced designs this usually does not cause problems, but in unbalanced cases the expected mean squares may not provide an easy solution to the proper construction of test statistics. In the latter situation it may be necessary to estimate the variance components and construct the test statistics from synthetic mean squares, which are linear combinations of the estimated variance components. Searle (1971) provided a thorough discussion of the issues and problems involved.

An example of a random effects model occurs in multicenter clinical trials, presented in Section II,D,2 of this chapter. Another situation in which random effects models traditionally have been applied is repeated measurements experiments. However, as discussed in Section IV, a more general approach to repeated measurements analysis is available.

## B. Checking Assumptions

Prior to conducting any analysis in the univariate linear model, we should examine the data for violations of assumptions. In particular, this consists of checking for independence, homogeneity of variance, and normality. The usual tool for investigating all of these is the set of residuals, defined by

$$\underline{R} = \underline{Y} - \underline{X}\hat{\underline{\beta}} = \{\underline{I}_n - \underline{X}(\underline{X}'\underline{X})^{-1}\underline{X}'\}\underline{Y} \tag{6}$$

Residual plots in terms of residuals versus (1) predicted values, (2) chronological order of measurement, (3) observed values, and (4) design variables are very helpful in spotting possible violations. Again, any standard regression text (e.g., Chapter 3 of Draper and Smith, 1981) addresses the use and interpretation of residual plots.

Certain procedures are available for formally testing some of the assumptions. In a one-way ANOVA classification, Bartlett's modified LR test (Bartlett, 1937) attempts to test the hypothesis of variance homogeneity (univariate and multivariate versions are available in PROC DISCRIM of SAS 1985). It should be used cautiously, however, because it is very sensitive to nonnormality of the data. Robust versions of this test have been proposed by Levene (1960), as discussed in Chapter 5, and Brown and Forsythe (1974), based on absolute deviations from the median or 10% trimmed mean. The robust tests do not exhibit this sensitivity to nonnormality (a test based on absolute deviations from the mean is available in 7D of BMDP 1983).

It is wise to subject the residuals to a normality test and/or examine normal probability plots. Mardia (1980) has reviewed the numerous tests of univariate normality that have appeared in the literature and the power comparisons that have been conducted via computer simulation. The Shapiro-Wilk test (Shapiro and Wilk, 1965) appears to be the best test with respect to a broad range of alternative hypotheses, especially for sample sizes $\leqslant 50$. This test is constructed to detect the departure of $\tilde{\sigma}^2/\hat{\sigma}^2$ from unity, where $\tilde{\sigma}^2$ is the estimate of $\sigma^2$ based on the order statistics from an underlying normal distribution, and $\hat{\sigma}^2$ is the sample variance. For sample sizes $>50$ many of the proposed tests, such as tests based on skewness and kurtosis and the Kolmogorov-Smirnov test, perform nearly as well as the Shapiro-Wilk test. Skewness is defined as the third central moment divided by the second central moment to the 3/2 power, and kurtosis is defined as the fourth central moment divided by the square of the second central moment. For a normal distribution, skewness equals zero and kurtosis equals three, so a test of normality can be constructed by comparing the sample counterparts to these numbers. The Kolmogorov-Smirnov test examines the goodness-of-fit of the empirical distribution function to the hypothesized normal distribution. (UNIVARIATE in SAS 1985 performs the Kolmogorov-Smirnov test for sample sizes $>50$ and the Shapiro-Wilk test for sample sizes $\leqslant 50$. Normal probability plots are available in this procedure and in 5D of BMDP 1983.)

There is a problem with applying the normality tests or constructing the normal probability plots in the univariate linear model, however, because the residuals given in (6) are not independent random variables. Theil (1965, 1968) introduced BLUS residuals (best linear unbiased among those with scalar covariance matrix $\sigma^2\underline{I}_n$), which are defined by

$$\underline{R}^* = (\underline{H}'\underline{H})^{-1/2}\underline{H}'\underline{Y} \qquad \text{with } \underline{R}^* \sim N_{n-r}(\underline{0}, \sigma^2\underline{I}_{n-r}) \tag{7}$$

where $\underline{H}$ is an $n \times (n - r)$ matrix formed from any $n - r$ independent columns of $\underline{I}_n - \underline{X}(\underline{X}'\underline{X})^{-1}\underline{X}'$ and $(\underline{H}'\underline{H})^{-1/2}$ denotes any square-root decomposition of the inverse of $\underline{H}'\underline{H}$. Notice that the elements of $\underline{R}^*$ are independent and have null expected value. A test of normality can be applied legitimately to the BLUS residuals in $\underline{R}^*$, assuming that the original observations in $\underline{Y}$ are independent. In terms of graphical displays, it is not clear how the n independent observations should be matched to the n − r independent components of $\underline{R}^*$. However, Anscombe and Tukey (1963) claimed that the correlations among the residuals in $\underline{R}$ have little, if any, effect on the graphical displays.

## C. Analysis with Nonnormal Data

There are three basic approaches to remedying nonnormal data. The first is to assume that the distribution function of $Y_i$, $1 \leq i \leq n$, is expressed as $F_i(y) = F(y - \beta_0 - \underline{x}_i'\underline{\beta})$, where $\beta_0$ and $\underline{\beta} = [\beta_1, \beta_2 \cdots \beta_r]'$ are unknown parameters, $\underline{x}_i$ is an r-vector of known constants, and F is a continuous distribution function. Instead of requiring that F be the univariate normal distribution, some other distribution function, such as the exponential or the Weibull, can be invoked. However, estimation and hypothesis testing can be very complicated, depending on the chosen F, and an appropriate choice of F is not usually clear.

A second approach is to conduct a nonparametric analysis using linear rank or aligned rank statistics (signed rank statistics are also available, but are mostly important in one-sample problems when dealing with symmetric distributions). In the one-way ANOVA classification (Hájek and Šidák, 1967), the test based on linear rank statistics reduces to the Kruskal-Wallis test if the Wilcoxon scores are used. Linear rank and aligned rank statistics also can be applied to more complicated univariate linear models, as presented by Puri and Sen (1971, 1985) and Hettmansperger (1984). However, as discussed below, the nonparametric analog of the F-test in (3) can be computationally difficult.

The following is a brief discussion of hypothesis testing with linear rank statistics in the univariate linear model, taken from Puri and Sen (1985, Section 5.3). Suppose that the distribution function of $Y_i$, $1 \leq i \leq n$, is expressed as $F_i(y) = F(y - \beta_0 - \underline{x}_i'\underline{\beta})$, where $\beta_0$ and $\underline{\beta} = [\beta_1\ \beta_2 \cdots \beta_r]'$ are unknown parameters, $\underline{x}_i$ is an r-vector of known constants, and F is a continuous distribution function. The null hypothesis of interest is $H_0$: $\{\underline{\beta} = \underline{0}\}$. Let $R_{n1}, R_{n2}, \ldots, R_{nn}$ represent the ranks (from smallest to largest) of the observations $Y_1, Y_2, \ldots, Y_n$, and let $a_n(i)$ be some function defined on the positive integers. This score function is derived from the generating function $\phi(u)$, $0 < u < 1$, as

$$a_n(i) = E\{\phi(U_{ni})\} \quad \text{or} \quad a_n(i) = \phi\left(\frac{i}{n+1}\right), \qquad 1 \leq i \leq n \tag{8}$$

where $U_{n1} \leq U_{n2} \leq \cdots \leq U_{nn}$ represent the order statistics of a sample from the uniform (0, 1) distribution. It is necessary that $\phi$ be square-integrable for the asymptotic theory to hold, i.e., $\int_0^1 \phi^2(u)\,du < \infty$. The vector of linear rank statistics is defined as

$$\underline{L}_n = \sum_{i=1}^{n} (\underline{x}_i - \bar{\underline{x}}_n)a_n(R_{ni}) = [L_{n1}\ L_{n2} \cdots L_{nr}]' \tag{9}$$

and under $H_0$, $E(\underline{L}_n) = \underline{0}$ and $Var(\underline{L}_n) = b_n\underline{X}_n$, where

$$b_n = (n-1)^{-1}\sum_{i=1}^{n}\{a_n(i) - \bar{a}_n\}^2, \qquad \underline{X}_n = \sum_{i=1}^{n}(\underline{x}_i - \bar{\underline{x}}_n)(\underline{x}_i - \bar{\underline{x}}_n)' \tag{10a}$$

and

$$\bar{\underline{x}}_n = n^{-1}\sum_{i=1}^{n}\underline{x}_i \quad \text{and} \quad \bar{a}_n = n^{-1}\sum_{i=1}^{n}a_n(i) \tag{10b}$$

Popular choices of generating functions are $\phi(u) = u$ for the Wilcoxon scores and $\phi(u) = \Phi^{-1}(u)$, the inverse of the standard normal distribution function, for the normal scores. The test statistic for $H_0$ is then given by

$$U_n = b_n^{-1}\underline{L}_n'\underline{X}_n^{-}\underline{L}_n \tag{11}$$

where $\underline{X}_n^{-}$ denotes any generalized inverse of $\underline{X}_n$. There are n! realizations of $U_n$ in (11), so its exact distribution can be tabulated and $H_0$ is rejected for large values. Assuming that $\underline{X}_n$ is of full rank r, and that the necessary regularity conditions hold, $U_n$ has an approximate central $\chi_r^2$ distribution for large values of n under the null hypothesis. Under a sequence of contiguous (local) alternatives, an asymptotic noncentral $\chi_r^2$ distribution for $U_n$ can be established. This noncentral distribution of $U_n$ leads to expressions for the asymptotic relative efficiency of $U_n$ when compared to its parametric counterpart, which is a function of the score function $a_n$ and the underlying distribution function F (Chapter 5 of Hettmansperger, 1984, and Chapter 5 of Puri and Sen, 1985).

For testing the null hypothesis $H_0$: $\underline{C}\,\underline{\beta} = \underline{0}$, where $\underline{C}$ is a known, full-rank, $c \times r$ matrix ($c \leqslant r$), we must resort to aligned rank statistics (see Chapter 7 of Puri and Sen, 1985). This involves a reparameterization as follows. Let $\underline{C}^*$ be a $(r - c) \times r$ matrix such that $[\underline{C}'\,\underline{C}^{*\prime}]'$ is of full rank r and let $[\underline{D}\,\underline{D}^*]$ be its inverse. Next, $F_i(y) = F(y - \beta_0 - \underline{x}_i'\underline{\beta})$ is rewritten as $F(y - \beta_0 - \underline{x}_i'\underline{D}\,\underline{C}\,\underline{\beta} - \underline{x}_i'\underline{D}^*\underline{C}^*\underline{\beta}) = F(y - \beta_0 - \underline{z}_{i1}'\underline{\gamma}_1 - \underline{z}_{i2}'\underline{\gamma}_2)$, where $[\underline{z}_{i1}'\,\underline{z}_{i2}'] = (\underline{x}_i'\underline{D}\,\underline{x}_i'\underline{D}^*]$, $1 \leqslant i \leqslant n$, and $[\underline{\gamma}_1'\,\underline{\gamma}_2']' = [\underline{\beta}'\underline{C}'\;\underline{\beta}'\underline{C}^{*\prime}]'$. Then testing $H_0$: $\underline{C}\,\underline{\beta} = 0$ is equivalent to testing $H_0$: $\gamma_1 = 0$ in the presence of the nuisance parameter $\underline{\gamma}_2$. Aligned rank statistics provide a means of estimating $\underline{\gamma}_2$, say by $\hat{\underline{\gamma}}_2$, so that the usual linear rank tests can be applied to the residuals $Y_i^* = Y_i - \underline{z}_{i2}'\hat{\underline{\gamma}}_2$, $1 \leqslant i \leqslant n$. Unfortunately, statistical software for nonparametric analysis within the univariate linear model is not well developed (PROC NPAR1WAY of SAS 1985, PROC MRANK of the 1985 SAS Supplement, and 3S of BMDP 1983 have some capabilities).

In terms of multiple comparisons, Sen (1980) and Miller (1981) reviewed the literature for nonparametric procedures in the univariate linear model based on rank statistics.

The third approach to nonnormal data involves transforming the data so that the transformed data are approximately normally distributed. If the response variable is restricted to the positive real line, the Box-Cox family of transformations (Box and Cox, 1964, 1982; Bickel and Doksum, 1981; Atkinson, 1986) can be invoked. This is a power transformation of the form

$$z = \begin{cases} (y^{\lambda} - 1)/\lambda, & \lambda \neq 0 \\ \log(y), & \lambda = 0 \end{cases} \qquad (12)$$

and in terms of the model in (1) each component $Y_i$ of $\underline{Y}$ is transformed to yield the vector $\underline{Z}$ such that $\underline{Z} = \underline{X}\,\underline{\beta} + \underline{\varepsilon}$, where the rows of $\underline{\varepsilon}$ are assumed to be independent, identically distributed, normal random variables with null expectation. Notice that the family of transformations is defined in this way so that it is a continuous function of $\lambda$, i.e., the limit of $(y^{\lambda} - 1)/\lambda$ is $\log(y)$ as $\lambda \to 0$. An added benefit of the transformation approach is that sometimes it can remove heterogeneity of variance.

The most straightforward approach to simultaneously estimating the set of parameters $(\underline{\beta}, \sigma^2, \lambda)$ is ML estimation with an iterative algorithm, such as a Newton-Raphson approach (Hinkley and Runger, 1984) or a nested minimization algorithm (Rode, 1986). The parameter estimates usually are highly correlated among themselves, which is analogous to the multicollinearity problem in multiple regression and can lead to unstable estimates.

As Hinkley and Runger (1984) discussed, there is some controversy in the literature as to whether the random nature of $\lambda$ should be accounted for when conducting inference on the parameter vector $\underline{\beta}$. In a one-way classification, for instance, if $\lambda$ is considered unknown and is estimated simultaneously along with the treatment group means, then the treatment group means are being compared on an unknown scale. Bickel and Doksum (1981) showed that the asymptotic distribution of the estimated parameter vector differs according to whether $\lambda$ is assumed to be known or unknown (and estimated by $\hat{\lambda}$). The general consensus in terms of data analysis (Carroll and Ruppert, 1981; Carroll, 1982; Taylor, 1985) appears to be to estimate $\lambda$ by $\hat{\lambda}$, transform the data according to $\hat{\lambda}$, and conduct the ANOVA conditional on $\hat{\lambda}$. Results can be interpreted in terms of the original scale via the inverse transformation of that defined in (12), and the penalty for estimating $\hat{\lambda}$ is minimal.

A program is provided in the Appendix (MULTBXCX) which conducts a multivariate Box-Cox transformation (Rode, 1986), of which the univariate version is a special case. An illustration of its application to a repeated measurements experiment is given in Example 2 at the end of Section IV.

## D. Issues in Clinical Trials

### 1. Pre- and Post-stratification of Prognostic Variables

Usually a set of prognostic variables is measured at the baseline period on all patients entering a clinical trial. Some of these may (1) correlate with the severity and progression of the disease and/or (2) interact with the treatments in some manner and affect the measurement of the response variables. The response variables themselves, measured at baseline, may fit into either category. Because the purpose of the analysis is to compare treatments with respect to the response variables, it is obvious that the statistical analysis must adjust for the important prognostic factors.

There are two general approaches to the adjustment of prognostic factors. The first, called prestratification, occurs at the randomization phase of the trial. A number of strata are created according to the levels of a subset of prognostic variables. When a patient is enrolled in the study, he/she is assigned to the appropriate stratum and then randomized to treatment within that stratum. Therefore, within each stratum, where the subset of prognostic variables is fixed, patients are equally distributed among the treatments. It is desirable to prestratify as many prognostic factors as possible, because this balances the treatment groups with respect to these factors. However, there is a trade-off between the number of strata, k, that can be employed and the total sample size, n. For example, in a two-armed trial k cannot be greater than n/2. In fact, it should be much less than n/2 to ensure that each stratum contains two or more patients.

The second approach to adjustment of prognostic factors is called post-stratification and occurs at the analysis phase after all the data have been collected. Usually some form of regression or ANCOVA provides a means of analysis in which the prognostic factors are incorporated as covariates in the univariate linear model. In this setting the effects of the covariates are partialled out (removed) from the treatment sum of squares (and, hence, the term partial sum of squares). This also has a tendency to improve the precision of the analysis because usually it decreases the amount of measured variability in the response, which is reflected in the error mean square.

Some authors provide strong evidence for one or the other approach (see Section 6.3 of Fleiss, 1985, for a discussion). Actually, a combination of the two methods is recommended. Usually there are a few prognostic factors which are known to play an important role and should be regarded as prestratifiers. (An important example of this is the factor of clinical center in multicenter clinical trials, discussed in more detail below.) The importance of some of the remaining factors can be determined in the post-stratification phase. Also, stratum should be incorporated as a block effect in the univariate linear model because the precision of the analysis is improved when the variability due to stratum is removed.

Sometimes it is the case that the baseline measurement of the response variable is an important prognostic variable. The question that arises is how to handle this in the statistical analysis. Viable arguments can be made for either of the following two approaches: (1) the dependent variable in the analysis should be the change or relative change in the response from baseline; (2) the dependent variable is unadjusted, but the baseline value serves as a covariate in an ANCOVA model. A disadvantage of the first approach is that the patients with extreme baseline values naturally exhibit the highest degree of change and change may act independently of treatment. A disadvantage of the second is that the effect of the baseline value in the ANCOVA model is assumed to be linear and that the slope is constant across treatment groups. If the latter analysis is applied, then the assumptions of linearity and homogeneity of slopes require investigation.

The appropriate choice between approaches 1 and 2 appears to vary from clinical trial to clinical trial. Some analysts have even used the two approaches simultaneously; i.e., the change from baseline is the dependent variable and the baseline is a covariate in an ANCOVA model.

## 2. Multicenter Clinical Trials

Multicenter clinical trials have grown in importance and popularity. A multicenter clinical trial consists of a number of clinical centers cooperating under a common clinical protocol to investigate the safety and efficacy of certain therapies. If the disease of interest is relatively rare, a multicenter trial can achieve the desired number of enrolled patients in a short period of time. For a discussion of the logistical problems in the conduct of a multicenter trial, the reader is referred to Friedman et al. (1981) and Pocock (1983). The ensuing discussion concentrates on the statistical issues in the analysis of multicenter trials.

The largest source of variability in a multicenter clinical trial is the set of clinical centers itself, even though the centers follow the same clinical protocol. Therefore, an essential component of the analysis is the inclusion of a stratifying variable for center. The difficulties in the analysis are those encountered when dealing with any multifactorial ANOVA. Let us assume that there are k centers in a multicenter clinical trial, r treatments to be compared, and $n_{ij}$ patients in the jth treatment group of the ith center, $1 \leqslant i \leqslant k$ and $1 \leqslant j \leqslant r$. For simplicity in presentation, we assume that there are no other effects that should be considered in the ANOVA model, so the modeled effects consist only of center and treatment main effects and center × treatment interactions. The first problem that arises is whether the center and center × treatment effects should be regarded as fixed or random effects. Theoretically, it makes sense that center is a random effect because centers are (1) randomly selected from a population of centers and (2) not fixed a priori in terms of it being important to compare different centers (in contrast to treatments, which are fixed a priori). However, statisticians have not reached any consensus as to whether fixed or random effects are more appropriate in multicenter trials. If center is regarded as random, and hence center × treatment also, the construction of the F-tests for the center and treatment main effects are altered. The denominator mean square of the F-tests is the mean square for the center × treatment interactions rather than the error mean square (as it would be if center were regarded as a fixed effect). The consequences of this can be drastic if k is small, because the denominator degrees of freedom for the F-test are greatly reduced, as illustrated in the following ANOVA table:

| Source | Degrees of freedom | Mean square | Fixed eff. F-Test | Random eff. F-Test |
|---|---|---|---|---|
| Center | $k - 1$ | $MS_C$ | $MS_C/MS_E$ | $MS_C/MS_{C\times T}$ |
| Treatment | $r - 1$ | $MS_T$ | $MS_T/MS_E$ | $MS_T/MS_{C\times T}$ |
| Center × treatment | $(k - 1)(r - 1)$ | $MS_{C\times T}$ | $MS_{C\times T}/MS_E$ | $MS_{C\times T}/MS_E$ |
| Error | $\Sigma_i\Sigma_j(n_{ij} - 1)$ | $MS_E$ | | |

(13)

Thus, one could be penalized with a very weak statistical test of treatment equivalence when applying a random effects approach in a small multicenter trial. In fact, one could get a more powerful test with fewer patients in a one-center trial. For this reason, some statisticians avoid a random effects approach.

The second problem is one of interpretation. If the estimated center × treatment interactions are significant, then the relative ordering of the estimated treatment means is not consistent across centers. A large degree of inconsistency, e.g., a two-armed trial with the treatment A mean larger than the treatment B mean in half the centers and vice versa in the other half, could negate the pooling of the data across centers. In such a situation, a one-way ANOVA is conducted within each of the k centers, but an attempt to pool and interpret the results of the k analyses could be difficult. One possibility in a two-armed trial is to calculate the binomial probability $p = \binom{k}{s}(0.05)^s(0.95)^{k-s}$, where s is the number of centers in which treatment A is significantly better than treatment B at the 0.05 significance level. Then a value of $p < 0.05$ would indicate that s out of k significant results itself is significant. Another possibility is to perform a meta-analysis, discussed below, for combining the statistical evidence from dissimilar clinical trials.

Some researchers, regardless of whether fixed or random effects are used, drop the center × treatment interaction terms from the ANOVA if these estimates are not significant. Johnson and Gunn (1984) reported that when fixed-effect interaction estimates are regarded as insignificant in a preliminary test, the interaction terms should remain in the ANOVA so that the tests for main effects will be based on partial sums of squares. If the interactions are removed from the ANOVA and unadjusted sums of squares are used, then the tests for main effects could be biased and less powerful when the true but unknown interactions are nonnull.

Sometimes it is desirable to combine the results of similar clinical trials which do not follow the same clinical protocol. Meta-analysis refers to that branch of statistics in which the results from independent but related experiments are combined. There are two basic approaches, namely combining p-values and combining estimators. Fisher (1932) introduced an approach to combining the p-values from k independent studies. The k p-values $p_1, p_2, \ldots, p_k$ represent an independent sample from the uniform (0, 1) distribution. Then the statistic

$$-2\sum_{i=1}^{k} \log(p_i) \sim \chi^2_{2k} \qquad \text{under } H_0 = \bigcap_{i=1}^{k} (H_{0i}) \tag{14}$$

where $H_{0i}$ is the null hypothesis for the ith experiment, $1 \leqslant i \leqslant k$, and $H_0$ is the combined null hypothesis. $H_0$ is rejected for large values of this statistic.

Hedges and Olkin (1985) reviewed some of the other tests that have been proposed for combining p-values from independent experiments. The obvious advantage of this type of approach is that the results from experiments with very different designs can be combined. The major disadvantage is the lack of robustness, as one very small p-value in the presence of $k - 1$ large p-values could lead to rejection of the combined null hypothesis $H_0$.

The second general approach in meta-analysis is to construct a weighted combination of the estimators $T_1, T_2, \ldots, T_k$, where the weights are usually taken as inverses of the respective estimated standard errors $S_1$, $S_2, \ldots, S_k$. Without loss of generality, it is assumed that $T_i$, $1 \leqslant i \leqslant k$, is the statistic for testing $H_0$: $\{\theta_i = 0\}$. Then under the homogeneity assumption that $\theta_1 = \theta_2 = \cdots = \theta_k$, the test of $H_0$ is based on the asymptotically normal statistic

$$T^* = \frac{\Sigma_{i=1}^{k} w_i T_i}{\Sigma_{i=1}^{k} w_i}, \qquad w_i = 1/S_i \text{ for } 1 \leqslant i \leqslant k \tag{15}$$

Hedges and Olkin (1985) derived the conditions under which asymptotic normality is attained and they examined more elaborate combinations.

The homogeneity assumption that $\theta_1 = \theta_2 = \cdots = \theta_k$ is necessary, otherwise the weighted combination in (15) is invalid. This is analogous to requiring the absence of significant estimated center × treatment interactions in the analysis of multicenter trials. Hedges and Olkin (1985) suggested the statistic

$$H_T = \sum_{i=1}^{k} w_i (T_i - T^*)^2 \tag{16}$$

as a test of homogeneity, where $H_T$ has an asymptotic $\chi^2_{k-1}$ distribution.

## 3. Combination Drug Trials

Occasionally clinical investigators wish to evaluate combinations of drugs for safety and efficacy. Two situations in particular come to mind. In the first case, drugs A and B are to be used in combination, but the investigators conduct a dose-ranging experiment to ascertain the optimal levels of combination. The best design for this would be an $a \times b$ factorial, where a and b represent the number of doses of drugs A and B, respectively. Response surface methodology (RSM) is ideally suited for this situation, because (1) a regression model is fit to the data and (2) an interpolation is performed within the fitted response surface when searching for the optimal combination. Actually, a wide variety of designs, other than factorials, have been proposed in RSM. The reader is referred to Mead and Pike (1975) and Myers (1976) for reviews on the design and analysis of response surface experiments. Carter et al. (1984) displayed a technique for determining whether the response at the estimated optimal combination is significantly better than each drug's optimal estimated response.

A more common situation that arises is that the individual drug's optimal dose has already been determined, and a combination drug is formed from these optimal doses. Then investigators might conduct a three-armed trial consisting of the therapies A, B, and AB. RSM cannot be applied in this instance, unless a $2 \times 2$ factorial design is employed, in which a fourth arm of placebo is added to the trial, but fitting and interpolating in a regression framework is not recommended.

Rather, statisticians have resorted to an ANOVA framework instead. This is also due to the Food and Drug Administration (FDA) requirement that the combination must be shown to be significantly better than each of the individual drugs. For a discussion of the ramifications of this requirement, see Pledger (1986). In terms of null and alternative hypotheses, if $\mu_A$, $\mu_B$, and $\mu_{AB}$ represent the respective population means for a particular response variable, then we need to test

$$H_0: \{\mu_A \geqslant \mu_{AB} \cup \mu_B \geqslant \mu_{AB}\} \qquad \text{against}$$

$$\text{against} \qquad H_1: \{\mu_A < \mu_{AB} \cap \mu_B < \mu_{AB}\} \tag{17}$$

Laska and Meisner (1986) showed that under normality and $\sigma^2$ known,

$$Z = [Z_A \; Z_B] = [\hat{\mu}_{AB} - \hat{\mu}_A \; \hat{\mu}_{AB} - \hat{\mu}_B] \tag{18}$$

is sufficient for $[\mu_{AB} - \mu_A \; \mu_{AB} - \mu_B]$. Also, $Z^* = \min\{Z_A, Z_B\}/(2\sigma^2)^{1/2}$ is the uniformly most powerful test of $H_0$ against $H_1$. They suggested extensions of this to the situations in which (1) $\sigma^2$ is unknown and (2) the underlying distribution function is symmetric but nonnormal, so that rank statistics can be applied.

## 4. Active- (or Positive-) Controlled Trials

For ethical reasons, an investigator may employ an active (or positive) control in a clinical trial rather than a placebo control. Suppose treatment A is the new experimental therapy and treatment B is the standard and accepted therapy which serves as the active control. For a discussion of the issues involved in active-controlled trials, see Lamborn (1983) and Temple (1983).

In this setting, it is necessary to show that treatment A is at least equivalent to treatment B, so that the test of interest is $H_0$: $\{\mu_A < \mu_B\}$ versus $H_1$: $\{\mu_A \geqslant \mu_B\}$. Unfortunately, a valid statistical approach analogous to the hypothesis-testing approach in (17) has not been established. The main problem is that the above hypothesis-testing situation, because of the monotone likelihood ratio property of exponential family distributions, reduces to the usual test of $H_0$: $\{\mu_A = \mu_B\}$ versus $H_1$: $\{\mu_A > \mu_B\}$ and acceptance of $H_0$ does not necessarily prove equivalence, especially for small samples. In some sense, this situation is analogous to establishing the bioequivalence of two drugs. Westlake (1972, 1976, 1979) and Metzler (1974) proposed the use of confidence intervals, while Selwyn et al. (1981) developed a Bayesian approach for the bioequivalence problem.

Blackwelder (1982) suggested testing $H_0$: $\{\mu_B - \mu_A \geqslant \Delta\}$ versus $H_1$: $\{\mu_B - \mu_A < \Delta\}$ for a positive-controlled trial, although a reasonable fixed value of $\Delta$ must be determined a priori. Hsu (1983) reviewed the statistical work in the area of positive-controlled trials and suggested visual examinations of the plots of the empirical cumulative distribution functions for the two groups.

Example 1. Ten investigators were enrolled in a randomized, double-blind, multicenter study to examine the effectiveness of two competing drugs (noted as drug A, the standard, and drug B, the experimental) in the control of menopausal symptoms. All women eligible for the study were administered drug A during the first 3 weeks at one of two levels. Week four was a washout period, after which patients were randomized (within each center and level of drug A) to either a continuation of that level of drug A or to drug B during weeks 5–7. The important efficacy variable measured was weekly average of hot flush frequency. For this efficacy variable the change in response was calculated by subtracting the average of weeks 1–3 from the average of weeks 5–7. A nonparametric two-way ANOVA model (three treatments and 10 investigators) was applied to the 124 observations because the Kolmogorov-Smirnov test on the BLUS residuals, discussed in Section II,B, revealed that the sample data deviated significantly ($p < 0.05$) from normality. The nonnormality appeared to be caused by the large degree of negative skewness (some women experienced a large decrease in their weekly average score). Otherwise, the residual plots did not indicate any problems.

The rank test for center × treatment interaction was not significant ($p > 0.50$). The three groups (level 1 of drug A, level 2 of drug A, and drug B) were not significantly different from one another with respect to either response variable ($p > 0.50$). The observed median changes were 0 for level 1 of drug A, −0.1667 for level 2 of drug A, and 0 for drug B. Because this is a variation of a positive-controlled study, it would be of interest to examine simultaneous confidence intervals of the group median differences. However, it is not clear how to proceed because of the nonnormality problem. Sen (1980) and Miller (1981) discussed simultaneous inference based on linear rank statistics for the two-way model with only one observation per cell, but this example contains multiple observations per cell. If the data were normal, Scheffé's method could be applied. Simultaneous inference within the framework of the Box-Cox transformation has not been examined in the statistical literature, although conditioning on the estimated transformation parameter might be a reasonable approach to the statistical analysis of this problem (see Example 2 at the end of Section IV).

## III. MULTIVARIATE LINEAR MODELS

In this section we review multivariate linear models, in which some of the results in the previous section are generalized to situations with more than one response variable. The multivariate linear model serves as a framework for multivariate regression, multivariate analysis of variance (MANOVA), and multivariate analysis of covariance (MANCOVA). The models presented in Sections IV through VI are members of this class of models, but are concerned with repeated measurements of the same set of response variables. The latter scenarios often are encountered in clinical research.

Statisticians often question the necessity of applying a multivariate linear model to clinical trial data. For instance, suppose in a clinical trial four treatment groups are to be compared with respect to three response variables. Invoking a one-way MANOVA model might reveal that the four groups differ significantly with respect to the response variables, but the clinical investigators will want to know exactly how the groups differ and which variables are responsible. Multivariate multiple comparison procedures can be constructed to answer these questions, but the sample size will need to be large for any semblance of statistical power. Therefore, statisticians might shy away from a MANOVA model so that power is not compromised. However, as is often the case, clinical investigators identify many more response variables than is necessary. Reducing the set of response variables would alleviate the problem, so that multivariate linear models could be used more frequently (and rightfully so).

Analogous to Section II, we express the multivariate linear model in terms of a full-rank design matrix $\underline{X}$, whose components can represent main effects, interactions, block effects, regressors, and/or covariates:

$$\underline{Y} = \underline{X}\,\underline{\beta} + \underline{\varepsilon} \tag{19}$$

In this model, $\underline{Y}$ is the $n \times p$ vector of observed responses, $\underline{X}$ is the $n \times r$ design matrix of known values, $\underline{\beta}$ is the $r \times p$ vector of unknown parameters, and $\underline{\varepsilon}$ is the $n \times p$ vector of unobservable random errors. In the standard set of parametric assumptions, the rows of $\underline{\varepsilon}$ are independent and identically distributed as $N_p(\underline{0}, \underline{\Sigma})$, where $\underline{\Sigma}$ is an unknown, $p \times p$, positive definite, variance-covariance matrix. Also, we need to assume that $n \geqslant p + r$ in order for the ML estimate of $\underline{\Sigma}$ to be positive definite in probability.

Notice how the multivariate linear model generalizes the univariate linear model. The columns of $\underline{Y}$ and the corresponding columns of $\underline{\beta}$ represent p separate, but dependent, univariate linear models, in which the design matrix $\underline{X}$ is the same for each of the p response variables. Although of interest and not discussed here, it is possible to allow a different design matrix for each of the p response variables. This is known as the multiple design multivariate (MDM) linear model (called seemingly unrelated regressions in the econometrics literature). Zellner (1962) derived a two-stage generalized least-squares estimator of the unknown parameters for this model, and McDonald (1975) developed exact methods of hypothesis testing under the assumption of normality. Schwab (1984) extended the results of Puri and Sen (1969) in constructing multivariate linear rank statistics for the MDM linear model.

As can be found in any standard text on multivariate analysis (e.g., Muirhead, 1982; Anderson, 1984; Morrison, 1976; Johnson and Wichern, 1982), the ML estimator of $\underline{\beta}$ is the same as the LS estimate under these assumptions and is given by

$$\hat{\underline{\beta}} = (\underline{X}'\underline{X})^{-1}\underline{X}'\underline{Y} \tag{20a}$$

and the ML estimator of $\underline{\Sigma}$ is revised slightly to attain the unbiased estimator

$$\underline{S} = (n - r)^{-1}\underline{Y}'\{\underline{I}_n - \underline{X}(\underline{X}'\underline{X})^{-1}\underline{X}'\}\underline{Y} \tag{20b}$$

Any testable hypothesis about the treatment groups in terms of all or some of the response variables can be generalized to the form $H_0\colon \underline{C}\,\underline{\beta}\,\underline{U} = \underline{0}$, where $\underline{C}$ is a known, full-rank, $c \times r$ matrix ($c \leqslant r$) and $\underline{U}$ is a known, full-rank, $p \times u$ matrix ($u \leqslant p$). Under the same set of assumptions, the likelihood ratio test of this null hypothesis rejects for small values of

$$\prod_{l=1}^{u}(1 + \omega_l)^{-1} \tag{21}$$

where $\omega_1 \geqslant \omega_2 \geqslant \cdots \geqslant \omega_u \geqslant 0$ are the eigenvalues of the $u \times u$ random matrix $\underline{S}_H\underline{S}_E^{-1}$. The random matrices $\underline{S}_H$ and $\underline{S}_E$ represent the hypothesis and error sums of squares and cross-products matrices, respectively, and have independent Wishart distributions given by

$$\underline{S}_H = (\underline{C}\,\hat{\underline{\beta}}\,\underline{U})'\{\underline{C}(\underline{X}'\underline{X})^{-1}\underline{C}'\}^{-1}(\underline{C}\,\hat{\underline{\beta}}\,\underline{U}) \sim W_u(c, \underline{U}'\underline{\Sigma}\,\underline{U}, \underline{\Delta}) \tag{22}$$

and

$$\underline{S}_E = \underline{U}'\underline{Y}'\{\underline{I}_n - \underline{X}(\underline{X}'\underline{X})^{-1}\underline{X}'\}\underline{Y}\,\underline{U} \sim W_u(n - r, \underline{U}'\underline{\Sigma}\,\underline{U}) \tag{23}$$

where the noncentrality parameter matrix $\underline{\Delta}$ is a null matrix when $H_0$ is true.

Other multivariate tests of the null hypothesis $H_0$: $\underline{C}\,\underline{\beta}\,\underline{U} = \underline{0}$ have been proposed, all of which are based on the eigenvalues of $\underline{S}_H\underline{S}_E^{-1}$. These include Roy's largest root $(\omega_1)$, the Hotelling-Lawley trace $(\Sigma\omega_1)$, and Pillai's trace $(\Sigma\omega_i(1 + \omega_i)^{-1})$. All four of the test criteria are invariant tests, but none is uniformly most powerful invariant (see Muirhead, 1982, Section 10.6.5, for further discussion).

Roy's largest root criterion leads to a method of simultaneous inference for all double linear combinations of the form $\underline{a}'\underline{C}\,\underline{\beta}\,\underline{U}\,\underline{b}$, where $\underline{a} \in R^c$ and $\underline{b} \in R^u$ are nonnull vectors (Roy and Bose, 1953; Section 5.5 of Morrison, 1976). If $\underline{\theta} = \underline{C}\,\underline{\beta}\,\underline{U}$ and $\hat{\underline{\theta}} = \underline{C}\,\hat{\underline{\beta}}\,\underline{U} = \underline{C}(\underline{X}'\underline{X})^{-1}\underline{X}'\underline{Y}$, the simultaneous confidence region is given by the following probability statement:

$$P\left[\underline{a} \in R^c,\ \underline{b} \in R^u:\ \underline{a}'\hat{\underline{\theta}}\underline{b} - \left\{\frac{x_\alpha(\underline{b}'\underline{S}_E\underline{b})\underline{a}'\underline{C}(\underline{X}'\underline{X})^{-1}\underline{C}'\underline{a}}{(1 + x_\alpha)}\right\}^{1/2} \leqslant \underline{a}'\underline{\theta}\underline{b}\right.$$
$$\left.\leqslant \underline{a}'\hat{\underline{\theta}}\underline{b} + \left\{\frac{x_\alpha(\underline{b}'\underline{S}_E\underline{b})\underline{a}'\underline{C}(\underline{X}'\underline{X})^{-1}\underline{C}'\underline{a}}{(1 + x_\alpha)}\right\}^{1/2}\right] = 1 - \alpha \tag{24}$$

where $x_\alpha$ is the upper $\alpha$ percentage point from Roy's largest root distribution and $\underline{S}_E$ is defined in (23). This method of simultaneous inference (comparable to Scheffé's method in the univariate model) yields very wide regions because it must allow for all possible linear combinations.

When only a few of the double linear combinations are of interest, say k, a Bonferroni correction might lead to narrower intervals. Because each $\underline{a}_i'\hat{\underline{\theta}}\underline{b}_i \sim N_1(\mu_i = \underline{a}_i'\underline{\theta}\underline{b}_i,\ \sigma_i^2 = \{\underline{a}_i'\underline{C}(\underline{X}'\underline{X})^{-1}\underline{C}'\underline{a}_i\}\{\underline{b}_i'\underline{U}'\underline{\Sigma}\,\underline{U}\,\underline{b}_i\})$, $1 \leqslant i \leqslant k$, we have the following conservative probability statement:

$$P[\underline{a}_i'\hat{\underline{\theta}}\underline{b}_i - \hat{\sigma}_i t_{n-r,1-\alpha/2k} \leqslant \underline{a}_i'\underline{\theta}\underline{b}_i \leqslant \underline{a}_i'\hat{\underline{\theta}}\underline{b}_i + \hat{\sigma}_i t_{n-r,1-\alpha/2k},$$
$$1 \leqslant i \leqslant k] = 1 - \alpha \tag{25}$$

where $t_{n-r,1-\alpha/2k}$ denotes the $1 - \alpha/2k$ critical point from the t-distribution with $n - r$ degrees of freedom.

Other approaches to simultaneous inference in the multivariate linear model have been proposed, such as Roy's step-down procedure, the $T^2_{max}$ procedure, and Krishnaiah's finite intersection tests. Unfortunately, these have not been implemented much in practice and the interested reader is referred to the review papers by Krishnaiah (1979) and Krishnaiah et al. (1980).

Numerical examples for applying the MANOVA model in regression, analysis of variance, and analysis of covariance settings are provided in any applied multivariate text, such as Morrison (1976) and Johnson and Wichern (1982).

Prior to conducting any multivariate analysis with the model in (19), we should examine the data for violations of assumptions. Analogous to the univariate model, this consists of checking for independence, homogeneity of variance matrices, and multivariate normality. The usual tool for investigating all of these situations is the set of p-variate residuals, defined by

$$\underline{R} = \underline{Y} - \underline{X}\hat{\underline{\beta}} = \{\underline{I}_n - \underline{X}(\underline{X}'\underline{X})^{-1}\underline{X}'\}\underline{Y} \tag{26}$$

Gnanadesikan (1977, Section 6.4) discussed some of the uses of multivariate residuals as a diagnostic tool in the multivariate linear model.

In the one-way MANOVA classification, Bartlett's modified LR test (Bartlett, 1937) can be generalized to the multivariate case to test the hypothesis of variance homogeneity. The statistical details are available in Morrison (1976, Section 7.4) or Muirhead (1982, Section 8.2), and DISCRIM of SAS 1985 is available for the computing aspects. As in the univariate case, this test should be used cautiously as it is very sensitive to nonnormality of the data. Although extensions to the MANOVA model of the robust tests proposed by Levene (1960) and Brown and Forsythe (1974) should be straightforward, such tests are not well known.

Gnanadesikan (1977, Section 5.4.2) and Mardia (1980) have reviewed the various tests of multivariate normality. Some of these are direct extensions from the univariate case. For instance, Mardia (1970) developed multivariate measures of skewness and kurtosis, while Malkovich and Afifi (1973) applied Roy's union-intersection principle to arrive at (1) multivariate measures of skewness and kurtosis (different from those of Mardia, 1970), and (2) a multivariate version of the Shapiro-Wilk test. A goodness-of-fit test, such as the Kolmogorov-Smirnov test, can be constructed for a large sample by forming $V_i = \underline{R}_i'\underline{S}^{-1}\underline{R}_i$, $1 \leq i \leq n$, for the ith residual p-vector. Then $V_i$ has an approximate chi-square distribution with p degrees of freedom because $\underline{S}$ in (20b) converges strongly to $\underline{\Sigma}$. This means that the empirical distribution function can be compared to the hypothesized chi-square distribution.

Again, because the residuals given by $\underline{R}$ in (26) are not independent random variables, it is best to work with the p-variate BLUS residuals from the MANOVA model, which are defined by

$$\underline{R}^* = (\underline{H}'\underline{H})^{-1/2}\underline{H}'\underline{Y} \qquad \text{with } \underline{R}^* \sim N_{p(n-r)}(\underline{0}, \underline{\Sigma}\otimes\underline{I}_{n-r}) \tag{27}$$

where $\underline{H}$ is an $n \times (n - r)$ matrix formed from any $n - r$ independent columns of the $n \times n$ matrix $\underline{I}_n - \underline{X}(\underline{X}'\underline{X})^{-1}\underline{X}'$, $(\underline{H}'\underline{H})^{-1/2}$ denotes any square-root decomposition of $(\underline{H}'\underline{H})^{-1}$, and $\underline{\Sigma}\otimes\underline{I}_{n-r}$, denotes the Kronecker or direct product of $\underline{\Sigma}$ with the $n - r \times n - r$ identity matrix. The elements of $\underline{R}^*$ are independent with null expected value, so a test of normality can be applied legitimately. To apply Mardia's (1970) test to the BLUS residuals, define the rows of $\underline{R}^*$ as $\underline{R}^* = [\underline{R}_1^*, \underline{R}_2^* \cdots \underline{R}_{n-r}^*]'$, $\bar{\underline{R}}^* = (n - r)^{-1} \Sigma_{i=1}^{n-r} \underline{R}_i^*$, and $r_{ij} = (\underline{R}_i^* - \bar{\underline{R}}^*)'\underline{S}^{-1}(\underline{R}_j^* - \bar{\underline{R}}^*)$, for $1 \leqslant i,j \leqslant n - r$, and where $\underline{S}$ is the $p \times p$ matrix version of $s^2$ in (4). Multivariate measures of skewness and kurtosis are

$$b_{1,p} = (n - r)^{-2} \sum_{i=1}^{n-r} \sum_{j=1}^{n-r} r_{ij}^3 \qquad \text{and} \qquad b_{2,p} = (n - r)^{-1} \sum_{i=1}^{n-r} r_{ii}^2 \tag{28}$$

respectively. Although the residuals within each column of $\underline{R}$ in (26) sum to zero, the same is not necessarily true for the BLUS residuals in (27). Asymptotically, $(n - r)b_{1,p}/6$ has a chi-square distribution with $p(p + 1) \times (p + 2)/6$ degrees of freedom and $b_{2,p}$ has a normal distribution with mean $p(p + 2)$ and variance $8p(p + 2)/(n - r)$. The large-sample goodness-of-fit test is formed from the random variables $r_{ii}$. $1 \leqslant i \leqslant n - r$, which are independent and identically distributed as $\chi_p^2$.

Programs for conducting tests of multivariate normality on the BLUS residuals, namely Mardia's (1970) test based on multivariate skewness and kurtosis and the Kolmogorov-Smirnov goodness-of-fit test, are provided in the Appendix (MULTNORM).

There are multivariate versions of linear rank, signed rank, and aligned rank statistics and Box-Cox transformations for situations in which the data do not approximate multivariate normality. Puri and Sen (1971, 1985) and Hettmannsperger (1984) have examined in detail multivariate rank statistics. The general formulation is analogous to that discussed in Section II,C, and the reader is referred to the above texts. Unfortunately, software for their use is lacking at present (PROC MRANK in the 1983 SAS Supplement has some capabilities). The Appendix lists a program (NPARMULT) for calculating a one-way multivariate linear rank statistic, which allows for the presence of incomplete responses. This formulation is taken from Koziol et al. (1981).

The multivariate Box-Cox transformation introduces a vector of parameters $\underline{\lambda} = [\lambda_1 \; \lambda_2 \cdots \lambda_p]'$ corresponding to p positive-valued response variables, such that the ith response variable is transformed as in (12) according to $\lambda_i$, $1 \leqslant i \leqslant p$. Then the transformed $n \times p$ observation matrix $\underline{Z}$ is assumed to satisfy $\underline{Z} = \underline{X}\,\underline{\beta} + \underline{\varepsilon}$, with the rows of $\underline{\varepsilon}$ being independent, identically distributed p-variate normal random variables. Multivariate Box-Cox transformations have been suggested by Gnanadesikan (1977, Section 5.3), Mardia (1980), and Hernandez and Johnson (1980), while Rode (1986) has provided the statistical details for finding the ML estimate of $(\underline{\beta}, \underline{\Sigma}, \underline{\lambda})$ via a combined Newton-Raphson algorithm and nested minimization approach. A version of his program is contained in the Appendix (MULTBXCX), and its use is illustrated in Example 2 for a repeated measurements experiment.

As Hinkley and Runger (1984) discussed, there is some controversy in the literature as to whether the random nature of $\underline{\lambda}$ should be accounted for when conducting inference on the parameter vector $\underline{\beta}$. The discussion in Section II,C for the univariate case applies here for the multivariate case. A reasonable approach appears to be to estimate $\underline{\lambda}$ by $\hat{\underline{\lambda}}$, transform the data according to $\hat{\underline{\lambda}}$, and conduct the MANOVA conditional on $\hat{\underline{\lambda}}$. Results can be interpreted in terms of the original scale via the inverse transformation of that defined in (12) for the ith response variable, $1 \leq i \leq p$.

## IV. REPEATED MEASURES EXPERIMENTS

Many clinical trials are designed as repeated measures experiments. In the one-way repeated measures design, patients are randomized to r treatment groups and the response variable is measured on p different occasions. The crossover design, in which a patient receives different treatments on different occasions, is considered in Section V. Because the p measurements for each patient are not independent, the univariate linear model in Section II is not appropriate. The following discussion concentrates on the one-way repeated measures design, but generalizations to more complicated designs are not much more difficult and examples are considered in PROC GLM of SAS 1985 and 2V and 4V of BMDP 1983.

Let $Y_{ijk}$ denote the response of the jth individual in the ith treatment group at the kth time point, $1 \leq i \leq r$, $1 \leq j \leq n_i$, and $1 \leq k \leq p$. Using a reference cell parameterization, the model is given by

$$Y_{ijk} = \mu + \tau_i + \gamma_k + (\tau\gamma)_{ik} + \varepsilon_{ijk}, \tag{29}$$

where $\tau_1 = 0$, $\gamma_1 = 0$, and $(\tau\gamma)_{1k} = 0$ and $(\tau\gamma)_{i1} = 0$ for all i and k. The parameters $\tau_i$ represent treatment effects, the $\gamma_k$ represent time effects, and the $(\tau\gamma)_{ik}$ represent the treatment × time interactions. We assume that individuals are independent, but within each individual the random errors $[\varepsilon_{ij1} \cdots \varepsilon_{ijp}]'$ are correlated and follow a p-variate normal distribution with an unknown variance-covariance matrix $\underline{\Sigma}$, as in (19).

Ordinarily, one would proceed with a one-way MANOVA for comparing the r groups. However, because of the repeated measures design, the model is expressed as a two-factor design with respect to treatment and time as in (29). In repeated measures terminology, the set of sample means over time for each treatment group comprises that treatment's profile. The goal of the analysis, of course, is to compare the treatment profiles. Just as in a univariate two-way ANOVA, the comparison is straightforward if the profiles are parallel or, equivalently, if the treatment × time interactions are null.

Huynh and Feldt (1970) and Rouanet and Lepine (1970) discovered that the multivariate tests in a one-way repeated measures analysis reduce to F-tests if the variance covariance matrix $\underline{\Sigma}$ is of type H, described as

$$\underline{\Sigma} = \begin{bmatrix} 2\alpha_1 + \gamma & \alpha_1 + \alpha_2 & \cdots & \alpha_1 + \alpha_p \\ \alpha_1 + \alpha_2 & 2\alpha_2 + \gamma & \cdots & \alpha_2 + \alpha_p \\ \vdots & \vdots & \cdots & \vdots \\ \alpha_1 + \alpha_p & \alpha_2 + \alpha_p & \cdots & 2\alpha_p + \gamma \end{bmatrix} \tag{30}$$

There are three important equivalent conditions to (30). Notationally, let $J_{r,s}$ denote the $r \times s$ matrix of unit values, let $\lambda_1 \geqslant \cdots \geqslant \lambda_{p-1} > 0$ denote the $p - 1$ nonnull eigenvalues of $\underline{\Sigma}(\underline{I}_p - p^{-1}\underline{J}_{p,p})$, let

$$\varepsilon = \frac{(p-1)^{-1}(\Sigma_{j=1}^{p-1} \lambda_j)^2}{(\Sigma_{j=1}^{p-1} \lambda_j^2)} \tag{31}$$

and let $\underline{C}$ be the remaining $p - 1$ rows of an orthogonal matrix whose first row is proportional to $\underline{J}_{1,p}$. Then $\underline{\Sigma}$ is of type H iff $\varepsilon = 1$ iff $\lambda_1 = \cdots = \lambda_{p-1}$ iff $\underline{C}\,\underline{\Sigma}\,\underline{C}' = \lambda \underline{I}_{p-1}$. A special case of the type H matrix in (30) is the matrix of compound symmetry, also called the equivariance/equicovariance matrix, in which all the diagonal elements are equal and all the off-diagonal elements are equal ($\alpha_1 = \alpha_2 = \cdots = \alpha_p$). The matrix of compound symmetry arises from random effects assumptions, and earlier authors thought that compound symmetry was necessary and sufficient for the validity of univariate F-tests. If $\underline{\Sigma}$ is of type H, then only $p + 1$ parameters are estimated for $\underline{\Sigma}$, while $p(p + 1)/2$ are estimated for an unspecified $\underline{\Sigma}$.

The assumption of a type H matrix is an important one, and with any repeated measures data set it is desirable to conduct a preliminary test that the variance-covariance matrix is of this form. It can be shown that an equivalent hypothesis to $H_0$: $\{\underline{\Sigma}$ of type H$\}$ is $H_0$: $\{\underline{C}\,\underline{\Sigma}\,\underline{C}' = \lambda \underline{I}_{p-1}\}$, so the sphericity test (Mauchly, 1940) can be adapted to this testing situation. Unfortunately, the sphericity test is not very powerful for the sample sizes usually encountered in clinical trials, and its sensitivity to departures from normality has been noted (see Thomas, 1983, for a brief discussion of this).

Assuming that $\underline{\Sigma}$ is of type H, the univariate analysis is summarized in the following ANOVA table ($N = \Sigma_{i=1}^{r} n_i$):

| Source | | Degrees of freedom | Mean square | F-test |
|---|---|---|---|---|
| Between subjects | Treatment | $r - 1$ | $MS_{Tr}$ | $MS_{Tr}/MS_{S(Tr)}$ |
| | Subject (treatment) | $N - r$ | $MS_{S(Tr)}$ | |
| Within subjects | Time | $p - 1$ | $MS_{Ti}$ | $MS_{Ti}/MS_E$ |
| | Treatment × time | $(r - 1)(p - 1)$ | $MS_{Tr \times Ti}$ | $MS_{Tr \times Ti}/MS_E$ |
| | Error | $(p - 1)(N - r)$ | $MS_E$ | |

(32)

In case $\underline{\Sigma}$ is not quite of type H, Greenhouse and Geisser (1959) suggested an adjustment of the numerator and denominator degrees of freedom in the F-tests within subjects by the factor $\hat{\varepsilon}_{GG}$, which is found by substituting

$\underline{S}$ in (20b) for $\underline{\Sigma}$ in (31). Huynh and Feldt (1976) recommended the alternative scaling factor

$$\hat{\varepsilon}_{HF} = \min\left[1, \frac{\{N(p-1)\hat{\varepsilon}_{GG} - 2\}}{(p-1)\{N - r - (p-1)\hat{\varepsilon}_{GG}\}}\right] \tag{33}$$

However, there is no real guidance in the literature as to how "large" $\hat{\varepsilon}_{GG}$ and $\hat{\varepsilon}_{HF}$ must be in order for these adjusted F-tests to serve as reliable tests.

Under the assumption of parallel profiles, the multivariate test of equal treatment effects is the same as the univariate test (see Morrison, 1976). Most of the available software for repeated measures analysis provides the sphericity test, the univariate tests along with Greenhouse-Geisser and Huynh-Feldt corrections, and the multivariate tests under the assumption of parallelism (PROC GLM of SAS 1985, 2V and 4V of BMDP 1983).

For situations in which the sample size is too small for multivariate tests, the time points are evently spaced, and the variance-covariance matrix does not appear to be "close" to a type H matrix, Hearne et al. (1983) derived a repeated measures analysis under the following first-order autocorrelation structure:

$$\underline{\Sigma} = \sigma^2/(1-\rho^2)\begin{bmatrix} 1 & \rho & \rho^2 & \cdots & \rho^{p-1} \\ \rho & 1 & \rho & \cdots & \rho^{p-2} \\ \rho^2 & \rho & 1 & \cdots & \rho^{p-3} \\ \vdots & \vdots & \vdots & \cdots & \vdots \\ \rho^{p-1} & \rho^{p-2} & \rho^{p-3} & \cdots & 1 \end{bmatrix} \tag{34}$$

They also derived a likelihood ratio test of this autocorrelation structure.

Aside from determining an appropriate structure for $\underline{\Sigma}$, the next major step in a repeated measures analysis is comparing the treatments when the profiles are not parallel. This is analogous to the problem of significant estimated interactions faced in a univariate two-way ANOVA. In the latter situation, some statisticians prefer to conduct a series of one-way ANOVAs within the levels of one of the factors. For the one-way repeated measures design, this corresponds to a set of one-way ANOVAs within each time point. However, an alternative approach in the repeated measures design is to construct a set of $p - 1$ orthogonal contrasts based on the $p - 1 \times p$ matrix $\underline{C}$, discussed above, in the presence of a type H matrix. For instance, if $\underline{C}$ represents the orthogonal polynomials, then one can test whether the treatment groups differ significantly over time with respect to linear, quadratic, cubic, etc. effects (PROC GLM of SAS 1985 has this capability).

Assuming that the profiles are parallel, simultaneous inference with the treatment groups usually is of interest. Surprisingly, this topic also has received little attention in the literature. In a preliminary report, Jensen (1986) stated that there is no real problem when the variance-covariance matrix is of compound symmetry. In this case, simultaneous inference is conducted as one would expect, with the $MS_{S(Tr)}$ from the table in (32) serving as the denominator mean square in any of the multiple comparison procedures discussed in Section II. However, multiple comparisons for the case of a type H matrix may not generalize in the same manner. In fact, Jensen (1986) felt that only Dunnett's procedure for comparing each treatment to control is appropriate for this situation. Since it is not clear whether the usual procedures are legitimate, the safe and conservative approach of simultaneous inference is to apply Roy's union-intersection principle for the multivariate case, as discussed in Section III.

The assumptions of normality and homogeneity of variance might not be appropriate in a repeated measures data set. As in the multivariate linear model, the statistician can resort to distribution-free procedures or transformations to approximate normality. However, there is no comparable distribution-free approach for a repeated measures analysis with linear rank statistics, so only the multivariate linear rank statistics discussed in Section III are applicable.

The added flexibility of commensurable response variables in a repeated measures data set, in contrast to any multivariate data set, allows for the wide variety of parametric analyses based on different assumptions about the covariance structure. Therefore, a multivariate Box-Cox power transformation for a repeated measures data set, under the constraint that the transformation parameters satisfy $\lambda_1 = \lambda_2 = \cdots = \lambda_p$, is worth investigation. The response variables under this type of transformation would remain commensurable and the usual parametric analyses could be applied. However, the analyst should consider conducting a preliminary test of $\lambda_1 = \lambda_2 = \cdots = \lambda_p$. Users of the program MULTBXCX, listed in the Appendix, can select the usual multivariate transformation or the constrained transformation for a repeated measures data set. The program can be executed under both conditions so that the LR test of $H_0$: $\{\lambda_1 = \lambda_2 = \cdots = \lambda_p\}$ is simply twice the difference of the log-likelihood under the full and constrained models, respectively, and this has an approximate chi-square distribution with $p - 1$ degrees of freedom.

Example 2. In a study conducted at Virginia Commonwealth University by Victoria C. Strider, mean arterial blood pressure was measured in patients twice before surgery, once during surgery, and every 10 minutes for 1 hour during recovery from surgery. Thirty-three patients were randomly allocated to three treatment groups prior to the study, 11 per group. Group 1 patients served as controls and received minimal verbal communication and no touch during recovery, group 2 patients received minimal verbal communication and touch, while group 3 patients received frequent verbal communication and touch. A baseline value was calculated as a weighted average of the three measurements prior to recovery. The raw data are listed in Table 1, and the responses of interest are those measured postsurgery.

Before applying any repeated measures analysis, the MANOVA model in (19) is fit with $n = 33$, $r = 4$ (three treatment groups and the baseline value serving as a covariate), and $p = 6$ in order to investigate nonnormality of the BLUS residuals in (27). Although it is always necessary to check the adequacy of the one-slope assumption in an ANCOVA model, for the sake of simplicity in presentation such checks are not performed here for this example. Mardia's estimates of multivariate skewness and kurtosis, defined in (28), are $b_{1,p} = 22.27$ ($p < 0.0001$) and $b_{2,p} = 60.43$ ($p = 0.066$), respectively, indicating significant departure from multivariate

Table 1 Raw Data Used in Examples 2 and 3[a]

| Group | MAP0 | MAP1 | MAP2 | MAP3 | MAP4 | MAP5 | MAP6 |
|---|---|---|---|---|---|---|---|
| 1 | 78.33 | 70.00 | 64.00 | 58.67 | 65.67 | 68.33 | 78.00 |
| 1 | 111.56 | 110.67 | 115.00 | 109.33 | 104.67 | 108.00 | 101.67 |
| 1 | 105.67 | 126.67 | 119.00 | 115.67 | 115.67 | 123.67 | 118.33 |
| 1 | 86.56 | 86.33 | 79.67 | 81.33 | 76.67 | 96.00 | 90.00 |
| 1 | 78.89 | 84.33 | 65.00 | 80.00 | 97.67 | 78.67 | 76.00 |
| 1 | 95.56 | 98.67 | 97.00 | 85.33 | 92.67 | 88.67 | 90.33 |
| 1 | 98.67 | 100.00 | 105.00 | 94.67 | 94.67 | 94.33 | 90.67 |
| 1 | 85.11 | 105.67 | 106.67 | 100.67 | 97.00 | 96.67 | 85.33 |
| 1 | 83.89 | 78.00 | 91.33 | 80.00 | 97.33 | 86.67 | 82.67 |
| 1 | 93.89 | 102.67 | 95.67 | 97.00 | 96.33 | 99.67 | 97.67 |
| 1 | 79.78 | 89.00 | 97.67 | 90.33 | 94.33 | 91.67 | 100.00 |
| 2 | 109.33 | 110.00 | 87.33 | 86.00 | 82.00 | 89.00 | 80.33 |
| 2 | 115.78 | 104.00 | 93.33 | 96.00 | 89.00 | 85.67 | 85.67 |
| 2 | 74.44 | 100.00 | 100.00 | 105.00 | 100.00 | 108.33 | 97.67 |
| 2 | 98.78 | 193.00 | 148.00 | 105.33 | 105.67 | 106.33 | 102.33 |
| 2 | 89.33 | 80.33 | 91.33 | 87.00 | 83.67 | 87.00 | 87.00 |
| 2 | 90.33 | 115.00 | 81.67 | 120.67 | 118.33 | 109.67 | 113.33 |
| 2 | 77.44 | 90.67 | 93.67 | 86.00 | 87.33 | 86.00 | 87.00 |
| 2 | 104.78 | 116.33 | 115.00 | 114.67 | 106.67 | 110.33 | 114.33 |
| 2 | 106.33 | 119.67 | 124.33 | 109.33 | 100.00 | 99.00 | 98.67 |
| 2 | 83.11 | 85.67 | 99.00 | 102.33 | 113.00 | 111.33 | 100.00 |
| 2 | 76.44 | 86.67 | 93.00 | 82.00 | 107.33 | 64.67 | 64.67 |
| 3 | 97.56 | 102.33 | 127.33 | 105.33 | 83.00 | 90.67 | 86.33 |
| 3 | 90.00 | 111.00 | 119.33 | 107.00 | 112.67 | 113.00 | 112.00 |
| 3 | 98.22 | 95.33 | 79.67 | 79.33 | 83.33 | 74.00 | 86.00 |
| 3 | 86.33 | 80.67 | 81.00 | 82.67 | 83.33 | 84.67 | 72.33 |
| 3 | 118.89 | 119.00 | 113.33 | 105.00 | 117.33 | 111.67 | 100.33 |
| 3 | 83.89 | 79.00 | 79.33 | 78.33 | 93.00 | 73.67 | 94.33 |
| 3 | 89.00 | 114.00 | 110.67 | 111.67 | 100.33 | 110.00 | 99.67 |
| 3 | 125.00 | 105.33 | 103.00 | 105.33 | 97.67 | 98.33 | 100.67 |
| 3 | 86.11 | 84.00 | 75.33 | 79.67 | 69.67 | 79.67 | 106.67 |

Table 1 (continued)

| Group | MAP0 | MAP1 | MAP2 | MAP3 | MAP4 | MAP5 | MAP6 |
|---|---|---|---|---|---|---|---|
| 3 | 104.78 | 108.00 | 106.67 | 99.67 | 108.67 | 97.33 | 95.33 |
| 3 | 84.00 | 76.33 | 90.67 | 90.33 | 81.67 | 89.00 | 93.00 |

[a]Taken from the experiment by Victoria C. Strider, consisting of mean arterial pressure measured at baseline (MAP0) and six more times at 10-minute intervals (MAP1–MAP6) for three treatment groups.

normality. Not surprisingly, Bartlett's test indicates that there is significant heterogeneity among the estimated variance matrices for the three groups ($p = 0.03$), which could be due to the sensitivity of this test to nonnormality.

In either case, the data exhibit nonnormality and heterogeneity of variance matrices. Application of the multivariate Box-Cox transformation reveals that the estimated transformation parameters are

$$\hat{\underline{\lambda}} = [-1.4725 \ 0.175564 \ 1.38842 \ 1.52092 \ 1.00627 \ 1.09126]'$$

The transformed data, based on $\hat{\underline{\lambda}}$, do not reveal the same assumption violations as the untransformed data. Mardia's estimates of multivariate skewness and kurtosis are $b_{1,p} = 12.3253$ ($p = 0.3470$) and $b_{2,p} = 44.8613$ ($p = 0.8058$), respectively, and the hypothesis of homogeneous variance-covariance matrices cannot be rejected ($p = 0.2513$) when Bartlett's test is applied.

However, the wide range of power transformations causes a few problems. Intuitively, it is not appealing that such different power transformations are needed for the same measurements taken within 1 hour. A transformation under the constraint $\lambda_1 = \lambda_2 = \cdots = \lambda_6$ was attempted, but the program MULTBXCX in the Appendix did not converge after 100 iterations with this constrained model, which could be due to the fact that the individual parameter estimates are so different that a fit of this model is not possible.

A more important concern is that after the full multivariate transformation the response variables are no longer commensurable over time. Thus, a repeated measures analysis, as discussed in this section, cannot be applied conditionally to the transformed data. Therefore, the analysis of the transformed data proceeds as that for the multivariate linear model discussed in Section III, with no effects for time and group × time interactions. The LR test indicated that the treatment groups did not differ significantly ($p = 0.609$), whereas the baseline value was significant ($p = 0.027$).

## V. CROSSOVER EXPERIMENTS

The standard version of a crossover experiment consists of a repeated measures design with each experimental unit subjected to each treatment.

Crossovers have great appeal as an experimental design in clinical trials because usually fewer patients are needed and each patient serves as his/her own control. However, the problems associated with poorly designed crossovers can outweigh the benefits, and Brown (1980) has investigated the efficiency aspects of crossover versus parallel designs. The order in which a patient receives the treatments can be extremely important, especially if effects of treatments from previous time periods carry over into the current time period. These carryover effects could invalidate the statistical analysis in certain crossover designs. Incorporating lengthy washout periods and/or baseline measurements between treatment administrations can eliminate the impact of carryover or residual effects on the analysis, but researchers do not always heed this advice. Due to the confounding of treatment and carryover effects in some crossover designs, many clinical trials have proved to be worthless in terms of statistically comparing the treatments. In such situations, only the data from the first period of treatment can by analyzed.

For the moment, we consider the r-treatment, r-period crossover design; i.e., each patient receives each of the r treatments during r periods of treatment administration. The order of treatment administration for a patient is called the sequence of treatments for that patient. Within the jth period, $1 \leqslant j \leqslant r$, it is possible that there are carryover effects from treatments administered during periods 1, 2, . . ., $j - 1$. In practice, we usually only consider the first-order carryover effects from period $j - 1$. There are two reasons for this: (1) if the first-order carryover effects have a negligible impact on the analysis, then ordinarily the higher-order carryover effects will do likewise; (2) the designs needed for eliminating the confounding between higher-order carryover effects and treatment effects are very cumbersome and not practical. Of course, when $r = 2$, we can have at most first-order carryover effects.

Designs for the r-treatment, r-period crossover are conveniently expressed in terms of Latin squares. For example, if $r = 4$ and A, B, C, and D represent the treatments, then

| | | Period | | | |
|---|---|---|---|---|---|
| | | 1 | 2 | 3 | 4 |
| Sequence | 1 | A | B | C | D |
| | 2 | B | D | A | C |
| | 3 | C | A | D | B |
| | 4 | D | C | B | A |

(35)

is a design employing four sequences of treatment (patients are randomized directly to sequences). The Latin square in (35) is specifically constructed so that each treatment precedes every other treatment the same number of times (once). When this occurs, the crossover design is said to be balanced with respect to first-order carryover effects. When r is an even number, only one Latin square is needed to achieve balance (r sequences); when r is odd, two Latin squares are needed (2r sequences). An example of this when $r = 3$ is as follows:

| | | Period 1 | Period 2 | Period 3 | | | Period 1 | Period 2 | Period 3 |
|---|---|---|---|---|---|---|---|---|---|
| Sequence | 1 | A | B | C | Sequence | 4 | A | C | B |
| | 2 | C | A | B | | 5 | B | A | C |
| | 3 | B | C | A | | 6 | C | B | A |

(36)

The design in (36) has each treatment preceding every other treatment twice, so it is balanced with respect to first-order carryover effects.

For the two-period crossover, Grizzle (1965) and Hills and Armitage (1979) considered parametric analyses, and Koch (1972) and Taulbee (1982) discussed nonparametric analyses. Fisher and Wallenstein (1981) and Laska et al. (1983) have investigated the statistical analysis for a larger class of crossover designs. Typically, a random effects model is employed for the r-treatment, r-period, s-sequence crossover experiment:

$$Y_{ijk} = \mu + \alpha_i + \beta_{j(i)} + \gamma_k + \tau_{f(i,k)} + \rho_{f(i,k-1)} + \varepsilon_{ijk} \tag{37}$$

where $\mu$ is the grand mean, $\alpha_i$ the ith sequence effect, $1 \leqslant i \leqslant s$, $\beta_{j(i)}$ the random effect of the jth subject within the ith sequence, $1 \leqslant j \leqslant n_i$, $\gamma_k$ the effect of the kth period, $\tau_{f(i,k)}$ the effect of the treatment administered during the kth period of the ith sequence, $\rho_{f(i,k-1)}$ the carryover effect from the treatment administered during the $(k - 1)$st period of the ith sequence, and $\varepsilon_{ijk}$ the random error. The $\beta_{j(i)}$ are assumed to be independent and identically distributed null mean random variables, as are the $\varepsilon_{ijk}$, and the $\beta_{j(i)}$ and the $\varepsilon_{ijk}$ are assumed to be independent. Not all of the effects described in this model are estimable, because the treatment and carryover effects are functions of the sequence $\times$ period interactions. The classical approach has been to assume that the sequence effects are null and test for equal carryover effects by summing each subject's observations and comparing these sums via a one-way ANOVA. If the test of equal carryover effects cannot be rejected, then the analysis proceeds in a straightforward manner with a revision of the model in (37) because the treatment effects become estimable. If the test of equal carryover effects is rejected, then only the data from the first period can be used to test for treatment effects. As noted by Brown (1980), however, this test of equal carryover effects is not very powerful, and failure to reject the null hypothesis is no guarantee that the treatment effects are estimable.

The assumptions of the model in (37) lead to a variance-covariance matrix $\underline{\Sigma}$ which is of compound symmetry. Appropriate multivariate methods for the analysis of treatment and carryover effects for an unrestricted $\underline{\Sigma}$ have not been well developed except for the asymptotic case.

Because of the problems associated with the statistical analysis of the r-treatment, r-period crossover experiment, researchers have started to investigate more elaborate crossover designs with (1) fewer treatments than periods, (2) baseline measurements between treatment administrations, and/or (3) repeated measurements within a period. The model in (37) can be modified easily to reflect these situations. For instance, Ebbutt (1984) examined some three-period, two-treatment crossover designs.

Designs which deserve particular attention are (1) strongly balanced uniform designs and (2) balanced uniform designs augmented with another period, discussed by Laska et al. (1983). A crossover design is said to be strongly balanced if each treatment precedes every other treatment, including itself, the same number of times. Such a design is constructed by repeating the last period in a balanced design. An example of a strongly balanced design, based on the balanced design in (35), is given by

| | | Period | | | | |
|---|---|---|---|---|---|---|
| | | 1 | 2 | 3 | 4 | 5 |
| Sequence | 1 | A | B | C | D | D |
| | 2 | B | D | A | C | C |
| | 3 | C | A | D | B | B |
| | 4 | D | C | B | A | A |

(38)

A crossover design is said to be uniform if (1) within each period, the same number of subjects are assigned to each treatment, and (2) within each subject, each treatment appears in the same number of periods. So although the design in (38) may be strongly balanced, it is not uniform. An example of a strongly balanced uniform design for two treatments is the set of sequences {ABBA, BAAB, AABB, BBAA} in which n patients are assigned to each sequence.

For convenience, let $\Omega_{t,N,p}$ denote the class of crossover designs with t treatment, N subjects, and p periods. Laska et al. (1983) claimed that for $p > t$, a strongly balanced uniform design in $\Omega_{t,N,p}$ is universally optimal (Kiefer, 1975) in $\Omega_{t,N,p}$, and that regardless of the presence of baseline measurements, treatment and carryover effects are estimable. Also, the universally optimal property holds for a design in $\Omega_{t,N,p}$ with $p > t$ in which (1) the first $p - 1$ periods are univorm and balanced, and (2) the pth period is a replicate of the $(p - 1)$st period. An example of an optimal design in $\Omega_{2,2n,5}$ is obtained from the latter situation as n patients assigned to each of the two sequences ABBAA and BAABB (because {ABBA, BAAB} is uniform and balanced).

## VI. MULTIVARIATE GROWTH AND DOSE-RESPONSE MODELS

In a repeated measures experiment, it might be of interest to fit a functional curve to the responses at the p time points. For example, a linear or quadratic curve could describe (1) height and/or weight in children, measured over time, or (2) physiologic response in patients, measured over increasing doses of the same drug (assuming negligible carryover effects). The standard approach is to model polynomial curves, but any function which is linear in terms of the unknown parameters will suffice (e.g., Bryant and Gillings, 1985, and Schwiderski, 1986, fit splines instead of polynomials).

The multivariate model which accommodates this experimental situation is the generalized multivariate analysis of variance (GMANOVA) model. As the name suggests, it is a generalization of the MANOVA model discussed in Section III. The current version was introduced by Potthoff and Roy (1964) and is given by

$$\underline{Y} = \underline{X}\,\underline{\beta}\,\underline{P} + \underline{\varepsilon} \tag{39}$$

where $\underline{Y}$ is the n × p matrix of observations, $\underline{X}$ the n × r between-units design matrix of full rank r, $\underline{\beta}$ the r × q unknown parameter matrix, $\underline{P}$ the q × p within-units design matrix of full rank q ($q \leqslant p$), and $\underline{\varepsilon}$ the n × p matrix of unobservable random errors. The usual assumptions about $\underline{\varepsilon}$ are in place; i.e., the rows of $\underline{\varepsilon}$ are independent and identically distributed as $N_p(\underline{0}, \underline{\Sigma})$, where $\underline{\Sigma}$ is an unknown, p × p, positive definite variance-covariance matrix. It is necessary that $n \geqslant p + r$ for the ML estimate of $\underline{\Sigma}$ to be positive definite in probability.

Rao (1967) and Khatri (1966) found the ML estimator of $\underline{\beta}$ to be

$$\hat{\underline{\beta}} = (\underline{X}'\underline{X})^{-1}\underline{X}'\underline{Y}\,\underline{S}^{-1}\underline{P}'(\underline{P}\,\underline{S}^{-1}\underline{P}')^{-1} \tag{40}$$

where $\underline{S}$ is defined as in (20b). Originally, Potthoff and Roy (1964) derived a similar estimator with a matrix $\underline{G}$, either nonstochastic or independent of $\underline{Y}$, replacing $\underline{S}$ in equation (40). Rao (1965, 1967) proved the inefficiency of the Potthoff-Roy estimator.

The ML estimator $\hat{\underline{\beta}}$ in (40) does not have a normal distribution unless $\underline{P}$ is a square matrix ($q = p$), in which case the ML estimator in the GMANOVA model reduces to the ML estimator in (20a) postmultiplied by $\underline{P}^{-1}$. Under the assumption of the patterned covariance matrix given in (41), however, the ML estimator of $\hat{\underline{\beta}}$ is given in (42) and is normally distributed:

$$\underline{\Sigma} = \sigma^2\underline{I}_p + \underline{P}'\underline{\Sigma}_1\underline{P} + \underline{Q}'\underline{\Sigma}_2\underline{Q} \tag{41}$$

$$\hat{\underline{\beta}} = (\underline{X}'\underline{X})^{-1}\underline{X}'\underline{Y}\,\underline{P}'(\underline{P}\,\underline{P}')^{-1} \tag{42}$$

In (41), $\underline{\Sigma}_1$ is a full-rank, q × q, positive definite matrix, $\underline{\Sigma}_2$ is a full-rank p − q × p − q, positive definite matrix, and $\underline{Q}$ is a known, full-rank, p − q × q matrix with $\underline{P}'\underline{Q} = \underline{0}$. Lee and Geisser (1972) and Chinchilli and Carter (1984) derived the LR test of this particular variance structure for the cases of one response variable and more than response variable, respectively, and Reinsel (1982) examined inference on the parameter matrix $\underline{\beta}$ under this patterned structure. Grizzle and Allen (1969) labeled the ML estimators in (40) and (42) the fully weighted and unweighted estimators, respectively. Actually, there are situations between these two extremes, called partially weighted estimators, based on more general types of variance structures. The interested reader is referred to Rao (1967), Grizzle and Allen (1969), and Verbyla (1986).

Grizzle and Allen (1969) derived a goodness-of-fit test for the GMANOVA model based on the LR criterion for testing $H_0$: {the model is given by (39)} versus $H_1$: {the model is given by (19)}. Prior to any inference on the parameter matrix $\underline{\beta}$, it is extremely important that the selected dimension (q) of the GMANOVA model is appropriate. Otherwise, rejection of a null general linear hypothesis about $\underline{\beta}$ could be due to lack of fit (Schwiderski, 1986). In terms of fitting polynomials, Grizzle and Allen (1969) suggested a sequential approach of trying q = 1, 2, 3, etc. until a good fit is attained.

The LR test is analogous to that described in Section III, in which the model and error random matrices are given by

$$\underline{S}_H = \underline{Q}\,\underline{Y}'\underline{X}(\underline{X}'\underline{X})^{-1}\underline{X}'\underline{Y}\,\underline{Q}' \sim W_{p-q}(r, \underline{Q}\,\underline{\Sigma}\,\underline{Q}', \underline{\Delta}) \tag{43}$$

and

$$\underline{S}_E = \underline{Q}\,\underline{Y}'\{\underline{I}_n - \underline{X}(\underline{X}'\underline{X})^{-1}\underline{X}'\}\underline{Y}\,\underline{Q}' \sim W_{p-q}(n - r, \underline{Q}\,\underline{\Sigma}\,\underline{Q}') \tag{44}$$

Tests of $H_0$: $\underline{C}\,\underline{\beta}\,\underline{U} = \underline{0}$ are conducted in the same manner as in Section III for the multivariate linear model (where now $\underline{U}$ is q × u instead of p × u), even though the ML estimator of $\underline{\beta}$ in (40) is not normally distributed. The only difference between the MANOVA and GMANOVA models in this instance is that the $\underline{S}_H$ matrix in the GMANOVA model has a Wishart distribution under the null hypothesis but not under the alternative hypothesis (Grizzle and Allen, 1969). The $\underline{S}_H$ and $\underline{S}_E$ matrices are

$$\underline{S}_H = (\underline{C}\,\hat{\underline{\beta}}\,\underline{U})'\{\underline{C}\,\underline{R}\,\underline{C}'\}^{-1}(\underline{C}\,\hat{\underline{\beta}}\,\underline{U}) \sim W_u(c, \underline{U}'(\underline{P}\,\underline{\Sigma}^{-1}\underline{P}')^{-1}\underline{U}) \tag{45}$$

and

$$\underline{S}_E = \underline{U}'(\underline{P}\,\underline{S}^{-1}\underline{P}')^{-1}\underline{U} \sim W_u(n - r, \underline{U}'(\underline{P}\,\underline{\Sigma}^{-1}\underline{P}')^{-1}\underline{U}) \tag{46}$$

where

$$\underline{R} = (\underline{X}'\underline{X})^{-1} + (\underline{X}'\underline{X})^{-1}\underline{X}'\underline{Y}\{\underline{S}^{-1} - \underline{S}^{-1}\underline{P}'(\underline{P}\,\underline{S}^{-1}\underline{P}')^{-1}\underline{P}\,\underline{S}^{-1}\}\underline{Y}'\underline{X}(\underline{X}'\underline{X})^{-1} \tag{47}$$

Roy's largest root criterion leads to a method of simultaneous inference for all double linear combinations of the form $\underline{a}'\underline{C}\,\underline{\beta}\,\underline{U}\,\underline{b}$, where $\underline{a} \in R^c$ and $\underline{b} \in R^u$ are nonnull vectors. Khatri (1966) applied this criterion (see Roy and Bose, 1953) to the GMANOVA model. If $\underline{\theta} = \underline{C}\,\underline{\beta}\,\underline{U}$ and $\hat{\underline{\theta}} = \underline{C}\,\hat{\underline{\beta}}\,\underline{U}$, where $\hat{\underline{\beta}}$ is given in (40), the simultaneous confidence region is described by the following probability statement:

$$P\Bigg[\underline{a} \in R^c, \underline{b} \in R^u: \underline{a}'\hat{\underline{\theta}}\underline{b} - \left\{\frac{x_\alpha(\underline{b}'\underline{S}_E\underline{b})(\underline{a}'\underline{C}\,\underline{R}\,\underline{C}'\underline{a})}{(1 + x_\alpha)}\right\}^{1/2} \leqslant \underline{a}'\underline{\theta}\underline{b}$$

$$\leqslant \underline{a}'\hat{\underline{\theta}}\underline{b} + \left\{\frac{x_\alpha(\underline{b}'\underline{S}_E\underline{b})(\underline{a}'\underline{C}\,\underline{R}\,\underline{C}'\underline{a})}{(1 _ x_\alpha)}\right\}^{1/2}\Bigg] = 1 - \alpha \tag{48}$$

where $x_\alpha$ is the upper α percentage point from Roy's largest root distribution, $\underline{S}_E$ is defined in (46), and $\underline{R}$ is defined in (47). This method of simultaneous inference (comparable to Scheffé's method in the univariate model) yields very wide regions because it must allow for all possible linear combinations.

The Appendix lists two separate programs (GCCOVPAT and GMANOVA) for statistical analysis within the GMANOVA framework. GCOVPAT provides the LR test that the variance-covariance matrix is of the structure given in (41). Acceptance of the null hypothesis in this instance means that the analysis proceeds as that for the multivariate linear model after a transformation from $\underline{Y}$ to $\underline{Y}\,\underline{P}'(\underline{P}\,\underline{P}')^{-1}$. GMANOVA conducts weighted estimation and hypothesis testing along with the goodness-of-fit test proposed by Grizzle and Allen (1969).

Puri and Sen (1985, Section 8.4) applied multivariate linear rank statistics to conduct hypothesis testing in the GMANOVA model.

Example 3. Consider the repeated measures data set of Example 2. The investigator was interested in describing the decrease of mean arterial pressure over time, so that the GMANOVA model could be applied with only a linear component for the design within experimental units. However, as discussed in Example 2, a multivariate Box-Cox transformation was needed and this destroyed the commensurability of the response variables. Therefore, fitting a growth curve over time to the transformed data is meaningless.

If the constrained transformation had been adequate, the GMANOVA model with a linear polynomial could have been fit and the Grizzle-Allen (1969) goodness-of-fit test conducted to determine whether a higher-degree polynomial was needed. Once the appropriate degree has been determined, the hypothesis of treatment equivalence can be tested. Incidentally, the covariate of baseline mean arterial pressure is modeled with the same degree polynomial as the growth curves of the treatment groups. An alternative approach to this is the mixed MANOVA-GMANOVA model considered by Chinchilli and Elswick (1985).

## VII. INCOMPLETE DATA

Incomplete data, in terms of patient withdrawals and missed patient visits, are an unfortunate aspect of clinical trials conducted longitudinally. In addition to the increased complexity this causes in the statistical analysis, if the data are not missing "at random," the analysis can be biased (Rubin, 1976). For instance, suppose that treatments A and B are compared in a controlled, parallel design clinical trial and that patient visits are scheduled once a month for eight consecutive months. If treatment A is effective after a few months and patients start feeling better with this treatment, then some of them might discontinue their treatment and withdraw after a few months. This will bias the results of the study against treatment A. On the other hand, suppose treatment B is just as effective as treatment A, but it has undesirable side effects. Patients on this treatment might not adhere to their dose regimen and withdraw from the study, so their condition does not improve. This will bias the efficacy analysis in favor of treatment A. Practically speaking, it may never be possible to remove entirely these types of bias from clinical trials. However, clinicians and statisticians should be aware of such potential biases in longitudinal clinical trials and cope with them in the best manner possible.

This section reviews incomplete data problems within the context of missing values on the response variables in the MANOVA and GMANOVA models. It is always assumed that the design matrix $\underline{X}$, regardless of the model, is complete. Of course, the easiest approach to this problem is to delete any observation vector of p responses in which at least one of the responses is missing. This process, known as listwise deletion, is adequate if only a few of the observation vectors are deleted. Obviously, it will not be satisfactory for situations in which many of the observation vectors are missing one or more responses. Another practice, common among repeated measurements designs in clinical trials, is to estimate the missing responses for a withdrawn patient by the most recent observation on that patient. This also is not entirely satisfactory in terms of conducting an appropriate statistical analysis, because the bias introduced could be very large.

McDonald (1971) proposed estimating the missing data in the univariate and multivariate linear models through the known design matrix $\underline{X}$. The missing values are estimated by those values which minimize the trace of the error matrix $\underline{S} = \underline{Y}'\{\underline{I}_n - \underline{X}(\underline{X}'\underline{X})^{-1}\underline{X}'\}\underline{Y}$. Now these estimates replace the missing values in the observation matrix $\underline{Y}$ and the ML estimation of $(\underline{\beta}, \underline{\Sigma})$ is conducted with the "complete" observation matrix $\underline{Y}$. There is no gain in efficiency for the univariate case, as the ML estimator of $\underline{\beta}$ is exactly the same as if only the nonmissing values were employed in the likelihood function. The same is true in the multivariate case if $\underline{Y}$ is partitioned into the $m \times p$ and $n - m \times p$ matrices $\underline{Y}_1$ and $\underline{Y}_2$, respectively, where $\underline{Y}_1$ is unobserved and $\underline{Y}_2$ is observed completely. For the typical incomplete-data situation in the MANOVA model, however, this will not be true and McDonald's procedure can be very difficult to apply.

Hartley and Hocking (1971) proposed an exact, but iterative, likelihood approach for the multivariate one-sample problem with incomplete data, and Srivastava (1985) extended it to the multivariate linear model. It involves partitioning the n observations into K groups of similar patterns of complete and incomplete data.

The likelihood function for the entire sample is written as the product of likelihoods based on the K groups of observations. An iterative algorithm, such as the Newton-Raphson algorithm or the scoring method, can be invoked in an attempt to find the ML estimators of $\underline{\beta}$ and $\underline{\Sigma}$, but convergence problems can arise if the number of observations within each pattern is very small. In the following simple example with $n = 10$ and $p = 4$, there are $K = 4$ different patterns:

| | | Response | | | |
|---|---|---|---|---|---|
| Pattern | i | 1 | 2 | 3 | 4 |
| 1 | 1 | √ | √ | √ | √ |
| | 2 | √ | √ | √ | √ |
| 2 | 3 | √ | √ | | √ |
| | 4 | √ | √ | | √ |
| | 5 | √ | √ | | √ |
| 3 | 6 | √ | | √ | √ |
| | 7 | √ | | √ | √ |
| | 8 | √ | | √ | √ |
| 4 | 9 | √ | | | √ |
| | 10 | √ | | | √ |

(49)

The EM algorithm (Dempster et al., 1977), and its prototype the missing information principle (Orchard and Woodbury, 1972), is also an exact but iterative procedure. There are two steps within each iteration:

the E-step (expectation) and the M-step (maximization). In the general framework for the exponential family of distributions, let $\hat{\underline{\phi}}_{(h)}$ denote the parameter estimate of $\underline{\phi}$ at the hth iteration and $\underline{T}(\underline{Y})$ denote a sufficient statistic for $\underline{\phi}$ based on the observation matrix $\underline{Y}$. In the E-step $\underline{T}(\underline{Y})$ is estimated by $\underline{T}_{(h)}$, which is its conditional expectation given the nonmissing part of $\underline{Y}$ and $\hat{\underline{\phi}}_{(h)}$. In the M-step $\hat{\underline{\phi}}_{(h+1)}$ is found as that value for which the conditional expectation of $\underline{T}(\underline{Y})$, given $\hat{\underline{\phi}}_{(h+1)}$, equals $\underline{T}_{(h)}$. Convergence is attained when $\hat{\underline{\phi}}_{(h)} \approx \hat{\underline{\phi}}_{(h+1)}$ for some h, which yields that $\hat{\underline{\phi}}_{(h+1)}$ is taken as the ML estimate of $\underline{\phi}$. In terms of the MANOVA model, $\underline{\phi} = (\underline{\beta}, \underline{\Sigma})$ and $\underline{T}(\underline{Y}) = (\hat{\underline{\beta}}, \hat{\underline{\Sigma}})$, where $\hat{\underline{\beta}}$ is defined in (20) and $\hat{\underline{\Sigma}} = (n - r)^{-1}\underline{S}$. Although the EM algorithm displays slow convergence in comparison to other iterative routines (Wu, 1983), it has some appealing statistical and computational properties. Liski (1985) applied the EM algorithm to the incomplete data problem in the GMANOVA model.

Kleinbaum (1973) derived a genralized least-squares method for incomplete responses within the GMANOVA model framework, which he called the generalized growth curve model. He converted the GMANOVA model to a univariate model of the form $\underline{Z} = \underline{A}\,\underline{\gamma} + \underline{\varepsilon}$, where $\underline{Z}$ is the N-vector collection of nonmissing values in $\underline{Y}$, $\underline{A}$ is the $N \times qr$ design matrix constructed from $\underline{X}$ and $\underline{P}$ of the GMANOVA model in (39), $\underline{\gamma} = \text{vec}(\underline{\beta})$ is the qr-vector of parameters constructed from stacking the columns of $\underline{\beta}$ in (39) under one another, and $\underline{\varepsilon}$ is the unobservable random error N-vector with $\underline{\varepsilon} \sim N_N(\underline{0}, \underline{\Omega})$. The dimension $N = n_1p_1 + \cdots + n_Kp_K$, where $n_g$ is the sample size for the gth pattern of missing data and $p_g$ is the number of nonmissing responses for this pattern, $1 \leq g \leq K$. $\underline{\Omega}$ is a block diagonal matrix and a function of the unknown $\underline{\Sigma}$. Then the generalized least-squares estimator of $\underline{\gamma}$ is given by

$$\hat{\underline{\gamma}} = (\underline{A}'\hat{\underline{\Omega}}^{-1}\underline{A})^{-1}\underline{A}'\hat{\underline{\Omega}}^{-1}\underline{Z} \tag{50}$$

where $\hat{\underline{\Omega}}$ is any consistent estimator of $\underline{\Omega}$, defined via some function of a consistent estimator $\hat{\underline{\Sigma}}$ of $\underline{\Sigma}$. Although Kleinbaum has proved that the two-stage estimator is a best asymptotically normal (BAN) estimator, it might be even better to iterate between estimates of $\hat{\underline{\gamma}}$ and $\hat{\underline{\Sigma}}$ until covergence is attained. Kleinbaum's approach has been extended to incomplete data problems in larger classes of models: (1) Elswick (1985) developed it for the mixed MANOVA-GMANOVA model of Chinchilli and Elswick (1985); (2) Fairclough and Helms (1983) developed it for the MDM linear model discussed in Section III.

To apply Kleinbaum's approach, however, a satisfactory method for calculating a consistent estimator of $\underline{\Sigma}$ is needed. Engelman (1980) surveyed the various options in this respect, such as estimating $\sigma_{jj'}$ from all nonmissing pairs of observations (covpair covariance) or from all possible nonmissing observations (allvalue covariance), $1 \leq j, j' \leq p$. In either approach, the resultant $\hat{\underline{\Sigma}}$ could be an indefinite matrix. In a different context, Bock and Peterson (1975) suggested a smoothing technique for $\hat{\underline{\Sigma}}$ in which every negative eigenvalue of $\hat{\underline{\Sigma}}$ is replaced by zero, and then $\hat{\underline{\Sigma}}$ is recalculated from the original eigenvectors and the revised eigenvalues. The revised estimate is at least nonnegative definite. All of the above methods are available in 8D and AM of BMDP 1983, but only for the one-way MANOVA design. Schwertman and Allen (1979) and Leeper and Woolson (1982) investigated the smoothed estimator within the context of Kleinbaum's approach.

For each of the likelihood procedures discussed in this section, it is straightforward to construct a Wald-type test for $H_0$: $\underline{C}\,\underline{\beta}\,\underline{U} = \underline{0}$, which will have an asymptotic chi-squared distribution with cu degrees of freedom. This is because $(\underline{U}' \otimes \underline{C})\hat{\underline{\gamma}}$ is the estimator of vec($\underline{\beta}$) and has the following approximate multivariate normal distribution:

$$(\underline{U}' \otimes \underline{C})\hat{\underline{\gamma}} \sim N_{cu}[(\underline{U}' \otimes \underline{C})\underline{\gamma}, \{(\underline{U}' \otimes \underline{C})(\underline{A}'\underline{\Omega}^{-1}\underline{A})^{-1}(\underline{U} \otimes \underline{C}')\}] \tag{51}$$

If the data are complete, the Wald test statistic reduces to the Hotelling-Lawley trace.

The Appendix lists a program (GGCM) for calculating the iterated version of Kleinbaum's generalized least-squares estimates. The program uses a smoothed estimate of the variance-covariance matrix.

Hosking (1981) ran a simulation comparing the Hartley-Hocking approach, the EM algorithm, and Kleinbaum's approach. He claimed that the EM algorithm and Kleinbaum's technique were roughly equivalent in convergence rates and performance, whereas the Hartley-Hocking approach did not always converge.

The methods for handling incomplete data in multivariate models which have been presented here are very general in their application. Some special methods are available for specific types of incomplete-data problems. For instance, Mudholkar and Subbaiah (1980) proposed a step-down procedure when the missing data form a hierarchical structure by design. Eaton and Kariya (1983) examined situations in which locally most powerful invariant tests are possible in the multivariate one-sample problem with incomplete data.

Another class of techniques for the analysis of incomplete data in the GMANOVA model arises by making random effects assumptions about $\underline{\beta}$, the parameter matrix in (39), in a manner analogous to that in Rao (1965). This results in a structured covariance matrix, which may not always be appropriate, but it does permit irregular time points of observations for individuals. The latter situation is a frequent occurrence in repeated measures clinical trials, as patient visits do not always correspond to the scheduled times of the clinical protocol. Laird and Ware (1982) and Ware (1985) used the EM algorithm as a solution to this problem, whereas Hui (1984) applied iteratively reweighted least squares and Vonesh and Carter (1987) and Vonesh and Story (1986) took an estimated generalized least-squares approach.

Example 4. Eckberg et al. (1986) examined the parasympathetic responses of electrocardiographic R-R intervals as affected by increasing doses of nitroprusside and phenylephrine in 10 diabetic and 12 nondiabetic young adults. Only the analysis of the phenylphrine portion of the study is presented here. After the baseline measurement, phenylphrine was administered in 10-minute infusion periods with increasing concentrations of 0.2, 0.4, 0.8, and 1.6 μg/kg per minute. Two readings of the R-R interval were taken during each 10-minute period. The raw data are presented in Table 2. The investigators were interested in comparing the two groups with respect to the change from baseline by applying a MANOVA model as in (19). Notice that only 8 of the 22 subjects provided a complete set of observations (infusion ceased when certain clinical signs appeared, such as a change of more than 15 mm Hg in diastolic pressure).

Table 2 Raw Data Used in Example 4[a]

| | | 0.2 μg/kg | | 0.4 μg/kg | | 0.8 μg/kg | | 1.6 μg/kg | |
|---|---|---|---|---|---|---|---|---|---|
| Diabetic | RR0 | RR1 | RR2 | RR3 | RR4 | RR5 | RR6 | RR7 | RR8 |
| Yes | 664 | 657 | 671 | 684 | 702 | 727 | 769 | 868 | 942 |
| Yes | 987 | 1054 | 1148 | 1090 | 1130 | | | | |
| Yes | 745 | 723 | 748 | 883 | 911 | | | | |
| Yes | 991 | 981 | 1021 | 1050 | 1001 | | | | |
| Yes | 1263 | 1254 | 1283 | 1319 | 1235 | 1143 | 1179 | 1321 | |
| Yes | 1015 | 983 | 1019 | 1042 | 1037 | 1092 | 1176 | | |
| Yes | 638 | 637 | 652 | 705 | 665 | 784 | 821 | | |
| Yes | 932 | 917 | 962 | 991 | 995 | 1046 | 1088 | | |
| Yes | 868 | 846 | 898 | 981 | 994 | 1110 | 1157 | 1226 | 1342 |
| Yes | 740 | 731 | 762 | 789 | 799 | 846 | 828 | 954 | |
| No | 1029 | 1086 | 1041 | 1046 | 1041 | 1054 | 1193 | 1351 | 1479 |
| No | 1188 | 1067 | 1148 | 1144 | 1165 | 1180 | 1262 | 1414 | |
| No | 893 | 996 | 1022 | 1090 | 1084 | 1099 | 1088 | | |
| No | 995 | 1019 | 1066 | 1070 | 1031 | 1133 | 1257 | 1318 | |
| No | 883 | 887 | 887 | 921 | 903 | 914 | 947 | 1014 | 1061 |
| No | 1319 | 1214 | 1280 | 1243 | 1280 | 1397 | 1452 | 1567 | |
| No | 1128 | 1137 | 1194 | 1201 | 1236 | 1281 | 1296 | 1421 | 1500 |
| No | 752 | 744 | 881 | 905 | 1005 | 1134 | 1189 | 1251 | |
| No | 854 | 847 | 918 | 923 | 986 | 942 | 866 | 1036 | 1119 |
| No | 764 | 780 | 884 | 871 | 929 | 94± | 985 | 1117 | 1198 |
| No | 839 | 858 | 870 | 999 | 1065 | 1080 | 1130 | 1187 | 1284 |
| No | 1147 | 1063 | 1242 | 1327 | 1338 | 1293 | 1410 | | |

[a]Taken from Eckberg et al. (1986), consisting of a baseline measurement (RR0) and two measurements each at concentrations of 0.2, 0.4, 0.8, and 1.6 μg/kg per minute (RR1–RR8).

Therefore, in order to employ as much of the data as possible, the iterated version of Kleinbaum's procedure, discussed above, is applied with $n = 22$, $r = 2$, and $p = q = 8$. The iterated version of the estimator in (50) is

$$\hat{\underline{\beta}} = \begin{bmatrix} -6.36 & 31.58 & 70.25 & 62.28 & 106.92 & 139.09 & 271.85 & 338.54 \\ -7.60 & 53.72 & 78.60 & 106.13 & 138.92 & 191.96 & 295.07 & 371.80 \end{bmatrix}$$

The rows of $\hat{\underline{\beta}}$ correspond to the diabetic and nondiabetic groups, respectively, and the columns correspond to the measurements RR1 through RR8 (the $8 \times 8$ estimated variance matrix is not listed, but the procedure required only three iterations to attain convergence). The null hypothesis that the change from baseline is the same for the diabetic and nondiabetic groups is rejected ($p < 0.0001$) using the asymptotic test statistic in (51).

## ACKNOWLEDGMENTS

Thanks are extended to Dr. Dwain L. Eckberg for the use and publication of his data in Example 4, and likewise to Victoria C. Strider for her data in Examples 2 and 3. Thanks are extended to Dr. Richard A. Rode for the use and publication of his multivariate Box-Cox algorithm.

## APPENDIX

Six programs written in PROC MATRIX of SAS 1985 by the author are listed here. These programs are written as SAS MACROS, i.e., they function as subroutines. The user must input the MACRO at the beginning of the job, supply the necessary data sets and parameter values, then call the MACRO by its name.

An important note of caution is that PROC MATRIX is being phased out by SAS Institute, to be replaced by IML (Interactive Matrix Language) In Version 6. There are some minor enhancements and changes from PROC MATRIX to IML, and also within BASE SAS (such as in the use of arrays). In the future the user of the programs listed in this appendix will have to make his/her own changes to allow for these revisions.

The six programs are as follows:

1. NPARMULT: one-way nonparametric multivariate analysis of variance
2. MULTNORM: Tests of multivariate normality
3. MULTBXCX: multivariate Box-Cox transformation
4. GCCOVPAT: LR test of a patterned variance matrix in the BMANOVA model
5. GMANOVA: weighted inference under the GMANOVA model
6. GGCM: inference under Kleinbaum's generalized growth curve model

```
MACRO NPARMULT
***********************************************************************
* THIS MACRO CALCULATES AN ASYMPTOTIC CHI-SQUARED TEST FOR THE        *
* P-VARIATE, K-SAMPLE LOCATION PROBLEM.  THE NULL HYPOTHESIS STATES   *
* THAT THE P-VARIATE DISTRIBUTION FUNCTIONS ARE THE SAME FOR THE K    *
* GROUPS, WHILE THE ALTERNATIVE HYPOTHESIS STATES THAT THEY DIFFER IN *
* LOCATION.  MISSING DATA ARE PERMISSIBLE, ALTHOUGH FOR EACH OF THE P *
* RESPONSES AT LEAST ONE EXPERIMENTAL UNIT IN EACH GROUP MUST HAVE A  *
* VALID OBSERVATION.  THE FORMULATION IS TAKEN FROM KOZIOL, ET AL.    *
* (1981) BIOMETRICS, VOLUME 37, PAGES 383-390.                        *
*                                                                     *
* IN ORDER TO USE THIS MACRO, THE USER MUST SUPPLY THE FOLLOWING:     *
* (1) THE MACRO VARNAMES, WHICH CONTAINS THE SAS VARIABLE NAMES OF THE*
*     P RESPONSES IN A ROW VECTOR.                                    *
* (2) THE MACRO SCORES, WHICH CONTAINS EITHER THE NAME 'WILCOXON' OR  *
*     'NORMAL' FOR THE DESIRED SCORE FUNCTION.                        *
* (3) THE MACRO PERMTEST, WHICH CONTAINS EITHER THE NAME 'FULL',      *
*     'PARTIAL', OR 'NONE' FOR REQUESTING A PERMUTATION TEST.  A      *
*     PERMUTATION TEST IS NOT VALID IF THE TOTAL NUMBER OF MISSING    *
*     VALUES FOR ANY RESPONSE IS GREATER THAN OR EQUAL TO THE SMALLEST*
*     GROUP SIZE.  ALSO, THE FULL PERMUTATION TEST SHOULD NOT BE      *
*     REQUESTED WHEN THE TOTAL NUMBER OF PERMUTATIONS IS LARGE (IF IT *
*     EXCEEDS 200, THEN EXTRA SPACE AND TIME MAY BE NEEDED).          *
* (4) THE SAS DATA SET NAMED FIRST, WHICH CONSISTS OF THE VARIABLE    *
*     GROUP FOLLOWED BY THE RESPONSE VARIABLES.  GROUP MUST BE NUMERIC*
*     AND NUMBERED 1 THROUGH K TO INDICATE THE K GROUPS.              *
***********************************************************************;
PROC SORT DATA=FIRST;BY GROUP;
DATA SECOND;SET FIRST;BY GROUP;ARRAY M VARNAMES;
DO OVER M;IF M=. THEN M=0;ELSE M=1;END;KEEP VARNAMES;
DATA THIRD;SET FIRST;BY GROUP;IF FIRST.GROUP THEN N=0;N+1;
KEEP N;IF LAST.GROUP THEN OUTPUT;
PROC MATRIX FUZZ;FETCH G DATA=FIRST(KEEP=GROUP);NTOT=NROW(G);
FETCH A DATA=FIRST(KEEP=GROUP);NTOT=NROW(G);
FETCH T DATA=SECOND(KEEP=VARNAMES);
FETCH N DATA=THIRD;K=NROW(N);PK=P#K;DF=PK-P;
DO L=1 TO P BY 1;A(LOC(T(,L)),L)=RANKTIE(A(LOC(T(,L)),));END;
NM=(DESIGN(G))'*T;GG=G;
A=(A'@||((NM(+,)+J(1,P,1))'##-1))';
S=SCORES;SS='NORMAL';IF S=SS THEN DO;
   A=A+(0.5#(J(NTOT,P,1)-T));A=PROBIT(A);
END;FREE S SS;
T=T#((A+,)#(NM(+,)##-1)@J(NTOT,1,1));A=A-T;FREE T;CALCULAT;
CH=CHISQ;PVALUE=1-PROBCHI(CH,DF);RESULTS=CH||DF||PVALUE;
R=' ';C='CHISQ' 'DF' 'PVALUE';
NOTE 'THE RESULTS OF THE ASYMPTOTIC CHI-SQUARE TEST:';
PRINT RESULTS COLNAME=C ROWNAME=R;
FULL='FULL';PARTIAL='PARTIAL';TYPE='PERMTEST';
IF TYPE=FULL OR TYPE=PARTIAL THEN DO;
   FETCH T DATA=SECOND(KEEP=VARNAMES);T=NTOT-MIN(T(+,));
   IF T>=MIN(N) THEN DO;
      NOTE 'THE PERMUTATION TEST IS NOT VALID FOR THIS DATA SET.';
   END;ELSE DO;I1=J(1,K,1);I2=J(1,K,N(1,1));
      DO I=2 TO K BY 1;
         I1(1,I)=SUM(N(1:(I-1),1))+1;I2(1,I)=SUM(N(1:I,1));
      END;NUMPERM=1;NT=NTOT+1;
      DO J=1 TO K BY 1;
         NJ=N(J,1);NJ1=NJ+1;
         DO JJ=1 TO NJ BY 1;
            NT=NT-1;NJ1=NJ1-1;NUMPERM=(NUMPERM#NT)#/NJ1;
         END;
      END;FREE J JJ NJ NJ1 NT;NN=1:NTOT;PVALUE=0;
      IF TYPE=PARTIAL THEN DO;
         NITER=MIN(NUMPERM//500);NP=NN;
         DO I=1 TO NITER BY 1;
            DO J=1 TO NTOT BY 1;U=U||(UNIFORM(0));END;
            NP(,RANK(U))=NN;FREE U;COMPARE;
         END;
```

```
      END;
      IF TYPE=FULL THEN DO;
         NP=NN;NPALL=NN;NITER=NUMPERM;
         DO I=1 TO NITER BY 1;
            DO J=1 TO NTOT BY 1;U=U||(UNIFORM(0));END;
            NP(,RANK(U))=NN;FREE U;
            DO J=1 TO K BY 1;
               L1=I1(1,J);L2=I2(1,J);NPP=NP(1,(L1:L2));
               NP(,(RANK(NP(1,(L1:L2)))+J(1,N(J,1),(L1-1))))=NPP;
            END;FREE NPP;FLAG=1;
            IF I>1 THEN DO II=1 TO I-1 BY 1;
               IF NP=NPALL(II,) THEN DO;FLAG=0;I=I-1;END;
            END;IF FLAG=1 THEN DO;NPALL=NPALL//NP;COMPARE;END;
         END;
      END;
      PVALUE=PVALUE#/NITER;RESULTS(1,3)=PVALUE;
      NOTE SKIP=6 'THE TYPE AND RESULTS OF THE PERMUTATION TEST:';
      CC='NITER' 'NUMPERM';RR=TYPE;TYPE=NITER||NUMPERM;
      PRINT TYPE COLNAME=CC ROWNAME=RR;
      PRINT RESULTS COLNAME=C ROWNAME=R;
   END;
END;PROC DELETE DATA=SECOND THIRD;
%
MACRO COMPARE FETCH NM DATA=SECOND(KEEP=VARNAMES);
G(NP,)=GG;NM=(DESIGN(G))'*NM;CALCULAT;
IF CH<=CHISQ THEN PVALUE=PVALUE+1;%
MACRO CALCULAT S=((DESIGN(G))'*A)#(NM##-1);S=SHAPE(S,1);
NM=SHAPE(NM,1);V=((NM*NM')##-1)#(N@J(P,PK,1));FREE NM;
V=V#((I(K)@J(P,P,1))-(J(PK,P,1)@(N'#/NTOT)));
V=(V#(J(K,K,1)@(A'*A)))#/(NTOT-1);
CHISQ=S'*GINV(V)*S;FREE S V;
%


MACRO MULTNORM PROC MATRIX FUZZ;
*******************************************************************
* THIS MACRO CALCULATES TESTS OF MULTIVARIATE NORMALITY ON THE BLUS   *
* RESIDUALS FROM A MULTIVARIATE LINEAR MODEL.  THE TESTS PROVIDED ARE *
* MARDIA'S MEASURES OF MULTIVARIATE SKEWNESS AND KURTOSIS AND A       *
* KOLMOGOROV-SMIRNOV GOODNESS-OF-FIT TEST.  THE FORMULATION IS TAKEN  *
* FROM MARDIA (1980) HANDBOOK OF STATISTICS, VOLUME I, NORTH-HOLLAND, *
* NEW YORK, PAGES 279-320.                                            *
*                                                                     *
* THE USER MUST SUPPLY THE SAS DATA SETS Y AND X, WHERE Y IS THE N BY *
* P RESPONSE MATRIX AND X IS THE N BY R DESIGN MATRIX OF FULL RANK R. *
*******************************************************************;
FETCH Y DATA=Y;FETCH X DATA=X;N=NROW(Y);P=NCOL(Y);R=NCOL(X);
XPXINV=INV(X'*X);XPY=X'*Y;BETA=XPXINV*XPY;
A=I(N)-(X*XPXINV*X');S=(Y'*A*Y)#/(N-R);INDEX=1;
DO I=2 TO N BY 1;
   IF ALL(EIGVAL(A(,INDEX||I)'*A(,INDEX||I)>0.00000001)) THEN
   INDEX=INDEX||I;
END;A=A(,INDEX);
AA=A'*A;EIGEN L E AA;AA=E*(DIAG(L##-0.5))*E';RSTAR=AA*A'*Y;
FREE A AA E L;RBAR=RSTAR(+,)#/(N-R);S=INV(S);
RSTAR=RSTAR-(J(N-R,1,1)*RBAR);RSTAR=RSTAR*S*RSTAR';
SKEWNESS=SUM(RSTAR##3)#/((N-R)##2);
RSTAR=VECDIAG(RSTAR);KURTOSIS=SUM(RSTAR##2)#/(N-R);
DF=P#(P+1)#(P+2)#/6;B1=(N-R)#SKEWNESS#/6;PB1=1-PROBCHI(B1,DF);
B2=(KURTOSIS-(P#(P+2)))#/((8#P#(P+2)#/(N-R))##0.5);PB2=1-PROBNORM(B2);
RESULTS=SKEWNESS||PB1||KURTOSIS||PB2;
CN='SKEWNESS' 'SK-PVAL' 'KURTOSIS' 'KU-PVAL';RN=' ';
NOTE 'THE MEASURES OF MULTIVARIATE SKEWNESS AND KURTOSIS:';
PRINT RESULTS COLNAME=CN ROWNAME=RN;
RR=RSTAR;RSTAR(RANK(RR),)=RR;FREE RR;
FN=(1:(N-R))'#/(N-R);FO=PROBCHI(RSTAR,P);
D=((N-R)##0.5)#MAX(ABS(FN-FO));P=0;FLAG=0;J=0;
```

```
DO WHILE (FLAG=0);J=J+1;
   T=((-1)##(J+1))#EXP(-2#J#J#D#D);
   IF ABS(T)<=0.00000001 THEN FLAG=1;P=P+T;
END;P=2#P;
RESULTS=D||P;CN='KOL-SMIR' 'P-VALUE';
NOTE SKIP=6 'THE KOLMOGOROV-SMIRNOV TEST:';
PRINT RESULTS COLNAME=CN ROWNAME=RN;
%

MACRO MULTBXCX PROC MATRIX FUZZ;
************************************************************************
* THIS MACRO CALCULATES THE ESTIMATED VECTOR OF TRANSFORMATION         *
* PARAMETERS IN THE MULTIVARIATE BOX-COX LINEAR MODEL YY = X*BETA + E,*
* WHERE YY IS THE N BY P MATRIX OF TRANSFORMED OBSERVATIONS, X IS THE *
* N BY R KNOWN DESIGN MATRIX OF FULL RANK R (< N), BETA IS THE UNKNOWN*
* R BY P MATRIX OF PARAMETERS, AND E IS THE UNOBSERVABLE N BY P MATRIX*
* OF RANDOM ERRORS, WHOSE ROWS ARE INDEPENDENT AND IDENTICALLY         *
* DISTRIBUTION AS P-VARIATE NORMAL RANDOM VECTORS WITH NULL MEAN AND  *
* VARIANCE-COVARIANCE MATRIX SIGMA.  THE ORIGINAL OBSERVATION MATRIX  *
* Y IS TRANSFORMED TO YY ACCORDING TO THE BOX-COX POWER TRANSFORMATION*
* APPLIED TO EACH COLUMN OF Y.  THE FORMULATION IS TAKEN FROM RODE     *
* (1986), UNPUBLISHED PH.D. DISSERTATION, VIRGINIA COMMONWEALTH        *
* UNIVERSITY.                                                          *
*                                                                      *
* THE USER MUST PROVIDE THE SAS DATA SETS Y AND X, AND THE MACRO       *
* RMANAL, WHICH CONTAINS THE VALUE ZERO OR ONE.  RMANAL CONTAINS THE  *
* VALUE OF ONE IF THE DATA SET CONSISTS OF REPEATED MEASURES AND ONLY *
* ONE TRANSFORMATION PARAMETER IS DESIRED FOR ALL P RESPONSES.         *
* OTHERWISE, RMANAL CONTAINS THE VALUE ZERO.                           *
*                                                                      *
* THE OUTPUT CONSISTS OF THE MLES OF LAMDA (THE TRANSFORMATION         *
* PARAMETERS), BETA, AND SIGMA, AND THE VALUE OF THE LOG-LIKELIHOOD   *
* FUNCTION EVALUATED AT THE MLES.                                      *
************************************************************************;
FETCH Y DATA=Y;FETCH X DATA=X;N=NROW(Y);P=NCOL(Y);R=NCOL(X);
RM=RMANAL;PT=P;PP=1;NT=N;
IF RM=1 THEN DO;N=N#P;PP=P;P=1;Y=SHAPE(Y',1);END;
LNY=LOG(Y);Z=(LNY(+,))';XPXINV=INV(X'*X);EPS1=0.000001;EPS2=0.00000001;
LAMDA=J(P,1,0);YY=J(N,P,0);
DO C=1 TO P BY 1;LOWER=-3;UPPER=3;L=-3;DL=0;FLAG=0;
   DO WHILE (FLAG=0);CHANGE=0;
      DO J=1 TO 26 BY 1;
         LOLD=L;DLOLD=DL;L=LOWER+((J-1)#(UPPER-LOWER)#/25);
         IF ABS(L)<EPS1 THEN DO;YT=LNY(,C);L1=(LNY(,C)##2)#/2;END;
         ELSE DO;YL=Y(,C)##L;YT=(YL-1)#/L;L1=((YL#LNY(,C))-YT)#/L;END;
         XPYT=(I(PP)@X')*YT;BETA=(I(PP)@XPXINV)*XPYT;
         SIGMA=((YT'*YT)-(XPYT'*BETA))#/N;
         YL=(YT-((I(PP)@X)*BETA))#/SIGMA;DL=Z(C,)-SUM(L1#YL);
         IF (J>=2) & (SIGN(DL) NE SIGN(DLOLD) THEN DO;
            CHANGE=1;LOWER=LOLD;UPPER=L;J=26;
         END;
      END;IF CHANGE=0 THEN DO; LOWER=LOWER-1;UPPER=UPPER+1;END;
      ELSE IF ABS(DL)<EPS1 THEN FLAG=1;
   END;LAMDA(C,)=L;YY(,C)= YT;
END;L=LAMDA;IF RM=1 THEN YY=SHAPE(YY,NT)';
FREE YT XPYT UPPER LOWER LOLD DLOLD;
XPYY=X'*YY;BETA=XPXINV*XPYY;SIGMA=((YY'*YY)-(XPYY'*XPXINV*XPYY))#/N;
NOTE 'THE INITIAL ESTIMATES OF LAMDA, BETA, AND SIGMA ARE:';
PRINT L BETA SIGMA;
IF RM=1 THEN DO;B=J(PT,1,L);Y=SHAPE(Y,NT)';LNY=SHAPE(LNY,NT)';END;
ELSE B=L;MXN=100;IT=0;YL=J(NT,PT,0);L1=YL;L2=YL;
START:SI=INV(SIGMA);DO C=1 TO PT BY 1;
   IF ABS(B(C,))<EPS1 THEN DO;
      YY(,C)=LNY(,C);L1(,C)=(LNY(,C)##2)#/2;L2(,C)=(LNY(,C)##3)#/3;
   END;ELSE DO;
      YL(,C)=Y(,C)##B(C,);YY(,C)=(YL(,C)-1)#/B(C,);
      L1(,C)=((YL(,C)#LNY(,C))-YY(,C))#/B(C,);
      L2(,C)=((YL(,C)#(LNY(,C)##2))-(2#L1(,C)))#/B(C,);
   END;
```

```
END;YL=(YY-(X*BETA))*SI;
IF RM=1 THEN DO;
   DL=Z-SUM(L1#YL);DLL=TRACE(L1*SI*L1')+SUM(L2#YL);B=B(1,1);
END;ELSE DO;
DL=Z-(((L1#YL)')(,+));DLL=((L1'*L1)#SI)+DIAG(((L2#YL)')(,+));
END;NEW=B+SOLVE(DLL,DL);
CHECK=MAX((ABS(NEW-B))#/((ABS(B))+EPS1));
IF CHECK<EPS2 THEN GO TO OK;IT=IT+1;
IF IT>=MXN THEN GOT TO NC;B=NEW;IF RM=1 THEN B=J(PT,1,NEW);
XPYY=X'*YY;BETA=XPXINV*XPYY;SIGMA=((YY'*YY)-(XPYY*XPXINV*XPYY))#/N;
GO TO START;
NC:NOTE PAGE 'FAILURE TO CONVERGE';PRINT MXN B;STOP;
OK:L=B;NOTE PAGE 'THE MLES OF LAMDA, BETA, AND SIGMA ARE:';
PRINT L BETA SIGMA;
LLIKE=((-NT#/2)#LOG(DET(SIGMA)))+SUM(Z#(L-1))-(TRACE(YL*SIGMA*YL'#/2));
NOTE 'THE LOG LIKELIHOOD EVALUATED AT THE MLES IS:';PRINT LLIKE;
%


MACRO GCCOVPAT PROC MATRIX FUZZ;
*******************************************************************
* THIS MACRO CONDUCTS A LIKELIHOOD RATIO TEST FOR DETERMINING WHETHER *
* THE MP BY MP COVARIANCE MATRIX FOR THE MP-VARIATE GMANOVA MODEL IS  *
* OF THE STRUCTURE WHICH PERMITS AN UNWEIGHTED ANALYSIS.  THE MODEL   *
* CONSISTS OF FITTING ORTHOGONAL POLYNOMIALS TO THE P REPEATED        *
* MEASURES OF M RESPONSE VARIABLES, I.E., E(Y) = A*BETA*(QQ@I), WHERE *
* Y IS THE N BY MQ RESPONSE MATRIX, A IS THE N BY R DESIGN MATRIX,    *
* BETA IS THE R BY MQ UNKNOWN PARAMETER MATRIX, QQ IS THE Q BY P      *
* MATRIX OF (Q-1)-DEGREE ORTHOGONAL POLYNOMIALS, AND I IS THE M BY M  *
* IDENTITY MATRIX.  THE FORMULATION IS TAKEN FROM * CHINCHILLI AND    *
* CARTER (1984) BIOMETRICS, VOLUME 40, PAGES 151-156.                 *
*                                                                     *
* THE USER NEEDS TO SUPPLY THE SAS DATA SET CALLED FIRST AND THE TWO  *
* MACROS CALLED LEVELS AND DMENSION.  FIRST CONTAINS N OBSERVATIONS   *
* AND MP+R VARIABLES CORRESPONDING TO THE N BY MP RESPONSE MATRIX Y   *
* AND THE N BY R, FULL-RANK DESIGN MATRIX A.  THE MP+R VARIABLES IN   *
* THIS DATA SET SHOULD BE ARRANGED SUCH THAT THE M RESPONSES ARE      *
* GROUPED TOGETHER FOR EACH OF THE P TIME POINTS, AND THEN THE R      *
* VARIABLES FROM THE DESIGN MATRIX FOLLOW.  THE MACRO LEVELS LISTS THE*
* P TIME POINTS IN A ROW VECTOR.  THE MACRO DMENSION LISTS THE VALUES *
* M, P, R, AND Q-1 IN A ROW VECTOR.                                   *
*******************************************************************
T=LEVELS;T=T';Q=DMENSION;M=Q(1,1);P=Q(1,2);R=Q(1,3);Q=Q(1,4)+1;
QQ=ORPOL(T);MP=M#P;MQ=M#Q;
FETCH A DATA=FIRST;Y=A(,1:MP);A=A(,(MP+1):(MP+R));N=NROW(A);
Y=Y*(QQ@I(M));FREE T QQ;YPY=Y'*Y;YPA=Y'*A;FREE Y;
A=INV(A'*A);S=YPY-(YPA*A*YPA');FREE A;
S11=S(1:MQ,1:MQ);S12=S(1:MQ,(MQ+1):MP);S22=S((MQ+1):MP,(MQ+1):MP);
S=I(MQ)-(INV(S11)*S12*INV(S22)*S12');FREE S11 S12 S22;
L=DET(S);FREE S;RHO=N-R-((MP+1)#/2);T=M#MQ#(P-Q);
G2=T#((MQ##2)+((M#(P-Q))##2)-5)#/48;
G4=((G2##2)#/2)+((T#/1920)#((3#(MQ##4))+(3#((M#(P-Q))##4))+
   (10#(M##4)#((Q#(P-Q))##2))-(50#(MQ##2))-(50#((M#(P-Q))##2))+159));
LL=-1#RHO#LOG(L);P1=PROBCHI(LL,T);
T4=PROBCHI(LL,T+4);T8=PROBCHI(LL,T+8);
P2=(G2#(RHO##-2))#(T4-P1);P3=(RHO##-4)#((G4#(T8-P1)-((G2##2)#(T4-P1)));
PVALUE=1-(P1+P2+P3);RESULTS=L||PVALUE;C='TESTSTAT' 'P-VALUE';R=' ';
NOTE 'THE RESULTS OF THE LIKELIHOOD RATIO TEST FOR A PATTERNED';
NOTE 'COVARIANCE MATRIX IN THE GMANOVA MODEL';
PRINT RESULTS COLNAME=C ROWNAME=R;
%
```

```
MACRO GMANOVA PROC MATRIX FUZZ;
**********************************************************************
* THE GENERALIZED MANOVA (GMANOVA) MODEL IS DESCRIBED BY             *
*           E(Y) = X*BETA*P AND COV(VEC(Y)) = SIGMA@I(N),            *
* WHERE Y IS THE N BY PP OBSERVATION MATRIX, X IS THE N BY R DESIGN  *
* MATRIX, BETA IS THE UNKNOWN R BY Q PARAMETER MATRIX, P IS THE Q BY *
* PP WITHIN-UNITS DESIGN MATRIX, I(N) IS THE N BY N IDENTITY MATRIX, *
* AND SIGMA IS THE UNKNOWN, POSITIVE-DEFINITE, PP BY PP COVARIANCE   *
* MATRIX.  THE MATRICES X AND P ARE ASSUMED TO BE OF FULL RANK R AND *
* Q, RESPECTIVELY, AND N >= PP+R.                                    *
*                                                                    *
* NOTE THAT IF P IS SET EQUAL TO I(PP), THEN THE GMANOVA MODEL REDUCES*
* TO THE CLASSICAL MANOVA MODEL.  THIS MACRO ALSO CALCULATES A       *
* GOODNESS-OF-FIT TEST WHENEVER THE MATRIX P IS NOT SQUARE.  A       *
* SIGNIFICANT RESULT INDICATES THAT THE WITHIN-UNITS MODELLING IS    *
* INADEQUATE, I.E., P NEEDS MORE ROWS.                               *
*                                                                    *
* BEFORE CALLING THIS MACRO, THE USER NEEDS TO SUPPLY THE SAS DATA   *
* SETS Y, X, AND PPRIME, WHICH CONTAIN THE MATRICES Y, X, AND P',    *
* RESPECTIVELY.                                                      *
*                                                                    *
* AFTER COMPLETION OF THIS MACRO, THE SAS DATA SETS BETA AND SIGMA ARE*
* AVAILABLE TO THE USER.  THESE DATA SETS CONTAIN THE MLES OF THESE  *
* VALUES.                                                            *
*                                                                    *
* IT IS ALSO POSSIBLE FOR THE USER TO TEST ANY NUMBER OF HYPOTHESES  *
* OF THE FORM C*BETA*U = 0, WHERE C IS A FULL RANK CC BY R MATRIX (CC *
* <= R) AND U IS A FULL RANK PP BY UU MATRIX (UU <= PP).  IN ORDER TO *
* DO THIS, THE USER NEEDS TO PROVIDE THE MATRICES C' AND U IN THE SAS *
* DATA SETS CALLED CPRIME AND U, RESPECTIVELY, BEFORE CALLING THE    *
* MACRO HYPTEST.                                                     *
**********************************************************************;
FETCH Y DATA=Y COLNAME=CY;FETCH X DATA=X COLNAME=CX;
PP=NCOL(Y);R=NCOL(X);N=NROW(X);
FETCH P DATA=PPRIME COLNAME=CP;P=P';Q=NROW(P);INV_XX=INV(X'*X);
IF Q<PP THEN DO;
   U=I(PP)-(P'*INV(P*P')*P);INDEX=1;
   DO I=2 TO PP;
      IF ALL(EIGVAL(U(,INDEX||I)'*U(,INDEX||I))>0.00000001) THEN
      INDEX=INDEX||I;
   END;U=U(,INDEX);SH=X'*Y;SH=U'*SH'*INV_XX*SH*U;
   SE=HALF(INV(U'*(Y'*Y)*U)-SH));LR=1#/DET(I(PP-Q)+(SE*SH*SE'));
   DF=(PP-Q)#R;RHO=1-((PP-Q+R+1)#/(2#N));
   GAMMA=(DF#(((PP-Q)##2)+(R##2)-5))#/48;LRO=-N#RHO#LOG(LR);
   PVALUE=1-PROBCHI(LRO,DF)+((GAMMA#/((RHO#N)##2))#
          (PROBCHI(LRO,DF)-PROBCHI(LRO,DF+4)));
   RESULTS=LR||(PP-Q)||R||(N-R)||PVALUE;
   CTITLE='LR_STAT' 'NROW_SH' 'DF_SH' 'DF_SE' 'P-VALUE';RTITLE=' ';
   NOTE 'THE RESULTS OF THE GOODNESS-OF-FIT TEST:';
   PRINT RESULTS COLNAME=CTITLE ROWNAME=RTITLE;NOTE PAGE;
END;BETA=INV_XX*(X'*Y);S=(Y'*Y)-((Y'*X)*BETA);INV_S=INV(S);
PSP=INV(P*INV_S*P');W=I(PP)-(INV_S*P'*PSP*P);
T=N-R-(PP-Q);T=(N-R-1)#/(T#(T-1));
SIGMA=(S+(W'*(Y'*X)*BETA*W))#T;BETA=BETA*INV_S*P'*PSP;
CDF=NROW(C);UDF=NCOL(U);
XY=X'*Y;INV_XX=INV(X'*X);S=(Y'*Y)-(XY'*INV_XX*XY);
INV_S=INV(S);PSP=INV(P*INV_S*P');
R=C*INV_XX*XY;R=R'*INV(C*INV_XX*C')*R;
SE=U'*PSP*U;SH=(U'*INV(P*INV(S+R)*P')*U)-SE;
SE=HALF(INV(SE));LR=1#/DET(I(UDF)+(SE*SH*SE');
DF=CDF#UDF;RHO=1-((RR+PP-Q-CDF+.5#((UDF+CDF+1)))#/N);
GAMMA=(DF#((UDF##2)+(CDF##2)-5))#/48;LRO=-N#RHO#LOG(LR);
PVALUE=1-PROBCHI(LRO,DF)+((GAMMA#/((RHO#N)##2))#
       (PROBCHI(LRO,DF)-PROBCHI(LRO,DF+4)));
```

```
NOTE 'THE ESTIMATES OF BETA AND SIGMA:';
OUTPUT BETA OUT=BETA COLNAME=CP ROWNAME=CX;
PRINT BETA COLNAME=CP ROWNAME=CX;
OUTPUT SIGMA OUT=SIGMA COLNAME=CY ROWNAME=CY;
PRINT SIGMA COLNAME=CY ROWNAME=CY;
%
MACRO HYPTEST PROC MATRIX FUZZ;
FETCH Y DATA=Y COLNAME=CY;FETCH X DATA=X COLNAME=CC;
PP=NCOL(Y);RR=NCOL(X);N=NROW(X);
FETCH P DATA=PPRIME COLNAME=RU;P=P';Q=NROW(P);
FETCH C DATA=CPRIME COLNAME=RC;C=C';FETCH U DATA=U COLNAME=CU;
RESULTS=LR||UDF||CDF||(N-RR)||PVALUE;
CTITLE='LR_STAT' 'NROW_SH' 'DF_SH' 'DF_SE' 'P-VALUE';RTITLE=' ';
NOTE 'THE RESULTS OF TESTING HO:C*BETA*U = 0';
PRINT C COLNAME=CC ROWNAME=RC;PRINT U COLNAME=CU ROWNAME=RU;
PRINT RESULTS COLNAME=CTITLE ROWNAME=RTITLE;
%


MACRO GGCM PROC MATRIX FUZZ;
***********************************************************************
* THIS MACRO CALCULATES A BEST ASYMPTOTICALLY NORMAL (BAN) ESTIMATE   *
* AND ITS ESTIMATED COVARIANCE MATRIX FOR THE PARAMETER MATRIX OF THE *
* GENERALIZED GROWTH CURVE MODEL, AS DESCRIBED BY KLEINBAUM (1973),   *
* JOURNAL OF MULTIVARIATE ANALYSIS, VOLUME 3, PAGES 117-124.  THE     *
* MODEL ALLOWS FOR MISSING DATA AND/OR INCOMPLETE MULTIRESPONSE       *
* DESIGNS.                                                            *
*                                                                     *
* THE GENERALIZED MANOVA (GMANOVA) MODEL IS DESCRIBED BY              *
*             E(Y) = X*BETA*P AND COV(VEC(Y)) = SIGMA@I(N),           *
* WHERE Y IS THE N BY PP OBSERVATION MATRIX, X IS THE N BY R DESIGN   *
* MATRIX, BETA IS THE UNKNOWN R BY Q PARAMETER MATRIX, P IS THE Q BY  *
* PP WITHIN-UNITS DESIGN MATRIX, I(N) IS THE N BY N IDENTITY MATRIX,  *
* AND SIGMA IS THE UNKNOWN, POSITIVE-DEFINITE, PP BY PP COVARIANCE    *
* MATRIX.  THE MATRICES X AND P ARE ASSUMED TO BE OF FULL RANK R AND  *
* Q, RESPECTIVELY, AND N >= PP+R.  UNDER THE ASSUMPTION OF NORMALITY, *
* THE ESTIMATES OF BETA AND SIGMA CALCULATED IN THIS MACRO ARE THE    *
* MAXIMUM LIKELIHOOD ESTIMATES BECAUSE THE ITERATED VERSION OF        *
* KLEINBAUM'S TECHNIQUE IS USED.  THE INITIAL ESTIMATE OF SIGMA IS    *
* FOUND BY THE ALLVALUE METHOD WITH SMOOTHING.                        *
*                                                                     *
* NOTE THAT IF P IS SET EQUAL TO I(PP), THEN THE GMANOVA MODEL REDUCES*
* TO THE CLASSICAL MANOVA MODEL.  THIS MACRO ALSO CALCULATES A        *
* GOODNESS-OF-FIT TEST WHENEVER THE MATRIX P IS NOT SQUARE.  A        *
* SIGNIFICANT RESULT INDICATES THAT THE WITHIN-UNITS MODELLING IS     *
* INADEQUATE, I.E., P NEEDS MORE ROWS.                                *
*                                                                     *
* BEFORE CALLING THIS MACRO, THE USER NEEDS TO SUPPLY THE SAS DATA    *
* SETS Y, X, AND PPRIME, WHICH CONTAIN THE MATRICES Y, X, AND P',     *
* RESPECTIVELY.                                                       *
*                                                                     *
* AFTER COMPLETION OF THIS MACRO, THE SAS DATA SETS BETA, SIGMA, AND  *
* COVBETA, THE ESTIMATED VARIANCE MATRIX OF VEC(BETA), ARE AVAILABLE  *
* TO THE USER.                                                        *
*                                                                     *
* IT IS ALSO POSSIBLE FOR THE USER TO TEST ANY NUMBER OF HYPOTHESES   *
* OF THE FORM C*BETA*U = 0, WHERE C IS A FULL RANK CC BY R MATRIX (CC *
* <= R) AND U IS A FULL RANK PP BY UU MATRIX (UU <= PP).  IN ORDER TO *
* DO THIS, THE USER NEEDS TO PROVIDE THE MATRICES C' AND U IN THE SAS *
* DATA SETS CALLED CPRIME AND U, RESPECTIVELY, BEFORE CALLING THE     *
* MACRO HYPTEST.                                                      *
***********************************************************************;
```

```
FETCH Y DATA=Y COLNAME=CY;N=NROW(Y);PP=NCOL(Y);PATNO=J(N,1,1);
DO I=1 TO N BY 1;DO J=1 TO PP BY 1;
   IF Y(I,J)=. THEN Y(I,J)=0;ELSE Y(I,J)=1;
   PATNO(I,)=PATNO(I,)+(Y(I,J)*(2**(J-1)));
END;END;
Y=Y||(1:N)'||PATNO;CYY='I' 'PATNO';CY=CY||CYY;
OUTPUT Y OUT=Y2 COLNAME=CY;FREE Y PATNO I;
PROC SORT DATA=Y2;BY PATNO;
DATA Y3;SET Y2;BY PATNO;LENGTH NUM $ 8;
IF FIRST.PATNO THEN N_UNITS=0;N_UNITS+1;
IF LAST.PATNO THEN DO;N+1;NUM=PUT(N,8.);DROP ROW PATNO N I;OUTPUT;END;
PROC MATRIX FUZZ;EPS1=0.000001;EPS2=0.00000001;
FETCH Y DATA=Y COLNAME=CY;FETCH X DATA=X COLNAME=CX;
PP=NCOL(Y);R=NCOL(X);N=NROW(X);
FETCH I DATA=Y2(KEEP=I);Y=Y(I,);X=X(I,);FREE I;
FETCH PATTERN DATA=Y3 COLNAME=CTITLE ROWNAME=NUM;
NOTE PAGE 'THE FOLLOWING PRINTOUT LISTS THE PATTERNS OF OBSERVED AND';
NOTE 'MISSING DATA IN THE Y MATRIX AND THE NUMBER OF UNITS WITH';
NOTE 'EACH PATTERN.  IN THE ANALYSIS WHICH FOLLOWS, ALL UNITS ARE';
NOTE 'DELETED IN WHICH EVERY RESPONSE VARIABLE IS UNOBSERVED.';
PRINT PATTERN COLNAME=CTITLE ROWNAME=NUM;NOTE PAGE;
NI=PATTERN(,PP+1);PATTERN=PATTERN(,1:PP);K=NROW(PATTERN);
IF SUM(PATTERN(1,))=0 THEN DO;
   NKEEP=(NI(1,)+1):N;X=X(NKEEP,);Y=Y(NKEEP,);FREE NKEEP;
   KKEEP=2:K;NI=NI(KKEEP,);PATTERN=PATTERN(KKEEP,);FREE KKEEP;
   N=N-NI(1,);K=K-1;
END;
FETCH P DATA=PPRIME COLNAME=CP;P=P';Q=NROW(P);
IF Q<PP THEN DO;
   SIGMA=I(PP);PT=I(PP);QT=PP;BETA=J(R,PP,0);ITERATE=0;FLAG=0;
   DO WHILE (FLAG=0);
      B=BETA;S=SIGMA;BETA_AV;SIGMA_AV;
      IF (SSQ(BETA-B)<(EPS1*(SSQ(B)+EPS2)))&(SSQ(SIGMA-S)<
         (EPS1*(SSQ(S)+EPS2))) THEN FLAG=1;
      ITERATE=ITERATE+1;IF ITERATE>50 THEN FLAG=1;
   END;
   U=I(PP)-(P'*INV(P*P')*P);INDEX=1;
   DO I=2 TO PP;
      IF ALL(EIGVAL(U(,INDEX||I)'*U(,INDEX||I))>0.00000001) THEN
      INDEX=INDEX||I;
   END;U=U(,INDEX);U=U'@I(R);
   W=U*SHAPE(BETA',1);W=W'*GINV(U*COVBETA*U')*W;
   DF=R#(PP-Q);PVALUE=1-PROBCHI(W,DF);RESULTS=W||DF||PVALUE;
   CTITLE='CHI_SQ' 'DEG_FREE' 'P-VALUE';RTITLE=' ';
   NOTE 'THE RESULTS OF THE GOODNESS-OF-FIT TEST:';
   PRINT RESULTS COLNAME=CTITLE ROWNAME=RTITLE;NOTE PAGE;
END;SIGMA=I(PP);PT=P;QT=Q;BETA=J(R,Q,0);ITERATE=0;FLAG=0;
DO WHILE (FLAG=0);B=BETA;S=SIGMA;BETA_AV;SIGMA_AV;
   IF (SSQ(BETA-B)<(EPS1*(SSQ(B)+EPS2)))&(SSQ(SIGMA-S)<
         (EPS1*(SSQ(S)+EPS2))) THEN FLAG=1;
   ITERATE=ITERATE+1;IF ITERATE>50 THEN DO;FLAG=1;
      NOTE 'CONVERGENCE NOT ATTAINED, SO THE ESTIMATES OF THE LAST';
      NOTE 'ITERATION ARE ASSUMED TO BE APPROPRIATE.';NOTE SKIP=10;
   END;
END;
NOTE 'THE NUMBER OF ITERATIONS AND THE MLES OF BETA AND SIGMA:';
RTITLE=' ';CTITLE='NUMBER';
PRINT ITERATE COLNAME=CTITLE ROWNAME=RTITLE;
OUTPUT BETA OUT=BETA COLNAME=CP ROWNAME=CX;
PRINT BETA COLNAME=CP ROWNAME=CX;
OUTPUT SIGMA OUT=SIGMA COLNAME=CY ROWNAME=CY;
```

```
PRINT SIGMA COLNAME=CY ROWNAME=CY;
OUTPUT COVBETA OUT=COVBETA;
%
MACRO BETA_AV BETA=J(R#QT,1,0);COVBETA=BETA*BETA';UL=0;LL=0;
DO KK=1 TO K BY 1;
   LL=UL+1;UL=UL+NI(KK,);IN=LOC(PATTERN(KK,));
   SKK=GINV(SIGMA(IN,IN))*PT(,IN)';XX=X(LL:UL,);YY=Y(LL:UL,IN);
   BETA=BETA+SHAPE((XX'*YY*SKK)',1);SKK=PT(,IN)*SKK;
   XX=XX'*XX;COVBETA=COVBETA+(SKK@XX);
END;FREE IN SKK XX YY;
COVBETA=GINV(COVBETA);BETA=SHAPE(COVBETA*BETA,R)';
%
MACRO SIGMA_AV SIGMA=J(PP,PP,0);
DO J=1 TO PP BY 1;DO JJ=J TO PP BY 1;
   UL=0;LL=0;J2=J||JJ;
   DO KK=1 TO K BY 1;LL=UL+1;UL=UL+NI(KK,);
      IF PATTERN(KK,J)#PATTERN(KK,JJ)=1 THEN DO;
         YY=YY||Y(LL:UL,J2);XX=XX||((X(LL:UL,)*BETA*PT)(,J2));
      END;
   END;NN=NROW(YY);XY=XX'*YY;YY=YY(,1)'*YY(,2);XX=XX(,1)'*XX(,2);
   SIGMA(J,JJ)=(YY+XX-XY(1,2)-XY(2,1))#/NN;SIGMA(JJ,J)=SIGMA(J,JJ);
   FREE XX YY XY;
END;END;EIGEN EE E SIGMA;
IF ANY(EE<0) THEN DO;
   JJ=LOC(EE<0);EE(JJ,)=J(NROW(JJ),1,0);SIGMA=E*DIAG(EE)*E';
END;
%
MACRO HYPTEST PROC MATRIX FUZZ;
FETCH BETA DATA=BETA;FETCH COVBETA DATA=COVBETA;
FETCH C DATA=CPRIME COLNAME=RC;C=C';FETCH U DATA=U COLNAME=CU;
FETCH X 1 DATA=X COLNAME=CC;FETCH P 1 DATA=PPRIME COLNAME=RU;
W=SHAPE((C*BETA*U)',1);FREE X P;
NOTE 'THE RESULTS OF TESTING HO:C*BETA*U = 0';
PRINT C COLNAME=CC ROWNAME=RC;PRINT U COLNAME=CU ROWNAME=RU;
U=U'@C;W=W'*GINV(U*COVBETA*U')*W;
DF=NROW(U);PVALUE=1-PROBCHI(W,DF);
RESULTS=W||DF||PVALUE;CTITLE='CHI_SQ' 'DEG_FREE' 'P-VALUE';RTITLE=' ';
PRINT RESULTS COLNAME=CTITLE ROWNAME=RTITLE;
%
```

## REFERENCES


Anderson, T. W. (1984). *An Introduction to Multivariate Statistical Analysis*, 2nd ed., Wiley, New York.

Anscombe, F. J. and Tukey, J. W. (1963). The Examination and Analysis of Residuals, *Technometrics*, 5: 141–160.

Atkinson, A. C. (1986). Diagnostic Tests for Transformations, *Technometrics*, 28: 29–37.

Bartlett, M. S. (1937). Some Examples of Statistical Methods of Research in Agriculture and Biology, *J. R. Stat. Soc. Suppl.*, 3: 137–170.

Bickel, P. J. and Doksum, K. A. (1981). An Analysis of Transformations Revisited, *J. Am. Stat. Assoc.*, 76: 296–311.

Blackwelder, W. C. (1982). Proving the Null Hypothesis in Clinical Trials, *Controlled Clin. Trials*, 3: 345–353.

Bock, D. R. and Peterson, A. C. (1975). A Multivariate Correction for Attenuation, *Biometrika*, 62: 673–678.

Box, G. E. P. and Cox, D. R. (1964). An Analysis of Transformations, *J. R. Stat. Soc. Ser. B*, 26: 211–243.

Box, G. E. P. and Cox, D. R. (1982). An Analysis of Transformations Revisited, Rebutted, *J. Am. Stat. Assoc.*, 77: 209–210.

Brown, B. W. (1980). The Crossover Experiment for Clinical Trials, *Biometrics*, 36: 69–79.

Brown, M. B. and Forsythe, A. B. (1974). Robust Tests for the Equality of Variance, *J. Am. Stat. Assoc.*, 69: 364–367.

Bryant, E. and Gillings, D. (1985). Statistical Analysis of Longitudinal Repeated Measures Designs, *Biostatistics: Statistics in Biomedical Public Health and Environmental Sciences* (P. K. Sen, ed.), North-Holland, New York, pp. 251–282.

Carroll, R. J. (1982). Tests for Regression Parameters in Power Transformation Models, *Scand. J. Stat.*, 9: 217–222.

Carroll, R. J. and Ruppert, D. (1981). Prediction and the Power Transformation Family, *Biometrika*, 68: 609–616.

Carter, W. H., Jr., Chinchilli, V. M., Campbell, E. D., and Wampler, G. L. (1984). Confidence Interval About the Response at the Stationary Point of a Response Surface with an Application to Pre-Clinical Cancer Therapy, *Biometrics*, 40: 1125–1130.

Chinchilli, V. M. and Carter, W. H., Jr. (1984). A Likelihood Ratio Test for a Patterned Covariance Matrix in a Multivariate Growth-Curve Model, *Biometrics*, 40: 151–156.

Chinchilli, V. M. and Elswick, R. K. (1985). A Mixture of the MANOVA and GMANOVA Models, *Commun. Stat.—Theory Methods*, 14: 3075–3089.

Dempster, A. P., Laird, N. M., and Rubin, D. B. (1977). Maximum Likelihood from Incomplete data via the EM Algorithm, *J. R. Stat. Soc. Ser. B*, 39: 1–38.

Draper, N. R. and Smith, H. (1981). *Applied Regression Analysis*, 2nd ed., Wiley, New York.

Eaton, M. and Kariya, T. (1983). Multivariate Tests with Incomplete Data, *Ann. Stat.*, 11: 654–665.

Ebbutt, A. F. (1984). Three-Period Crossover Designs for Two Treatments, *Biometrics*, 40: 219–224.

Eckberg, D. L., Harkins, S. W., Fritsch, J. M., Musgrave, G. E., and Gardner, D. F. (1986). Baroflex Control of Plasma Norepinephrine and Heart Period in Healthy Subjects and Diabetic Patients, *J. Clin. Invest.*, 78: 366–374.

Elswick, R. K. (1985). The Missing Data Problem as Applied to the Extended Version of the GMANOVA Model, Ph.D. dissertation, Virginia Commonwealth University, Richmond, Va.

Engelman, L. (1980). An Efficient Algorithm for Computing Covariance Matrices from Data with Missing Values, *BMDP Tech. Rep. No. 76*, BMDP Statistical Software, Los Angeles.

Fairclough, D. L. and Helms, R. W. (1983). "PROC MATRIX Macros for Generalized Incomplete (GIM) and Multiple design (MDM) Multivariate Models," Proc. 8th Annual SAS Users Group International Converence, SAS Institute, Cary, N.C., pp. 883–887.

Fisher, R. A. (1932). *Statistical Methods for Research Workers*, 4th ed., Oliver & Boyd, London.

Fisher, A. C. and Wallenstein, S. (1981). Crossover Designs in Medical Research, *Statistics in the Pharmaceutical Industry* (C. R. Buncher and J. Y. Tsay, eds.), Marcel Dekker, New York, pp. 139–156.

Fleiss, J. L. (1985). *The Design and Analysis of Clinical Experiments*, Wiley, New York.

Friedman, L. M., Furberg, C. D., and DeMets, D. L. (1981). *Fundamentals of Clinical Trials*, John Wright, PSG, Boston.

Gnanadesikan, R. (1977). *Methods for Statistical Data Analysis of Multivariate Observations*, Wiley, New York.

Greenhouse, S. W. and Geisser, S. (1959). On Methods in Analysis of Profile Data, *Psychometrika*, 24: 95–112.

Grizzle, J. E. (1965). The Two-Period Change-Over Design and Its Use in Clinical Trials, *Biometrics*, 21: 467–480. [Corrigenda: (1974). *Biometrics*, 30: 727.]

Grizzle, J. E. and Allen, D. M. (1969). Analysis of Growth and Dose Response Curves, *Biometrics*, 25: 357–381.

Hájek, J. and Šidák, Z. (1967). *Theory of Rank Tests*, Academic Press, New York.

Hartley, H. O. and Hocking, R. R. (1971). The Analysis of Incomplete Data, *Biometrics*, 27: 783–823.

Hearne, E. M., III, Clark, G. M., and Hatch, J. P. (1983). A Test for Serial Correlation in Univariate Repeated-Measures Analysis, *Biometrics*, 39: 237–243.

Hedges, L. V. and Olkin, I. (1985). *Statistical Methods for Meta-Analysis*, Academic Press, New York.

Hernandez, F. and Johnson, R. A. (1980). The Large Sample Behavior of Transformations to Normality, *J. Am. Stat. Assoc.*, 75: 855–861.

Hettmansperger, T. P. (1984). *Statistical Inference Based on Ranks*, Wiley, New York.

Hills, M. and Armitage, P. (1979). The Two-Period Cross-Over Clinical Trial, *Br. J. Clin. Pharmacol.*, 8: 7–20.

Hinkley, D. V. and Runger, G. (1984). The Analysis of Transformed Data, *J. Am. Stat. Assoc.*, 79: 302–309.

Hosking, J. D. (1981). "Missing Data in Multivariate Linear Models: A Comparison of Several Estimation Techniques," Proc. 6th Annual SUGI Conference, SAS Institute, Cary, N.C., pp. 46–51.

Hsu, J.-P. (1983). "The Assessment of Statistical Evidence from Active Control Clinical Trials," ASA 1983, Proc. of the Biopharmaceutical Section, pp. 12–18.

Hui, S. L. (1984). Curve Fitting for Repeated Measurements Made at Irregular Time-Points, *Biometrics*, 40: 691–697.

Huynh, H. and Feldt, L. S. (1970). Conditions Under Which Mean Square Ratios in Repeated Measurements Designs Have Exact F-Distributions, *J. Am. Stat. Assoc.*, 65: 1582–1589.

Huynh, H. and Feldt, L. S. (1976). Estimation of the Box Correction for Degrees of Freedom from Sample Data in the Randomized Block and Split-Plot Designs, *J. Educ. Stat.*, 1: 69–82.

Jensen, D. R. (1986). "Topics in the Analysis of Repeated Measurements," presented at the Virginia Academy of Science, 64th Annual Meeting, Harrisonburg, Va., May 16.

Johnson, R. A. and Wichern, D. W. (1982). *Applied Multivariate Statistical Analysis*, Prentice-Hall, Englewood Cliffs, N.J.

Johnson, R. E. and Gunn, R. D. (1984). "A Comparison of Error Probabilities for Some Nonorthogonal Analyses of Variance," presented at the Joint Statistical Meetings, Philadelphia, August 14.

Khatri, C. G. (1966). A Note on a MANOVA Model Applied to Problems in Growth Curves, *Ann. Inst. Stat. Math.*, 18: 75–86.

Kiefer, J. (1975). Construction and Optimality of Generalized Youden Designs, *A Survey of Statistical Design and Linear Models* (J. N. Srivastava, ed.), North-Holland, New York, pp. 333–353.

Kleinbaum, D. G. (1973). A Generalization of the Growth Curve Model Which Allows for Missing Data, *J. Multivariate Analy.*, 3: 117–124.

Koch, G. G. (1972). The Use of Nonparametric Methods in the Statistical Analysis of the Two-Period Changeover Design, *Biometrics*, 28: 577–584.

Koziol, J. A., Maxwell, D. A., Fukushima, M., Colmerauer, M. E. M., and Pilch, Y. H. (1981). A Distribution-Free Test for Tumor-Growth Curve Analyses with Application to an Animal Tumor Immunotherapy Experiment, *Biometrics*, 37: 383–390.

Krishnaiah, P. R. (1979). Some Developments on Simultaneous Test Procedures, *Developments in Statistics*, vol. 2 (P. R. Krishnaiah, ed.), Academic Press, New York, pp. 157–210.

Krishnaiah, P. R., Mudholkar, G. S., Subbaiah, P. (1980). Simultaneous Test Procedures for Mean Vectors and Covariance Matrices, *Handbook of Statistics*, vol. 1 (P. R. Krishnaiah, ed.), North-Holland, New York, pp. 631–671.

Laird, N. M. and Ware, J. H. (1982). Random-Effects Models for Longitudinal Data, *Biometrics*, 38: 963–974.

Lamborn, K. R. (1983). "Some Practical Issues and Concerns in Active Control Clinical Trials," ASA 1983, Proc. of the Biopharmaceutical Section, pp. 8–11.

Laska, E. and Meisner, M. (1986). "Testing Whether an Identified Treatment Is Best: The Combination Problem," presented at the Biometrics Society (ENAR), ASA, and IMS Meetings, Atlanta, March 18.

Laska, E., Meisner, M., and Kushner, H. B. (1983). Optimal Crossover Designs in the Presence of Carryover Effects, *Biometrics*, 39: 1087–1091.

Lee, J. C. and Geisser, S. (1972). Growth Curve Prediction, Sankhyā Ser. A, 34: 393–412.

Leeper, J. D. and Woolson, R. F. (1982). Testing Hypotheses for the Growth Curve Model When the Data Are Incomplete, *J. Stat. Comput. Simulation*, 15: 97–107.

Levene, H. (1960). Robust Tests for Equality of Variance, *Contributions to Probability and Statistics* (I. Olkin, ed.), Stanford University Press, Palo Alto, Calif., pp. 278–292.

Liski, E. P. (1985). Estimation from Incomplete Data in Growth Curve Models, *Comm. Stat.—Simulation Comput.*, 14: 13–27.

Malkovich, J. F. and Afifi, A. A. (1973). On Tests for Multivariate Normality, *J. Am. Stat. Assoc.*, 68: 176–179.

Mardia, K. V. (1970). Measures of Multivariate Skewness and Kurtosis with Applications, *Biometrika*, 57: 519–530.

Mardia, K. V. (1980). Tests of Univariate and Multivariate Normality, *Handbook of Statistics*, vol. 1 (P. R. Krishnaiah, ed.), North-Holland, New York, pp. 279–320.

Mauchly, J. W. (1940). Significance Test for Sphericity of a Normal n-Variate Distribution, *Ann. Math. Stat.*, 29: 204–209.

McDonald, L. L. (1971). On the Estimation of Missing Data in the Multivariate Linear Model, *Biometrics*, 27: 535–543.

McDonald, L. L. (1975). Tests for the General Linear Hypothesis Under the Multiple Design Multivariate Linear Model, *Ann. Stat.*, 3: 461–466.

Mead, R. and Pike, D. J. (1975). A Review of Response Surface Methodology from a Biometric Viewpoint, *Biometrics*, 31: 803–851.

Metzler, C. M. (1974). Bioavailability—A Problem in Equivalence, *Biometrics*, 30: 309–317.

Miller, R. G., Jr. (1981). *Simultaneous Statistical Inference*, 2nd ed., Springer-Verlag, New York.

Morrison, D. F. (1976). *Multivariate Statistical Methods*, 2nd ed., McGraw-Hill, New York.

Mudholkar, G. S. and Subbaiah, P. (1980). A Review of Step-Down Procedures for Multivariate Analysis of Variance, *Multivariate*

*Statistical Analysis* (R. P. Gupta, ed.), North-Holland, New York, pp. 161–178.

Muirhead, R. J. (1982). *Aspects of Multivariate Statistical Theory*, Wiley, New York.

Myers, R. H. (1976). *Response Surface Methodology*, Edwards Brothers, Ann Arbor, Mich.

Orchard, T. and Woodbury, M. A. (1972). A Missing Information Principle: Theory and Applications, *Proceedings of the 6th Berkeley Sumposium on Mathematical Statistics and Probability*, vol. 1, University of California Press, Berkeley, pp. 697–715.

Pledger, G. (1986). "Thoughts on the Design, Analysis and Interpretation of Combination Drug Trials," presented at the Biometric Society (ENAR), ASA, and IMS Meetings, Atlanta, March 18.

Pocock, S. J. (1983). *Clinical Trials: A Practical Approach*, Wiley, New York.

Potthoff, R. F. and Roy, S. N. (1964). A Generalized Multivariate Analysis of Variance Model Useful Especially for Growth Curve Problems, *Biometrika*, 51: 313–326.

Puri, M. L. and Sen, P. K. (1969). A Class of Rank Order Tests for a General Linear Hypothesis, *Ann. Math. Stat.*, 40: 610–618.

Puri, M. L. and Sen, P. K. (1971). *Nonparametric Methods in Multivariate Analysis*, Wiley, New York.

Puri, M. L. and Sen, P. K. (1985). *Nonparametric Methods in General Linear Models*, Wiley, New York.

Rao, C. R. (1965). Theory of Least Squares When the Parameters Are Stochastic and Its Application to the Analysis of Growth Curves, *Biometrika*, 52: 447–458.

Rao, C. R. (1967). Least Squares Theory Using an Estimated Dispersion Matrix and Its Application to Measurement of Signals, *Proceedings of the 5th Berkeley Symposium on Mathematical Statistics and Probability*, vol. 1 (L. M. LeCam and J. Neyman, eds.), University of California Press, Berkeley, pp. 355–372.

Reinsel, S. G. (1982). Multivariate Repeated Measurement or Growth Curve Models with Multivariate Random-Effects Covariance Structure, *J. Am. Stat. Assoc.*, 77: 190–195.

Rode, R. A. (1986). The Use of Box-Cox Transformations in the Development of Multivariate Tolerance Regions with Applications to Clinical Chemistry, Ph.D. dissertation, Virginia Commonwealth University, Richmond.

Rouanet, H. and Lepine, D. (1970). Comparison Between Treatments in a Repeated-Measures Design: ANOVA and Multivariate Methods, *Br. J. Math. Stat. Psychol.*, 23: 147–163.

Roy, S. N. and Bose, R. C. (1953). Simultaneous Confidence Interval Estimation, *Ann. Math. Stat.*, 24: 513–536.

Rubin, D. P. (1976). Inference and Missing Data, *Biometrika*, 63: 581–592.

Schwab, B. H. (1984). Parametric and Non-Parametric Analysis of the Multiple Design Multivariate Linear Model, Ph.D. dissertation, Virginia Commonwealth University, Richmond.

Schwertman, N. C. and Allen, D. M. (1979). Smoothing an Indefinite Variance-Covariance Matrix, *J. Stat. Comput. Simulation*, 9: 183–194.

Schwiderski, U. E. (1986). *A Unified MANOVA-GMANOVA Theory and Splines*, Ph.D. dissertation, Virginia Commonwealth University, Richmond.

Searle, S. R. (1971). *Linear Models*, Wiley, New York.

Selwyn, M. R., Dempster, A. P., and Hall, N. R. (1981). A Bayesian Approach to Bioequivalence for the 2 × 2 Changeover Design, *Biometrics*, 3: 11–21.

Sen, P. K. (1980). Nonparametric Simultaneous Inference for Some MANOVA Models, *Handbook of Statistics*, vol. 1 (P. R. Krishnaiah, ed.), North-Holland, New York, pp. 673–702.

Shapiro, S. S. and Wilk, M. B. (1965). An Analysis of Variance Test for Normality (Complete Samples), *Biometrika*, 52: 591–611.

Srivastava (1985). Multivariate Data with Missing Observations, *Comm. Stat.—Theory Methods*, 14: 775–792.

Taulbee, J. D. (1982). A Note on the Use of Nonparametric Methods in the Statistical Analysis of the Two-Period Changeover Deisgn, *Biometrics*, 38: 1053–1055.

Taylor, J. M. G. (1985). Measures of Location of Skew Distributions Obtained Through Box-Cox Transformations, *J. Am. Stat. Assoc.*, 80: 427–432.

Temple, R. (1983). "Difficulties in Evaluating Positive Control Trials," ASA 1983, Proc. of the Biopharmaceutical Section, pp. 1–7.

Theil, H. (1965). The Analysis of Disturbances in Regression Analysis, *J. Am. Stat. Assoc.*, 60: 1067–1079.

Theil, H. (1968). A Simplification of the BLUS Procedure for Analyzing Regression Disturbances, *J. Am. Stat. Assoc.*, 63: 242–251.

Thomas, D. R. (1983). Univariate Repeated Measures Techniques Applied to Multivariate Data, *Psychometrika*, 48: 451–464.

Verbyla, A. P. (1986). Conditioning in the Growth Curve Model, *Biometrika*, 73: 475–483.

Vonesh, E. F. and Story, K. O. (1986). "A Generalized Growth Curve Procedure for the Analysis of Incomplete Longitudinal Data," Proc. of the 11th Annual SAS Users Group International Conference, SAS Institute, Cary, N.C., pp. 889–894.

Vonesh, E. F. and Carter, R. L. (1987). Efficient Inference for Random Coefficient Growth Curve Models with Unbalanced Data, to appear in *Biometrics*.

Ware, J. H. (1985). Linear Models for the Analysis of Longitudinal Studies, *Am. Stat.*, 39: 95–101.

Westlake, W. J. (1972). Use of Confidence Intervals in Analysis of Comparative Bioavailability Trials, *J. Pharm. Sci.*, 61: 1340–1341.

Westlake, W. J. (1976). Symmetrical Confidence Intervals for Bioequivalence Trials, *Biometrics*, 32: 741–744.

Westlake, W. J. (1979). Statistical Aspects of Comparative Bioavailability Trials, *Biometrics*, 35: 273–280.

Wu, C. F. J. (1983). On the Convergence Properties of the EM Algorithm, *Ann. Stat.*, 11: 95–103.

Zellner, A. (1962). An Efficient Method of Estimating Seemingly Unrelated Regressions and Tests for Aggregration Bias, *J. Am. Stat. Assoc.*, 57: 348–368.

# 9 Clinical Efficacy Trials with Categorical Data

GARY G. KOCH *Biostatistics Department, School of Public Health, University of North Carolina, Chapel Hill, North Carolina*

SUZANNE EDWARDS *Clinical Statistics Department, Burroughs Wellcome Co., Research Triangle Park, North Carolina*

## I. INTRODUCTION

Efficacy response measures in clinical trials often take the form of categorical variables. Such response variables may be dichotomous (e.g., healed vs. not healed), ordinal (e.g., symptom severity of none, mild, moderate, or severe), discrete counts (e.g., number of occurrences of some symptom), or grouped survival times (e.g., time interval of recurrence of symptoms). An overview of the primary methods used in the analysis of categorical efficacy response measures is shown in Table 1. Examples of both nonparametric randomization-based methods, such as Fisher's exact test, and model-based methods, such as logistic regression, appear in this table. The rationale for choosing one of these methods of inference over the other lies in the assumptions one is willing to make about the connection between the sample of patients enrolled in the trial and the target population to which results are to be generalized.

In most cases, the patients in a clinical trial are a convenience sample, chosen by their ability to satisfy the study protocol, and the investigators are a judgment sample, chosen by their expertise in the area of study. Given this sampling process, the patients in a clinical trial may not be statistically representative of any well-defined target population. If assumptions which involve random sampling from a larger target population are not tenable, then nonparametric randomization-based methods may be more appropriate than model-based methods. Nonparametric randomization-based methods require no assumptions other than the randomization of patients to treatment groups. However, these methods have limited scope in that the conclusions drawn on a statistical basis apply only to the patients who were actually randomized; generalization to a larger population requires nonstatistical arguments about the representativeness of the observed patients (see Koch et al., 1982; Koch and Gillings, 1983). Also, nonparametric randomization-based methods are useful mainly in hypothesis testing; they must be supplemented by other methods for estimation purposes.

Table 1 Overview of Statistical Methods for Categorical Efficacy Response Variables

| Response[a] variable | Randomization-based nonparametric methods | | Model-based methods (single or combined investigators) |
|---|---|---|---|
| | Single investigator | Combined investigators | |
| Dichotomous | Fisher's exact test | Mantel-Haenszel test | Maximum likelihood logistic regression, weighted least squares for correlated proportions from repeated measures |
| Ordinal or discrete counts | Wilcoxon rank sum or Kruskal-Wallis test | Extensions of the Mantel-Haenszel procedure | Extensions of logistic regression (e.g., the proportional odds model), weighted least squares for rank measures of association or mean scores |
| Grouped survival times | Logrank test | Stratified logrank test | Maximum likelihood fitting of piecewise exponential models |

[a]Polytomous nominal variables are not included in the table because they are rarely used as efficacy variables in clinical trials. Such variables can be analyzed with chi-square tests, multivariate forms of the Mantel-Haenszel test, or log-linear model extensions of logistic regression.

Model-based methods are appropriate if it can be assumed that the patients selected for the trial are equivalent to a (possibly stratified) random sample of some larger population. Assumed probability models can then be used to describe the relationship of the response variable to explanatory variables such as treatment, investigator, and pretreatment patient characteristics. As noted by Koch and Sollecito (1984), model-based methods have several advantages over nonparametric randomization-based methods: (1) the capacity to assess the homogeneity of effects across strata (e.g., center-by-treatment interactions), (2) greater flexibility with respect to adjustment for demographic or pretreatment variables which are not equivalently distributed for the treatment groups, and (3) more powerful analysis through the framework of reduced variability provided by adjustment for explanatory variables which are strongly associated with the response variable. To clarify advantages (2) and (3) further, it must be noted that equivalent distributions for demographic and pretreatment variables are expected as a consequence of the randomized assignment of treatments. When noteworthy imbalances in important explanatory variables do occur, they can be adjusted for with randomization-based methods via stratification. The advantage of model-based methods for such adjustments is that a larger number of explanatory variables may be adjusted for by incorporating them in the model, and these variables may be either categorical or continuous. The disadvantages of model-based methods are that they may be more difficult to implement and interpret than randomization-based methods. Also, since the assumptions necessary for model-based methods cannot be proved, the results of such analyses are more likely to be subject to debate. A reasonable strategy is thus to use both types of methods in combination.

In this chapter, a variety of model-based and nonparametric randomization-based methods for analysis of categorical data are reviewed, using examples from clinical trials with different data structures (i.e., univariate, multivariate) and measurement scales for the response and explanatory variables.

## II. ANALYSIS OF 2 × 2 TABLES

The data in Table 2 are from a randomized double-blind placebo-controlled clinical trial in patients with rheumatoid arthritis. Patients evaluated the effectiveness of treatment according to a three-point scale: no improvement, some improvement, or marked improvement. The explanatory variables of interest are treatment (test drug vs. placebo), sex, and age.

If one reduces the response variable to a dichotomous variable and considers treatment as the only explanatory variable of interest, the data may be presented as follows:

| | Improvement | | |
|---|---|---|---|
| Treatment | None | Some or marked | Total |
| Test drug | $n_{11} = 13$ | $n_{12} = 28$ | $n_{1+} = 41$ |
| Placebo | $n_{21} = 29$ | $n_{22} = 14$ | $n_{2+} = 43$ |
| Total | $n_{+1} = 42$ | $n_{+2} = 42$ | $n = 84$ |

(1)

Table 2 Rheumatoid Arthritis Data

| Test drug treatment | | | | Placebo treatment | | | |
|---|---|---|---|---|---|---|---|
| Pt. # | Sex | Age | Improvement[a] | Pt. # | Sex | Age | Improvement[a] |
| 57 | M | 27 | 1 | 9 | M | 37 | 0 |
| 46 | M | 29 | 0 | 14 | M | 44 | 0 |
| 77 | M | 30 | 0 | 73 | M | 50 | 0 |
| 17 | M | 32 | 2 | 74 | M | 51 | 0 |
| 36 | M | 46 | 2 | 25 | M | 52 | 0 |
| 23 | M | 58 | 2 | 18 | M | 53 | 0 |
| 75 | M | 59 | 0 | 21 | M | 59 | 0 |
| 39 | M | 59 | 2 | 52 | M | 59 | 0 |
| 33 | M | 63 | 0 | 45 | M | 62 | 0 |
| 55 | M | 63 | 0 | 41 | M | 62 | 0 |
| 30 | M | 64 | 0 | 8 | M | 63 | 2 |
| 5 | M | 64 | 1 | 80 | F | 23 | 0 |
| 63 | M | 69 | 0 | 12 | F | 30 | 0 |
| 83 | M | 70 | 2 | 29 | F | 30 | 0 |
| 66 | F | 23 | 0 | 50 | F | 31 | 1 |
| 40 | F | 32 | 0 | 38 | F | 32 | 0 |
| 6 | F | 37 | 1 | 35 | F | 33 | 2 |
| 7 | F | 41 | 0 | 51 | F | 37 | 0 |
| 72 | F | 41 | 2 | 54 | F | 44 | 0 |
| 37 | F | 48 | 0 | 76 | F | 45 | 0 |
| 82 | F | 48 | 2 | 16 | F | 46 | 0 |
| 53 | F | 55 | 2 | 69 | F | 48 | 0 |
| 79 | F | 55 | 2 | 31 | F | 49 | 0 |
| 26 | F | 56 | 2 | 20 | F | 51 | 0 |
| 28 | F | 57 | 2 | 68 | F | 53 | 0 |
| 60 | F | 57 | 2 | 81 | F | 54 | 0 |
| 22 | F | 57 | 2 | 4 | F | 54 | 0 |
| 27 | F | 58 | 0 | 78 | F | 54 | 2 |
| 2 | F | 59 | 2 | 70 | F | 55 | 2 |
| 59 | F | 59 | 2 | 49 | F | 57 | 0 |
| 62 | F | 60 | 2 | 10 | F | 57 | 1 |
| 84 | F | 61 | 2 | 47 | F | 58 | 1 |

**Table 2** (continued)

| Test drug treatment | | | | Placebo treatment | | | |
|---|---|---|---|---|---|---|---|
| Pt. # | Sex | Age | Improvement[a] | Pt. # | Sex | Age | Improvement[a] |
| 64 | F | 62 | 1 | 44 | F | 59 | 1 |
| 34 | F | 62 | 2 | 24 | F | 59 | 2 |
| 58 | F | 66 | 2 | 48 | F | 61 | 0 |
| 13 | F | 67 | 2 | 19 | F | 63 | 1 |
| 61 | F | 68 | 1 | 3 | F | 64 | 0 |
| 65 | F | 68 | 2 | 67 | F | 65 | 2 |
| 11 | F | 69 | 0 | 32 | F | 66 | 0 |
| 56 | F | 69 | 1 | 42 | F | 66 | 0 |
| 43 | F | 70 | 1 | 15 | F | 66 | 1 |
| | | | | 71 | F | 68 | 1 |
| | | | | 1 | F | 74 | 2 |

[a]Improvement: 0, none; 1, some; 2, marked.

Although a few patients may have been excluded from each group due to protocol violations, the row marginal totals ($n_{1+}$, $n_{2+}$) can be considered fixed by the treatment allocation process. The column marginal totals ($n_{+1}$, $n_{+2}$) can also be considered as fixed under the null hypothesis $H_0$ of no treatment difference for each patient, $n_{+1}$ and $n_{+2}$ being respectively the number with no improvement and the number with at least some improvement, regardless of treatment. Given that all marginal totals are fixed, the probability model implied by randomization is given by the hypergeometric distribution, i.e.,

$$\Pr\{n_{ij} \mid H_0\} = \frac{n_{1+}!\ n_{2+}!\ n_{+1}!\ n_{+2}!}{n!\ n_{11}!\ n_{12}!\ n_{21}!\ n_{22}!} \tag{2}$$

Thus, the expected value of $n_{11}$ is

$$E\{n_{11} \mid H_0\} = \frac{n_{1+}n_{+1}}{n} = m_{11} = 20.50 \tag{3}$$

and the variance is

$$V\{n_{11} \mid H_0\} = \frac{n_{1+}n_{2+}n_{+1}n_{+2}}{n^2(n-1)} = v_{11} = 5.31 \tag{4}$$

If the sample size is sufficiently large, $n_{11}$ has approximately a normal distribution by central limit theory (Hannan and Harkness, 1963; Puri and Sen, 1971; Plackett, 1981), and so

$$Q = \frac{(n_{11} - m_{11})^2}{v_{11}} \tag{5}$$

approximately has a chi-square distribution with 1 degree of freedom (d.f.). The value of Q does not depend on which of the four cells is used in the calculations because for the ij-th cell, $(n_{ij} - m_{ij}) = \pm(n_{11} - m_{11})$ and $v_{ij} = v_{11}$, where $m_{ij}$ and $v_{ij}$ are the expected value and variance of $n_{ij}$ under $H_0$. The statistic Q can also be written as

$$Q = \frac{\{(n_{1+}n_{2+}/n)(p_{11} - p_{21})\}^2}{v_{11}} \tag{6}$$

where $p_{i1} = (n_{i1}/n_{i+})$ denotes the proportion of patients in the ith group with no improvement. This expression shows how larger values of Q can be interpreted as indicating larger differences between treatments with respect to the proportions of patients with no improvement; in this sense, $(p_{11} - p_{21})$ describes the association between treatment and response.

The relationship between the randomization chi-square statistic, Q, and the well-known Pearson chi-square statistic, $Q_P$, where

$$Q_P = \sum_{i=1}^{2}\sum_{j=1}^{2} \frac{(n_{ij} - m_{ij})^2}{m_{ij}} = \frac{n(n_{11}n_{22} - n_{12}n_{21})^2}{n_{1+}n_{2+}n_{+1}n_{+2}} \tag{7}$$

for 2 × 2 tables is $Q = [(n - 1)/n]Q_P$. Thus, in large samples the two statistics are nearly identical. In our example, $Q = 10.59$ ($p < 0.01$) and $Q_P = 10.72$ ($p < 0.01$).

A commonly used rule for the suitability of chi-square tests such as Q or $Q_P$ to 2 × 2 contingency tables is that the expected value, $m_{ij} = n_{i+}n_{+j}/n$, for all cells should exceed 5.0. This is true in our example. A more appropriate method than chi-square tests when cell counts are small is Fisher's exact test. The p value for Fisher's (two-sided) exact test is found by enumerating all tables with the same marginal totals as the observed table, calculating the probability of each table using (2), and then summing the probabilities of those tables which are as likely as or less likely than the observed table. For the one-sided Fisher's exact test, summation is applied to the probabilities of tables with association at least as strong as that for the observed table in the direction specified by the one-sided alternative (here, tables in which the test drug has the more favorable response). In our example, the p value from a two-sided Fisher's exact test is 0.002, and the p value from a one-sided Fisher's exact test is 0.001. When the sample sizes in the two treatment groups are nearly equal, as in our example, the p value from a two-sided Fisher's exact test is approximately twice the p value from a one-sided test; when they are actually equal, this approximation becomes an identity. However, when the sample sizes differ, the

Fisher's exact test is usually nonsymmetric, and the two-sided p value can be notably less than twice the one-sided p value.

Some authors (Haber, 1986; Overall and Starbuck, 1983; Salama et al., 1984) have advocated the use of exact tests which are more powerful than Fisher's exact test. However, Fisher's exact test has the advantage of following directly from randomization, whereas other tests require that one assumes the data for each treatment have a binomial distribution. Justification of this assumption requires that the patients in each group can be viewed as equivalent to a random sample from some larger target population. For situations in which the assumption of binomial distributions is reasonable, a numerical study by Upton (1982) suggests that chi-square approximations for Q are somewhat better than those for $Q_P$.

A continuity correction is often applied to both the randomization and the Pearson chi-square statistics to make the results agree more closely with results of an exact test. The continuity-corrected randomization statistic is

$$Q_C = \frac{\{|n_{11} - m_{11}| - 0.5\}^2}{v_{11}} \qquad (8)$$

and the continuity-corrected Pearson statistic (Yates, 1934) is

$$Q_{PC} = \sum_{i=1}^{2}\sum_{j=1}^{2} \frac{\{|n_{ij} - m_{ij}| - 0.5\}^2}{m_{ij}} \qquad (9)$$

In our example, the continuity-corrected Pearson statistic, $Q_{PC} = 9.34$ ($p = 0.002$), does more closely approximate the two-sided Fisher's exact test than the uncorrected statistic. Fleiss (1981) recommends that the Pearson chi-square statistic always be corrected for continuity, although there is some controversy on this point (see Grizzle, 1967; Conover, 1974). One aspect of this controversy concerns whether evaluation is based on the hypergeometric distribution induced by randomization or on the assumption of binomial distributions for each treatment. Another consideration is that in nonsymmetric situations, the one-sided p value from a continuity-corrected chi-square statistic tends to approximate the one-sided Fisher's exact test p value well, but its two-sided counterpart (which is twice as large) then tends to be larger than the two-sided Fisher's exact test p value to an overly conservative extent.

In summary, for the analysis of 2 × 2 tables from clinical efficacy trials, Fisher's exact test is always applicable and easily obtained by computer. Approximations for it (i.e., all chi-square tests, continuity-corrected or not) should be used cautiously, particularly in nonsymmetric situations, unless the sample sizes are sufficiently large (e.g., all $m_{ij} \geq 10$).

## III. ANALYSIS OF s × r TABLES

Another way to view the data in Table 2 is to construct a 2 × 3 contingency table for the cross-classification of treatment and the ordinal response variable.

| | Improvement | | | |
|---|---|---|---|---|
| Treatment | None | Some | Marked | Total |
| Test drug | 13 | 7 | 21 | 41 |
| Placebo | 29 | 7 | 7 | 43 |
| Total | 42 | 14 | 28 | 84 |

(10)

Let $n_{ij}$ denote the number of patients who received the ith treatment ($i = 1, 2$) and had the jth response ($j = 1, 2, 3$). Again, the row marginals, $n_{i+}$, may be considered fixed by the treatment allocation process, and the column marginals, $n_{+j}$, may be considered fixed under the null hypothesis $H_0$ of no treatment effect for each patient by considering the $n_{+j}$ as the number with the jth level of improvement, regardless of treatment. The probability model implied by randomization is

$$\Pr\{n_{ij} \mid H_0\} = \frac{\Pi_{i=1}^{2} n_{i+}! \ \Pi_{j=1}^{3} n_{+j}!}{n! \ \Pi_{i=1}^{2} \Pi_{j=1}^{3} n_{ij}!} \tag{11}$$

which is sometimes called the multivariate hypergeometric model. The frequencies $n_{ij}$ have expected values

$$E\{n_{ij} \mid H_0\} = \frac{n_{i+} n_{+j}}{n} \tag{12}$$

and covariance structure

$$\text{Cov}\{n_{ij}, n_{i'j'} \mid H_0\} = \frac{m_{ij}(n\delta_{ii'} - n_{i'+})(n\delta_{jj'} - n_{+j'})}{n(n-1)} \tag{13}$$

where $\delta_{kk'} = 1$ if $k = k'$ and $\delta_{kk'} = 0$ if $k \neq k'$.

The form of the Pearson chi-square statistic for $s \times r$ tables is the same as (7) except that the summation for i is from 1 to s, where s is the number of groups (e.g., treatments; here $s = 2$), and the summation for j is from 1 to r, where r is the number of response levels (here $r = 3$). For sufficiently large samples (e.g., all $m_{ij} \geq 5$), $Q_P$ approximately has the chi-square distribution with $(r - 1)(s - 1)$ degrees of freedom. In this example, $Q_P = 13.06$, d.f. = 2, and $p < 0.01$.

The general form of the randomization chi-square statistic counterpart of $Q_P$ is given by the quadratic form

$$Q = (\underline{n} - \underline{m})' \, \underline{A}' \, (\underline{A}\,\underline{V}\,\underline{A}')^{-1} \underline{A}(\underline{n} - \underline{m}) \tag{14}$$

where $\underline{n} = (13, 7, 21, 29, 7, 7)'$ is the compound vector of observed frequencies, $\underline{m}$ is the corresponding $6 \times 1$ vector of expected frequencies from (12), $\underline{V}$ is the $6 \times 6$ covariance matrix from (13), and $\underline{A}$ is any $2 \times 6$ matrix such that $\underline{A}\,\underline{V}\,\underline{A}'$ is nonsingular. For $\underline{A}\,\underline{V}\,\underline{A}'$ to be nonsingular, $\underline{A}$

must be linearly independent of the restrictions that sums of the $(n_{ij} - m_{ij})$ in the same row or in the same column are identically zero; thus, Rank $[\underline{A}', \underline{R}']$ must equal Rank$[\underline{A}']$ + Rank$[\underline{R}']$, where $\underline{R}$ is a basis for the restrictions, e.g.,

$$\underline{R} = \begin{bmatrix} 1 & 1 & 1 & 0 & 0 & 0 \\ 0 & 0 & 0 & 1 & 1 & 1 \\ 1 & 0 & 0 & 1 & 0 & 0 \\ 0 & 1 & 0 & 0 & 1 & 0 \end{bmatrix} \tag{15}$$

One convenient choice is $\underline{A} = [\underline{I}_2, \underline{0}_{2,4}]$, where $\underline{I}_2$ is the $2 \times 2$ identity matrix and $\underline{0}_{2,4}$ is the $2 \times 4$ matrix of zeros. As noted in Koch et al. (1982) and other references, $Q = (n - 1)Q_P/n$ for any $\underline{A}$ satisfying the above criteria, so Q has the same approximate chi-square distribution as $Q_P$ with d.f. = $(r - 1)(s - 1)$. In this example, Q = 12.90, d.f. = 2, and $p < 0.01$.

Although Q and $Q_P$ are useful for detecting general types of departures from $H_0$, they are not as effective as other methods for alternatives involving location shifts across ordinal response levels (e.g., the more favorable response categories being more likely for some treatments than others). A class of randomization statistics which accounts for ordinal response levels can be constructed by changing the form of the $\underline{A}$ matrix in (14). Let $\underline{a} = \{a_j\} = (a_1, a_2, a_3)'$ be a set of scores reflecting response levels. Then the mean score for the test drug group is

$$\bar{f}_1 = \sum_{j=1}^{3} \frac{a_j n_{1j}}{n_{1+}} \tag{16}$$

which has expected value

$$E\{\bar{f}_1 \mid H_0\} = \sum_{j=1}^{3} \left( a_j \frac{n_{1+} n_{+j}}{n_{1+} n} \right) = \sum_{j=1}^{3} a_j \frac{n_{+j}}{n} = \mu_{\underline{a}} \tag{17}$$

and variance

$$\begin{aligned} \text{Var}\{\bar{f}_1 \mid H_0\} &= \sum_{j=1}^{3} \sum_{j'=1}^{3} a_j a_{j'} \frac{n_{1+}(n - n_{1+}) n_{+j} (n\delta_{jj'} - n_{+j'})}{n^2 n_{1+}^2 (n - 1)} \\ &= \frac{n - n_{1+}}{n_{1+}(n - 1)} \sum_{j=1}^{3} (a_j - \mu_{\underline{a}})^2 \left( \frac{n_{+j}}{n} \right) \\ &= \frac{(n - n_{1+}) v_{\underline{a}}}{n_{1+}(n - 1)} \end{aligned} \tag{18}$$

where $\mu_{\underline{a}}$ and $v_{\underline{a}}$ are the finite population mean and variance of scores $\underline{a}$ for the 84 patients on study. In large-sample situations, $\bar{f}_1$ approximately has a normal distribution by randomization central limit theory, and so the mean score statistic

$$Q_S = \frac{(\bar{f}_1 - \mu_{\underline{a}})^2}{[(n - n_{1+})/n_{1+}(n-1)]v_{\underline{a}}}$$

$$= \left(\frac{n-1}{n}\right)\frac{(\bar{f}_1 - \bar{f}_2)^2}{[(1/n_{1+}) + (1/n_{2+})]v_{\underline{a}}} \tag{19}$$

approximately has the chi-square distribution with d.f. = 1. The general expression of this statistic is

$$Q_S = (\underline{n} - \underline{m})'\underline{A}_S'(\underline{A}_S\underline{V}\underline{A}_S')^{-1}\underline{A}_S(\underline{n} - \underline{m}) \tag{20}$$

where $\underline{A}_S = [\underline{a}', -\underline{a}']$ for the comparison of s = 2 treatments.

When the overall comparison among s > 2 treatments is of interest for a response with r levels, a convenient choice for $\underline{A}_S$ is the (s − 1) × sr matrix

$$\begin{bmatrix} \underline{a}' & \underline{0}' & \cdots & \underline{0}' & -\underline{a}' \\ \underline{0}' & \underline{a}' & \cdots & \underline{0}' & -\underline{a}' \\ & & \cdots & & \\ \underline{0}' & \underline{0}' & \cdots & \underline{a}' & -\underline{a}' \end{bmatrix} \tag{21}$$

As noted in Koch and Bhapkar (1982) and elsewhere, $Q_S$ in (20) can be simplified to the one-way analysis of variance form

$$Q_S = \frac{(n-1)\Sigma_{i=1}^{s} n_{i+}(\bar{f}_i - \mu_{\underline{a}})^2}{v_{\underline{a}}} \tag{22}$$

where $\bar{f}_i = (\Sigma_{j=1}^{r} a_j n_{ij}/n_{i+})$ is the mean score for the ith group. The statistic $Q_S$ approximately has the chi-square distribution with d.f. = (s − 1) in large-sample situations.

A variety of possible scoring systems are described in Landis et al. (1979) and Koch et al. (1985a). In the analysis of clinical efficacy trials, the scores which may be of most interest are as follows:

1. *Integer scores*, defined as $a_j = j$ for j = 1, 2, . . . ., r. In our example, $\underline{a} = (1, 2, 3)'$ and $Q_S = 12.86$. Integer scores are useful when the response levels are discrete counts or ordinal categories which can be viewed as equally spaced. A rationale for their use in more general situations lies in the fact that $Q_S$ can be expressed as a function of

$$(\bar{f}_1 - \mu_{\underline{a}}) = \sum_{j=1}^{r} j\left(\frac{n_{1j}}{n_{1+}} - \frac{n_{+j}}{n}\right) = \frac{1}{n_{1+}} \sum_{k=2}^{r} (N_{1k} - M_{1k}) \tag{23}$$

where $N_{1k} = \Sigma_{j=k}^{r} n_{1j}$ is the number of patients on test drug with responses at least as favorable as the kth and $M_{1k} = \Sigma_{j=k}^{r} m_{1j}$ is its corresponding expected value under $H_0$; thus, integer scores enable evaluation of the extent to which the $\{N_{1k}\}$ are consistently larger (or smaller) than the $\{M_{1k}\}$.

2. *Standardized midranks*, defined as

$$a_j = \frac{2[\Sigma_{k=1}^{j} n_{+k}] - n_{+j} + 1}{2(n + 1)} \tag{24}$$

In our example, $\underline{a}$ = (43/170 = 0.253, 99/170 = 0.582, 141/170 = 0.829)' and $Q_S$ = 12.73. Standardized midranks lie in the interval (0, 1). In situations with no ties in the sense that all $n_{+j} = 1$, they represent expected values of order statistics for the uniform distribution. The use of $Q_S$ with standardized midranks provides the contingency table counterpart of the Wilcoxon rank sum statistic (see Koch and Bhapkar, 1982). An advantage of standardized midranks over integer scores is that they involve no scaling of the response categories other than that implied by their relative ordering. Also, they will be noted in Section V to have favorable power relative to actual midranks [i.e., $(n + 1)a_j$] in situations involving a set of $2 \times r$ tables. In the SAS (1985) procedure FREQ, standardized midrank scores are referred to as modified ridits.

3. *Logrank scores*, defined as

$$a_j = 1 - \sum_{k=1}^{j} \left(\frac{y_{+k}}{\Sigma_{m=k}^{r} y_{+m}}\right) \tag{25}$$

In our example, $\underline{a}$ = (0.500, 0.167, −0.833)' and $Q_S$ = 12.61. When there are no ties in the sense that all $n_{+j} = 1$, the quantities $\{(1 - a_j)\}$ represent expected values of order statistics from the exponential distribution with unit mean. The use of $Q_S$ with logrank scores is of interest when the data have L-shaped distributions and there is greater interest in treatment differences for higher value response categories than lower value ones; see Koch et al. (1985a, 1985b) for further discussion and examples. In the SAS procedure NPAR1WAY, the Savage statistic is based on scores which are identical to the $\{-a_j\}$ when there are no ties, but on somewhat different quantities when there are ties.

4. *Binary scores*, $a_j = 1$ for some levels of response and $a_j = 0$ for the other levels. The use of the binary scores $\underline{a}$ = (0, 1, 1)' yields a statistic identical to (5), i.e., $Q_S$ = 10.59.

When a study is concerned with comparisons of ordinal response distributions for s treatments which are also ordinally scaled with respect to some factor like dose, potential trends in the mean scores $\{\bar{f}_i\}$ are often of interest. A class of randomization statistics which are effective for addressing such alternative hypotheses can be constructed in terms of a linear function

$$\bar{f} = \sum_{i=1}^{s} c_i \bar{f}_i \left(\frac{n_{i+}}{n}\right) = \sum_{i=1}^{s} \sum_{j=1}^{r} \frac{c_i a_j n_{ij}}{n} \tag{26}$$

where $\underline{c} = (c_1, c_2, \ldots, cs)'$ is a vector of scores which account for the ordinal scaling of the subpopulations. Under the hypothesis $H_0$,

$$E\{\bar{f} \mid H_0\} = \sum_{i=1}^{s} c_i \left(\frac{n_{i+}}{n}\right) \sum_{j=1}^{r} a_j \left(\frac{n_{+j}}{n}\right) = \mu_{\underline{c}} \mu_{\underline{a}} \tag{27}$$

and

$$\mathrm{Var}\{\bar{f} \mid H_0\} = \left\{ \sum_{i=1}^{s} (c_i - \mu_{\underline{c}})^2 \left(\frac{n_{i+}}{n}\right) \sum_{j=1}^{r} \frac{(a_j - \mu_{\underline{a}})^2 (n_{+j}/n)}{(n-1)} \right\} \tag{28}$$

$$= \frac{v_{\underline{c}} v_{\underline{a}}}{(n-1)}$$

The linear function $\bar{f}$ approximately has a normal distribution in large samples, and so for these situations,

$$Q_{CS} = \frac{[\bar{f} - E\{\bar{f} \mid H_0\}]^2}{\mathrm{Var}\{\bar{f} \mid H_0\}} \tag{29}$$

$$= \frac{(n-1)[\Sigma_{i=1}^{s} \Sigma_{j=1}^{r} (c_i - \mu_{\underline{c}})(a_j - \mu_{\underline{a}}) n_{ij}]^2}{[\Sigma_{i=1}^{s} (c_i - \mu_{\underline{c}})^2 n_{i+}][\Sigma_{j=1}^{r} (a_j - \mu_{\underline{a}})^2 n_{+j}]}$$

$$= (n-1) r_{ac}^2$$

approximately has the chi-square distribution with d.f. = 1. The statistic $r_{ac}$ represents the correlation coefficient between group scores and response scores, and so $Q_{CS}$ is often called a correlation chi-square statistic. Its use can involve any of the types of scores (1)-(4) for either the groups or the response categories.

Since $Q_S$ and $Q_{CS}$ are based on linear combinations of the $\{n_{ij}\}$, moderate sample sizes are usually sufficient to support chi-square approximations for their distributions (e.g., all $n_{i+} \geq 20$ for $Q_S$ and $n \geq 25$ for $Q_{CS}$).

Additional discussion of the relative advantages and limitations of randomization chi-square statistics like $Q$, $Q_S$, and $Q_{CS}$ is given in Landis et al. (1978) and Koch et al. (1982, 1985a).

In some instances (e.g., several $m_{ij} < 5$) an exact test for $s \times r$ tables may be preferable to a chi-square test. An exact test for $s \times r$ tables may be calculated using the same principle as the Fisher's exact test, but with the probabilities to be summed calculated from the multivariate hypergeometric distribution. For the example, the result for this test was $p = 0.0014$. Algorithms for calculating the exact p values are described in Pagano and Halvorsen (1981) and Mehta and Patel (1983). Also, Mehta et al. (1984) describe an algorithm for obtaining exact p values when the mean score statistic $Q_S$ is used with standardized midranks.

## IV. THE MANTEL-HAENSZEL METHOD

In the analysis of efficacy trials, one may wish to compare the treatments by combining the results from a set of strata (e.g., centers in a multicenter trial) and adjusting for the effect of the stratification variable(s). Postrandomization stratification on baseline or other explanatory variables which have at least moderately strong association with response may also be of interest. When the treatment groups are unbalanced by chance with respect to the distribution of important explanatory variables, stratification provides an adjusted framework for undertaking comparisons so that observed differences between the treatment groups can be more clearly interpreted as actually due to treatment.

In the rheumatoid arthritis example, sex has a moderately strong relationship with response (58% of the females had favorable response compared with 32% of the males). Although the treatment groups are only slightly unbalanced with respect to the distribution of sex (66% female in the test drug group versus 74% female in the placebo group), it is still of interest to use sex as a stratification variable in order to conduct an overall test of treatment effectiveness, adjusted for sex.

Let $h = 1, 2$ index the two strata (sexes) in our example, and for simplicity let us consider the response as dichotomous. The data may be presented as follows:

| | | Improvement | | |
|---|---|---|---|---|
| Sex | Treatment | None | Some or marked | Total |
| Female | Test drug | $n_{111} = 6$ | $n_{112} = 21$ | $n_{11+} = 27$ |
| Female | Placebo | $n_{121} = 19$ | $n_{122} = 13$ | $n_{12+} = 32$ |
| Female total | | $n_{1+1} = 25$ | $n_{1+2} = 34$ | $n_1 = 59$ |
| Male | Test drug | $n_{211} = 7$ | $n_{212} = 7$ | $n_{21+} = 14$ |
| Male | Placebo | $n_{221} = 10$ | $n_{222} = 1$ | $n_{22+} = 11$ |
| Male total | | $n_{2+1} = 17$ | $n_{2+2} = 8$ | $n_2 = 25$ |

(30)

Among the females, 78% of the test drug group had a favorable response, as compared to 41% of the placebo group. The males were less responsive to both treatments; 50% of the males in the test drug group had favorable response, as compared to 9% of those in the placebo group.

When a set of 2 × 2 tables arises from stratification by center in a multicenter study, the separate treatment randomizations at each center induce independent hypergeometric distributions for the within-center frequencies, and so the distribution for the full table is the product of these hypergeometric distributions. The same distribution (i.e., the product hypergeometric) applies via conditional distribution arguments to data from a single center where there is postrandomization stratification on demographic or pretreatment variables. Thus, under the null hypothesis $H_0$ of no treatment difference for each patient of each sex, the expected value of $n_{h11}$ is

$$E\{n_{h11} \mid H_0\} = \frac{n_{h1+}n_{h+1}}{n_h} = m_{h11} \tag{31}$$

and its variance is

$$\text{Var}\{n_{h11} \mid H_0\} = \frac{n_{h1+}n_{h2+}n_{h+1}n_{h+2}}{n_h^2(n_h - 1)} = v_{h11} \tag{32}$$

Note that in the 2 × 2 table for males, two of the $m_{hij}$ are less than 5.0, so a chi-square statistic is not appropriate for testing $H_0$ for males separately. The two-sided Fisher's exact test p value for males is 0.042, and that for females is 0.008.

A method for evaluating the overall association of treatment and response, adjusted for sex, is the Mantel-Haenszel (1959) statistic:

$$Q_{MH} = \frac{\{\Sigma_{h=1}^{2} n_{h11} - \Sigma_{h=1}^{2} m_{h11}\}^2}{\Sigma_{h=1}^{2} v_{h11}} \tag{33}$$

$$= \frac{\{\Sigma_{h=1}^{2}(n_{h1+}n_{h2+}/n_h)(p_{h11} - p_{h21})\}^2}{\Sigma_{h=1}^{2} v_{h11}}$$

$$= 12.59$$

where $p_{hi1} = (n_{hi1}/n_{hi+})$ is the proportion of patients of the hth sex who received the ith treatment and had no improvement. Even when the within-strata sample sizes are small, the Mantel-Haenszel statistic approximately has the chi-square distribution with d.f. = 1 as long as the combined strata sample sizes,

$$n_{+1+} = \sum_{h=1}^{2}\sum_{j=1}^{2} n_{h1j} = 41 \quad \text{and} \quad n_{+2+} = \sum_{h=1}^{2}\sum_{j=1}^{2} n_{h2j} = 43 \tag{34}$$

are sufficiently large. Thus, the result $Q_{MH} = 12.59$ is interpreted as significant with $p < 0.01$. Here, it is appropriate to note that some references (e.g., Fleiss, 1981; Breslow and Day, 1980) apply a continuity correction for $Q_{MH}$ to improve the quality of the chi-square approximation.

Mantel and Fleiss (1980) proposed the following criterion for considering the chi-square approximation suitable for the distribution of the Mantel-Haenszel statistic in the general setting of q strata:

$$\min\left\{\left[\sum_{h=1}^{q} m_{h11} - \sum_{h=1}^{q} (n_{h11})_L\right], \left[\sum_{h=1}^{q} (n_{h11})_U - \sum_{h=1}^{q} m_{h11}\right]\right\} \geqslant 5 \tag{35}$$

where $(n_{h11})_L = \max(0, n_{h1+} - n_{h+2})$ and $(n_{h11})_U = \min(n_{h+1}, n_{h1+})$ are respectively the lowest and highest possible values for $n_{h11}$ across all possible randomizations with the marginal frequencies $\{n_{hi+}\}$ and $\{n_{h+j}\}$ fixed. Thus, the criterion specifies that the across-strata sum of expected values for a particular cell should be at least 5.0 from both the minimum possible sum and the maximum possible sum of observed values. Any of the four cells may be used in the calculations. In our example,

$$\sum_{h=1}^{2} m_{h11} = 11.4 + 9.5 = 20.9$$

$$\sum_{h=1}^{2} (n_{h11})_L = 0 + 6 = 6 \tag{36}$$

and

$$\sum_{h=1}^{2} (n_{h11})_U = 25 + 14 = 39$$

Since $(20.9 - 6) \geqslant 5$ and $(39 - 20.9) \geqslant 5$, the Mantel-Fleiss criterion is satisfied and the use of the chi-square distribution with d.f. = 1 for the Mantel-Haenszel statistic is appropriate for this example. When the Mantel-Fleiss criterion is not met for a set of $2 \times 2$ tables, an exact test may be carried out using the algorithm of Thomas (1975). The one-sided p value provided by this method is $p = 0.0003$. A computationally more efficient algorithm for exact inference in sets of $2 \times 2$ tables is given in Mehta et al. (1985).

The Mantel-Haenszel statistic $Q_{MH}$ is effective for detecting patterns of treatment differences across the respective strata when there is a strong tendency for the $\{(p_{h11} - p_{h21})\}$ to have the same sign. For this reason, it is sometimes called a test of average partial association in order to distinguish it from the total partial association statistic

$$Q_T = \sum_{h=1}^{q} \frac{(n_{h11} - m_{h11})^2}{v_{h11}} \tag{37}$$

When there are sufficiently large within-stratum sample sizes (e.g., all $m_{hij} \geq 5$), $Q_T$ has approximately the chi-square distribution with d.f. = q. For our example, $Q_T$ = 12.69, d.f. = 2, but as noted previously, the sample sizes for males may not be large enough to support the use of the chi-square approximation for $Q_T$.

Because $Q_{MH}$ algebraically cannot exceed $Q_T$ and because $Q_{MH}$ tends to increase with greater similarity among strata with respect to strength and direction of the association of treatment and response, the difference between $Q_T$ and $Q_{MH}$ has been used as a statistic for assessing the homogeneity of the strata with respect to the association of treatment and response. Given that the sample size conditions stated for $Q_T$ are met, $(Q_T - Q_{MH})$ has approximately a chi-square distribution with d.f. = (q − 1) under $H_0$. However, this statistic must be interpreted cautiously because $H_0$ involves a framework of no association rather than one of homogeneous nonnull association. For this reason, $(Q_T - Q_{MH}) = Q_{PH}$ needs to be viewed as a pseudohomogeneity statistic for detecting departures from $H_0$ which involve substantial variation across the respective strata in the magnitude and direction of treatment differences. For the example, $Q_{PH}$ = 12.69 − 12.59 = 0.10 is a relatively small component of $Q_T$, and so the pattern of treatment differences is sufficiently consistent across strata for the Mantel-Haenszel statistic to provide the principal basis for contradicting $H_0$. Additional discussion of the interpretation of the pseudohomogeneity statistic $Q_{PH}$ is given in Koch et al. (1985a). In Section VI, an appropriate method for assessing the across-strata homogeneity of nonnull association of treatment and response is presented in the setting of a logistic regression model.

## V. EXTENSIONS OF THE MANTEL-HAENSZEL METHOD

Mantel (1963) proposed an extension of the Mantel-Haenszel procedure to the analysis of a set of (2 × r) tables with ordinal response categories. This method involves a combination of the principles underlying the mean score statistic $Q_S$ in (19) and the Mantel-Haenszel statistic $Q_{MH}$ in (33). Its application to the rheumatoid arthritis data is described here.

After stratification by sex, the frequencies for the cross-classification of treatment with the ordinal response variable are as follows:

| | | Improvement | | | |
|---|---|---|---|---|---|
| Sex | Treatment | None | Some | Marked | Total |
| Female | Test drug | 6 | 5 | 16 | 27 |
| Female | Placebo | 19 | 7 | 6 | 32 |
| Female total | | 25 | 12 | 22 | 59 |

| | | Improvement | | | |
|---|---|---|---|---|---|
| Sex | Treatment | None | Some | Marked | Total |
| Male | Test drug | 7 | 2 | 5 | 14 |
| Male | Placebo | 10 | 0 | 1 | 11 |
| Male total | | 17 | 2 | 6 | 25 |

(38)

Let $n_{hij}$ denote the number of patients of the hth sex (h = 1, 2) who received the ith treatment (i = 1, 2) and had the jth response (j = 1, 2, 3). Under the null hypothesis of no treatment difference for each patient, the applicable probability model is the product multivariate hypergeometric distribution:

$$P\{n_{hij} \mid H_0\} = \prod_{h=1}^{2} \frac{[\Pi_{i=1}^{2} n_{hi+}! \, \Pi_{j=1}^{3} n_{h+j}!]}{[n_h! \, \Pi_{i=1}^{2} \Pi_{j=1}^{3} n_{hij}!]} \tag{39}$$

Let $\{a_{hj}\}$ be a set of scores for the response levels in the hth stratum. The sum of across-strata scores for the test drug group is

$$f_{+1+} = \sum_{h=1}^{2} \sum_{j=1}^{3} a_{hj} n_{h1j} = \sum_{h=1}^{2} n_{h1+} \bar{f}_{h1} \tag{40}$$

where $\bar{f}_{h1} = \Sigma_{j=1}^{3}(a_{hj} n_{h1j}/n_{h1+})$ is the mean score for the test drug group in the hth stratum. Under $H_0$, $f_{+1+}$ has expected value

$$E\{f_{+1+} \mid H_0\} = \sum_{h=1}^{2} n_{h1+} \mu_h = \mu_* \tag{41}$$

and variance

$$\mathrm{Var}\{f_{+1+} \mid H_0\} = \sum_{h=1}^{2} \frac{n_{h1+}(n_h - n_{h1+})}{(n_h - 1)} v_h = v_* \tag{42}$$

where $\mu_h = \Sigma_{j=1}^{3}(a_{hj} n_{h+j}/n_h)$ and $v_h = \Sigma_{j=1}^{3}(a_{hj} - \mu_h)^2 (n_{h+j}/n_h)$ are the finite subpopulation mean and variance of scores for the hth stratum.

When the across-strata sample sizes $\{n_{+i+}\}$ are large, $f_{+1+}$ approximately has a normal distribution by central limit theory, and so

$$Q_{EMH} = \frac{\{f_{+1+} - \mu_*\}^2}{v_*} \tag{43}$$

approximately has a chi-square distribution with d.f. = (s − 1) = 1. In the case of two treatments, $Q_{EMH}$ can be shown to be a linear function of the differences in the mean scores of the two treatments for the respective strata, i.e.,

$$Q_{EMH} = \frac{[\Sigma_{h=1}^{2} n_{h1+}(\bar{f}_{h1} - \mu_h)]^2}{[\Sigma_{h=1}^{2} n_{h1+} n_{h2+} v_h/(n_h - 1)]}$$

$$= \frac{[\Sigma_{h=1}^{2} (n_{h1+} n_{h2+}/n_h)(\bar{f}_{h1} - \bar{f}_{h2})]^2}{\Sigma_{h=1}^{2} (n_{h1+} n_{h2+}/n_h)^2 \bar{v}_h} \tag{44}$$

where the $\{\bar{v}_h = (1/n_{h1+} + 1/n_{h2+})[n_h/(n_h - 1)]v_h\}$ are the variances of the mean score differences $\{(\bar{f}_{h1} - \bar{f}_{h2})\}$ for the respective strata. From the structure of (44), it is apparent that $Q_{EMH}$ is effective for detecting consistent patterns of treatment differences where the $(\bar{f}_{h1} - \bar{f}_{h2})$ strongly tend to have the same sign.

In our example, the results of the extended Mantel-Haneszel procedure for the scores defined in 1–4 of Section III were as follows:

1. Integer scores: $Q_{EMH} = 14.63$
2. Within-stratum standardized midrank scores: $Q_{EMH} = 15.00$
3. Within-stratum logrank scores: $Q_{EMH} = 13.89$
4. Binary scores for at least some improvement: $Q_{EMH} = 12.59$

Relative to the chi-square distribution with d.f. = 1, all of these results are significant with $p < 0.01$. It is of interest to note that $Q_{EMH}$ with standardized midranks represents the categorical data counterpart of the procedure proposed by van Elteren (1960) for combining Wilcoxon rank sum tests across a set of strata. As discussed in Lehmann (1975), the van Elteren statistic has a locally most powerful property for continuous data situations where treatment effects are similar for the respective strata.

Extensions of the Mantel-Haenszel procedure to sets of (s × r) tables can also be constructed. These have the general form

$$Q_{EMH} = \left\{\sum_{h=1}^{q} (\underline{n}_h - \underline{m}_h)'\underline{A}'_h\right\}\left\{\sum_{h=1}^{q} \underline{A}_h\underline{V}_h\underline{A}'_h\right\}^{-1}\left\{\sum_{h=1}^{q} \underline{A}_h(\underline{n}_h - \underline{m}_h)\right\} \tag{45}$$

where $\underline{n}_h$, $\underline{m}_h$, and $\underline{V}_h$ are the observed frequency vector, the expected frequency vector, and the covariance matrix for the hth stratum and have definitions analogous to $\underline{n}$, $\underline{m}$, and $\underline{V}$ in (14); the $\{\underline{A}_h\}$ are (u × sr) matrices with full rank $u \leqslant (s - 1)(r - 1)$ and structure such that

$\{\Sigma_{h=1}^{q} \underline{A}_h \underline{V}_h \underline{A}'_h\}$ is nonsingular. For purposes of interpretation, the $\{\underline{A}_h\}$ specify the linear functions of the $\{(\underline{n}_h - \underline{m}_h)\}$ at which the test statistic is directed. Alternative choices of the $\{\underline{A}_h\}$ along the lines discussed in Section III provide the stratification-adjusted counterparts to the overall randomization chi-square statistic Q, the mean score statistic $Q_S$, and the correlation statistic $Q_{CS}$. These test statistics can be computed with the SAS (1985) procedure FREQ or the program PARCAT documented in Landis et al. (1979).

It is also possible to construct counterparts to the total partial association statistic $Q_T$ and the pseudohomogeneity statistic $Q_{PH}$ for sets of $s \times r$ tables. Issues concerning their application are similar to those outlined in Section IV for sets of $2 \times 2$ tables. For additional discussion and references concerning the construction and properties of alternative types of randomization-based test statistics for sets of $s \times r$ tables, see Landis et al. (1978) and Koch et al. (1982, 1985a).

## VI. LOGISTIC REGRESSION

Logistic regression can be used to describe the relationship between a dichotomous response variable and a set of explanatory variables (e.g., treatment, age, sex, pretreatment status). Such analysis is of interest for the evaluation of treatment effects as well as interaction effects of treatment with other explanatory variables. Confirmation of no interaction between treatment and any particular explanatory variable is desirable because treatment effects can then be interpreted as homogeneous across the levels of that variable, thus supporting the generalizability of treatment effects.

Let us return to the data structure of Section IV as shown in Table 3. If we assume that the patients of each sex are statistically representative of some larger target population (i.e., the sample of patients enrolled is equivalent to a stratified simple random sample), then the overall sex-by-treatment-by-response cross-classification has the product binomial distribution:

$$P\{n_{hij}\} = \prod_{h=1}^{2} \prod_{i=1}^{2} \frac{n_{hi+}!}{(n_{hi1}!n_{hi2}!)} (1 - \theta_{hi})^{n_{hi1}} (\theta_{hi})^{n_{hi2}} \tag{46}$$

where $\theta_{hi}$ is the probability that a patient of sex h who receives treatment i will have some or marked improvement and $n_{hi1}$ and $n_{hi2}$ are the numbers of patients of the hth sex and ith treatment who had no improvement or improvement, respectively. The logistic model for describing the variation among the $\{\theta_{hi}\}$ has the specification:

$$\theta_{hi} = \{1 + \exp[-(\alpha + \underline{x}'_{hi}\underline{\beta})]\}^{-1} \tag{47}$$

where $\alpha$ is the intercept parameter, $\underline{\beta}$ is a vector of regression parameters, and $\underline{x}'_{hi}$ is a row vector of explanatory variables corresponding to the (h, i)th subpopulation. If we let $\underline{x}'_{hiA} = [1, \underline{x}'_{hi}]$, then the $\underline{x}'_{hiA}$ are the rows of a model specification matrix $\underline{X}_A$. An important property of the

**Table 3** Observed Frequencies, Observed Percentages, and Predicted Percentages from Logistic Regression Analysis of Rheumatoid Arthritis Data

| | | Observed frequencies | | Percentage of patients with some or marked improvement | |
|---|---|---|---|---|---|
| Sex | Treatment | No improvement | Some or marked improvement | Observed % (s.e.) | Model-predicted % (s.e.) |
| Female | Test drug | 6 | 21 | 77.8 ( 8.0) | 79.4 ( 7.2) |
| | Placebo | 19 | 13 | 40.6 ( 8.7) | 39.3 ( 8.2) |
| Male | Test drug | 7 | 7 | 50.0 (13.4) | 47.0 (11.9) |
| | Placebo | 10 | 1 | 9.1 ( 8.7) | 13.0 ( 6.8) |

logistic model is that all possible values of $(\alpha + \underline{x}'_{hi}\underline{\beta})$ in $(-\infty, +\infty)$ correspond to values of $\theta_{hi}$ in (0, 1). With a logistic transformation, the model becomes

$$\text{logit}(\theta_{hi}) = \log_e \frac{\theta_{hi}}{(1 - \theta_{hi})} = \alpha + \underline{x}'_{hi}\underline{\beta} \tag{48}$$

The logit of $\theta_{hi}$ is thus the logarithm of the odds of some or marked improvement to no improvement for the (h, i)th subpopulation.

The parameters $\alpha$ and $\underline{\beta}$ are usually estimated by maximum likelihood. By replacing $\theta_{hi}$ in the likelihood equation (46) with its model expression in (47), then differentiating $\log_e$ of (46) with respect to $\alpha$ and $\underline{\beta}$, and setting the results to 0, one obtains the following equation:

$$\sum_{h=1}^{2} \sum_{i=1}^{2} (n_{hi1} - n_{hi+} \hat{\theta}_{hi})\underline{x}'_{hiA} = 0 \tag{49}$$

where $\hat{\theta}_{hi} = \{1 + \exp[-(\hat{\alpha} + \underline{x}'_{hi}\hat{\underline{\beta}})]\}^{-1}$ is the model predicted maximum likelihood estimate of $\theta_{hi}$ based on $\underline{x}_{hi}$, and $\hat{\alpha}$ and $\hat{\underline{\beta}}$ are the maximum likelihood estimates of $\alpha$ and $\underline{\beta}$. Since this equation is nonlinear, $\hat{\alpha}$ and $\hat{\underline{\beta}}$ must usually be calculated with an iterative procedure such as the Newton-Raphson method [see Cox (1970), Koch and Edwards (1985), or McCullagh and Nelder (1983) for further discussion]. The estimates $\hat{\alpha}$ and $\hat{\underline{\beta}}$ are approximately normal when $\Sigma_{h=1}^{2}\Sigma_{i=1}^{2} n_{hi1}\underline{x}'_{hiA}$ is approximately normal. If we let $\hat{\underline{\beta}}_A = (\hat{\alpha}, \hat{\underline{\beta}}')'$, then a consistent estimate for the covariance matrix of $\hat{\underline{\beta}}_A$ is

$$\underline{V}(\hat{\underline{\beta}}_A) = \left[ \sum_{h=1}^{2} \sum_{i=1}^{2} \{n_{hi+}\hat{\theta}_{hi}(1 - \hat{\theta}_{hi})\underline{x}_{hiA}\underline{x}'_{hiA}\} \right]^{-1} \tag{50}$$

A model of interest for the data in Table 3 is the main effects model

$$\underline{X}_A = \begin{bmatrix} 1 & 1 & 1 \\ 1 & 1 & 0 \\ 1 & 0 & 1 \\ 1 & 0 & 0 \end{bmatrix} \tag{51}$$

for which the parameter estimates, their standard errors, and p values for tests of zero values are as follows:

| Parameter | Estimate | Standard error | p value | Interpretation |
|---|---|---|---|---|
| $\alpha$ | −1.904 | 0.598 | <0.010 | $\text{Log}_e$ odds of improvement for males receiving placebo |

| Parameter | Estimate | Standard error | p value | Interpretation |
|---|---|---|---|---|
| $\beta_1$ | 1.469 | 0.576 | 0.011 | Increment in $\log_e$ odds due to female sex |
| $\beta_2$ | 1.782 | 0.519 | <0.010 | Increment to $\log_e$ odds due to test drug |

(52)

The p values are obtained from the Wald statistics based on the squares of the ratios of the parameter estimates to their estimated standard errors, i.e., $\{\text{est.}/\text{se(est.)}\}^2$. These test statistics approximately have chi-square distributions with d.f. = 1.

The goodness of fit of model $\underline{X}_A$ can be evaluated either by the Pearson chi-square statistic

$$Q_P = \sum_{h=1}^{2}\sum_{i=1}^{2}\sum_{j=1}^{2} \frac{(n_{hij} - \hat{m}_{hij})^2}{\hat{m}_{hij}} = 0.26 \tag{53}$$

where $\hat{m}_{hi1} = n_{hi+}(1 - \hat{\theta}_{hi})$ and $\hat{m}_{hi2} = n_{hi+}\hat{\theta}_{hi}$, or by the log-likelihood ratio chi-square statistic

$$Q_L = \sum_{h=1}^{2}\sum_{i=1}^{2}\sum_{j=1}^{2} 2n_{hij} \log_e \frac{n_{hij}}{\hat{m}_{hij}} = 0.28 \tag{54}$$

[in which, for $n_{hij} = 0$, $n_{hij} \log_e(n_{hij}/\hat{m}_{hij})$ is defined to be 0]. For large-sample situations (e.g., all $m_{hij} \geqslant 5$), the statistics $Q_P$ and $Q_L$ are asymptotically equivalent, and both approximately have chi-square distributions with d.f. = $(qs - t) = 1$, where $q = 2$ is the number of sexes, $s = 2$ the number of treatments, and $t = 3$ the rank of $\underline{X}_A$.

A third equivalent way to test the goodness of fit of model $\underline{X}_A$ is to fit an expanded model which includes a parameter for sex-by-treatment interaction and then to verify that the Wald statistic for this parameter is nonsignificant. If we fit the model $[\underline{X}_A, \underline{W}]$ where $\underline{W} = (1\ 0\ 0\ 0)'$, the Wald statistic for this interaction term is $Q = 0.26$. All three of the statistics $Q_P$, $Q_L$, and $Q_W$ are nonsignificant with $p > 0.50$, so the goodness of fit of the model is well supported. In this regard, one can note that numerical studies (e.g., Larntz, 1978) tend to support the use of $Q_P$ because of the applicability of chi-square approximations to its distribution for many types of situations with small or moderate sample sizes (e.g., most $\hat{m}_{hij} > 2$ and few $< 1$).

Values of the logits predicted by model $\underline{X}_A$ for each of the four subpopulations are the respective elements of $\underline{X}_A\hat{\underline{\beta}}_A$, as shown below. Their estimated standard errors are the square roots of the diagonal elements of $\underline{X}_A\hat{\underline{V}}(\hat{\underline{\beta}}_A)\underline{X}'_A$. The predicted odds of improvement for each subpopulation are found from exponentiating the predicted logit.

| Sex | Treatment | Estimates from model $\underline{X}_A$: logit | s.e. (logit) | Odds of improvement |
|---|---|---|---|---|
| F | Test drug | $\hat{\alpha} + \hat{\beta}_1 + \hat{\beta}_2 = 1.347$ | 0.437 | $\exp(1.347) = 3.846$ |
| F | Placebo | $\hat{\alpha} + \hat{\beta}_1 = -0.435$ | 0.345 | $\exp(-0.435) = 0.647$ |
| M | Test drug | $\hat{\alpha} + \hat{\beta}_2 = -0.122$ | 0.480 | $\exp(-0.122) = 0.885$ |
| M | Placebo | $\hat{\alpha} = -1.904$ | 0.598 | $\exp(-1.904) = 0.149$ |

(55)

From the model-estimated odds of improvement, one can calculate the model-estimated probability of improvement. For example, the estimated probability of improvement for females receiving test drug ($\hat{\theta}_{11}$) is found from $\hat{\theta}_{11}/(1 - \hat{\theta}_{11}) = 3.846$, i.e., $\hat{\theta}_{11} = 0.794$. The standard error of $\hat{\theta}_{11}$ is found from $(\hat{\theta}_{11})(1 - \hat{\theta}_{11})[\text{s.e.}(\underline{x}'_{11A}\hat{\underline{\beta}}_A)] = (0.794)(0.206)(0.437) = 0.072$. The estimated probabilities of improvement for the four groups are shown alongside the observed probabilities in Table 3. Note that the standard errors for the model-estimated probabilities are somewhat smaller than those for the observed proportions. This gain in precision is due to the elimination of extraneous variability for effects not in the model $\underline{X}_A$ (i.e., the variability due to sex-by-treatment interaction). The results from the maximum likelihood logistic regression analysis of this example were calculated with the SAS (1985) procedure CATMOD; they could also be obtained with the SAS procedure LOGIST documented in Harrell (1986) or the BMDP procedure PLR documented in Engelman (1983).

Since the fit of the model $\underline{X}_A$ in (51) is supported by the nonsignificance of $Q_P$, $Q_L$, and the Wald statistic $Q_W$ for the interaction term in the model $[\underline{X}_A, \underline{W}]$, the association of treatment and response can be interpreted as being relatively similar for the two sexes. A measure of the association between treatment and response which can be estimated directly from the logistic model is the odds ratio, i.e., the ratio of the odds of improvement for the test drug to the odds of improvement for the placebo treatment. Since $\underline{X}_A$ is a main effects model, the estimated odds ratio for males, $\exp(\hat{\alpha} + \hat{\beta}_2)/\exp(\hat{\alpha}) = \exp(\hat{\beta}_2)$, is equal to the estimated odds ratio for remales, $\exp(\hat{\alpha} + \hat{\beta}_1 + \hat{\beta}_2)/\exp(\hat{\alpha} + \hat{\beta}_1) = \exp(\hat{\beta}_2) = \exp(1.782) = 5.94$. A 95% normal approximation confidence interval for the common odds ratio, $\exp(\beta_2)$, is

$$\exp\{\hat{\beta}_2 \pm 1.96[\text{s.e.}(\hat{\beta}_2)]\} = (2.15, 16.43) \tag{56}$$

An exact confidence interval for the common odds ratio can be obtained through the product noncentral hypergeometric distribution (see Breslow and Day, 1980; Gart, 1971; Kleinbaum et al., 1982) from the computing procedure of Thomas (1975). The exact 95% confidence interval obtained by this method is (1.94, 18.78). A more efficient algorithm for obtaining this type of exact confidence interval is described in Mehta et al. (1985).

Logistic regression, as previously described, can be generalized to include continuous as well as categorical explanatory variables. The main difference between a logistic model with a small number of categorical explanatory variables and a model with both categorical and continuous explanatory variables is in the evaluation of goodness of fit. For example, in the previously described model, there were four subpopulations corresponding to the cross-classification of sex and treatment. If we expand this model to include age also, so that the number of subpopulations corresponds to the cross-classification of sex by treatment by age, then methods [such as the Pearson chi-square (53) or the log-likelihood ratio chi-square (54)] which depend on having a sufficiently large number of patients per subpopulation (e.g., $n_{hi+} \geqslant 10$) are no longer applicable.

The strategy of fitting an expanded model and then verifying the nonsignificance of effects not in the original model is still applicable, however. If the specification matrix $\underline{X}_A$ for the original model has rank t, then the expanded model $[\underline{X}_A, \underline{W}]$ must have rank t + w, where w is the rank of $\underline{W}$. The significance of the contribution of $\underline{W}$ may be evaluated by the difference of the log-likelihood ratio chi-square statistics for the models $\underline{X}_A$ and $[\underline{X}_A, \underline{W}]$, i.e.,

$$Q_{LR} = \sum_{i=1}^{s} \sum_{j=1}^{2} 2n_{ij} \left[ \log_e \left( \frac{\hat{m}_{ij,w}}{\hat{m}_{ij}} \right) \right] \tag{57}$$

where s is the total number of subpopulations with at least one subject for the cross-classification of explanatory variables, $n_{ij}$ the number of patients in the ith subpopulation with the jth response, $\hat{m}_{ij}$ the predicted value of $n_{ij}$ for model $\underline{X}_A$, and $\hat{m}_{ij,w}$ the predicted value of $n_{ij}$ for model $[\underline{X}_A, \underline{W}]$. $Q_{LR}$ has an approximate chi-square distribution with d.f. = w.

An alternative statistic which does not require fitting the expanded model is the Rao score statistic for assessing the association of the residuals $(\underline{n}_{*1} - \hat{\underline{m}}_{*1})$ with $\underline{W}$ through the linear functions $\underline{g} = \underline{W}'(\underline{n}_{*1} - \hat{\underline{m}}_{*1})$. A computational expression for this criterion is

$$Q_{RS} = \underline{g}'\{\underline{W}'[\underline{D}_{\hat{v}}^{-1} - \underline{D}_{\hat{v}}^{-1}\underline{X}_A(\underline{X}'_A\underline{D}_{\hat{v}}^{-1}\underline{X}_A)^{-1}\underline{X}'_A\underline{D}_{\hat{v}}^{-1}]\underline{W}\}^{-1}\underline{g} \tag{58}$$

where $\underline{n}_{*1} = (n_{11}, n_{21}, \ldots, n_{s1})'$, $\hat{\underline{m}}_{*1} = (\hat{m}_{11}, \hat{m}_{21}, \ldots, \hat{m}_{s1})'$, and $\underline{D}_{\hat{v}}$ is a diagonal matrix with diagonal elements $\hat{v}_i = [n_{i+}\hat{\theta}_i(1 - \hat{\theta}_i)]^{-1}$. Both $Q_{LR}$ and $Q_{RS}$ approximately have the chi-square distribution with d.f. = w when the overall sample size is sufficiently large to support an approximately multivariate normal distribution for the linear functions $[\underline{X}'_A, \underline{W}']\underline{n}_{*1}$; see Imrey et al. (1981, 1982) and Koch et al. (1985a) for further discussion.

A logistic regression model with explanatory variables of sex, treatment, and age was fit to the data of Table 2 with the SAS procedure LOGIST of Harrell (1986). The estimated model parameters and their standard errors were as follows:

| Parameter | Estimate | Standard error | p value | Interpretation |
|---|---|---|---|---|
| $\alpha$ | −4.503 | 1.307 | <0.010 | $\log_e$ odds of improvement for a hypothetical male of age 0 who received placebo |
| $\beta_1$ | 1.488 | 0.595 | 0.012 | Increment due to female sex |
| $\beta_2$ | 1.760 | 0.536 | <0.010 | Increment due to test drug |
| $\beta_3$ | 0.049 | 0.021 | 0.018 | Increment per year of age |

(59)

Thus, sex, treatment, and age are all significant predictors of improvement. It is important to note that the model is useful for predicting the relationship between age and response only for patients with ages similar to the ages of patients in the study (i.e., ages 23 to 74). Also, the model assumes a linear relationship between age and response, which may not be the case. One way of assessing potential departures from linearity is through the contribution of the square of age to the model.

The goodness of fit of the model was tested by evaluating the potential contribution of the following four additional variables: the sex × treatment interaction, the age × treatment interaction, the sex × age interaction, and the square of age (where interaction variables are defined as products of their components). The Rao score statistic (58) for the joint contribution of these four variables was $Q_{RS} = 4.03$, which is nonsignificant when compared to the chi-square distribution with d.f. = 4 ($p = 0.402$). However, when attention was directed at the individual components of this overall test, the sex × age interaction was found to be nearly significant ($Q_{RS} = 3.69$, d.f. = 1, $p = 0.055$). This result can be interpreted either as a chance event for the situation with no sex × age interaction or as an indicator that the model needs to be expanded to include the sex × age interaction. The former interpretation is supported by the sex × age interaction not seeming strongly evident in view of the multiplicity of goodness-of-fit assessments involved in the overall $Q_{RS}$ and its separate components. Alternatively, the latter interpretation is compatible with the theme of confirming conclusions about treatment effects by considering settings which account for all other possibly relevant sources of variation. When the expanded model with age, sex, age × sex interaction, and treatment effects was applied, the test drug was still found to be associated with significantly ($p < 0.010$) more favorable response, and (treatment × sex) and (treatment × age) interactions were found to be nonsignificant. The sex × age interaction corresponded to the disparity between a significant linear relationship with age for females and an essentially null relationship for males. For additional discussion and references concerning logistic regression methods, see Breslow and Day (1980), Cox (1970), Kleinbaum et al. (1982),

Koch et al. (1982, 1985a), Koch and Edwards (1985), and McCullagh and Nelder (1983).

## VII. EXTENSIONS OF LOGISTIC REGRESSION

For ordinal variables with three levels (e.g., no improvement, some, or marked), two logits for more favorable versus less favorable response can be constructed:

1. The logit for some or marked improvement versus no improvement. If, for the ith subpopulation, we let $\pi_{i1}$ denote the probability of no improvement, $\pi_{i2}$ denote the probability of some improvement, and $\pi_{i3}$ denote the probability of marked improvement, this logit can be expressed as $\log_e\{(\pi_{i2} + \pi_{i3})/(\pi_{i1})\} = \text{logit}(\theta_{i1})$. This is the logit considered in Section VI.
2. The logit for marked improvement versus some or no improvement, i.e., $\log_e\{(\pi_{i3})/(\pi_{i1} + \pi_{i2})\} = \text{logit}(\theta_{i2})$.

The assumed probability structure for the frequencies $\{n_{ij}\}$ of response outcomes j = 1, 2, . . ., r (here, r = 3) for the samples of $\{n_{i+}\}$ subjects from the i = 1, 2, . . ., s subpopulations is the product multinomial distribution:

$$P\{n_{ij}\} = \prod_{i=1}^{s}\left[n_{i+}! \prod_{j=1}^{r} \frac{\pi_{ij}^{n_{ij}}}{n_{ij}!}\right] \tag{60}$$

As noted in Section VI, a supportive basis for the structure (60) is the perspective that the subjects under study are conceptually representative of the respective subpopulations in a sense equivalent to stratified simple random sampling.

We can consider a model for both of the logits in 1 and 2 simultaneously with the specification

$$\text{logit}(\theta_{ik}) = \alpha_k + \underline{x}_i'\underline{\beta}_k \tag{61}$$

where k = 1, 2 indexes the two logits. With this model, there are separate intercept parameters $\{\alpha_k\}$ and vectors of regression parameters $\{\underline{\beta}_k\}$ for the two types of logits. This model does not account for the ordinal nature of the response because comparisons between the ith and i'th subpopulations with respect to it have (r − 1) = 2 components

$$\{\text{logit}(\theta_{ik}) - \text{logit}(\theta_{i'k})\} = (\underline{x}_i - \underline{x}_{i'})'\underline{\beta}_k \qquad \text{for } k = 1, 2 \tag{62}$$

which express general association. However, under the condition that the respective logits have the same regression parameter vector $\underline{\beta}$ (i.e., all $\underline{\beta}_k = \underline{\beta}$), the structure (62) simplifies to

$$\{\text{logit}(\theta_{ik}) - \text{logit}(\theta_{i'k})\} = (\underline{x}_i - \underline{x}_{i'})'\underline{\beta} \tag{63}$$

which involves only one component. Since $\exp\{(\underline{x}_i - \underline{x}_{i'})'\underline{\beta}\}$ expresses the extent to which more favorable response categories are more likely in the ith subpopulation than the i'th subpopulation, it is potentially indicative of location shifts. Thus, the corresponding model

$$\mathrm{logit}(\theta_{ik}) = \alpha_k + \underline{x}_i'\underline{\beta} \tag{64}$$

provides a framework which accounts for ordinal response categories; it is called the proportional odds model by McCullagh (1980) as well as other authors. Maximum likelihood estimation for the parameters of the proportional odds model (64) is discussed in McCullagh (1980), McCullagh and Nelder (1983), and Walker and Duncan (1967). Computations can be readily undertaken with the SAS procedure LOGIST of Harrell (1986).

For the rheumatoid arthritis data, the model with sex and treatment as explanatory variables in a form analogous to the last two columns of (51) was fit. The resulting maximum likelihood estimates, their standard errors, and p values were as follows:

| Parameter | Estimate | Standard error | p value | Interpretation |
|---|---|---|---|---|
| $\alpha_1$ | −1.813 | 0.565 | <0.010 | $\log_e$ odds of some or marked improvement versus no improvement for males receiving placebo |
| $\alpha_2$ | −2.667 | 0.606 | <0.010 | $\log_e$ odds of marked improvement versus some or no improvement for males receiving placebo |
| $\beta_1$ | 1.319 | 0.538 | 0.014 | Increment for both types of $\log_e$ odds due to female sex |
| $\beta_2$ | 1.797 | 0.472 | <0.010 | Increment for both types of $\log_e$ odds due to test drug |

(65)

For situations with categorical explanatory variables and contingency table data structure, like that shown in Section V for this example, the goodness of fit of the model (64) can be evaluated with counterparts to the Pearson chi-square $Q_P$ in (53) or the log-likelihood ratio chi-square $Q_L$ in (54). The model-predicted frequencies for these methods are obtained as follows:

$$\hat{m}_{i1} = n_{i+}[1 + \exp(\hat{\alpha}_1 + \underline{x}_i'\hat{\underline{\beta}})]^{-1} \tag{66}$$

$$\hat{m}_{ij} = n_{i+}\{[1 + \exp(\hat{\alpha}_j + \underline{x}_i'\hat{\underline{\beta}})]^{-1} - [1 + \exp(\hat{\alpha}_{(j-1)} + \underline{x}_i'\hat{\underline{\beta}})]^{-1}\}$$

$$\text{for } j = 2, \ldots, (r-1)$$

$$\hat{m}_{ir} = n_{i+}[1 + \exp(-\hat{\alpha}_{(r-1)} - \underline{x}_i'\hat{\underline{\beta}})]^{-1}$$

When the sample sizes for each subpopulation are sufficiently large (e.g., most $\hat{m}_{ij} > 2$ and few $< 1$), $Q_P$ approximately has the chi-square distribution with d.f. = $\{(r - 1)s - t\}$, where t is the dimension of $\underline{\beta}_A = (\alpha_1, \alpha_2, \underline{\beta}')'$. The result $Q_P = 4.80$ supports the use of the model illustrated here by its nonsignificance ($p = 0.308$) with respect to the chi-square distribution with d.f. = 4.

Maximum likelihood estimation for the parameters of the proportional odds model is also applicable to situations with categorical and continuous explanatory variables. An example of such analysis is the fit of the model with age, sex, and treatment explanatory variables to the rheumatoid arthritis data in Table 2. Results from the SAS procedure LOGIST of Harrell (1986) are as follows:

| Parameter | Estimate | Standard error | p value | Interpretation |
|---|---|---|---|---|
| $\alpha_1$ | −3.784 | 1.144 | <0.010 | $\log_e$ odds of some or marked improvement versus no improvement for males of age o receiving placebo |
| $\alpha_2$ | −4.683 | 1.187 | <0.010 | $\log_e$ odds of marked improvement versus some or no improvement for males of age o receiving placebo |
| $\beta_1$ | 1.252 | 0.546 | 0.022 | Increment for both types of $\log_e$ odds due to female sex |
| $\beta_2$ | 1.745 | 0.476 | <0.010 | Increment for both types of $\log_e$ odds due to test drug |
| $\beta_3$ | 0.038 | 0.018 | 0.038 | Increment for both types of $\log_e$ odds per year of age |

(67)

As discussed for logistic models in Section V, the evaluation of goodness of fit of proportional odds models with categorical and continuous variables needs to be undertaken in terms of the potential contribution of additional explanatory variables. For this purpose, either extensions of the log-likelihood ratio chi-square statistic (57) or the Rao score statistic (58) can be used. For the model expansion consisting of sex × treatment interaction, age × treatment interaction, sex × age interaction, and the square of age, the Rao score statistic $Q_{RS} = 3.53$ was nonsignificant ($p = 0.473$) relative to the chi-square distribution with d.f. = 4. On this basis, the model with age, sex, and treatment as the explanatory variables is found to be satisfactory. Treatment effects are thereby interpreted as homogeneous across sex and age. The nature of treatment effects is expressed through $\exp(\hat{\underline{\beta}}_2) = \exp(1.745) = 5.73$. This quantity represents the model-predicted ratio of test drug to placebo for both the odds of some or marked improvement versus no improvement and the odds of marked improvement versus some or no improvement in the setting where sex and age are held constant; i.e., it is an odds ratio which describes the multiplicative extent to which more favorable response is more likely for test drug patients than placebo patients. Thus, the odds of some or marked improvement versus no improvement and the odds of marked improvement versus some or no improvement are 5.73 times greater for test drug patients than placebo patients. Additional discussion of the proportional odds model and methods to assess its goodness of fit are given in Harrell (1986), Koch et al. (1985c), McCullagh and Nelder (1983), and Peterson (1986).

Another extension of the logistic model which is useful for the analysis of data for responses with $r > 2$ levels is the log-linear model with the structure

$$\pi_{ij} = \frac{\exp(\alpha_j + \underline{x}_i'\underline{\beta}_j)}{\{1 + \Sigma_{j=1}^{(r-1)} \exp(\alpha_j + \underline{x}_i'\underline{\beta}_j)\}} \tag{68}$$

where $j = 1, 2, \ldots, (r-1)$ for a set of explanatory variables $\{\underline{x}_i\}$ for the respective subpopulations; for $j = r$, the model is $\pi_{ir} = 1 - \Sigma_{j=1}^{(r-1)} \pi_{ij}$. Since (68) implies that

$$\log_e \left\{ \frac{\pi_{ij}}{\pi_{ir}} \right\} = \alpha_j + \underline{x}_i'\underline{\beta}_j \tag{69}$$

for $j = 1, 2, \ldots, (r-1)$, the $\{\alpha_j\}$ represent intercept parameters and the $\{\underline{\beta}_j\}$ represent vectors of regression parameters. As was the case for the model (61), the model (68) describes general association between subpopulations and response in the sense that comparisons between the ith and i'th subpopulations have $(r-1)$ components

$$\log_e \left\{ \frac{\pi_{ij}\pi_{i'r}}{\pi_{ir}\pi_{i'j}} \right\} = (\underline{x}_i - \underline{x}_{i'})'\underline{\beta}_j \tag{70}$$

for $j = 1, 2, \ldots, (r-1)$. When response outcomes are nominal rather than ordinal, the evaluation of such general association would be of interest. Maximum likelihood methods for estimating model parameters and assessing goodness of fit for the log-linear model (68) in these situations are discussed in such references as Andersen (1980), Bishop et al. (1975), Fienberg (1980), Haberman (1978), and Imrey et al. (1981, 1982). Computations for this type of analysis can be undertaken with the SAS (1985) procedure CATMOD or the BMDP procedure P4F documented in Brown (1983).

A refinement of the log-linear model (68) which can account for ordinal response categories is the additional condition of equality for odds ratios involving adjacent response categories, i.e.,

$$\left( \frac{\pi_{ij}\pi_{i',(j+1)}}{\pi_{i,(j+1)}\pi_{i'j}} \right) = (\underline{x}_i - \underline{x}_{i'})'\underline{\beta} \tag{71}$$

for $j = 1, 2, \ldots, (r-1)$. Since (71) also implies that

$$\log_e \left\{ \frac{\pi_{ij}\pi_{i'r}}{\pi_{ir}\pi_{i'j}} \right\} = (r-j)(\underline{x}_i - \underline{x}_{i'})'\underline{\beta} \tag{72}$$

the comparison of the ith and i'th subpopulations with respect to the log-linear model with equal adjacent odds ratio structure has only one component. A specification for this model is

$$\pi_{ij} = \left\{ \frac{\exp\{\alpha_j + (r-j)\underline{x}_i'\underline{\beta}\}}{\{1 + \Sigma_{j=1}^{(r-1)} \exp\{\alpha_j + (r-j)\underline{x}_i'\underline{\beta}\}} \right\} \quad \text{for } j = 1, 2, \ldots, (r-1)$$

$$\pi_{ir} = 1 - \sum_{j=1}^{(r-1)} \pi_{ij} \tag{73}$$

for i = 1, 2, . . ., s. Aspects of the application of maximum likelihood methods to the model (73) are similar to those for the general model (68) and are discussed in the references cited previously for it. Results for the rheumatoid arthritis data from the fit of the equal adjacent odds ratio log-linear model with sex and treatment as explanatory variables are as follows:

| Parameter | Estimate | Standard error | p value | Interpretation |
|---|---|---|---|---|
| $\alpha_1$ | 2.607 | 0.707 | <0.010 | $\text{Log}_e$ odds of no improvement versus marked improvement for males receiving placebo |
| $\alpha_2$ | 0.566 | 0.515 | 0.271 | $\text{Log}_e$ odds of some improvement versus marked improvement for males receiving placebo |
| $\beta_1$ | −0.741 | 0.325 | 0.023 | Increment in adjacent odds of less favorable versus more favorable response due to female sex |
| $\beta_2$ | −1.076 | 0.293 | <0.010 | Increment in adjacent odds of less favorable versus more favorable response due to test drug |

(74)

The goodness of fit of this model can be assessed by application of counterparts to the Pearson chi-square statistic $Q_P$ in (53) or the log-likelihood ratio chi-square statistic $Q_L$ in (54) to the contingency table data structure shown in Section V. For these test statistics, the model-predicted frequencies are given by $\hat{m}_{ij} = n_{i+}\hat{\pi}_{ij}$, where $\hat{\pi}_{ij}$ is the maximum likelihood estimate for $\pi_{ij}$ as obtained from substitution of the maximum likelihood estimates $\{\hat{\alpha}_j\}$ and $\hat{\underline{\beta}}$ for model parameters into (73). Since $Q_P = 2.22$ and $Q_L = 3.36$ are both nonsignificant ($p = 0.695$ and $p = 0.499$, respectively) relative to their approximately chi-square distribution for which d.f. = $\{(r-1)s - t\} = 4$, where t is the dimension of $\underline{\beta}_A = (\alpha_1, \alpha_2, \underline{\beta}')'$, the use of this model is supported. Accordingly, $\exp(\hat{\beta}_2) = \exp(-1.076) = 0.341$ expresses the nature of treatment effects in the sense of the extent to which both the odds of no response versus some response and the odds of some response versus marked response are less likely for test drug patients than placebo patients. The corresponding reciprocal, $\exp(-\hat{\beta}_2) = 2.93$, expresses the extent to which the odds of more favorable response in adjacent pairs versus less favorable response are greater for test drug patients than placebo patients.

For applications such as the rheumatoid arthritis data analyzed in this chapter, the choice between the proportional odds model and the equal adjacent odds model mainly involves philosophical and computational considerations since both had satisfactory goodness of fit. The proportional odds model has two noteworthy advantages. One is that the pooling of adjacent response categories only influences its specification through reduction of the number of logit functions which it encompasses, whereas such pooling necessitates complete respecification of the equal adjacent odds ratio model due to the creation of a new set of adjacent outcomes. A second advantage of the proportional odds model is the availability of the SAS procedure LOGIST of Harrell (1986), which can be used to implement models with both categorical and continuous explanatory variables. Readily available computing procedures for the equal adjacent odds ratio model are currently limited to situations with categorical explanatory variables. However, the equal adjacent odds ratio model has the important advantage of being in the general class of log-linear models, and therefore the methods for the evaluation of its goodness of fit are more straightforward than those for the proportional odds model. Also, the equal adjacent odds model has somewhat greater flexibility for extensions which account for potential lack of fit. Of course, in some situations, only one of these two models might provide a satisfactory fit, so the choice of model would be based on practical considerations. If neither of these types of models was appropriate, the applicability of other reasonable structures would need to be explored. References which deal with alternative methods for the analysis of ordinal data or provide illustrative examples include Agresti (1983, 1984), Andrich (1979), Clogg (1982), Cox and Chuang (1984), Goodman (1979), Koch et al. (1982, 1985a, 1985c), McCullagh (1980), and McCullagh and Nelder (1983).

## VIII. RANK MEASURES OF ASSOCIATION

The application of the extended Mantel-Haenszel statistic to a set of $2 \times r$ tables with ordinal response categories was described in Section V. A relevant consideration for the effectiveness of that method was the consistency across strata for the patients with one treatment to have more favorable responses than those with the other treatment. One way to address this issue of homogeneity of treatment effects is through tests of treatment $\times$ strata interaction effects in a statistical model for ordinal data such as the proportional odds model or the equal adjacent odds ratio model discussed in Section VII. A limitation of this approach, however, is that these types of models may not have sufficiently satisfactory goodness of fit for their use to be appropriate. In such situations, an alternative approach involving the across-strata comparison of measures of association between treatment and response is of interest.

One useful rank measure of association between a dichotomous treatment classification and an ordinal response variable for the hth stratum is

$$g_h = \left\{ \sum_{j=1}^{3} p_{h1j} \left[ \left( \sum_{k=1}^{j} p_{h2k} \right) - 0.5 p_{h2j} \right] \right\} \tag{75}$$

where $p_{hij} = (n_{hij}/n_{hi+})$ denotes the proportion of subjects with the jth response for the hth stratum and the ith treatment. The function $g_h$ is related to the Mann-Whitney (1947) statistic in the sense of being an hth stratum estimator for the probability $\xi_h$ with which a randomly selected test drug patient has more favorable response than a randomly selected placebo patient when ties are randomly broken with probability 1/2. Since the hypothesis of no treatment effects on the responses of subjects in the hth stratum implies $\xi_h = 0.5$, the values of $(\xi_h - 0.5)$ describe the extent of treatment differences for the respective strata.

Given that the frequencies $\{n_{hij}\}$ have a product multinomial distribution like that shown in (60), a consistent estimator $\underline{V}_{\underline{g}}$ for the covariance matrix of $\underline{g} = (g_1, g_2)'$ can be constructed by linear Taylor series methods. This estimator has the form

$$\underline{V}_{\underline{g}} = [\underline{H}(\underline{p})]\underline{V}_{\underline{p}}[\underline{H}(\underline{p})]' \tag{76}$$

where $\underline{p} = (\underline{p}_{11}', \underline{p}_{12}', \underline{p}_{21}', \underline{p}_{22}')'$ is the compound vector with components $\underline{p}_{hi}' = (p_{hi1}, p_{hi2}, p_{hi3})$ for the respective subpopulations; $\underline{V}_{\underline{p}}$ is the block diagonal matrix with diagonal blocks $[\underline{D}_{\underline{p}_{hi}} - \underline{p}_{hi}\underline{p}_{hi}']/n_{hi+}$ and corresponds to the estimated covariance matrix of $\underline{p}$; and $\underline{H}(\underline{p}) = [\partial \underline{g}/\partial \underline{p}]$. Also, if $\underline{g}$ is expressed as the compound function

$$\underline{g} = \underline{A}_3\{\underline{\exp}[\underline{A}_2 \, \underline{\log}_e(\underline{A}_1 \underline{p})]\} \tag{77}$$

with $\underline{\exp}$ and $\underline{\log}_e$ being operations that respectively exponentiate and compute natural logarithms of the elements of a vector and

$$\underline{A}_1 = \begin{bmatrix} 1&0&0 &&& &&& &&& \\ 0&1&0 &&& &&& &&& \\ 0&0&1 &&& &&& &&& \\ &&& .5&0&0 &&& &&& \\ &&& 1&.5&0 &&& &&& \\ &&& 1&1&.5 &&& &&& \\ &&& &&& 1&0&0 &&& \\ &&& &&& 0&1&0 &&& \\ &&& &&& 0&0&1 &&& \\ &&& &&& &&& .5&0&0 \\ &&& &&& &&& 1&.5&0 \\ &&& &&& &&& 1&1&.5 \end{bmatrix} \tag{78}$$

$$\underline{A}_2 = \begin{bmatrix} 1&0&0&1&0&0 &&&&&& \\ 0&1&0&0&1&0 &&&&&& \\ 0&0&1&0&0&1 &&&&&& \\ &&&&&& 1&0&0&1&0&0 \\ &&&&&& 0&1&0&0&1&0 \\ &&&&&& 0&0&1&0&0&1 \end{bmatrix} \quad \text{and} \quad \underline{A}_3 = \begin{bmatrix} 1&1&1&0&0&0 \\ 0&0&0&1&1&1 \end{bmatrix}$$

then $\underline{\Pi}(\underline{p}) = \underline{A}_3 \underline{D}_{\underline{a}_2} \underline{A}_2 \underline{D}_{\underline{a}_1}^{-1} \underline{A}_1$ with $\underline{a}_1 = \underline{A}_1 \underline{p}$ and $\underline{a}_2 = \underline{\exp}[\underline{A}_2 \underline{\log}_e \underline{a}_1]$. Additional discussion concerning the use of compound functions to represent rank measures of association $\underline{g}$ and the construction of their estimated covariance matrix $\underline{V}_{\underline{g}}$ is given in Forthofer and Lehnen (1981), Koch et al. (1982, 1985a), and Semenya et al. (1983).

For the example ,

$$\underline{g} = \begin{bmatrix} 0.733 \\ 0.698 \end{bmatrix} \quad \text{and} \quad \underline{V}_{\underline{g}} = \begin{bmatrix} 0.3814 & 0 \\ 0 & 0.6704 \end{bmatrix} \times 10^{-2} \tag{79}$$

As a consequence of central limit theory for Mann-Whitney statistics (see Puri and Sen, 1971), the estimators $\underline{g}$ approximately have a multivariate normal distribution when the sample sizes for the respective (sex × treatment) subpopulations are sufficiently large (e.g., $n_{hi+} \geqslant 10$). Thus, the variation among the $\{g_h\}$ can be analyzed by weighted least-squares methods and Wald statistics as discussed in Grizzle et al. (1969) and Koch et al. (1977, 1985a). The Wald statistic for comparing the $\{g_h\}$ for the two strata is

$$Q_W = \underline{g}'\underline{W}'[\underline{W}\underline{V}_{\underline{g}}\underline{W}']^{-1}\underline{W}_{\underline{g}} = 0.12 \tag{80}$$

where $\underline{W} = [1, -1]$. This result is nonsignificant with p = 0.732 relative to the approximately chi-square distribution with d.f. = 1 for $Q_W$, and so the two sexes are interpreted as having similar patterns of association between treatment and response relative to the measures $\{\xi_h\}$. On this basis, the $\{g_h\}$ can be described with the linear model

$$\underline{E}_{\underline{A}}\{\underline{g}\} = \underline{\xi} = \begin{bmatrix} \xi_1 \\ \xi_2 \end{bmatrix} = \begin{bmatrix} 1 \\ 1 \end{bmatrix} \beta = \underline{X}\beta \tag{81}$$

where $E_A\{\ \}$ denotes expected value for large samples, $\underline{X}$ is the model specification matrix, and $\beta$ is the parameter for the common value of the $\{\xi_h\}$ for the two strata under the model. The weighted least-squares estimator b for $\beta$ and a consistent estimator $V_b$ for its variance are given by

$$b = (\underline{X}'\underline{V}_{\underline{g}}^{-1}\underline{X})^{-1}\underline{X}'\underline{V}_{\underline{g}}^{-1}\underline{g} = 0.720 \tag{82}$$

and

$$V_b = (\underline{X}'\underline{V}_{\underline{g}}^{-1}\underline{X})^{-1} = 0.002431 \tag{83}$$

The goodness of fit of the model specified by $\underline{X}$ is supported by the statistic

$$Q_W = (\underline{g} - \underline{X}b)'\underline{V}_{\underline{g}}^{-1}(\underline{g} - \underline{X}b) = 0.12 \tag{84}$$

which approximately has the chi-square distribution with d.f. = (u − t) = 1, where u denotes the number of elements in $\underline{g}$ and t denotes the number of elements in b. $Q_W$ is identical to the Wald statistic shown in (80) when

$\underline{X}$ and $\underline{W}'$ are orthocomplements of one another (i.e., $\underline{W}\,\underline{X} = \underline{0}_{(u-t),t}$ and Rank[$\underline{X}$, $\underline{W}'$] = u).

Relative to the model $\underline{X}$ in (81), the hypothesis of no treatment effects can be expressed as $H_0$: $\beta = 0.5$. Since b approximately has a normal distribution,

$$Q_b = \frac{(b - 0.5)^2}{V_b} = 19.91 \tag{85}$$

approximately has the chi-square distribution with d.f. = 1. This result is significant with $p < 0.01$, and so a difference between treatments is evident. The estimator b = 0.720 enables this difference to be interpreted as involving 0.72 probability for a randomly selected test drug patient having more favorable response than a randomly selected placebo patient. The computations for obtaining $\underline{g}$ and $\underline{V}_g$ and carrying out weighted least-squares analysis were undertaken with the program GENCAT documented in Landis et al. (1976); with some modifications in the operations for specifying g, the SAS (1985) procedure CATMOD could have been used.

Since the statistic $Q_b$ in (85) is based on the combination of the rank measures of association $\{g_h\}$, it is analogous to the extended Mantel-Haenszel statistic $Q_{EMH}$ with standardized midrank scores in Section V. The main advantage it has over $Q_{EMH}$ is its determination from a framework through which the across-strata homogeneity of measures of association between treatment and response can be evaluated. On the other hand, the application of $Q_b$ is limited to situations where the $\{g_h\}$ are homogeneous because this condition is an assumption of its underlying model (81), whereas $Q_{EMH}$ can be used more broadly. Also, since $Q_{EMH}$ is based on linear statistics, chi-square approximations for its distribution tend to be reasonable for somewhat smaller sample sizes than the methods for the nonlinear statistics considered in this section. Further discussion of analyses of rank measures of association and additional examples illustrating their comparison with other methods are given in Forthofer and Lehnen (1981), Koch et al. (1982, 1985a, 1986b), and Semenya et al. (1983).

## IX. A GROUPED SURVIVAL TIMES EXAMPLE

The data in Table 4 are from a multicenter trial to compare a test drug to placebo with respect to time to healing of a gastrointestinal condition. Patients were examined at 2 weeks to determine whether healing had occurred. If healing had not occurred, they were reexamined at 4 weeks. All patients were considered to have completed the study at 4 weeks regardless of whether their condition had been healed.

A variety of useful preliminary analyses can be performed for these data without giving specific attention to their time to healing (or ulcer survival) structure. For example, the Mantel-Haenszel method (Section IV) was applied to each of two binary response variables: healed at 2 weeks versus not healed at 2 weeks, and healed by 4 weeks (i.e., healed at 2 weeks or healed at 4 weeks) versus not healed. There was no significant difference between the treatments with respect to healing by 2 weeks, controlling for center ($Q_{MH}$ = 1.94, d.f. = 1, p = 0.164). Also, no lack of homogeneity between centers is suggested with respect to this variable

**Table 4** Data from a Multicenter Trial for the Comparison of Treatments with Respect to Healing of a Gastrointestinal Condition

| | | Contingency table format | | | |
|---|---|---|---|---|---|
| Center | Treatment | Number healed at 2 weeks | Number healed at 4 weeks but not at 2 weeks | Number not healed | Total |
| 1 | Test drug | 15 | 17 | 2 | 34 |
| | Placebo | 15 | 17 | 7 | 39 |
| 2 | Test drug | 17 | 17 | 10 | 44 |
| | Placebo | 12 | 13 | 15 | 40 |
| 3 | Test drug | 7 | 17 | 16 | 40 |
| | Placebo | 3 | 17 | 18 | 38 |

($Q_{PH} = Q_T - Q_{MH} = 2.51 - 1.94 = 0.57$, d.f. = 2, p = 0.752). However, there was a significant treatment difference with respect to healing by 4 weeks, controlling for center ($Q_{MH} = 4.01$, d.f. = 1, p = 0.045), although none of the treatment differences at the individual centers was significant at the 0.05 level. The pseudohomogeneity statistic was nonsignificant for this response variable ($Q_{PH} = 0.99$, d.f. = 2, p = 0.610).

Logistic regression analyses were also undertaken for each of the previously specified dichotomous response variables in order to compare the results with those from the Mantel-Haenszel method and to obtain a more convincing assessment of the homogeneity of the association between treatment and response at the three centers. Computations were performed with the SAS procedure LOGIST. Two indicator variables were formed to represent the effect of center (1 if center 1, 0 otherwise; and 1 if center 2, 0 otherwise). The treatment variable was coded as 1 if test drug, 0 if placebo. Two indicator variables were also formed to represent center-by-treatment interaction (1 if center 1 and test drug, 0 otherwise; and 1 if center 2 and test drug, 0 otherwise). The indicator variables for center and treatment were forced to be included in the model, and the two interaction indicator variables were considered as candidate variables for entry. The Rao score statistic for testing the combined contribution of the interaction variables was nonsignificant for both response variables, so the models with main effects for center and treatment were considered to have satisfactory goodness of fit.

In the main effects logistic regression model for healing by 2 weeks, the indicator variables for center 1 and center 2 were both significant ($p < 0.01$ for both Wald statistics), and the parameter estimates were both positive, indicating that healing was significantly faster at centers 1 and 2 than at center 3. The Wald statistic for treatment effect was nonsignificant ($Q_W = 1.95$, p = 0.163) and was very similar to the corresponding Mantel-Haenszel statistic ($Q_{MH} = 1.94$). In the logistic regression for healing by 4 weeks, the indicator variable for center 1 was significant

($p < 0.001$) and the indicator variable for center 2 was nearly significant ($p = 0.069$), with both parameter estimates being positive. The Wald statistic for treatment effect ($Q_W = 4.01$, $p = 0.045$) was again virtually identical to the corresponding Mantel-Haenszel statistic ($Q_{MH} = 4.01$).

Other possible methods which could be applied to these data include extensions of the Mantel-Haenszel method (Section V) and extensions of logistic regression (Section VII). In these analyses, the three responses (healed at 2 weeks, healed between 2 and 4 weeks, and not healed) would be considered as an ordinal response variable with three levels.

A survival data extension of the Mantel-Haenszel procedure was suggested by Mantel (1966) and later derived by Cox (1972) from likelihood theory under the Cox regression model for essentially continuous time to event response variables. The Mantel-Cox test $Q_{MC}$ involves restructuring the contingency table data as a set of $2 \times 2$ tables with a life table format for each center, and then applying the Mantel-Haenszel procedure to this set of tables. The data in Table 4 would be restructured as shown in Table 5. The frequencies in this table satisfy the Mantel-Fleiss (1980) criterion (Section IV), so a chi-square approximation is appropriate for the evaluation of the Mantel-Haenszel (or Mantel-Cox) statistic. There is a significant difference between the treatments according to the Mantel-Cox statistic, $Q_{MC} = 4.25$ (d.f. = 1, $p = 0.039$). If the Mantel-Fleiss criterion had not been satisfied, the computing procedure of Thomas (1975) could have been used to provide an exact probability for $Q_{MC}$.

The Mantel-Cox statistic is closely related to the mean score statistic $Q_S$ in (20) with logrank scores or its stratified extension $Q_{EMH}$ in (44). For this reason, tests based on $Q_{MC}$ are often referred to as logrank tests. Both $Q_S$ and $Q_{MC}$ have the same numerator. The denominator of $Q_{MC}$ is like that shown in (33) for sets of $2 \times 2$ tables, whereas the denominator of the $Q_S$ or $Q_{EMH}$ is like that shown in (44) for sets of $2 \times r$ tables [see Lee (1980) and Koch et al. (1985b) for additional discussion of the difference between the two statistics].

One could also approach the data in Table 5 from a model-based survival analysis standpoint. One model which can be used to describe the relationship between a grouped survival time response variable and a set of categorical explanatory variables is the piecewise exponential model. This model requires the assumption that within each of the intervals, 0 to 2 weeks and 2 to 4 weeks, the events of healing have independent exponential distributions. The piecewise exponential likelihood function for our example may be written as

$$L_{PE} = \prod_{i=1}^{6} \prod_{j=1}^{2} \lambda_{ij}^{n_{ij}} [\exp(-\lambda_{ij} N_{ij})] \tag{86}$$

where i indexes the six subpopulations formed from the cross-classification of center and treatment; j indexes the two time intervals; and $n_{ij}$, $N_{ij}$, and $\lambda_{ij}$ are respectively the number of patients with healing, the number of person-weeks at risk to heal, and the hazard for the ith subpopulation in the jth interval. The quantities $n_{ij}$ and $N_{ij}$ are shown in the fourth and last columns, respectively, of Table 5. The $N_{ij}$ were calculated under the assumption that healing occurred uniformly throughout the interval and

**Table 5** Life Table Format for Data from a Multicenter Trial for the Comparison of Treatments with Respect to Healing of a Gastrointestinal Condition

| | | | Life table format | | | |
|---|---|---|---|---|---|---|
| Center | Time interval (weeks) | Treatment | Number healed during interval | Number not healed during interval | Total | Estimated person-weeks at risk |
| 1 | 0–2 | Test drug | 15 | 19 | 34 | 53 |
| | | Placebo | 15 | 24 | 39 | 63 |
| | 2–4 | Test drug | 17 | 2 | 19 | 21 |
| | | Placebo | 17 | 7 | 24 | 31 |
| 2 | 0–2 | Test drug | 17 | 27 | 44 | 71 |
| | | Placebo | 12 | 28 | 40 | 68 |
| | 2–4 | Test drug | 17 | 10 | 27 | 37 |
| | | Placebo | 13 | 15 | 28 | 43 |
| 3 | 0–2 | Test drug | 7 | 33 | 40 | 73 |
| | | Placebo | 3 | 35 | 38 | 73 |
| | 2–4 | Test drug | 17 | 16 | 33 | 49 |
| | | Placebo | 17 | 18 | 35 | 53 |

therefore can be viewed as occurring at the midpoint of the interval; thus, $N_{ij} = (n_{ij} \times 1) + [(\text{number in subpopulation i not healed in interval j}) \times 2]$.

We can express the hazard rates $\lambda_{ij}$ in terms of a log-linear model with specification matrix $\underline{X}$ and parameter vector $\underline{\beta}$ as follows:

$$\lambda_{ij} = \exp(\underline{x}'_{ij}\underline{\beta}) \tag{87}$$

where $\underline{x}'_{ij}$ is the row of $\underline{X}$ corresponding to the ith subpopulation and jth interval. The matrix $\underline{X}$ is required to have full rank $t \leqslant sd$, where $s = 6$ is the number of subpopulations and $d = 2$ the number of time intervals. Maximum likelihood estimation of the regression parameters for this model can be undertaken using log-linear Poisson regression computing procedures because, as shown by Holford (1980) and Laird and Olivier (1981), the likelihoods for the piecewise exponential and Poisson frameworks are proportional. If we consider the numbers healed $\{n_{ij}\}$, conditional on their exposures to treatment $\{N_{ij}\}$, as having independent Poisson distributions with means $\mu_{ij} = N_{ij}\lambda_{ij}$, then the Poisson counterpart to (86) is

$$L_{PO} = \prod_{i=1}^{6}\prod_{j=1}^{2} \frac{(N_{ij}\lambda_{ij})^{n_{ij}}\,[\exp(-N_{ij}\lambda_{ij})]}{n_{ij}!} \tag{88}$$

$$= L_{PE}\left[\prod_{i=1}^{6}\prod_{j=1}^{2} \frac{N_{ij}^{n_{ij}}}{n_{ij}!}\right]$$

It is not necessary to actually assume that the $\{n_{ij}\}$ have Poisson distributions. The Poisson likelihood (88) is presented only to substantiate the proportionality of $L_{PO}$ and $L_{PE}$. Since the two likelihood functions are proportional, the $\underline{\beta}$ which maximize $L_{PO}$ also maximize $L_{PE}$, and thus maximum likelihood Poisson regression computing procedures may be used for estimation of piecewise exponential model parameters. This is of practical as well as theoretical importance because computing procedures for Poisson regression are more widely available than those for piecewise exponential modeling.

The maximum likelihood Poisson regression approach to parameter estimation involves substituting $\exp(\underline{x}'_{ij}\underline{\beta})$ for $\lambda_{ij}$ in (88), differentiating the $\log_e$ of (88) with respect to $\underline{\beta}$, and equating the result to zero. The set of nonlinear equations resulting from this process has the form

$$\underline{X}'\underline{n} = \underline{X}'\hat{\underline{\mu}} = \underline{X}'\underline{D}_{\underline{N}}[\exp(\underline{X}\,\hat{\underline{\beta}})] \tag{89}$$

where $\underline{n} = \{n_{ij}\} = (n_{11}, n_{12}, \ldots, n_{61}, n_{62})'$, $\hat{\underline{\mu}}$ is the corresponding vector of predicted values (where $\hat{\mu}_{ij} = N_{ij}\hat{\lambda}_{ij}$), and $\underline{D}_{\underline{N}}$ is a diagonal matrix with the elements of $\underline{N} = (N_{11}, N_{12}, \ldots, N_{61}, N_{62})'$ down the main diagonal. Since these equations usually do not have an explicit solution, an iterative method, such as the Newton-Raphson procedure, is necessary to solve for $\underline{\beta}$. The maximum likelihood estimator $\hat{\underline{\beta}}$ approximately has a multivariate normal distribution with a covariance matrix which can be consistently estimated by

$$\underline{V}(\hat{\beta}) = (\underline{X}'\underline{D}_{\hat{\mu}}\underline{X})^{-1} \tag{90}$$

when the observed $n_{ij}$ are sufficiently large for the linear functions $\underline{X}'\underline{n}$ to have approximately a multivariate normal distribution from central limit theory; a relatively conservative guideline for this purpose is all $n_{ij} \geqslant 5$.

Two measures of goodness of fit for the model $\underline{X}$ are the Pearson chi-square criterion

$$Q_P = \sum_{i=1}^{6} \sum_{j=1}^{2} \frac{(n_{ij} - \hat{\mu}_{ij})^2}{\hat{\mu}_{ij}} \tag{91}$$

and the log-likelihood ratio chi-square criterion

$$Q_L = \sum_{i=1}^{6} \sum_{j=1}^{2} 2n_{ij} \log_e \left[ \frac{n_{ij}}{\hat{\mu}_{ij}} \right] \tag{92}$$

Both of these statistics have d.f. equal to the number of rows of $\underline{X}$ minus the number of columns of $\underline{X}$; in this example, d.f. = (sd − t) = (12 − t).

As in Section VI, the goodness of fit of the model may also be evaluated by assessing the contribution of columns $\underline{W}$ (a matrix with rank w) as an expansion to the model $\underline{X}$ via a log-likelihood ratio statistic $Q_{LR}$ with structure analogous to (57) or by the Rao score statistic

$$Q_{RS} = (\underline{n} - \hat{\underline{\mu}})'\underline{W}[\underline{W}'\{\underline{D}_{\hat{\mu}} - \underline{D}_{\hat{\mu}}\underline{X}(\underline{X}'\underline{D}_{\hat{\mu}}\underline{X})^{-1}\underline{X}'\underline{D}_{\hat{\mu}}\}\underline{W}]^{-1}\underline{W}'(\underline{n} - \hat{\underline{\mu}}) \tag{93}$$

Both $Q_{LR}$ and $Q_{RS}$ approximately have chi-square distributions with w degrees of freedom.

A model of interest for the data in Table 5 is

$$\underline{X} = \begin{bmatrix} 1 & 1 & 1 & 1 & 1 & 1 & 1 & 1 & 1 & 1 & 1 & 1 \\ 0 & 0 & 0 & 0 & 1 & 1 & 1 & 1 & 0 & 0 & 0 & 0 \\ 0 & 0 & 0 & 0 & 0 & 0 & 0 & 0 & 1 & 1 & 1 & 1 \\ 1 & 0 & 1 & 0 & 1 & 0 & 1 & 0 & 1 & 0 & 1 & 0 \\ 0 & 0 & 1 & 1 & 0 & 0 & 1 & 1 & 0 & 0 & 1 & 1 \end{bmatrix}' \tag{94}$$

for which the parameter estimates and their standard errors are as follows:

| Parameter | Estimate | Standard error | p value | Interpretation |
|---|---|---|---|---|
| $\beta_1$ | −1.515 | 0.172 | <0.010 | Reference value of the $\log_e$ hazard rate for placebo group patients in center 1 during weeks 0–2 |
| $\beta_2$ | −0.421 | 0.181 | 0.020 | Increment due to center 2 |

| Parameter | Estimate | Standard error | p value | Interpretation |
|---|---|---|---|---|
| $\beta_3$ | −0.887 | 0.197 | <0.010 | Increment due to center 3 |
| $\beta_4$ | 0.306 | 0.156 | 0.049 | Increment due to test drug |
| $\beta_5$ | 0.964 | 0.158 | <0.010 | Increment for interval 2–4 weeks |

(95)

Computations for this example were undertaken with the SAS macro CATMAX documented in Stokes and Koch (1983).

The goodness of fit of model $\underline{X}$ is supported by the nonsignificance of the Pearson and log-likelihood criteria with respect to the chi-square distribution with d.f. = 7 ($Q_P = 6.29$, p = 0.506 and $Q_L = 6.66$, p = 0.466). A Rao score statistic was used to explicitly test the effect of center-by-treatment interaction; the result was nonsignificant ($Q_{RS} = 0.14$, d.f. = 2, p = 0.931).

The model $\underline{X}$ is said to have a proportional hazards structure because the effects of the explanatory variables are specified to be the same in both time intervals. This proportionality assumption was also tested with Rao score statistics. The center-by-time interaction was assessed by the Rao score statistic for the contribution of the matrix

$$\underline{W}_1 = \begin{bmatrix} 0 & 0 & 0 & 0 & 0 & 0 & 1 & 1 & 0 & 0 & 0 & 0 \\ 0 & 0 & 0 & 0 & 0 & 0 & 0 & 0 & 0 & 0 & 1 & 1 \end{bmatrix}' \qquad (96)$$

Its result approached significance ($Q_{RS} = 5.08$, d.f. = 2, p = 0.078). The contribution of the treatment-by-time interaction

$$\underline{W}_2 = [0\ 0\ 1\ 0\ 0\ 0\ 1\ 0\ 0\ 0\ 1\ 0]' \qquad (97)$$

was nonsignificant ($Q_{RS} = 0.01$, d.f. = 1, p = 0.931). Also, the joint contribution of the center-by-time and treatment-by-time interactions, $[\underline{W}_1, \underline{W}_2]$, was nonsignificant ($Q_{RS} = 5.09$, d.f. = 3, p = 0.165). Thus, the overall interpretation is that the effects of center and treatment during the first 2 weeks of the study are similar to their effects during the second 2 weeks, although it is recognized that there is some suggestion of possible departure of center effects from this structure. If this were considered important, the model could be expanded to include center-by-time interaction; such analysis is not illustrated here. For further discussion of the application of Poisson regression methods to the analysis of grouped survival data like those shown in Table 4, see Frome (1983), Holford (1980), Koch et al. (1985a, 1986a), and Laird and Olivier (1981).

## X. A LONGITUDINAL STUDY EXAMPLE

Many clinical trials have longitudinal designs in which the response status of each patient in the respective treatment groups is determined at two or more visits. For the resulting repeated measurements data structure of these studies, the objectives of statistical analysis are as follows:

1. Comparisons among groups for the response distributions at each visit separately and for their average across visits
2. Comparisons among visits for the response distributions within each group and for their average across subpopulations
3. Comparisons among groups for differences between visits in response distributions (i.e., group × visit interaction)
4. Description of the relationship of response distributions with group and visit

Some methods for addressing these objectives are discussed here in the context of an example involving two treatments (test drug and placebo) in a randomized clinical trial for a skin disorder. The response variables are the dichotomous response status, satisfactory (S) or unsatisfactory (U), at each of two visits (7 days and 14 days). The data for this example are summarized in Table 6 in terms of a 2 × 4 contingency table. The rows of this table correspond to the $s = 2$ treatments, and the columns correspond to the $r = 2^2 = 4$ possible response profiles [i.e., (S, S), (S, U), (U, S), and (U, U)].

For the data in Table 6, the proportion of patients with satisfactory classifications for each (group × visit) describe the corresponding response distributions. These quantities indicate that 0.50 of the patients with test drug had satisfactory response at 7 days, and 0.73 of them had this status at 14 days. In contrast, the proportions of placebo patients with satisfactory response were 0.28 at 7 days and 0.23 at 14 days. Thus, the differences between treatments in the proportions of patients with satisfactory response status favor test drug at both visits, and this tendency appears much stronger at 14 days than at 7 days. Subsequent analysis provides more formal confirmation of these conclusions.

The comparison of the two treatments at each visit separately can be undertaken by applying the methods in Section II to the corresponding 2 × 2 marginal tables. More specifically, let $n_{ijk}$ denote the number of patients in the ith treatment group with the jth response outcome at 7 days and the kth response outcome at 14 days, where i = 1, 2 for test drug and placebo, and j, k = 1, 2 for satisfactory and unsatisfactory. Then,

**Table 6** Data from a Longitudinal Clinical Trial for the Comparison of Treatments for a Skin Disorder

| | Frequencies of response at 7 days and 14 days[a] | | | | |
|---|---|---|---|---|---|
| Treatment | (S, S) | (S, U) | (U, S) | (U, U) | Number of patients |
| Test drug | 24 | 6 | 20 | 10 | 60 |
| Placebo | 8 | 9 | 6 | 37 | 60 |

[a]S, Satisfactory; U, unsatisfactory.

the $\{n_{ij+} = \Sigma_{k=1}^{2} n_{ijk}\}$ and the $\{n_{i+k} = \Sigma_{j=1}^{2} n_{ijk}\}$ represent the response distributions for the two treatments at 7 and 14 days, respectively. For these 2 × 2 tables, the two-sided Fisher's exact tests yield $p = 0.024$ for the treatment comparison at 7 days and $p < 0.001$ for that at 14 days, and so patients with test drug have significantly more favorable experience than their placebo counterparts at both visits.

Since the proportions of patients with satisfactory response at 7 days and at 14 days are given by $p_{i1+} = (n_{i11} + n_{i12})/n_{i++}$ and $p_{i+1} = (n_{i11} + n_{i21})/n_{i++}$, respectively, their average is given by the mean score

$$\bar{f}_i = \frac{p_{i1+} + p_{i+1}}{2} = \frac{2n_{i11} + n_{i12} + n_{i21}}{2n_{i++}} = \sum_{j=1}^{2} \sum_{k=1}^{2} \frac{a_{jk} n_{ijk}}{n_{i++}} \tag{98}$$

where $(a_{11}, a_{12}, a_{21}, a_{22}) = (1, 0.5, 0.5, 0) = \underline{a}'$. It follows from Section III that the mean score statistic $Q_S$ in (19) is applicable for the comparison of the $\{\bar{f}_i\}$ for the two treatment groups. A convenient computational strategy to obtain $Q_S$ is to apply the integer scores (1, 2, 3) to the following transformation of Table 6:

| Treatment | (U, U) | {(U, S) or (S, U)} | (S, S) | Total |
|---|---|---|---|---|
| Test drug | 10 | 26 | 24 | 60 |
| Placebo | 37 | 15 | 8 | 60 |

(99)

The justification for this approach is the invariance of $Q_S$ with respect to linear transformations of $\underline{a}$. The result $Q_S = 23.77$ has $p < 0.01$ relative to the chi-square distribution with d.f. = 1, and so the test drug is interpreted as significantly better than placebo with respect to the average proportion of patients with satisfactory response across the two visits.

Information concerning variation between visits is provided by the differences

$$\bar{g}_i = (p_{i+1} - p_{i1+}) = \frac{n_{i21} - n_{i12}}{n_{i++}} = \sum_{j=1}^{2} \sum_{k=1}^{2} \frac{\bar{a}_{jk} n_{ijk}}{n_{i++}} \tag{100}$$

where $(\bar{a}_{11}, \bar{a}_{12}, \bar{a}_{21}, \bar{a}_{22}) = (0, -1, 1, 0) = \bar{\underline{a}}'$. Their comparison between groups through $Q_S$ in (19) provides an assessment of group × visit interaction. A convenient way to obtain $Q_S$ for this situation is to apply the integer scores (1, 2, 3) to the following transformation of Table 6:

| Treatment | (S, U) | {(S, S) or (U, U)} | (U, S) | Total |
|---|---|---|---|---|
| Test drug | 6 | 34 | 20 | 60 |
| Placebo | 9 | 45 | 6 | 60 |

(101)

Such computation yields $Q_S = 7.17$, which is significant ($p < 0.01$) relative to the chi-square distribution with d.f. = 1. This significant group × visit interaction corresponds to the much larger difference in favor of test drug at 14 days than at 7 days.

The comparison of visits within each group separately is addressed in terms of the extent to which the $\{\bar{g}_i\}$ are different from 0, or equivalently the extent to which the ratios $\{(n_{i21}/(n_{i21} + n_{i12})\}$ are different from 0.5. This latter type of hypothesis can be evaluated exactly with the binomial distribution in a spirit similar to the sign test. Such analysis is usually known as McNemar's test. For test drug, $n_{121} = 20$ and $(n_{112} + n_{121}) = 26$. The two-sided exact probability of outcomes at least this extreme for the binomial distribution with 26 trials and probability parameter 0.5 is $p = 0.0094$. Similarly, for placebo, where $n_{221} = 6$ and $(n_{212} + n_{221}) = 15$, the two-sided exact probability for outcomes at least as extreme is $p = 0.607$. Thus, for active treatment patients, the proportion of patients with satisfactory response increases significantly between 7 days and 14 days, whereas for placebo it tends to remain unchanged. This substantial difference in the pattern of change across visits for the two treatment groups corresponds to an alternative description of the group × visit interaction. Its nature implies that any comparison between visits for the pooled groups would need to be viewed cautiously. The application of McNemar's test to this setting would involve outcomes at least as extreme as $(n_{121} + n_{221}) = 26$ relative to $(n_{112} + n_{121} + n_{212} + n_{221}) = 41$ binomial trials. Since the number of binomial trials under consideration exceeds 40, approximation with the chi-square distribution with d.f. = 1 is applied to the continuity-corrected statistic

$$Q_M = \frac{\{|26 - (41/2)| - 0.5\}^2}{(41/4)} = 2.44 \tag{102}$$

Since the two-sided $p = 0.118$, it appears that the change between visits for the pooled groups is nonsignificant. As noted previously, the nature of the significant group × visit interaction tends to imply that the separate groups provide a better framework for interpreting comparisons between visits than the combined groups.

Although the previously considered methods for longitudinal studies are useful for specific statistical tests concerning groups, visits, and group × visit interaction, they do not provide an overall description of the variation among groups and visits for the proportions of patients with satisfactory response. One way to address this latter objective of analysis is to fit linear regression models by the weighted least-squares (WLS) procedures discussed in Grizzle et al. (1969). This general methodology has three stages: (1) construction of appropriate functions of the responses and a corresponding consistent estimate of their covariance structure, (2) estimation of model parameters, and (3) tests for hypotheses concerning model parameters. Its principal assumption is the availability of sufficient sample size for the functions in (1) to have an approximately multivariate normal distribution for which the covariance matrix can be viewed as known. This requirement tends to limit the use of WLS methods to situations with categorical explanatory variables for which each of the cross-classified subpopulations has at least moderate sample size (e.g., $\geq 25$). A noteworthy advantage of WLS methods is their applicability to other types of functions than those involved in logistic regression or its extensions (as discussed in Sections VI and VII). These include linear functions as well as more complex functions like the rank measures of association discussed in Section VIII. This capability of WLS methods is particularly relevant to longitudinal studies in which linear functions like $p_{i1+}$ and $p_{i+1}$ for describing the marginal distributions of response at each visit are often of interest.

In the subsequent discussion of the example summarized in Table 6, the patients in each treatment group are assumed to be conceptually representative of corresponding large target subpopulations in a sense equivalent to simple random sampling. On this basis, the $\{n_{ijk}\}$ have the product multinomial distribution

$$P\{n_{ijk}\} = \prod_{i=1}^{2}\left[n_{i++}! \prod_{j=1}^{2}\prod_{k=1}^{2}\left(\pi_{ijk}^{n_{ijk}} \Big/ n_{ijk}!\right)\right] \tag{103}$$

where the $\{\pi_{ijk}\}$ denote the probabilities of the jth response outcome at 7 days and the kth response outcome at 14 days for randomly selected patients who received the ith treatment. Let $\underline{\pi}_i = (\pi_{i11}, \pi_{i12}, \pi_{i21}, \pi_{i22})'$, and let $\underline{p}_i = (p_{i11}, p_{i12}, p_{i21}, p_{i22})'$, where $p_{ijk} = (n_{ijk}/n_{i++})$ denotes the sample estimator of $\pi_{ijk}$. Also, define the compound vectors $\underline{\pi} = (\underline{\pi}_1', \underline{\pi}_2')'$ and $\underline{p} = (\underline{p}_1', \underline{p}_2')'$. A set of linear functions of $\underline{p}$ can be expressed in the form $\underline{F} = \underline{A}\,\underline{p}$. For the proportions $p_{i1+}$ and $p_{i+1}$ of patients with satisfactory response at 7 days and 14 days,

$$\underline{A} = \begin{bmatrix} 1 & 1 & 0 & 0 & 0 & 0 & 0 & 0 \\ 1 & 0 & 1 & 0 & 0 & 0 & 0 & 0 \\ 0 & 0 & 0 & 0 & 1 & 1 & 0 & 0 \\ 0 & 0 & 0 & 0 & 1 & 0 & 1 & 0 \end{bmatrix} \tag{104}$$

A consistent estimate for the covariance matrix of the linear functions $\underline{F}$ is given by

$$\underline{V}_{\underline{F}} = \underline{A}\,\underline{V}_{\underline{p}}\underline{A}' \tag{105}$$

where $\underline{V}_{\underline{p}}$ is the estimated covariance matrix for $\underline{p}$ and has the block diagonal structure

$$\underline{V}_{\underline{p}} = \begin{bmatrix} (\underline{D}_{\underline{p}_1} - \underline{p}_1\underline{p}_1')/n_{1++} & \underline{0}_{4,4} \\ \underline{0}_{4,4} & (\underline{D}_{\underline{p}_2} - \underline{p}_2\underline{p}_2')/n_{2++} \end{bmatrix} \tag{106}$$

In (106), $\underline{D}_{\underline{y}}$ denotes a diagonal matrix with elements of $\underline{y}$ on the diagonal, and $\underline{0}_{4,4}$ denotes the $4 \times 4$ matrix of 0's. The specific forms of $\underline{F}$ and $\underline{V}_{\underline{F}}$ for the data in Table 6 are as follows:

$$\underline{F} = \begin{bmatrix} p_{11+} \\ p_{1+1} \\ p_{21+} \\ p_{2+1} \end{bmatrix} = \begin{bmatrix} 0.500 \\ 0.733 \\ 0.283 \\ 0.233 \end{bmatrix} \quad \text{and} \quad \underline{V}_{\underline{F}} = \begin{bmatrix} 41.67 & 5.56 & 0.00 & 0.00 \\ 5.56 & 32.59 & 0.00 & 0.00 \\ 0.00 & 0.00 & 33.84 & 11.20 \\ 0.00 & 0.00 & 11.20 & 29.82 \end{bmatrix} \times 10^{-4} \tag{107}$$

The functions $\underline{F}$ are viewed here as approximately having a multivariate normal distribution since the sample sizes for the two treatment groups are moderately large (i.e., $n_{1++} = n_{2++} = 60$ and all $n_{ijk} \geq 5$). On this basis, the application of weighted least-squares methods to fit a linear model to $\underline{F}$ is appropriate. Linear models for a $(u \times 1)$ vector $\underline{F}$ are expressed as

$$\underline{E}_A\{\underline{F}\} = \underline{X}\,\underline{\beta} \tag{108}$$

where $\underline{E}_A\{\ \}$ denotes expected value for large samples, $\underline{X}$ is the $(u \times t)$ model specification matrix with full rank $t \leq u$, and $\underline{\beta}$ is the $(t \times 1)$ vector of unknown parameters. The weighted least-squares estimators $\underline{b}$ for $\underline{\beta}$ and their estimated covariance matrix are given by

$$\underline{b} = (\underline{X}'\underline{V}_{\underline{F}}^{-1}\underline{X})^{-1}\underline{X}'\underline{V}_{\underline{F}}^{-1}\underline{F} \qquad \text{and} \qquad \underline{V}_{\underline{b}} = (\underline{X}'\underline{V}_{\underline{F}}^{-1}\underline{X})^{-1} \tag{109}$$

Goodness of fit of the model (108) can be assessed with the Wald statistic

$$Q_W = (\underline{F} - \underline{X}\,\underline{b})'\underline{V}_{\underline{F}}^{-1}(\underline{F} - \underline{X}\,\underline{b}) = \underline{F}'\underline{W}'(\underline{W}\,\underline{V}_{\underline{F}}\underline{W}')^{-1}\underline{W}\,\underline{F} \tag{110}$$

where $\underline{X}$ and $\underline{W}'$ are orthocomplements to one another (i.e., $\underline{W}\,\underline{X} = \underline{0}$ and Rank $[\underline{X}, \underline{W}'] = u$). When the model applies, $Q_W$ approximately has the chi-square distribution with d.f. $= (u - t)$. For models which are considered to have satisfactory fit, linear hypotheses $H_0$: $\underline{C}\,\underline{\beta} = \underline{0}$ (where $\underline{C}$ is a $c \times t$ specification matrix) can be tested with the Wald statistic

$$Q_C = \underline{b}'\underline{C}'(\underline{C}\,\underline{V}_{\underline{b}}\underline{C}')^{-1}\underline{C}\,\underline{b} \tag{111}$$

$Q_C$ approximately has the chi-square distribution with d.f. $= c$. A useful preliminary framework for assessing the sources of variation that pertain to the functions $\underline{F}$ in (107) for the example is the cell mean (or identity model). It has the structure

$$\underline{E}_A\{\underline{F}\} = \underline{X}\,\underline{\beta} = \underline{I}_4\underline{\beta} \tag{112}$$

where $\underline{I}_4$ is the $(4 \times 4)$ identity matrix. For this model, $\underline{b} = \underline{F}$ and $\underline{V}_b = \underline{V}_F$. The goodness-of-fit statistic $Q_W$ is not applicable for the cell mean model because the model involves no reduction in dimension (i.e., $u = t$, so d.f. $= 0$ for $Q_W$). For this reason, the principal use of this model is for testing hypotheses concerning $\underline{F}$ with the $Q_C$ statistic in (111). The specifications and results for some hypotheses of interest concerning group effects, visit effects, and group $\times$ visit interaction are as follows:

| | | | |
|---|---|---|---|
| Comparison of groups | $\underline{C} = [1\ \ 1\ -1\ -1]$ | $Q_C = 29.96$ | $p < 0.010$ |
| Comparison of visits | $\underline{C} = [1\ -1\ \ 1\ -1]$ | $Q_C = 3.22$ | $p = 0.073$ |
| Group $\times$ visit interaction | $\underline{C} = [1\ -1\ -1\ \ 1]$ | $Q_C = 7.69$ | $p < 0.010$ |

(113)

The conclusions from this analysis agree with those from the previously described methods involving explicit construction of statistical tests. Both analyses indicate that the difference between groups (in the sense of an average over visits) and the group × visit interaction are significant.

From the elements of $\underline{F}$ in (107), one can see that the group × visit interaction is due to the disparity between the test drug group having a substantial increase in the proportion of patients with satisfactory response over the time from 7 days to 14 days and the placebo group having essentially no change. This interpretation can be expressed with the reduced model

$$E\{\underline{F}\} = \underline{X}_R \underline{\beta}_R = \begin{bmatrix} 1 & 1 & 0 \\ 1 & 1 & 1 \\ 1 & 0 & 0 \\ 1 & 0 & 0 \end{bmatrix} \begin{bmatrix} \beta_{R,1} \\ \beta_{R,2} \\ \beta_{R,3} \end{bmatrix} \tag{114}$$

For this model, $\beta_{R,1}$ corresponds to the probability of satisfactory response for placebo patients at either 7 days or 14 days, $\beta_{R,2}$ is the effect for test drug at 7 days, and $\beta_{R,3}$ is the interaction increase in the effect of test drug between 7 days and 14 days.

The goodness of fit of the model $\underline{X}_R$ is evaluated with the Wald statistic in (110). Since $Q_W = 0.61$ is nonsignificant ($p = 0.436$) with respect to the chi-square distribution with d.f. = 1, use of the model is supported. The weighted least-squares estimates $\underline{b}_R$ of the parameters $\underline{\beta}_R$ and their estimated covariance matrix $\underline{V}_{\underline{b}_R}$ are

$$\underline{b}_R = \begin{bmatrix} 0.256 \\ 0.244 \\ 0.233 \end{bmatrix} \quad \text{and} \quad \underline{V}_{\underline{b}_R} = \begin{bmatrix} 21.42 & -21.42 & 0.00 \\ -21.42 & 63.08 & -36.11 \\ 0.00 & -36.11 & 63.15 \end{bmatrix} \times 10^{-4} \tag{115}$$

Model-predicted values for the proportions of patients with satisfactory response are given by $\hat{\underline{F}} = \underline{X}_R \underline{b}_R$. Their estimated standard errors are obtainable as square roots of the diagonal elements of $\underline{V}_{\hat{\underline{F}}} = \underline{X}_R \underline{V}_{\underline{b}_R} \underline{X}'_R$. These results were as follows:

| | Test drug | | Placebo | |
|---|---|---|---|---|
| | 7 days | 14 days | 7 days | 14 days |
| Predicted proportion with satisfactory response | 0.500 | 0.733 | 0.256 | 0.256 |
| Estimated standard error | 0.065 | 0.057 | 0.046 | 0.046 |

(116)

Thus, the use of models like $\underline{X}_R$ provides an effective way of describing the relationship of response distributions with group and visit.

The methods described in this section can be extended to more complex studies with longitudinal or repeated measurements designs, e.g., those involving more than two visits, ordinal data, or several groups which encompass both treatments as well as other explanatory variables. Discussion of some analysis strategies for such situations is given in Koch et al. (1977, 1985a, 1986b, 1987), Landis and Koch (1978, 1979), Landis et al. (1987), and Wei et al. (1985).

## XI. A CHANGEOVER STUDY EXAMPLE

Clinical trials pertaining to the relief of symptoms of chronic or recurrent conditions often make use of changeover designs. In such studies, each patient is randomly assigned to a sequence of treatments which are to be received during successive evaluation periods. The resulting data have a repeated measurements structure for which the methods of analysis are similar to those discussed for the longitudinal study example in Section X. Some aspects of their application are considered here.

The data in Table 7 are from a two-period changeover study concerned with the comparison of a test drug and placebo for relief of a recurrent pain condition. There are two sequence groups, to each of which 50 patients were randomly assigned. The patients in the T:P group received test drug for the first episode of the pain condition during the course of the study and placebo for the second episode; the P:T group received the opposite regimen. Also, the two evaluation periods for the two episodes were separated by several weeks so that the treatment for the first would tend not to influence the response during the second (i.e., to minimize treatment × period interaction due to potential carryover effects). For each evaluation period, the response of each patient was classified as either favorable (F) or unfavorable (U), so for periods jointly, there are $r = 2^2 = 4$ possible response profiles [i.e., (F, F), (F, U), (U, F), (U, U)].

As was the case for the longitudinal study example in Section X, the response distributions for the data in Table 7 for the respective (group × period) combinations are described by the corresponding proportions with favorable response. These quantities are the following linear functions of the frequencies $\{n_{ijk}\}$ for the joint occurrence of the jth response outcome during evaluation period 1 and the kth response outcome during evaluation period 2 for the $n_{i++} = \Sigma_{j=1}^{2} \Sigma_{k=1}^{2} n_{ijk} = 50$ patients in the ith sequence group:

**Table 7** Data from a Changeover Clinical Trial for the Comparison of Treatments for Relief of a Recurrent Pain Condition

| Treatment sequence | Frequencies of response at evaluation periods 1 and 2[a] | | | | Number of patients |
|---|---|---|---|---|---|
| | (F, F) | (F, U) | (U, F) | (U, U) | |
| Test drug : placebo | 20 | 16 | 5 | 9 | 50 |
| Placebo : test drug | 16 | 6 | 18 | 10 | 50 |

[a]F, Favorable; U, unfavorable.

| Sequence group | Period | Treatment | Proportion favorable |
|---|---|---|---|
| T:P | 1 | T | $p_{11+} = (n_{111} + n_{112})/n_{1++} = 0.720$ |
| T:P | 2 | P | $p_{1+1} = (n_{111} + n_{121})/n_{1++} = 0.500$ |
| P:T | 1 | P | $p_{21+} = (n_{211} + n_{212})/n_{2++} = 0.440$ |
| P:T | 2 | T | $p_{2+1} = (n_{211} + n_{221})/n_{2++} = 0.680$ |

(117)

Thus, about 70% of the patients have favorable response during the period with test drug, whereas about 50% have favorable response during the period with placebo. Also, the tendency of about 20% more patients to have favorable response during test drug treatment than during placebo is relatively homogeneous across both periods and sequence groups. Results which support these general conclusions are presented in the following discussion.

A two-sided Fisher's exact test is used to compare the treatments for each of the two periods separately. In both periods the proportion of patients with favorable response is higher in the test drug group. In period 1 the result of Fisher's exact test is significant ($p = 0.008$), and in period 2 it is suggestive ($p = 0.103$). As discussed in Koch et al. (1983), the extent of carryover effects (i.e., treatment × period interaction) can be assessed through the comparison of the sequence groups with respect to the averages $\{\bar{f}_i = (p_{i1+} + p_{i+1})/2\}$. From computations like those for the analogous $\{\bar{f}_i\}$ in (98) for the longitudinal study example in Section X, the test statistic $Q_S = 0.475$ is obtained. This result is nonsignificant ($p = 0.491$) relative to the chi-square distribution with d.f. = 1. Similarly, the comparison of sequence groups with respect to the differences $\{\bar{g}_i = (p_{i+1} - p_{i1+})\}$ provides a test for treatment effects (in an average sense across periods). The result from computations like those for the analogous $\{\bar{g}_i\}$ in (100) of Section X is $Q_S = 11.64$, which is significant ($p < 0.010$) with respect to the chi-square distribution with d.f. = 1. Thus, when the combined information for both periods is taken into account, a strong difference in favor of test drug is detected.

For changeover studies in which carryover effects are expected to be negligible (on the basis of knowledge about the clinical condition under investigation) and are confirmed to be nonsignificant by a method like that described previously, an exact test is available for the assessment of treatment effects. It is undertaken by applying Fisher's exact test to the two sequence groups for the subtable of Table 7 corresponding to the (F, U) and (U, F) response profiles; the rationale for this method is discussed in Gart (1969). A relevant consideration is that the (F, F) and (U, U) response profiles do not provide information useful for discriminating between the two treatments since the patients with either of these profiles had the same response to both treatments. The result from this application of Fisher's two-sided exact test is $p = 0.001$; it significantly favors test drug because (F, U) was more prevalent for the T:P sequence (where the F corresponded to test drug) and (U, F) was more prevalent for the P:T sequence (where the F also corresponded to test drug).

The fitting of linear regression models by weighted least-squares methods provides a way of describing the variation among the proportions of patients with favorable response for the respective (treatment × period) combinations. The considerations involved in applying this analysis strategy are essentially the same as outlined in (103)–(111) for the longitudinal data example in Section X. A model of interest for the data in Table 7 has the specification

$$\underline{E}_A\{\underline{F}\} = \underline{E}_A \begin{bmatrix} p_{11+} \\ p_{1+1} \\ p_{21+} \\ p_{2+1} \end{bmatrix} = \begin{bmatrix} 1 & 0 & 1 \\ 1 & 1 & 0 \\ 1 & 0 & 0 \\ 1 & 1 & 1 \end{bmatrix} \begin{bmatrix} \beta_1 \\ \beta_2 \\ \beta_3 \end{bmatrix} = \underline{X}\,\underline{\beta} \qquad (118)$$

for which $\beta_1$ represents the probability of favorable response for placebo during period 1, $\beta_2$ represents the effect of period 2, and $\beta_3$ represents the effect of test drug. The goodness of fit of this model is assessed with the Wald statistic (110). The result, $Q_W = 0.48$, is nonsignificant ($p = 0.489$) with respect to the chi-square distribution with d.f. = 1; thus, use of the model is supported. Moreover, this finding can be interpreted as indicating that carryover effects are negligible since the structure of $\underline{X}$ expresses additive effects for treatment and period. The weighted least-squares estimates $\underline{b}$ of the parameters $\underline{\beta}$, their estimated standard errors [from square roots of the diagonal elements of $\underline{V}_b$ in (109)], and p-values from the test statistics $Q_C$ in (111) for hypotheses $H_0$: $\beta_h = 0$ are as follows:

| Parameter | Estimate | Standard error | p value |
|---|---|---|---|
| $\beta_1$ | 0.466 | 0.060 | <0.010 |
| $\beta_2$ | 0.006 | 0.063 | 0.920 |
| $\beta_3$ | 0.231 | 0.063 | <0.010 |

(119)

These results indicate that the proportion of patients with favorable response is significantly greater by 0.231 for test drug than for placebo. Also, for each treatment, there is essentially no difference between the two periods for the proportions of patients with favorable response. A description of the pattern of variation corresponding to these conclusions is provided by the following model-predicted values $\hat{\underline{F}} = \underline{X}\underline{b}$ and their estimated standard errors (from square roots of diagonal elements of $\underline{X}\underline{V}_b\underline{X}'$):

| | Test drug Period 1 | Placebo Period 2 | Placebo Period 1 | Test drug Period 2 |
|---|---|---|---|---|
| Predicted proportion with favorable response | 0.697 | 0.472 | 0.466 | 0.703 |
| Estimated standard error | 0.054 | 0.058 | 0.060 | 0.057 |

(120)

Since neither period effects nor carryover effects are evident for this example, one additional test for treatment effects is of interest. It is based on the application of McNemar's test to determine whether the favorable components of (F, U) or (U, F) responses are equally split between test drug and placebo. For the data in Table 7, (16 + 18) = 34 of the (16 + 5 + 6 + 18) = 45 responses of this type favored test drug. The continuity-corrected McNemar statistic for this is $Q_M = \{|34 - 22.5| - 0.5\}^2/(45/4) = 10.76$, which is significant ($p < 0.010$) with respect to the chi-square distribution with d.f. = 1.

Other types of repeated measurements data structures which have research designs analogous to those of change-over studies can be analyzed by the methods described here. A noteworthy example is the bilateral study where one treatment is applied to the right side of the body (e.g., skin on arms or legs, eye, mouth) and the other to the left side. Also, extensions to deal with greater complexity, such as more than two evaluation periods or more than two treatments, are available or feasible to develop, particularly for situations where the sample size for each sequence group is moderately large. For additional discussion of statistical methods for the analysis of categorical data from change-over design studies, see Fleiss (1981), Gart (1969), and Koch et al. (1977, 1983, 1985a).

## XII. SOME COMMENTS ON POWER AND SAMPLE SIZE ISSUES

This chapter has been primarily concerned with describing statistical methods for the analysis of clinical efficacy trials with categorical data. However, two related considerations which often require attention are the power of statistical tests for treatment comparisons and the sample size needed to achieve desired levels of power. With model-based methods, such as logistic regression or its extensions (Sections VI and VII), piecewise exponential modeling (Section IX), or weighted least-squares procedures (Section VIII, X, and XI), an indication of the a posteriori power for an already completed study can be obtained by making use of the approximately normal distribution of the estimated model parameter $\hat{\beta}_T$ for treatment effects. More specifically, for the null hypothesis $H_0$: $\beta_T = 0$, an approximation of the power to detect the alternative, $H_A$: $\beta_T = \delta$, through the test statistic $Z = \hat{\beta}_T/\{s.e.(\hat{\beta}_T)\}$ in the setting with two-sided type I error $\alpha$ is given by

$$\text{Power} = 1 - \Phi\left\{Z_{(\alpha/2)} - \left[\frac{\delta}{s.e.(\hat{\beta}_T)}\right]\right\} \tag{121}$$

where $\Phi$ is the cumulative distribution function for the standard normal distribution with expected value 0 and variance 1 and $Z_{(\alpha/2)}$ is the corresponding $100\{1 - (\alpha/2)\}$th percentile.

Methods for determining sample sizes to have specific levels of power in studies which are to be initiated also often involve the use of normal approximations. For the comparison of two treatments with respect to a dichotomous response, Fleiss (1981) provides extensive tables of sample sizes. Further discussion of sample size issues which is relevant to this and more general situations with categorical data is given in Donner (1984) and Lachin (1981).

## ACKNOWLEDGMENTS

The research for this chapter was supported in part by the U.S. Bureau of the Census through Joint Statistical Agreement JSA-84-5. The authors would also like to thank Ingrid Amara, Susan Atkinson, Gregory Carr, and Amanda Sullivan for assistance with computations.

## REFERENCES

Agresti, A. (1983). A Survey of Strategies for Modeling Classifications Having Ordinal Variables, *J. Am. Stat. Assoc.*, 78: 184–197.

Agresti, A. (1984). *Analysis of Ordinal Categorical Data*, Wiley, New York.

Andersen, E. B. (1980). *Discrete Statistical Models with Social Science Applications*, North Holland, Amsterdam.

Andrich, D. (1979). A Model for Contingency Tables Having an Ordered Response Classification, *Biometrics*, 35: 403–415.

Bishop, Y. M. M., Fienberg, S. E., and Holland, P. W. (1975). *Discrete Multivariate Analysis: Theory and Practice*, MIT Press, Cambridge, Mass.

Breslow, N. E. and Day, N. E. (1980). *Statistical Methods in Cancer Research*, vol. I: *The Analysis of Case-Control Studies*, International Agency for Research on Cancer, Lyon.

Brown, M. B. (1983). P4F: Frequency Tables, *BMDP Statistical Software* (W. J. Dixon et al., eds.), University of California Press, Los Angeles, pp. 143–206.

Clogg, C. C. (1982). Some Models for the Analysis of Association in Multiway Cross-Classifications Having Ordered Categories, *J. Am. Stat. Assoc.*, 77: 803–815.

Conover, W. J. (1974). Some Reasons for Not Using the Yates Continuity Correction on 2 × 2 Contingency Tables, *J. Am. Stat. Assoc.*, 69: 374–382.

Cox, D. R. (1970). *The Analysis of Binary Data*, Methuen, London.

Cox, D. R. (1972). Regression Models and Life Tables, *J. R. Stat. Soc. Ser. B*, 34: 187–220.

Cox, C. and Chuang, C. (1984). A Comparison of Chi-Square Partitioning and Two Logit Analyses of Ordinal Pain Data from a Pharmaceutical Study, *Stat. Med.*, 3: 273–285.

Donner, A. (1984). Approaches to Sample Size Estimation in the Design of Clinical Trials—a Review, *Stat. Med.*, 3: 199–214.

Engelman, L. (1983). PLR: Stepwise Logistic Regression, *BMDP Statistical Software* (W. J. Dixon et al., eds.), University of California Press, Los Angeles, pp. 330–344.

Fienberg, S. E. (1980). *The Analysis of Cross-Classified Categorical Data*, 2nd edition, MIT Press, Cambridge, Mass.

Fleiss, J. L. (1981). *Statistical Methods for Rates and Proportions*, 2nd Edition, Wiley, New York.

Forthofer, R. N. and Lehnen, R. G. (1981). *Public Program Analysis: A New Categorical Data Approach*, Wadsworth, Belmont, Calif.

Frome, E. L. (1983). The Analysis of Rates Using Poisson Regression Models, *Biometrics*, 39: 665–674.

Gart, J. J. (1969). An Exact Test for Comparing Matched Proportions in Crossover Designs, *Biometrika*, 56: 75–80.

Gart, J. J. (1971). The Comparison of Proportions: A Review of Significance Tests, Confidence Intervals, and Adjustments for Stratification, *Int. Stat. Rev.*, 39: 148–169.

Goodman, L. A. (1979). Simple Models for the Analysis of Association in Cross-Classifications Having Ordered Categories, *J. Am. Stat. Assoc.*, 74: 537–552.

Grizzle, J. E. (1967). Continuity Correction in the $\chi^2$-Test for 2 × 2 Tables, *Am. Stat.*, 21: 28–32.

Grizzle, J. E., Starmer, C. F., and Koch, G. G. (1969). Analysis of Categorical Data by Linear Models, *Biometrics*, 25: 489–504.

Haber, M. (1986). An Exact Unconditional Test for the 2 × 2 Comparative Trial, *Psychol. Bull.*, 99: 129–132.

Haberman, S. J. (1978). *Analysis of Qualitative Data: Vol. 1 - Introductory Topics; Vol. 2 - New Developments*, Academic Press, New York.

Hannan, J. and Harkness, W. L. (1963). Normal Approximation to the Distribution of Two Independent Binomials, Conditional on Fixed Sum, *Ann. Math. Stat.*, 34: 1593–1595.

Harrell, F. E. (1986). The LOGIST Procedure, *SUGI Supplemental Library User's Guide, Version 5 Edition*, SAS Institute, Cary, N.C., pp. 269–293.

Holford, T. R. (1980). The Analysis of Rates and Survivorship Using Log-Linear Models, *Biometrics*, 36: 299–305.

Imrey, P. B., Koch, G. G., Stokes, M. E., Darroch, J. N., Freeman, D. H., and Tolley, H. D. (1981). Categorical Data Analysis: Some Reflections on the Log Linear Model and Logistic Regression, Part I, *Int. Stat. Rev.*, 49: 265–283.

Imrey, P. B., Koch, G. G., Stokes, M. E., Darroch, J. N., Freeman, D. H., and Tolley, H. D. (1982). Categorical Data Analysis: Some Reflections on the Log Linear Model and Logistic Regression, Part II, *Int. Stat. Rev.*, 50: 35–64.

Kleinbaum, D. G., Kupper, L. L., and Morgenstern, H. (1982). *Epidemiologic Research: Principles and Quantitative Methods*, Lifetime Learning Publications, Belmont, Calif.

Koch, G. G. and Bhapkar, V. P. (1982). Chi-Square Tests, *Encyclopedia of Statistical Sciences*, vol. 1 (N. L. Johnson and S. Kotz, eds.), Wiley, New York, pp. 442–457.

Koch, G. G. and Edwards, S. (1985). Logistic Regression, *Encyclopedia of Statistical Sciences*, vol. 5 (N. L. Johnson and S. Kotz, eds.), Wiley, New York, pp. 128–133.

Koch, G. G. and Gillings, D. B. (1983). Inference, Design Based vs. Model Based, *Encyclopedia of Statistical Sciences*, vol. 4 (N. L. Johnson and S. Kotz, eds.), Wiley, New York, pp. 84–88.

Koch, G. G. and Sollecito, W. A. (1984). Statistical Considerations in the Design, Analysis, and Interpretation of Comparative Clinical Studies: An Academic Perspective, *Drug Inf. J.*, 18: 131–151.

Koch, G. G., Landis, J. R., Freeman, J. L., Freeman, D. H., and Lehnen, R. G. (1977). A General Methodology for the Analysis of Experiments with Repeated Measurement of Categorical Data, *Biometrics*, 33: 133–158.

Koch, G. G., Amara, I. A., Davis, G. W., and Gillings, D. B. (1982). A Review of Some Statistical Methods for Covariance Analysis of Categorical Data, *Biometrics*, 38: 563–595.

Koch, G. G., Gitomer, S. L., Skalland, L., and Stokes, M. E. (1983). Some Non-Parametric and Categorical Data Analyses for a Change-Over Design Study and Discussion of Apparent Carry-Over Effects, *Stat. Med.*, 2: 397–412.

Koch, G. G., Imrey, P. B., Singer, J. M., Atkinson, S. S., and Stokes, M. E. (1985a). *Analysis of Categorical Data*, Les Presses de l'Université de Montreal, Montreal.

Koch, G. G., Sen, P. K., and Amara, I. A. (1985b). Logrank Scores, Statistics and Tests, *Encyclopedia of Statistical Sciences*, vol. 5 (N. L. Johnson and S. Kotz, eds.), Wiley, New York, pp. 136–142.

Koch, G. G., Singer, J. M., and Amara, I. A. (1985c). A Two-Stage Procedure for the Analysis of Ordinal Categorical Data, *Biostatistics: Statistics in Biomedical, Public Health and Environmental Sciences, the Bernard G. Greenberg Volume* (P. K. Sen, ed.), North-Holland, New York, pp. 357–387.

Koch, G. G., Atkinson, S. S., and Stokes, M. E. (1986a). Poisson Regression, *Encyclopedia of Statistical Sciences*, vol. 7 (N. L. Johnson and S. Kotz, eds.), Wiley, New York, pp. 32–41.

Koch, G. G., Singer, J. M., Stokes, M. E., Carr, G. J., Cohen, S. B., and Forthofer, R. N. (1986b). "Some Aspects of Weighted Least Squares Analysis for Longitudinal Categorical Data," Proc. of the Workshop on Longitudinal Methods in Health Research, Berlin (forthcoming).

Koch, G. G., Elashoff, J. D., and Amara, I. A. (1987). Repeated Measurements Studies, Design and Analysis, *Encyclopedia of Statistical Sciences*, vol. 8 (N. L. Johnson and S. Kotz, eds.), Wiley, New York, forthcoming.

Lachin, J. M. (1981). Introduction to Sample Size Determination and Power Analysis for Clinical Trials, *Controlled Clin. Trials*, 2: 93–113.

Laird, N., and Olivier, D. (1981). Covariance Analysis of Censored Survival Data Using Log-Linear Analysis Techniques, *J. Am. Stat. Assoc.*, 76: 231–240.

Landis, J. R. and Koch, G. G. (1978). The Measurement of Observer Agreement for Categorical Data, *Biometrics*, 33: 159–174.

Landis, J. R. and Koch, G. G. (1979). The Analysis of Categorical Data in Longitudinal Studies of Behavioral Development, *Longitudinal Research in the Study of Behavior and Development* (J. R. Nesselroade and P. B. Baltes, eds.), Academic Press, New York, Chap. 9, pp. 233–261.

Landis, J. R., Stanish, W. M., Freeman, J. L., and Koch, G. G. (1976). A Computer Program for the Generalized Chi-Square Analysis of Categorical Data Using Weighted Least Squares (GENCAT), *Comput. Programs Biomed.*, 6: 196–231.

Landis, J. R., Cooper, M. M., Kennedy, T., and Koch, G. G. (1979). A Computer Program for Testing Average Partial Association in Three-Way Contingency Tables (PARCAT), *Comput. Programs Biomed.*, 9: 223–246.

Landis, J. R., Heyman, E. R., and Koch, G. G. (1978). Average Partial Association in Three-Way Contingency Tables: A Review and Discussion of Alternative Tests, *Int. Stat. Rev.*, 46: 237–254.

Landis, J. R., Miller, M. E., Davis, C. S., and Koch, G. G. (1987). Some General Methods for the Analysis of Categorical Data in Longitudinal Studies, *Stat. Med.*, 6, forthcoming.

Larntz, K. (1978). Small Sample Comparisons of Exact Levels for Chi-Squared Goodness of Fit Statistics, *J. Am. Stat. Assoc.*, 73: 253–263.

Lee, E. T. (1980). *Statistical Methods for Survival Data Analysis*, Lifetime Learning Publications, Belmont, Calif.

Lehmann, E. L. (1975). *Nonparametrics: Statistical Methods Based on Ranks*, Holden-Day, San Francisco.

Mann, H. B. and Whitney, D. R. (1947). On a Test of Whether One of Two Random Variables Is Stochastically Larger Than the Other, *Ann. Math. Stat.*, 18: 50–60.

Mantel, N. (1963). Chi-Square Tests with One Degree of Freedom: Extensions of the Mantel-Haenszel Procedure, *J. Am. Stat. Assoc.*, 58: 690–700.

Mantel, N. (1966). Evaluation of Survival Data and Two New Rank Order Statistics Arising in Its Consideration, *Cancer Chemother. Rep.*, 50: 163–170.

Mantel, N. and Fleiss, J. (1980). Minimum Expected Cell Size Requirements for the Mantel-Haenszel One-Degree of Freedom Chi-Square Test and a Related Rapid Procedure, *Am. J. Epidemiol.*, 112: 129–134.

Mantel, N. and Haenszel, W. (1959). Statistical Aspects of the Analysis of Data from Retrospective Studies of Disease, *J. Nat. Cancer Inst.*, 22: 719–748.

McCullagh, P. (1980). Regression Models for Ordinal Data (with Discussion), *J. R. Stat. Soc. Ser. B*, 42: 109–142.

McCullagh, P. and Nelder, J. A. (1983). *Generalized Linear Models*, Chapman & Hall, New York.

Mehta, C. R. and Patel, N. R. (1983). A Network Algorithm for Performing Fisher's Exact Test in r × c Contingency Tables, *J. Am. Stat. Assoc.*, 78: 427–434.

Mehta, C. R., Patel, N. R., and Tsiatis, A. A. (1984). Exact Significance Testing to Establish Treatment Equivalence with Ordered Categorical Data, *Biometrics*, 40: 819–825.

Mehta, C. R., Patel, N. R., and Gray, R. (1985). Computing an Exact Confidence Interval for the Common Odds Ratio in Several 2 × 2 Contingency Tables, *J. Am. Stat. Assoc.*, 80: 969–973.

Overall, J. E. and Starbuck, R. R. (1983). F-Test Alternatives to Fisher's Exact Test and to the Chi-Square Test of Homogeneity in 2 by 2 Tables, *J. Educ. Stat.*, 8: 59–74.

Pagano, M. and Halvorsen, K. T. (1981). An Algorithm for Finding the Exact Significance Levels of r × c Tables, *J. Am. Stat. Assoc.*, 76: 931–934.

Peterson, B. L. (1986). Proportional Odds and Partial Proportional Odds Models for Ordinal Response Variables, Dissertation submitted to Department of Biostatistics, University of North Carolina, Chapel Hill.

Plackett, R. L. (1981). *The Analysis of Categorical Data*, Griffin, London.

Puri, M. L. and Sen, P. K. (1971). *Non-Parametric Methods in Multivariate Analysis*, Wiley, New York.

Salama, I. A., Quade, D., and Koch, G. G. (1984). Tables for Testing the Equality of Two Proportions when Prior Information on Their Common Value May Be Available, *Biometrical J.*, 25: 301–320.

SAS Institute, Inc. (1985). *SAS User's Guide: Statistics, Version 5 Edition*, SAS Institute, Cary, N.C.

Semenya, K. A., Koch, G. G., Stokes, M. E., and Forthofer, R. N. (1983). Linear Models Methods for Some Rank Function Analyses of Ordinal Categorical Data, *Commun. Stat.—Theory Methods*, 12: 1277–1298.

Stokes, M. E. and Koch, G. G. (1983). "A Macro for Maximum Likelihood Fitting of Log-Linear Models to Poisson and Multinomial Counts with Contrast Matrix Capability for Hypothesis Testing," Proc. of the 8th Annual SAS Users Group International Conference, SAS Institute, Cary, N.C., pp. 795–800.

Thomas, D. G. (1975). Exact and Asymptotic Methods for the Combination of 2 × 2 Tables, *Comput. Biomed. Res.*, 8: 423–446.

Upton, G. G. G. (1982). A Comparison of Alternative Tests for the 2 × 2 Comparative Trial, *J. R. Stat. Soc.*, 145: 86–105.

van Elteren, P. H. (1960). On the Combination of Independent Two-Sample Tests of Wilcoxon, *Bull. Int. Stat. Inst.*, 37: 351–361.

Walker, S. H. and Duncan, D. B. (1967). Estimation of the Probability of an Event as a Function of Several Independent Variables, *Biometrika*, 54: 167–179.

Wei, L. J., Stram, D., and Ware, J. H. (1985). Analysis of Repeated Ordered Categorical Outcomes with Possibly Missing Observations, Department of Biostatistics, Harvard School of Public Health, Boston, Technical Report No. 6.

Yates, F. (1934). Contingency Tables Involving Small Numbers and the $\chi^2$ Test. *J. R. Stat. Soc. Suppl.*, 1: 217–235.

# 10 Design, Analysis, and Reporting of Cancer Clinical Trials

RICHARD SIMON *Biometric Research Branch, Cancer Therapy Evaluation Program, Division of Cancer Treatment, National Cancer Institute, Bethesda, Maryland*

## I. INTRODUCTION

### A. Objectives

The basic objectives in the design and analysis of good cancer clinical trials are no different from those in other medical areas. These objectives are to ask an important question and to employ a design and method of analysis that yield a reliable answer. In this chapter I will discuss some aspects of how to achieve these goals.

### B. Kinds of Cancer Clinical Trials

Cancer clinical trials are generally classified somewhat differently from other trials. Phase I trials attempt to determine the relationship between toxicity and dose-schedule of treatment. The subjects are generally previously treated patients with various kinds of cancer for whom no further effective therapy is known. A result of a phase I trial of a chemotherapeutic agent should be a dose schedule which is safe for use but is close to a maximum tolerated dose (MTD). Reliable data on the antitumor activity of a drug are not available from a phase I trial because not enough patients with the same tumor type are treated at a dose close to the MTD. Also, patients included in phase I trials often have been previously treated with several other chemotherapy regimens. Consequently their tumors may be resistant. Lack of antitumor activity seen for a drug in such patients does not mean that the drug would be inactive in patients with less extensive disease.

The main objective of a single-agent phase II trial is to determine whether a drug has antitumor activity in patients having a given type of cancer. Since antitumor activity can be highly variable by tumor type, degree of prior treatment, and performance status of patient, phase II trials should be limited to patients with the same tumor type, the minimum amount of previous treatment ethically appropriate, and the best performance status. In this way the drug has the best chance of demonstrating antitumor activity and patients least likely to respond are spared toxic treatment (Simon, 1982c; Wittes et al., 1985).

Phase II trials of single agents generally include 25 to 50 patients. There is usually no control group because the question posed is simply whether the drug can cause tumor shrinkage. Although there are randomized Phase II trials (Simon et al., 1985), the objective of such trials is generally not comparative.

Reports of Phase II studies sometimes include a comparison of survivals between responders and nonresponders given the same treatment. This is an invalid analysis for many reasons (Tannock and Murphy, 1983; Anderson et al., 1983, 1985; Weiss et al., 1983; Simon and Makuch, 1984). There is a time bias because responders must live long enough for response to be observed. Also, responders may have more favorable prognostic factors than nonresponders, leading to a difference in survival regardless of treatment. It is even possible that treatment may shorten the survival of nonresponders rather than lengthen that of responders. Improving partial response rates for solid tumors may not result in improving survival. Demonstrating that a treatment prolongs survival requires comparing the survival of all patients receiving that treatment to an appropriate control group. This is generally not possible in a Phase II trial.

In addition to the many Phase II single-agent clinical trials in oncology, there are also many trials of combination therapy conducted without a concurrent control group. These trials also are often referred to as Phase II trials. Their objective is not, however, to determine whether the combination has any antitumor activity, for this is usually known in advance. Their objective is usually to determine whether the degree of antitumor activity is sufficient to warrant a randomized Phase III trial. Although such trials can be useful, they are sometimes reported as if they were adequate Phase III trials.

Finally, the objective of a Phase III trial is to determine the relative merits of two or more treatments. Such trials are generally most useful if one of the treatments is a standard therapy, which may be no treatment at all in some circumstances. Traditionally, a new experimental treatment has been of interest if it was more effective than the control. For example, the control treatment in a study of stage II breast cancer might be a simple mastectomy, whereas the experimental treatment might be simple mastectomy followed by one year of a specified chemotherapy regimen. Recently, "equivalence" or "positive control" Phase III trials have become important. With such trials we are interested in whether the experimental treatment is as good as or better than the control. The experimental treatment is considered of value if it is no worse than the control because the control is considered effective and the experimental treatment has some other advantage such as less toxicity or lower cost. To continue the example given above, if simple mastectomy is the control treatment, then breast-preserving tumor excision might be the experimental treatment in an equivalence trial. In both kinds of Phase III trials the objective is to determine the relative merits of the treatments, but the decision frameworks relating outcomes of the trial to disposition of the experimental treatment are quite different. Consequently, there are important differences in the design and reporting of the two types of trials.

## II. DESIGN OF SINGLE-AGENT PHASE II TRIALS

The purpose of most single-agent Phase II trials in oncology is to determine whether the drug has sufficient antitumor activity to warrant evaluation of the drug's effectiveness in a Phase III trial. Effectiveness or clinical usefulness is not evaluable in Phase II trials. The future disposition of a drug can depend on several factors in addition to the observed degree of antitumor activity—for example, its spectrum of toxicity, dose-response relationship and evidence for non-cross-resistance with other active agents. Nevertheless, the observed response rate and the durability of those responses are key endpoints for evaluation of Phase II studies.

There is frequently great variability in the response rates reported from different Phase II studies of the same agent. Some major factors that contribute to this variability are listed in Table 1 (Wittes et al., 1985; Simon et al., 1985). The first factor is patient selection (Moertel et al., 1974). Response rates generally decrease as the extent of prior therapy increases. Patients who have failed several prior regimens are more likely to have tumors composed of large numbers of resistant cells, and such patients are also less likely to be able to tolerate full doses of the investigational drug. Whereas it may be appropriate to screen for drugs cytotoxic to tumor cells that are resistant to standard agents, such screening generally cannot be effectively performed in patients with poor bone marrow and other organ system reserves. The steep dose-response relations observed in experimental tumors for some cytotoxic agents suggest that adequate Phase II evaluations should be performed at maximally tolerated doses. Probably the most frequent problem with Phase II studies is that the patients selected are so debilitated by disease and prior therapy that adequate evaluation of antitumor activity is impossible. Such patients are less likely to benefit from the drug, and the trial itself will not contribute meaningfully to evaluation of the drug. Debilitated patients are more likely to die or withdraw early in the course of treatment.

A second important factor is variability in response criteria among institutions and groups (Tonkin et al., 1985). Some reports of Phase II trials do not describe or reference the exact criteria used in sufficient detail to adequately interpret the results. A third factor is subjectivity in assessment of response. There have been few formal evaluations of the impact of measurement error on reported rates of tumor response (Gurland

**Table 1** Sources of Variability in Results of Phase II Trials

1. Patient selection
2. Response criteria
3. Interobserver variability in response assessment
4. Dosage modification and protocol compliance
5. Reporting procedures
6. Sample size

and Johnson, 1965; Moertel and Hanley, 1976; Lavin and Flowerdew, 1980). In a recent study Warr et al. (1984) compared measurements by several physicians on real or simulated malignant lesions and found that some commonly used criteria of response are subject to large errors. For this reason, they recommended requiring more extensive and longer-lasting evidence of tumor reduction as a basis for the definition of partial response. A fourth source of variability in results includes differences in dosage modification guidelines and differences in protocol compliance. These features reflect differences in aggressiveness of treatment and carefulness in performing clinical trials. They may also reflect differences in patient selection, as protocol compliance may be more difficult in a population of heavily pretreated patients. The variations in policies for handling exclusions in the calculation of response rates and the small sample sizes of Phase II trials also influence the variability of results. Certainly, the outcome of treatment for all patients entered in a trial should be reported regardless of whether the investigator considers some of the patients to be unevaluable for response.

Protocols for Phase II trials should attempt to deal with the problems listed in Table 1. Eligibility criteria should be limited to patients with the least amount of prior treatment consistent with good medical care. Protocols for Phase II trials should clearly define the response criteria used, and use criteria for which interobserver agreement is satisfactory. It is useful to have second-party review of response assessment.

There is a substantial literature on statistical designs for Phase II trials in oncology and this has been reviewed by Herson (1984). Too small a sample size may lead to erroneous conclusions. Too large a sample size may expose patients unnecessarily to inactive but toxic treatment. Although a variety of interesting designs have been proposed, Fleming (1982) has proposed one that seems particularly useful. This is a multistage design for testing a null hypothesis that the true response probability is less than some uninteresting level $H_0$: $p \leqslant p_0$ against an alternative hypothesis $H_1$: $p \geqslant p_1$ that the true response probability is at least as large as a target level $p_1$. Schultz et al. (1973) had previously studied a broad class of multistage designs containing the Fleming design.

If one specifies the number of stages desired, the target levels $p_0$ and $p_1$, and the error limits for rejecting $H_0$ when it is true and for rejecting $H_1$ when it is true, then designs of this class can be determined. Consider a design with $p_0 = .05$ and $p_1 = .20$ for which the error limits are to be less than .10. These constraints can be met with a two-stage design as reported by Fleming, with 15 patients in the first stage and 20 patients in the second stage. If no responses are observed in the first 15 patients, then the trial is terminated and $H_1$: $p \geqslant .20$ is rejected. If at least three responses are observed in the first 15 patients, then the trial is terminated and $H_0$: $p \leqslant .05$ is rejected. If 1 or 2 responses are observed in the first 15 patients, then 20 more patients are accrued. After all 35 patients are evaluated, $H_0$ is rejected if the response rate is $\geqslant 4/35$ (11.4%) and $H_1$ is rejected if the response rate is $\leqslant 3/35$ (8.6%).

The design reported by Fleming is attractive because it forces one to specify levels of activity of interest and to attempt to define how decisions will be influenced by results. Such quantification of objectives is even more important for "Phase II" trials of combinations or analogs of active agents. One must bear in mind, however, that rejection of the null hypothesis $p \leqslant .05$ does not imply acceptance of the alternative $p \geqslant .20$. For example, the null hypothesis is rejected if the observed response rate is 4/35 or 11.4%. With $P_0 = .05$, rejection of the null hypothesis merely indicates that the drug is active, i.e., produces responses above a background rate of 5%. Such rejection does not carry the implication that the response rate is greater than a target response rate of 20% that might define an interesting level of activity. If $p \leqslant p_0$ is the hypothesis rejected, the design does not itself specify how much further study of the drug is appropriate. It is, of course, quite possible that the true response probability is in the range between $p_0$ and $p_1$. In this case both hypotheses $H_0$ and $H_1$ are false, but only one will be "rejected" and one "not rejected." For interpreting results after use of Fleming's design it is useful to employ confidence intervals (Jennison and Turnbull, 1983; Atkinson and Brown, 1985; Chang and O'Brien, 1986). Such intervals make clear what is consistent with the results and what is not and avoid the potential misinterpretations that may result from accept-reject terminology. Some important practical aspects of the registration and monitoring of patient accrual in Phase II trials are discussed by Herson (1983).

In some cases the objective of a Phase II study involves selecting a promising agent from among a set of candidates in addition to or instead of evaluating each of the candidates. For example, we may wish to select the best dose level for Phase III trial of a biologic based on a randomized Phase II trial of two or more dose levels. Or we may wish to select for Phase III trial a preparation of a biologic from among many possible preparations. Although these objectives are comparative, the usual statistical

**Table 2** Number of Patients per Treatment for Phase II Selection Design[a]

| Smallest response rate (%) | Number of treatments | | |
|---|---|---|---|
| | 2 | 3 | 4 |
| 10 | 21 | 31 | 37 |
| 20 | 29 | 44 | 52 |
| 30 | 35 | 52 | 62 |
| 40 | 37 | 55 | 67 |
| 50 | 36 | 54 | 65 |
| 60 | 32 | 49 | 59 |
| 70 | 26 | 39 | 47 |
| 80 | 16 | 24 | 29 |

[a]Probability of correctly selecting the best treatment is 0.90 when it is superior by an absolute difference of 15% in response rate.

sample size formulas for Phase III trials are not always appropriate. For example, if we are interested only in selecting a promising dose rather than in really testing the hypothesis that there is no dose-response effect, then a selection formulation rather than a hypothesis testing formulation is appropriate. This distinction is important from a sample size standpoint, and one should carefully specify the objectives of a Phase II study involving comparison or selection. Some randomized Phase II studies are actually inadequate Phase III trials.

Simon et al. (1985) studied randomized Phase II trials for selecting the best from among a small set of candidate treatments. The selection formulation assumes that the treatment which produces the greatest response rate will be selected for a subsequent study regardless of how small that apparent superiority is. The sample size is based on the constraint that if one treatment is in fact superior by an amount D, then the probability of correctly selecting it must be P. Table 2 shows the required number of patients per treatment if D = 15% and P = .90. The sample size depends somewhat on the number of candidate treatments and the baseline response probabilities. The sample sizes are mostly in the range of 20 to 60 patients per treatment.

## III. DESIGN OF PHASE III TRIALS

### A. Endpoints

The term "endpoint" refers to the criterion by which patient benefit is measured. The major endpoints for evaluating the effectiveness of a treatment should be measures of patient welfare. Duration of survival and quality of life are two such endpoints. Quality of life is used infrequently because sufficiently simple and reproducible measures of important aspects of quality of life are only now being developed (Schipper and Levitt, 1985).

Survival is generally the most meaningful measure of benefit for Phase III studies (Simon, 1982c). In many cases, improving complete response rates, complete response durations and disease-free survival results in improved survival. These endpoints are, therefore, often appropriate surrogates of survival. This is not always the case, however, and the situation is much more tenuous with regard to partial responses. A treatment that causes partial responses is not necessarily beneficial to the patient. Even if one demonstrates that partial responders live longer than nonresponders, one cannot conclude that treatment is beneficial. An increase in partial response rates without corresponding improvements in survival or quality of life cannot in itself be viewed as a therapeutic improvement of benefit to the patient. Partial responses are, however, useful indicators of biologic activity for Phase II trials.

Peto has argued that cause-specific mortality is in some ways a better endpoint than total mortality. For example, one might distinguish cancer mortality from deaths due to other causes to try to obtain more precise estimates of the relative merits of the treatments. This viewpoint is controversial because of questions concerning the accuracy of cause-of-death assessments. Analyses based on cause-specific mortality would also most likely have to treat deaths from other causes as noninformative censoring.

This would, unfortunately, introduce an unverifiable assumption into the analysis.

Although "disease-free survival" is often used as an endpoint for surgical adjuvant studies, the term is somewhat ambiguous and often not defined. The main issue is whether a death without recurrence counts as a failure event or a censored observation. It seems natural that disease-free survival should denote survival as well as remaining disease-free, and this is the definition recommended here. If death without recurrence is treated as censoring, it seems more appropriate to call the endpoint time till recurrence, disease-free interval, or complete remission duration. One defect of this endpoint is the necessity of assuming that death without recurrence represents a noninformative censoring mechanism. Duration of local control is of even more concern in this regard.

## B. Controls for Phase III Trials

The interpretation of most Phase III studies involves some type of comparison of results. In some cases the basis of comparison will be the natural history of the disease, and in others it will be another treatment. We shall use the term "control" to represent the basis against which a treatment is to be evaluated. Rarely, if ever, do we just want to know whether a treatment is better or worse than the control. We want to estimate the degree of difference. All measurement ultimately is comparative, however, and the categorization of a treatment as "good" or "bad" involves an implicit comparison to the natural history of the disease.

To determine whether a new treatment cures any patients with a disease that is uniformly and rapidly fatal, history is a satisfactory control. In this situation the patient population is completely homogeneous with regard to cure in the absence of the new therapy. If 20% of patients are cured by conventional therapy and we can identify them by patient and tumor characteristics measured at diagnosis, we can restrict a study to the remaining 80% and have complete homogeneity. Once we leave the setting of complete homogeneity with regard to the chosen endpoint, the definition of an adequate nonrandomized control becomes problematic.

In many studies the controls are either numbers determined from publications or patients treated in nonexperimental settings in which the information is abstracted from tumor registries, data banks, or medical records. The meaningfulness of such controls is questionable. Often diagnostic and staging procedures, supportive care, secondary treatments, and methods of evaluation and follow-up are different for the controls and the current treatment group. There generally is differential bias in the selection of patients to be treated resulting from judgments by the physicians, self-selection by the patients, and differences in referral patterns. There may be bias in treatment ineligibility rates. Current patients are sometimes excluded from analysis for not meeting eligibility criteria, not receiving "adequate" treatment, refusing treatment, or a major protocol violation. The controls, on the other hand, generally contain all the patients. There may be differences in the distribution of known and unknown prognostic factors between the controls and the current treatment group. Often there is inadequate information to determine whether there are such differences, and current known prognostic factors may not have been measured or recorded for the controls. It generally is difficult to tell whether the controls would have been eligible for the current study and in what way they represent a selection of all eligible patients.

Formation of the control group by random assignment of treatment as an integral part of the planned study can avoid most of the systematic biases mentioned above (Byar et al., 1976; Peto et al., 1976, 1977; Simon, 1982a; Micciolo et al., 1985). The random assignment should not be performed until the patient is found eligible, and then a truly random or nondecipherable mechanism should be used. Alternation, day of the week, or other predictable procedures are not adequate because they allow bias in the decision of whether to enter a patient into a study based on knowledge beforehand of what treatment the patient will receive. Randomization does not ensure that the study will include a representative sample of all patients with the disease, but it does help to ensure an unbiased evaluation of the relative merits of the two treatments for the types of patients entered.

Some of the advantages of randomization are subtle and not widely understood. For example, it is sometimes said that randomization is unnecessary because matched historical or concurrent controls can be selected. But one can match only with regard to known prognostic factors, and these generally explain only a minor portion of the variability in prognosis among patients. Matching with regard to known factors gives no assurance that the distributions of unknown factors are similar between the treatment groups. It also is sometimes said that randomization is not effective in ensuring that the treatment groups are similar with regard to unknown prognostic factors unless the number of patients is large. This is true but reflects misunderstanding of randomization. Randomization does not ensure that the groups are medically equivalent, but it distributes the unknown biasing factors according to a known random distribution so that their effects can be rigorously allowed for in significance tests and confidence intervals (Simon, 1982a). This is true regardless of the study size. A significance level represents the probability that differences in outcome are due to random fluctuations. Without randomized treatment allocation, a "statistically significant difference" may be due to a nonrandom difference in the distribution of unknown prognostic factors.

### C. Randomization and Stratification Methods

The most important aspects of the randomization procedure are that randomization should generally take place as late as possible before effecting the treatment of the patient and that the results of the randomization should be unpredictable to those individuals entering patients. If physicians can predict treatment assignment, they can consciously or unconsciously bias the results of the trial by deciding whether or not to enter a patient based on this information. Assuring unpredictability requires that randomization be performed centrally by a statistical or operations office using lists, cards in envelopes, or computers unavailable to those caring for patients.

Except in very large studies, some form of restricted randomization is usually employed in order to provide blocking over time, institution, or prognostic factors. Blocking with regard to institution or prognostic factors is usually referred to as stratification. The traditional method of stratification is to prepare a separate randomization list for each distinct subset of patients (stratum). Each list is usually prepared using permuted blocks of length ranging from 4 to 10 if there are two treatments. That is, the sequence of treatment assignments within each stratum is random subject to the restriction that the numbers of assignments of each treatment must be equal after each block of the length used.

The number of strata increases multiplicatively with the number of stratification factors because the patient subsets are defined by combinations of these factors. Although limited stratification often is desirable, overstratification is detrimental to the trial (Simon, 1980). If there is extensive stratification, numerous strata will contain very few patients. Consequently, balance with regard to the most important factor or factors may be seriously impaired by the inclusion of factors of secondary importance. Even the total numbers of patients assigned to each of the treatments may be very unequal. Extensive overstratification becomes equivalent to randomization with no stratification at all.

It generally is best to limit stratification to factors definitely known to have important independent effects on response. If two factors are closely correlated, one, at most, should be included in the stratification. Some statisticians believe that stratification is an unnecessary complication because adjustment for imbalances of known factors can be made in the analysis. For small studies, however, such adjustments should not be relied on. Stratification may obviate the chance of gross imbalances that cannot be adjusted for and ensures that the treatment comparisons are not totally dependent on statistical adjustment methods. Simon (1979), Pocock (1979), Kalish and Begg (1985) have reviewed the various stratification methods available including the newer adaptive methods.

### D. Sample Size for Phase III Trials

An effective clinical trial must ask an important question and provide a reliable answer. A major determinant of the reliability of the answer is the size of the trial. Trials of inadequate size may cause contradictory and erroneous results and thereby lead to inappropriate treatment of patients (Freiman et al., 1978; George, 1983; Simon, 1985).

Most Phase III trials in oncology are planned as fixed sample size trials in the Neyman-Pearson framework. That is, it is assumed that at the end of the trial a statistical significance test of size $\alpha$ of the null hypothesis that the treatments are of equal efficacy will be performed. One plans the sample size so that if an alternative hypothesis is true and the relative treatment efficacy is $\Delta$, then the power for rejecting the null hypothesis will be a specified value $1 - \beta$.

Clearly, to determine sample size, one must define the endpoint and scale of measurement and then select values of power and $\Delta$. For example, survival or disease-free survival will generally be the endpoint in adjuvant studies, and the scale of $\Delta$ could be the difference in five-year survival proportions or ratio of median survivals. The technical aspects of scale of measurements are not of fundamental importance to the main issues here. The main point is that for a given endpoint, scale of measurement, and power for detecting a difference of size $\Delta$, the required number of patients depends critically on $\Delta$. This can be seen in Table 3, where the number of patients per treatment for detecting a difference in five-year survival rate is tabulated as a function of $\Delta$ and the expected five-year survival rate in the control group. In this table, a two-sided size of .05 and a power of .90 are used. It is assumed that analysis occurs after all patients are followed for five years. This table is based on the exact calculations of Casagrande et al. (1978a) for the power of the Fisher-Irwin exact test.

**Table 3** Number of Patients per Treatment to Have Power .90 for Detecting Difference in Survival Proportion[a]

| Survival proportion in control group | Difference in survival proportion | | |
|---|---|---|---|
| | .10 | .15 | .20 |
| .05 | 198 | 106 | 69 |
| .10 | 281 | 143 | 89 |
| .20 | 410 | 197 | 117 |
| .30 | 496 | 231 | 133 |
| .40 | 533 | 244 | 141 |
| .50 | 533 | 242 | 133 |
| .60 | 496 | 216 | 117 |
| .70 | 410 | 172 | 89 |
| .80 | 281 | 106 | — |

[a]Two-sided significance test of size .05.

Either their tables or calculations based on their approximations (Casagrande et al., 1978b; Fleiss et al., 1980; Ury and Fleiss, 1980) should be used for planning trials that compare proportions using the Fisher exact test or continuity corrected chi-squared test. Many other approximations that have been published and employed substantially misspecify the required sample sizes.

The parameter $\Delta$ does not represent the difference we hope exists; it represents the size of difference that the study will be able to detect (Simon, 1985). For major Phase III studies, it should generally represent the smallest difference that would be of clinical significance. For some pilot studies, it may be reasonable to have a larger $\Delta$. Many "negative" studies are misinterpreted due to the use of significance tests rather than confidence intervals and the use of unrealistically large $\Delta$'s (Simon, 1986). Lack of a statistically significant difference does not demonstrate that the treatments are equivalent. If $\Delta$ is chosen very large, then non-statistically significant results will be consistent both with no difference and with major differences that are less than a "home run." For example, a complete response rate of 25/50 (50%) is not significantly different from one of 20/50 (40%) ($P = .42$), but an approximate 95% confidence interval for the true difference ranges from 13% favoring one treatment to 33% favoring the other treatment!

Although there have been a few home runs in cancer therapeutics, they are often apparent regardless of the methodology employed. Randomized clinical trials are needed to identify important advances that are not home runs and to develop stepwise components of major breakthroughs. An absolute 15% increase in survival for a common type of cancer represents a major public health accomplishment that has rarely been achieved in oncology. Based on current concepts of the probability that a solid tumor has developed multiple resistance by the time of diagnosis, and based on the difficulty of substantially extending survival compared to temporarily shrinking a solid tumor, it may not often be plausible to expect survival differences of greater than 15% for adjuvant therapy studies of current cytotoxic therapy.

Differences in survival proportions of less than 10% are very difficult to seek in a single study. Therefore, it seems important to design at least some major trials to have 90% power for detecting differences in survival proportion of 10 to 15%. Using Table 3, this generally implies a sample size in the range of 150 to 500 patients per treatment group for each set of patients that will be separately analyzed.

For diseases in which survival is very short, it is not natural to speak in terms of differences in the proportion of patients who survive beyond a specified time. In general, for studies with survival or disease-free survival as endpoint, unless the curves show a clear plateau, then planning the trial based on comparing the entire curves is a better approach. Table 4 shows the approximate total number of deaths necessary to provide power .90 for detecting a specified ratio of median survivals for exponential distributions (George and Desu, 1974; Rubinstein et al., 1981). For example, to detect a 50% increase in median survival, it is required that the combination of number of patients accrued and duration of follow-up must be such that a total of about 256 deaths are observed by the time of analysis. The 50% increase in median survival corresponds to at most a 15% absolute difference in proportion surviving beyond a specified time (e.g., five-year survival proportion). Similarly, the 30% (1.3) and 75% (1.75) increases in medians correspond to maximum absolute differences in survival proportion of about 10 and 20%, respectively. To observe 256 deaths in a disease with a long median survival, many more than 256 patients need to be entered if the study is to be completed in a reasonable period of time. For diseases with very poor prognosis, where median survival is <1 year, it may be reasonable to target the smallest ratio of medians of clinical significance to be >75%. For example, if the median survival is only six months, it may be reasonable to target a doubling of median survival to be that of smallest clinical significance. Since it would be feasible to follow all patients until failure in this circumstance, a clinical trial of only 88 total patients might be appropriate (Table 4). But for other types of cancer, a 50% increase in median represents a very major effect, and larger differences may not be plausible. For example, newly diagnosed prostatic cancer patients with stage $D_2$ disease have a median survival of about 3 years. A 50% increase represents an increase of 1.5 years in median and 2.2 years in mean survival. To have 90% power for detecting an increase of this size, a study with about 256 total deaths would be required. With an accrual rate of 100 patients per year and a follow-up period of 3 years after the end of accrual, accrual for 4.25 years with entry of 425 patients is necessary.

In general for exponential distributions, the power for detecting a ratio of medians $\Delta$ is given by:

$$\text{Power} = \Phi^{-1}\left\{\frac{\ln \Delta}{\sqrt{\{1/d_1 + 1/d_2\}}} - Z_\alpha\right\}$$

**Table 4** Total Number of Deaths in Both Groups to Have Power .90 for Detecting Ratio of Median Survivals[a]

| Ratio of median survival | Number of total deaths to observe |
|---|---|
| 1.2 | 1268 |
| 1.3 | 612 |
| 1.4 | 372 |
| 1.5 | 256 |
| 1.6 | 190 |
| 1.7 | 150 |
| 1.8 | 122 |
| 1.9 | 102 |
| 2.0 | 88 |

[a]Two-sided significance test of size .05.

where

$$d_i = (nA/2)\left\{1 - \frac{\exp(-\lambda_i F)}{\lambda_i A}[1 - \exp(-\lambda_i A)]\right\}$$

A and F are the durations of the accrual and follow-up periods in years, n is the accrual rate per year, $Z_\alpha$ is the normal deviate corresponding to an $\alpha$ level type 1 error (1.96 for a two-sided .05 level test), $\Phi^{-1}$ denotes the inverse normal distribution function, $\lambda_1$ denotes the failure rate (ln 2/median) for the control group, and $\lambda_2 = \lambda_1/\Delta$.

### 1. Patient Subsets

The calculations above ignore the substantial heterogeneity in prognosis and biology seen in many kinds of human cancer. The calculations assume that treatment comparisons will be made only for all randomized patients as a whole. For many diseases, patient and tumor characteristics have more effect on prognosis than treatment (Simon, 1984). With heterogeneous diseases such as primary breast cancer, there is almost always interest in evaluating relative treatment efficacy within subsets defined by variables such as menopausal status, extent of nodal involvement, and presence of hormonal receptors. Unfortunately, clinical trials rarely include enough patients to perform subset analyses reliably. It is common for reports of trials to describe no overall significant difference but then to compare the treatments within many subsets and find one or more with a statistically significant difference. Although such analyses may provide important information, more frequently such claims are never confirmed. Unfortunately, the probability that a significant treatment difference within subsets will occur spuriously by chance alone is large and the power for detecting true treatment differences within subsets is small. For example, in comparing two equivalent treatments within 10 disjoint subsets, the probability of obtaining at least one significant difference by chance alone is 40%. If the subsets are of equal size and the trial was designed for an overall 90% power, then the power of treatment comparisons with subsets is only about 18%.

To perform reliable subset analyses, one must use statistical methods to control the probability of false-positive conclusions and must have sample sizes within subsets to provide adequate power for treatment comparisons (Simon, 1982b; Gail and Simon, 1985). For stage II breast cancer, this means at least a doubling of the numbers of patients described in the previous section. Studies without adequate numbers within subsets may suggest hypotheses for testing in subsequent studies, but such hypotheses should be expected to be frequently spurious. Without adequate numbers of patients within subsets, the only reliable inference that can be made from individual studies is usually the overall comparison of treatments. For a disease such as breast cancer, this is not satisfactory for major therapeutic questions (e.g., chemotherapy versus no chemothrrapy) or questions in which treatment by subset interactions seem likely (e.g., tamoxifen and hormonal receptors).

## 2. Demonstrating Treatment Equivalence

In most studies of a new treatment versus a standard treatment, it is implicit that the new treatment will be widely adopted only if it produces improved survival or disease-free survival. There are an increasing number of studies, however, in which the new treatment is less debilitating and thus will be accepted as long as it does not have inferior efficacy. With a finite number of patients, one can never demonstrate that two treatments are precisely equivalent with regard to outcome, but one can bound the size of difference which might exist. Such studies need to be large because it is important to demonstrate that the loss in survival rate caused by the new therapy is very small at worst. Studies that reliably identify or exclude small differences must be very large. In reporting the results of such studies, it is important to calculate confidence limits for the survival differences that are consistent with the observed data, because significance tests can be very misleading. Confidence intervals are almost always much more informative than significance tests, but the distinction is particularly important for equivalence studies because of the implications of a misinterpretation of nonsignificant differences. Makuch and Simon (1978) provided a simple formula for planning the size of such studies based on confidence limit analysis. Improved methods for calculating confidence limits have been described more recently by Santner and Snell (1980), Mee (1984), and Hauck and Anderson (1986). Although the use of confidence limits for reporting results is very important, sample sizes can be planned using conventional significance testing methods by requiring power at least 90% for the smallest difference of clinical significance. Assuming that the two treatments are equivalent and that inference will be based on a 90% two-sided confidence interval, Table 5 indicates the approximate number of patients per group needed to have a 90% probability of detecting 5 and 10% difference in survival rate. As can be seen from the table, to exclude 10% differences in T-year survival rates, one generally requires 300 to 500 patients per treatment.

**Table 5** Number of Patients per Treatment for Showing Equivalence in Order to Have 90% Probability for Detecting Differences of 5 or 10%

| Survival proportion in control group | Difference in survival proportions | |
|---|---|---|
| | .05 | .10 |
| .95 | 503 | 165 |
| .90 | 782 | 232 |
| .80 | 1232 | 338 |
| .70 | 1538 | 409 |
| .60 | 1762 | 445 |
| .50 | 1730 | 445 |

Excluding 5% differences generally requires at least 1000 patients per treatment. It is assumed that all patients are followed for at least T years.

## E. Interim Analysis

In addition to having a target sample size large enough to detect the treatment differences of clinical significance, it is generally important to perform some interim monitoring of results. In some circumstances, the treatment differences are larger than anticipated and accrual can be reliably terminated early. More frequently, however, large early differences disappear with further follow-up. Many unconfirmable positive reports are probably attributable to inappropriate early termination of studies and premature reporting of results. This was well illustrated in a computer simulation by Fleming et al. (1984a), who assumed that patients were accrued to a two-armed randomized study at a constant rate of 40 patients per year over three years. The endpoint was survival and the treatments were equivalent. If analyses were performed every three months, then the probability of obtaining a statistically significant ($p < .05$) result by chance alone at some point during the four-year trial was 26%. This illustrates how misleading interim results can be. The presentation of interim results to trial participants may hurt, rather than help, accrual because of the presence or absence of apparent differences. Such presentations may also provide unreliable information to the general medical community. For these reasons most cancer cooperative groups keep interim results blinded from participants and some have established data monitoring committees for Phase III studies.

Peto et al. (1977) described a very simple sequential design that permits early termination if interim results are extreme. They proposed termination if a chi-square with 1 degree of freedom for the treatment comparison reached a value of nine or greater. If this did not occur then the trial would end accrual at its preplanned fixed sample size. This monitoring rule was felt to be sufficiently conservative that an analysis based on a fixed sample size would be appropriate if early termination did not occur. This conservative approach is useful and was developed more fully by O'Brien and Fleming (1979) and by Fleming et al. (1984b). It represents a useful approach for several reasons. First, large differences are rather uncommon in cancer clinical trials. Hence, designs that substantially increase the maximum trial size in order to increase the probability of early termination if large differences occur are not desirable. Second, the decision to terminate a trial early is complex and sequential designs are based on oversimplified models of the true clinical trial. Consequently, conservative designs are desirable. Third, with survival data early differences may not be representative of later effects. Hence, conservative designs are again desirable. Rubinstein and Gail (1982) also introduced a useful design for monitoring accumlating data in survival studies. Pocock (1982) compared some of these sequential designs.

In some circumstances, interim results indicate that the new treatment is very unlikely to be better than control by a clinically significant amount. In such a case the study can be terminated early. This type of interim analysis can frequently be very important but it must be performed carefully with appropriate statistical methodology. One approach to this analysis is described by Ellenberg and Eisenberger (1985). Another method, due to Lan et al. (1982), is based on the notion of conditional power. At each interim analysis, one calculates the probability of rejecting the null hypothesis if the alternative hypothesis is true and the trial were continued to its target sample size. This power is calculated given the interim results already observed. If the conditional power becomes too small, say less than 0.2, the trial is terminated. Such monitoring can substantially reduce the expected sample size when the null hypothesis is true. Another approach to early termination without rejection of the null hypothesis is described by Fleming et al. (1984b).

In addition to using proper statistical methods, interim decision making must be based on data that have been uniformly updated and carefully reviewed. Even under these conditions, interim results may be transient. It is therefore important that published reports state the target sample size, the history of interim analyses, and the reason for terminating the trial (Simon and Wittes, 1985). For a more complete description of interim analysis, see Chapter 11 by Geller and Pocock.

### F. Factorial Designs

Consider a randomized trial to compare external radiotherapy to radioactive implants for the treatment of women with stage I breast cancer. One might contemplate a subrandomization of each treatment group with regard to whether or not to administer one year of CMF chemotherapy. This is an example of $2 \times 2$ factorial design. Each of the four treatment groups consists of one of the two levels of each of the two factors (Peto, 1978; Simon, 1982c; Byar and Piantadosi, 1985).

Factorial designs offer the possibility of answering two questions for about the same cost, in terms of number of patients, as one. In comparing the two types of radiotherapy, for example, one would stratify with regard to adjuvant chemotherapy if the randomizations were performed simultaneously. If the adjuvant chemotherapy randomization was performed later, then the analysis of radiotherapy would be pooled over patients who received and did not receive chemotherapy. The essential point is, however, that if a total of N patients have been placed on study, then evaluation of each factor involves comparing two treatment groups each with about N/2 patients.

One may want to increase the sample size somewhat to account for the fact that two comparisons are made. That is, one may wish to work at the $\alpha/2$ significance level rather than the $\alpha$ level one would use for a single comparison. This is a controversial point, however. Even if a multiple comparison correction is adopted, the saving in number of patients compared to two separate phase III studies is enormous.

If the relative efficacy of adjuvant chemotherapy depends seriously on the type of radiotherapy administered, then the factorial analysis described above may be very misleading. The estimated contrast between the two levels of one factor will be an average of the contrasts for each of the two levels of the other factor. The reasonableness of the factorial analysis is based on the assumption that there is no substantial interaction between the effects of the two factors. Unfortunately, if the study is planned to answer two questions for about the same number of patients as one, the sample size will be inadequate to test for interactions. Hence, finding no "significant" interactions will not really mean anything. The meaningfulness of the analysis will depend on an unverifiable assumption. In the above example, I believe that the assumption is reasonable and the study would be worthwhile. Reasonableness is an a priori judgment that will not, however, be subject to verification.

In many cases the factorial design will be obviously inappropriate or questionable. For example, suppose chemotherapy for patients receiving external radiotherapy must be given in lower doses than for those given implants. If one knows this from the outset and specifies two different chemotherapy regimens in the protocol, one does not really have a factorial design and should not plan sample sizes as if one did. If one plans to administer the same chemotherapy but suspects that it will not be possible, then a factorial design is risky and probably not advisable.

Although the factorial design is not applicable to most trials, there are situations where it can be very appropriate and useful. This is particularly true when dealing with treatments which are not very toxic or those for which no effect is likely.

### G. Prerandomization

Zelen (1979) proposed a design in which patients are randomized after being found eligible for the protocol but before being asked for consent to participate. With one version of his proposal, consent is sought only for patients randomized to an experimental therapy. This version was controversial with regard to ethical considerations. The other version of Zelen's proposal involves seeking consent for both randomized gorups of patients. Since the treatment assignment is known and can be presented at the time consent is sought, it was thought that physicians would approach more patients for participation in the trial and patients would be more likely to participate. To avoid the possibility of bias in analysis of results, patients must be compared "as randomized" rather than "as treated"; that is, a patient randomized to treatment A who refused and received treatment B must be considered in the A group for analysis. Otherwise, the treatment comparison can be biased by a possible relationship between prognosis, treatment, and refusal rate. Since the analysis must be performed "as

Table 6 Sample Size Inflation Factor According to Overall Refusal Rate

| Refusal rate | Inflation factor |
|---|---|
| 0.02 | 1.09 |
| 0.05 | 1.23 |
| 0.10 | 1.56 |
| 0.15 | 2.04 |
| 0.20 | 2.78 |
| 0.25 | 4.00 |
| 0.30 | 6.25 |
| 0.35 | 11.11 |
| 0.40 | 25 |
| 0.45 | 100 |

randomized," refusals dilute the differences in outcome that can be observed. To counteract this effect, the number of patients must be increased compared to a conventional trial in which most refusals occur before randomization and thus can be excluded from analysis. The relationship between refusal rate and increased sample size required for a prerandomization design is dramatic and is shown in Table 6 (Ellenberg, 1984). For a 15% refusal rate, the number of patients required is double that of a conventional randomized trial. Even though the informed consent process may be more comfortable for the physician and patient, one must question how frequently prerandomization would result in a doubling of accrual.

Two other concerns have been expressed about prerandomization (Ellenberg, 1984). First is the ethical issue of whether patients are really fully informed about possible alternative treatments when randomization has already occurred. The second issue is that there will be demand to analyze results both "as randomized" and "as treated." The results may differ, ruining credibility of the trial.

Prerandomization is attractive to many physicians, although there is lack of awareness of the problems mentioned above. The limited experience to date with this design has given mixed results. Refusal rates have ranged from 10 to 30%. Accrual generally has increased but not always by an amount sufficient to compensate for the inefficiency of the design. The prerandomization design is not an attractive alternative but may be useful for trials that cannot be performed otherwise.

## IV. ANALYSIS OF PHASE III TRIALS

### A. Exclusions

Excluding patients from analysis because of treatment deviations, early death, or patient withdrawal for other reasons may seriously bias the results. Excluded patients often have poorer outcomes than those not excluded. One can rationalize that patients not receiving treatment as specified in the protocol did worse because of that fact, but this is just a rationalization, which may be erroneous. Canner (1981) showed that patients with poor compliance to placebo pills in the Coronary Drug Project had much greater cardiovascular mortality than other placebo patients and the difference was not explainable on the basis of known prognostic factors. Excluding patients (or "analyzing them separately," which is equivalent to excluding them) for reasons other than that they did not satisfy the eligibility criteria of the study is a major defect in the analysis and reporting of Phase III trials. If the conclusions of a study depend on exclusions, then these conclusions are suspect. Eligibility criteria for both patients and collaborators should be established in such a way that there will be few protocol deviations. Generally, the treatment plan should be viewed as a policy to be evaluated. This policy cannot be applied completely to all patients, but all patients should generally be evaluable in Phase III studies.

When survival is the endpoint, the vital status of all patients can be traced and there is little justification for exclusion of eligible patients. The report of such a study should always include an analysis without exclusion of eligible patients.

Censoring patients who go "off-study" or receive a nonprotocol treatment is a form of exclusion. Because such censorship certainly cannot be assumed to be noninformative, it invalidates analysis of the data. Such censorship can introduce great bias and should not be employed. The trial should be designed and conducted so that a straightforward analysis of survival without exclusions or informative censoring is sufficient. Otherwise the trial has failed and cannot be reliably retrieved by statistical analysis. If survival is the important endpoint, then crossover treatment should not be part of the design.

## B. Confidence Intervals and Significance Tests

Although the Neyman-Pearson theory of hypothesis testing is generally used in planning the size of a clinical trial, this theory is not a satisfactory framework for analysis (Cutler et al., 1966). It is not usually good science to regard a significance level of 0.04 as "significant" but one of 0.06 as "not significant." This dichotomous approach to analysis is used to a surprising extent with many anamolous and erroneous results. Nonsignificant differences in small studies are also sometimes taken as evidence that the treatments are equivalent or that differences are not clinically important. If the study is small, however, results are often inconclusive, not negative. Confidence intervals are generally more informative than significance levels or the declarations of hypothesis tests (Rothman, 1978a; Detsky and Sackett, 1985; Gardner and Altman, 1986; Simon, 1986). Confidence intervals should be routinely used in reporting results as supplements or alternatives to significance levels. For small studies they clarify whether the results are inconclusive or negative. They are more appropriate than statements of statistical power because the latter do not take into consideration the results actually obtained. Small studies with results favoring the control treatment sometimes can conclusively rule out the possibility that the experimental treatment is substantially better than the control. Such results will be evident by reporting the confidence interval, but not by stating statistical power.

### C. Binary Response Methods

For comparing two proportions, the Fisher-Irwin exact test provides an exact significance level. When sample sizes are adequately large, the contingency chi-square test with continuity correction provides a good Fisher-Irwin exact test.

If treatment group i has $d_i$ responses in $r_i$ patients then the continuity corrected chi-square can be written in the form

$$X_c^2 = \frac{(|d_1 - E| - 0.5)^2}{V'}$$

where

$$E = \frac{r_1(d_1 + d_2)}{(r_1 + r_2)}$$

$$V' = \frac{r_1 r_2 (d_1 + d_2)(r_1 + r_2 - d_1 - d_2)}{(r_1 + r_2)^3} \tag{1}$$

Under the null hypothesis $X_c^2$ is asymptotically chi-square distributed with 1 degree of freedom. This test is easily generalized to the Mantel-Haenszel (1959) test for the comparison of two treatments while adjusting for other qualitative factors. Suppose that there are I disjoint strata of patients. Within each stratum calculate $E_i$ as in (1), but calculate $V_i$ from

$$V = \frac{r_1 r_2 (d_1 + d_2)(r_1 + r_2 - d_1 - d_2)}{(r_1 + r_2)^2 (r_1 + r_2 - 1)} \tag{2}$$

Let E equal the sum of the $E_i$'s over the strata. Similarly, let V be the sum of the $V_i$'s and let 0 denote the total number of responses to treatment 1. The Mantel-Haenszel test statistic is

$$X^2 = \frac{(0 - E)^2}{V} \tag{3}$$

and under the null hypothesis $X^2$ is asymptotically chi-square distributed with 1 degree of freedom. In summing over the strata, one should not use the continuity correction.

In some situations one may wish to adjust for baseline covariates by a regression rather than a stratified analysis approach. The most commonly used regression model for binary response clinical trials is the logit model (Cox, 1970)

$$\log \frac{p(x)}{q(x)} = bx$$

where $q(x) = 1 - p(x)$, x is a vector of variables, and b is a vector of unknown regression coefficients. One of the variables is coded to represent the treatment assigned and the other variables represent baseline covariates (it is invalid to adjust for covariates measured after treatment assignment). Model development, assessment, and validation is a much more complex procedure than most statistical packages would lead one to believe. Discussion of these complexities is beyond the scope of this chapter, but some aspects are discussed by Tsiatis (1980), Pregibon (1981), Simon (1983), and Landwehr et al. (1984).

If one observes d responses in r patients, it is easy to calculate approximate confidence intervals for the true response probability p. The Clopper-Pearson (1934) interval $(p_L, p_U)$ is obtained from

$$B(d; r, p_U) = \frac{\alpha}{2}$$

$$B(d - 1; r, p_L) = 1 - \frac{\alpha}{2} \tag{4}$$

where $B(x; r, p)$ is the cumulative binomial probability of x or fewer responses in r patients if the response probability is p. Although exact binomial probabilities are used here, the intervals do not have coverage probabilities exactly equal to the nominal confidence levels. This is due to discreteness. The coverage probabilities, however, are guaranteed to be at least as great as the nominal levels.

Another useful approximate confidence interval for the true response probability is given by (Ghosh, 1979).

$$\frac{\hat{p} + A/2 \pm Z\sqrt{\hat{p}(1 - \hat{p})/r + A/4r}}{1 + A} \tag{5}$$

where $A = Z^2/r$ and $\hat{p} = d/r$. In this formula $Z = 1.96$, 1.645, or 1.28 for 95, 90, or 80% confidence intervals, respectively.

An approximate confidence interval for the difference in two response probabilities can be calculated in the usual way from

$$(\hat{p}_1 - \hat{p}_2) \pm Z\sqrt{\left\{\frac{\hat{p}_1(1 - \hat{p}_1)}{r_1} + \frac{\hat{p}_2(1 - \hat{p}_2)}{r_2}\right\}} \tag{6}$$

where Z is identified as above. Easily calculated confidence intervals analogous to (4) or (5) are not available for the two-sample problem, but methods more accurate than (6) are presented by Mee (1984) and by Hauck and Anderson (1986).

The output of logit model programs will include the maximum likelihood estimate, say $\hat{b}_0$, of the regression coefficient associated with treatment and estimate s of the standard error of $\hat{b}_0$. If the treatment variable was coded 1 for treatment 1 and 0 for treatment 2, then $\hat{b}_0 \pm Zs$ is an approximate confidence interval for the logarithm of the odds ratio of treatment 1 relative to treatment 2.

## D. Censored Data Methods

Censored data methods are generally needed for the analysis of survival, disease-free survival, or time-till-tumor progression. We shall use the term "survival" here in the generic sense to apply to all such data with right censoring. The generally used methods all assume "noninformative" censoring (Kalbfleisch and Prentice, 1980). This means roughly that the survival and censoring times are independent. We have already discussed situations for which censoring is likely to be informative. In the following discussion, however, we shall assume that censoring is noninformative.

The product-limit method of Kaplan and Meier (1958) is commonly used to estimate the survival distribution. Let $t_1, t_2, \ldots, t_n$ denote distinct times at which deaths occur and let $d_1, \ldots, d_n$ denote the number of deaths at each time. Let $r_i$ denote the number of patients whose survivals are equal to or greater than $t_i$. The Kaplan-Meier estimator is

$$S(t_1) = \frac{(r_1 - d_1)}{r_1}$$

$$S(t_i) = S(t_{i-1}) \frac{(r_i - d_i)}{r_i}$$

for $i = 2, \ldots, n$. The points $(t_i, S(t_i))$, should be plotted as a step function, the value $S(t_{i-1})$ drawn horizontally to $t_i$ where it drops to $S(t_i)$. It is useful either to indicate censored observations by tick marks on the graph or to indicate the number of patients at risk at various points of the graph.

The Kaplan-Meier estimator is asymptotically normal and the usually encountered estimate of the variance of $S(t_i)$ is

$$[S(t_i)]^2 \Sigma \frac{d_j}{r_j(r_j - d_j)}$$

where the summation is over all failure times up to and including $t_i$. Dividing this expression for the variance into $[S(t_i)][1 - S(t_i)]$, one obtains the effective sample size at $t_i$

$$ESS(t_i) = \frac{1 - S(t_i)}{S(t_i) \Sigma d_j/r_j(r_j - d_j)} \qquad (7)$$

An approximate confidence interval for the probability of survival beyond $t_i$ can be obtained by using expression (5) with $S(t_i)$ replacing $\hat{p}$ and $ESS(t_i)$ replacing r (Rothman, 1978b). It is useful to show approximate confidence intervals for the Kaplan-Meier estimate in the right tail of the curve, where clinicians sometimes overinterpret the estimate.

Median survival is sometimes used as a descriptive statistic instead of displaying the entire Kaplan-Meier curve. If the sample size is small or if the curve is flat around the median, then the estimate may be very imprecise. It is generally useful to report a confidence interval when the median is used in order to avoid gross misinterpretations. One method of obtaining an approximate $1 - \alpha$ level confidence interval for the median is to determine the set of failure times $t_i$ for which 0.5 is included in the $1 - \alpha$ level confidence interval for the probability of surviving beyond $t_i$ (Simon and Lee, 1982). This involves repeated calculation of the type of confidence interval described above. A simpler method, called the reflection method, is as follows (Slud et al., 1984).

Calculate

$$p_U = 0.5\left(1 + \frac{Z}{\sqrt{ESS}}\right)$$

$$p_L = 0.5\left(1 - \frac{Z}{\sqrt{ESS}}\right) \tag{8}$$

where Z is a standard normal percentile described following expression (5) and ESS is the effective sample size at the estimated median (a failure time). The confidence interval for the median is obtained by drawing horizontal lines at levels $p_U$ and $p_L$ on the y axis. The times on the x axis where these lines intersect the Kaplan-Meier curve represent the endpoints of the confidence interval.

The Mantel-Haenszel test (Mantel, 1966), also called the log-rank test, is one of the most commonly used significance tests for comparing survival distributions. The test statistic is given by expression (3). 0 represents the total number of deaths for treatment 1; E and V are calculated from

$$E = \sum_{i=1}^{n} \frac{r_{1i} d_i}{r_i} \tag{9}$$

$$V = \sum_{i=1}^{n} \frac{d_i(r_i - d_i) r_{1i} r_{2i}}{r_i^2 (r_i - 1)} \tag{10}$$

In these calculations $r_{ji}$ represents the number of patients in treatment group j with survival greater than or equal to $t_i$, $d_{ji}$ represents the number of deaths in treatment group j at time $t_i$, $r_i = r_{1i} + r_{2i}$, $d_i = d_{1i} + d_{2i}$, and the summation is over distinct death times $t_1, \ldots, t_n$. Under the null hypothesis, $X^2$ has a chi-square distribution with 1 degree of freedom.

This test is easily modified to account for an important prognostic factor. One may wish to do this because of an imbalance in the groups being compared with regard to this factor. It is important to recognize that the fact that an imbalance is not "statistically significant" does not mean it is not distorting the survival comparison; lack of "statistical significance" merely means that the imbalance may have arisen by chance. One may also wish to incorporate into the analysis an important prognostic factor for which there is no imbalance. One sometimes obtains a more powerful survival comparison by accounting in this way for known sources of variability (Simon, 1984). To incorporate such information one subdivides each of the two treatment groups into distinct strata determined by the prognostic factors. Considering each stratum entirely separately one calculates $0_j$, $E_j$, and $V_j$ for group one, in which the subscript j is used to denote stratum. One then calculates 0, E, and V by summing over the strata, and performs the test $X^2 = (0 - E)^2/V$ as before. Only a few strata can be accommodated in this type of analysis unless there are many patients. The resulting test is an overall comparison of the two groups adjusted for the influences of the prognostic factors.

There are a number of ways in which significance tests described here are commonly misused. If there are more than two groups to be compared, an overall test comparing all groups simultaneously should be employed before contrasting all pairs. It is also not valid to compare one treatment to another within subsets without first demonstrating that the relative treatment efficacy varies among the subsets. These more complex analyses require more sophisticated methods.

The quantities 0, E, and V can be used to obtain an approximate confidence limit for the hazard ratio of two treatments:

$$\exp\left[\frac{(0 - E)}{V} \pm \frac{Z}{\sqrt{V}}\right] \tag{11}$$

It can be shown that $(0 - E)/V$ is a first-order Taylor series approximation to the maximum likelihood estimator of the log hazard ratio and that the asymptotic variance of $(0 - E)/V$ is $1/V$ (Haybittle, 1979). This is the statistical basis for (11).

A number of other methods for obtaining approximate confidence intervals for the hazard ratio have been studied. Several of these are reviewed by Bernstein et al. (1981). We shall present here an easily calculated approximate confidence interval for the ratio of median survivals when the survival distributions are exponential. The confidence interval is (Rubinstein et al., 1981)

$$R \exp(\pm Zs) \tag{12}$$

where $s = \sqrt{(1/d_1 + 1/d_2)}$. Here the d's denote the total number of deaths in the two treatment groups and R denotes the observed ratio of median survivals. For an exponential distribution the best estimate of the median is the total follow-up time (summation of times till death or censoring) divided by the number of deaths. R is the ratio of the two medians calculated in this way.

Approximate confidence intervals for the difference in probability of surviving beyond specified time t (e.g., five years) can be obtained from (6) using the Kaplan-Meier estimates for the p's and the effective sample sizes at t for the r's.

Many regression models have been developed for censored data problems. One of the most commonly used models is the semiparametric proportional hazards model of Cox (1972). The hazard function for an individual with covariate vector x is assumed to be of the form $\lambda_0(t) \exp(bx)$, where b is a vector of unknown regression coefficients and $\lambda_0(t)$ is an unknown nonnegative function that does not depend on b or x. The ratio of hazard functions for two individuals with different covariate vectors is constant over time; hence, this is called a proportional hazards model. The attractiveness of this approach is that inference about b can be made without making parametric assumptions about the underlying distribution $\lambda_0(t)$. The proportional hazards assumption is itself a rather strong assumption, however, and so evaluation of model adequacy is very important. The class of proportional hazards models is not closed under the omission of covariates, and hence biased estimates of regression coefficients may result from such model misspecification (Gail et al., 1984). The literature of censored data regression models is vast. A few good references are Kay (1977), Kalbfleish and Prentice (1980), Lee (1980), Elandt-Johnson and Johnson (1980), Miller (1981), Schoenfeld (1982), Lawless (1982), and Cox and Oakes (1984).

### E. Subset Analysis

Subset analysis is a common source of "false positive" results in the analysis of clinical trials. As stated previously, unless sample size is established for adequate separate analysis of predefined subsets, the power for detecting subset-specific effects will be unsatisfactorily small and the probability of having a "significant" treatment difference within some subset by chance alone will be unsatisfactorily large. When major treatment by subset interactions seem likely a priori, sample size should be established so that the subsets of interest can be analyzed separately with adequate statistical power. In other cases, subset analysis should be viewed as hypothesis generation. Even in this case the strength of evidence in favor of treatment by subset interactions can be assessed by formal tests of significance. Such tests are described for binary and survival data by Simon (1982b). Gail and Simon (1985) have introduced a test for qualitative treatment by subset interactions. Let relative efficacy of treatment 2 versus treatment 1 be measured on an accepted scale and let $\Delta_{min}$ denote the smallest difference in rlative efficacy of clinical importance. In some cases we may take $\Delta_{min} = 0$. A qualitative interaction exists if the relative efficacy is $>\Delta_{min}$ for some subsets but $<\Delta_{min}$ for others. The relevance of this concept is discussed by Gail and Simon. Unless the protocol is designed for separate analysis of subsets, claims of subset-specific effects should be documented by the demonstration of significant treatment by subset interactions.

## V. REPORTING RESULTS

Effective reporting of results is an integral part of good research. Unfortunately, statistical methodology is often used poorly in reports of medical studies. By "statistical methodology" I do not mean merely the detailed technical aspects of calculating significance levels and confidence intervals, but more generally the methods for determining who to analyze, of summarizing evidence, and of reaching conclusions. The nine items listed below were developed by Simon and Wittes (1985) and have been adopted by most major cancer journals as methodologic guidelines for reports of clinical trials.

1. Authors should discuss briefly the quality control methods used to ensure that the data are complete and accurate. A reliable procedure should be cited for ensuring that all patients entered on study are actually reported on. If no such procedures are in place, their absence should be noted. Any procedures employed to ensure that assessment of major endpoints is

reliable should be mentioned (e.g., second-party review of responses) or their absence noted.

2. All patients registered on study should be accounted for. The report should specify for each treatment the number of patients who were not eligible, who died, or who withdrew before treatment began. The distribution of follow-up times should be described for each treatment, and the number of patients lost to follow-up should be given.

3. The study should not have an inevaluability rate for major endpoints of greater than 15%. Not more than 15% of eligible patients should be lost to follow-up or considered inevaluable for response due to early death, protocol violation, missing information, etc.

4. In randomized studies, the report should include a comparison of survival and/or other major endpoints for all eligible patients as randomized, that is, with no exclusions other than those not meeting eligibility criteria.

5. The sample size should be sufficient to either establish or conclusively rule out the existence of effects of clinically meaningful magnitude. For "negative" results in therapeutic comparisons, the adequacy of sample size should be demonstrated by either presenting confidence limits for true treatment differences or calculating statistical power for detecting differences. For uncontrolled phase II studies, a procedure should be in place to prevent the accrual of an inappropriately large number of patients, when the study has shown the agent to be inactive.

6. Authors should state whether there was an initial target sample size and, if so, what it was. They should specify how frequently interim analyses were performed and how the decisions to stop accrual and report results were arrived at.

7. All claims of therapeutic efficacy should be based on explicit comparison with a specific control group, except in special circumstances where each patient is his own control. If nonrandomized controls are used, the characteristics of the patients should be presented in detail and compared to those of the experimental group. Potential sources of bias should be adequately discussed. Comparison of survival between responders and nonresponders does not estalbish efficacy and should not generally be included. Reports of Phase II trials which draw conclusions about antitumor activity but not therapeutic efficacy generally do not require a control group.

8. The patients studied should be adequately described. Applicability of conclusions to other patients should be carefully dealt with. Claims of subset-specific treatment differences must be carefully documented statistically as more than the random results of multiple-subset analyses.

9. The methods of statistical analysis should be described in detail sufficient that a knowledgeable reader could reproduce the analysis if the data were available.

## REFERENCES

Anderson, J. R., Cain, K. C., and Gelber, R. D. (1983). Analysis of Survival by Tumor Response, *J. Clin. Oncol.*, 1: 710–719.

Anderson, J. R., Cain, K. C., Gelber, R. D., and Gelman, R. S. (1985). Analysis and Interpretation of the Comparison of Survival by Treatment Outcome Variables in Cancer Clinical Trials, *Cancer Treat. Rep.*, 69: 1139–1146.

Atkinson, E. N. and Brown, B. W. (1985). Confidence Limits for Probability of Response in Multistage Phase II Clinical Trials, *Biometrics*, 41: 741–744.

Bernstein, L., Anderson, J., and Pike, M. C. (1981). Estimation of the Proportional Hazard in Two Treatment Group Clinical Trials, *Biometrics*, 37: 513–519.

Byar, D. P. and Piantadosi, S. (1985). Factorial Designs for Randomized Clinical Trials, *Cancer Treat. Rep.*, 69: 1055–1064.

Byar, D. P., Simon, R. M., Friedewald, W. T., Schlesselman, J. J., DeMets, D. L., Ellenberg, J. H., Gail, M. H., and Ware, J. S. (1976). Randomized Clinical Trials: Perspectives on Some Recent Ideas, *N. Engl. J. Med.*, 295: 74–80.

Canner, P. L. (1981). Influence of Treatment Adherence in the Coronary Drug Project (letter), *N. Engl. J. Med.*, 304: 612–613.

Casagrande, J. J., Pike, M. C., and Smith, P. G. (1978a). The Power Function of the "Exact" Test for Comparing Two Binomial Distributions, *Appl. Stat.*, 27: 176–201.

Casagrande, J. T., Pike, M. C., and Smith, P. G. (1978b). An Improved Formula for Calculating Sample Sizes for Comparing Two Binomial Distributions, *Biometrics*, 34: 483–486.

Chang, M. and O'Brien, P. (1986). Confidence Intervals Following Group Sequential Tests, *Controlled Clin. Trials*, 7: 18–26.

Clopper, C. J. and Pearson, E. S. (1934). The Use of Confidence or Fiducial Limits Illustrated in the Case of the Binomial, *Biometrika*, 26: 406–413.

Cox, D. R. (1970). *Analysis of Binary Data*, Methuen, London.

Cox, D. R. (1972). Regression Models and Life Tables, *J. R. Stat. Soc. Ser. B*, 34: 187–220.

Cox, D. R. and Oakes, D. (1984). *Analysis of Survival Data*, Chapman & Hall, London.

Cutler, S. J., Greenhouse, S. W., Cornfield, J., Schneiderman, M. A., Zelen, M., Shaw, L. W., and Beebe, G. W. (1966). The Role of Hypothesis Testing in Clinical Trials, *J. Chronic Dis.*, 19: 857–882.

Detsky, A. S. and Sackett, D. (1985). When Was a "Negative" Clinical Trial Big Enough? *Arch. Intern. Med.*, 145: 704–712.

Elandt-Johnson, R. C. and Johnson, N. L. (1980). *Survival Models and Data Analysis*, Wiley, New York.

Ellenberg, S. S. (1984). Randomization designs in Comparative Clinical Trials, *N. Engl. J. Med.*, 310: 1404.

Ellenberg, S. S. and Eisenberger, M. A. (1985). An Efficient Design for Phase III Studies of Combination Chemotherapies, *Cancer Treat. Rep.*, 69: 1147–1152.

Fleiss, J. L., Tytun, A., and Ury, H. K. (1980). A Simple Approximation for Calculating Sample Size for Comparing Independent Proportions, *Biometrics*, 36: 343–346.

Fleming, T. R. (1982). One Sample Multiple Testing Procedure for Phase II Clinical Trials, *Biometrics*, 38: 143–151.

Fleming, T. R., Green, S. J., and Harrington, D. P. (1984a). Considerations for Monitoring and Evaluating Treatment Effects in Clinical Trials, *Controlled Clin. Trials*, 5: 55–66.

Fleming, T. R., Green, S. J., and Harrington, D. P. (1984b). Designs for Group Sequential Tests, *Controlled Clin. Trials*, 5: 348–361.

Freiman, J. A., Chalmers, T. C., Smith, H., Jr., and Kuebler, R. R. (1978). The Importance of Beta, the Type II Error, and Sample Size in the Design and Interpretation of the Randomized Controlled Trial. Survey of 71 "Negative" Trials, *N. Engl. J. Med.*, 299: 690–694.

Gail, M. and Simon, R. (1985). Testing for Qualitative Interactions between Treatment Effects and Patient Subsets, *Biometrics*, 41: 361–372.

Gail, M. H., Wieand, S., and Piantadosi, S. (1984). Biased Estimates of Treatment Effect in Randomized Experiments with Nonlinear Regressions and Omitted Covariates, *Biometrika*, 71: 431–444.

Gardner, M. J. and Altman, D. G. (1986). Confidence Intervals Rather Than p Values: Estimation Rather Than Hypothesis Testing, *Br. Med. J.*, 292: 746–750.

George, S. L. (1983). The Required Size and Length of a Phase III Clinical Trial, *Cancer Clinical Trials: Design, Practice and Analysis* (M. E. Buyse, M. J. Staquet, and R. J. Sylvester, eds.), Oxford University Press, New York.

George, S. L. and Desu, M. M. (1974). Planning the Size and Duration of a Clinical Trial Studying the Time to Some Critical Event, *J. Chronic Dis.*, 27: 15–24.

Ghosh, B. K. (1979). A Comparison of Some Approximate Confidence Intervals for Binomial Parameter, *J. Am. Stat. Assoc.*, 74: 894–900.

Gurland, J. and Johnson, R. O. (1965). How Reliable Are Tumor Measurements, *J. Amer. Medical Assoc.*, 194: 125–130.

Hauck, W. W. and Anderson, S. (1986). A Comparison of Large-Sample Confidence Interval Methods for the Difference of Two Binomial Probabilities, *The American Statistician*, 40: 318–322.

Haybittle, J. I. (1979). The Reporting of Non-Significant Results in Clinical Trials, *Clinical Trials in "Early" Breast Cancer* (B. R. Scheurlen, G. Weckesser, and I. Armbruster, eds.) Springer-Verlag, Berlin, pp. 28–39.

Herson, J. (1983). Practical Side of Multistage Clinical Trials for Screening New Agents, *Cancer Treat. Rep.*, 67: 71–75.

Herson, J. (1984). Statistical Aspects in the Design and Analysis of Phase II Clinical Trials, *Cancer Clinical Trials: Methods and Practice* (M. E. Buyse, M. J. Staquet, and R. J. Sylvester, eds.), Oxford University Press, London.

Jennison, C. and Turnbull, B. W. (1983). Confidence Intervals for a Binomial Parameter Following a Multistage Test with Application to MIL-STD 105D and Medical Trials, *Technometrics*, 25: 49–58.

Kalbfleisch, J. D. and Prentice, R. L. (1980). *The Statistical Analysis of Failure Time Data*, Wiley, New York.

Kalish, L. A. and Begg, C. B. (1985). Treatment Allocation Methods in Clinical Trials: A Review, *Stat. Med.*, 4: 129–144.

Kaplan, E. L. and Meier, P. (1958). Nonparametric Estimation from Incomplete Observations, *J. Am. Stat. Assoc.*, 53: 457–481.

Kay, R. (1977). Proportional Hazard Regression Models and the Analysis of Censored Survival Data, *Appl. Stat.*, 26: 227–237.

Lan, G. K. K., Simon, R., and Halperin, M. (1982). Stochastically Curtailed Tests in Long-Term Clinical Trials, *Commun. Stat.—Sequential Anal.*, 1: 207–219.

Landwehr, J. M., Pregibon, D., and Shoemaker, A. C. (1984). Graphical Methods for Assessing Logistic Regression Models, *J. Am. Stat. Assoc.*, 79: 61–83.

Lavin, P. T. and Flowerdew, G. (1980). Studies in Variation Associated with the Measurement of Solid Tumors, *Cancer*, 46: 1286–1290.

Lawless, J. F. (1982). *Statistical Models and Methods for Lifetime Data*, Wiley, New York.

Lee, E. T. (1980). *Statistical Methods for Survival Data Analysis*, Lifetime Learning Publications, Belmont, Calif.

Makuch, R. and Simon, R. (1978). Sample Size Requirements for Evaluating a Conservative Therapy, *Cancer Treat. Rep.*, 62: 1037–1040.

Mantel, N. (1966). Evaluation of Survival Data and Two New Rank Order Statistics Arising in Its Consideration, *Cancer Chemother. Rep.*, 50: 163–170.

Mantel, N. and Haenszel, W. (1959). Statistical Aspects of the Analysis of Data from Retrospective Studies of Disease, *J. Natl. Cancer Inst.*, 22: 719–748.

Mee, R. W. (1984). Confidence Bounds for the Difference between Two Probabilities, *Biometrics*, 40: 1175–1176.

Micciolo, R., Valagussa, P., and Marubini, E. (1985). The Use of Historical Controls in Breast Cancer: An Assessment in Three Consecutive Trials, *Controlled Clin. Trials*, 6: 259–270.

Miller, R. G., Jr. (1981). *Survival Analysis*, Wiley, New York.

Moertel, C. G. and Hanley, J. A. (1976). The Effect of Measuring Error on the Results of Therapeutic Trials in Advanced Cancer, *Cancer*, 38: 388–394.

Moertel, C. G., Schutt, A. J., Hahan, R. G., et al. (1974). Effects of Patient Selection on Results of Phase II Chemotherapy Trials in Gastrointestinal Cancer, *Cancer Chemother. Rep.*, 59: 257.

O'Brien, P. C. and Fleming, T. R. (1979). A Multiple Testing Procedure for Clinical Trials, *Biometrics*, 35: 549–556.

Peto, R. (1978). Clinical Trial Methodology, *Biomedicine*, 28: 24–36.

Peto, R., Pike, M. C., Armitage, P., Breslow, N. E., Cox, D. R., Howard, S. V., Mantel, N., McPherson, K., Peto, J., and Smith, P. G. (1976).

Design and Analysis of Randomized Clinical Trials Requiring Prolonged Observation of Each Patient. 1. Introduction and Design, *Br. J. Cancer*, 34: 585–612.

Peto, R., Pike, M. C., Armitage, P., Breslow, N. E., Cox, D. R., Howard, S. V., Mantel, N., McPherson, K., Peto, J., and Smith, P. G. (1977). Design and Analysis of Randomized Clinical Trials Requiring Prolonged Observation of Each Patient. 2. Analysis and Examples, *Br. J. Cancer*, 35: 1–39.

Pocock, S. J. (1979). Allocation of Patients to Treatment in Clinical Trials, *Biometrics*, 35: 183–197.

Pocock, S. J. (1982). Interim Analyses for Randomized Clinical Trials: The Group Sequential Approach, *Biometrics*, 38: 153–162.

Pregibon, D. (1981). Logistic Regression Diagnostics, *Ann. Stat.*, 9: 705–724.

Rothman, K. J. (1978a). A Show of Confidence, *N. Engl. J. Med.*, 299: 1362–1363.

Rothman, K. J. (1978b). Estimation of Confidence Limits for the Cumulative Probability of Survival in Life Table Analysis, *J. Chronic Dis.*, 31: 557–560.

Rubinstein, L. V. and Gail, M. H. (1982). Monitoring Rules for Stopping Accrual in Comparative Survival Studies, *Controlled Clin. Trials*, 3: 325–343.

Rubinstein, L. V., Gail, M. H., and Santner, T. J. (1981). Planning the Duration of a Comparative Clinical Trial with Loss to Follow-up and a Period of Continued Observation, *J. Chronic Dis.*, 36: 469–474.

Santner, T. J. and Snell, M. A. (1980). Small-Sample Confidence Intervals for $P_1-P_2$ and $P_1/P_2$ in $2 \times 2$ contingency Tables, *J. Am. Stat. Assoc.*, 75: 386–394.

Schipper, H. and Levitt, M. (1985). Measuring Quality of Life: Risks and Benefits, *Cancer Treat. Rep.*, 69: 1115–1126.

Schoenfeld, D. (1982). Partial Residuals for the Proportional Hazards Regression Model, *Biometrika*, 69: 239–241.

Schultz, J. R., Nichol, F. R., Elfring, G. L., and Weed, S. D. (1973). Multiple-Stage Procedures for Drug Screening, *Biometrics*, 29: 293–300.

Simon, R. (1979). Restricted Randomization Designs in Clinical Trials, *Biometrics*, 35: 503–512.

Simon, R. (1980). Patient Heterogeneity in Clinical Trials, *Cancer Treat. Rep.*, 64: 405–410.

Simon, R. (1982a). Randomized Clinical Trials and Research Strategy, *Cancer Treat. Rep.*, 66: 1083–1087.

Simon, R. (1982b). Patient Subsets and Variation in Therapeutic Efficacy, *Br. J. Clin. Pharmacol.*, 14: 473–482.

Simon, R. M. (1982c). Design and Conduct of Clinical Trials, *Cancer: Principles and Practice of Oncology* (V. T. DeVita, Jr., S. Hellman, and S. A. Rosenberg, eds.), Lippincott, Philadelphia.

Simon, R. (1983). Use of Regression Models: Statistical Aspects, *Cancer Clinical Trials: Design, Practice and Analysis* (M. E. Buyse, M. J.

Staquet, and R. J. Sylvester, eds.), Oxford University Press, London, p. 445–466.

Simon, R. (1984). Importance of Prognostic Factors in Cancer Clinical Trials, *Cancer Treat. Rep.*, 68: 185–192.

Simon, R. (1985). The Size of Phase III Cancer Clinical Trials, *Cancer Treat. Rep.*, 69: 1087–1092.

Simon, R. (1986). Confidence Intervals for Reporting Clinical Trial Results, *Ann. Intern. Med.*, 105: 429–435.

Simon, R. and Lee, Y. J. (1982). Nonparametric Confidence Limits for Survival Probabilities and Median Survival Time, *Cancer Treat. Rep.*, 66: 37–42.

Simon, R. and Makuch, R. W. (1984). A Nonparametric Graphical Representation of the Relationship between Survival and the Occurrence of an Event: Application to Responder Versus Nonresponder Bias, *Stat. Med.*, 3: 35–44.

Simon, R. and Wittes, R. E. (1985). Methodologic Guidelines for Reports of Clinical Trials, *Cancer Treat. Rep.*, 69: 1–3.

Simon, R., Wittes, R. E., and Ellenberg, S. S. (1985). Randomized Phase II Clinical Trials, *Cancer Treat. Rep.*, 69: 1375–1381.

Slud, E. V., Byar, D. P., and Green, S. B. (1984). A Comparison of Reflected versus Test-Based Confidence Intervals for the Median Survival Time Based on Censored Data, *Biometrics*, 40: 587–600.

Tannock, I. and Murphy, K. (1983). Reflections on Medical Oncology: An Appeal for Better Clinical Trials and Improved Reporting of Their Results, *J. Clin. Oncol.*, 1: 66–70.

Tonkin, K., Tritchler, D., and Tannock, I. (1985). Criteria of Tumor Response Used in Clinical Trials of Chemotherapy, *J. Clin. Oncol.*, 3: 870–875.

Tsiatis, A. A. (1980). A Note on a Goodness of Fit Test for the Logistic Regression Model, *Biometrika*, 67: 250–251.

Ury, H. K. and Fleiss, J. L. (1980). On Approximate Sample Sizes for Comparing Two Independent Proportions with the Use of Yates' Correction, *Biometrics*, 36: 347–351.

Warr, D., McKinney, S., and Tannock, I. (1984). Influence of Measurement Error on Assessment of Response to Anti-Cancer Chemotherapy and a Proposal for New Criteria of Tumor Response, *J. Clin. Oncol.*, 2: 1040–1046.

Weiss, G. B., Bunce, H., and Hokanson, J. A. (1983). Comparing Survival of Responders and Nonresponders after Treatment: A Potential Source of Confusion in Interpreting Cancer Clinical Trials, *Controlled Clin. Trials*, 4: 43–52.

Wittes, R. E., Marsoni, S., Simon, R., and Leyland-Jones, B. (1985). The Phase II Trial, *Cancer Treat. Rep.*, 69: 1235–1239.

Zelen, M. (1979). A New Design for Randomized Clinical Trials, *N. Engl. J. Med.*, 300: 1242–1245.

# 11 Design and Analysis of Clinical Trials with Group Sequential Stopping Rules

NANCY L. GELLER *Division of Biostatistics, Department of Epidemiology and Biostatistics, Memorial Sloan-Kettering Cancer Center, New York, New York*

STUART J. POCOCK *Department of Clinical Epidemiology and General Practice, The Royal Free Hospital School of Medicine, London, England*

## I. INTRODUCTION

Until recently, clinical trials were designed with the implicit assumption that a single analysis of the data would occur, after the trial was completed. Despite design, it was commonplace for analyses to be requested by investigators while the trial was ongoing. Statisticians usually complied, although it was generally known that with repeated analyses of accumulating data, it would become more and more likely to find "significant" results, even if the null hypothesis were true (Armitage, 1971; Armitage et al., 1969; McPherson, 1974; McPherson and Armitage, 1971).

From a clinical viewpoint, there are practical reasons for undertaking interim analyses. The trial should be stopped on grounds of efficiency and ethics if evidence of one treatment's superiority emerges in a randomized trial or if, in a single-armed trial, the agent under test is either clearly ineffective or so effective that further development should proceed. From a statistical viewpoint, aside from problems arising from multiple hypothesis testing on accumulating data, another objection to unplanned interim analysis is that the stopping rule is ill-defined. The decision following an interim analysis is quite complex (DeMets and Lan, 1984) and need not be either statistical or logical, but the view is taken here that there always is a decision.

It thus comes as a logical consequence of clinical trial reality that statisticians should work out clinical trial designs which allow for interim data analysis. Since the late 1970s, planned interim analyses for clinical trials have been theoretically justified through a variety of group sequential designs. These allow a limited number of analyses and yet maintain prespecified overall type I and type II errors.

This chapter focuses on the design and analysis of clinical trials with planned interim analysis, with emphasis on their practical statistical aspects. Several recent review papers from various perspectives are recommended to those wishing further details (DeMets and Lan, 1984; Gail, 1984; Geller and

Pocock, 1987). After a brief historical background and a description of multiple testing plans for single-armed trials, a number of group sequential designs for randomized trials are reviewed. Tables of nominal significance levels and required sample sizes for several plans are given. Some aspects of implementing such designs are then discussed, and examples of three clinical trials which use these procedures are presented.

### A. Sequential Procedures

Because patients enter clinical trials serially rather than simultaneously, their responses become available serially and so it is possible to analyze cumulative data sequentially, as they become available. Applicability to medical trials of the sequential analysis methods first developed by Wald (1947) was advocated by Armitage (1975) and more recently by Whitehead and Jones (1979) and Whitehead (1982). A recent example of a completely sequential trial was given by Jones et al. (1982).

Armitage mentioned several limitations of sequential trials. These were that response should be available soon after treatment is started, that monitoring is restricted to one endpoint (and early stopping would imply sacrificing a high degree of precision for any secondary comparisons), and that there would be organizational problems, such as coordination in multicenter trials and a much greater amount of work for the statistician. In addition, in much of the theory, an analysis occurs after each pair of observations, one in each treatment arm. While this appears to imply inherent pairing of observations, which many consider unrealistic, only balance overall is assumed.

Perhaps the early advocates of sequential trials were ahead of their time; for the advantage of group sequential trials over completely sequential trials is that the disadvantages described above arise only a few times, rather than after every pair of responses is recorded. In addition, the group sequential methodology blends well with clinical reality: most investigators do not want to evaluate results every time new data arrive, but do want to know every few months if the trial is worth the time and effort they are putting into it. Whitehead recently considered the analysis of secondary endpoints after a sequential trial has stopped (Whitehead, 1986) and his methods should be applicable also in the group sequential setting.

## II. MULTIPLE TESTING FOR SINGLE-ARMED TRIALS

A design which permits early stopping for single-armed trials which have a dichotomous endpoint, say response, was proposed by Schultz et al. (1973) and developed by Fleming (1981). In these procedures, the null hypothesis that the response proportion is $p_0$ or less is tested against the alternative hypothesis that the response proportion is $p_A$ or more. An indifference region, that is, the region between $p_0$ and $p_A$, of .15 or .2 in length allows plans to be tabulated (Fleming, 1981) which accrue between 10 and 50 patients in two or three stages and maintain an overall significance level close to .05 and overall power close to .8. In a k-stage procedure, $n_1$ patients initially would be evaluated and if there were either $a_1$ or fewer responses, the trial would stop, rejecting $H_A$. If there were $b_1$ ($>a_1$) or more responses, the trial would stop, rejecting $H_0$. If there were between $a_1$ and

$b_1$ responses, a second group of $n_2$ patients would be evaluated and a similar decision rule would apply. In the final stage of this procedure, a decision to reject either $H_0$ or $H_A$ has to be made, that is, $b_k = a_k + 1$. Thus, the values $a_i$ and $b_i$, $i = 1, 2, \ldots, k$, constitute the stopping boundaries for the trial. A biostatistical considerations section for a protocol for a trial designed according to this procedure is given in Geller (1984). The first example in Section IV,C is a trial designed according to this procedure.

Following hypothesis testing, a confidence interval for the proportion of responses is often of interest. If a multiple testing procedure has been used, the confidence interval calculation should use the information obtained from the earlier analyses. A method to obtain confidence intervals for the proportion of responses after termination of a single-armed trial designed according to Schultz et al. (1973) and Fleming (1981) was first proposed by Jennison and Turnbull (1983). Their confidence intervals depend on which boundary was crossed, the number of stages completed, and the number of responses. The interval coincides with the single-stage procedure confidence interval if the trial is stopped after the first analysis or if there are no responses. Otherwise, the farther from the boundary the trial outcome is, the greater the difference between the single-stage confidence intervals and Jennison and Turnbull's. Chang and O'Brien (1986) proposed an alternate ordering of the sample space using the likelihood principle, which yields shorter confidence intervals than the Jennison and Turnbull intervals. For both procedures, computer programs are needed and are available from the authors (Jennison and Turnbull, 1983; Chang and O'Brien, 1986).

Multiple testing procedures are especially recommended for "extended" Phase II trials, when it is reasonable to enter 15 or 20 patients initially and then to decide whether the trial should be continued. Such a strategy would be appropriate in a single-armed trial of efficacy of a combination chemotherapy regimen for cancer treatment, when each component of the combination is known to be active.

## III. GROUP SEQUENTIAL PROCEDURES FOR RANDOMIZED TRIALS

Many proposals have been forwarded for planned interim data analysis in randomized trials and a number of them will be described in this section. While such group sequential procedures differ in detail, they have certain common features. A group sequential design first gives a schedule which relates patient accrual to when the interim analyses will occur. This schedule is conveniently expressed in terms of the proportion of the maximal possible number of patients that the trial could accrue. Second, such designs give a sequence of statistics used to test the null hypothesis, and third, they give a stopping rule defined in terms of a monotone increasing sequence of nominal significance levels at which each test will be conducted. This sequence of significance levels is carefully chosen to maintain the overall type I error at some desired level (e.g., .05 or .10). Either the number of analyses is prespecified or the rate at which the overall significance level is "used up" is fixed in advance (Lan and DeMets, 1983). Thus, undertaking group sequential trials implies a willingness to perform hypothesis testing at nominal significance levels less than a prestated overall significance level, as well as recognition that if results are extreme enough, the trial should be stopped.

### A. Basic Group Sequential Plans

A number of group sequential plans which may be used with any asymptotically normal test statistic have been proposed (Geller and Pocock, 1986; Lan and DeMets, 1983; DeMets and Ware, 1980, 1982; Fleming et al., 1984; Haybittle, 1971; Jennison and Turnbull, 1984; Kim and DeMets, 1985b; Lan et al., 1984; O'Brien and Fleming, 1979; Peto et al., 1976; Pocock, 1977, 1982). Tables 1a and 1b show a number of sequences of nominal significance levels for two-, three-, four-, or five-stage group sequential trials planned to maintain an overall (two-sided) significance level of .05 (Table 1a) or .10 (Table 1b). Table 1b (1a) may also be used to generate nominal significance levels for group sequential trials with overall one-sided significance level .05 (.025), simply by halving the tabulated nominal significance level and looking up the corresponding normal critical value. It is useful to note that these nominal significance levels will always add up to more than the overall significance level, because with multiple significance testing, the probability of rejecting the null hypothesis does not accumulate additively.

The first column of Tables 1 a and 1b gives the number of stages, that is, the maximal number of analyses to be undertaken, here between two and five. The second column gives nominal significance levels of the original group sequential plans suggested by Pocock (1977). These use up the overall significance level equally each time an analysis is undertaken, in the sense of maintaining an equal conditional probability of rejecting the null hypothesis at each analysis, given that the null hypothesis is true and that it has not been rejected previously. The third column in Table 1a is based on an early observation of Haybittle (1971) and Peto et al. (1976), that if repeated analyses are undertaken at the .001 nominal significance level, the final analysis still can be undertaken at almost the .05 level, since repeated testing at the .001 level uses up very little of the overall significance level. The fourth column of Tables 1a and 1b, due to O'Brien and Fleming (1979), was motivated by the possibility that a trial designed to maintain overall significance level .05 (.10) using the original group sequential plan could run to completion, have a final p value between .016 and .05 (.039 and .10), and still not reject the null hypothesis. To overcome this disadvantage, O'Brien and Fleming (1979) suggested starting with a very stringent nominal significance level, making it very difficult to stop the trial early on, and ending with a nominal significance level very close to the overall significance level. The fifth column is an example of nominal significance levels which increase less steeply than O'Brien and Fleming's (1979) but are not totally flat like Pocock's (1977). A proposal of Fleming et al. (1984), the fifth column selected a final nominal significance level somewhat close to the overall significance level (in Table 1a, close to .04, and in Table 1b, close to .085) and used up the remainder of the overall significance level so that there is an equal unconditional probability of rejecting the null hypothesis (when it is true) at each prior test. The last column gives examples of nominal significance levels which are optimal, in the sense that they minimize the average sample number for the alternative hypothesis that could be detected with the given power (Pocock, 1982).

**Table 1a** Standardized Normal Deviates and Corresponding Nominal Significance Levels for Two-Sided Group Sequential Trials Which Maintain an Overall Significance Level of .05

| Stages | Pocock | Haybittle/ Peto | O'Brien/ Fleming | Fleming/ Harrington/ O'Brien | Optimal for .75 power |
|---|---|---|---|---|---|
| 2 | 2.178<br>.029 | 3.290<br>.001 | 2.797<br>.005 | 2.432<br>.015 | 2.280<br>.023 |
| | 2.178<br>.029 | 1.962<br>.050 | 1.977<br>.048 | 2.036<br>.042 | 2.101<br>.036 |
| 3 | 2.289<br>.022 | 3.290<br>.001 | 3.471<br>.0005 | 2.576<br>.010 | 2.520<br>.012 |
| | 2.289<br>.022 | 3.290<br>.001 | 2.454<br>.014 | 2.493<br>.013 | 2.306<br>.021 |
| | 2.289<br>.022 | 1.964<br>.050 | 2.004<br>.045 | 2.059<br>.039 | 2.134<br>.033 |
| 4 | 2.361<br>.018 | 3.290<br>.001 | 4.049<br>.0001 | 2.713<br>.007 | 2.769<br>.006 |
| | 2.361<br>.018 | 3.290<br>.001 | 2.863<br>.004 | 2.641<br>.008 | 2.418<br>.016 |
| | 2.361<br>.018 | 3.290<br>.001 | 2.338<br>.019 | 2.567<br>.013 | 2.329<br>.020 |
| | 2.361<br>.018 | 1.967<br>.049 | 2.024<br>.043 | 2.051<br>.040 | 2.146<br>.032 |
| 5 | 2.413<br>.016 | 3.290<br>.001 | 4.562<br>.00001 | 2.801<br>.005 | 2.988<br>.003 |
| | 2.413<br>.016 | 3.290<br>.001 | 3.226<br>.001 | 2.743<br>.006 | 2.537<br>.011 |
| | 2.413<br>.016 | 3.290<br>.001 | 2.634<br>.008 | 2.684<br>.007 | 2.407<br>.016 |
| | 2.413<br>.016 | 3.290<br>.001 | 2.281<br>.023 | 2.617<br>.009 | 2.346<br>.019 |
| | 2.413<br>.016 | 1.967<br>.049 | 2.040<br>.041 | 2.051<br>.040 | 2.156<br>.031 |

**Table 1b** Standardized Normal Deviates and Corresponding Nominal Significance Levels for Two-Sided Group Sequential Trials Which Maintain an Overall Significance Level of .10

| Stages | Pocock | Haybittle/ Peto | O'Brien/ Fleming | Fleming/ Harrington/ O'Brien | Optimal for .75 power |
|---|---|---|---|---|---|
| 2 | 1.875<br>.061 | 3.090<br>.002 | 2.373<br>.018 | 2.170<br>.030 | 1.990<br>.047 |
| | 1.875<br>.061 | 1.647<br>.10 | 1.678<br>.093 | 1.717<br>.086 | 1.789<br>.074 |
| 3 | 1.992<br>.046 | 3.090<br>.002 | 2.961<br>.003 | 2.326<br>.020 | 2.213<br>.027 |
| | 1.992<br>.046 | 3.090<br>.002 | 2.094<br>.036 | 2.220<br>.026 | 2.021<br>.043 |
| | 1.992<br>.046 | 1.649<br>.099 | 1.710<br>.087 | 1.739<br>.082 | 1.830<br>.067 |
| 4 | 2.067<br>.039 | 3.090<br>.002 | 3.466<br>.0005 | 2.475<br>.013 | 2.429<br>.015 |
| | 2.067<br>.039 | 3.090<br>.002 | 2.455<br>.014 | 2.383<br>.017 | 2.132<br>.033 |
| | 2.067<br>.039 | 3.090<br>.002 | 2.001<br>.045 | 2.294<br>.022 | 2.044<br>.045 |
| | 2.067<br>.039 | 1.651<br>.099 | 1.733<br>.083 | 1.731<br>.083 | 1.845<br>.065 |
| 5 | 2.122<br>.034 | 3.090<br>.002 | 3.912<br>.0001 | 2.570<br>.010 | 2.638<br>.008 |
| | 2.122<br>.034 | 3.090<br>.002 | 2.766<br>.006 | 2.493<br>.013 | 2.241<br>.025 |
| | 2.122<br>.034 | 3.090<br>.002 | 2.258<br>.026 | 2.418<br>.016 | 2.124<br>.034 |
| | 2.122<br>.034 | 3.090<br>.002 | 1.955<br>.054 | 2.337<br>.019 | 2.057<br>.040 |
| | 2.122<br>.034 | 1.653<br>.098 | 1.752<br>.084 | 1.728<br>.084 | 1.854<br>.064 |

Tables 1a and 1b were generated using numerical integration (Pocock, 1977, 1982). The reader should realize that there is a certain amount of selection required in the preparation of short tables of nominal significance levels, such as these, and that many other plans are available for consideration when a group sequential trial is being designed. Extensive tables of the nominal significance levels exemplified by the fifth column of Tables 1a and 1b are given by Fleming et al. (1984) and a number of sequences of nominal significance levels of the type in the sixth column of Tables 1a and 1b are given by Pocock (1982). Table 1a overlaps with Table 1 of Geller and Pocock (1986).

Tables 2a and 2b give the sample size per arm per stage required for group sequential designs corresponding to Tables 1a and 1b, in order to maintain overall power of .75, .80, and .90. For normally distributed endpoints, it is convenient to express the mean treatment difference to be detected, $\delta$, in units of the common standard deviation, $\sigma$. Thus, to calculate the sample size per arm per stage to be able to detect a mean treatment difference $\delta$ in the two arms with a certain power, the tabulated value must be multiplied by $\sigma^2/\delta^2$. In the case of a dichotomous endpoint, $\delta$ is the difference in proportions to be detected and $\sigma^2 = \pi(1 - \pi)$, where $\pi$ is the anticipated overall proportion.

For designing a one-sided group sequential trial at the overall .05 (.025) level, Table 2b (2a) will give an approximation for the required sample size, although sometimes a slight underestimate. For example, in order to detect a .125 decrease in the response proportion, that is, from .90 response to .775 response, with power .80, when undertaking a five-stage group sequential trial at the .05 overall significance level, using the plan of O'Brien and Fleming (fourth column of Table 1b, with nominal significance levels divided by 2), the required sample size per arm per stage, when actually computed (DeMets and Ware, 1982), would be 24. Use of Table 2b would give 23 patients per arm per stage gotten by calculating the difference to be detected, $\delta = .125$, $\pi = (.775 + .9)/2 = .8375$, $\sigma^2 = .8375\ (1 - .8375) = .13609375$, and multiplying $\sigma^2/\delta^2 = 8.71$ by 2.57, the entry in Table 2b (fifth column, second row from the bottom), and rounding up. The third example in Section IV,C makes use of these results.

In designing a trial, the reader may wonder how one would make a choice of a sequence of nominal significance levels and of sample size. One might argue in favor of the two types of plans in the rightmost columns in Tables 1a and 1b because one has some chance of stopping early, yet does not "use up" so much of the nominal significance level early on that the final analysis is at a nominal significance level too far from the overall significance level. Correspondingly, analyzing with few patients per arm can hardly be recommended: an early analysis may lead to early termination of the trial, and with few patients there will be very low power for secondary comparisons. The second example in Section IV,C illustrates some of the difficulties that can arise.

### B. Plans with Flexible Boundaries

The plans given in Tables 1a and 1b were devised assuming that analysis will occur at most k times (k = 2, 3, 4 or 5), exactly after a proportion 1/k of the maximal number of patients is evaluated. Since it is often preferred to analyze the data at other times, Lan and DeMets (1983) generalized

**Table 2a** Sample Size per Arm per Stage[a] for Designs Chosen from Table 1a, Which Have Power .75, .80, .90 to Detect a Mean Treatment Difference $\delta$

| Stages | Power | Pocock | Haybittle/ Peto | O'Brien/ Fleming | Fleming/ Harrington/ O'Brien | Optimal for .75 power |
|---|---|---|---|---|---|---|
| 2 | .75 | 7.74 | 6.95 | 7.00 | 7.20 | 7.44 |
| | .80 | 8.72 | 7.86 | 7.91 | 8.13 | 8.40 |
| | .90 | 11.56 | 10.52 | 10.58 | 10.88 | 11.17 |
| 3 | .75 | 5.43 | 4.64 | 4.71 | 4.86 | 5.02 |
| | .80 | 6.10 | 5.25 | 5.33 | 5.48 | 5.66 |
| | .90 | 8.06 | 7.02 | 7.12 | 7.31 | 7.53 |
| 4 | .75 | 4.20 | 3.48 | 3.56 | 3.63 | 3.78 |
| | .80 | 4.72 | 3.94 | 4.02 | 4.10 | 4.26 |
| | .90 | 6.22 | 5.27 | 5.37 | 5.47 | 5.66 |
| 5 | .75 | 3.44 | 2.79 | 2.86 | 2.90 | 3.03 |
| | .80 | 3.86 | 3.16 | 3.23 | 3.28 | 3.42 |
| | .90 | 5.07 | 4.22 | 4.32 | 4.37 | 4.54 |

[a]Multiply each entry by $\sigma^2/\delta^2$.

**Table 2b** Sample Size per Arm per Stage[a] for Designs Chosen from Table 1b, Which Have Power .75, .80, .90 to Detect a Mean Treatment Difference $\delta$

| Stages | Power | Pocock | Haybittle/ Peto | O'Brien/ Fleming | Fleming/ Harrington/ O'Brien | Optimal for .75 power |
|---|---|---|---|---|---|---|
| 2 | .75 | 6.06 | 5.38 | 5.18 | 5.58 | 5.79 |
| | .80 | 6.93 | 6.82 | 6.28 | 6.40 | 6.64 |
| | .90 | 9.51 | 8.57 | 8.69 | 8.84 | 9.13 |
| 3 | .75 | 4.27 | 3.59 | 3.69 | 3.76 | 3.93 |
| | .80 | 4.88 | 4.13 | 4.25 | 4.32 | 4.50 |
| | .90 | 6.65 | 5.72 | 5.85 | 5.95 | 6.18 |
| 4 | .75 | 3.32 | 2.70 | 2.79 | 2.81 | 2.96 |
| | .80 | 3.78 | 3.10 | 3.20 | 3.23 | 3.39 |
| | .90 | 5.14 | 4.29 | 4.42 | 4.45 | 4.66 |
| 5 | .75 | 2.72 | 2.16 | 2.24 | 2.25 | 2.37 |
| | .80 | 3.10 | 2.48 | 2.57 | 2.58 | 2.72 |
| | .90 | 4.21 | 3.44 | 3.55 | 3.56 | 3.73 |

[a]Multiply each entry by $\sigma^2/\delta^2$.

nominal significance levels to accommodate this need. In their scheme, only the maximal number of patients and not the total number of analyses needs to be fixed in advance. They introduce a "use function," a*(t), such that a*(0) = 0, t is the proportion of the maximal number of patients evaluated thus far, a*(t) is monotonically increasing, and a*(1) = $\alpha$, the overall significance level. The function a*(t) generates the cumulative significance level "used up" after the current analysis and is related to the sequence of nominal significance levels through a Brownian motion process. When k is fixed and $t_i = i/k$, appropriate Lan and DeMets use functions could be used to generate Tables 1a and 1b. As examples, for overall one-sided significance level $\alpha$, Lan and DeMets show that

$$a^*(0) = 0$$

$$a^*(t) = 2 - 2\,N(z_{\alpha/2}t^{-1/2}) \qquad (0 < t \leqslant 1)$$

approximates the O'Brien and Fleming (1979) boundary, where N is the cumulative standard normal distribution function, and that

$$a^*(t) = \alpha \ln[1 + (e - 1)t]$$

approximates the Pocock (1977) boundary. They also suggest the uniform use function

$$a^*(t) = \alpha t \qquad (0 < t \leqslant 1)$$

Conversely, when k is fixed, Lan and DeMets (1983) and Kim and DeMets (1987) convert sequences of nominal significance levels which were generated for equal accrual between analyses into corresponding sequences of nominal significance levels for analyses at other times. For example, for overall significance level .05, when five analyses will be at t = .3, .6, .8, .9, 1 (late analyses), the O'Brien and Fleming nominal normal deviates would become 3.93, 2.67, 2.29, 2.19, 2.08. Conversions of this type require a computer program obtainable from the authors (Lan and DeMets, 1983). Kim and DeMets (1987) observe that converted boundaries are not always monotonically decreasing and so violate the definition of use function. Of course, it would not be wise to design a trial in which it is harder to stop as further analysis is undertaken.

### C. Survival Analysis

The main difficulty in applying the above methods to survival data is staggered entry. With staggered entry, each time the data are reviewed, an observation that was censored at the least analysis may become a death, causing a reordering of the death times and necessitating a recalculation of any rank statistic (such as the logrank test) on which hypothesis testing was based. If all patients enter the trial simultaneously, death times could be viewed as accumulating data and the ordinary group sequential theory would apply.

Gail (1984) provides a good summary of group sequential methods for survival analysis. Here only the major results are summarized. The prespecified boundaries of group sequential theory can be applied to survival data if the analyses are undertaken after fixed increments of statistical information are obtained and the test statistic in use produces uncorrelated increments from one analysis to the next. In particular, the logrank test calculated after equal numbers of deaths can be used (Tsiatis, 1982), whereas Gehan's modified Wilcoxon test does not have asymptotically uncorrelated increments (Slud and Wei, 1982). For analyses undertaken at fixed calendar times, Gail, DeMets and Slud (1982) and DeMets and Gail (1985) show via simulations that the logrank test with the Pocock (1977) or O'Brien and Fleming (1979) boundaries yields approximately correct overall size and power. Using simulations, Canner (1977, 1984) provides nominal critical values and power for analysis of survival data at fixed calendar times with fixed or staggered entry of a preset number of patients, using a variety of boundaries and an exponential mortality model.

In monitoring survival data, a procedure which allows a treatment decision to be made early on does not consider that later follow-up could see a reversal in treatment trend, as would be the case if the hazard functions crossed. Instead, one may monitor merely to stop accrual to the trial ("accrual monitoring"), with the understanding that a treatment decision will be made once, and only after a prespecified amount of statistical information is obtained (Rubinstein and Gail, 1982). Assuming that patients remain on their assigned treatment until the required number of deaths is seen, Rubinstein and Gail (1982) show that it is possible to stop accrual early when using a test statistic with uncorrelated increments if a large treatment difference begins to emerge and yet maintain the fixed sample properties of the final inference.

### D. Estimation

To estimate treatment differences following a group sequential trial, one must consider that the observed treatment difference and its unadjusted confidence interval will be biased upward and that the bias will be greater when more looks have been taken and when the outcome is more extreme. The upward bias occurs because the information that the trial has stopped now and not at a prior analysis is ignored. A numerical method to calculate exact confidence intervals following group sequential tests with Pocock (1977) and O'Brien and Fleming (1979) boundaries is given by Tsiatis et al. (1984). By ordering the sample space in a manner similar to that used by Jennison and Turnbull (1984), these authors derive confidence intervals based on normal distribution theory which pull the naive confidence intervals toward zero and are no longer symmetric about the sample mean. They comment that their method is applicable to any asymptotically normal test statistic which has uncorrelated increments and for which the variance can be estimated consistently. As with Jennison and Turnbull's intervals, there is no adjustment if the trial is stopped after the first analysis. The Chang and O'Brien (1986) approach to confidence intervals is also applicable to this problem.

Rather than consideration of confidence intervals after the trial stops, Jennison and Turnbull (1984) calculate confidence intervals to be used as the trial proceeds. Because their confidence intervals are simultaneously valid for each stage of analysis, Jennison and Turnbull's intervals (1984) can be used as another approach to group sequential analysis. A trial would be stopped as soon as the treatment difference falls outside the repeated confidence intervals. The simultaneous nature of these intervals makes them unduly wide if one is interested in a single confidence interval after conduct of a group sequential trial.

Concerning point estimation, Whitehead (1982) observes that with a group sequential procedure, the maximum likelihood estimate of treatment difference is biased. As an alternative, he advocates median unbiased estimation, since the property of median unbiasedness is maintained no matter when the trial is stopped. He proposes median unbiased confidence intervals and tabulates them for his test with triangular boundaries.

Kim and DeMets (1985) use the confidence intervals of Tsiatis et al. (1984) and those of Whitehead (1982) along with the flexible boundaries of Lan and DeMets (1983) to recommend the following group sequential procedure. At each interim analysis, for whatever proportion of the maximum sample size available for analysis, determine a group sequential boundary and perform a significance test. When the boundary is crossed or the last analysis is completed, compute the upper and lower confidence limits and a point estimate of the treatment difference.

### E. Other Procedures: Stochastic Curtailment

Curtailed procedures were originally designed for trials without planned interim analysis (Lan et al., 1982). As a clinical trial is taking place, investigators might ask whether the outcome of the major hypothesis test has already been determined. If this were the case, it would not be sensible to continue the trial and so the trial would be curtailed. Of course, to consider whether a trial should be curtailed, investigators had to undertake an unplanned analysis, and this was ignored.

Recently, it was realized that it might well be worthwhile to curtail a clinical trial if it was extremely unlikely, but not necessarily certain, to either reject or not reject the null hypothesis. Halperin and co-workers (Lan et al., 1982; Halperin et al., 1982) suggested that a less conservative view than curtailing per se would be stochastic curtailing. This would mean that investigators would reject the null hypothesis prior to accumulating all of the data if, under the null hypothesis, the probability of rejecting the null hypothesis at the end of the trial were sufficiently large, say greater than $\gamma$; similarly, they would accept the null hypothesis early if, under an alternative hypothesis of interest, the probability of accepting the null hypothesis were sufficiently large, say greater than $\gamma'$. If a trial is designed at significance level $\alpha$ and power $1 - \beta$, then during the entire monitoring process, the probability of rejecting the null hypothesis when it is indeed true will not exceed $\alpha/\gamma$ and the probability of accepting the null hypothesis when the alternative of interest is true will not exceed $\beta/\gamma'$ (Lan et al., 1982). The computation for stochastic curtailing is generally complex and has been done for the difference in the proportion of responses using the approximate normality of the test statistic (Halperin et al., 1982).

It is only natural to combine stochastic curtailing with group sequential procedures. A proposal of this type has been made by Lan and Friedman (1986), who combine asymmetric group sequential testing (DeMets and Ware, 1980, 1982) with stochastic curtailing. Such a procedure would be useful if one were testing for a one-sided treatment effect but also would wish to stop if the new treatment did not have much chance of being superior. It is clear that further theoretical work is needed in this area.

# IV. PRACTICAL CONSIDERATIONS AND EXAMPLES

## A. How Extensive Should Interim Analyses Be?

The planned interim analyses should be simple, often involving only one major comparison. Without this precaution, an interim analysis can become a full-blown analysis and all control of the overall significance level is lost.

Stratification on known prognostic factors, preferably using the method of Pocock and Simon (1975) and White and Freedman (1978), can facilitate this. Otherwise, balance on known prognostic factors should be checked before the primary endpoint is compared in the two treatment arms. It is best that a list of prognostic factors be specified before any analysis is undertaken; if an imbalance is found, the test statistic for the primary comparison should be modified to account for the imbalance(s). Without such poststratification, results may be completely misleading. For example, one could make an incorrect decision about stopping a trial.

## B. Who Should Decide Whether a Trial Should Stop?

When a clinical trial has a group sequential stopping rule, the question of who monitors the trial, that is, who knows the results of the interim analyses and who makes the decision of whether or not the trial should stop, arises. The "N.I.H. model" (DeMets, 1984), which assembles a monitoring committee consisting of individuals with diverse backgrounds—statistical, ethical, medical—and various degrees of involvement with the trial, may seem overly cumbersome, especially in single-institution trials. On the other hand, leaving the decision to stop the trial to the principal investigator alone or to the principal investigator and the statistician is not optimal. The principal investigator may be eager to end the trial and report results and the statistician may be eager to apply the stopping rule literally. These two are often too close to the data to be objective. It is wise to have some individuals not directly involved in the trial as members of the monitoring committee to which results should be presented. The committee may be small: two clinicians in addition to the principal investigator and statistician may suffice in a single-institution trial. The members of the monitoring committee who are not involved in the trial should be convinced by those involved in the trial that their current decision on stopping or continuing is justified. This committee can also ensure that interim results of a trial are not reported publicly, which can influence future accrual and trial outcome in ways which are completely uncontrollable (Geller and Pocock, 1987). To ensure objectivity, the committee should be selected before any analysis is undertaken. The third example in Section IV,C sets up such a committee within the biostatistical considerations section of a clinical trial protocol.

## C. Examples

Three examples are presented. The first two are of trials already completed. The first is a single-armed cancer trial designed according to Fleming (1981) and the second is a randomized trial in an infectious disease, in which interim analysis took place when only a few patients were enrolled. The third is an example of a design for a randomized trial in cancer which has not yet begun. Other examples are given by Geller and Pocock (1987) and DeMets (1984).

### 1. 4'-Deoxydoxorubicin for Advanced Breast Cancer (Leitner et al., 1985)

A single-arm Phase II trial of 4'-deoxydoxorubicin, a structural analogue of doxorubicin, the best single chemotherapeutic agent known for the treatment of advanced breast cancer, was designed according to Fleming's (1981) multiple testing procedure. Because the proportion of patients responding to doxorubicin was believed to be .25 and it was considered important not to reject the drug if it was active, the trial was designed to test the null hypothesis that the proportion responding to the new agent was at most .05 against the alternative that the response proportion was at least .20. At most three groups of 10 patients each were to be enrolled. If, among the first 10 patients, there were at least 3 responses, the trial would stop, rejecting the null hypothesis. Otherwise, 10 more patients would be evaluated. If, among the 20 patients, there were 0 or 1 responses, the trial would stop, rejecting the alternative hypothesis. If, on the other hand, there were at least 4 responses among the 20 patients, the trial would stop, rejecting the null hypothesis. If there were 2 or 3 responses among the first 20 patients, a final 10 patients would be evaluated. If at least 4 of the 30 patients responded the null hypothesis would be rejected, and otherwise the alternative hypothesis would be rejected. This design had overall power of .86 to distinguish between the null and alternative hypotheses and an overall significance level of .06.

This specific design was chosen because the only possibility of stopping after 10 patients was if the new agent yielded at least 3 responses. The investigators rejected an alternative design which would allow stopping the trial if there were no responses in the first 10 patients, since with 0 responses out of 10 patients, the 95% confidence limits for response would have been 0 to .26. Further, since it was considered important not to conclude that an active agent was inactive, a design with high power was chosen.

In this trial, there were no responses in the first 10 patients, so another 10 patients were accrued. Again there were no responses. The conclusion was to reject the alternative hypothesis: the proportion responding to deoxydoxorubicin was at most .20. A 95% confidence interval for the proportion of responses was 0 to .16 (Blyth and Still, 1983). It follows that there would be limited further interest in this agent for advanced breast cancer.

### 2. FIAC versus Ara-A for Herpes Zoster (Leyland-Jones et al., 1986)

To compare the efficiacy of FIAC, a new treatment for herpes zoster, to conventional treatment, Ara-A, a double-blind, randomized trial was designed to enroll 40 patients per arm in order to be able to detect a 60% response in one arm versus a 90% response in the other arm with power .9 when a two-sided $\chi^2$ test was undertaken at the .05 level. No interim analyses were planned. Initially, "response" was defined as improvement in subcutaneous disease. For this example, suppose that response is taken to mean crusting of lesions within 72 hours and a decreased number of days until the appearance of the last new lesions.

Because accrual was lagging, an unplanned interim analysis was undertaken after eight FIAC patients and nine Ara-A patients were evaluated. FIAC was found to be superior to Ara-A with respect to both crusting of lesions within 72 hours (P = .011) and number of days to last new lesion (logrank test, P = .00004). Although results were striking, no stopping rule was in place and, because of the small sample size, a decision was made to continue the trial.

At this time, the trial was redesigned as a group sequential trial, allowing up to four analyses, including the one described. The O'Brien and Fleming procedure was chosen because of its stringency at the first interim analysis (Geller and Pocock, 1986): a P-value of .0001 would have been required at the first interim analysis (Fleming, 1981). Since this was not met for the crusting endpoint, proceeding to a second analysis appeared appropriate.

After 15 FIAC patients and 16 Ara-A patients were evaluated, a second interim analysis was undertaken. To stop the trial at that analysis, a P-value of .004 was required. A higher proportion of crusting within 72 hours was seen on the FIAC arm (P = .0009) as well as a shorter number of days to last new lesion (logrank P = .00004) and the trial was stopped.

Several comments about this trial can be made in retrospect. Redesign of trials in progress to allow interim analyses is sensible and is discussed further in Geller and Pocock (1986). This trial was redesigned before Lan and DeMets (1983) appeared; however, the slight inequality in numbers of patients accrued between analyses would have had only a sight effect on the nominal significance levels required. Last, a trial of this type did not have a single, obvious endpoint and so additional problems of multiple testing arose. In requiring the nominal significance level to be achieved by each endpoint, the reported analysis was somewhat stringent, since the endpoints were positively correlated.

### 3. Combination Chemotherapy for Advanced Germ Cell Tumors

A group sequential trial of two combination chemotherapy regimens for the treatment of "good risk" patients with disseminated germ cell tumors was designed in order to compare the conventional in-hospital treatment with a new combination which could be administered on an outpatient basis. The biostatistical considerations section written for the protocol is given as an example because many of the recommendations described in this chapter were included.

#### SAMPLE BIOSTATISTICAL CONSIDERATIONS FOR A GROUP SEQUENTIAL TRIAL

1. Major endpoint, power and significance level: The major endpoint of this trial is complete response to chemotherapy (CR). In good risk patients, the current CR to the conventional therapy (Arm A) is approximately 90%. We will consider the new treatment (Arm B) to be comparable if its CR is no lower than 77.5%. Thus, the alternative hypothesis will be that the CR to treatment B is lower than the CR to treatment A. The trial will be designed as a group sequential trial, so that early stopping, before the maximum number of patients is accrued, is possible. A sufficient number of

patients will be entered in order to be able to detect the stated difference in CR with overall power .8 when hypothesis testing is undertaken to maintain an overall significance level of .05.

2. Group sequential design: This trial is designed according to a one-sided asymmetric group sequential plan (DeMets and Ware, 1982). As the data accumulate, at most five analyses will be undertaken. Assuming that these take place after successive groups of 24 patients per arm have been entered, the sequence of nominal significance levels will be .00005, .003, .013, .027 and .042. The first hypothesis test (based on 24 patients per arm) will be undertaken at the .00005 level, and if the difference in CRs in the two arms is not extreme enough to stop the trial, a second group of 24 patients per arm will be accrued and a second analysis undertaken (based on 48 patients per arm) at the .003 level, etc. Accrual of the required number of patients is anticipated within three years.

3. Randomization: After eligibility is established, patients will be randomized via a telephone call to the randomization office. Randomization will be by the method of randomly permuted blocks, stratified for patient institution (Simon, 1979).

4. Interim analyses: Fisher's exact test will be used to compare the proportion of CRs in each arm. Balance in hCG, AFP, LDH and number of sites of metastasis will be checked at each interim analysis. If any imbalance is found, a stratified Fisher's exact test (Zelen, 1971) will be used to compare the proportion of CRs in each arm. Toxicities in the two arms will be tabulated. Beginning at the third interim analysis, time to relapse will also be monitored.

5. Timing of interim analyses: The interim analyses may be rescheduled at 6, 12, 18, 30 months and at the end of the accrual period and then the sequence of nominal significance levels will be modified (since an unequal number of patients will have been accrued between analyses) using the computer program of Lan and DeMets (1983).

6. Decision to stop the trial: The interim analyses will be reported to a data monitoring committee, consisting of the principal investigator, the statistician and two other clinicians (not involved in this trial). After an interim analysis is completed, the data monitoring committee will decide if the trial should stop. Results of the trial will not be reported at a professional meeting until a decision to stop the trial has been made.

7. Final analysis: Time to relapse, survival time and "event-free survival" (time to relapse or death, whichever comes first) will be estimated using the method of Kaplan and Meier (1958) and comparisons of the two treatment arms will be made using the logrank test, stratified if appropriate (Mantel, 1966).

## V. CONCLUSIONS

A number of methods for planned interim analyses for trials of various types have been described. Although several unsolved problems remain and this area is still in rapid development, at this time a sufficient variety of techniques are available that planned interim analyses should be considered whenever a clinical trial is designed. Since few large trials do not require an analysis of the data while the trial is in progress, planning for such eventualities is in order.

## ACKNOWLEDGMENT

We appreciate the assistance of Eva Y. W. Chan in generating Tables 1a, 1b, 2a, and 2b.

## REFERENCES

Armitage, P. (1971). *Statistical Methods in Medical Research*, Blackwell, Oxford.

Armitage, P. (1975). *Sequential Medical Trials*, 2nd ed., Blackwell, Oxford.

Armitage, P., McPherson, C. K., and Rowe, B. C. (1969). Repeated Significance Tests on Accumulating Data, *J. R. Stat. Soc. Ser. A*, 132: 235.

Blyth, C. R. and Still, H. A. (1983). Binomial Confidence Intervals, *J. Am. Stat. Assoc.*, 78: 108.

Canner, P. L. (1977). Monitoring Treatment Differences in Long-Term Clinical Trials, *Biometrics*, 33: 235.

Canner, P. L. (1984). Monitoring Long-Term Clinical Trials for Beneficial and Adverse Treatment Effects, *Commun. Stat.—Theory Methods*, 13: 2374.

Chang, M. N. and O'Brien, P. C. (1986). Confidence Intervals Following Group Sequential Tests, *Controlled Clin. Trials*, 7: 18.

DeMets, D. L. (1984). Stopping Guidelines Versus Stopping Rules: A Practitioner's Point of View, *Commun. Stat.—Theory Methods*, 13: 2395.

DeMets, D. L. and Gail, M. H. (1985). Use of Logrank Tests and Group Sequential Methods at Fixed Calendar Times, *Biometrics*, 41: 1039.

DeMets, D. L. and Lan, K. K. G. (1984). An Overview of Sequential Methods and Their Application in Clinical Trials, *Commun. Stat.—Theory Methods*, 13: 2315.

DeMets, D. L. and Ware, J. H. (1980). Group Sequential Methods for Clinical Trials with a One-Sided Hypothesis, *Biometrika*, 67: 651.

DeMets, D. L. and Ware, J. H. (1982). Asymmetric Group Sequential Boundaries for Monitoring Clinical Trials, *Biometrika*, 69: 661.

Fleming, T. R. (1981). One-Sample Multiple Testing Procedure for Phase II Trials, *Biometrics*, 38: 143.

Fleming, T. R., Harrington, D. P., and O'Brien, P. C. (1984). Designs for Group Sequential Tests, *Controlled Clin. Trials*, 5: 348.

Gail, M. H. (1984). Nonparametric Frequentist Proposals for Monitoring Comparative Survival Studies, *Handbook of Statistics*, vol. 14 (P. Krishnaiah, ed.)

Gail, M. H., DeMets, D. L., and Slud, E. V. (1982). Simulation Studies on Increments of the Two-Sample Logrank Score Test for Survival Time Data, with Application to Group Sequential Boundaries, *Survival Analysis* (J. Crowley and R. A. Johnson, eds.), Institute of Mathematic Statistics Lecture Notes-Monograph Series, Hayward, Calif., pp. 287–301.

Geller, N. L. (1984). Design of Phase I and II Clinical Trials in Cancer: A Statistician's View, *Cancer Invest.*, 2: 483.

Geller, N. L. and Pocock, S. J. (1987). Interim Analyses in Randomized Clinical Trials: Ramifications and Guidelines for Practitioners, *Biometrics*, 43: 213.

Halperin, M., Lan, K. K. G., Ware, J. H., Johnson, N. J., and DeMets, D. M. (1982). An Aid to Data Monitoring in Long-Term Clinical Trials, *Controlled Clin. Trials*, 3: 311.

Haybittle, J. L. (1971). Repeated Assessment of Results in Clinical Trials in Cancer Treatment, *Br. J. Radiol.*, 44: 793.

Jennison, D. and Turnbull, B. W. (1983). Confidence Intervals for a Binomial Parameter Following a Multistage Test with Application to MIL-STD 105D and Medical Trials, *Technometrics*, 25: 49.

Jennison, C. and Turnbull, B. (1984). Repeated Confidence Intervals for Group Sequential Trials, *Controlled Clin. Trials*, 5: 33.

Jones, D. R., Newman, C. E., and Whitehead, J. (1982). The Design of a Sequential Clinical Trial for the Comparison of Two Lung Cancer Treatments, *Stat. Med.*, 1: 73.

Kim, K. and DeMets, D. L. (1985a). Estimation Following Group Sequential Tests in Clinical Trials, Techn. Rep. No. 32, Wisconsin Clinical Cancer Center, University of Wisconsin-Madison.

Kim, K. and DeMets, D. L. (1987). Design and Analysis of Group Sequential Tests Based on the Type I Error Spending Rate Function, *Biometrika*, 74: 149.

Kaplan, E. L. and Meier, P. (1958). Nonparametric Estimation from Incomplete Observations, *J. Am. Stat. Assoc.*, 53: 457.

Lan, K. K. G. and DeMets, D. L. (1983). Discrete Sequential Boundaries for Clinical Trials, *Biometrika*, 70: 659.

Lan, K. K. G. and Friedman, L. (1986). Monitoring Boundaries for Adverse Effects in Long-Term Clinical Trials, *Controlled Clin. Trials*, 7: 1.

Lan, K. K. G., Simon, R., and Halperin, M. (1982). Stochastically Curtailed Tests in Long-Term Clinical Trials, *Commun. Stat.—Sequential Analysis*, 1: 207.

Lan, K. K. G., DeMets, D., and Halperin, M. (1984). More Flexible Sequential and Non-Sequential Designs in Long-Term Clinical Trials, *Commun. Stat.—Theory Methods*, 13: 2339.

Leitner, S. P., Casper, E. S., Hakes, T. B., Kaufman, R. J., Winn, R. J., Scoppetuolo, M., Raymond, V., Geller, N. L., and Young, C. W. (1985). Phase II Trial of 4'-Deoxydoxorubicin in Patients with Advanced Breast Cancer, *Cancer Treat. Rep.*, 69: 1319.

Leyland-Jones, B., Donnelly, H., Groshen, S., Myskowski, P., Donner, A. L., Fanucchi, M., Fox, J., and the Memorial Sloan-Kettering Antiviral Working Group. (1986). 2'-Fluoro-5-iodoarabinosylcytosine, a New Potent Antiviral Agent: Efficacy in Immunosuppressed Individuals with Herpes Zoster, *J. Infect. Dis.*, 154: 430.

Mantel, N. (1966). Evaluation of Survival Data and Two New Rank Order Statistics Arising in Its Consideration, *Cancer Chemother. Rep.*, 50: 163.

McPherson, C. K. (1974). Statistics: The Problem of Examining Accumulating Data More Than Once, *N. Engl. J. Med.*, 290: 501.

McPherson, C. K. and Armitage, P. (1971). Repeated Significance Tests on Accumulating Data When the Null Hypothesis Is Not True, *J. R. Stat. Soc. Ser. A*, 134: 15.

O'Brien, P. C. and Fleming, T. R. (1979). A Multiple Testing Procedure for Clinical Trials, *Biometrics*, 35: 549.

Peto, R., Pike, M. C., Armitage, P., Breslow, N. E., Cox, D. R., Howard, S. V., Mantel, N., McPherson, K., Peto, J., and Smith, P. G. (1976). Design and Analysis of Randomized Clinical Trials Requiring Prolonged Observation of Each Patient. Part I. Introduction and Design, *Br. J. Cancer*, 34: 585.

Pocock, S. J. (1977). Group Sequential Methods in the Design and Analysis of Clinical Trials, *Biometrika*, 64: 191.

Pocock, S. J. (1982). Interim Analyses for Randomized Clinical Trials: The Group Sequential Approach, *Biometrics*, 38: 153.

Pocock, S. J. and Simon, R. (1975). Sequential Treatment Assignment with Balancing for Prognostic Factors in the Controlled Clinical Trial, *Biometrics*, 31: 103.

Rubinstein, L. V. and Gail, M. H. (1982). Monitoring Rules for Stopping Accrual in Comparative Survival Studies, *Controlled Clin. Trials*, 3: 325.

Schultz, J. R., Nichol, F. R., Elfring, G. L., and Weed, S. D. (1973). Multiple Stage Procedures for Drug Screening, *Biometrics*, 29: 293.

Simon, R. (1979). Restricted Randomization Designs in Clinical Trials, *Biometrics*, 35: 503.

Slud, E. and Wei, L. J. (1982). Two-Sample Repeated Significance Tests Based on the Modified Wilcoxon Statistic, *J. Am. Stat. Assoc.*, 77: 862.

Tsiatis, A. A. (1982). Repeated Significance Testing for a General Class of Statistics Used in Censored Survival Analysis, *J. Am. Stat. Assoc.*, 77: 855.

Tsiatis, A. A., Rosner, G. L., and Mehta, C. R. (1984). Exact Confidence Intervals Following a Group Sequential Test, *Biometrics*, 40: 797.

Wald, A. (1947). *Sequential Analysis*, Wiley, New York.

White, S. J. and Freedman, L. S. (1978). Allocation of Patients to Treatment Groups in a Controlled Clinical Study, *Br. J. Cancer*, 37: 849.

Whitehead, J. (1982). *The Design and Analysis of Sequential Clinical Trials*, Ellis Horwood, Chichester.

Whitehead, J. (1986). Supplementary Analysis at the Conclusion of a Sequential Clinical Trial, *Biometrics*, 42: 461.

Whitehead, J. and Jones, D. (1979). The Analysis of Sequential Clinical Trials, *Biometrika*, 66: 443.

Whitehead, J. and Stratton, I. (1983). Group Sequential Clinical Trials with Triangular Continuation Regions, *Biometrics*, 39: 227.

Zelen, M. (1971). The Analysis of Several Two by Two Contingency Tables, *Biometrika*, 58: 129.

# 12 Applications of Interval Inference

**A. LAWRENCE GOULD** *Biostatistics and Research Data Systems, Merck Sharp & Dohme Research Laboratories, West Point, Pennsylvania*

## I. INTRODUCTION

### A. Preamble

Mainland (1960, 1967) categorized investigations as *experiments* or *surveys*. Experiments deal with questions or hypotheses identified in advance. Surveys seek associations or relationships that lead to the particular questions or hypotheses addressed by analyses of data from subsequent experiments. As a rule, investigations related to pharmaceutical development have both survey and experimental aspects.

These distinctions are important for statisticians concerned with the development of pharmaceuticals for use in humans. The development of any pharmaceutical agent aims first to identify, then to establish, its claim structure. Market and medical needs and the results from early (clinical pharmacology, Phase I, Phase II) trials identify a provisional claim structure that is confirmed and refined by the results from late Phase II and Phase III trials. Especially during these latter stages, it is essential to distinguish between the findings providing answers to questions identified in the design of the trial, and the findings reflecting quirks about the behavior of the agent unsuspected from the results of the earlier trials. These two kinds of findings have different implications for the claim structure: as a rule, only the former can provide the support for particular claims (e.g., identifying the population and conditions for which the agent is indicated); the latter usually will be expressed only as negative findings, e.g., as warnings about events that have been observed in the trials but are not confirmed as associated with the agent. Effective analysis of the results of any trial requires a clear understanding of the questions that the trial is intended to answer (which may include, in the early stages, questions about the existence of associations to be confirmed or denied by later trials). How the protocols for the sequence of trials in the development of any pharmaceutical agent should address these questions is discussed elsewhere (e.g., Chapter 6), although ensuring that they are addressed thoroughly and clearly is an important responsibility of pharmaceutical statisticians. The need to provide answers to carefully specified questions defines (or should) the analytic approach. Usually, but not always, standard methods will be appropriate. Often, however, modifications of these methods will be needed; many standard methods and modifications are described in this book. Interval inference methods (e.g., confidence intervals) comprise a body of techniques that are well known, but not often described in the context of pharmaceutical trial design and analysis. This chapter describes some applications of interval inference methods to issues that have arisen in practice, with the aim of illustrating their applicability and encouraging their use in answering relevant questions.

## B. Interval Inference

Pharmaceutical data are analyzed most commonly in practice by hypothesis-testing or point estimation procedures, many of which are described elsewhere in this book. These procedures answer questions such as "do the data support rejection of the null hypothesis at some specified level of significance (e.g., 5%)?" or "what is the best estimate of the population mean value of some clinically relevant parameter (e.g., the reduction in supine diastolic blood pressure after 8 weeks of treatment with some agent)?". The simple answer ("yes/no" or a single computed value) these approaches give often will not answer adequately the complex questions that commonly arise in practice. The following examples illustrate situations for which hypothesis-testing and point estimation procedures would be inappropriate.

Example 1. The following table summarizes the results after 4 weeks of treatment from several placebo-controlled trials of anti-ulcer agents in the treatment of acute gastric ulcer. Are the various agents equally effective?

| | Trial 1 | | Trial 2 | | Trial 3 | | Trial 4 | |
|---|---|---|---|---|---|---|---|---|
| | Act. | Pbo. | Act. | Pbo. | Act. | Pbo. | Act. | Pbo. |
| No. at risk | 149 | 145 | 36 | 33 | 20 | 20 | 123 | 75 |
| No. healed | 70 | 45 | 19 | 9 | 11 | 6 | 80 | 33 |
| % healed | 47% | 31% | 53% | 27% | 55% | 30% | 65% | 44% |

Example 2. In a published trial of two anti-inflammatory agents, 8 of 19 (42%) of the patients on a new agent and 5 of 20 (25%) of the patients on a standard agent felt that their osteoarthritis had improved after 8 weeks of treatment. How large a sample would be required for a definitive trial (i.e., one that had 80% power for detecting the difference that appears to exist)?

Example 3. More than one-third of the patients in a trial comparing an active agent against placebo in their ability to prevent cystoid macular edema following cataract extraction surgery failed to provide follow-up angiograms. What results might have been expected for these patients, and how might these results have affected the conclusions?

Example 4. A new agent for the treatment of congestive heart failure was evaluated in two trials; one compared the agent to placebo, the other to a standard agent. The latter trial contained no placebo group, but if it had, how confident can one be that the outcome for the new agent would have been significantly superior to the placebo group outcome?

Explicit incorporation of the underlying uncertainty is the point of interval inference techniques in general and the applications described in this chapter in particular. Confidence intervals are the examples of interval inference techniques most familiar to most statisticians. Other examples include prediction intervals, many Bayesian techniques, etc.

### C. Contents of Chapter

This chapter describes the following methods that have been developed in response to particular data analytic problems that have arisen in practice: (1) response rate ratio methods, (2) sample size estimation incorporating prior uncertainty, (3) assessing robustness of conclusions with many missing values, and (4) inferring the likelihood of a significant difference between the outcome observed for a treatment in an active-control study and the outcome that might have been observed for a placebo group had one been included.

## II. RESPONSE RATE RATIO METHODS

When the key response variables in a trial take a finite number of discrete values (e.g., two or three), the outcomes of two or more treatment groups can be compared usefully in terms of the *ratio* of the response rates or some function of them. These comparisons lead directly to statements such as "A response to treatment A is at least $1\frac{1}{2}$ times as likely as a response to treatment B." Comparisons of this sort often are of central interest in the development of therapeutic agents, especially in anti-inflammatory agent development, where they provide a basis for assessing the outcome for a new agent relative to aspirin, a recognized standard treatment. They also provide a simple, direct way to assess treatment equivalence; for example, an analysis revealing that the ratio of the response rates for two treatments is very likely to lie between 0.9 and 1.1 certainly strongly supports an assertion that the treatments are essentially equivalent with respect to the response. The development of response rate ratio analyses also leads to technically interesting and operationally useful generalizations, including some applications to trial design.

### A. Dichotomous Responses, Two Populations

Let X and Y each have binomial distributions, $X \sim B(m, p)$ and $Y \sim B(n, q)$. Differences between the true values of p and q may be expressed in various ways, e.g., the arithmetic difference $p - q$, the ratio $p/q$, and the odds ratio $p(1 - q)/q(1 - p)$. Of these, the ratio seems to have the greatest practical appeal: it is easy to explain to nonstatisticians, the conclusions resulting from its analysis readily can be expressed in plain language and directly related to relevant hypotheses, the necessary computations are easily programmed and numerically stable, etc.

The confidence distribution of the response rate ratio can be derived easily using a Bayesian argument (Aitchison and Bacon-Shone, 1981) that adapts readily to the confidence case. Suppose that p has a beta(a, b) prior distribution, and that q has a beta(c, d) prior distribution. Then

$$\Pr\{p \leqslant p^*\} = I_{p^*}(a, b) \qquad \text{and} \qquad \Pr\{q \leqslant q^*\} = I_{q^*}(c, d)$$

where, for computational convenience, b and d are integers (not an important restriction, since the prior distribution rarely has to be specified exactly), and $I_p$ denotes the incomplete beta function. If x denotes a realization of X and y a realization of Y, then the posterior density of the ration $R = p/q$ can be determined easily:

$$f_R(r;\, m, n, x, y, a, b, c, d) \tag{II.1}$$

$$= C \sum_{j=0}^{b+m-x-1} \binom{b+m-x-1}{j} B(a + c + x + y,\, d + j + n - y) r^{a+j+x-1}$$

$$\times (1 - r)^{b+m-x-j-1} \qquad r \leqslant 1$$

where $C = [B(a+x,\, b+m-x)B(c+y,\, d+n-y)]^{-1}$ and $B(a, b) = \Gamma(a)\Gamma(b)/\Gamma(a+b)$. For $r > 1$, the density of R can be obtained by using (II.1) to compute the density of $U = R^{-1}$ and then performing the variable transformation $U \to U^{-1} = R$. The distribution function of R follows immediately from (II.1):

$$\Pr\{R \leqslant r \mid m, n, x, y, z, b, c, d\} \tag{II.2}$$

$$= C \sum_{j=0}^{b+m-x-1} \binom{b+m-x-1}{j} B(a + c + x + y,\, d + j + n - y)$$

$$\times B(a + x + j,\, b + m - x - j)\, I_r(a + x + j,\, b + m - x - j) \qquad r \leqslant 1$$

$$= 1 - C \sum_{j=0}^{d+n-y-1} \binom{d+n-y-1}{j} B(a + c + x + y,\, b + j + m - x)$$

$$\times B(c + y + j,\, d + n - y - j)\, I_{1/r}(c + y + j,\, d + n - y - j) \qquad r \geqslant 1$$

The sums in (II.1) and (II.2) will be finite only if b and d are integers, which is the computational reason for assuming that they are. The result (II.2) also applies in a regular confidence context, since confidence statements about p or q (separately) can be approximated reasonably well by assuming that the confidence density of p or q is a beta density as above with $a = b = c = d = 1$ (Gould, 1983). The calculations required for (II.2) can be programmed easily with a reliable subroutine for calculating values of the incomplete Beta function (e.g., Majumdar and Bhattacharjee, 1973). Because all of the terms of (II.1) and (II.2) are positive, the calculations for the confidence (or posterior) density and distribution function of the response rate ratio R are numerically stable.

Example. The following table summarizes the results after 4 weeks of treatment from four placebo-controlled trials of anti-ulcer agents in the treatment of acute gastric ulcer.

| | Trial 1 | | Trial 2 | | Trial 3 | | Trial 4 | |
|---|---|---|---|---|---|---|---|---|
| | Act. | Pbo. | Act. | Pbo. | Act. | Pbo. | Act. | Pbo. |
| No. at risk | 149 | 145 | 36 | 33 | 20 | 20 | 123 | 75 |
| No. healed | 70 | 45 | 19 | 9 | 11 | 6 | 80 | 33 |
| % healed | 47% | 31% | 53% | 27% | 55% | 30% | 65% | 44% |

The question, "Are the various agents equally effective?" can be answered easily and unambiguously with the 95% confidence intervals for the ratios of the healing rates (active/placebo) for each trial:

| | Rate ratio (active/placebo) | 95% confidence interval |
|---|---|---|
| Trial 1 | 1.51 | 1.13, 2.04 |
| Trial 2 | 1.94 | 1.06, 3.66 |
| Trial 3 | 1.83 | 0.87, 3.96 |
| Trial 4 | 1.48 | 1.13, 1.99 |

The answer is evident: the healing rates (relative to placebo) in the four trials are essentially identical.

When, as here, the response rate ratios are nearly identical, it may be useful to provide overall point and interval estimates for the response rate ratio. This can be accomplished in at least two ways. First, one could start with the joint likelihood expressed as the product of the likelihoods from the constituent trials,

$$\prod_i \binom{m_i}{x_i}\binom{n_i}{y_i} p^{x_i}(1-p)^{m_i-x_i} q^{y_i}(1-q)^{n_i-y_i}$$

and proceed as described above; the result would be expressions (II.1) and (II.2) with (m, x, n, y) replaced by ($\Sigma_i m_i$, $\Sigma_i x_i$, $\Sigma_i n_i$, $\Sigma_i y_i$), respectively. This approach assumes that common values of p and q apply for each trial, i.e., that there is no trial-to-trial variation in the response rates, resulting in a misleadingly narrow interval estimate of the common response ratio.

Constancy of the response rate ratio across trials does not require constancy of the response rates themselves across trials. The response rate ratio will not be affected if the response rates in any trial are multiplied by a trial-specific constant that might reflect the circumstances peculiar to that trial. For example, the response rate ratios in trials 1 and 4 above are very nearly the same (about 1.5), yet the response rates in trial 4 are about 1.4 times the response rates in trial 1.

An alternative method for obtaining overall estimates arises from replacing the assumption of response rate constancy with the weaker assumption that a common process generates the response rates for each trial; this assumption allows for trial-specific response rates and is discussed more fully in Section V. Under this assumption the information about the overall value of the response rate ratio is expressed in terms of the process that generates response rates according to the density

$$g(p, q; a, b, c, d) = B^{-1}(a, b)\, B^{-1}(c, d)\, p^{a-1}(1-p)^{b-1} q^{c-1}(1-q)^{d-1}$$

It follows immediately from the development leading to (II.1) and (II.2) that the process generates response rate ratios according to a density given by (II.1) with $x = y = m = n = 0$. Therefore, given values of a, b, c, and d, (II.2) can be used to provide an overall prediction or confidence interval for the response rate ratio. An overall point estimate of the ratio may be obtained by substituting the values of a, b, c, and d into the expression for the response rate ratio expectation (Hora and Kelly, 1983),

$$\varepsilon(R) = a(c + d - 1)/(a + b)(c - 1) \tag{II.3}$$

The outcomes from the various trials can be used to estimate the parameters (a, b, c, d) of this process, since the probability function of the outcomes can be expressed as a function of these parameters,

$$f(x, y \mid m, n, a, b, c, d) = B^{-1}(a, b)B^{-1}(c, d) \binom{m}{x}\binom{n}{y} \times B(a + x, b + m - x)B(c + Y, d + n - y) \tag{II.4}$$

This expression is the marginal likelihood of the outcome for a single trial. The marginal likelihood of the outcomes of (say) k trials is the product of k factors like (II.4).

This formulation differs from the development leading to (II.1) and (II.2), where the values of the parameters of the prior distribution were assumed known. Here, the values of a, b, c, and d are not assumed known a priori, but are estimated using the data from the various trials. The difference in approach reflects the difference between the problems. The development leading to (II.1) and (II.2) determined the distribution of the ratio of a given pair of "true" response rates on the basis of data from a single trial (or set of trials with the same "true" response rates) and information available before the data were obtained. The aim here is to characterize a property of the *process* that generates trial-specific "true" rates using the outcomes (observed rate ratios) from several trials.

The computations are straightforward. Maximum likelihood estimates of a and b (c and d) may be obtained using the outcomes for the active agent (placebo) as described in Section V,C below. Substituting the parameter estimates into (II.3) yields a point estimate of the overall response rate ratio, and substituting the estimates into (II.2) with $m = n = x = y = 0$ yields an overall interval estimate.

Applying these two ways of obtaining overall estimates to the data given above illustrates that they differ primarily in the width of the confidence intervals they generate:

| | Common response rate ratio | | |
|---|---|---|---|
| Method | Expectation | Median | 95% conf. int. |
| "Pooling" | 1.62 | 1.61 | (1.34, 1.96) |
| "Common process" | 1.63 | 1.60 | (1.18, 2.12) |

The "pooling" method yields a narrower interval than the "common process" method. In view of the results presented separately above for each trial, the confidence interval generated by the "common process" method seems more realistic.

The paper by Aitchison and Bacon-Shone (1981) appears to be the first to give computationally stable forms for the density and distribution function [Aitchison and Bacon-Shone actually consider the distribution of the quantity $R/(R + 1)$, from which the distribution of R may be obtained readily]. Weisberg (1972) studied the corresponding problem for multinomial populations and expressed the distribution function of R as an alternating series which may be sensitive to roundoff errors as m and n become large.

One of the advantages of formulating the problem in terms of the response rate ratio (besides ease of interpretation) is the fact that its density and distribution function involve only finite series of positive terms, which simplifies the computations. Similar expressions (involving infinite series) can be derived for the density and distribution function of the odds ratio $p(1 - q)/q(1 - p)$ (Altham, 1969; Latorre, 1982; Hora and Kelley, 1983). Hora and Kelley also provide expressions for the moments of the distributions of the response rate and odds ratios. The assumption of independent prior distributions for p and q is not necessary: one may start with processes that are interrelated a priori (see Antelman, 1972) and work through the (more complicated) algebra to derive results analogous to (II.1) and (II.2).

Finally, the Bayesian approach used above is not the only way to determine confidence intervals for response rate ratios, although it seems one of the simplest. Noether (1957) describes confidence intervals for the ratio $p/q$ based on asymptotic normal theory. Thomas and Gart (1977) provide a table of "exact" confidence limits for the ratio $p/q$, arithmetic difference $p - q$, and odds ratio $p(1 - q)/q(1 - p)$ based on Fisher's conditional "exact" test. Santner and Snell (1980) describe geometric algorithms for determining confidence intervals for the ratio $p/q$ and the arithmetic difference $p - q$. These methods appear to be most useful when the confidence level is fixed and the confidence limits are to be determined; they are less convenient when the limits are fixed and the level is to be determined.

### B. Determining Sample Sizes

The methods described above can be used for trial design (i.e., determining sample sizes) as well as for analysis. The sample sizes are determined so as to make the probability of satisfying the relation $G(r) \leq \theta$, where $G(r)$ is given by (II.2), at least equal to a specified value, $\gamma$, using what one knows or believes at the outset about p and q. That is, the sample sizes must be large enough so that (given the prior information) the probability is at least $\gamma$ that the data will support stating that $R = p/q$ exceeds r at a confidence level of at least $100\theta\%$.

Let n denote the number of observations to be drawn from each binomial population. The set of possible outcomes (x, y) consists of the $(n + 1)^2$ points (0, 0), (0, 1), . . ., (n, n); G(r) can be computed for any member of this set, using (II.2). Let

$$\delta_{r,\theta}(x, y) = \begin{cases} 1 & \text{if } G(r) \leq \theta \\ 0 & \text{otherwise} \end{cases}$$

The marginal density of the points (x, y) with respect to the prior distribution is given by (II.4). The probability a priori of an outcome for which $G(r) \leq \theta$ therefore is

$$S(r, n, a, b, c, d) = \sum_{x=0}^{n} \sum_{y=0}^{n} \delta_{r,\theta}(x, y)\, f(x, y; n, n, a, b, c, d) \tag{II.5}$$

The sample size required for each group is the least value of n for which $S(r, n, a, b, c, d) \geq \gamma$. Not all of the $(n + 1)^2$ possible terms of S need to be calculated. The sequence of terms (x, 0), (x, 1), . . . corresponds to monotone increasing values of G(r) so that, given x, the range of y is $0 \leq y \leq y^*$, where $y^*$ is such that G(r) calculated at $(x, y^*)$ is $\leq \theta$, but G(r) calculated at $(x, 1 + y^*)$ exceeds $\theta$. Also, the range of x is $x^* \leq x \leq n$, where $(x^*, 0)$ gives $G(r) \leq \theta$, but $(x^* - 1, 0)$ gives $G(r) > \theta$. The value of S(r, n, a, b, c, d) will increase with n as n becomes large, but will not in general go to 1 as n becomes very large because the marginal distribution of $x_1/n$ and $x_2/n$ converges to the joint prior distribution for the parameters p and q. Since the prior probability content of the region of (p, q) values for which $\delta = 0$ does not vanish as n increases, neither does the marginal probability content. Sample size determination therefore requires some balancing among n, r, and $\theta$.

Example. A belief that the active response rate is about 55%, with 80% confidence a priori that it is between 45 and 65%, implies a beta prior distribution for the active treatment response rate with parameters (22, 18). A corresponding belief that the placebo response rate is about 35%, with 80% confidence a priori that it is between 25 and 45%, implies a beta prior distribution for the placebo response rate with parameters (13, 24). How large must the per-group sample size, n, be to have a reasonable chance that the sample outcome will support the statement "With at least 90% confidence, the active treatment response rate is at least 20% better than the placebo response rate?" The probability, $\gamma$, that the trial outcome will support this statement depends upon n, the number of patients assigned to each treatment:

| n | 10 | 20 | 40 | 80 | 120 | $\infty$ |
|---|---|---|---|---|---|---|
| $\gamma$ | 0.42 | 0.52 | 0.59 | 0.66 | 0.69 | 0.85 |

Given the prior information, there is a chance that the statement will not be supported by the trial outcome regardless of the sample size.

### C. Polytomous Responses, Two Populations

This situation arises when the observations from each of two treatment groups consist of the counts in several ordered or unordered categories. For example, the categories might denote "response to treatment" evaluated as "good," "no change," or "worse," and one might wish to calculate the confidence that the proportion of patients with a "no change" response *and* the proportion of patients with a "worse" response *both* are less for a test agent than for a control. This calculation cannot be accomplished by determining the confidence that a patient on the test agent is more likely than a patient on the control to have a "good" response. This latter quantity expresses the confidence only that the proportion of patients with a "no change" *or* "worse" response is less for the active agent than for the control.

Let X denote an observation from a multinomial distribution with parameters M and $p_1, p_2, \ldots, p_k$ where $\Sigma\, p_i = 1$. That is,

$$\Pr\{X = x \mid M, p\} = \frac{M!}{x_1!\, x_2! \cdots x_k!}\, p_1^{x_1} p_2^{x_2} \cdots p_k^{x_k}$$

where $\Sigma\, x_i = M$. Likewise, let Y denote an observation from a multinomial distribution with parameters N and $q_1, q_2, \ldots, q_k$ where $\Sigma\, q_i = 1$ and $\Sigma\, y_i = N$. The tractability of the analysis depends on how the ratios are defined, i.e.,

$$U_i = \frac{p_1 + \cdots + p_i}{q_1 + \cdots + q_i} \qquad \text{or} \qquad V_i = \frac{p_i}{q_i}$$

The $U_i$ correspond essentially to ratios of the underlying distribution function if the categories are ordered, and the $V_i$ correspond to ratios of the category probabilities.

The analysis of the U ratios has been studied by Altham (1969) and Weisberg (1972). Both present expressions for the joint distribution function evaluated at $U_1 = U_2 = \cdots = 1$, which amounts to evaluating the confidence that the X and Y distributions are stochastically ordered. The complexity of the algebraic development and the computational algorithm is beyond the scope of this discussion.

Analysis of the V ratios is somewhat more tractable. Suppose that the $p_i$ have a Dirichlet prior distribution with positive integer parameters $a_1, a_2, \ldots, a_k$ and the $q_i$ have a Dirichlet prior with positive integer parameters $b_1, \ldots, b_k$:

$$f(p \mid a) = \Gamma(a_1 + \cdots + a_k)[\Gamma(a_1) \cdots \Gamma(a_k)]^{-1}$$

$$\times p_1^{a_1 - 1} \cdots (1 - p_1 - \cdots - p_{k-1})^{a_k - 1}$$

and similarly for $f(q \mid b)$. Let $a_i' = a_i + x_i$ and $b_i' = b_i + y_i$, $i = 1, \ldots, k$. Then the joint distribution function of $V_1, \ldots, V_{k-1}$ in the unit hypercube $0 \leqslant v_i \leqslant 1$, $i = 1, \ldots, k-1$, has the form

$$G(v \mid a, b, x, y) = \binom{a_.' + b_.' - 2}{b_.' - 1}^{-1} \sum_{j=0}^{a_k - 1} \binom{a_k' + b_k' - j - 2}{b_k' - 1}$$

$$\times \sum_{j_1} \cdots \sum_{j_{k-1}} \prod_{i=1}^{k-1} \binom{a_i' + b_i' + j_i - 1}{b_i' - 1} I_{v_i}(a_i', j_i + 1) \qquad \text{(II.6)}$$

where the multiple summation is over nonnegative integer values of $j_1, \ldots, j_{k-1}$ such that $\Sigma j_i = j$ and $a_.' = a_1' + \cdots + a_k'$, etc. The derivation of (II.6) is almost the same as that of (II.2), except for more cumbersome algebra. Note that (II.6) provides a joint confidence statement and that (II.2) provides marginal confidence statements.

Example. A trial comparing two nonsteroidal anti-inflammatory agents in osteoarthritis of the hip (Dieppe et al., 1976) provided the following outcomes for the patients' evaluations of their therapeutic outcomes (withdrawals were for lack of efficacy):

| | Improved | No change | Worse | Withdrew | Total |
|---|---|---|---|---|---|
| Sulindac | 9 | 8 | 0 | 2 | 19 |
| Ibuprofen | 6 | 7 | 3 | 4 | 20 |

Assuming a uniform prior distribution for the response parameter ($a_i = b_i = 1$, $i = 1, 2, 3$) means, in analogy to the case when $k = 2$, that the quantity computed from (II.6) can be interpreted as a confidence level. Then the confidence from (II.2) is 96% that the proportion of patients withdrawing *or* rating themselves as worse on sulindac is less than the corresponding proportion on ibuprofen. From (II.6), the joint confidence is 72% that the proportion of patients withdrawing on sulindac is less than the proportion withdrawing on ibuprofen *and* the proportion on sulindac rating themselves as worse is less than the proportion on ibuprofen rating themselves as worse; this is a stronger statement then the one implied by (II.2), so the confidence level is lower.

## D. Dichotomous Responses, Several Populations

In this case there are k binomial populations such that the response rate parameter from the i-th, $p_i$, has a beta prior distribution with positive integer parameters $(a_i, b_i)$. Let $X_i$ denote the random variable for the i-th population, so that $X_i \sim B(n_i, p_i)$. The ratios for this case may be defined in various ways, e.g.,

$$U_i = \frac{p_i}{p_k} \quad \text{or} \quad V_i = \frac{p_i}{p_{i+1}}, \qquad i = 1, 2, \ldots, k-1$$

The $U_i$ values might correspond to multiple comparisons against a control, depending on the context of the analysis, while the $V_i$ express stochastic ordering among the populations.

It follows straightforwardly from algebra similar to that used to establish (II.2), omitted here, that the joint distribution function of the U ratios has the form

$$G(u \mid a, b, n, x) = B^{-1}(a_k + x_k, b_k + n_k - x_k) \times \sum_{j_1=1}^{s_1} \cdots \sum_{j_{k-1}=1}^{s_{k-1}} B\left(\sum_1^k (a_i + x_i), \sum_1^k (b_i + n_i - x_i) + \sum_1^{k-1} j_i\right) \times \prod_{i=1}^{k-1} \binom{a_i + b_i + n_i - j_i - 1}{j_i - 1} I_{u_i}(a_i + b_i + n - j_i, j_i) \tag{II.7}$$

where, as before, $0 \leqslant u_i \leqslant 1$, $i = 1, \ldots, k-1$ and $s_i = b_i + n_i - x_i$. Note that (II.7) provides the *joint* confidence that *each* of the first $k-1$ response rates is no more than a specified fraction of the k-th.

The analysis of the V ratios is almost as straightforward. Approximate solutions have been described in the literature: Leonard (1972) developed a procedure for comparing several binomial distributions that involves reparametrization and a normal approximation; Bratcher and Bland (1975) described a decision theoretic method for ordering binomial distributions.

An "exact" solution can be obtained readily from the product of the confidence distributions for the individual populations by writing $1 - p_i = (1 - p_{i+1}) + p_{i+1}(1 - v_i)$, where $v_i = p_i/p_{i+1}$, $i = 1, \ldots, k-1$. If $v_k = p_k$, then the joint distribution of the V ratios follows after making the transformation $p_1, \ldots, p_k \rightarrow v_1, \ldots, v_k$, integrating out $v_k$ (assuming the remaining $v_i$ all are between 0 and 1), and integrating with respect to $v_1, \ldots, v_{k-1}$. The result is

$$G(v \mid a, b, n, x) = \{\Pi\, B(a_i + x_i, b_i + n_i - x_i)\}^{-1}$$

$$\times \sum_{i_1=1}^{s_1} \binom{s_1 - 1}{i_1 - 1} B(a_1 + x_1, i_1) I_{v_1}(a_1 + x_1, i_1)$$

$$\times \sum_{i_2=1}^{s_2} \binom{s_2 - 1}{i_2 - 1} B(a_2^+ + x_2^+ - 1, i_2) I_{v_2}(a_2^+ + x_2^+ - 1, i_2)$$

$$\times \cdots \times \sum_{i_{k-1}=1}^{s_{k-1}} \binom{s_{k-1} - 1}{i_{k-1} - 1} B(a_{k-1}^+ + x_{k-1}^+$$

$$+ 2 - k, i_{k-1}) I_{v_{k-1}}(a_{k-1}^+ + x_{k-1}^+ + 2 - k, i_{k-1})$$

$$\times B(a_k^+ + i_{k-1}^+ + 1 - k, b_k^+ - i_{k-1}^+) \qquad \text{(II.8)}$$

where

$$a_j^+ = a_1 + \cdots + a_j \qquad \text{(for instance)}$$

and

$$s_1 = b_1 + n_1 - x_1 \qquad \text{and} \qquad s_j = b_j^+ + n_j^+ - x_j^+ - i_{j-1}^+,$$

$$j = 2, \ldots, k - 1$$

This algebraically complicated expression turns out to be fairly simple to compute by recursion.

Examples. A study of the effect of alcohol combined with aspirin or ibuprofen on the gastroduodenal mucosa of normal volunteers (Lanza et al., 1985), yielded the following gastric endoscopy scores for the aspirin, aspirin + alcohol, and ibuprofen + alcohol groups (0/1 = no effect or 1 submucosal hemorrhage or superficial ulceration; 2/4 = more than 1 submucosal hemorrhage or superficial ulceration):

| Score | Aspirin | Aspirin + alcohol | Ibuprofen + alcohol |
|---|---|---|---|
| 0/1 | 1 | 2 | 6 |
| 2/4 | 9 | 8 | 4 |

How confident do these results allow us to be that the ulceration rate for ibuprofen + alcohol is less than the ulcertaion rate for aspirin *and* for aspirin + alcohol? Assume uniform priors for the ulceration rates in each group (so that $a_i = b_i = 1$, $i = 1, 2, 3$). Evaluating (II.7) at $u_1 = u_2 = 1$ yields the value of 95.1% for this confidence, suggesting that the risk of material gastric ulceration with aspirin or with aspirin + alcohol is significantly greater than with ibuprofen + alcohol.

To determine the confidence that the ulceration rate for aspirin is less than the ulceration rate for aspirin + alcohol *and* the ulceration for ibuprofen + alcohol is less than the ulceration rate for aspirin, evaluate (II.8) with $(n_1, x_1) = (4, 10)$, $(n_2, x_2) = (9, 10)$, $(n_3, x_3) = (8, 10)$, and $a_i = b_i = 1$, $i = 1, 2, 3$. The calculated confidence value is 28.3%, which suggests that the ulceration rate for aspirin is not less than the rate for aspirin + alcohol or the ulceration rate for ibuprofen + alcohol is not less than the rate for aspirin, or neither presumed inequality is true. Since the confidence that the ulceration rate for ibuprofen + alcohol is less than the rate for aspirin + alcohol is 96%, the conclusion from this analysis is that adding alcohol to aspirin does not increase the ulceration rate, but that the combination of ibuprofen + alcohol poses less of an ulceration risk than the combination of aspirin + alcohol (or aspirin alone).

## III. SAMPLE SIZE ESTIMATES WITH UNCERTAINTY

Sample sizes usually are determined in practice by specifying (1) the null and alternative hypotheses, (2) the statistical procedure that will be used to distinguish between these hypotheses, (3) the type I error rate, and (4) the type II error rate. Once these are specified, along with a reliable estimate of the response variability, the required sample size can be determined straightforwardly. There seldom is anything uncertain about the error rates or the particular statistical method that will be used for the analysis. Nor, as a rule, is there any doubt as to the null hypothesis, which usually is specified as a zero value of some quantity or equivalence in some sense among two or more underlying distributions.

However, there invariably is a great deal of uncertainty associated with the specification of the alternative hypothesis. For example, in designing a trial to determine if the likelihood of a response to a test agent differs from the likelihood of a response to placebo, an investigator might wish the trial to distinguish between a placebo response rate of (say) 25% and an active response rate of 60% with given type I and type II error rates. But what is the basis for specifying "25%" and "60%"? If the placebo response rate is *known* to be 25%, then it is necessary only to test the null hypothesis that the active response rate is 25% against the alternative that it exceeds 25%. If the possibility of positive bias due to the knowledge that nobody is receiving placebo can be ruled out, then in principle the placebo group really is not needed. The fact is, the placebo response rate is *not* known to be 25%; at best, one might be very confident that it is between 20 and 30%, or information from previous studies might indicate that it is 25% ± some quantity that depends on the sample sizes of these studies. A similar consideration applies for the active response rate. Indeed, the choice of its expected value may be based only on the response rates observed for other agents with similar properties. The principle of inherent uncertainty associated with the hypotheses, described here for the case of a simple binomial trial, applies as well for more complex trials. Arbitrarily specifying the null and alternative hypothesis values allows the sample size to be determined by a convenient formula, table, or nomogram (e.g., Schneiderman, 1964; Gail and Gart, 1973; Haseman, 1978; Feigl, 1978; Casagrande et al., 1978; Aleong and Bartlett, 1979, but may be misleading because doing so ignores this inherent uncertainty.

This uncertainty can be incorporated explicitly in a simple way into the determination of the sample size for simple binomial trials and for trials where the response variable is normally distributed. In both cases, the uncertainty is expressed by confidence intervals for the sample size.

### A. Binomial Trials (Gould, 1983)

Let X and Y have binomial distributions $X \sim B(n, p)$ and $Y \sim B(n, q)$. Lachin (1977) has shown that for sufficiently large samples, the noncentrality parameter of the usual $2 \times 2$ chi-square test for comparing two binomial proportions is $\lambda = 2nt$, where n denotes the per-group sample size,

$$t = \frac{(p - q)^2}{(p + q)(2 - p - q)}$$

and p and q represent the response rate parameters. As long as $\lambda \geqslant \lambda_{1-\beta}$ for a test of a given size (see Table 1), the power of the test will be at least $1 - \beta$. Uncertainty about p and q (e.g., due to sampling variation of their estimators from previous or pilot trials) implies uncertainty about t and, therefore, about power, sample size estimates, etc., that depend on t. Thus, for a fixed level of power (fixed $\lambda$), uncertainty about t corresponds to uncertainty about the required sample size, since $\mathrm{Conf}(t \geqslant t_0) = \mathrm{Conf}(n \leqslant n_0)$ and $\lambda = nt = n_0t_0$. Similarly, if n is fixed (e.g., by resource limitations) then the uncertainty in t implies a corresponding uncertainty in the sensitivity of the trial, since $\mathrm{Conf}(t \geqslant t_0) = \mathrm{Conf}(\lambda \leqslant \lambda_0) = \mathrm{Conf}(\beta \leqslant \beta_0)$. Finally, if the power and sample size are specified, then one may determine the confidence that the trial indeed will have the expected power [because specifying the power and sample size specifies a value, $t_0$, for t, and the desired confidence is $\mathrm{Conf}(t \geqslant t_0)$].

The set of (p, q) values for which $t \geqslant t_0$ turns out be

$$r = \frac{2t_0}{(1 + t_0)} \leqslant p \leqslant 1, \qquad 0 \leqslant q \leqslant U - V \qquad (p > q)$$

$$0 \leqslant p \leqslant 1 - r, \qquad U + V \leqslant q \leqslant 1 \qquad (p < q) \qquad \text{(III.1)}$$

**Table 1** Values of Noncentrality Parameter, $\lambda$, Required to Yield Specified Power for Testing a Null Hypothesis at a Specified Significance Level Using a Chi-Square Test with 1 d.f.[a]

| | | Power | | | |
|---|---|---|---|---|---|
| | $\alpha$ | 0.50 | 0.80 | 0.90 | 0.95 |
| One-tail | 0.05 | 2.707 | 6.184 | 8.567 | 10.827 |
| | 0.01 | 5.414 | 10.038 | 13.021 | 15.777 |
| Two-tail | 0.05 | 3.841 | 7.849 | 10.507 | 12.995 |
| | 0.01 | 6.635 | 11.679 | 14.879 | 17.814 |

[a]From Gould, 1983.

where $U = 2[t_0 + p(1 - t_0)]/(1 + t_0)$ and $V = [t_0^2 + 4t_0p(1 - p)]^{0.5}/(1 + t_0)$. The computation of $\mathrm{Conf}(t \geq t_0)$ therefore requires an expression for the confidence distribution of p, which may be approximated closely by a beta distribution: if $x_1$ responses have been observed among $n_1$ independend observations from a binomial distribution with parameter p, then $\mathrm{Conf}(p \leq p^*) \cong I_{p^*}(x_1 + 1, n_1 - x_1 + 1)$, so the confidence distribution of p is a beta distribution with parameters $(x_1 + 1, n_1 - x_1 + 1)$. Likewise, the confidence distribution of q is a beta distribution with parameters $(x_2 + 1, n_2 - x_2 + 1)$. This approximation permits computation of the confidence associated with the set of (p, q) values defined by (III.1):

$$C_1 = c\int_r^1 z^{x_1}(a - z)^{n_1-x_1} I_{U-V}(x_2+1, n_2-x_2+1)dz \qquad (p > q)$$

$$C_2 = c\int_0^{1-r} z^{x_1}(a - z)^{n_1-x_1} I_{1-U-V}(n_2-x_2+1, x_2+1)dz \qquad (p > q) \qquad \text{(III.2)}$$

where $c = B^{-1}(x_1 + 1, n_1 - x_1 + 1)$ and B denotes the usual beta function. From (III.2), $\mathrm{Conf}(t \geq t_0) = C_1 + C_2$. Numerical integration of $C_1$ and $C_2$ is required and easily accomplished, e.g., by Gaussian quadrature (Abramowitz and Stegun, 1965, p. 887).

Example. In a trial comparing two nonsteroidal anti-inflammatory agents in osteoarthritis of the hip (Dieppe et al., 1976), 8 of 19 (42%) patients on sulindac and 5 of 20 (25%) patients on ibuprofen felt that they had improved after 8 weeks of treatment. These findings may be regarded as pilot information that can indicate what resources a definitive trial would require. The confidence density for the sulindac response rate is a beta(9, 12) distribution, and the density for the ibuprofen response rate is a beta(6, 16) distribution. Performing the computations with these data yields a variety of results, of which the following are typical (using a 5% two-tail test). In all cases, the term "power" refers to the probability of detecting a response rate difference as large as these findings *suggest* exists. Since the actual response rate difference is unknown and may be more or less than 17 percentage points, the true power will be greater or less than the power corresponding to a 17 percentage point difference. The uncertainty about the true value of the difference is a consequence of the finiteness of the available information and implies a corresponding uncertainty about the actual power level.

1. The available information supports a 50% (67%, 80%) level of confidence that 132 (294, 800) observations on each treatment will be enough for the proposed trial to have 80% power.
2. If at most 200 patients can be studied (because of limited time, money, etc.) then with 50% (67%, 80%) confidence, the power of the trial will be at least 72% (40%, 22%).
3. If 120 patients are on each treatment, one can be 48% confident that the trial will have power at least 80% (120 per group would be required to differentiate between response rates of 0.25 and 0.42 with 80% power).
4. If 120 patients are on each treatment, the expected power of the trial is 94%.

In this particular example the distribution of t is almost L-shaped, which helps to explain the apparent disparity between the average sample size and the sample sizes obtained by considering the quantiles of the distribution. With this degree of skewness, it is more realistic to base the sample size recommendations on the quantiles than on the mean.

Choosing a sample size on the basis of the computations described above involves making a trade-off between the confidence level and the precision of the prior information. If the prior information is relatively imprecise, then a confidence level of 50 to 60% would be appropriate; with greater assurance about the true values of the response rates a confidence level of 70 to 80% would be appropriate.

The computations just described can be carried out when the prior information reflects what the investigator believes about the response rates by replacing $x_i$ and $n_i - x_i$ in (III.2) with parameters $a_i$ and $b_i$ reflecting this belief. For example, a beta distribution with parameters a = 30 and b = 90 would reflect a prior belief that the placebo response rate is about 25% and 80% a priori confidence that it is between 20 and 30%.

### B. Continuous Measurements

A test of the hypothesis that the mean of a normal distribution equals zero against the alternative that it equals $\mu \neq 0$ (say, $\mu > 0$) conventionally uses Student's t statistic,

$$t = \frac{(X - \mu)}{C_X s} \tag{III.3}$$

where $C_X$ is defined by $\mathrm{Var}(X) = C_X^2\sigma^2$ and $C_X^2 s^2$ estimates $C_X^2\sigma^2$. This formulation covers one- and two-sample tests, tests of regression coefficients, etc. The noncentrality parameter of (III.3) is

$$\lambda = \frac{(\mu/\sigma)}{C_X} = \frac{\delta}{C_X} \tag{III.4}$$

The key part of (III.4) for determining sample size is $\delta$, since $C_X$ will be determined by the design of the experiment. Suppose that information regarding the values of $\mu$ and $\sigma^2$ is available from a previous trial, say X (to estimate $\mu$) and $s^2$ (to estimate $\sigma^2$). Given values of X and $s^2$, confidence statements about $\delta$ and $\sigma^2$ can be made by integrating the joint density of $U = (X - \sigma\delta)/C_X\sigma$ and $V = ms^2/\sigma^2$, considered as a function of $\delta$ and $\sigma^2$, over the regions of $(\delta, \sigma^2)$ values implied by the confidence statements. Here, m denotes the degrees of freedom associated with $s^2$. The value of $\text{Conf}\{\delta \geqslant \delta^*\}$, where $\delta^*$ is known, can be computed easily. A simple transformation $U \to \delta$ and $V \to V$ yields the joint confidence density of $\delta$ and $ms^2/\sigma^2$ (= V):

$$f(\delta, V \mid C_X, X, s^2) = C_X^{-1}(2\pi)^{-1/2}\exp(-1/2C_X^2)\left[\delta - X\left(\frac{V}{ms^2}\right)^{1/2}\right]^2$$

$$\times [2^{m/2}\Gamma(m/2)]^{-1}V^{(m/2 - 1)}\exp\left\{\frac{-V}{2}\right\} \tag{III.5}$$

Direct integration of (III.5) with respect to V to obtain the marginal confidence density of $\delta$ is possible, but difficult. However, since (1) the conditional distribution of $\delta$ given V is normal with mean $X(V/ms^2)^{1/2}$ and variance $C_X^2$, (2) the marginal density of V is a chi-square density with m degrees of freedom, and (3) $\text{Conf}\{\delta \geqslant \delta^*\} = \int \text{Conf}\{\delta \geqslant \delta^* \mid V\}f(V)\,dV$, some simple algebra yields

$$\text{Conf}\{\delta \geqslant \delta^*\} = \int \Phi\left[\frac{(X/(Vms^2)^{1/2} - \delta^*)}{C_X}\right]$$

$$\left[2^{m/2}\Gamma\left(\frac{m}{2}\right)\right]^{-1} V^{-(m/2 + 1)}\exp\left\{\frac{-1}{2V}\right\}dV \tag{III.6}$$

which is easy to evaluate numerically; $\Phi$ denotes the cumulative distribution function (CDF) of a standard normal distribution. The density of $\delta$ evaluated at $\delta^*$ may be obtained by replacing $\Phi$ in (III.6) with $\phi$, the standard normal density. Note that (III.6) is the product of a normal CDF and the density of the *inverse* of a chi-square variate; using the inverse of a chi-square variate simplifies the numerical computation of the integral by quadrature.

This method is similar in principle to a procedure described by Dudewicz (1972), which is concerned with accommodating the uncertainty associated with the estimate of $\sigma^2$ rather than $\delta$. As with Dudewicz's method, the method described here (and the method described above for binomially distributed data) can be used to analyze a completed trial with respect to power, or to design a future trial. Also as with Dudewicz's method, the method described here can be extended to more complex situations (e.g., analysis of variance tests).

Example. From Table 3, in Section V,C, the outcomes from a trial comparing a new active agent against a placebo with respect to improvement in exercise tolerance time after 12 weeks of treatment were:

| | N | Mean | Std. dev. |
|---|---|---|---|
| Active agent | 78 | 169.1 | 180.26 |
| Placebo | 36 | 62.3 | 101.18 |

How large a trial should be undertaken to confirm these findings? An answer to this question depends on the sensitivity desired for the trial, i.e., its power, which in turn requires a reasonably precise estimate of the noncentrality parameter value. These findings provide an estimate of the noncentrality parameter that may be expected to apply for a confirmatory trial, but this estimate is subject to error because the findings are based on finite samples. For any presumed value, $\delta^*$, of the noncentrality parameter, the best one can do (given these findings) is to calculate the value of the confidence that the unknown true noncentrality parameter value is no less than $\delta^*$. Since the sensitivity (power) of the trial depends monotonically on the value of the noncentrality parameter once the per-group sample size is fixed, this confidence also is the confidence that the sensitivity of the trial is at least equal to that implied by the value of $\delta^*$. From the findings above, $m = 112$, $X = 169.1 - 62.3 = 106.8$, $s^2 = (35 \times 101.18^2 + 77 \times 180.26^2)/112 = 25{,}538.58$, and $C_X{}^2 = (1/36 + 1/78) = 0.0406$; a point estimate of the noncentrality parameter is provided as $X/s = 0.668$. Calculating (III.6) with these values yields the following values of $\text{Conf}(\delta \geqslant \delta^*)$ as a function of $\delta^*$:

| $\delta^*$ | 0.2 | 0.324 | 0.4 | 0.515 | 0.6 | 0.695 | 0.8 | 1.0 |
|---|---|---|---|---|---|---|---|---|
| $\text{Conf}(\delta \geqslant \delta^*)$ | 0.977 | 0.95 | 0.91 | 0.8 | 0.67 | 0.5 | 0.31 | 0.07 |

The actual noncentrality parameter for the confirmatory study is given by (III.4), with $C_X$ determined by the design of that study. For example, if the active:placebo allocation will be about 2:1 as in the completed study, then $C_X{}^2 = 2/3n$; if the allocation is 1:1, then $C_X{}^2 = 1/2n$.

## IV. ROBUSTNESS OF CONCLUSIONS WHEN THERE ARE MANY MISSING VALUES

### A. Introduction

It may not always be possible to get follow-up measurements on every patient in a clinical trial, especially if the measurement process is uncomfortable or (perceived by the patient as) hazardous. Basing the conclusions only on the observations provided by the patients with follow-up measurements may be misleading, especially if observations are missing for a substantial fraction of the patients: the missing observations, if they were available, could influence the conclusions materially. In such a situation, when the follow-up measurements cannot be obtained, the best that can be done is to assess the impact on the conclusions that the missing observations are likely to have. One way to do this is to assign the worst possible response to each missing datum and then proceed with an appropriate analysis. This is an unduly conservative strategy, since one of the reasons for the missing response may be that the treatment has been so effective that the patient is content and does not want to return to undergo an uncomfortable or potentially hazardous measurement procedure.

Alternatively, one might determine the extent to which the observations that were obtained could be predicted from data available for all (or almost all) of the patients, and then use this information to predict the likely responses for the patients who did not return for a follow-up measurement. This approach is not as conservative as assuming the worst case for all missing data, but does require assuming that the responses of the patients without follow-up on a treatment are related to their observable data in the same way as the responses of the patients who do provide follow-up data. Although it seldom will be possible in practice to verify this assumption, it often will be possible to assess the extent to which apparent violations of it will affect the validity of the conclusions. The purpose of this section is to illustrate the principle behind this approach by describing its application to a situation that arose in practice. The measurement in this example happens to be dichotomous, but the basic principle can be extended straightforwardly to polytomous or continuous data.

### B. Illustration of Method

A clinical trial conducted to evaluate the effectiveness of an active agent (versus placebo) in minimizing the likelihood of cystoid macular edema following cataract surgery required periodic angiograms to assess retinal status. Angiography (see, e.g., Berkow et al., 1984) is not a pleasant procedure and, indeed, about one-third of the patients in the trial did not provide follow-up angiograms. The following table summarizes the outcomes; a "negative" angiogram is a good clinical outcome:

| | Total patients | Angiogram (% negative) | No angiogram |
|---|---|---|---|
| Active | 327 | 208 (90) | 119 |
| Placebo | 170 | 110 (81) | 60 |

If one considers only the patients providing a follow-up angiogram, the usual chi-square statistic for comparing the proportions of negative responses equals 5.08, which is significant at the 0.025 level (two-tail), so the observed responses clearly imply that the active agent is more effective than placebo. However, 36% of the patients on the active agent and 35% of the patients on placebo did not return for their follow-up angiograms, and it is possible that the conclusion might be different if the results for these patients were known. For example, if all of the missing angiograms had been "positive," then the chi-square statistic value would have been 1.06; if all had been "negative," the value would have been 6.52; and if half the missing angiograms had been "positive," then the chi-square statistic value would have been 1.77. Clearly, distributing the missing observations among the response categories arbitrarily can affect the conclusions profoundly.

Assigning the missing observations among the response categories in proportion to the distribution of the observed outcomes yields a chi-square value of 7.50, suggesting very strongly that the active agent is more effective than placebo. However, this approach assumes as a minimum that (1) the patients on a given treatment without angiograms are similar with respect to all factors that might effect their response to the patients on the treatment who did provide angiograms, and (2) the angiogram outcome depends on the predictive factors in the same way for the patients with and the patients without angiograms. While there does not seem to be any way to get around assumption (2) in the absence of relevant ancillary information (the usual situation), assumption (1) is not necessary. Moreover, in addition to removing the dependence on assumption (1), one also can avoid the unsatisfyingly deterministic aspect of proportional assignment and incorporate the uncertainty about the responses that the patients without angiograms might have given.

The basic strategy started by using the data for the patients with angiograms to estimate the parameters of logistic regression models expressing the probability of a "negative" angiogram as a function of measurements available for all (or almost all) of the patients, whether they had angiograms or not. This was done separately within each treatment group to ensure that the event probability estimates were based only on information from the patients with angiograms who were as nearly as possible like the patients without angiograms. The regression models for different treatment groups need not have the same parameter values, or even the same parameters. The regression model used has the generic form ($\underline{x}$ denotes a vector of explanatory variable values):

$$\Pr\{\text{"negative" angiogram} \mid \underline{x}\} = \frac{e^{\underline{b}'\underline{x}}}{1 + e^{\underline{b}'\underline{x}}} \qquad \text{(IV.1)}$$

where $\underline{b}$ denotes a vector of parameter values to be estimated. Exploration of different models using stepwise regression and other techniques led to the predictive models for the two treatment groups given in Table 2; all x values are 0 or 1. Thus, for example, the estimated probability of a "negative" response for a 60-year-old man in the placebo group who underwent extracapsular extraction and had a treated visual acuity of 20/20 is $0.756 = e^{1.131}/(1 + e^{1.131})$, and the estimated probability of a "negative" response for a woman in the active group who underwent phacoemulsification and had a treated visual acuity of 20/40 is $0.973 = e^{3.581}/(1 - e^{3.581=-(-1.771 + -1.810)})$. These estimated probabilities were computed only for the patients not providing angiograms; for the patients who did provide angiograms, the probability was 0 ("positive" angiogram) or 1 ("negative" angiogram) and did not have to be estimated. Each patient now had a probability score: 0 or 1 for the patients with angiograms, or an estimated probability between 0 and 1 for the patients without angiograms.

The next step was to perform a series of Monte Carlo simulations using these probability scores. For each simulation, a patient was assigned a 1 with probability equal to that patient's probability score, or a 0. Given these assignments, a series of test statistics were computed, e.g., Mantel-Haenszel tests stratifying on various factors such as age, age + sex, and surgery type. The simulation steps can be repeated as often as desired. Fifty repetitions were performed with the following results (without stratifying, and stratifying by age, sex, and type of surgery):

| | Mean p | Median p | Range of p values |
|---|---|---|---|
| Stratified | | | |
| Act. vs. Pbo. | 0.01 | 0.005 | 0.001–0.039, 0.07 |
| Not stratified | | | |
| Act. vs. Pbo. | 0.008 | 0.005 | 0.001–0.049 |
| Heterogeneity | 0.46 | 0.45 | 0.13–0.79 |

**Table 2** Logistic Regression Models for Each Treatment Group

| | Independent Variable (x = 1) | b |
|---|---|---|
| | Active Treatment Group | |
| $x_0$ | Intercept (always = 1) | −1.771 |
| $x_1$ | Women undergoing extracapsular extraction | 1.177 |
| $x_2$ | Age between 60 and 69 | 0.458 |
| $x_3$ | Age between 60 and 69 undergoing phacoemulsification | −1.482 |
| $x_4$ | Treated visual acuity = 20/20 | −0.455 |
| $x_5$ | Treated visual acuity = 20/25−30 | 0.688 |
| $x_6$ | Women, treated visual acuity = 20/40 | −1.810 |
| $x_7$ | Women, treated visual acuity = 20/50−60 | 0.984 |
| | Placebo Group | |
| $x_0$ | Intercept (always = 1) | −1.131 |
| $x_1$ | Gender (= 1 for women) | 0.468 |
| $x_2$ | Surgery type (= 1 for phacoemulsification | −0.858 |
| $x_3$ | Men, age between 70 and 79 | 0.960 |
| $x_4$ | Women, age less than 70 | −1.148 |
| $x_5$ | Women, age greater than 79 | 1.278 |
| $x_6$ | Men, treated visual acuity = 20/25−30 | 0.415 |
| $x_7$ | Men, treated visual acuity = 20/40 | 1.756 |
| $x_8$ | Women, treated visual acuity = 20/25−30 | −1.306 |

These results suggest that the missing angiograms probably would not have influenced the conclusions materially.

This technique, while useful for providing a perspective on the range of outcomes that might be expected had data been available for patients not followed up, has a number of limitations. Only the results based on the outcomes actually observed can be regarded as definitive; the results obtained by taking account of what might have happened for the patients not followed up at best reinforce and provide perspective on the definitive results. The validity of the procedure depends vitally on the assumption that the observable variables are predictive for the target response in the same way for the patients in whom the response was observed and for the patients in whom the response was not observed. The fact that this assumption generally cannot be verified does not render the method worthless, but does imply the need to be cautious in its use.

## C. Other Cases

The principle illustrated by the preceding discussion applies as well to data that are not dichotomous. For example, if the response can be one of c ordered categories, then McCullagh's (1980) method could be used to relate the distribution among the categories of the patients with responses to observations obtained on all of the patients. Any patient lacking a response would receive c scores, all between 0 and 1 and adding to 1, that would govern the random assignment of the patient to a response category. A patient with a response also would receive c scores, the score corresponding to the observed response category being 1 and the remaining $c - 1$ scores, 0. The simulation otherwise would proceed as described above, with different test statistics.

If the response is continuous (say, normally distributed), then one could regress the responses for those patients with responses on the generally observed variables. The regression equation(s) thus obtained would be used to predict the response of a patient lacking a response; this value, along with the estimate of the residual variance provided by the patients with responses, would be taken as the parameters of a normal distribution from which a response would be drawn at random for that patient. The simulation would proceed as described above, using the appropriate test statistics.

## V. PLACEBO COMPARISONS IN ACTIVE-CONTROLLED TRIALS

### A. Introduction

Clinical drug trials often include only active agents because ethical considerations preclude treatment with placebo. Typical indications are long-term control of hypertension, prophylactic treatment of patients recovering from myocardial infarction, duodenal or gastric ulcers, etc. Evaluating the outcomes of such trials often requires determining the likelihood that the observed outcomes would have differed significantly from the outcome for a placebo group had one been included in the trial. Because these trials lack placebo groups, one cannot *know* that the differences would have been significant; the best one can do is to determine the likelihood (or confidence) of a significant difference. Although cast here in the context of assessing placebo outcomes, this discussion applies as well for outcomes of active treatments (e.g., alternative standard therapies) not included in the trial.

The information available for making this assessment consists of the trial's outcome and the outcomes observed for similar kinds of patients on placebo in similarly designed trials that included a placebo group. The nature of this information determines how it is used. When the placebo outcome information comes from a single trial, confidence principles can be applied to predict the distribution of future or hypothetical placebo outcomes. When the placebo outcome information comes from several trials, an analogous, but not identical, principle can be used to predict the distribution of the placebo outcomes. Either situation requires assuming that the process that generated the placebo group outcomes in the previous trials also would have generated the outcomes for a placebo group in the active-control trial if it had included a placebo group.

This section describes methods for determining the likelihood that observed outcomes in active-control trials would have differed from the outcome of a placebo group, had one been included when the information about the placebo outcome comes from a single trial and from several trials. Meta-analysis (see, e.g., Light and Pillemer, 1984, or, for a more mathematical treatment, Hedges and Olkin, 1985) techniques can provide a convenient and coherent way to accumulate this information.

## B. Placebo Information from a Single Trial

Two studies addressed the efficacy of a new agent for the treatment of congestive heart failure (CHF); one compared the new agent with a standard agent, and the other compared the new agent with placebo. The protocols for these two studies were essentially the same, aside from the choice of control. Because the active-control study contained no placebo group, it was necessary to determine the likelihood that the outcome for the new active agent in that study would have been superior to the outcome for a placebo group had one been included. Table 3 summarizes the relevant outcomes.

Preliminary examination of the relationships (regardless of treatment) between change in exercise tolerance and (1) baseline exercise tolerance, (2) age, and (3) weight at baseline revealed no material relationships in either study; the distributions of the changes from baseline also appeared to be very nearly normally distributed. Therefore, a normal model was used to determine the likelihood that the outcome of a sample of (say) 60 observations from a placebo group (had it been available) would have been significantly inferior to the observed outcome for the new active agent.

Suppose that a sample of $n_1$ observations from an active treatment group yields a mean $\bar{x}_1$ and a standard deviation $s_1$; assume that this sample comes from a normal parent distribution with mean $\mu_1$ and variance $\sigma^2$. Now suppose that one might draw a sample of n observations from a normal distribution with mean $\mu$ and variance $\sigma^2$ in order to test $H_0$: $\mu \geqslant \mu_1$ against $H_1$: $\mu < \mu_1$. $H_0$ will be rejected when the test statistic

$$t = \frac{\bar{x}_1 - \bar{x}}{\left[\frac{(m_1 s_1^2 + ms^2)}{m_1 + m}\left(\frac{1}{n_1} + \frac{1}{n}\right)\right]^{1/2}}$$

satisfies $t > t^*$, the appropriate critical value from a central t distribution with $m_1 + m$ degrees of freedom ($m = n - 1$; $m_1 = n_1 - 1$). The event $t > t^*$ is equivalent to the event

**Table 3** Change in Exercise Tolerance (seconds) After 12 Weeks of Treatment

| Group[a] | No. of patients | Mean | Std. dev. |
|---|---|---|---|
| Placebo | 36 | 62.3 | 101.18 |
| New active (P) | 78 | 169.1 | 180.26 |
| New active (A) | 59 | 151.6 | 121.39 |

[a](A) refers to the group from the active-control study; (P) refers to the group from the placebo-controlled study.

$$\bar{x} < x^* = \bar{x}_1 - t^* \left[ \frac{(m_1 s_1^2 + m s^2)}{m_1 + m} \left( \frac{1}{n_1} + \frac{1}{n} \right) \right]^{1/2} \qquad (V.1)$$

If $n_1$, $\bar{x}_1$, and $s_1$ are fixed then, given $\mu$, $\sigma^2$, and n, the probability that (V.1) will occur is

$$\pi_n(\mu, \sigma) = \int_{s^2=0}^{\infty} \int_{\bar{x}=-\infty}^{x^*} f(\bar{x}, s^2 \mid \mu, \sigma^2, n)\, d\bar{x}\, ds^2 \qquad (V.2)$$

On making the transformation $U = n^{1/2}(\bar{x} - \mu)/\sigma$, $V = \sigma^2/ms^2$, and noting that U and $V^{-1}$ are independently distributed with, respectively, N(0, 1) and central $\chi^2$ distributions, (V.2) becomes

$$\pi_n(\mu, \sigma) = \int_{v=0}^{\infty} \Phi \left[ n^{1/2} \left\{ \bar{x}_1 - \mu - t^* \frac{(m_1 s_1^2 + \sigma^2/v)}{m_1 + m} \left( \frac{1}{n_1} + \frac{1}{n} \right) \right\}^{1/2} \Big/ \sigma \right] f_m(v)\, dv \qquad (V.3)$$

where $\Phi$ denotes the CDF of a standard normal distribution, and $f_m$ denotes the density of a variate whose inverse has a central $\chi^2$ distribution with m degrees of freedom. The use of V instead of $V^{-1}$ means that the non-negligible values of $f_m$ will be between 0 and 1, which simplifies evaluating the integral in (V.3) by Gaussian quadrature (Abramowitz and Stegun, 1965, p. 887).

The quantity $\pi_n(\mu, \sigma)$ is the likelihood that $H_0$ would have been rejected had the outcome of a sample of n observations from an $N(\mu, \sigma^2)$ distribution been compared with the given outcome for the active treatment group. For values of $(\mu, \sigma)$ near $(\bar{x}_1, s_1)$, $\pi_n(\mu, \sigma)$ will be small; for values of $(\mu, \sigma)$ far from $(\bar{x}_1, s_1)$, $\pi_n(\mu, \sigma)$ will be large. Given $\gamma$, a number between 0 and 1, (V.3) can be used to identify a region $R_\gamma$ of $(\mu, \sigma)$ values:

$$R_\gamma = \{\mu, \sigma \mid \pi_n(\mu, \sigma) > \gamma\} \qquad (V.4)$$

The boundary of $R_\gamma$ (i.e., the values of $\mu$ and $\sigma$ for which $\pi_n = \gamma$) will be a curve $\mu_\gamma(\sigma)$, so that $R_\gamma$ also can be defined by

$$R_\gamma = \{\mu, \sigma \mid 0 < \sigma < \infty, \mu < \mu_\gamma(\sigma)\} \qquad (V.5)$$

The curve $\mu_\gamma(\sigma)$ often may be well approximated by a simple polynomial, e.g., a quadratic; with this approximation, (V.5) provides a more convenient definition for computational purposes (see below) than (V.4). If $\bar{x}_1 = 151.6$ and $s_1 = 121.39$, the values for the new active group from the active-control study, then $\mu_\gamma(\sigma)$ almost coincides with a quadratic curve for $\gamma = 0.8$, 0.9, 0.95, and 0.99; the coefficients are:

| $\gamma$ | Coefficient of $\sigma^0$ | $\sigma^1$ | $\sigma^2$ |
|---|---|---|---|
| 0.8 | 126.32 | −0.1544 | −0.0004164 |
| 0.9 | 126.33 | −0.2113 | −0.0004183 |
| 0.95 | 126.33 | −0.2583 | −0.0004198 |
| 0.99 | 126.33 | −0.3464 | −0.0004228 |

In practice, $\mu$ and $\sigma$ will be unknown. However, information about their values from a previous study (say $n_0$ observations with mean $\bar{x}_0$ and standard deviation $s_0$) may be used to set confidence limits on $\pi_n(\mu, \sigma)$. Confidence limits are the best that can be done in this situation, in view of the uncertainty associated with the estimates from the previous study due to sampling variation. Simply computing $\pi_n(\bar{x}_0, s_0)$ may provide a misleading assessment of the likelihood of rejecting $H_0$ by ignoring this uncertainty.

The uncertainty about the values of $\mu$ and $\sigma$ due to the sampling variation of $\bar{x}_0$ and $s_0$ is reflected in the joint confidence distribution of $\mu$ and $\sigma$, with density $g_0(\mu, \sigma) = g(\mu, \sigma | n_0, \bar{x}_0, s_0)$. The function $g_0(\mu, \sigma)$ is the confidence density of $\mu$ and $\sigma$ in the sense that confidence statements about regions of $(\mu, \sigma)$ values are obtained by integrating $g_0(\mu, \sigma)$ over these regions. Since U and V (see above) are jointly sufficient statistics for $\mu$ and $\sigma$, it follows that the joint confidence density for $\mu$ and $\sigma$ can be expressed essentially as the product of a normal density and the inverse of a central $\chi^2$ density,

$$g_0(\mu, \sigma) = 2\left(\frac{n_0^{1/2} m_0 s_0^2}{\sigma^3}\right) \phi\left(\frac{n_0^{1/2}(\mu - \bar{x}_0)}{\sigma}\right) f_{m_0}\left(\frac{m_0 s_0^2}{\sigma^2}\right)$$

where $m_0 = n_0 - 1$, $\phi$ represents a standard normal density, and $f_m$ represents a central $\chi^2$ density with m degrees of freedom.

Given values of $n_i$, $\bar{x}_i$, and $s_i$ ($i = 0, 1$), various useful quantities can be computed, e.g.,

1. The expected likelihood of rejection of $H_0$ is

$$\int_\mu \int_\sigma g_0(\mu, \sigma) \pi_n(\mu, \sigma)\, d\mu\, d\sigma \qquad \text{(V.6)}$$

2. The confidence that the likelihood of rejecting $H_0$ is at least $\gamma$ (specified) is

$$\int_\mu \int_\sigma g_0(\mu, \sigma) d\mu\, d\sigma \qquad \text{(V.7)}$$
$$\mu, \sigma \in R_\gamma$$

With $\bar{x}_0 = 62.3$ and $s_0 = 101.18$ (the outcome for the placebo group) and the $R_\gamma$ region tabulated above, (V.6) and (V.7) can be used to support the needed confidence statements. For example, (V.6) would lead to the statement "The expected likelihood of rejecting the null hypothesis that the new active is no more effective than placebo if 60 observations from a placebo group had been available in the active-control study exceeds 99%." Here is what this statement means. For any particular values of $\mu$ and $\sigma^2$, the likelihood that a sample of 60 observations drawn from an $N(\mu, \sigma^2)$ parent distribution will cause (V.1) to be satisfied may be more or less than 0.99, depending on the values of $\mu$ and $\sigma^2$; the expected value of these likelihoods with respect to the confidence distribution of $\mu$ and $\sigma^2$ exceeds 0.99.

Alternatively, (V.7) would lead to the statement "One can be 95% confident that with probability at least 0.99, a sample of 60 observations from a placebo group would have yielded an outcome significantly inferior to that observed for the new active group in the active-control study." This statement expresses two levels of uncertainty: the confidence of 95% and the probability of at least 0.99. The probability of 0.99 reflects the uncertainty associated with the sample outcome *for particular (unknown) "true" values of the mean and standard deviation from the parent distribution generating the placebo outcome*, and therefore is a conditional uncertainty. The 95% confidence reflects the uncertainty about these "true" values arising from the sampling variation *of the outcome actually observed for the placebo group in the placebo-controlled trial*, and so is a marginal uncertainty. The statement implied by (V.6) is less complex, but generally will have an associated confidence level of 50 to 60%.

### C. Placebo Information from Several Trials

A multiclinic trial compared three dosage regimens of a new active agent with a standard agent in the treatment of acute duodenal ulcer. The key response variable was the proportion of patients whose ulcer was healed after 4 weeks of treatment; healing was defined as replacement of the ulceration by scar tissue or normal mucosa. The results were:

| | New active | | | |
|---|---|---|---|---|
| | Regimen 1 | Regimen 2 | Regimen 3 | Standard |
| No. of patients | 240 | 247 | 247 | 246 |
| No. with healed ulcer | 164 | 191 | 201 | 186 |
| Healing rate (%) | 68 | 77 | 81 | 76 |

Although no placebo group was included in the trial, it still was necessary to determine the likelihood that the outcomes for these treatments would have been superior to the outcome for a placebo group if one had been included.

A number of trials, some fairly large, have been carried out comparing various active agents against placebo in the treatment of acute duodenal ulcer. Table 4 summarizes the 4-week duodenal ulcer healing rates from 23 trials (some multiclinic) that included a placebo group. In general, the trials were double-blinded and included outpatients with duodenal ulcers endoscopically proven within 2 to 3 days before starting their (randomly

**Table 4** Numbers and Percentages of Patients with Healed Duodenal Ulcers after 4 Weeks on Placebo in Double-Blind Controlled Trials[a]

| | | Healed at 4 weeks | | | | Healed at 4 weeks | |
|---|---|---|---|---|---|---|---|
| Site | N | N | (%) | Site | N | N | (%) |
| Belgium/Holland | 73 | 29 | (40) | Switzerland | 106 | 61 | (58) |
| Czechoslovakia | 55 | 33 | (60) | U.K. | 143 | 41 | (29) |
| France | 87 | 49 | (56) | U.K. | 151 | 47 | (31) |
| France | 99 | 54 | (55) | U.S.A. | 340 | 167 | (49) |
| Hong Kong | 24 | 4 | (17) | U.S.A. | 195 | 80 | (41) |
| Hungary | 20 | 12 | (60) | W. Germany | 123 | 74 | (60) |
| Ireland | 35 | 14 | (40) | W. Germany | 101 | 63 | (62) |
| Italy | 758 | 242 | (32) | | | | |
| Italy | 166 | 60 | (36) | Italy[1] | 17 | 5 | (29) |
| Norway | 72 | 36 | (50) | Italy[2] | 80 | 23 | (29) |
| Norway | 20 | 12 | (60) | Norway[3] | 24 | 11 | (46) |
| Switzerland | 164 | 93 | (57) | U.S.A.[4] | 168 | 76 | (45) |

[a]All results are from Wormsley (1981) except as noted: [1]Dobrilla et al. (1981); [2]Porro et al. (1983); [3]Berstad et al. (1980); [4]Hirschowitz (1983).

assigned) treatment. As far as could be determined, the published trials and the present trial were conducted similarly (see Gould, 1985, for details). The findings displayed in Table 4 appear to represent reasonably the spectrum of outcomes that might be expected from groups of patients with acute duodenal ulcer receiving placebo for 4 weeks in double-blind trials (see also the discussion at the end of this section). The information provided in Table 4 can be used to determine the likelihood that the outcomes from the present trial would have been significantly superior to the outcome for a placebo group had one been included.

The considerable variation in the healing rates displayed in Table 4 (to a greater extent than can be accounted for by sampling variation alone) suggests that a universal "true" placebo healing rate is not an operationally useful concept even if it exists, since (quasi-) random variations in the trial populations' demographics, diets, stress levels, medical characteristics, etc. will lead to variations in the "true" rates that will apply for the trials and, as a consequence, to variations in the observed healing rates. It is conceivable that a detailed knowledge of the characteristics of the various trial populations and of the relationship between these characteristics and the likelihood of healing while on placebo could be used to express the observed (or the "true") healing rate in each trial in terms of the universal "true" value with adjustments for the particular constellation of characteristics of the patients in each trial. This is not a practicable approach, however, since this knowledge does not exist. Instead, one must assume that there is some process generating trial-specific "true" rates at random. With this assumption, confidence statements about upper bounds on the likely value of the placebo healing rate that may apply in any trial can be determined using estimates of the parameters of the process based on the data in Table 4.

Suppose that for any trial, the "true" placebo healing rate is a random realization from a distribution with density

$$f(p) = B^{-1}(a, b)p^{a-1}(1-p)^{b-1} \qquad 0 \leqslant p \leqslant 1,\ a, b > 0 \tag{V.8}$$

where B(a, b) represents the (complete) beta function and a and b are the parameters of theprocess to be estimated using the available information. Once a value of p has been "generated" for a trial via (V.8), the probability that there will be x healed patients out of n entered has the form

$$g(x; n, p) = \binom{n}{x} p^x (1-p)^{n-x} \tag{V.9}$$

The joint probability function for the outcome and the value of p is the product of (V.8) and (V.9):

$$h(x, p; n, a, b) = \binom{n}{x} B^{-1}(a, b)p^{a+x-1}(1-p)^{b+n-x-1} \tag{V.10}$$

The relationship between the trial outcome (x) and the process generating the "true" response rate for the trial follows from integrating (V.10) with respect to p,

$$k(x; n, a, b) = \binom{n}{x} B^{-1}(a, b)B(a + x, b + n - x) \tag{V.11}$$

This is the a priori likelihood function for the trial, i.e., the probability function of the outcome (x) in terms of the process that generates the (trial-specific) "true" response rates. The quantity (V.11) would apply for each trial, so the joint likelihood for m trials is the product of quantities like (V.11),

$$L(\underline{x}; \underline{n}, a, b) = B^{-m}(a, b) \prod_{i=1}^{m} \binom{n_i}{x_i} B(a + x_i, b + n_i - x_i)$$

The values of a and b that maximize L, given the outcomes $\underline{x} = (x_1, x_2, \ldots, x_m)$ from m trials, turn out to be the solutions of the equations

$$\psi(a) - \psi(a+b) - \frac{1}{m}\Sigma[\psi(a + x_i) - \psi(a + b + n_i)] = 0$$

$$\psi(b) - \psi(a+b) - \frac{1}{m}\Sigma[\psi(b + n_i - x_i) - \psi(a + b + n_i)] = 0 \tag{V.12}$$

where $\psi$ denotes the derivative of the logarithm of the gamma function. The solution of these equations proceeds in two steps with the use of an approximation to $\psi$ provided by Gnanadesikan et al. (1967). Since values of a and b satisfying the difference between the two equations in (V.12),

$$\psi(a) - \psi(b) - \frac{1}{m}\Sigma[\psi(a + x_i) - \psi(b + n_i - x_i)] = 0 \qquad (V.13)$$

turn out to be very nearly linearly related, one solves the first equation in (V.12) for a after expressing b as a linear function of a using the coefficients determined numerically from (V.13).

For the 23 studies in Table 4, a = 9.3 and b = 11.2. The expected "true" response rate is $\varepsilon(p) = a/(a + b) = 0.453$, about 45%. Substituting these values of a and b into (V.11) gives the estimated prior likelihood of the observed number of responses (to placebo) among n patients treated. This likelihood is used as the confidence distribution for the outcomes that might have been observed had a placebo group been present.

The active treatment group outcomes are known, so one can identify the (potential placebo) outcomes that would be judged significantly different from them. Suppose that y patients responded out of m treated with an active agent. Also, let c denote the critical value for a chi-square variable with 1 degree of freedom corresponding to a two-tail test at some significance level (e.g., for a 5% level test, c = 3.84). Given y and m, an outcome of x responses out of n patients treated will result in a chi-square statistic value exceeding c if

$$x \geqslant x_U = \frac{mn(2Ny - 2cy + Nc) + N\{m^2n^2c^2 + 4mncNy(m - y)\}^{0.5}}{2m(mN + nc)}$$

or

$$x \leqslant x_L = \frac{mn(2Ny - 2cy + Nc) - N\{m^2n^2c^2 + 4mncNy(m - y)\}^{0.5}}{2m(mN + nc)}$$

Let $x_U^*$ = smallest integer $\geqslant x_U$ and let $x_L^*$ = largest integer $\leqslant x_L$. Then the confidence that a sample of n patients on placebo would yield an outcome significantly different from that observed for the active agent (y responses out of m tested) is

$$\text{Conf\{Sign. Diff.} \mid m, y, n\} = \sum_{x \leqslant x_L^*} k(x; n, a, b) + \sum_{x \geqslant x_U^*} k(x; n, a, b) \qquad (V.14)$$

Table 5 displays the outcomes from the present trial and the results of performing the calculations described above. The confidence level exceeds 95% for three of the four groups even if a placebo group of less than half the size of any of the active treatment groups is used. Therefore, if a placebo group had been included, its healing rate very likely would have been significantly less than the healing rate for at least three of the four active treatment groups (and probably for all of them).

The assertion that the outcomes presented in Table 4 represent realizations from some process generating "true" healing rates can be justified on statistical grounds (C. L. Mallows, personal communication). Each of the outcomes in Table 4 is generated by some distribution. Replacing each

**Table 5** Confidence of Significant Active-Placebo Differences

| | Outcomes from present trial | | | Confidence that a significant difference will be observed if the placebo group sample size is | | | | |
|---|---|---|---|---|---|---|---|---|
| | | Healed | | | | | | |
| Treatment group | N | No. | (5) | 100 | 200 | 220 | 240 | 260 |
| Test, reg. 1 | 240 | 164 | (68) | .844 | .888 | .897 | .898 | .899 |
| Test, reg. 2 | 247 | 191 | (77) | .968 | .983 | .983 | .984 | .986 |
| Test, reg. 3 | 247 | 201 | (81) | .987 | .995 | .995 | .995 | .996 |
| Standard | 246 | 186 | (76) | .954 | .973 | .975 | .976 | .978 |

outcome by the value of its distribution's CDF evaluated at the outcome value yields realizations from a uniform distribution. These CDF values will fall on or near the line connecting the points (0, 0) and (1, 1) when plotted against their relative ranks [corresponding values of the CDF for a uniform (0, 1) distribution]. The degree to which the values resulting from using assumed, rather than known, distributions coincide with this straight line reflects the degree to which the assumptions are consistent with the data. If the CDF values based on the assumed distribution(s) deviate greatly from the line (which can be assessed by a Kolmogorov goodness-of-fit test, for example), then the assumptions ought not to be regarded as being supported by the data.

If each outcome in Table 4 is assumed to come from a distribution of the form (V.11), the maximum absolute difference between the (ordered) calculated CDFs and the corresponding uniform (0, 1) order statistics turns out to be 0.137, which is less than the 20% critical value for a Kolmogorov test based on 23 observations (Miller, 1956). This suggests that the findings in Table 4 can be regarded as having been generated by a common process.

The validity of the analysis also rests on the assumption that the results used to identify the process generating the placebo outcomes represent accurately the spectrum of outcomes that might be generated. Compiling these results properly requires attention to a number of considerations addressed in recent books by Light and Pillemer (1984) and Hedges and Olkin (1985).

## REFERENCES

Abramowitz, M. and Stegun, I. A. (1965). *Handbook of Mathematical Functions*, Dover, New York.

Aitchison, J. and Bacon-Shone, J. (1981). Bayesian Risk Ratio Analysis, *Am. Stat.*, 35: 254–257.

Aleong, J. and Bartlett, D. E. (1979). Improved Graphs for Calculating Sample Sizes When Comparing Two Independent Binomial Distributions, *Biometrics*, 35: 875–888.

Altham, P. M. E. (1969). Exact Bayesian Analysis of a 2 × 2 Contingency Table, and Fisher's "Exact" Significance Test, *J. R. Stat. Soc. Ser. B*, 31: 261–269.

Antelman, G. R. (1972). Interrelated Bernoulli Processes, *J. Am. Stat. Assoc.*, 67: 831–841.

Berkow, J. W., Kelley, J. S., and Orth, D. H. (1984). *Fluorescein Angiography*, American Academy of Ophthalmology, San Francisco.

Berstad, A., Kett, K., Aadland, E., Carlsen, E., Frislid, K., Saxhaug, K., and Kruse-Jensen, A. (1980). Treatment of Duodenal Ulcer with Ranitidine, a New Histamine H2-Receptor Antagonist, *Scand. J. Gastroenterol.*, 15: 637–639.

Bratcher, T. L. and Brand, R. P. (1975). On Comparing Binomial Probabilities from a Bayesian Viewpoint, *Commun. Stat.*, 4(10): 975–985.

Casagrande, J. T., Pike, P. C., and Smith, P. G. (1978). An Improved Approximate Formula for Calculating Sample Sizes for Comparing Two Binomial Distributions, *Biometrics*, 34: 483–486.

Dieppe, P. A., Burry, H. C., Graham, R. C., and Perera, T. (1976). Sulindac in Osteoarthritis of the Hip, *Rheumatol. Rehabil.*, 15: 112–115.

Dobrilla, G., de Pretis, G., Felter, M., and Chilovi, F. (1981). Endoscopic Double-Blind Clinical Trial on Ranitidine vs. Placebo in the Short-Term Treatment of Duodenal Ulcer, *Hepatogastroenterology*, 28: 49–52.

Dudewicz, E. J. (1972). Confidence Intervals for Power with Special Reference to Medical Trials, *Aust. J. Stat.*, 14: 211–216.

Feigl, P. (1978). A Graphical Aid for Determining Sample Sizes When Comparing Two Independent Proportions, *Biometrics*, 34: 111–122.

Gail, M. and Gart, J. J. (1973). The Determination of Sample Sizes for Use with the Exact Conditional Test in 2 × 2 Comparative Trials, *Biometrics*, 29: 441–448.

Gnanadesikan, R., Pinkham, R. S., and Hughes, L. P. (1967). Maximum Likelihood Estimation of the Parameters of the Beta Distribution from Smallest Order Statistics, *Technometrics*, 9: 607–620.

Gould, A. L. (1983). Sample Sizes Required for Binomial Trials When the True Response Rates Are Estimated, *J. Stat. Planning Inference*, 8: 51–58.

Gould, A. L. (1985). Would the Active Have Differed from Placebo? Proc. Biopharmaceutical Section, Am. Stat. Assoc., pp. 96–99.

Haseman, J. (1978). Exact Sample Sizes for Use with the Fisher-Irwin Test for 2 × 2 Tables, *Biometrics*, 34: 106–109.

Hedges, L. V. and Olkin, I. (1985). *Statistical Methods for Meta-Analysis*, Academic Press, New York.

Hirschowitz, B. I. (1983). Lessons from the U.S. Multicenter Trial of Ranitidine Treatments for Duodenal Ulcer, *J. Clin. Gastroenterol.*, 5(Suppl. 1): 115–122.

Hora, S. C. and Kelley, G. D. (1983). Bayesian Inference on the Odds and Risk Ratios, *Commun. Stat.—Theory Methods*, 12(6): 725–738.

Lachin, J. M. (1977). Sample Size Determination for r × c Comparative Trials, *Biometrics*, 33: 315–324.

Lanza, F. L., Royer, G. L., Jr., Nelson, R. S., Rack, M. F., and Seckman, C. S. (1985). Ethanol, Aspirin, Ibuprofen, and the Gastroduodenal Mucosa: An Endoscopic Assessment, *Am. J. Gastroenterol.*, 80: 767–769.

Latorre, G. (1982). The Exact Posterior Distribution of the Cross-Ratio of a 2 × 2 Contingency Table, *J. Stat. Comput. Simulation*, 16: 19–24.

Leonard, T. (1972). Bayesian Methods for Binomial Data, *Biometrika*, 59: 581–589.

Light, R. J. and Pillemer, D. B. (1984). *Summing Up: The Science of Reviewing Research*, Harvard University Press, Cambridge, Mass.

Mainland, D. (1960). The Use and Misuse of Statistics in Medical Publications, *Clin. Pharmacol. Ther.*, 1: 411–422.

Mainland, D. (1967). Statistical Ward Rounds—1, *Clin. Pharmacol. Ther.*, 8: 139–146.

Majumder, K. L. and Bhattacharjee, G. P. (1973). The Incomplete Beta Integral, *Appl. Stat.*, 22: 409–404.

McCullagh, P. (1980). Regression Models for Ordinal Data, *J. R. Stat. Soc. Ser. B*, 42: 109–142.

Miller, L. H. (1956). Table of Percentage Points of Kolmogorov Statistics, *J. Am. Stat. Assoc.*, 51: 111–121.

Noether, G. E. (1957). Two Confidence Intervals for the Ratio of Two Probabilities and Some Measures of Effectiveness, *J. Am. Stat. Assoc.*, 52: 36–45.

Porro, G. B., Petrillo, M., and Lazzaroni, M. (1983). Ranitidine in the Short-Term Treatment of Duodenal Ulcer: A Multicenter Endoscopic Double-Blind Trial, *The Clinical Use of Ranitidine. Proceedings of the Second International Symposium on Ranitidine* (J. J. Misiewicz and K. G. Wormsley, eds.), Medicine Publishing Foundation, Oxford, pp. 136–142.

Santner, T. J. and Snell, M. K. (1980). Small-Sample Confidence Intervals for p1 − p2 and p1/p2 in 2 × 2 Contingency Tables, *J. Am. Stat. Assoc.*, 75: 386–394.

Schneiderman, M. A. (1964) The Proper Size of a Clinical Trial: "Grandma's Strudel" Method, *J. New Drugs*, 4: 3–11.

Thomas, D. G. and Gart, J. J. (1977). A Table of Exact Confidence Limits for Differences and Ratios of Two Proportions and Their Odds Ratios, *J. Am. Stat. Assoc.*, 72: 73–76.

Weisberg, H. I. (1972). Bayesian Comparison of Two Ordered Multinomial Populations, *Biometrics*, 28: 859–867.

Wormsley, K. G. (1981). Short-Term Treatment of Duodenal Ulceration, *Cimetidine in the 80s* (J. G. Baron, ed.), Churchill-Livingstone, Edinburgh, pp. 3–8.

# 13 Assessment of Safety

**ROBERT T. O'NEILL** *Division of Biometrics, Center for Drugs and Biologics, Food and Drug Administration, Rockville, Maryland*

## I. INTRODUCTION

The assessment of the safety of a new drug is a continuing process that takes place throughout the development and marketing of a drug. It begins in the premarketing stage with preclinical animal studies and with early Phase I studies which examine the absorption, excretion, dose ranging, tolerance, and other pharmacokinetic properties of the drug in humans. It continues with the collection of safety data on larger numbers of subjects in controlled clinical trials in Phase II and Phase III and in the uncontrolled clinical trials which supplement the evidence for safety and efficacy from the comparative trials. In the experimental stages of drug development the conditions under which a drug is administered are relatively controlled and the size and disease severity of the population and the durations of exposure are different from the situations characterizing the marketing stage of the drug. During the postmarketing phase of a drug's lifetime, safety information is obtained in broader patient populations by means of voluntary reports, monitoring systems, uncontrolled patient follow-up, and formal epidemiological studies. The assessment of drug safety may focus on both how that drug behaves in patients to whom it is administered and how offspring are affected, particularly the offspring of women exposed to drugs during pregnancy.

The goal of this chapter is to describe approaches to the assessment of safety of a drug product during its developmental and marketing stages, emphasizing appropriate statistical considerations for planning, analysis, and interpretation where useful. This chapter will treat considerations related to clinical trials, both controlled and uncontrolled, as well as considerations derived from the closely allied field of drug epidemiology, which provides the structure for observational studies necessary for the surveillance of drug safety in the postmarketing setting. Drug-induced adverse events can occur in various types and patterns depending on the drug and disease treated, and a number of examples will be provided to illustrate some of these patterns, as well as measures that are best able to describe the event rates. Statistical issues will be discussed concerning the estimation of rates of occurrence of adverse events associated with drug exposure in single cohorts or treatment groups and the comparative evaluation of rates and differences between rates observed in two or more exposure groups or cohorts. Of particular interest will be the role of time and duration of drug exposure in the estimation of adverse event rates. Measures common to the epidemiology field, such as ratios of rates as employed in both the relative risk and odds ratio, will also be discussed, particularly with respect to how misclassification of safety outcomes affects their interpretation.

In general, the assessment and characterization of drug safety in clinical trials has not received the same level of attention as the assessment of efficacy. Indeed, this is not so much a criticism of the drug development process as it is a statement about the level of sophistication and attention that drug safety has received in the medical literature. A study in the *British Medical Journal* (Venulet et al., 1982) described an evaluation of the quality and completeness of published articles on adverse drug reactions. The study reported that of 5737 articles from 80 countries published between 1972 and 1979, only 61% of the articles included information on the number of patients treated and the number with adverse drug reactions. In only 55% could the incidence rate of a particular adverse reaction be calculated, and only 35% of the articles provided any information on duration of treatment. Though recent medical literature has improved on this record, it is for the most part a sporadic and heterogeneous improvement characterized by confusion in the way rates of adverse reactions are reported and compared. Rates of side effects of drugs observed in clinical trials or in observational settings are often compared to determine whether side effect profiles differ, especially from rates observed in actual use situations (Coles et al., 1983) and seldom are the data in a form to make meaningful comparisons. Stephens (1985) describes some of the problems involved with the detection of adverse reactions both before and after marketing and methods used to overcome these problems.

Experience with statistical approaches to the evaluation of drug safety in clinical trials is also limited. Methods applicable to the comparative analysis of efficacy may need to be modified for comparative safety assessment since a priori hypotheses and safety outcomes are less well defined. With safety assessment a multiplicity of potential outcomes can be unexpected in both type and frequency, making ordinary hypothesis-testing concepts designed for planned efficacy comparisons less helpful. Sample sizes of clinical studies designed to demonstrate effictiveness are likely to be inadequate to simultaneously satisfy acceptable statistical power criteria for safety assessment. Safety data from clinical trials, both controlled and uncontrolled, are more often used to estimate rates of occurrence of adverse events in exposed groups and for exploratory analysis to examine subgroups of patients experiencing differential rates of adverse events. Descriptive statistics are useful for this purpose.

On the other hand, the field of drug epidemiology has evolved over the past two decades to address the surveillance of drug safety in the postmarketing stage. Statistical methods have played a major role in the design and analysis of cohort and case-control studies used to evaluate drug safety. Statistical aspects of the design of drug monitoring systems received attention (Finney, 1965, 1971a, 1971b, 1974) in the early 1970s and recent federal regulations [21 CFR 314.80(b)] may stimulate interest in this area.

It is important to develop a sound statistical structure to assess drug safety since safety is often summarized in terms of rates or relative risks, the magnitude and patterns of which can influence whether and how a physician practitioner uses a drug with a patient. The magnitude, accuracy, and uncertainty of these rates can influence whether a drug that appears to offer no more effectiveness than another is ascribed the reputation as a drug of second choice because of its estimated safety profile. Any drug, once on the market, can be removed or relabeled more restrictively on the basis of new safety information. A recent example concerns a request to the Food and Drug Administration (Petition filed by Public Citizen, January 8, 1986) to ban, as an imminent hazard to the public health, the use of a widely marketed antiarthritic drug in people aged 60 and older because of alleged gastrointestinal toxicity more common and severe with the drug than with similar arthritis drugs, particularly in elderly patients.

Estimates of risk associated with use of a drug are described in the professional instructions for how to use the drug, called the drug label, which, by law, must be accurate and not misleading and must provide adequate directions for using the drug in patients. One has only to read advertisements in medical journals that describe the safety of a drug in terms of its adverse reaction rate profile to realize that the statistical basis for such statements needs to be improved. In addition to inferential statistical considerations related to the planning, analysis, and interpretation of safety studies, the practical considerations of how safety data are summarized, displayed, and provided to the user is very much a descriptive process. This chapter will try to illustrate the role that statistical concepts play in this process.

## II. FEDERAL DRUG REGULATIONS PERTAINING TO SAFETY ASSESSMENT

### A. Premarketing

Federal drug regulations have a major role in influencing the quantity and quality of safety information on drugs marketed to the general population. The Food, Drug, and Cosmetic Act requires that in order for a new drug to be marketed in the United States evidence of its safety and effectiveness be compiled and submitted by the drug's sponsor to the Food and Drug Administration prior to its availability to the general public. The safety data available during the premarketing stage are derived from controlled and uncontrolled clinical trials, the majority of which are used to support the claims for the effectiveness of the drug product. The required evidence is packaged in the form of a New Drug Application, the requirements for which have recently (May 21, 1985) been revised to emphasize the summarization and analysis of safety information. In particular, section 21 CFR 314.50(d)(5)(ii) of the Code of Federal Regulations requires an applicant for a new drug to submit an integrated summary of all available information about the safety of the drug product, including a description of any statistical analyses performed in analyzing safety data [21 CFR 314.50(d)(6)]. Such a summary will generally include estimates of rates of adverse reactions in controlled and uncontrolled trials, as well as analysis of any major differences in rates related to dose, duration of use, or patient characteristics.

Biological products, such as vaccines and allergenic products, must also be demonstrated to be safe, pure, potent, and effective (21 CFR 601.25). Safety of some biologicals, such as vaccines, is assessed in large field trials.

### B. Postmarketing

After a drug is approved and available in the marketplace, information on its safety is accrued from many sources. Section 21 CFR 314.80(b) of the Code of Federal Regulations requires a holder of an approved application for a new drug to promptly review all adverse drug experience information obtained or otherwise received by the applicant from any source, foreign or domestic, including information derived from commercial marketing experience, postmarketing clinical investigations, postmarketing epidemiological/surveillance studies, reports in the scientific literature, and unpublished scientific papers. Sometimes, special observational studies are carried out by the drug sponsor to collect additional safety data. A major source of safety information is generated by reports from health practitioners who observe adverse drug experiences in patients to whom they have prescribed a drug in the course of professional practice. A holder of a new drug application is required to periodically review the frequency of adverse drug experience reports which they receive that are both serious and expected (already in the labeling instructions for using the drug), and report to the Food and Drug Administration any significant increase in frequency [21 CFR 314.80(c)(1)(ii)] which might suggest a drug-related incidence that is higher than previously observed or expected.

Patients often are exposed to several drugs at the same time, and postmarketing surveillance is intended to gather information on interaction between two or more drugs.

### C. Drug Labeling: The Mechanism for Informing the Physician and Patient

The information on adverse reactions collected and evaluated from clinical studies, postmarketing studies, and voluntary spontaneous reports is summarized in the professional labeling of a drug so that the drug can be used in a safe and effective manner. The drug label is the vehicle which instructs the prescribing physician about the benefits and risks associated with the drug and aids the physician in instructing the patient about such benefits and risks. Two sections of the federal regulations on drug labeling are relevant to the statistical characterization of adverse reaction rates.

Section 21 CFR 201.57(e) on warnings states when the label should describe serious adverse reactions and potential safety hazards, limitations in use imposed by them, and steps that should be taken if these serious reactions occur. Drug labels are revised to include a warning as soon as there is reasonable evidence of an association of a serious hazard with a drug. A causal relationship need not have been proved. The frequency of the serious adverse reactions and, if known, the approximate mortality and morbidity rates for patients sustaining the reaction are expressed under the "Adverse Reactions" section of the labeling [21 CFR 201.57(g)(2)]. In this section of the labeling, adverse reactions may be categorized by organ system, severity of the reaction, frequency, toxicological mechanism, or a combination of these as appropriate. The approximate frequency of each adverse reaction is usually expressed in rough orders of magnitude, such as 1 in 30 patients, or less than 1 in 100 patients. Sometimes the rates are reported for a specific clinical trial. Sometimes, the results of a postmarketing study might be described in the labeling including adverse event rates or relative risk estimates if applicable. However, any claim comparing one drug to another in terms of frequency, severity, or character of adverse reactions should be based on adequate and well-controlled studies. There are few examples of comparative randomized clinical trials statistically demonstrating differences in risks between two drugs used for the same indication.

## III. TYPES OF ADVERSE EVENTS AND THEIR PATTERN OF OCCURRENCE

As a convention, the term" adverse event" will be used throughout most of this chapter to denote adverse health events associated with exposure to a drug. No causality judgment is being made for adverse events because small study sizes, ascertainment methods, or lack of comparison groups often preclude statistical demonstration of a drug-induced relationship. Adverse events associated with exposure to a drug can take many different forms depending on the dose, duration of exposure, preexisting conditions of the subject, or presence of concomitant medications taken for a concomitant disease. An adverse event can be serious, causing withdrawal of a patient from treatment, hospitalization, and sometimes death. Or an adverse event can be of a less serious nature yet annoying and undesirable to the patient, such as nausea, headache, dry mouth, or constipation. Less serious adverse events may reoccur because a subject remains under treatment even after the adverse event occurs, while serious events may be associated with withdrawal from treatment. The duration of drug exposure may also influence the occurrence of an adverse event. Diseases of a self-limiting or acute nature, such as an infection or allergic rhinitis, are usually treated with a relatively short regimen of drug therapy, say of 2 weeks duration. Diseases of a chronic nature, such as arthritis, are usually treated with longer-term drug therapy, sometimes for many years or a lifetime. To appropriately analyze and characterize adverse event rates one must be concerned with which of these situations applies.

The pattern of occurrence of adverse events can also differ according to drug, disease treated, and use of the drug. For example, a nonsteroidal anti-inflammatory drug was recently withdrawn from the market because of an anaphylactic-type reaction which was suspected of occurring only after prior sensitization involving intermittent use of the drug. Tardive dyskinesia, a type of involuntary movement disorder, has been reported (Joint Statement on Tardive Dyskinesia by the Adverse Drug Reactions Advisory Committee, 1983) to be observed with 12 months or more use of antipsychotic drugs in the treatment of patients with anxiety, depression, or personality disorders. The first of these two serious reactions might be considered an acute event occurring relatively soon after drug exposure, while the latter reaction is a delayed event. Many adverse events observed in clinical studies or postmarketing studies occur in subjects who remain under treatment, possibly at reduced dosages, and thus are at risk for subsequent adverse events of the same or different types. Certain types of milder adverse events occurring to a patient may predict more serious adverse events later on.

Less serious adverse events which do not censor a patient or subject from treatment probably require a different method of presentation and analysis than serious events associated with censoring from treatment. The former events are more likely to be influenced by study design and methods of ascertainment, whereas the latter, when they do occur, may be detected and less influenced by methods of ascertainment.

In statistically characterizing adverse event rates it is necessary to examine which of these situations applies because the measures used to express a rate of occurrence depend on assumptions that may or may not be satisfied by the adverse event patterns. The next two sections will describe different measures used to characterize event rates and show a number of examples of drug-associated reactions which follow different patterns.

## IV. MEASURES OF ADVERSE EVENT RATES

Given the variety of scenarios that describe the occurrence of adverse events, what measures should be used to estimate their rate of occurrence and to convey the concept of risk to nonstatisticians whose use or choice of a drug may depend on this information? Different subject matter disciplines have employed different measures. Several authors (Morgenstern et al., 1980; Elandt-Johnson, 1975) have also considered measures of disease incidence and various definitions of rates. For example, clinical trial data are seldom summarized in terms of average number of occurrences per unit of time, whereas the epidemiological field frequently uses such a measure to describe or summarize event rates. The following measures have been used to estimate rates of occurrence of adverse events and may be appropriate depending on the situation.

### A. The Crude Rate

The crude incidence rate $R = X/N$ is defined as the number of subjects or patients X experiencing a certain event, divided by the number of subjects or patients N initially exposed to the drug, regardless of duration of use. The crude rate is usually considered to be statistically distributed as a binomial variable, the theory for which is well known (Fleiss, 1981) and serves as the basis for tests of significance and confidence interval estimation. The crude rate is the measure most frequently used in the medical literature to estimate adverse event rates. Probably because of its simplicity, it is also the most misused measure as it is used to characterize event rates with short-term and long-term exposure, with different time-specific patterns (hazard rates), and with differential follow-up of exposed patients. The crude rate is most appropriate in the situation where all subjects or patients are treated and followed for the same period of time or for very short-term drug use (e.g., 2 weeks) or for acute events following close in time after exposure. In other situations, especially when subjects are exposed for longer periods of time and for differential durations of time, the crude rate is inappropriate. It can overestimate the true occurrence rate early after exposure and underestimate the true occurrence rate later in the exposure period. The examples in Section V illustrate the differences between several estimates of risk including the crude rate when duration of exposure is an important factor to consider.

### B. Occurrences per Unit Time of Exposure

The measure "occurrences per unit time of exposure" is frequently used in drug epidemiology to summarize the risk associated with exposure to a drug. Inman in the *PEM News* (1983) used a similar measure to compare three nonsteroidal anti-inflammatory (NSAID) drugs with regard to rates of reported occurrences of skin, gastrointestinal, central nervous system (CNS), cardiovascular, respiratory, and genitourinary events while on treatment with a specific drug. In this context, numbers of prescriptions were a surrogate for total exposure time. Jick (1985) used a similar measure for calculating events per numbers of drug exposures from a computerized patient record of prescriptions to characterize the rate of gastrointestinal bleeding and ulcers associated with NSAID use.

The assumption involved in such a measure is that the risk per unit of time exposed is constant during the entire period of treatment exposure (i.e., a patient's risk in the first month of exposure is the same as the risk in the ninth month of exposure, given that the patient is unaffected through month eight) and is the same for each patient or subject. An exponential time to occurrence of an adverse event is assumed when the measure "average events per unit of exposure" is used. As illustrated in Section V, some adverse events are delayed and take several months to develop and others do not follow a constant risk per unit of time (hazard function). For such adverse events it is inappropriate to use a per-unit-of-time measure because, for example, the risk associated with the total exposure time of five patients each exposed for 1 year is likely not the same as the risk associated with one patient exposed for 5 years. In particular, for a delayed event that may take a year to develop, none of the five patients exposed for 1 year is likely to experience the event but the patient exposed for a longer period of time is more likely to experience the event. For delayed events, more patient time exposure is being counted with this measure than is appropriate, thus diluting the overall estimate of cumulative risk.

Particularly in postmarketing observational studies, which often utilize prescription data as surrogates for total drug exposure, there is a tendency to rely on a measure like "events per 1000 patient prescriptions" which masks the actual numbers of patients exposed. Although it is often difficult in postmarketing surveillance to obtain data on the exact numbers of patients exposed for various durations of time, the numbers of prescriptions for a given drug is not an acceptable substitute for numbers of patients exposed unless something is known or assumed about the time pattern of the adverse event occurrence.

### C. Life Table Rates or Cumulative Rates

Another measure useful in describing adverse events is the cumulative life table rate P(t) defined as the probability that an adverse event of a specified type occurs by time t. The cumulative rate can account for differential duration of drug exposure among an exposed cohort of patients or subjects as well as the time pattern of when, during exposure, the events occur relative to the number of subjects or patients at risk. Because the exact time of occurrence of some adverse events may not be known, the life table method groups events and subjects at risk into intervals of time, the choices of which should depend on the study designs, the data available, and the purposes for which the data will be used. The life table method makes no assumption about the functional form of the underlying risk per unit of time. The measure is appropriate for increasing or decreasing risk-per-unit-of-time functions (hazard rates) and is reported with reference to a defined time interval, e.g., a 3-month or 6-month rate. The cumulative rate assumes that subjects who withdraw from treatment follow-up (censored subjects) would have exhibited the same event rate as the noncensored subjects had they not been censored and that censoring is not related to the effects of the drug. Both of these assumptions are probably violated in many situations and need to be carefully assessed in any particular application. Several good references discuss and give methods for calculation of the life table estimates (Lee, 1980; Gross and Clark, 1975; Kalbfleisch and Prentice, 1980).

When subjects or patients are at risk for an adverse event for different durations of time, particularly for chronically used drugs, a disparity can exist between the crude incidence rate unadjusted for exposure time and the cumulative life table rate adjusted for exposure time. To illustrate this point, consider the following situation. Assume that 3000 patients are initially exposed to a drug and followed for up to 1 year but that the numbers of patients exposed for various durations is as described in Figure 1. Assume further that 30 patients among the 3000 initially exposed experience serious adverse events but that they occur in three different patterns as

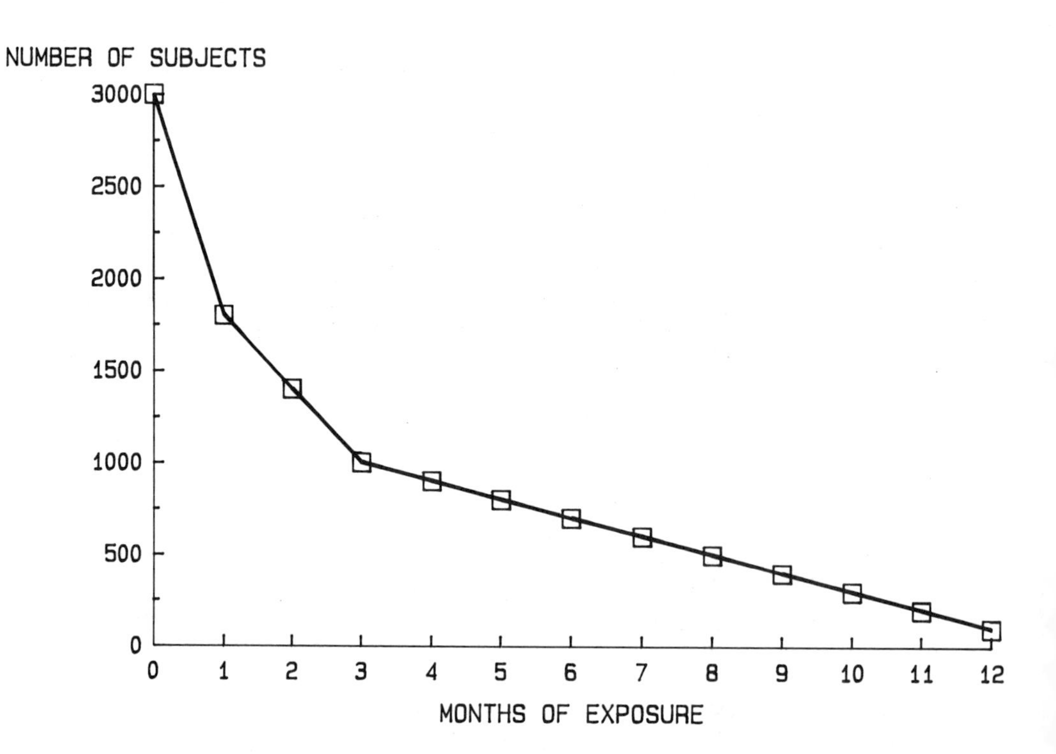


Fig. 1 Duration of exposure pattern for subjects followed up.

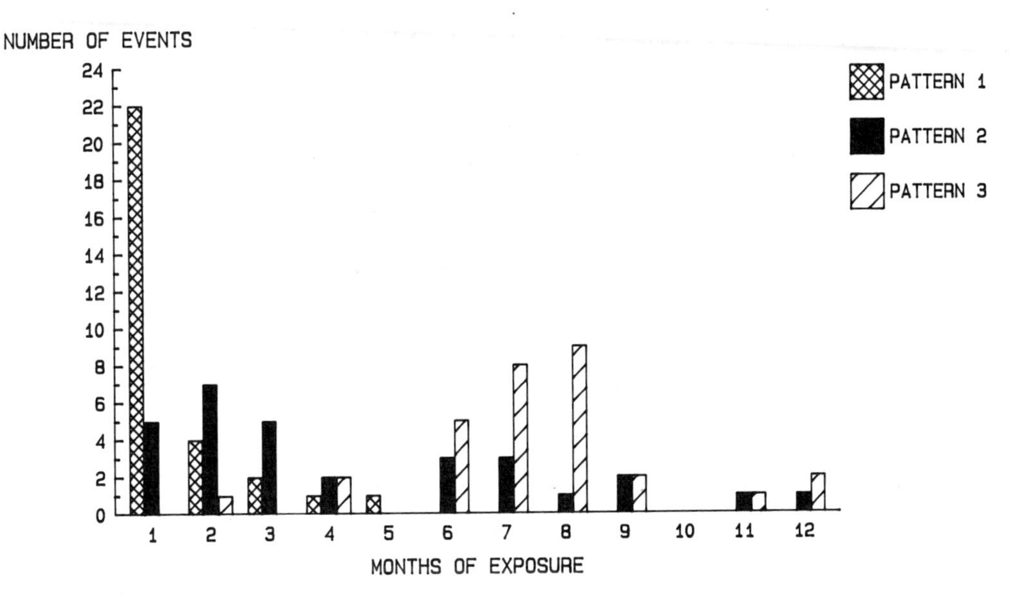


Fig. 2 Frequency of occurrence of 30 adverse events according to three different patterns.

depicted in Figure 2. In pattern one, most of the events occur early; in pattern two the events are spread out over the entire 1-year period of follow-up; in pattern three, most of the events occur later on with increased duration of exposure. Although 30 out of 3000 patients experienced these adverse events, the different times at which the events occurred relative to the number of patients at risk at that time differed. This difference has an impact on the estimation of the adverse event rates when using a time-adjusted estimation method such as the life table described above. For example, the crude rate is 30/3000 = .01, and the 12-month cumulative rates for patterns one, two, and three are .015, .036, and .055, respectively. Figure 3 illustrates the cumulative rate for each of these three patterns of occurrence. Thus, solely because of different numbers of patients at risk over time, the same 30 events provide three markedly different point estimates of adverse event rates.

It is clear that for patterns where the event occurs early in the exposure interval the crude and cumulative rates have better agreement. But when events take place later in time as fewer patients are at risk, the crude rate underestimates the longer-term risk and overestimates the shorter-term risk.

Table 1 provides a basic data layout for the life table and illustrates the life table calculations with data from pattern three. The examples in Section V illustrate the difference in cumulative and crude rates for a variety of data sets describing the pattern of occurrence of some adverse events.

## D. Hazard Rates and Functions

The hazard function or rate h(t) is a measure that complements the information provided by the cumulative rate in that the hazard rate is

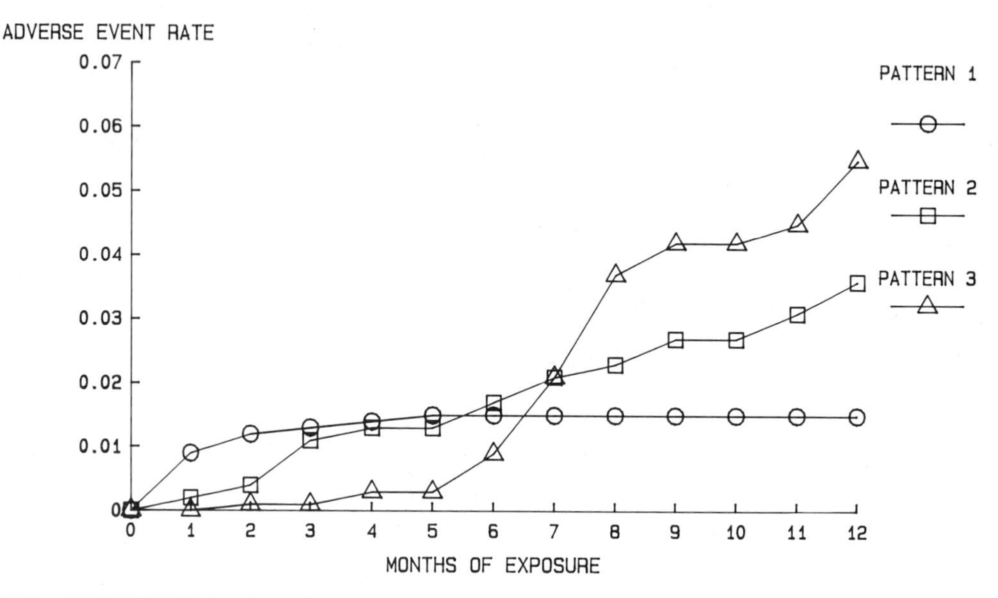


Fig. 3 Estimated cumulative adverse event rates for three patterns of adverse event occurrence.

conditional on the time of exposure. It is defined as the probability of an adverse event during a small but specified interval of time, assuming that the subject or patient has been exposed and followed to the beginning of the interval. In practice, the hazard function is usually estimated as the proportion of patients with the adverse event in an interval of unit time (say a week or a month) given that they h/ave been exposed and followed to the beginning of the interval. The actuarial approach uses the average hazard rate of the interval, whereby the number of subjects with the adverse event in the interval is divided by the average number of subjects followed at the midpoint of the interval. The hazard function is a measure of the risk of an adverse event per unit of time over the entire exposure interval. It thus complements the cumulative adverse event rate or life table rate, which is always a nondecreasing function of time. The hazard rate may increase, decrease, remain constant, or have other more complex patterns. In particular, use of the measure "average occurrences per unit time of exposure" assumes that the adverse event follows a constant hazard rate.

The following example illustrates the marked differences in estimates and interpretations of risk when using the per-unit-of-time measure under two assumptions and compares these measures of adverse event occurrence when the event is assumed to have a 1-month delay in occurrence.

Example 1. Assume an adverse event follows a constant risk per unit of time (hazard rate) of .05 per patient month of exposure. Assume, however, that the event is delayed and takes at least 1 month of exposure to manifest itself. The following data illustrate the effect of ignoring the

**Table 1** Format of Life Table

| Interval (months) | Midpoint | Width | Number entering interval | Number lost to follow-up | Number with adverse event | Number at risk | Conditional proportion with adverse event | Conditional proportion without event | Cumulative proportion having the adverse event | Hazard $h(t_{mi})$ |
|---|---|---|---|---|---|---|---|---|---|---|
| 0–1 | .5 | 1 | 3000 | 1200 | 0 | 2400 | .0000 | 1.0 | .0000 | .0000 |
| 1–2 | 1.5 | 1 | 1800 | 399 | 1 | 1600.5 | .0006 | .9994 | .0006 | .0006 |
| 2–3 | 2.5 | 1 | 1400 | 400 | 0 | 1200 | .0000 | 1.0 | .0006 | .0000 |
| 3–4 | 3.5 | 1 | 1000 | 98 | 2 | 951 | .0021 | .9979 | .0027 | .0021 |
| 4–5 | 4.5 | 1 | 900 | 100 | 0 | 850 | .0000 | 1.0 | .0027 | .0000 |
| 5–6 | 5.5 | 1 | 800 | 95 | 5 | 752.5 | .0066 | .9934 | .0094 | .0067 |
| 6–7 | 6.5 | 1 | 700 | 92 | 8 | 654 | .0122 | .9878 | .0215 | .0123 |
| 7–8 | 7.5 | 1 | 600 | 91 | 9 | 554.5 | .0162 | .9838 | .0374 | .0164 |
| 8–9 | 8.5 | 1 | 500 | 98 | 2 | 451 | .0044 | .9956 | .0416 | .0044 |
| 9–10 | 9.5 | 1 | 400 | 100 | 0 | 350 | .0000 | 1.0 | .0416 | .0000 |
| 10–11 | 10.5 | 1 | 300 | 99 | 1 | 250.5 | .0040 | .9960 | .0454 | .0040 |
| 11–12 | 11.5 | 1 | 200 | 0 | 2 | 200 | .0100 | .9900 | .055 | .0000 |

time of occurrence of the event and counting exposure time of those subjects exposed for less than 1 month when using a measure like occurrences per unit time of exposure.

| Months exposed | Patients followed at least this long | Events occurring in interval |
|---|---|---|
| 1 | 300 | 0 |
| 2 | 200 | 10 |
| 3 | 100 | 5 |

The total exposure time for the 300 patients followed is $100 \times 1 + 100 \times 2 + 100 \times 3 = 600$ months, so the estimated risk per patient month of exposure is $15/600 = .025$ per month when the delay in the time of occurrence of the event is ignored. Because use of this measure assumes that the adverse event follows a constant hazard rate and therefore the cumulative rate of occurrence by time t follows an exponential distribution, one can estimate the cumulative rate at the end of each month as follows: $P(1) = 1 - \exp(-.025 \times 1) = .025$; $P(2) = 1 - \exp(-.025 \times 2) = .05$; $P(3) = 1 - \exp(-.015 \times 3) = .072$. However, since the biologic or pharmacologic assumption is that the adverse event could not occur for at least a month, the appropriate calculation for the delayed event does not utilize the exposure time in the first month on patients not followed longer than a month. Rather, the per-month risk is zero for the first month and the total exposure time for the delayed event is $100 \times 2 + 100 \times 1 = 300$ yielding a per-patient-month risk of $15/300 = .05$. Using this estimate of the per-month risk and the functional form $1 - \exp(-.05 \times (t - 1))$ for the cumulative probability of an adverse reaction which is delayed at least 1 month, the cumulative rate at 1 month is $1 - \exp(- .05 \times (1 - 1)) = 0$; at 2 months, $1 - \exp(-.05 \times (2 - 1)) = .05$; and at 3 months, $1 - \exp(-.05 \times (3 - 1)) = .10$. These estimates are consistent with life table cumulative estimates, which would be $P(1) = 0/300 = 0$; $P(2) = 1 - (1 - 10/200) = .05$; and $P(3) = 1 - (1 - .05) \times (1 - 5/100)) = 1 - (.95 \times .95) = .10$. The crude rate is estimated to be $15/300 = .05$. Figure 4 illustrates the cumulative risk for these two sets of assumptions.

Thus, an assumption that the per-month risk is constant regardless of the delay in the event results in estimates that are markedly different from each other and from the crude rate estimate.

## E. Competing Risks

With the previous measures the occurrence of an individual adverse event has been characterized in a marginal sense. That is, life table estimates consider all other types of adverse events as censoring mechanisms and therefore provide a cumulative adverse event rate of a particular type in the presence of other adverse events. Adverse event data are more general in two important respects. First, the adverse event can be one of several types such that termination of exposure can result from any one of these adverse events occurring. Methodology to handle these problems is called competing risk theory. The second is that more than one adverse event of the same or different type may be experienced by an individual, and this

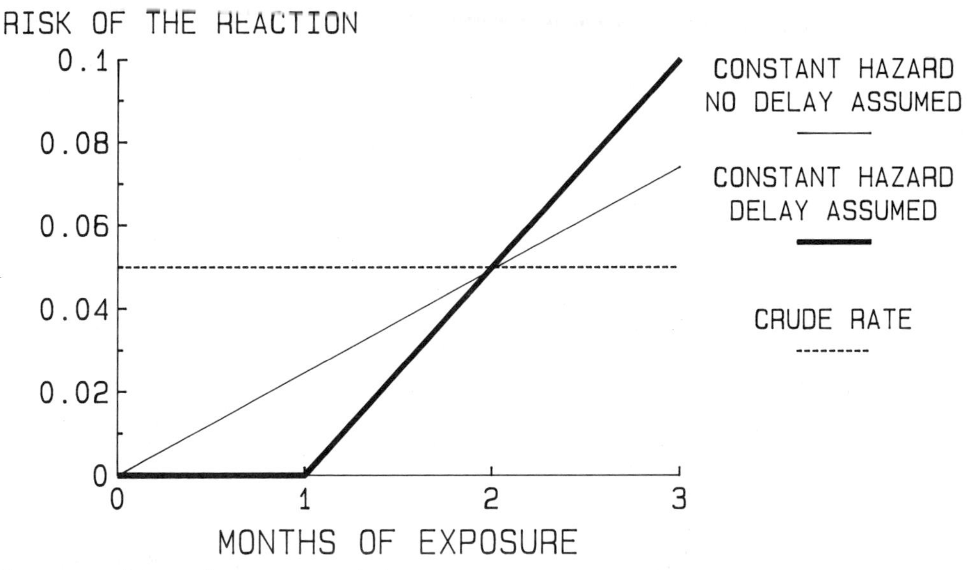


Fig. 4 Cumulative risk of a delayed adverse reaction using two sets of assumptions for a constant hazard rate.

involves methodologies for recurrent events and multivariate failure time data. Some recurrent events are not necessarily serious enough to be associated with treatment withdrawal, and it may be of no real practical benefit to characterize these in a statistically rigorous manner. On the other hand, serious reinfections, seizures, or sensitivity reactions may require this level of attention. To date, there is little experience and few examples to illustrate this problem. Kalbfleisch and Prentice (1980, Chapter 7) provide an excellent discussion and list of additional references on both of these topics.

From the patient's perspective it may be useful to describe the chance of occurrence of any one of some very serious adverse events when exposed chronically for longer periods of time. For example, there may be three serious adverse reactions associated with a particular drug, which follow different shape hazard rates. The first adverse reaction, if it occurs, does so relatively close to initial exposure such as an anaphylactic reaction; the second adverse reaction may follow a constant hazard rate such as gastrointestinal bleeding; and the third type of reaction may be delayed for 6 months with an increasing hazard rate thereafter. What are the chances that any one of these three events may occur by time t? If the short-term event does not occur by time $t_1$, then what are the chances that either of the other two reactions might occur by time $t_2$ ($t_2 > t_1$)?

The actual estimation of such probabilities may not be simple, particularly with the type of follow-up data and loss to follow-up patterns in many clinical trials and observational cohort studies. Further, thr usually low rates of occurrence of the serious events may preclude rigorous statistical treatment. Again, there is not much statistical literature in this area.

### F. More on the Role of Time in Estimating Adverse Event Rates

Accounting for differential exposure time is routine practice in the evaluation of drug efficacy but not so in safety assessment, particularly in the clinical trial setting. For chronically used drugs, experience with clinical trials and observational epidemiological studies shows that seldom do all patients in a study have comparable follow-up times and seldom will the patterns of adverse events be similar from one disease to another.

Not only does the crude rate in many situations underestimate the cumulative rate but also it provides a false sense of the precision with which the true rate is estimated because confidence intervals for the crude rate are based on methods for the binomial distribution. When large disparities in exposure durations exist among patients and events occur throughout the follow-up time, confidence limits for binomial rates will be too narrow. The appropriate confidence limits for cumulative life table rates will, in many situations where heavy censoring is present, be quite wide and therefore include a larger range of possible true cumulative rates consistent with the data. The standard errors associated with life table rates in which fewer patients are exposed for longer durations reflect the weight given to these estimates. Standard errors based on the crude rate estimator (binomial rate) will usually falsely suggest a higher degree of precision in the estimate than is warranted, particularly for longer-term studies with heavy censoring.

All the examples in Section V have been provided to illustrate these points for a number of different drug-disease associations.

## V. EXAMPLES OF ADVERSE EVENTS FOLLOWING DIFFERENT PATTERNS OF OCCURRENCE

This section provides examples of different patterns of occurrence of adverse events observed in subjects in clinical trials or in specially designed postmarketing studies. Because the relationship of time to occurrence of the adverse events is a major focus of the chapter, many of the examples concern longer-term exposure to drugs used to treat chronic conditions. Some of the data sets have been compiled from information collected in more than one study but the rarity of the adverse event was such that a trial of a smaller size would not be useful to illustrate the points. Much of these data were subject to heavy censoring in that subjects were exposed for various durations of time and the observed patterns should be judged with the influence of censoring in mind. That is, the numbers of subjects exposed for longer duration may have been too small to observe any adverse events with reasonable certainty given the rarity of the interval-specific event rate. Thus, estimated decreasing hazard rates should be judged against this fact as well as the likelihood that subjects who were censored earlier might have been at higher risk than subjects remaining under treatment. To estimate these time patterns, life table methods (BMDP program P1L; Dixon, 1984) were utilized to calculate the cumulative adverse event rates and hazard rates, where appropriate, both of which are compared with the crude rate.

Example 2: Skin rash associated with an NSAID. Figure 5 displays the pattern of occurrence of skin rash associated with a nonsteroidal anti-inflammatory drug (NSAID). It suggests that skin rash occurs relatively

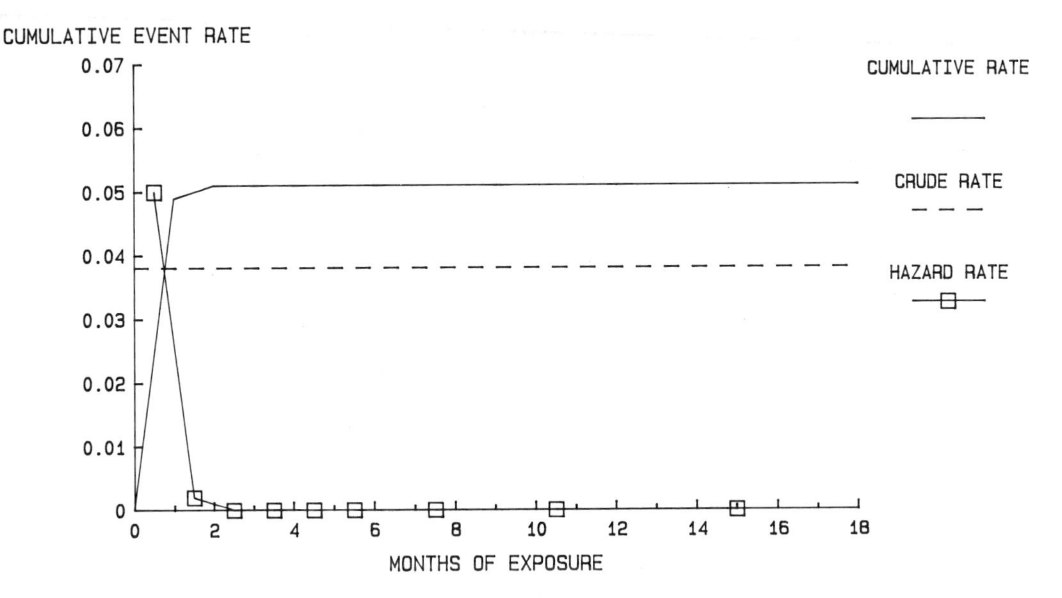


**Fig. 5** Occurrence of skin rash in subjects treated with an NSAID.

soon (e.g., within the first month of exposure) after initial exposure to the drug and that after this initial period of time, if the event has not been observed, subjects on longer durations of treatment are at a decreased level of risk relative to the initial period. This is an example of a decreasing hazard function.

Example 3: Photosensitivity associated with an NSAID. Figure 6 displays the pattern of occurrence of photosensitivity, a reaction in which the skin is sensitive to sunlight exposure, associated with an NSAID. These data suggest that the risk of the reaction is greatest within the first several months, with a gradual decrease (but not a disappearance) of the risk per unit of time with longer exposure. A decreasing hazard function is suggested by these data.

Example 4: Gastric ulcers and bleeding associated with an NSAID. Figure 7 displays the pattern of occurrence of gastric ulcer and bleeding associated with an NSAID. These data suggest that, after an initial lower risk period during the first month of exposure, the risk per unit of time remains relatively constant over the remaining period of exposure. A constant hazard rate is suggested by these data with a possible slight increase early after the first month.

Example 5: Gynecomastia associated with an antiulcer drug. Figure 8 displays the pattern of occurrence of gynecomastia, a reaction which enlarges the breasts, associated with an antiulcer drug. These data were collected in a postmarketing study and illustrate several points. The first is that the hazard rate appears to be relatively constant, with events being observed throughout the entire follow-up period. Second, these data were summarized in a promotional brochure to show the risk associated with each of two intervals of exposure time. Table 2 displays the data as summarized in the promotional brochure and illustrates how these data are presented to the prescribing physician in an advertisement.

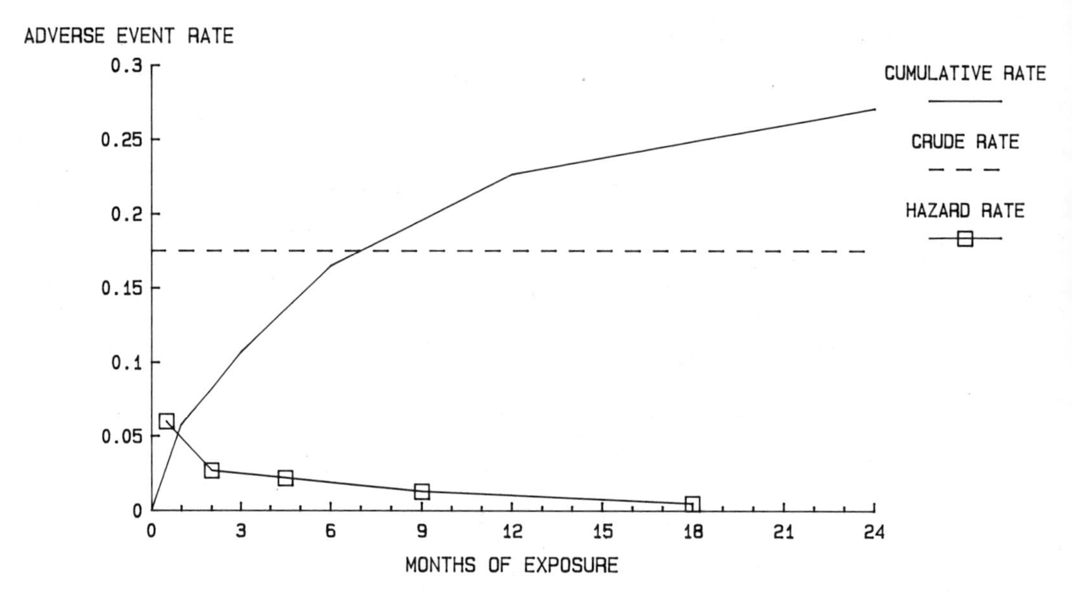


Fig. 6 Occurrence of photosensitivity reactions in subjects treated with an NSAID.

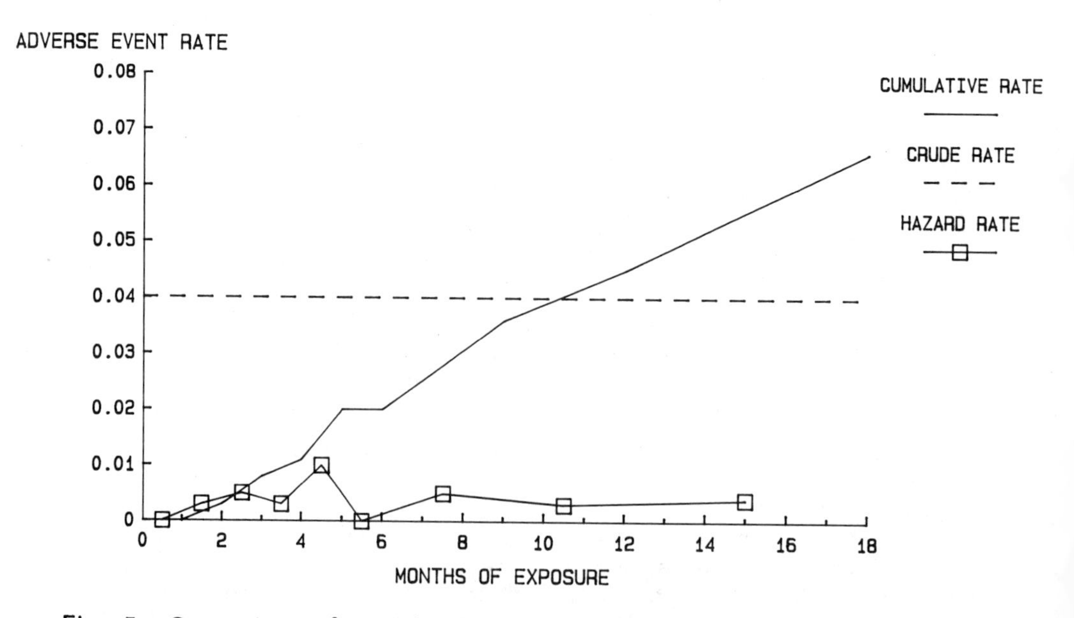


Fig. 7 Occurrence of gastric ulcers and bleeding in subjects treated with an NSAID.

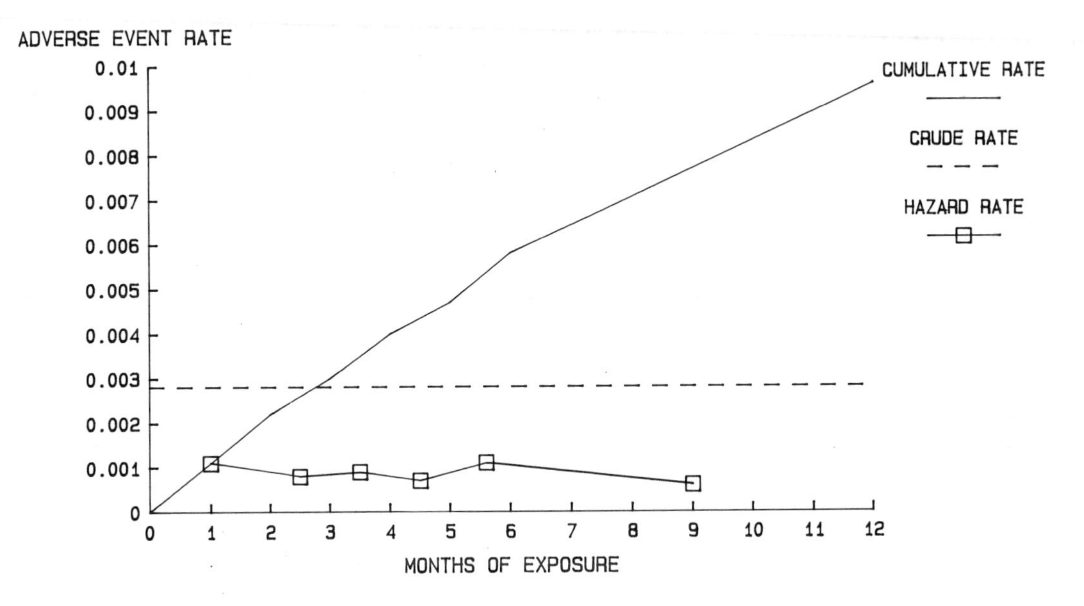


SOURCE: GIFFORD ET AL. 1980, ROSSI ET AL.1983

**Fig. 8** Occurrence of gynocomastia in subjects treated with an antiulcer drug.

Example 6: Six adverse reactions associated with an antihypertensive drug. Figure 9 displays the occurrence of six adverse effects, namely rash, taste disturbance, neutropenia, proteinuria, hypotension, and miscellaneous, that resulted in discontinuation of an antihypertensive drug. Of particular interest are the patterns of two of the adverse reactions, neutropenia and proteinuria. All of the neutropenia reactions were observed in the first 3 months and none thereafter, suggesting a decreasing hazard rate. The majority of proteinuria reactions were observed in the 7- to 12-month period, suggesting an increasing hazard function. Figure 10 illustrates the estimated hazard rate for proteinuria associated with this drug. These data are interesting in that they provide evidence of differing hazard functions for different adverse reactions as well as competing risks posed by each reaction on the same drug in the same patient population.

**Table 2** Frequency of Gynocomastia Expressed as a Percentage of Courses of Therapy[a]

| Duration of courses of therapy | | | |
|---|---|---|---|
| (4800 courses) | <8 weeks–6 months (3095 courses) | <6 months (1827 courses) | Overall average (9763 courses) |
| .0% | .2% | .8% | .2% |

[a]From promotional brochure, SK&F Lab. Co. (1981).

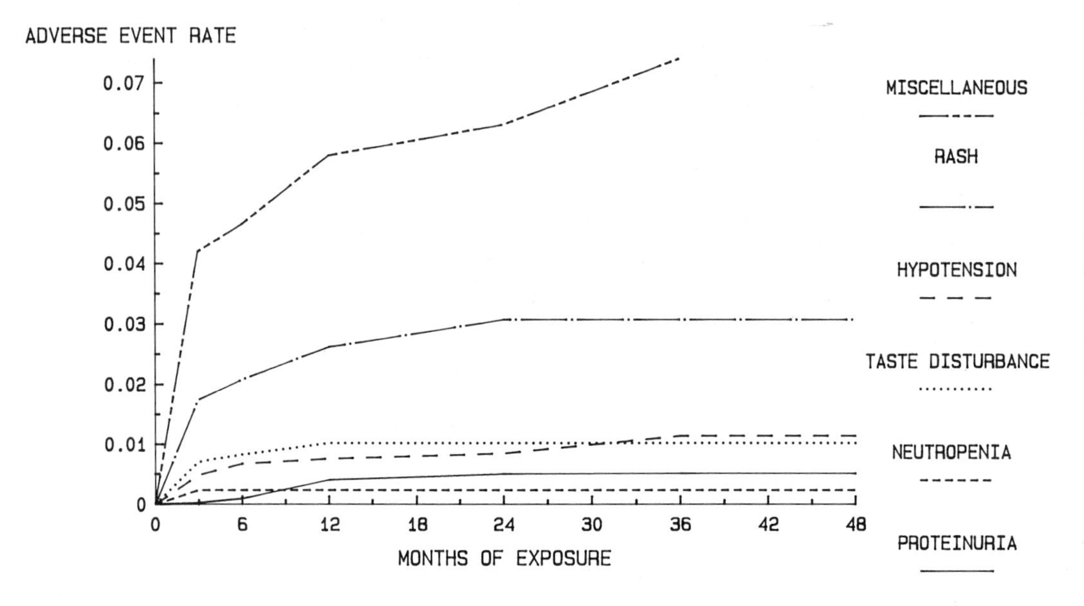


SOURCE: GROEL ET AL. (1983)

Fig. 9 Cumulative rates of six adverse reactions associated with treatment withdrawal of an antihypertensive drug.

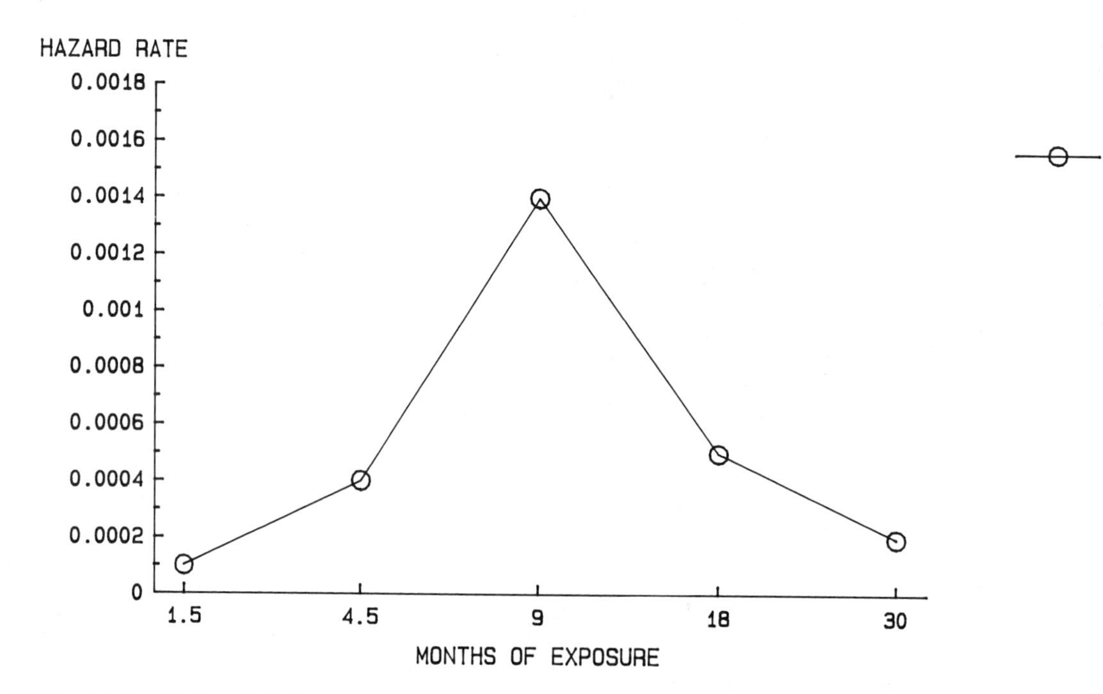


SOURCE: GROEL ET AL. (1983)

Fig. 10 Estimated hazard rate for proteinuria associated with antihypertensive drug.

Example 7: Seizures associated with an antidepressant—relationship with dose and duration of use. Figure 11 displays the occurrence of seizures associated with an antidepressant drug in which both the duration of exposure and the highest dose of the drug given to the subject are simultaneously accounted for. This is an interesting example because frequently adverse reaction rates are displayed by dose but not simultaneously by duration of exposure so that it is not clear whether the estimated rates reflect effects of dose or of longer duration of exposure. However, subjects often are titrated upward on dose so that they may be exposed to lower doses prior to the dose on which they may have an event. Whether a subject's duration of use at a lower dose should be counted as part of the duration at the higher dose is arguable, and no one way of displaying or analyzing the data is generally the best.

Example 8: Discontinuation rates associated with an antihypertensive drug. Sometimes, the reported results of a study describe the discontinuation rates of subjects under study. Generally, the discontinuations may be for lack of effectiveness of a treatment or for a side effect experienced by a subject. With longer-term treatment, the temporal pattern of the discontinuations is informative and may shed light on the role of efficacy as well as safety of the drug under long-term use. Figure 12 presents data on the cumulative discontinuation rates for treatment failure or side effects associated with an antihypertensive drug. When subjects are not all followed for the same duration of time, a binomial discontinuation rate with no reference to time is not totally informative.

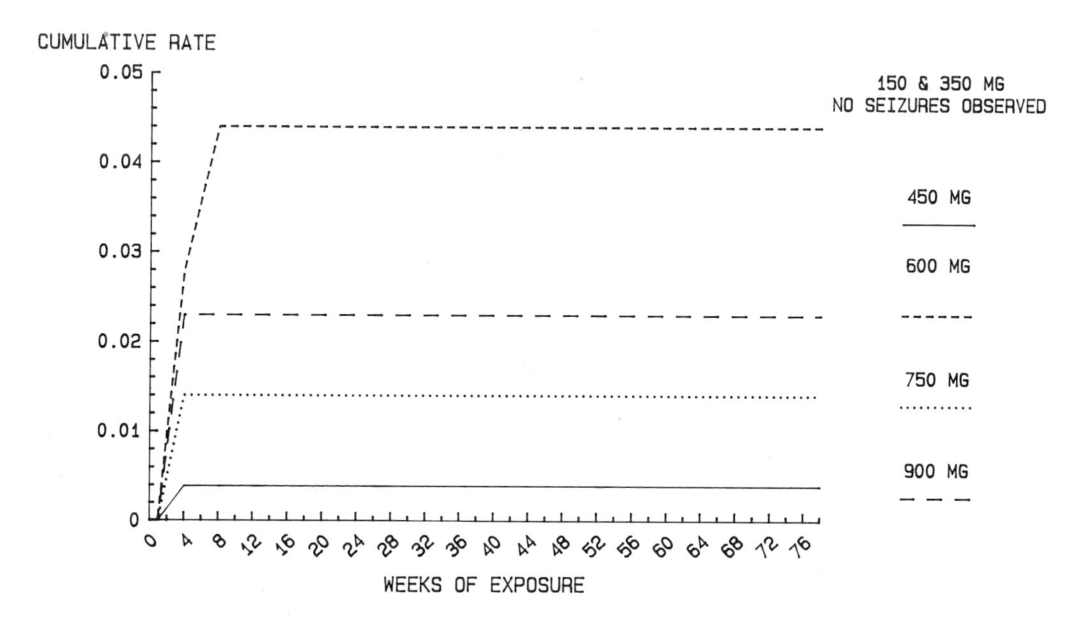


Fig. 11 Cumulative incidence rates of seizures by highest dose of antidepressant attained.

| MONTHS | EXPOSED THIS LONG | TREATMENT FAILURE | CUM. RATE | SIDE EFFECT | CUM. RATE |
|---|---|---|---|---|---|
| 0 - 3 | 7,103 | 141 | 0.025 | 267 | 0.046 |
| 4 - 6 | 4,397 | 31 | 0.033 | 46 | 0.057 |
| 7 - 12 | 3,614 | 38 | 0.046 | 62 | 0.085 |
| 13 - 24 | 1,704 | 20 | 0.066 | 17 | 0.102 |
| 25 - 36 | 511 | 0 | 0.066 | 5 | 0.116 |
| 37 - 48 | 163 | 0 | 0.066 | 0 | 0.116 |

SOURCE: GROEL ET AL. (1983)

**Fig. 12** Cumulative discontinuation rates for side effects or treatment failure (no efficacy).

Example 9: Side effects associated with increasing dose levels. Table 3 shows the relationship between increasing doses of an NSAID and the rate of occurrence of gastrointestinal (GI) side effects, and Table 4 shows the relationship between increasing doses of an antidepressant and the occurrence rate of seizures. Sometimes it is also of interest to take account simultaneously of pretreatment or preexposure status of the patient as well as dose. Table 5 shows the relationship between increasing dose and the existence of pretreatment renal disease to the rate of occurrence of proteinuria in patients treated with an antihypertensive drug. Although these data suggest an increasing dose-response relationship, their interpretation is sometimes difficult because studies will include patients exposed to a number of different dosage levels of a given drug. Usually patients may be titrated upward on increasing doses dependent on the effectiveness

**Table 3** Incidence Rates of Gastrointestinal Effects by Dose[a]

| Dose (mg/day) | Patients | % with GI effects | % with ulcers | % discontinued |
|---|---|---|---|---|
| 10 | 438 | 9.6 | .5 | 1.6 |
| 20 | 2544 | 18.4 | .9 | 3.4 |
| 30 | 717 | 22.3 | 2.6 | 4.3 |
| 40 | 288 | 29.9 | 6.9 | 6.6 |

[a]From Pitts (1980).

**Table 4** Incidence Rates of Seizure by Dose[a]

| Dose (mg/day) | Rate | 95% Confidence limits |
|---|---|---|
| 150 | 0/74 = 0 | (0, .040) |
| 300 | 0/334 = 0 | (0, .009) |
| 450 | 1/393 = .003 | (.0001, .014) |
| 600 | 5/199 = .025 | (.008, .058) |
| 750 | 1/107 = .009 | (.0002, .051) |
| 900 | 1/45 = .022 | (.0006, .118) |

[a]From Peck et al. (1983).

response achieved or on the occurrence of tolerance at a given dose. Thus, there may not exist mutually exclusive cohorts of patients exposed to only one dose level. Unless data are presented in a manner to do so, it may not be possible to distinguish risks associated with increasing dose from risk associated with increasing or longer duration of exposure. For this purpose, it is useful to compare Table 4 with Figure 10, which shows the same data but with duration of exposure displayed in the figure. This display suggests that, regardless of dosage level, the occurrence of the seizure was observed within the first month and therefore longer time of exposure and follow-up to reach the higher dose would not be an explanation for the observed rates.

**Table 5** Incidence Rates of Proteinuria by Dose and Prior Renal Disease[a]

| Preexisting disease | Dose (mg) | Rate | 95% confidence limits |
|---|---|---|---|
| No | ≤150 | 4/2126 = .002 | (.0005, .0048) |
| No | >150 | 15/1447 = .010 | (.0058, .017) |
| Yes | ≤150 | 12/1180 = .010 | (.0053, .018) |
| Yes | >150 | 36/1016 = .035 | (.025, .049) |

[a]From Groel et al. (1983).

## VI. PREMARKETING ASSESSMENT OF SAFETY

Prior to the marketing of a drug its safety is evaluated in clinical trials, some of which are controlled and some of which are uncontrolled. A drug will usually be studied in 5 to 10 or more clinical trials of varying durations and designs comprising a total of 1000 to 2000 exposed subjects in the premarketing stage. Adverse-event information is obtained in controlled studies usually in a blinded manner, whereas unblinded methods of ascertainment are usually employed in the uncontrolled studies. This is an important consideration in deciding how to compare or combine rates from two different studies with different designs.

A clinical trial is generally characterized by a predefined period of treatment and predefined frequencies of subject follow-up visits during this treatment period. A protocol will contain questions related to various conditions of the subject categorized by body system, on which a subject will be systematically assessed to determine whether the condition has or has not been experienced during the treatment period or since the last visit. Some events that are recorded may be elicited while others may be volunteered by the subject. Table 6 presents one classification scheme for categorizing the body system events that might be observed.

### A. Controlled Trials

Randomized controlled trials (RCTs) are the primary mechanism for assessing the effectiveness of a new drug. RCTs also provide a comparison of the rate of occurrence of adverse events in both test and control subjects

**Table 6** Adverse Events by Body System

| Body system | Number of patients with event (%) |
|---|---|
| Cardiovascular | |
| Digestive | |
| Endocrine | |
| Hemic and lymphatic | |
| Metabolic and nutritional | |
| Musculoskeletal | |
| Nervous | |
| Respiratory | |
| Skin and appendages | |
| Special senses | |
| Urogenital | |
| Body as a whole | |

to determine whether increased rates are associated with drug exposure. The principal limitation of RCTs as a mechanism for comparative safety evaluation is their size and duration. Usually, the sample size of an RCT is based on detecting certain levels of efficacy, and sample sizes based on efficacy considerations will be inadequate to statistically demonstrate increased adverse event rates except in unusual situations where the drug-associated rates are dramatically increased over control rates. Further, the effectiveness of a drug can usually be demonstrated in a relatively short period of time, say 3 to 6 months, as is the case with drugs for ulcers, hypertension, depression, arthritis, and arrhythmias, and this time frame may be inadequate for safety evaluation.

Despite these limitations, RCTs are a major source of detailed and planned safety information not usually available to postmarketing or observational studies. For example, during planned periodic visits subject data are collected on changes in dosing patterns, changing clinical chemistry measurements, changing efficacy measurements, and changing concomitant conditions or drug use, all of which are available for analysis. By use of specifically designed instruments to measure and record subtle changes or treatment emergent signs and symptoms, otherwise undetectable differences between treatment and control groups might be demonstrated. At least one such instrument has been under development and tested in the psychopharmacology field (Levine and Schooler, 1986). Information related to reasons for differential duration of subject drug exposure or withdrawal from treatment is also likely to be better captured in the clinical trial setting and useful in analysis.

RCTs provide for unbiased statistical comparisons of rates of adverse events in test and control groups, but it should first be established that any differences, or lack of differences, are not due to pretreatment differences in baseline characteristics of the subjects studied, that comparable methods of event ascertainment are used for all groups, that definitions of the adverse event are standardized for both groups, and that censoring mechanisms, if present, are not responsible for observed differences or lack of differences. Randomization and blinding in RCTs can contribute a great deal to ensuring the unbiasedness of these comparisons, but limited sample sizes are likely to preclude further gains.

For purposes of organization, RCTs will be categorized into short-term and long-term studies. Short-term studies will be assumed to have all subjects followed for the same duration of exposure. Some studies are short-term because the disease being treated is self-limiting or because one course of therapy, such as with antibiotics, is sufficient to treat the disease. Long-term studies will be defined as those in which not all subjects are exposed and followed for the same duration of time. Some long-term studies are carried out on chronic diseases, where it can take months or years for an effect to be demonstrated or where many patients must be followed to demonstrate moderate reductions in low-incidence outcomes such as morbidity or mortality. Many large multicenter studies conducted by the National Institutes of Health are examples of longer-term trials. However, studies of 3 to 6 months duration can be considered long-term when they are characterized by numerous dropouts and differential subject exposure. The focus on long-term studies is to point out that the estimation and testing of differences in adverse event rates are dependent on examination of patterns of differential drug exposure, censoring mechanisms, and time patterns that adverse events follow.

### 1. Short-Term Trials

In short-term RCTs, the statistical comparison of observed adverse event rates in the treated group and control group is of major interest. Statistical methods based on the binomial distribution can be used for tests of hypotheses and confidence intervals on the difference of two rates. As an example, for each treatment group the proportion of subjects with selected adverse event terms within each body system category, as in Table 6, can be compared. Other rates of interest are the proportion of subjects experiencing any adverse event, or experiencing any adverse event in a body system category, or being discontinued from treatment for an adverse event, for a hospitalization, or for death while on treatment. Subjects can experience more than one type of adverse event, so multiple events characteristic of a severe outcome sometimes are combined. Terminology of adverse events is often loose, and caution should be exercised in splitting up events into categories of terms which have very low rates. Conversely, terms referring to events that are not related should not be combined.

Confidence interval estimation for the difference between rates and for each individual group treatment rate is informative in describing the range of true differences of rates consistent with the data and because it reflects the imprecision in the estimates of rates when sample sizes are small, as they usually will be in many RCTs. When planning formal statistical comparisons for many different categories of events which may not have been a priori comparisons of interest, some consideration should be given to statistical adjustments for multiple testing, multiple correlated outcomes, and related issues.

### 2. Long-Term Trials

Most longer-term RCTs of drugs are in chronic diseases, where the drug effect takes time to exhibit itself or where the accrual of large sample sizes and/or exposure time is necessary to demonstrate statistically the benefit of the drug. Such studies can be either placebo-controlled or active treatment-controlled. In chronic disease conditions, such as arthritis, where an effective outcome may be evident within several months, it becomes difficult to keep placebo subjects under study for prolonged periods of time without withdrawals for lack of effectiveness. Thus, placebo-controlled studies of diseases for which palliative treatment is the objective will generally be of more limited duration. This is a basic difference between long-term placebo-controlled studies of drugs which do not otherwise symptomatically affect the subjects (e.g., lipid-lowering agent or platelet-thinning agents) and in which the efficacy outcome may be reduction in mortality or morbidity.

In long-term studies not all subjects are followed for the same period of time because some become censored for administrative reasons, some for lack of effectiveness, some for adverse events, some for a combination of both, and some for formal study termination. The methods to account for such differential exposure in adverse reaction assessment have been inconsistently applied and reported in the medical literature. We will describe three long-term placebo-controlled RCTs that illustrate different methods of analysis of adverse event data.

In a report of the Coronary Drug Project Research Group (1977), on 2680 placebo-treated subjects and 1051 clofibrate-treated subjects, a statistically significant increase over placebo was demonstrated in the incidence of cholecystitis and cholelithiasis in subjects treated with clofibrate in a 6-year follow-up period. Because not all subjects were treated for the same duration, the life table analysis was the method employed to compare the groups with respect to the occurrence of gallbladder disease. Data on the adverse event were displayed according to its interval of occurrence, allowing calculation of the curves in the next two figures.

Figure 13 illustrates the cumulative occurrence of gallbladder disease in the placebo and clofibrate gorups and Figure 14 displays the estimated hazard functions for gallbladder disease for both groups. This example is of interest in that the background rate over a long interval of time is available for the placebo group to assess the natural course and pattern of occurrence of gallbladder disease.

In another large-scale placebo-controlled RCT (Norwegian Multicenter Study Group, 1981) the drug Timolol was studied in patients surviving acute myocardial infarction. A total of 1884 patients (945 on Timolol and 939 on placebo) were treated and followed for 12 to 33 months. Adverse events were compared between the two groups and the numbers of subjects withdrawn for all reasons according to time of withdrawal was displayed, amounting to 219 in the placebo group and 275 in the Timolol group. The numbers of subjects in each treatment group with various adverse reactions were displayed and statistically compared between groups. However, whether the duration of exposure was accounted for in the safety comparisons, as it was in the efficacy analysis of cardiovascular death, was not clear from the report and was not discussed.

In another large-scale multiclinic placebo-controlled RCT (Anturane Reinfarction Trial Research Group, 1978, 1980) the drug sulfinpyrazone was studied in 1558 subjects followed for an average of 16 months after a documented myocardial infarction. Four hundred and fifteen subjects

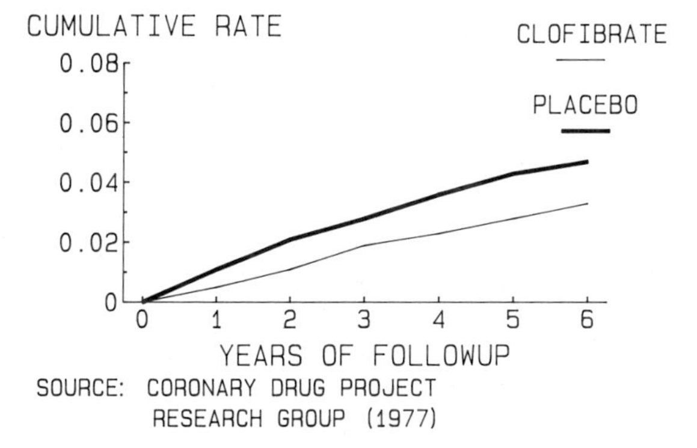


**Fig. 13** Cumulative incidence rate of new gallbladder disease.

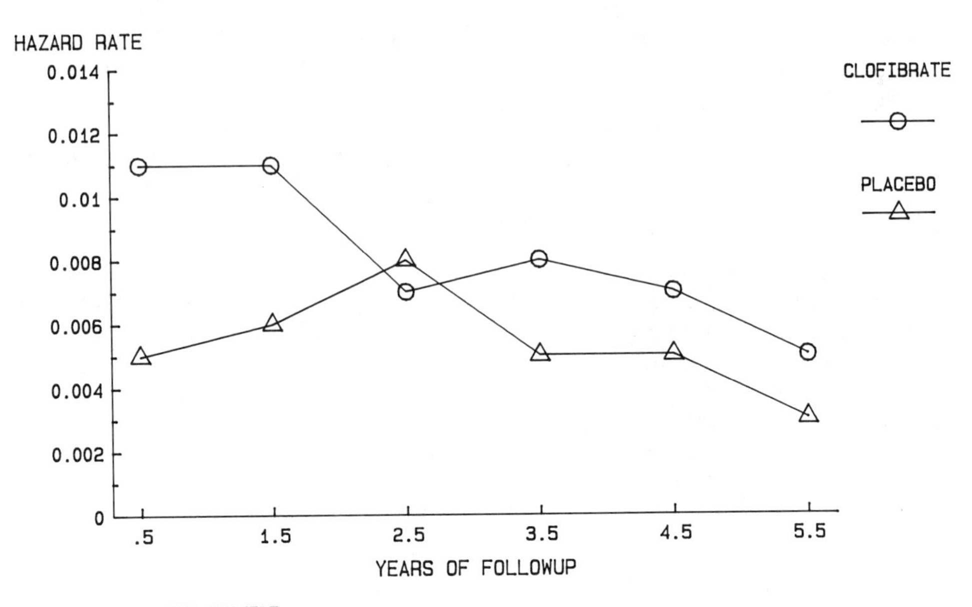


Fig. 14 Estimated hazard functions for new gallbladder disease.

withdrew prematurely from the study, having an average exposure period of 7.4 months. All other subjects not prematurely withdrawing were exposed and followed between 12 and 24 months. The numbers of subjects with signs and symptoms that emerged after entry into the trial were reported and compared between the two treatment groups, but the time of their occurrence was neither reported nor accounted for in the analysis as it was in the efficacy analysis of sudden death.

These three studies reported in the literature illustrate the nonuniform methods of reporting and analyzing adverse event data in long-term clinical trials. The clofibrate study reported occurrence of gallbladder disease by intervals of exposure time, analyzed the events accounting for differential exposure, and provided estimates of the cumulative rate of gallbladder disease. The Timolol and Anturane studies, while also having considerable differential exposure among subjects, did not report events by exposure time and essentially compared crude rates of adverse events among treatment groups.

Methods for the statistical comparison of adverse event rates derived from a test and control group in long-term RCTs are formally similar to those used in comparative evaluation of efficacy where the outcome of interest is a time-to-occurrence variable. Application of such methods requires some assumptions about the relationship of the hazard functions for the adverse event in each of the two treatment groups. These methods also require assumptions about the relationship of the censoring distribution to the outcome event and about the comparability of censoring between the two treatment groups. Many of these assumptions may be violated for adverse event data, particularly when a high degree of differential dropout, withdrawals from treatment, and other censoring mechanisms are operating. Tests for the comparison of survival curves (Lee, 1980; Kalbfleisch and Prentice, 1980) assume that there is a proportional relationship between the hazard rate for the control group and the test drug group, a weaker assumption than requiring each group to follow a constant but different hazard function. However, even a proportional hazard assumption may not be satisfied if the particular adverse event is delayed, increasing, or decreasing in time and the comparative group is placebo-treated. Different comparative survival analysis methods (Tarone and Ware, 1977) weight early and later events in different ways which may have some bearing on the interpretation of the comparative analysis. Because of the low statistical power of RCTs for adverse event evaluation, comparative hypothesis tests may not be as useful as reporting confidence intervals on the difference between cumulative rates, the ratio of rates, or the ratio of hazards.

Reporting of cumulative rates of adverse events where the time period is specified is standard practice in efficacy evaluation of long-term studies and should be the practice for safety evaluation for more serious events. The role of duration of exposure, the number of subjects exposed for various durations, and the time pattern of the adverse event all need to be addressed simultaneously when making a comparative evaluation of drug safety for chronically used drugs. The estimation of adverse event rates in single-sample treatment groups will be the major focus of the next section, dealing with uncontrolled studies.

### B. Uncontrolled Trials

Uncontrolled or open-label clinical trials of drugs provide a major source of adverse event data on subjects, particularly for serious events associated with long-term exposure. Some uncommon events such as liver injury, renal failure, and hematologic abnormalities, and events that reoccur on rechallenge, bear a close temporal relationship to drug use and resolve on discontinuation of treatment may be identified in uncontrolled studies. Serious adverse events associated with treatment withdrawal or dose reduction are the focus of this section. Subjects in uncontrolled studies usually know they are exposed to the test drug and therefore recording of some types of events can be influenced by the ascertainment methods. Uncontrolled studies are usually of two types. Either the trial is from the outset designed to collect safety experience in an unblinded and uncontrolled manner or subjects are continued on the test drug for longer periods of time in an unblinded manner after having first completed the double-blind randomized portion of a controlled study. Few clinical trials have placebo control groups followed beyond 3 to 6 months for direct comparison of adverse event rates because drug efficacy can be demonstrated in that interval of time and it is ethically problematic to extend placebo treatment in conditions or diseases like pain, ulcers, hypertension, depression, arthritis, arrhythmias, and other diseases for which palliative treatment is available. Thus, uncontrolled trials afford the opportunity of assessing the rates of occurrence of serious adverse events under longer-term exposure.

The characterization of serious adverse events derived from uncontrolled clinical trials is a descriptive effort in estimating patterns of occurrence and rates of serious adverse events and in determining whether there exists a relationship between these rates and relevant factors such as age, sex, disease severity, dose, and duration of exposure.

Most uncontrolled trials are characterized by major differences in the durations of follow-up and exposure of subjects in part because subjects will generally remain under follow-up only if they perceive an effective response. Figure 15 provides data on the average proportion of subjects remaining under treatment for varying periods of time in Phase III clinical trials based on duration of treatment with four nonsteroidal anti-inflammatory, three cardiovascular, and four psychopharmacological drugs (Idanpaan-Heikkila, 1983). While some of these data are from controlled trials with predetermined termination times, the point is that markedly different proportions of subjects are exposed for various durations of time and this needs to be taken into account when estimating rates from uncontrolled trials.

As there are many reasons for differential durations of drug exposure among subjects in an uncontrolled trial, it can be informative for interpretation of study results to quantify the rates of discontinuation and reasons for withdrawal or loss to follow-up. Subjects will have different patterns of follow-up visits, may change their dosing and compliance patterns, and may exit from the study for a number of reasons distinct from a controlled study. These censoring mechanisms can convey a considerable amount of information about which or how many subjects to consider at risk for the adverse events of interest. When a high degree of differential exposure among subjects exists, certain assumptions must be made in the statistical summary of the event rates. For example, in the life table method differential durations of drug exposure are handled in a manner that assumes subjects with shorter exposure time are similar to those with longer exposure and, had they not been censored, would have a risk similar to those with longer exposure. The assumption is that the censoring

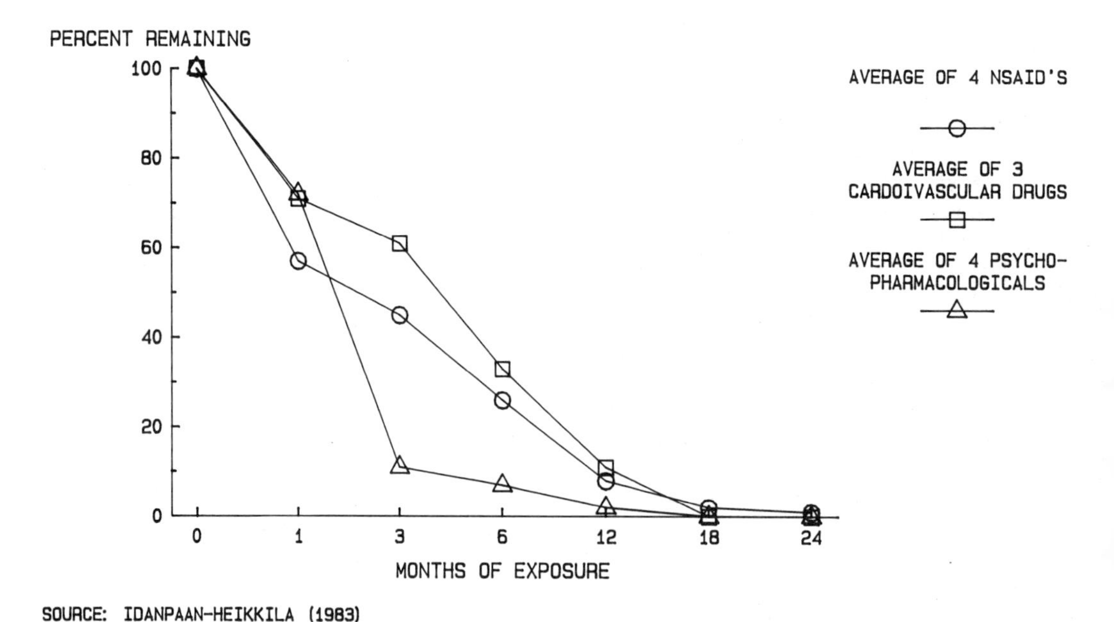


Fig. 15 Average proportion of subjects remaining in Phase III clinical studies according to duration of treatment.

distribution is independent of the outcome (the adverse event). Such an assumption can be questionable, but in the absence of information to assess the appropriateness of the assumption few options are left for a meaningful analysis.

One method displaying discontinuation rates and adverse event rates is illustrated in Figure 12. This example uses the life table method and displays cumulative rates as a function of exposure time. A display of data according to a complete life table layout, including different columns for each category of withdrawal, could provide more information on discontinuation patterns over time. In concept, separate life tables can be produced for relevant age, sex, dose, and other factor combinations and compared for differences in patterns or cumulative rates. Such a display might also be useful if one were combining cumulative rates of specific events from more than one study as it would provide insight into which studies may justifiably on a statistical basis be combined. However, this is a very exploratory and data-driven effort and is more useful for describing the data than for deriving any inferential conclusions.

### C. Combining Adverse Event Data from Several Clinical Trials

Serious adverse events like seizure, liver damage, or gastrointestinal ulcer are sometimes characteristic of a drug class and considered drug-induced. These serious adverse events may be observed in several clinical trials whose lengths of follow-up differ, and interest centers on obtaining an overall estimate of their rate of occurrence. The medical literature provides many examples of summary adverse event data from several trials or from many patients exposed to a drug for different durations. The examples of seizure, proteinuria, and neutropenia in Section V are from subjects exposed in several trials. The purpose of this section is to discuss strategies for combining adverse event data from several trials in order to estimate rates with improved precision or to examine patterns of occurrence of adverse events that might not be apparent in one study but might be if several studies are evaluated together. Recent changes in the FDA regulations [21 CFR 314.50(d)(5)(vii)] require an integrated summary of all safety data submitted in a New Drug Application, including the summary of estimated rates of adverse reactions from controlled and uncontrolled trials separately. Such a summary is intended to help in the assessment of the overall rates of adverse events which may be reflected in the drug labeling.

Since it may not be appropriate to combine data from studies with markedly different designs, the choice of which studies and which observed rates to combine deserves fuller examination than this section allows and therefore will not be discussed except to point out that such an exercise should be carried out cautiously. It will be assumed that the definition of the event and its methods of ascertainment are similar among the several trials chosen for combining.

Conceptually, assume that data on a specific serious adverse event is available from K separate trials of subjects who have been treated with the same drug and that the adverse event data can be displayed as K separate life tables with common follow-up intervals. The treated groups within each trial may have different numbers of subjects initially exposed, different durations of exposure, and different intensities of censoring over time. The issue is how best to characterize the adverse event rates from several studies which may differ in length. One strategy is to obtain an overall estimate of the cumulative adverse event rate by combining data from each trial within comparable intervals of exposure and follow-up and calculate a cumulative life table rate as if all the data came from a common source. Such an approach implicitly assumes that the pattern of occurrence of the adverse events in each group follows a common hazard rate function except that some groups are exposed for longer durations of time. Table 7 illustrates data from three trials of different durations where this assumption applies, because the 1-month rates in trials 1, 2, and 3 are the same while the interval specific rates in the second month are the same for trials 2 and 3. The 3-month cumulative rate is the same for all the data combined as it is for group 3 alone. The estimates of the cumulative rates of adverse events (calculated according to life table methods) at the end of each interval of exposure for the combined data are .015, .030, and .069, respectively. This example illustrates the impact of combining large short-term studies with smaller longer-term studies and the importance of taking the differential exposure in each trial into account when calculating an overall rate. The crude rates unadjusted for exposure time for trials 1, 2,

**Table 7** Combining Adverse Event Data from Three Trials

| Trial | Measure | Months of exposure | | |
|---|---|---|---|---|
| | | 1 | 2 | 3 |
| 1 | No. with event | 5 | | |
| | No. exposed | 500 | | |
| | Cumulative rate | .01 | | |
| | Crude rate | .01 | | |
| 2 | No. with event | 2 | 2 | |
| | No. exposed | 200 | 100 | |
| | Cumulative rate | .01 | .03 | |
| | Crude rate | .02 | | |
| 3 | No. with event | 1 | 1 | 1 |
| | No. exposed | 100 | 50 | 25 |
| | Cumulative rate | .01 | .03 | .069 |
| | Crude rate | .03 | | |
| Combined | No. with event | 8 | 3 | 1 |
| | No. exposed | 800 | 150 | 25 |
| | Cumulative rate | .01 | .03 | .069 |
| | Crude rate | .015 | | |

and 3 are 5/100 = .01, 4/200 = .02, and 3/100 = .03, respectively, and the crude rate for all three trials combined is 12/800 = .015. It is clear that the large number of subjects followed for a short period of time dominates the overall crude rate and provides a misleading estimate of the rate over time. An additional impact is that the confidence intervals for the crude rate will be much more narrow than those for the cumulative rate because the large number of subjects in trial 1, followed for a short period, heavily influences the standard error of the crude rate while it has less influence on the cumulative rate. Subjects not followed beyond 1 month do not contribute to the standard error of the cumulative estimates beyond 1 month. For example, the 95% confidence limits for the crude rate based on the combined data are (.004, .020), using standard asymptotic normal theory calculations for binomial rates (e.g., Kleinbaum et al., 1982). The corresponding 95% confidence limits for the cumulative rates (method in Kalbfleisch and Prentice, 1980, p. 15) at each of the three time points are (.008, .033), (.023, .107), and (.034, .216), indicating marked differences in the certainty of the true cumulative rates compared with the crude rate limits. Further, the increasing width of the confidence intervals on the cumulative rates illustrates the imprecision in estimates of cumulative rates as smaller numbers of subjects are followed in time.

In practice, preliminary statistical tests to determine whether it is justifiable to combine data from two or more studies can be carried out. Such statistical tests are applicable to the testing of the equality of K survival curves or cumulative life table curves (e.g., Breslow, 1970; Lee, 1980, p. 139). Because adverse event rates are usually small, it is unlikely that tests of the equality of K life table curves would have much statistical power to detect differences among the groups. On the other hand, it is possible that the shapes of the hazard functions which generate the observed cumulative adverse event rate curves might differ if some treatment groups consist of different age, sex, or severity factors which have a predictive association with the pattern of occurrence of the adverse event. When statistical differences among the trials, or among subgroups, are demonstrated, it may be useful to further explore the reasons for the differences.

Censoring patterns observed in clinical trials can differ markedly, as can the reasons for such censoring, so insight may be gained from examining the characteristics of subjects who are censored, subjects who remain under treatment, and subjects who contribute adverse events. These characteristics might be identified and compared across trials. Makuch (1982) and Hankey and Myers (1971) have dealt with the problem of estimating an overall survival curve when covariable associations with different survival curves exist. Duffy (1983) also has considered the merging of survival rates from several studies to obtain an overall rate.

## VII. POSTMARKETING SAFETY ASSESSMENT AND SURVEILLANCE

### A. Purposes, Goals, and Strategies

The amount of safety information that can practically be obtained on a drug prior to its marketing is limited by the number and types of subjects exposed to the drug during its development and testing and by the relatively controlled conditions under which the drug is studied in the premarketing stage. The patient population selected for study in the premarketing stage may have less concomitant disease and may be less exposed to concomitant drugs. Once available to the general population, a drug may be used in a manner not previously evaluated, or at different doses, or in populations with characteristics different from those of the subjects tested premarketing. Patient populations may experience safety profiles on the drug which are different from those observed premarketing.

Safety information of varying quality and utility is accrued from many sources after a drug is approved and available to the general public, and that information is used to update incidence rates of some adverse reactions, to add reports of serious adverse reactions to the professional labeling of the drug, or to modify the labeling directions for appropriate prescribing of the drug.

Postmarketing surveillance (PMS) refers to the process of obtaining that information on drug experience after a drug is approved for use by the general population. PMS will be defined as a collection of ad hoc and formal studies and monitoring systems used

1. To estimate or compare the incidence of adverse reactions, to explore a specific pharmacologic effect, or to obtain more information of a circumscribed nature about a drug,
2. To determine long-term effects of a drug on morbidity and mortality, and
3. To study patient populations not adequately studied in the premarketing phase, e.g., children, the aged, or women of childbearing ages.

Although PMS methods have been considered since the late 1960s, the past 10 years have seen an increasing and evolving experience with PMS by the Food and Drug Administration and the pharmaceutical industry. The Food and Drug Administration funded an extensive project titled "An Experiment in Early Postmarketing Surveillance of Drugs" (IMS America, Health Care Services Research Group, 1980) to provide the basis for development of an overall system of postmarketing surveillance. The report described methods of PMS existing at that time and evaluated the strengths and limitations of each. These methods include

1. The systematic collection and interpretation of voluntary spontaneouse adverse reaction reports, as in the United States (Faich, 1986) and the yellow card system in the United Kingdom.
2. Hospital-based monitoring programs such as the Boston Collaborative Drug Surveillance Program (Slone et al., 1966, 1969; Jick et al., 1970), which is geared primarily toward acute in-hospital drug effects, and the Drug Epidemiology Unit multiple case-control approach (Slone et al., 1977), which was designed to discover and quantify suspected drug-induced diseases associated with drugs in common use over long periods of time by outpatients but which result in hospitalization.
3. Drug use and disease or diagnosis record linkage systems such as those based on computerized records from health maintenance organizations like the Kaiser Permanente groups in California, the Seattle, Washington, Puget Sound Health Cooperative (Jick et al., 1984; Jick, 1979), the Oxford Record Linkage Study in the United

Kingdom, and more recently computerized third-party reimbursement record data bases such as Medicaid.

4. Special studies such as that of the Royal College of General Practitioners (1974), studies of oral contraceptives, and several U.S. industry-sponsored studies of specific newly marketed drugs (see Section VII,C). These industry-sponsored studies have usually obtained patient level data through physician or pharmacist recruitment of the patient in the normal course of health care.

More recently, the Drug Surveillance Research Unit at the University of Southampton, under the direction of Dr. William Inman, has set up a Prescription Event Monitoring (PEM) system (PEM News, 1983) to detect adverse events occurring during drug treatment. PEM uses a method which identifies cohorts of drug users through the Prescription Pricing Authority in the United Kingdom. Identified individuals are followed up by questionnaires mailed to the treating physician, who then records all events occurring to the patient while on the drug and after discontinuation. The incidence of these events (as opposed to adverse reactions) can then be estimated from the user cohort.

All of the above approaches may be used alone, or in combination, to develop further safety information on a drug. Of course, the methods and strategies used depend on the prevalence of exposure to the drugs, the rarity of the side effect or adverse reaction, the nature of the reaction (e.g., hospitalization, death, outpatient treatment), whether the reaction is known or suspected or is unknown, the time pattern of the reaction (e.g., acute or delayed), the data bases available, the methods for patient recruitment, and the choice of control groups if required. Jick et al. (1979) and Mitchell et al. (1980) provide insightful discussions for various scenarios under which different surveillance and follow-up methods might be useful.

The following scheme is useful for conceptually describing the strategies for studying drugs and adverse reactions in the postmarketing stage. These strategies are not considered fixed, and it should be noted that studies of newly marketed drugs (first 2 to 3 years after approval) may require data accrual methods different from those for drugs on the market for some time. The emphasis here is on comparative observational studies.

Conceptual Strategy for Studying Drug-Induced Diseases[a]

| Rate of illness[b] | | Prevalence of use | Acute or | Known or | Research |
|---|---|---|---|---|---|
| On drug | background | in population | delayed | unknown | strategy |
| High | Low | High | Either | Known | Case-control |
| High | Low | High or low | Either | Unknown | Cohort |
| Low | High | High | Either | Known | Questionable |
| Low | High | Low | Either | Unknown | Questionable |
| Low | Low | High | Either | Known | Case-control |
| Low | Low | High | Either | Unknown | Cohort |
| High | High | Either | Either | Either | Either |

| Rate of illness[b] | | Prevalence of use in population | Acute or delayed | Known or unknown | Research strategy |
|---|---|---|---|---|---|
| On drug | background | | | | |
| Intermediate | Intermediate | — | — | Known | Case-control |
| Intermediate | Intermediate | — | — | Unknown | Cohort |

[a]Adapted from Jick (1977, 1979).
[b]High prevalence rate defined as 1% and greater; low prevalence rate defined as .1% and lower; high exposure rate defined as 1% and greater; low exposure rate defined as .1% and lower.

These scenarios suggest that for some types of drug-induced reactions formal studies of any type are impractical and not likely to detect the events of interest. For these situations, spontaneous adverse reaction reports generated by practicing health professionals provide numerator data that can be useful in assessing the safety of a drug. In some cases, a comparison group may not be possible or practical and estimates of the rates are all that can be expected. This leads into more detailed discussions of some of the postmarketing methods in the next section.

## B. Analysis of Spontaneous Reports

Spontaneous reports are voluntary reports generated by health care providers who make a judgment that an association between an adverse reaction and a drug may exist. These reports are generated during the usual practice of medicine and can differ in quality and completeness. The primary mechanism for capturing information of this type is the 1639 Form [21 CFR 314.80(b)] which health practitioners fill out and which reaches the U.S. Food and Drug Administration either directly from the health care provider or from the manufacturer of the drug product, who is required by law and regulations to submit such reports to the FDA. During any one calendar year the FDA may receive more than 40,000 reports from all sources. The reports contain the identified health event and names of the drugs a patient was on when the health care provider observed the adverse health event in the patient.

The spontaneous report data base has a number of limitations, including underreporting, reporting artifacts, uncertain causality, inadequate information on the form, and misunderstanding of the adverse reaction terms by the provider. Elapsed time after initial marketing, publicity, initial rate of sales, patterns of use, and historical changes in rates of reporting by health care providers all affect the numbers of reports received in the system. These limitations represent biases that must be carefully assessed when using these data bases for analysis of adverse reactions.

Although it is not possible to determine incidence rates from spontaneous reports, several important uses of these data bases have been established. Rossi (1983, 1984) has shown that some new serious and unexpected adverse reactions have been detected by using spontaneous reports. Faich (1986) provides a current update of the use of adverse drug reaction monitoring, its impact on recent regulatory decisions, and several examples of drugs removed from the market or relabeled as a result of information provided by spontaneous reports. These spontaneous report data bases may be the only mechanism for uncovering very rare reactions that are characteristic of drug-induced illness such as bone marrow depression, phocomelia, acute renal and noninfectious liver disease, rare eye and skin reactions, and severe allergic reactions. Inman (1977) used the U.K. Committee on Safety of Drugs (now Medicine) registry of adverse reactions of drug-induced fatal aplastic anemia or agranulocytosis and prescription estimates to calculate mortality rates from oxyphenbutazone and phenylbutazone, two drugs whose safety was seriously questioned and which were the subject of a petition to the U.S. Department of Health and Human Services in 1984 to ban their use for safety reasons.

Recently, an additional area that has received attention is the monitoring and use of aggregate numbers of reports whose frequency is greater than expected. One example is the comparison of reports of adverse reactions associated with use of different drugs in the same therapeutic class. Stern and Bigby (1984) describe data submitted to a dermatology registry in which the relative numbers of prescriptions dispensed for each of the drugs was used in comparing the relative frequency of reports received for a large number of nonsteroidal anti-inflammatory drugs. Seigel (1971) employed an analogous approach to adverse reaction data for oral contraceptives to examine whether high- and low-dose contraceptives were equally associated with cardiovascular outcomes.

These concepts are similar to those proposed by Finney (1965, 1971a, 1971b) and used by Knapp et al. (1980). MacRae (1984), following Finney's idea, proposed a standardized event ratio to compare the observed to the expected number of reports, an index similar to the standardized mortality ratio commonly used in epidemiology. Comparative analysis of adverse reaction reports naming different drugs requires considerable understanding of the biases and reporting patterns and the use of statistical adjustments to control known or suspected sources of systematic bias.

The use of aggregate numbers of reports to signal an unsuspected association between a drug and some adverse reaction has been attempted with varying success over the past two decades. Finney (1965, 1971a, 1971b, 1974) and others (Brockett and Kemperman, 1980; Chen, 1978; Chen et al., 1982; Hill et al., 1968; Levine et al., 1977; Mandel et al., 1976; Moussa, 1978; Weatherall and Haskey, 1976) have addressed some of the conceptual and statistical issues in the analysis and signaling of adverse reaction reports, particularly within the context of the World Health Organization reporting systems. More recently, the use of spontaneous reports has been extended to the assessment of whether there is an increased frequency of reports in one time period relative to another comparable time period. The 1985 New Drug Regulations require that a manufacturer monitor the frequency of reports it receives and report when an increase in frequency occurs (21 CFR 314.80). Such an increase may signal that a drug is being used in a population at higher risk than those studied in the premarketing stage or that a delayed reaction may be occurring at a rate previously thought to be low. Usually, such signals will be evaluated to determine whether more formal epidemiological studies should be carried out to assess increased risks associated with a particular drug.

## C. Special Postmarketing Surveillance Studies

Pharmaceutical firms have conducted several large postmarketing studies relatively soon after initial marketing of their drug to obtain information about the effects of a newly approved drug in broader and larger patient populations. Most of these studies have been uncontrolled and nonrandomized and their purpose was usually to examine safety issues and to assess the kinds of patients that may be using the drug, their compliance and tolerance of the medication, the refill performance of a drug, and other marketing characteristics. Two uncontrolled surveys will be described briefly (Rossi et al., 1983).

The Cimetidine Postmarketing Outpatient Surveillance Program (Gifford et al., 1980) was a nonrandomized, prospective observational survey of users of Cimetidine for 6 weeks up to a year. Cimetidine, a drug for the treatment of gastrointestinal ulcers, was the subject of a physician-based prospective survey of almost 10,000 patients treated with the drug. Company representatives enrolled 1232 physicians (60% family practitioners, 30% internists, 10% gastroenterologists), each of whom was asked to provide specific case report data on at least 10 patients either already taking Cimetidine or for whom they prescribed the drug during a subsequent $3\frac{1}{2}$-month period. In the first phase of the survey, the case report data contained information on the patient (name, birth date, weight, height, sex, date of initial visit, diagnosis), on the drug (dates starting and stopping, doses), and results of therapy, adverse effects, and hospitalizations.

Questions asked included whether any adverse reactions occurred and whether any consultations, hospitalizations, changes in associated conditions, or new diagnoses occurred. No specific identified adverse effects were asked about (e.g., no focused or a priori hypotheses). A subsequent phase of the survey was to include longer-term follow-up data in a similar fashion. Of the physicians; 1049 responded (85.1%) with data on 9907 patients, of whom 881 were excluded from further tabulations because they did not fulfill the qualifications for study entrance. A total of 577 adverse effects were reported by 442 of the 9907 patients (4.4%), which were reported not to differ in type or incidence from those reported in premarketing clinical studies.

Comments on the study:

1. Adverse reactions occurring to patients in the cohort depended on a patient voluntarily returning to the same physician and voluntarily reporting or presenting with an adverse effect.
2. There was no systematic querying of the patient.
3. No specific monitoring and follow-up procedures were stated in the protocol. There were no rules for how long patients should be followed, what exit criteria would be, or how to report and analyze the data.
4. It was not possible to determine on the basis of reporting to what degree, if any, selective physician inclusion of patient case report forms or patient nonresponse influenced results.
5. The sample size criterion used to plan the study size stated that the cohort of 10,000 patients would provide a 90% probability of including an event with an incidence of 1/4348 and an 80% probability of including events occurring with an incidence of 1/6250 (see Section VIII for formula). This is not a very useful criterion

from the perspective of estimating a rate with specified precision, for comparisons of rates in different risk groups, or for assessing the effects of duration of treatment.

6. Results of the postmarketing study were compared to the results of the clinical trials without assessing the basic differences in design and methods of event ascertainment.

The Cyclobenzaprine (Flexeril) Postmarketing Surveillance Program (Rauch, 1979; Nibbelink and Streikland, 1979) was a nonrandomized, prospective observational survey of a cohort of short-term users (1 to 2 weeks) of Flexeril, a tricyclic drug indicated as an adjunct to rest and physical therapy for short-term relief of muscular spasm associated with acute painful musculoskeletal conditions. The objective of this surveillance program was to obtain clinical information on 8000 to 10,000 patients by asking 2000 to 2200 physicians in various specialties to treat five patients, record their finding on a patient report form, and at the end of a 2-week treatment period return it to a company representative for review of completeness and accuracy. In addition to the usual demographic data for each patient, the physician was asked to furnish information on etiology of muscle spasm, presence of concomitant illness, adjunctive medications, and dosage of Flexiril. Adverse reactions were classified as volunteered if offered by the patient to a general question on adverse effects or elicited when evoked by specific questioning. Specifically listed adverse reactions asked about were CNS-related hallucination, disorientation, confusion, seizures, tachycardia, and cardiac arrhythmias (focused part of the study). Reports on a total of 4749 patients were processed, of which 28% volunteered adverse experiences and 2.8% experienced elicited adverse reaction obtained by specific questioning of the patient. The drug was discontinued in 101 patients (2.1%) because of elicited adverse reactions.

An interesting statement in the report indicated that the highest frequency of volunteered adverse reactions (41%) was evident after the first 1000 reports were analyzed, and this incidence gradually diminished to an overall level of 28.2% at the end of 4749 patient reports. This suggests some type of bias operating at the later parts of the survey with regard to either a change in the types of patients chosen by physicians for treatment compared to earlier patients or a modification in the physician-patient exchange in filling out the case report form or in suggesting to the patient events which might be adverse reactions.

Comments on the study: The results of the survey were compared with rates observed in premarketing clinical trials indicating that the survey results were uniformly lower than rates observed in the clinical trials. This is not an unexpected result given that there is a potential bias in underreporting of some events in this type of study compared with premarket studies.

In both studies discussed above, the data were collected prospectively by physician interface with patients. Other studies have involved pharmacist recruitment of patients. Borden and Lee (1982) reported on a pharmacy study of orally administered antibacterial agents used short-term and dispensed in an ambulatory care setting. The study included the entire class of antibacterial agents and looked for health events occurring within 30 days subsequent to exposure. During the study over 20 different antibacterials were dispensed, but four drugs accounted for the bulk of the prescriptions. Registration and follow-up of patients on an entire class of therapeutic agents offered the opportunity of constructing comparison groups of patients on individual antibacterials. Patients were registered and recruited at the time of dispensing a medication from a participating pharmacy. Since the intent of the study was, in part, to characterize every health event occurring to the patient during the 30-day period independent of any perception of a relationship to drug, direct patient contact was made via a mail questionnaire with a telephone follow-up when needed. This procedure was an attempt to avoid missing any events because of selective inclusion or reporting of cases by the treating physician or because the patient did not advise the original treating physician of a subsequent illness. Data on health events occurring in outpatients were reported and analyzed by cohorts of drug exposures, age, sex, and recruitment clinic.

These three examples illustrate methods that have been specifically designed to collect and evaluate safety data. There are few examples of such special studies that have followed patients for longer-term exposure where patient follow-up is well accounted for.

Another class of studies have used data bases that were already collected for other reasons, such as for basic health care records and for financial reimbursement of health services provided. Because of the computerized linkage of diagnoses and drug usage these data bases have been used for adverse drug reaction surveillance.

### 1. Large-Scale Automated Data Bases

Some large-scale automated population-based data resources (Jick, 1977, 1985; Jick et al., 1984; Lawson, 1984; Maronde et al., 1971; Friedman and Ury, 1980) that include information on both drug exposures and diagnoses of illnesses, some requiring hospitalizations, are becoming available in the United States, Canada, and Britain and have been utilized for the study of drug effects after marketing. Jick (1985) provides a description of some of these data bases and a summary of some of their advantages and limitations. One advantage of these resources, he states, is their cost efficiency; another is their ability to provide continuous long-term follow-up studies (although few have demonstrated this capability to date), allowing the study of delayed drug effects as well as short-term effects. The limitations of automated data bases include the quality and extent of critical information that is likely to be needed to properly study a problem. Studies can have important distortion by misclassification when original clinical records are unavailable for review. Another limitation is that recent information on a large number of possibly important patient characteristics including smoking, alcohol habits, and family history, is unlikely to be available for a large number of people.

Medicaid, a federally supported and state-administered assistance program providing medical care for certain low-income individuals and families, is being considered as a potential source of computerized health records on millions of patients useful for adverse event surveillance (Strom et al., 1985; Shaffner et al., 1983). The U.S. Food and Drug Administration has been interested in developing this resource and currently has available longitudinal patient records of paid claims for outpatient and inpatient health care services and outpatient prescriptions for eligible recipients in several states (FDA contract 223-85-3020). Data on individual patients include sex, age, race, state, and county of residence, drug NDC code, strength, and amount dispensed, and the ICD9-CM code for the diagnosis or condition for which each service was rendered.

Medicaid data bases can be utilized for cohort and case-control studies of drug-induced diseases and for studies of patient prescriptions (Federspeil et al., 1976; Van de Carr et al., 1983). To utilize the data base, computerized records of all users of specified drugs can be selected and examined for the presence or absence of certain diagnoses indicative of the adverse event of interest. Since adverse reactions are not specifically recorded or identified as such on the record, an investigator must design criteria appropriate to the particular data base to create drug exposure cohorts, outcomes, and cases or controls. For example, a decision must be made on the class of diagnoses that represent the event of interest, on the temporal criteria for when subsequent to drug exposure the event (diagnosis) will be counted as drug-related, and on classification of events in patient records where the event is likely an incidence of a drug-induced illness rather than a prevalence which might have existed on the patient record prior to drug exposure. Usually, operational algorithms are developed and applied to a patient record in order to calculate duration of exposure or follow-up because records only include prescription dates which are not equally spaced in time. Prescriptions for drugs and diagnoses other than the study drug and outcome may also be used to categorize patient records or control for potential confounding drugs or diagnoses because the data base consists only of records of sick people who are being treated.

Automated data bases usually contain information on all drugs to which a patient is exposed, thus allowing examination of possible drug interaction and/or statistical assessment of multivariate factors associated with adverse outcomes.

Because of the susceptibility of observed rates of diagnoses to the definitions and inclusion and exclusion criteria employed, to the grouping of diagnoses, and to the data collection and coding procedures peculiar to that system, comparative cohort evaluations within the data base are the most informative. It is extremely difficult to interpret rates of diagnoses derived from a single cohort relative to rates derived from other data sets in the absence of comparative data within the same data base because of the inherent biases that exist.

Experience with the Medicaid data base is currently too limited to judge its ultimate utility for assessing drug-induced disease associations and for deciding what types of diagnoses may be best studied using the system. Many biases can be identified within the data base, but these may contribute to random noise, making any associations more difficult to detect. Although recall bias is not an issue with this data base as with many drug studies, there are questions about the quality of some of the diagnosis data [FDA contract 223-82-3021 (1984)] and the ability to easily access the patient clinical records to confirm the automated data record.

### D. Case-Control Studies

The case-control method has played a fundamental role in the postmarketing surveillance of drug safety, particularly in the study of drugs on the market for many years and in the study of drug-induced diseases which take a considerable number of years to develop after initial drug exposure or which occur after chronic long-term use. The observational case-control method emerged as an important epidemiological and investigational tool for study of the potential effects of marketed drugs in the late 1960s, when it was used effectively by several British investigators (Vessey and Doll, 1968, 1969; Inman and Vessey, 1968) to demonstrate a relationship between oral contraceptives and the occurrence of thromboembolic disease and death. In addition to being extensively used as a means of studying the postmarketing effects of oral contraceptives, the case-control method has been used in the assessment of carcinogenic, teratogenic (Mann et al., 1975; Herbst et al., 1971; Greenwald et al., 1971; Levy et al., 1973; Nora and Nora, 1973; Jannerich et al., 1974; Safra and Oakley, 1975; Greenberg et al., 1975; Gold, 1978; Bracken and Holford, 1981; Saxen, 1974, 1975; Saxen et al., 1974), and other delayed effects of marketed drugs, in particular estrogens (Smith et al., 1975; Ziel and Finkle, 1975; Mack et al., 1976).

Case-control designs are limited to assessing known or suspected drug-induced disease associations because the case disease must first be identified and then appropriate controls chosen for the cases selected. Usually, case-control studies are carried out after a suspicion is raised by clusters of spontaneous reports of drug side effects or other general concerns derived from monitoring systems. In particular, when the disease of interest is rare and the exposure to the suspected drug is relatively high, case control methods can be the only practical study design to detect drug-induced disease associations; prospective cohort studies required to address the problem would be difficult to mount because of the enormous sample sizes needed. Moreover, in an observational setting patients are often exposed to many drugs during the same interval of time so that multiple exposures to several drugs must be sorted out and statistically controlled in order to interpret the data meaningfully. Case-control designs can deal with multiple exposures to many different drugs more efficiently than cohort designs, particularly where small incidence rates of adverse reactions are likely. Case-control studies cannot estimate absolute risk associated with a drug, but only risk relative to some background rate of occurrence of the disease. For this purpose, the measure of association is the odds ratio which is an approximation for the relative risk defined as the ratio of two incidence rates (Cornfield, 1951).

The limitations of the case-control design have been argued in the literature by a number of authors (Feinstein, 1973; Ibrahim and Spitzer, 1979) and several recent texts provide excellent accounts of the design's use and applications (Schlesselman, 1982; Breslow and Day, 1980; Kleinbaum et al., 1982). Bias is one of the inherent criticisms of the case-contol method, and case-control studies of drug-induced disease associations must pay particular attention to these concerns. Kleinbaum et al., (1982) provide a detailed discussion of the problems of validity in estimating an effect of a drug when many sources of bias must be considered simultaneously. The particular example used to illustrate the issue was the Boston Collaborative Drug Surveillance Program's report on reserpine and breast cancer (1974), which, though reporting a positive association, was criticized on the grounds of potential biases in the study design. Schlesselman (1982) gives an excellent discussion of the design and analysis of case-control studies, using examples of studies of drug-induced diseases to illustrate many points. In particular, the study of Shapiro et al. (1979) on the relation of myocardial infarction to recent oral contraceptive use is extensively discussed and analyzed. Breslow and Day (1980) also give an excellent account of the analysis of case-control studies of cancer associated with various exposures and risk factors. These authors use the study of Mack et al. (1976) on estrogens and endometrial cancer to illustrate a number of methodological approaches to the analysis of case-control studies.

The design of case-control studies of drug-induced disease may utilize previously collected health care or hospital records, for which case definitions are created to extract cases which satisfy the definitions and for which the drug exposure data are expected to be captured in the data base. This is more likely to be the situation for drugs that have been on the market for a number of years. The majority of case-control studies to date have been of this type, where the data base already exists and the study can be mounted and completed in a relatively short period of time compared with de novo accrual of patient data. Another strategy for case-control studies has been to take advantage of data that are being accrued for other reasons, such as routine screening programs. One example of such a strategy is a case-control study of breast cancer and reserpine that followed the published study by the Boston Collaborative Drug Surveillance Program and included 481 breast cancer cases and 1268 controls obtained from a joint national mammography screening project of the National Cancer Institute and the American Cancer Society (Williams et al., 1978).

However, case-control studies can be designed prospectively according to a priori well-defined protocol criteria in the sense that cases can be accrued as they occur in the future, matched to appropriate controls, and analyzed according to either a fixed or sequential design strategy. There are no examples of case-control studies being analyzed according to a sequential strategy, though some issues relevant to such designs have been discussed (O'Neill and Anello, 1978; Pasternack and Shore, 1980, 1981). A recent study on the association between Reye's syndrome and certain medications including salicylates (Hurwitz et al., 1985) is an example of a case-control study that was carried out prospectively over the course of one flu season in which cases were identified and collected as they occurred, appropriate controls were identified and matched to the cases, and a battery of demographic and exposure information queried for each study participant according to protocol definitions.

The studies of the association between reserpine and breast cancer, oral contraceptives and myocardial infarction, conjugated estrogens and endometrial cancer in postmenopausal women, and aspirin and Reye's syndrome are a few examples of relatively rare diseases in which the background prevalence of use of the suspected drug was relatively high, say greater than 5%, making the case-control design the method of choice over cohort designs. In the first three examples the duration of exposure prior to the occurrence of the disease was relatively long, ranging up to many years for cancer. In the Reye's syndrome study, drug exposure was very closely temporally related to the outcome and therefore duration of exposure and delay of the outcome were not important issues.

Tables 8 and 9 contrast the study sizes required under a case-control design and those required under a cohort design to detect various levels of risk. Table 8 contains sample sizes for a case-control design extracted from Schlesselman (1980) that would be required to detect specified increased risks for selected control group prevalence rates of exposure to a drug and statistical type I and II errors. Table 9 provides sample sizes for a cohort design required to detect specified increased risks for selected control group disease rates and statistical type I and II errors.

Table 8 Case-Control Sample Size; $\alpha$ = .05 (Two-Sided), $\beta$ = .20[a]

| Relative risk | Rate of exposure in control group | | | | | |
|---|---|---|---|---|---|---|
| | .01 | .05 | .10 | .15 | .20 | .25 |
| 1.5 | 7954 | 1687 | 910 | 656 | 534 | 465 |
| 2.0 | 2394 | 515 | 282 | 207 | 171 | 152 |
| 2.5 | 1245 | 271 | 151 | 112 | 94 | 84 |
| 3.0 | 803 | 176 | 99 | 75 | 63 | 58 |
| 3.5 | 579 | 129 | 73 | 56 | 48 | 44 |
| 4.0 | 448 | 100 | 58 | 44 | 38 | 36 |
| 4.5 | 363 | 82 | 48 | 37 | 32 | 30 |
| 5.0 | 304 | 69 | 41 | 32 | 28 | 26 |

[a]From Schlesselman (1974).

Table 9 Cohort Sample Sizes; $\alpha$ = .05 (Two-Sided), $\beta$ = .20[a]

| Relative risk | Rate R of event in control group | | | | |
|---|---|---|---|---|---|
| | .001 | .005 | .01 | .05 | .10 |
| 1.5 | 78,301 | 15,581 | 7,741 | 1,469 | 685 |
| 2.0 | 23,484 | 4,668 | 2,316 | 434 | 199 |
| 2.5 | 12,173 | 2,417 | 1,197 | 221 | 99 |
| 3.0 | 7,823 | 1,551 | 767 | 140 | 62 |
| 3.5 | 5,631 | 1,115 | 551 | 99 | 43 |
| 4.0 | 4,343 | 859 | 423 | 75 | 31 |
| 4.5 | 3,509 | 693 | 341 | 60 | 24 |
| 5.0 | 2,930 | 578 | 284 | 49 | 19 |

[a]From Schlesselman (1974).

## VIII. STATISTICAL PLANNING CONSIDERATIONS

Comparison or estimation of adverse event rates from two or more groups is often carried out after a study is completed with no particular attention paid to how capable the study design was of assessing specified rates or differences in rates. Before undertaking a RCT, a comparative observational cohort study, or an uncontrolled trial, it is possible to assess the likelihood of being able to observe adverse events with specified rates of occurrence, to estimate rates with a specified degree of precision, or to statistically detect differences between group rates given the sample sizes available for study. An extensive literature exists for sample size planning applicable to efficacy evaluation both for the comparative two-sample problem employing tests of hypothesis and for estimation of proportions, cumulative rates, hazard rates, and differences of these between groups (Lachin, 1981; McHugh and Lee, 1984; Schlesselman, 1974). Many of the statistical methods developed for the comparative evaluation of efficacy in clinical trials, particularly in the survival analysis area, can be useful for safety evaluation (Halperin et al., 1968; Pasternack and Gilbert, 1971; Pasternack, 1972; George and Desu, 1974; Palta and McHugh, 1980; Taulbee and Symons, 1983).

RCTs are not usually designed with the primary hypothesis of interest involving testing the difference of two adverse event rates. More often direct comparison of adverse event rates is the primary objective of a comparative observational cohort study in the postmarketing setting. Sometimes confidence interval estimation of the difference between rates from two treatment groups is more useful in the evaluation of safety because sample sizes can be too small for moderate differences between rates to be statistically detected with any certainty or because interest rests in determining the closeness of safety profiles of two drugs. In situations where a single treatment group or cohort of users is being followed, the primary goal may be to estimate rates with a desired degree of precision. Thus, it is informative to examine the sample sizes that would be needed to estimate a rate or to detect or estimate differences of specified amounts between rates for two different treatment groups. The magnitude of rates that can feasibly be studied in most comparative clinical studies is about .01 and higher. Observational cohort studies usually can assess rates on the order of .001 and higher. The following considerations might be useful in planning or evaluating a study.

1. *Sample sizes to detect differences between two binomial event rates.* If $P_i$ is the true rate of occurrence of an adverse event in a specified interval of time for treatment group or cohort i, then the sample size required in each group to be able to conclude statistically that there is a significant difference between the observed rates in different groups when the type I error is .05 and the chance of statistically detecting that difference is $1 - \beta$ can be calculated using a formula in Lachin (1981, p. 100). Schlesselman (1974) provides formulas for relative risks in cohort and case-control studies.

2. *Sample sizes to estimate differences between two binomial event rates with specified level of precision.* The absolute precision with which it is desired to estimate the difference between rates can also be specified at the outset. If the estimate of $D = P_1 - P_2$, the true difference between rates, is to be within d units of the true value D, then d is defined as the absolute precision of the estimator. If the true difference between the adverse event rates in two groups is $P_1 - P_2 = D$, then the sample size in each group needed to estimate D with precision d with $1 - \alpha$ confidence for selected

values of $P_i$, D, and d can be calculated with formulas in McHugh and Lee (1984, p. 160). Cochran (1963) provides methods for estimating a single rate. O'Neill (1984) provides sample sizes for estimating the odds ratio with a specified precision.

3. *Sample sizes to detect differences between two hazard rates.* When the functional form of the time to occurrence of an adverse event is the exponential distribution, then the hazard rate is constant over time. The hazard rate is defined in Section IV and describes the risk per unit of time. Assuming that a constant hazard rate applies can be useful for planning purposes when adverse event occurrence is associated with differential follow-up or exposure time that is different among patients. If patients are uniformly entered into study over the time interval (0, T) or are entered at the same time and then uniformly censored over this time interval, then Lachin (1981, p. 103) provides a formula to calculate the sample size required in each group to test the difference between two hazard rates. In order to use this formula, the maximum duration T of follow-up must be specified beforehand and one must assume censoring occurs uniformly over that interval. For many adverse reactions the assumption of an exponential distribution may be unrealistic. However, the calculations are informative in examining the sensitivity of resources to simultaneous consideration of patient sample size and follow-up time.

4. *Sample sizes to estimate the difference between two hazard rates with specified precision.* To estimate under conditions similar to those above the difference between two hazard rates with a specified precision d, McHugh and Lee (1984, p. 162) provide a formula for calculating the sample sizes needed under various configurations of follow-up time T, hazard rates, and levels of confidence.

5. *Sample sizes for comparative survival analysis using the logrank test.* Freedman (1982) and Schoenfeld (1983) provide formulas for determining the number of adverse events that should be observed to test the equality of two cumulative adverse event curves when using the logrank test under the proportional hazards model. Freedman discusses how to arrive at the total number of patients to be entered in order to meet the objective.

6. *Smallest detectable risk.* When faced with a completed study and an estimate of the adverse event rate in a control group it is possible to estimate the smallest risk that could have been detected statistically for a specified type I error given the sample sizes used in the study. Schlesselman (1980, p. 153) provides a formula due to Walter (1977) that permits calculation of the smallest increase in risk detectable for a given statistical power.

7. *Single-sample issues.* Sometimes a single treatment group or cohort of users is followed to identify unique adverse events that suggest a drug effect such as jaundice or blood dyscrasias. This is usually the situation in uncontrolled observational studies. When independent information on the rate of occurrence of such an event is available, say from other studies, there may be interest in testing whether the event rate in the follow-up cohort is the same or higher. Makuch and Simon (1980) provide sample size formulas for testing rates observed in such cohorts against a historical control rate. For single treatment groups a major objective might be to estimate a rate with specified precision. Formulas in the reference cited above can be modified to estimate rates with specified precision in the single-sample situation.

At times one may not observe any specific adverse events in a treatment group and questions might arise about the likelihood of observing or not observing at least one adverse event outcome in a specified time period in a single-cohort study. Several formulas can be useful in interpreting outcomes of this type. If R is the true rate of experiencing an adverse event in a given time period and N is the number of subjects exposed and followed for the same period of time, then the probability of observing at least one adverse event out of N subjects in that fixed period of time is $P = 1 - (1 - R)^N$. This formula can be used in a number of different ways. Table 10 displays for selected sample sizes N the probability of observing at least one adverse event in a fixed time interval when the true rate of an adverse event is R. This table can be useful for determining the likelihood of observing an event in a specific time interval when the sample size has been decreasing under loss to follow-up.

For example, if an adverse event is assumed to have a constant hazard rate of .005 per month, and of a total of 1000 patients initially exposed to a drug, 600 are exposed for 2 months, 300 for 3 months, and 100 for 4 months, the following can be said. There is a .99 chance of observing at least one event in the first month in 1000 patients, a .92 chance of observing at least one event in the second month in 600 patients, a .78 chance of observing at least one event in 300 patients in the third month, and a .39 chance of observing at least one event in 100 patients in the fourth month. If the interval lengths were not of the same size, additional modifications to these probabilities would be needed. Sometimes the sample size may be so small in a given follow-up interval that the likelihood of observing an event is low even if the true risk per unit of time is constant.

Table 11 displays for selected probability levels P (degree of confidence) and adverse event rates R the sample sizes needed to observe at least one adverse event with probability P when the true adverse event rate is R. Sometimes no adverse event is observed in N patients in a given time interval. Table 12 provides upper $1 - \alpha$ percent confidence limits on the true adverse event rate R in a fixed time interval given that the adverse event has not been observed in N patients.

Although it is informative to know the probability of observing at least one adverse event when its true rate is R, in practice when only one event is observed it may be difficult to interpret in the absence of other events occurring in patients. This is where criteria for estimating the rate R with a specified precision d are useful.

Example 10. The purpose of this example is to illustrate the impact on planning and analysis of the time interval over which an adverse event is studied. Assume that two cohorts each exposed to a different drug are being compared for the purpose of assessing whether one drug increases the rate of an adverse event threefold. Assume further than an adverse event follows an exponential distribution with a constant hazard rate. The hazard rate for drug 1 is $k_1 = .0002$ per month and for drug 2 is $k_2 = .0006$ (i.e., 2 and 6 events per 10,000 person-months exposed are expected respectively in each cohort). The cumulative rate of the adverse event for drug 1 after 3 months of exposure is $P_1(3) = 1 - \exp(-.0002 \times 3) = .0006$ and for drug 2 it is $P_2(3) = 1 - \exp(-.0006 \times 6) = .0018$. The cumulative rate after 6 months of exposure is $P_1(6) = 1 - \exp(-.0002 \times 6) = .0012$ for drug 1 and $P_2(6) = 1 - \exp(-.0006 \times 6) = .0036$ for drug 2. The sample size required to detect the difference between .0006 and .0018 at 3 months is 12,194 in both groups for a two-sided test at $\alpha = .05$ and power = .8.

**Table 10** Probability P of Observing in a Fixed Time Interval at Least One Adverse Event Whose True Rate is R for Selected Sample Sizes N

| Sample size N | Event rate R | | | | | | |
|---|---|---|---|---|---|---|---|
| | .001 | .005 | .01 | .05 | .10 | .15 | .20 |
| 1500 | .777 | .999 | 1.000 | | | | |
| 1000 | .632 | .993 | 1.000 | | | | |
| 950 | .613 | .991 | 1.000 | | | | |
| 900 | .594 | .989 | 1.000 | | | | |
| 850 | .573 | .986 | 1.000 | | | | |
| 800 | .551 | .982 | 1.000 | | | | |
| 750 | .528 | .977 | .999 | 1.000 | | | |
| 700 | .504 | .970 | .999 | 1.000 | | | |
| 650 | .478 | .962 | .999 | 1.000 | | | |
| 600 | .451 | .951 | .998 | 1.000 | | | |
| 550 | .423 | .937 | .996 | 1.000 | | | |
| 500 | .394 | .918 | .993 | 1.000 | | | |
| 450 | .363 | .895 | .989 | 1.000 | | | |
| 400 | .330 | .865 | .982 | 1.000 | | | |
| 350 | .295 | .827 | .970 | 1.000 | | | |
| 300 | .259 | .778 | .951 | 1.000 | | | |
| [illegible]50 | .221 | .714 | .919 | 1.000 | | | |
| 200 | .181 | .633 | .866 | 1.000 | | | |
| 150 | .139 | .529 | .779 | 1.000 | | | |
| 100 | .095 | .394 | .634 | .994 | 1.000 | | |
| 90 | .086 | .363 | .595 | .990 | 1.000 | | |
| 80 | .077 | .330 | .552 | .983 | 1.000 | | |
| 70 | .068 | .296 | .505 | .972 | .999 | 1.000 | |
| 60 | .058 | .260 | .453 | .954 | .998 | 1.000 | |
| 50 | .049 | .222 | .395 | .923 | .995 | 1.000 | |
| 40 | .039 | .182 | .331 | .871 | .985 | .998 | |
| 30 | .030 | .140 | .260 | .785 | .958 | .992 | .999 |
| 20 | .020 | .095 | .182 | .642 | .878 | .961 | .988 |
| 10 | .010 | .049 | .096 | .401 | .651 | .803 | .893 |

**Table 11** Sample Sizes Needed to Observe with Probability P at Least One Adverse Event Whose True Rate is R in a Given Time Interval

| Event rate R | P | | | | | | |
|---|---|---|---|---|---|---|---|
| | .99 | .95 | .90 | .80 | .70 | .60 | .50 |
| .001 | 4603 | 2994 | 2301 | 1609 | 1203 | 916 | 693 |
| .005 | 919 | 598 | 459 | 321 | 240 | 183 | 138 |
| .010 | 458 | 298 | 229 | 160 | 120 | 91 | 69 |
| .050 | 90 | 58 | 45 | 31 | 23 | 18 | 14 |
| .060 | 74 | 48 | 37 | 26 | 19 | 15 | 11 |
| .070 | 63 | 41 | 32 | 22 | 17 | 13 | 10 |
| .080 | 55 | 36 | 28 | 19 | 14 | 11 | 8 |
| .090 | 49 | 32 | 24 | 17 | 13 | 10 | 7 |
| .100 | 44 | 28 | 22 | 15 | 11 | 9 | 7 |
| .150 | 28 | 18 | 14 | 10 | 7 | 6 | 4 |
| .200 | 21 | 13 | 10 | 7 | 5 | 4 | 3 |
| .250 | 16 | 10 | 8 | 6 | 4 | 3 | 2 |

**Table 12** Upper P% Confidence Limit on the True Probability R of an Adverse Event in a Fixed Time Interval Given the Event Is Not Observed in N Patients

| N | P | | | |
|---|---|---|---|---|
| | .99 | .95 | .90 | .80 |
| 5000 | .00092 | .00060 | .00046 | .00032 |
| 4000 | .00115 | .00075 | .00058 | .00040 |
| 3000 | .00153 | .00100 | .00077 | .00054 |
| 2000 | .00230 | .00150 | .00115 | .00080 |
| 1000 | .00459 | .00299 | .00230 | .00161 |
| 900 | .00510 | .00332 | .00256 | .00179 |
| 800 | .00574 | .00374 | .00287 | .00201 |
| 700 | .00656 | .00427 | .00328 | .00230 |
| 600 | .00765 | .00498 | .00383 | .00268 |
| 500 | .00917 | .00597 | .00459 | .00321 |
| 400 | .01145 | .00746 | .00574 | .00402 |

Table 12 (continued)

| N | P | | | |
|---|---|---|---|---|
| | .99 | .95 | .90 | .80 |
| 300 | .01523 | .00994 | .00765 | .00535 |
| 200 | .02276 | .01487 | .01145 | .00801 |
| 100 | .04501 | .02951 | .02276 | .01597 |
| 90 | .04988 | .03274 | .02526 | .01772 |
| 80 | .05594 | .03675 | .02837 | .01992 |
| 70 | .06367 | .04189 | .03236 | .02273 |
| 60 | .07388 | .04870 | .03765 | .02647 |
| 50 | .08799 | .05816 | .04501 | .03168 |
| 40 | .10875 | .07216 | .05594 | .03944 |
| 30 | .14230 | .09503 | .07388 | .05223 |
| 20 | .20567 | .13911 | .10875 | .07732 |
| 10 | .36904 | .25887 | .20567 | .14866 |

The sample size required to detect the difference between .0012 and .0036 at 6 months is 6090 in each group for a two-sided test at $\alpha = .05$ and power .8.

Similarly, the respective cumulative rates at 9 months are $P_1(9) = .0054$ and $P_2(9) = .0018$ and the required sample size to detect a difference is 4056 in each group. Notice that at each time point there is a threefold relative increase but the absolute magnitude of the rate changes as a function of time. This directly translates into changing sample size requirements to meet the objective. Had a study initially started with only 4056 subjects in each group, assuming the hazard rates given, the statistical power ($\alpha = .05$, two-sided) to detect a difference between the two rates after 3 months is .37. At 6 months the statistical power to detect a difference is .63. Thus, for an exponential-type adverse event both follow-up time and patient sample size simultaneously play a role in the statistical assessment of the difference between the rates. This issue is seldom given thorough consideration in adverse event assessment. With events that follow more complicated patterns than a constant hazard rate, the planning and analysis consideration will need to be additionally addressed.

## IX. IMPACT OF MISCLASSIFICATION OF ADVERSE EVENTS AND DRUG EXPOSURE STATUS

The frequency of recorded occurrences of adverse events in clinical trials and in observational postmarketing studies is influenced by the definitions of the terms and the algorithms used to categorize the adverse drug responses as well as by the methods of ascertainment of the adverse events and reactions. In both clinical studies and postmarketing studies the criteria used to ascertain and classify an event as an adverse event or adverse reaction influences whether true drug-induced adverse events are statistically detected or not. For example, the assessment criterion spelled out in the protocol of a clinical trial plays a major role in whether different physicians rate or score an event as an adverse reaction or whether health events are elicited or volunteered by study patients. Different clinicians will disagree on the rating for the same health event (Hutchinson et al., 1983) and therefore categorize it differently. It is known that computerized data bases with linkage of disease diagnoses and drug exposure information can contain substantial misclassification of prescription and/or diagnosis information used to construct case groups, exposure cohorts, or rates of certain diagnoses in different exposure cohorts.

A study (FDA Contract 223-82-3021, 1984) of a sample of Medicaid computerized paid claims files for two states indicated that agreement between the diagnosis in the computer record and that corresponding to the physician-patient record was in the range of 50 to 60% for all diagnoses studied. This study addressed the agreement level and not whether the physician's diagnosis itself was correct and did not assess agreement by the degree of objectivity of the diagnosis. Another sample comparing Medicaid computerized pharmacy claims with actual pharmacy records indicated a high degree of agreement (above 95%) between the computer record and the pharmacy record (FDA Contract 223-82-3021, 1984) suggesting that drug exposure data are more reliable and accurate within this data base.

Some observational studies depend on patient recall of drug usage and other health-related information, which has been shown in some studies to vary by type of drug studied and types of records used for comparison (Paganini-Hill and Ross, 1982). For example, health-related information from multiple sources was collected on 334 women living in two retirement communities as part of a case-control study of cancer of the breast conducted in 1977–1978. That study utilized interviews, medical charts, and pharmacy records and found that agreement between data sources varied from a low of 69% for use of barbituates to a high of 87% for antihypertensive medications. The correspondence observed between medical record and personal interview with the subject was better than between either the medical and pharmacy records or the interview and pharmacy records. Patient recall of drug use is also likely to be a function of acutely used drugs or chronically used drugs, which may be easier to recall. More generally, case-control studies have been criticized for having a number of potential biases related to ascertainment of exposure and case information (Ibrahim and Sptizer, 1979; Sackett, 1979) which affect study design and interpretation. To the extent that a true adverse reaction (as opposed to a health event) that is drug-induced did or did not occur and was judged by some algorithm to occur or not occur, or that drug exposure was accurately captured and categorized, there is misclassification.

The sensitivity of a classification scheme is defined as the proportion of correctly classified positives (e.g., true adverse reactions), and the specificity of a classification scheme is defined as the proportion of correctly classified negatives (e.g., true nonadverse reactions). Often the true state of nature is not known and it is not possible to say with certainty that an adverse event is a drug-induced adverse reaction. However, this discussion provides a conceptual framework for assessing the potential impact of misclassification. The sensitivity and specificity of the categorization or classification procedure determine the magnitude of the false positive or false negative classification rates. Misclassification of adverse reactions can have a dramatic effect on the bias in estimates of rates and on the power of statistical comparisons of rates of adverse reactions in two or more groups. This is particularly the case for rates of adverse events which are usually low and where the false positive or false negative rates of classification may be equal to or higher than the true adverse event rate.

The impact is both on study planning considerations, in that larger sample sizes than otherwise thought may be required to demonstrate differences in risk between exposure groups, and on the interpretation of findings from completed studies, which may be biased toward under- or overestimates depending on the circumstances.

The emphasis here is on the impact of misclassification of an observed event on the estimate of a rate of adverse reaction from a single cohort and on the estimate of the difference of rates in two cohorts as well as on the power of the statistical test of the equality of two rates. Much of the conceptual framework is taken from Quade et al. (1980) and Goldberg (1975). Similar issues related to the impact on estimates of odds ratios and relative risks have been considered by a number of authors, and the impact is equally dramatic and will be considered briefly. Quade and McClish (1985) describe strategies for enhancing the classification scheme for estimating the prevalence of a disease.

For the i-th group let $P_i$ denote the true incidence rate of a particular adverse drug reaction, $N_i$ the number of individuals in each cohort group evaluated, and $X_i$ the number of individuals in each cohort classified as having the adverse reaction. Then $\hat{P}_i = X_i/N_i$ is the apparent or observed adverse reaction rate. Now let U be the chance of declaring that an individual has the adverse reaction when, in fact, he does (sensitivity) and let $1 - V$ denote the chance of declaring an individual to have an adverse reaction when, in fact, he does not (V is specificity). Then, in general, the expected chance that an individual in the i-th group has an adverse reaction of a specific type is

$$E_i = P_i U + (1 - P_i)(1 - V)$$

The bias in the estimate $\hat{P}_i$ is

$$E_i - P_i = -P_i(1 - U) + (1 - P_i)(1 - V)$$

This bias can be evaluated for various values of the sensitivity, specificity, and true incidence rate $P_i$. Table 13 presents the magnitude and direction of the bias in estimated rates for a range of U, V, and P typical of adverse reaction rates seen in clinical studies. As Quade et al. (1980) point out, imperfect sensitivity U will lead to underestimates of the true rate P, imperfect specificity V will lead to overestimates of the true rate P, and various combinations of the two will lead to different magnitudes of bias. This formula shows that estimated rates of adverse reactions from a single uncontrolled cohort may have a bias which makes the cohort unsuitable for meaningful comparisons to rates obtained from other data sets having their own particular biases.

**Table 13** Expected Rate E for Selected Sensitivity U and Specificity V When the True Rate Is P

| U | V | P | | | |
|---|---|---|---|---|---|
| 1 | 1 | .001 | .01 | .05 | .10 |
| .95 | 1 | .00095 | .0095 | .0475 | .095 |
| .90 | 1 | .00090 | .0090 | .0450 | .090 |
| .85 | 1 | .00085 | .0085 | .0425 | .080 |
| 1 | .99 | .0110 | .020 | .059 | .109 |
| .95 | .99 | .0109 | .019 | .057 | .104 |
| .90 | .99 | .0109 | .019 | .054 | .099 |
| .85 | .99 | .0108 | .018 | .052 | .094 |
| 1 | .98 | .0210 | .030 | .069 | .118 |
| .95 | .98 | .0210 | .029 | .066 | .113 |
| .90 | .98 | .0210 | .029 | .064 | .108 |
| .85 | .98 | .0210 | .028 | .061 | .103 |
| 1 | .97 | .0310 | .040 | .078 | .127 |
| .95 | .97 | .0309 | .039 | .076 | .122 |
| .90 | .97 | .0309 | .039 | .073 | .117 |
| .85 | .97 | .0308 | .038 | .071 | .112 |

The impact on estimates of the difference between rates of adverse reactions from two cohorts is not only on the estimate of the difference in rates but also on the power of the statistical test of the equality of the two rates. Assuming the simplest situation, where there is nondifferential misclassification of adverse reactions in each cohort (i.e., the sensitivity is the same and the specificity is the same in both cohorts), the following can be said. If $d = P_2 - P_1$ and $\hat{d}$ is the estimate of this difference based on $\hat{P}_i$ then the expected value of $\hat{d}$ is $E = E_2 - E_1$, which reduces to

$$d(U + V - 1)$$

Quade et al. (1980) provide an expression for the approximate statistical power to detect a true difference d when sensitivity U and specificity V in the classification procedure exists (power is the probability of detecting a true difference between $P_2$ and $P_1$ when a true difference of a specified amount actually exists). This expression is

$$\text{Power} = \Phi[\Phi^{-1}(\alpha) - E/S]$$

where $\Phi$ is the standard normal cumulative distribution function, $\Phi^{-1}(\alpha)$ is the critical value at the stated cutoff within parentheses, $E = E_2 - E_1$ is the expectation of the difference $\hat{P}_2 - \hat{P}_1 = (X_2/N_2) - (X_1/N_1)$ of observed rates, $S = (S_1{}^2 + S_2{}^2)^{1/2}$ is the standard error of $\hat{P} - \hat{P}_1$, and $2S_i = [E_i(1 - E_i)/N_i]^{1/2}$. Table 14 provides an example of the expected difference and the power to detect that difference for selected U and V when $P_1 = .10$, $P_2 = .05$, and $N_1 = N_2 = 500$. It is clear that the true difference in rates is underestimated when any misclassification is present and that for a given sample size the power of the statistical test of any true difference is reduced from what might be planned or expected in the absence of misclassification. This makes any comparative inferences less sensitive in an already insensitive situation because of the small rates and sample sizes usually used to evaluate safety in RCTs and in observational cohort studies.

If the sensitivity and specificity of the classification rules in either cohort are different, then the situation becomes more complex, as does the prediction of the directions in which the biases will go. For equal misclassification in both groups the direction of the bias is always toward the null value.

Some drug epidemiology studies use third-party health care data bases and some case-control studies rely on patient recall or the quality of a patient record, each of which has considerable potential for misclassification of exposure or disease identifiers, as discussed earlier. It is possible for misclassification of both drug exposure and the disease categorization so that attempts to determine their relative impact may be necessary at the planning stage. Appropriate designs and sample sizes can then be planned to compensate for such misclassification. Further, because observational studies usually are characterized by considerable statistical adjustments to control for confounding, one cannot overlook that the covariables themselves may be misclassified. This introduces an additional complexity into the interpretation of the analysis. For example, in the Medicaid data base referred to earlier, prescription drug codes are more likely to be accurately reflected in the data than disease diagnosis codes (FDA Contract 223-82-3021, 1984), so that given the options of adjusting for certain confounding

**Table 14** Impact of Nondifferential Misclassifications on Estimates of Difference in Two Rates and on Power: $d = P_2 - P_1$; $\hat{d} = \hat{P}_2 - \hat{P}_1$ Has Bias $d(U + V - 1)$[a]

Example: $P_1 = .10$, $P_2 = .05$, $N_1 = N_2 = 500$

| U | V | Expected difference | Power |
|---|---|---|---|
| 1.0 | 1.0 | 0.05 | .91 |
| 1.0 | .90 | 0.045 | .60 |
| .80 | 1.0 | 0.04 | .85 |
| .80 | .90 | 0.035 | .46 |

[a]From Quade et al. (1980).

variables it may be more prudent to use drug codes as surrogates for the disease conditions when possible so that the analysis will be less influenced by misclassified confounding variables. Schlesselman (1982), Kleinbaum et al., (1982), Greenland (1980), Gladen and Rogan (1979), and Kupper (1984) provide good discussions of these issues.

### A. Impact on the Odds Ratio and Relative Risk

Case-control and cohort observational studies are major instruments for assessing comparative safety issues. Misclassification of disease diagnosis and drug exposure status may affect observed results from such studies to the extent that point estimates consistently appear to be close to the null value when nondifferential misclassification exists. Copeland et al. (1977) illustrate the impact of misclassification on the relative risk in a cohort study and on the odds ratio in a case-control study. They show that, generally, specificity is more important than sensitivity in determining the bias in the estimate of the relative risk from a comparative cohort study, suggesting that in the range of adverse event rates usually observed most of the bias occurs before the specificity drops below 85%. The rarer the incidence of disease the greater the bias, because the true adverse reactions become swamped with the false positives (false adverse reaction categorizations), producing a dilution of the relative risk (a problem when loose criteria exists). Figures 16 and 17 illustrate, for a sensitivity of .95 in the disease classification, the bias in the estimates of a true relative risk for two relative risks of different magnitudes and two sets of disease rates that are characteristic of those observed in drug epidemiology studies.

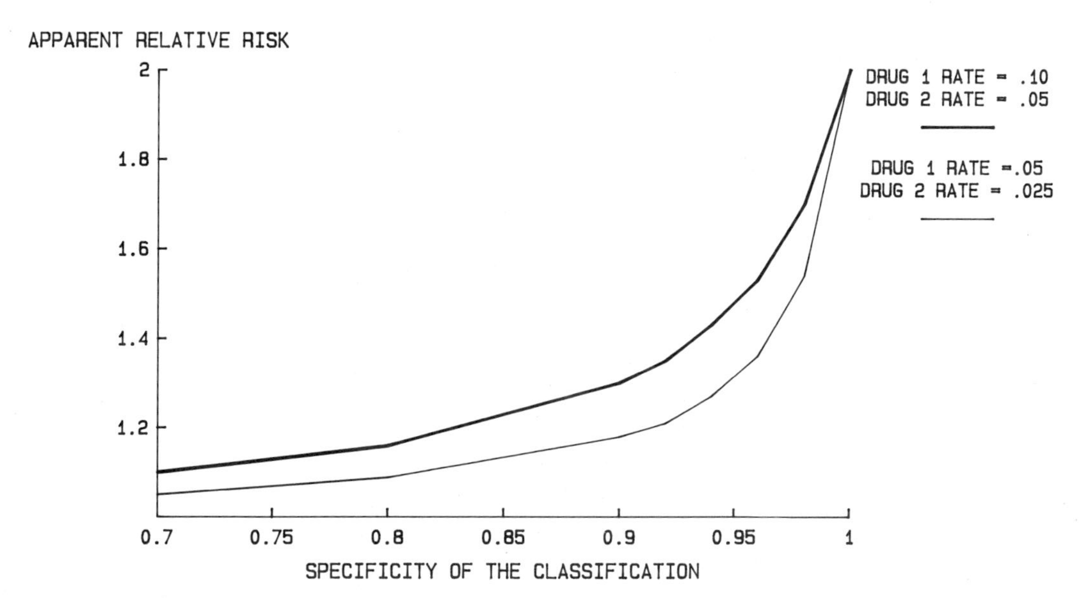


**Fig. 16** Bias in the estimate of a true relative risk equal to 2 as a function of specificity of the disease classification for a cohort study.

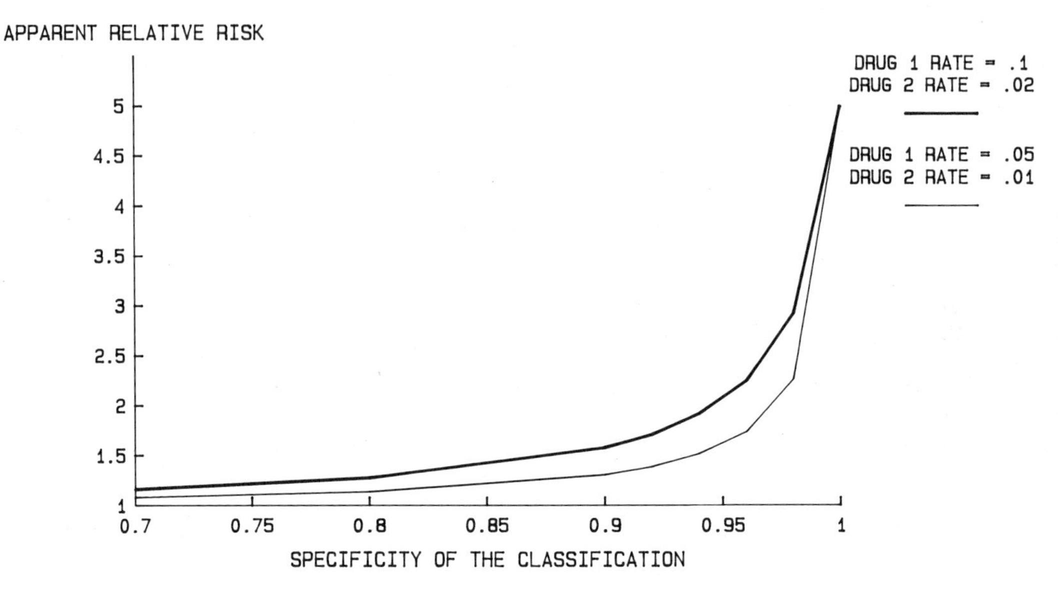


Fig. 17 Bias in the estimate of a true relative risk equal to 5 as a function of specificity of the disease classification for a cohort study.

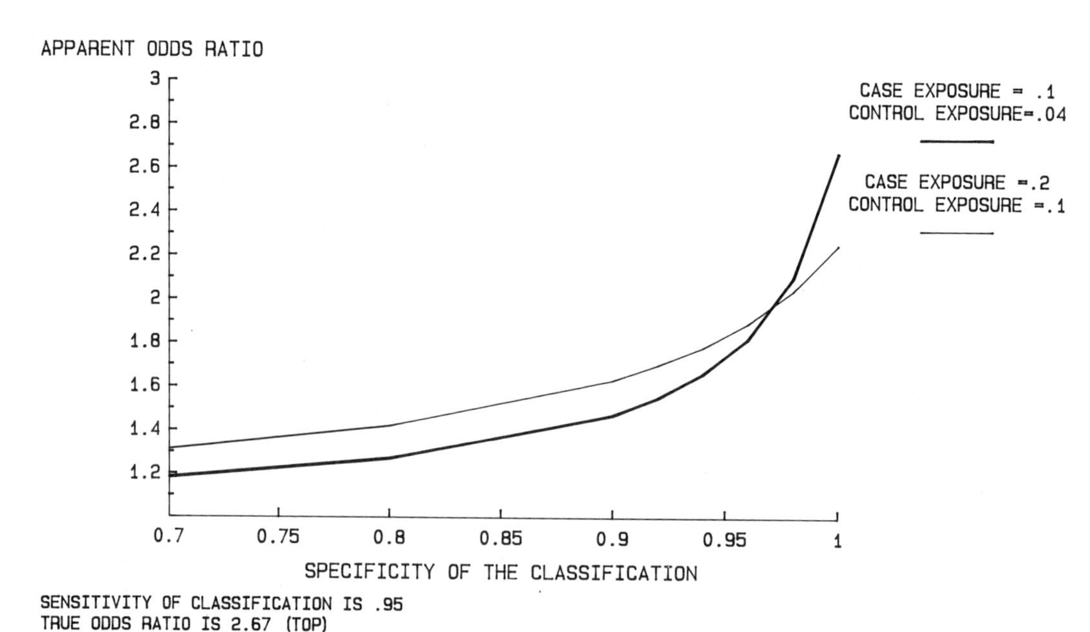


Fig. 18 Bias in the estimate of the true odds ratio as a function of specificity of exposure classification for a case-control study.

In each situation, the observed or apparent relative risk will be dramatically reduced from the true risk even before 95% specificity is reached.

For case-control studies, sensitivity of the exposure classification plays a larger role in the bias of the estimate of the odds ratio, and the less prevalent the exposure in the cases and controls the greater is the bias. Figure 18 illustrates, for a sensitivity of .95 in the drug exposure classification, the bias in the estimates of a true odds ratio for two odds ratios of different magnitudes that are characteristic of those observed in studies of drug-induced disease. Greenland (1982) illustrates similar effects of misclassification in matched pair case-control studies.

Thus the study of adverse reactions of low incidence can be substantially influenced by misclassification of the event and drug exposure which will make it more difficult to detect true differences or true relative risks. The interpretation of studies using data bases primarily designed for reasons other than drug epidemiology and whose quality may be less than desired should address these issues.

## REFERENCES

Anturane Reinfarction Trial Research Group. (1978). Sulfinpyrazone in the Prevention of Cardiac Death after Myocardial Infarction, *N. Engl. J. Med.*, 298(6): 289–295.

Anturane Reinfarction Trial Research Group. (1980). Sulfinpyrazone in the Prevention of Sudden Death after Myocardial Infarction, *N. Engl. J. Med.*, 302: 250–256.

Borden, E. K. and Lee, J. G. (1982). A Methodologic Study of Post-Marketing Drug Evaluation Using a Pharmacy-Based Approach, *J. Chronic Dis.*, 35: 803–816.

Boston Collaborative Drug Surveillance Program's Report on Reserpine and Breast Cancer. (1974). *Lancet*, 2: 669–671.

Bracken, M. and Holford, T. (1981). Exposure to Prescribed Drugs in Pregnancy and Association With Congenital Malformations. *Obstetrics and Gynecology*, 58(3): 336–344.

Breslow, N. E. (1970). A Generalized Kruskal-Wallis Test for Comparing K Samples Subject to Unequal Patterns of Censorship, *Biometrika*, 57: 579–594.

Breslow, N. E. and Day, N. E. (1980). *Statistical Methods in Cancer Research*, Vol. 1, *The Analysis of Case-Control Studies*, IARC Sci. Publ. No. 32, WHO, Lyon.

Brockett, P. L. and Kemperman, J. H. B. (1980). Statistical Recognition of Trends in Health Monitoring Systems, *Methods Inf. Med.*, 19(2): 106–111.

Chen, R., Mantel, N., Connelly, R. R., and Isacson, P. (1982). A Monitoring System for Chronic Disease. *Methods Inf. Med.*, 21: 86–90.

Chen, R. (1978). A Surveillance System for Congenital Malformations, *J. Am. Stat. Assoc.*, 73: 323–327.

Cochran, W. G. (1963). *Sampling Techniques*, Wiley, New York.

Code of Federal Regulations. (1986). Title 21, Revised April 1, 1986, U.S. Government Printing Office, Washington, D.C.

Coles, L. S., Fries, J. F., Kraines, R. G., and Roth, S. H. (1983). From Experiment to Experience: Side Effects of Nonsteroidal Anti-Inflammatory Drugs, *Am. J. Med.*, 74: 820–828.

Copeland, K. T., Checkoway, H., McMichael, A. J., and Holbrook, R. H. (1977). Bias Due to Misclassification in the Estimation of the Relative Risk, *Am. J. Epidemiol.*, 105: 488–495.

Cornfield, J. (1951). A Method of Estimating Comparative Rates from Clinical Data. Application to Cancer of the Lung, Breast and Cervix, *J. Natl. Cancer Inst.*, 11: 1269–1275.

Coronary Drug Project Research Group. (1977). Gallbladder Disease as a Side Effect of Drugs Influencing Lipid Metabolism, *N. Engl. J. Med.*, 296: 1185–1190.

Dixon, W. (1984). BMDP Statistical Software, 3rd ed., Los Angeles, Calif.

Duffy, S. W. (1983). On the Merging of Quoted Survival Rates to Obtain an Overall Rate, *Commun. Stat.—Theory Methods*, 12(23): 2753–2760.

Elandt-Johnson, R. C. (1975). Definition of Rates: Some Remarks on Their Use and Misuse, *Am. J. Epidemiol.*, 102: 267–271.

Faich, G. A. (1986). Adverse Drug Reaction Monitoring, *N. Engl. J. Med.*, 314: 1589–1592.

Federspeil, C. F., Ray, W. A., and Schaffner, W. (1976). Medicaid Records as a Valid Data Source, the Tennessee Experience, *Med. Care*, 14(2): 166–172.

Feinstein, A. R. (1973). Clinical Biostatistics: The Epidemiologic Trohoc, The Ablative Risk Ration, and "Retrospective" Research, *Clin. Pharmacol. Ther.*, 14: 291–307.

FDA Contract 223-82-3021. (1984). Final Report: Medicaid Data as a Source for Postmarketing Surveillance Information. Contract No. 223-82-3021, Food and Drug Administration, Washington, D.C.

Finney, D. J. (1965). The Design and Logic of a Monitor of Drug Use, *J. Chronic Dis.*, 18: 77–98.

Finney, D. J. (1971a). Statistical Aspects of Monitoring for Dangers in Drug Therapy, *Methods Inf. Med.*, 10: 1–8.

Finney, D. J. (1971b). Statistical Logic in the Monitoring of Reactions to Therapeutic Drugs, *Methods Inf. Med.*, 10(4): 237–245.

Finney, D. J. (1974). Systematic Signalling of Adverse Reactions to Drugs, *Methods Inf. Med.*, 13(1): 1–10.

Fleiss, J. L. (1981). *Statistical Methods for Rates and Proportions*, Wiley, New York.

Freedman, L. A. (1982). Tables of the Number of Patients Required in Clinical Trials Using the Logrank Test, *Stat. Med.*, 1: 121–129.

Friedman, G. D. and Ury, H. K. (1980). Initial Screening for Carcinogenicity of Commonly Used Drugs, *J. Natl. Cancer Inst.*, 65: 723–733.

George, S. L. and Desu, M. M. (1974). Planning the Size and Duration of a Clinical Trial Studying the Time to Some Critical Event, *J. Chronic Dis.*, 27: 15–24.

Gifford, L. M., Aeugle, M. E., Myerson, R. M., and Tannenbaum, P. J. (1980). Cimetidine Postmarket Outpatient Surveillance Program: Interim Report on Phase I, *J. Am. Med. Assoc.*, 243: 1532–1535.

Gladen, B. and Rogan, W. J. (1979). Misclassification and the Design of Environmental Studies, *Am. J. Epidemiol.*, 109: 607–616.

Gold, E., Gordis, L., Tonascia, J., and Szklo, M. (1978). Increased Risk of Brain Tumors in Children Exposed to Barbiturates. *J. Natl. Cancer Inst.*, 61: 1031–1034.

Goldberg, J. D. (1975). The Effects of Misclassification on the Bias in the Differences Between Two Proportions and the Relative Odds in the Fourfold Table, *J. Am. Stat. Assoc.*, 70: 561–567.

Greenberg, G., et al. (1975). Hormonal Pregnancy Tests and Congenital Malformations, *Br. Med. J.*, 191 (April 26).

Greenland, S. (1980). The Effect of Misclassification in the Presence of Covariates, *Am. J. Epidemiol.*, 112: 524–569.

Greenland, S. (1982). The Effect of Misclassification in Matched-Pair Case-Control Studies, *Am. J. Epidemiol.*, 116: 402–406.

Greenwald, P., et al. (1971). Vaginal Cancer with Maternal Treatment with Synthetic Estrogens, *N. Engl. J. Med.*, 285: 390–392.

Groel, J. T., Tadros, S. S., Dreslinski, G. R., and Jenkins, A. C. (1983). Long-Term Antihypertensive Therapy with Captopril, *Hypertension, Suppl. III*, 5(5): 145–151.

Gross, A. J. and Clark, V. (1975). *Survival Distributions: Reliability Applications in the Biomedical Sciences*, Wiley, New York.

Halperin, Rogot, Gurian, and Ederer. (1968). Sample Sizes for Medical Trials With Special Reference to Long-Term Therapy. *J. Chron. Dis.*, 21: 13–23.

Hankey, B. F. and Myers, M. H. (1971). Evaluating Differences in Survival between Two Groups of Patients, *J. Chron. Dis.*, 24: 523–531.

Herbst, A. L., et al. (1971). Adenocarcinoma of the Vagina: Association of Maternal Stilbestrol Therapy with Tumor Appearance in Young Women, *N. Engl. J. Med.*, 284: 878–881.

Hill, G. B., Spicer, C. C., and Weatherall, J. A. C. (1968). The Computer Surveillance of Congenital Malformations, *Br. Med. Bull.*, 24: 215–218.

Hurwitz, E. S., et al. (1985). Public Health Service Study on Reye's Syndrome and Medications, *N. Engl. J. Med.*, 313: 849–857.

Hutchinson, T. A., Flegel, K. M., Ho PingKing, H., Bloom, W. S., Kramer, M. S., and Trummer, E. G. (1983). Reasons for Disagreement in the Standardized Assessment of Suspected Adverse Drug Reactions, *Clin. Pharmacol. Ther.*, 34(1): 421–426.

Ibrahim, M. A. and Spitzer, W. O. (eds.). (1979). The Case-Control Study, Consensus and Controversy, Pergamon, New York; Special Issue, *J. Chronic Dis.*, 32(1): 2.

Idanpaan-Heikkila, J. (1983). A Review of Safety Information Obtained from Phase I-II and Phase III Clinical Investigations of Sixteen Selected

Drugs, U.S. Department of Health and Human Services, Office of New Drug Evaluation, National Center for Drugs and Biologics.

IMS America, Health Care Services Research Group. (1980). Final Report—Task A, B & C, An Experiment in Early Post-Marketing Surveillance of Drugs, FDA Contract 223-78-3007.

Inman, W. H. W. and Vessey, M. P. (1968). Investigation of Deaths from Pulmonary, Coronary and Cerebral Thrombosis and Embolism in Women of Childbearing Age, *Br. Med. J.*, 2: 193–200 (April 27).

Inman, W. H. W., Vessey, M. P., Westerholm, B., et al. (1970). Thromboembolic Disease and the Steroidal Content of Oral Contraceptives, a Report to the Committee on Safety of Drugs, *Br. Med. J.*, 2: 203–209.

Inman, W. H. W. (1977). Study of Fatal Bone Marrow Depression with Special Reference to Phenylbutazone and Oxyphenbutazone, *Br. Med. J.*, 1: 1500–1505.

Janerich, D. T., et al. (1974). Oral Contraceptives and Congenital Limb-Reduction Defects, *N. Engl. J. Med.*, 291(14): 697–700.

Jick, H., Miettinen, O. S., Shapiro, S., Lewis, G. P., Siskind, V., and Slone, D. (1970). Comprehensive Drug Surveillance, *J. Am. Med. Assoc.*, 213(9): 1455–1460.

Jick, H. (1977). The Discovery of Drug-Induced Illness, *N. Engl. J. Med.*, 296: 481–486.

Jick, H., Walker, A. M., and Spriet-Pourra, C. (1979). Postmarketing Follow-Up, *J. Am. Med. Assoc.*, 242(21): 2310–2314.

Jick, H. (1979). The Commission on Professional and Hospital Activities-Professional Activity Study: A National Resource for the Study of Rare Illnesses, *Am. J. Epidemiol.*, 109: 625–627.

Jick, H., Madsen, S., Nudelman, P. M., Perera, D. R., and Stergachis, A. (1984). Postmarketing Follow-Up at Group Health Cooperative of Puget Sound, *Pharmacotherapy*, 4: 99–100.

Jick, H. (1985). Use of Automated Data Bases to Study Drug Effects after Marketing, *Pharmacotherapy*, 5: 278–279.

Jick, H., Feld, A. D., and Perera, D. R. (1985). Certain Nonsteroidal Anti-Inflammatory Drugs and Hospitalization for Upper Gastrointestinal Bleeding, *Pharmacotherapy*, 5: 280–284.

Joint Statement on Tardive Dyskinesia by the Adverse Drug Reactions Advisory Committee and the Psychotropic Drug Committee of the Royal Australian and New Zealand College of Psychiatrists. (1983). *Med. J. Aust.*, (August 20).

Kalbfleisch, J. D. and Prentice, R. L. (1980). *The Statistical Analysis of Failure Time Data*, Wiley, New York.

Knapp, D. E., Zax, B. B., Rossi, A. C., and O'Neill, R. T. (1980). A Method for Postmarketing Screening of Adverse Reactions to Drugs—Initial Results, *Drug Intell. Clin. Pharm.*, 14: 23–27.

Kleinbaum, D. G., Kupper, L. L., and Morgenstern, H. (1982). *Epidemiologic Research, Principles and Quantitative Methods*, Lifetime Learning Publications, Belmont, Calif.

Kupper, L. (1984). Effects of the Use of Unreliable Surrogate Variables on the Validity of Epidemiologic Studies, *Am. J. Epidemiol.*, 120: 643–648.

Lachin, J. M. (1981). Introduction to Sample Size Determination and Power Analysis for Clinical Trials, *Controlled Clin. Trials*, 2: 93–113.

Lawson, D. H. (1984). Pharmacoepidemiology: A New Discipline, *Br. Med. J.*, 289: 940–941.

Lee, E. T. (1980). *Statistical Methods for Survival Analysis*, Lifetime Learning Publications, Belmont, Calif.

Levine, A., Mandel, S. P. H., and Santamaria, A. (1977). Pattern Signalling in Health Information Monitoring Systems, *Methods Inf. Med.*, 16(3): 138–144.

Levine, J., Schooler, N. R. (1986). SAFTEE: A Technique for the Systematic Assessment of Side Effects in Clinical Trials, *Psychopharmacol. Bull.*, 22: 343–381.

Levy, E. P., et al. (1973). Hormone Treatment during Pregnancy and Congenital Hearth Defects, *Lancet*, 611 (March 17).

Mack, T. M., Pike, M. C., Henderson, B. E., Pfeffer, R. I., Gerkins, V. R., Arthur, B. S., and Brown, S. E. (1976). Estrogens and Endometrial Cancer in a Retirement Community, *N. Engl. J. Med.*, 294: 1262–1267.

MacRae, K. D. (1984). The Statistical Analysis of Reports of Serious Adverse Reactions to the CSM, unpublished.

Makuch, R. W. and Simon, R. M. (1980). Sample Size Considerations for Non-Randomized Comparative Studies, *J. Chronic Dis.*, 33: 175–181.

Makuch, R. W. (1982). Adjusted Survival Curve Estimation Using Covariates, *J. Chronic Dis.*, 35: 437–443.

Mandel, S. P. H., Levine, A., and Beleno, G. E. (1976). Signalling Increases in Reporting in International Monitoring of Adverse Reactions to Therapeutic Drugs, *Methods Inf. Med.*, 15(1): 1–10.

Mann, J. I., et al. (1975). Myocardial Infarction in Young Women with Special Reference to Oral Contraceptive Practice, *Br. Med. J.*, (May 3), 241–245.

Maronde, R. F., Lee, P. V., McCarron, M. M., et al. (1971). A Study of Prescribing Patterns, *Med. Care*, 9: 383–395.

McHugh, R. B. and Lee, C. T. (1984). Confidence Estimation and the Size of a Clinical Trial, *Controlled Clin. Trials*, 5: 157–163.

Mitchell, A. A., Slone, D., Shapiro, S., and Goldman, P. (1980). Adverse Drug Effects and Drug Surveillance, *Pediatric Pharmacology: Therapeutic Principles in Practice* (S. J. Yaffe, ed.), Grune & Stratton, New York.

Morgenstern, H., Kleinbaum, D. G., and Kupper, L. L. (1980). Measures of Disease Incidence Used in Epidemiologic Research, *Int. J. Epidemiol.*, 9: 97–104.

Moussa, M. A. A. (1978). Statistical Problems in Monitoring Adverse Drug Reactions, *Methods Inf. Med.*, 17(2): 105–112.

Nibbelink, D. W. and Streikland, S. C. (1979). Cyclobenzaprine (Flexeril) Postmarketing Surveillance Program. Preliminary Report, *Current Ther. Res.*, 25: 564–570.

Nora, J. J. and Nora, A. H. (1973). Birth Defects and Oral Contraceptives, *Lancet*, 941–942 (April 28).

Norwegian Multicenter Study Group. (1981). The Timolal-Induced Reduction in Mortality and Reinfarction in Patients Surviving Acute Myocardial Infarction, *N. Engl. J. Med.*, 304: 801–807.

O'Neill, R. T. and Anello, C. (1978). Case-Control Studies: A Sequential Approach, *Am. J. Epidemiol.*, 108: 415–424.

O'Neill, R. T. (1984). Sample Sizes for Estimation of the Odds Ratio in Unmatched Case-Control Studies, *Am. J. Epidemiol.*, 120: 145–153.

Paganini-Hill, A. and Ross, R. K. (1982). Reliability of Recall of Drug Usage and Other Health-Related Information, *Am. J. Epidemiol.*, 116: 114–122.

Palta, M. and McHugh, R. (1980). Planning the Size of a Cohort Study in the Presence of Both Losses to Follow-Up and Non-Compliance, *J. Chronic Dis.*, 33: 501–512.

Pasternack, B. S. and Gilbert, H. S. (1971). Planning the Duration of Long-Term Survival Time Studies Designed for Accrual by Cohorts, *J. Chronic Dis.*, 24: 681–700.

Pasternack, B. S. (1972). Sample Sizes for Clinical Trials Designed for Patient Accrual by Cohorts, *J. Chronic Dis.*, 25: 673–681.

Pasternack, B. S. and Shore, R. E. (1980). Group Sequential Methods for Cohort and Case-Control Studies, *J. Chronic Dis.*, 33: 365–373.

Pasternack, B. S. and Shore, R. E. (1981). Sample Sizes for Group Sequential Cohort and Case-Control Study Designs, *Am. J. Epidemiol.*, 113: 182–191.

Peck, A. W., et al. (1983). Incidence of Seizures During Treatment with Tricyclic Antidepressant Drugs and Bupropion, *J. Clin. Psychiatry*, 44(5): 197–201.

PEM (Prescription Event Monitoring) News. (1983). Drug Research Unit, University of Southhampton, Botley, Southhampton S03 2BX No. 1.

Pitts, N. E. (1980). Review of Clinical Trial Experience with Piroxicam, *Piroxicam: A New Non-Steroidal Anti-Inflammatory Agent* (Proc. IXth European Congress of Rheumatology), Academy Professional Information Services, New York, pp. 48–66.

Prentice, R. L., Williams, B. J., and Peterson, A. V. (1981). On the Regression Analysis of Multivariate Failure Time Data, *Biometrika*, 68: 373–379.

Quade, D., Lackenbruch, P. A., Whaley, F. S., McClish, D. K., and Haley, R. W. (1980). Effects of Misclassification on Statistical Inferences in Epidemiology, *Am. J. Epidemiol.*, 111: 503–515.

Quade, D. and McClish, D. (1985). Improving Estimates of Prevalence by Repeated Testing, *Biometrics*, 41: 81–89.

Rauch, E. F. (1979). Data Entry Experience with the Flexeril Surveillance Program, *Drug. Inf. J.*, 75-83 (Sept.).

Rossi, A. C., et al. (1983). A Review of Selected Phase IV Studies and Their Contribution to Postmarketing Surveillance, Rep. FDA, CDB, Division of Drug Experience.

Rossi, A. C., et al. (1983). Discovery of Adverse Drug Reactions. A Comparison of Selected Phase IV Studies with Spontaneous Reporting Methods, *J. Am. Med. Assoc.*, 249(16): 2226–2228.

Rossi, A. C. (1984). Discovery of New Adverse Drug Reactions, A Review of the ,ood and Drug Administration's Spontaneous Reporting System, *J. Am. Med. Assoc.*, 252(8): 1030–1033.

Royal College of General Practitioners. (1974). *Oral Contraceptives and Health*, Pitman, London.

Sackett, D. L. (1979). Bias in Analytic Research, *J. Chronic Dis.*, 32: 51–63.

Safra, M. J. and Oakley, G. P. (1975). Association Between Cleft Lip with or without Cleft Palate and Prenatal Exposure to Diazepam, *Lancet*, 478–480 (Sept. 13).

Saxen, I. (1974). Cleft Lip and Palate in Finland: Parental Histories, Course of Pregnancy and Selected Environmental Factors. *Int. J. Epid.*, 3: 263–270.

Saxen, I., Klemetti, A., and Haro, A. S. (1974). A Matched-Pair Register for Studies of Selected Congenital Defects. *Am. J. Epidemiol.*, 100: 297–306.

Saxen, I. (1975). Associations Between Oral Clefts and Drugs Taken During Pregnancy, *Int. J. Epid.*, 4: 37–44.

Schlesselman, J. J. (1974). Sample Size Requirements in Cohort and Case-Control Studies of Disease, *Am. J. Epidemiol.*, 99: 381–384.

Schlesselman, J. J. (1982). *Case-Control Studies, Design, Conduct, Analysis*, Oxford University Press, New York.

Schoenfeld, D. (1983). Sample-Size Formula for the Proportionay Hazards Regression Model, *Biometrics*, 39: 499–503.

Seigel, D. G. (1971). The Use of Historical Data and Adverse Reaction Reporting Systems for Epidemiologic Study, *Am. J. Epidemiol.*, 94: 210–214.

Schaffner, W., Ray, W. A., Federspiel, F., and Miller, W. O. (1983). Improving Antibiotic Prescribing in Office Practice: A Controlled Trial of Three Education Methods, *J. Am. Med. Assoc.*, 250: 1728–1732.

Shapiro, S. and Slone, D. (0000). Post-Marketing Assessment of Drugs, *Post-Marketing Surveillance of Adverse Reactions to New Medicines*, Medico-Pharmaceutical Forum No. 7, 19–25.

Shapiro, S., Slone, )D., Rosenberg, L., Kaufman, D. W., Stolley, P. D., and Miettinen, O. S. (1979). Oral Contraceptive Use in Relation to Myocardial Infarction, *Lancet*, 1: 743–747.

SK&F Lab. Co. (1981). Promotional Brochure TG31, Tagamet Post-Marketing Surveillance Program Initial and Six Month Follow-Up Reports (May).

Slone, D., Jick, H., Borda, I., Chalmers, T., Feinleib, M., Muench, H., Lipworth, L., Bellotti, C., and Gilman, B. (1966). Drug Surveillance Utilizing Nurse Montors, *Lancet*, 7469: 901–903 (Oct. 22).

Slone, D., Gaetano, L. F., Lipworth, L., Shapiro, S., Lewis, G. P., and Jick, H. (1969). Computer Analysis of Epidemiologic Data: Effect of Drugs on Hospital Patients, *Public Health Rep.*, 34(1): 39–52.

Slone, D., Shapiro, S., and Miettinen, O. S. (1977). Case-Control Surveillance of Serious Illnesses Attributable to Ambulatory Drug Use, *Epidemiological Evaluation of Drugs* (D. Columbo, S. Shapiro, D. Slone, X. Tognoni, eds.), pp. 59–70.

Smith, D. C., et al. (1975). Association of Exogenous Estrogen and Endometrial Carcinoma, *N. Eng. J. Med.*, 293: 1164–1167.

Stephens, M. D. B. (1985). *The Detection of New Adverse Drug Reactions*, Macmillan,

Stern, R. S. and Bigby, M. (1984). An Expanded Profile of Cutaneous Reactions to Nonsteroidal Anti-Inflammatory Drugs. *J. Am. Med. Assoc.*, 252: 1433–1437.

Strom, B. L., Carson, J. L., Morse, M. L., and Leroy, A. A. (1985). The Computerized on Lone MEdicaid Pharmaceutical Analysis and Surveillance System: A New Resource for Postmarketing Drug Surveillance, *Clin. Pharmacol. Ther.*, 38: 359–364.

Tarone, R. E. and Ware, J. (1977). On Distribution-Free Tests of Equality of Survival Distributions, *Biometrika*, 64(1): 156–160.

Taulbee, J. D. and Symons, M. J. (1983). Sample Size and Duration for Cohort Studies of Survival Time with Covariables, *Biometrics*, 39: 351–360.

Van de Carr, S. W., Kennedy, D. L., Rosa, F. W., Anello, C., and Jones, J. K. (1983). Relationship of Oral Contraceptive Estrogen Dose to Age, *Am. J. Epidemiol.*, 117: 153–160.

Venulet, J., Blattner, R., Von Bulow, J., and Berneker, G. C. (1982). How Good are Articles on Adverse Drug Reactions? *Br. Med. J.*, 284: 252–254.

Vessey, M. P. and Doll, R. (1968). Investigation of Relation between Use of Oral Contraceptives and Thromboembolic Disease, *Br. Med. J.*, 2: 199–205.

Vessey, M. P. and Doll, R. (1969). Investigation of Relation between Use of Oral Contraceptives and Thromboembolic Disease, *Br. Med. J.*, 2: 651–657.

Walter, S. D. (1977). Determination of Significant Relative Risks and Optimal Sampling Procedures in Prospective and Retrospective Comparative Studies of Various Sizes, *Am. J. Epidemiol.*, 105: 387–397.

Weatherall, J. A. C. and Haskey, J. C. (1976). Surveillance of Malformations, *Br. Med. Bull.*, 32(1): 39–44.

Williams, R. R., Feinleib, M., Connor, R. J., and Stegens, N. L. (1978). Case-Control Study of Anti-Hypertensive and Diuretic Use by Women with Malignant and Benign Breast Lesions Detected in a Mammography Screening Program, *J. Natl. Cancer Inst.*, 61: 327–335.

Ziel, H. K. and Finkle, W. D. (1975). Increased Risk of Endometrial Carcinoma among Users of Conjugated Estrogens, *N. Engl. J. Med.*, 293: 1167–1170.

# 14 Manufacturing and Quality Control

ROBERT C. KOHBERGER *Quality Control Support Operations, Lederle Laboratories, Pearl River, New York*

## I. INTRODUCTION

Drug development does not end with the assessment of clinical safety and efficacy of a dosage form. It continues with postmarketing surveillance studies and activities in manufacturing and quality control to ensure that the product continues to meet the high standards required by the patient. The product evolves throughout its life to improve quality as delivered to the patient and to increase the efficiency of the manufacturing process.

Quality control in the pharmaceutical industry is involved in many different activities ranging from the receipt and storage of raw materials through the distribution of packaged final product to wholesalers and pharmacies. The quality of a pharmaceutical product includes the quality of

Design
Facilities
Raw Materials
Manufacturing
Testing Procedures
Labeling
Packaging
Distribution
Marketing
Sales

These areas form links of a quality chain and all are important if the patient is to receive a product that is safe and effective. This chapter will concentrate on the areas of design, manufacturing, and testing where the primary quality characteristics of interest are the chemical and physical attributes of the product. The chapter will first discuss the most common pharmaceutical products and their quality characteristics. Two of the major statistical application areas will then be discussed—validation, both assay and process validation, and the analysis of long term stability studies. There are other important application areas in manufacturing and quality control such as acceptance sampling and statistical process control, but the two areas to be discussed have characteristics unique to pharmaceutical dosage forms.

Pharmaceutical products place a unique responsibility on a manufacturer. The ultimate consumer—the patient—has little ability to judge the quality of the product. To be sure, the consumer can make some judgements concerning quality, primarily in the physical appearance of the product, but the key concerns—efficacy and safety—can be made only on an ex post facto basis: the product is consumed and the consumer can judge whether the product worked with minimal side effects. In as critical an area as health care such judgments occur too late for the consumer's benefit. Because of this, as in other areas of drug development, manufacturers have very strict internal control procedures and the industry itself is strongly regulated by the government.

## II. DOSAGE FORMS AND QUALITY

The quality concerns and quality measurements of a pharmaceutical product depend on the particular dosage form of the product.

### A. Tablets

The most common form of pharmaceutical dosage is the compressed tablet. Tablets consist of an active ingredient plus various additional components that have different functions in the tablet. Diluents or bulking agents add volume or bulk to the tablets. The therapeutic dose of most drugs is so small that manufacturing into a tablet the active ingredient alone is impossible with current technology. Bulking agents add sufficient volume to make compression possible. Binders make the various powders cohesive and hold them together after compression. Disintegrants are substances that help the tablet break up and release the active ingredient after the patient has ingested the tablet. Glidants are added to the tablet to aid the powder to flow from a storage bin or hopper into the compression punch. After a tablet has been compressed, lubricants are necessary to have the compressed tablets release from the compression punches. Colorants are added to the tablet primary to aid in product identification by the patient and to minimize the risk of product mixup at the manufacturing or dispensing site.

Prior to the compression into a tablet, the various powders usually undergo some intermediate processing. From an ease of processing viewpoint, an ideal tablet formulation may be directly compressed from a dry powder state. With these types of formulations the dry powders making up the tablet are mixed together and the resulting dry powder compressed on a tablet compression machine. Very often pharmaceutical products cannot be directly compressed. There are many different reasons, such as too fine a powder or inability to disintegrate with the level of binder necessary for compression. To overcome these compression difficulties granulation techniques are used. Whether the granulation is wet or dry, the basic idea behind granulation is to product coarse granular aggregates which have a greater cohesiveness than the original powders. In dry granulation the powder is held under compression for a longer timo than is usual for tablet making which strengthens the bonds holding the powder together. In wet granulation a solvent is added to the powders to produce a wet mass. This mass is then dryed and milled to produce aggregates with the desired characteristic.

Once the powders are in a form amenable to compression the powder or granulation enters a tablet press for compression into tablets. The granulation is loaded into hoppers, where, either by gravity or force feeding, the granulation enters the die cavity between two punches, where it is compressed and a tablet produced. Usual speeds of tablet presses are from 2000-5000 tablets/minute with compression forces of 2-6 tons.

The three major quality characteristics of tablets are potency, dose uniformity, and dissolution. Potency is a measure of the average amount of active ingredient in a tablet. Content uniformity is a measure of the variability in amount of active ingredient between tablets. Dissolution is a measure of the amount of active ingredient dissolved in a solution after a specified amount of time.

Potency is usually determined by compositing 20 tablets and performing either a chemical or microbiologic assay on the composite. The composite is made by grinding the tablets with a mortar and pestle or some other suitable means to obtain a uniform powder mixture. The specification on tablets is usually in the form of percent label claim (% LC) and often allows a variation in the assayed amount of 90 to 110% LC. Specific products may have different specifications.

Dose uniformity is a measure of the amount of active ingredient in an individual tablet. The amount of active ingredient may be calculated by two different methods:

1. Assay the individual tablet.
2. Assay a composite mixture of tablet powder and obtain the concentration of active on a weight per weight basis. Weigh an individual tablet and obtain an estimate of the amount of active ingredient by multiplying tablet weight by concentration.

The United States Pharmacopeia, 21st edition, (USP XXI) states that method 2 may be used for uncoated tablets that contain more than 50 mg of active ingredient and where the active ingredient comprises more than 50% of the tablet's weight. The USP XXI specifications for a tablet whose potency limits average 100% LC or less are given in Table 1. These limits apply whether the amount of active is determined by method 1 or method 2.

Dissolution is determined by placing the tablet in a vessel filled with dissolution medium, stirring the solvent, and after a specified time assaying the amount of active ingredient expressed as % LC dissolved in the solvent. The USP XXI specifications are given in Table 2. The intent of the dissolution test is to model in vitro the bioavailability of a pharmaceutical product. It is a matter of some contention whether the dissolution test is actually correlated with bioavailability, but at the least it provides a quality control measure on the consistency of production on a surrogate measure.

Other quality parameters are important and are often measured during the production process to control the tablet-making operations. These parameters include tablet thickness, hardness, weight, disintegration, and friability. The measurements are quickly and easily made on the production

**Table 1** USP XXI Dose Uniformity Specifications: Tablets with Potency Limits Average 100% LC

| Stage | Sample size | Pass stage, if: |
|---|---|---|
| 1 | 10 | All tablets between 85 and 115% LC and relative standard deviation <6.0% |
| 2 | 20 | For all tablets (stages 1 and 2) not more than 1 tablet outside range of 85–115% LC and no tablets outside range of 75–125% LC and relative standard deviation <7.8% |

floor, whereas assay measurements must be done under more specialized conditions and most often are too time-consuming for effective production control.

### B. Capsules

Capsule pharmaceutical products consist of a gelatin shell enclosing either a powder or a liquid containing the active ingredient. The shell itself may be a hard or soft shell capsule with one or two pieces. Capsules are preferred by many consumers because they have a smooth, slippery surface and are easy to swallow. Disadvantages include product cost and the fact that the product (the powder) is, in a sense, hidden, allowing for an opportunity for product tampering.

The primary manufacturing difference between tablets and capsules is that for capsules, binders and lubricants in the capsule powder are not necessary. The prime consideration is to obtain a capsule powder that is free-flowing and does not interact with the gelatin shell. The capsule powder is loaded into a hopper and, by either gravity or force feeding, is loaded into the individual capsules. Capsule filling equipment is capable of producing several thousand capsules per minute. Care must be taken to

**Table 2** USP XXI Dissolution Specifications

| Stage | Sample size | Pass stage, if: |
|---|---|---|
| 1 | 6 | Each tablet is not less than Q + 5%[a] |
| 2 | 6 | Average of 12 tablets (stages 1 and 2) is equal to or greater than Q and no tablet is less than Q − 15% |
| 3 | 12 | Average of 24 tablets (stages 1, 2, and 3) is equal to or greater than Q and not more than 2 tablets are less than Q − 15% |

[a]Q is set to some specified label claim, often 80% LC.

demonstrate that over time the ingredients in the capsule powder do not interact with the gelatin capsule to produce either brittle or soft capsules.

The primary quality characteristics for capsules are the same as those for tablets: potency, dose uniformity, and dissolution. The specifications for dose uniformity of capsules given by USP XXI differ from those for tablets and are given in Table 3. It should be noted that the dose uniformity requirements for capsules are less stringent than those for tablets because of the greater variability inherent in the manufacturing process. The method for determining the amount of active ingredient in a capsule is the same as that for tablets, and may be either the assay content method or the weight method, depending on which is appropriate.

In-process controls for capsules consist of capsule weight checks and physical appearance of the capsules. It is important that the two halves of a capsule mate correctly and that the capsule itself present a smooth surface without any pinhole leaks.

### C. Sustained Action Dose Forms

Sustained action dose forms are designed to slowly release active ingredient into the body over a period of time. They may be formulated as tablets, capsules, or patches applied to the skin.

A common method of obtaining sustained action in a tablet formulation is to disperse the active ingredient in a solid which is less soluble than the drug. The solid then impedes the dissolution of the active ingredient. Sustained release capsules are often obtained by coating particles of different sizes, thereby obtaining differential dissolution characteristics. The coated particle mix cannot withstand compression forces and is filled into capsules. For tablets and capsules, the usual quality characteristics apply with an important distinction for dissolution. Rather than measuring the amount dissolved at a single time point, multiple time points are observed to ensure that an adequate release profile is obtained. It is important to determine that no dumping of active ingredient occurs early in time and that the full therapeutic dose is released over time. Because transdermal patches are a relatively recent development and are not yet common, they are still under study by the USP to determine the most suitable dose release test specifications to cover a wide range of applications.

**Table 3** USP XXI Dose Uniformity Specifications: Capsules with Potency Limits Average 100% LC

| Stage | Sample size | Pass stage, if: |
|---|---|---|
| 1 | 10 | Not more than 1 capsule outside range of 85–115% LC and no capsule outside range of 75–125% LC and relative standard deviation <6.0% |
| 2 | 20 | For all capsules (stages 1 and 2) not more than 3 capsules outside range of 85–115% LC and no capsule outside range of 75–125% LC and relative standard deviation <7.8% |

### D. Sterile Products

Sterile or parenteral products are intended for use by injection through the skin. The products themselves may be solutions, suspensions, or sterile solids for reconstitution. The product dose units may be vials, syringes, or bags and may be either single use or multiple use units.

Because parenteral products bypass the usual body defense mechanisms, the primary quality concern for them is purity, both chemical and microbiologic. Sterility or absence of microorganisms must be guaranteed with a high degree of confidence. Although sterility concepts are couched in such statistical terms as "probability of organism kill" and "probability of nonsterility," statisticians have had little to do with the development of sterility concepts in the pharmaceutical industry. One reason is that, because of the high degree of assurance required, 100% inspection would be required if the concepts of acceptance sampling were applied. Furthermore, the actual sterility test is a destructive test. The approach taken to control sterility is through the development of theoretical models, empirically validated, that relate the probability that a microorganism is killed to the processing variables of temperature (heat) and time. The processing variables are then controlled on this basis and lot acceptance sampling is not necessary. Although such procedures, called parametric release, have been proposed, most companies still do batch acceptance tests to detect gross violations of established process control parameters.

Other quality aspects of parenterals, in addition to sterility, include potency, dose uniformity, and volume of fill. For all parenteral products potency is a quality consideration. For single-use containers potency is expressed on a dose unit basis and the USP XXI dose uniformity requirements apply. For multiple-use containers potency is expressed on a concentration basis and volume of fill requirements, rather than dose uniformity, apply.

### E. Other Products

There are many other pharmaceutical dose forms. A partial listing includes:

Ointments
Creams
Aerosols
Suppositories
Syrups
Elixirs
Suspensions

Each has its own use and its own specific quality characteristics. However, all entail the same concerns of potency, dose uniformity, and, in the case of multiuse containers, adequate volume or weight of contents.

## III. REGULATORY CONSIDERATIONS

Pharmaceutical manufacturers are regulated by the Food and Drug Administration (FDA) and also their products, if so labeled, must meet the requirements of the United States Pharmacopeial Convention (USP). The USP is a volunteer organization composed of individuals from medical, pharmaceutical, legal, and chemical fields. It provides uniform standards for pharmaceutical products, including standards for potency, purity, packaging and storage, and assay methods. The USP standard for dose uniformity has already been discussed. The USP standards have legal authority as written by the Food, Drug, and Cosmetic Act of 1938.

In addition to the review of new drug entities, the FDA has the authority to regulate and supervise drug product quality. This function is performed by random sampling of actual production lots and by plant inspections. Random sampling occurs at the manufacturing site or anywhere in the distribution chain. The samples are examined at FDA analytical laboratories to determine conformance with established specifications. In-plant inspections are designed to determine conformance with regulations called Current Good Manufacturing Practices (CGMPs). These regulations are designed to direct manufacturers into high-quality production practices. Some of the areas addressed in CGMPs are

Production and facility design
Equipment
Personnel training
Raw material components
Record keeping
Packaging
Quality control test procedures
Product stability
Product complaints

Failure to meet regulations and specifications can result in product recall, product seizure, fines, or plant closings.

## IV. VALIDATION

Validation is an important part of the drug development process because it is one of the final steps in the development of a drug from the discovery of a new chemical entity to the production of a dosage form that may be sold to the consumer. Validation is defined as establishing documented evidence that a process does what it purports to do (Chapman, 1983). Pharmaceutical manufacturers have been practicing validation since the first pharmaceutical manufacturing operations began, but recent FDA regulations and compliance activities have emphasized its importance. Under CGMP regulations adequate validation documentation is required, and firms not meeting the regulations are subject to legal remedies available to the FDA.

The major impact of validation regulations has been to examine process and assay development as a system with benchmarks along the way to demonstrate that the system works as it should with documentation to justify such conclusions. There is considerable confusion and controversy over what constitutes an adequate validation program. The controversies center around defining what a system is supposed to do and the criteria for determining whether the system has met the objectives. Because pharmaceutical manufacturing and assays system exhibit variation, statistical considerations form a key part of the validation process.

It is beyond the scope of this section to present a recommended validation program. The more common statistical techniques used in the execution of a validation program will be presented. The exact program requires multi-disciplinary expertise in the areas of pharmaceutical manufacturing, analytical chemistry, microbiology, regulation, quality control, and statistics.

The major statistical applications are in the validation of test methodology (assay validation) and the validation of manufacturing methods (process validation).

### A. Assay Validation

Assay validation is the demonstration that an assay reliably and accurately measures what it is supposed to measure.

In a tablet product, an active ingredient is surrounded by a matrix of inactive ingredients. Accuracy of an assay method depends on whether the assay obtains a value equal to the amount of active ingredient actually in the tablet. Experiments to answer this question are called recovery studies, the idea being that an assay recovers the active from the excipients. In a recovery study an exact amount of active is added to a placebo containing the tablet excipients. The amount of active added is at varying levels, from below the potency specification to above the potency specification. Often three levels are added: 80, 100, and 120% of label claim. At each level several replicate assays are obtained.

One form of analysis proceeds along the lines of simple linear regression analysis. The model to be fit is

$$y(\text{found}) = a + b * x(\text{added}) + e$$

where e is normally distributed error with mean = 0 and variance = $\sigma^2$. The regression estimates are given by

$$\hat{b} = \frac{\Sigma(y_i - \bar{y})}{\Sigma(x_i - \bar{x})^2}$$

$$\hat{a} = \bar{y} - b\bar{x}$$

$$s^2 = \frac{\Sigma(y_i - \hat{a} - \hat{b}x_i)^2}{n - 1}$$

$$se^2(b) = \frac{s^2}{\Sigma(x_i - \bar{x})^2}$$

$$se^2(a) = \frac{s^2 \Sigma x_i^2}{n\Sigma(x_i - \bar{x})^2}$$

where n is the number of observations. An accurate assay will obtain values of a = 0 and b = 1, so that the model will be

$$y(\text{found}) = 0 + 1 * x(\text{added}) = x$$

The joint hypothesis may be tested using a simultaneous confidence ellipse for the parameters a and b. The equation for this ellipse is given by

$$2s^2F = n(a - \hat{a})^2 + \Sigma x_i^2(b - \hat{b})^2 + 2n\bar{x}(a - \hat{a})(b - \hat{b})$$

where F = the 1 − α upper point of the F(2, n − 2) distribution. If the value a = 0, b = 1 falls outside this ellipse there is a statistically significant inaccuracy in the assay at the α significance level. Table 4 gives an example of the calculations and Figure 1 shows the relevant portion of the confidence ellipse. As can be seen from the figure, there is no evidence to suggest inaccuracy of the assay. If inaccuracy is demonstrated, then individual confidence limits could be placed on the regression parameters to determine the kind of inaccuracy (whether absolute or relative).

A different form of the analysis is sometimes favored by analytical chemists. In this analysis the found values are divided by added to obtain percent recovered. In terms of the original recovery model, this obtains observations of the form

$$\% \text{ recovery} = \left(\frac{a}{x} + b\right) * 100$$

If the assay is accurate (a = 0, b = 1) then this transformed observation will be 100%. The analysis proceeds by first comparing the percent recovered across added levels (via analysis of variance techniques); then, if no significant difference is found, pooling the observations to determine the average percent recovered. The average could then be tested against 100% to determine whether there is significant inaccuracy. Table 5 shows

Table 4 Recovery Experiment[a]

| Added (% LC) | Found (% LC) |
|---|---|
| 80 | 78 |
| 80 | 82 |
| 100 | 100 |
| 100 | 103 |
| 120 | 118 |
| 120 | 125 |

[a] $\hat{a} = -2.75$, $se(\hat{a}) = 7.782$, $\hat{b} = 1.0375$, $se(\hat{b}) = 0.077$, $s^2 = 9.438$. Confidence ellipse:

$$\begin{aligned} 131.00 = {} & 6(a + 2.75)^2 \\ & + 61600(b - 1.0375)^2 \\ & + 1200(a + 2.75)(b - 1.0375) \end{aligned}$$

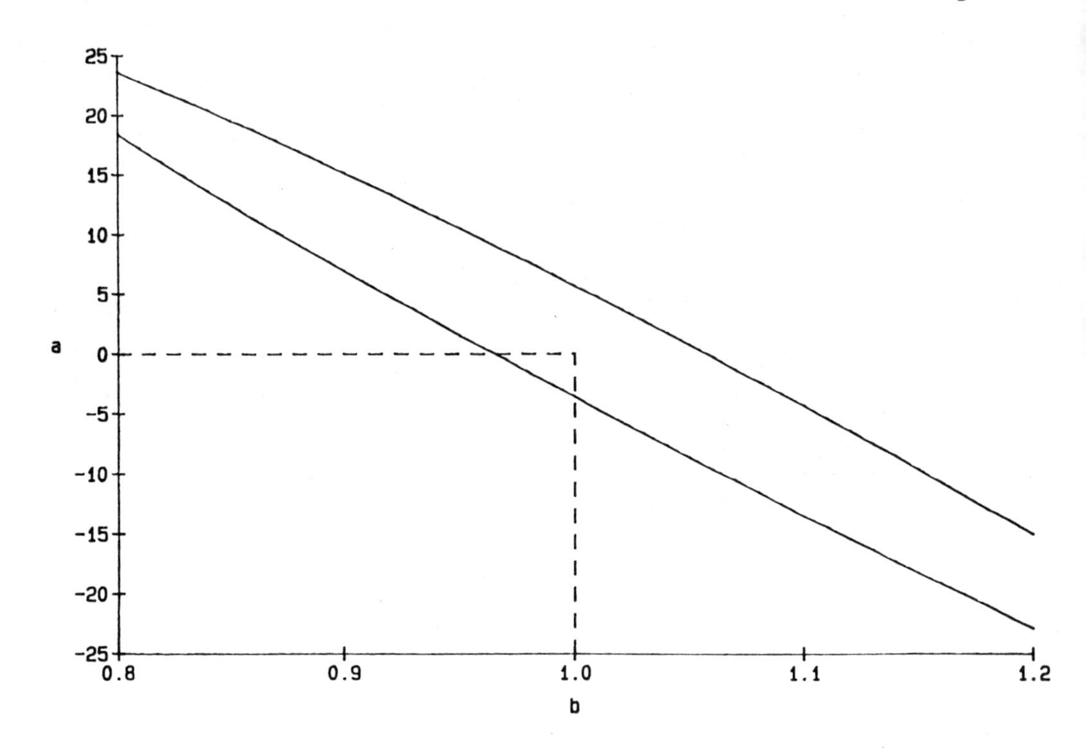


Fig. 1 Joint confidence ellipse—95% confidence a, b.

Table 5 Recovery Experiment

| Added (% LC) | Found (% LC) | % Recovery |
|---|---|---|
| 80 | 78 | 97.5 |
| 80 | 82 | 102.5 |
| 100 | 100 | 100.0 |
| 100 | 103 | 103.0 |
| 120 | 118 | 98.3 |
| 120 | 125 | 104.2 |

| Added | Average | Analysis of variance | | |
|---|---|---|---|---|
| 80 | 100.0 | Source | MS | F |
| 100 | 101.5 | Level | 1.3 | 0.11[a] |
| 120 | 101.3 | Error | 11.5 | |
| Overall | 100.9[b] | | | |

[a]Not significant, p > .05.
[b]95% confidence limit: 100.9 ± 2.9 = 98.0, 103.8

such an analysis. Although this method uses variables easily understood by a chemist, it suffers from some disadvantages: because a conditional procedure is used the p-level of the initial test is critical, and if inaccuracy is found it does not aid in determining whether this is due to absolute or relative errors.

In addition to the accuracy of an assay, the variability of the test method is of interest. Assay methods consist of many steps, each of which contributes to the overall variability of the method. Consider a method with three steps:

Separation
↓
Extraction
↓
Instrument reading

A typical design to measure the variability of the method would be:

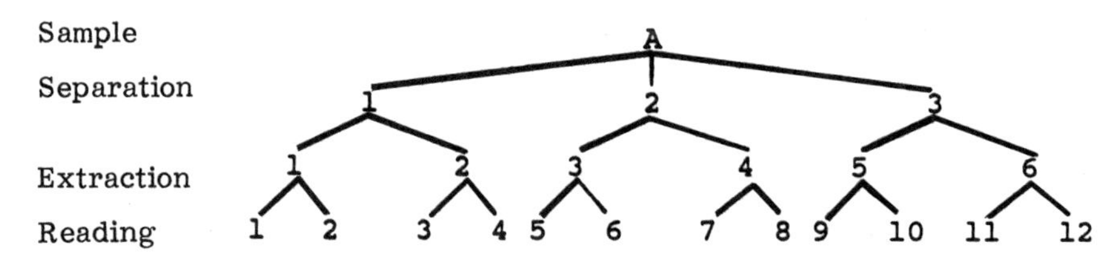


The data would be modeled as a nested variance components model of the form

$$y = M + S + E(S) + R(S, E)$$

where

M = constant, level in Sample A

S = Separation effect—normal $(0, \sigma_S^2)$

E = Extraction effect—normal $(0, \sigma_E^2)$

R = Reading effect—normal $(0, \sigma_R^2)$

The term R(S, E) is the usual "error" term in the analysis of variance model and is usually assumed to represent the instrument reading variability, although the variability does include any variability remaining after accounting for the previous terms. The total variation is given by

$$\sigma^2(\text{total}) = \sigma_S^2 + \sigma_E^2 + \sigma_R^2$$

Such an analysis is quite useful in the development of an assay in identifying the steps that have the greatest contribution to the total variation.

To illustrate the variance components model another common experiment will be discussed—the reproducibility/repeatability experiment of interlaboratory tests. In this experiment a common sample is sent to several laboratories. Each laboratory obtains replicates of the sample which are true replicates; i.e., each assay value is obtained from independent steps in the assay. The variability within a laboratory is then an estimate of the previously discussed "total" variation. The design is:

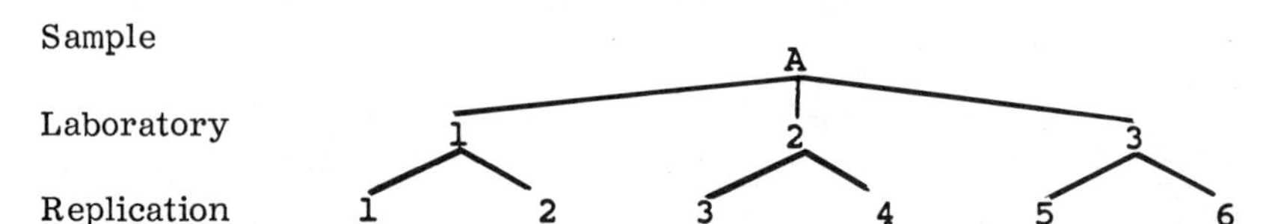


The two way nested analysis of variance model is given by

$$y_{ij} = M + L_i + e_{j(i)} \qquad i = 1, \cdots, I \quad j = 1, \cdots, J$$

where

$y_{ij}$ = j-th replication in i-th laboratory
M = constant, level of A
L = laboratory effect—normal $(0, \sigma_L^2)$
e = replication effect—normal $(0, \sigma^2)$
I = number of laboratories
J = number of replications within laboratory

The two variances may be estimated through the analysis of variance table:

| Source | SS | df | MS | EMS |
|---|---|---|---|---|
| Laboratory | $J\Sigma(\bar{y}_i - \bar{y})^2$ | $I - 1$ | $SS_L/I - 1$ | $\sigma^2 + J\sigma_L^2$ |
| Replication | $\Sigma\Sigma(y_{ij} - \bar{y}_i)^2$ | $I(J - 1)$ | $SS_R/I(J - 1)$ | $\sigma^2$ |

$$\bar{y}_i = \sum_{j=1}^{J} y_{ij}/J$$

$$\bar{y} = \sum_{i=1}^{I}\sum_{j=1}^{J} y_{ij}/IJ$$

The variance estimates are obtained by equating MS (mean squares) with their expectations, obtaining

$$\sigma^2 = MS_R$$
$$\sigma_L^2 = (MS_L - MS_R)/J$$
$$\sigma^2(\text{total}) = \sigma_L^2 + \sigma^2$$

The variation within a laboratory ($\sigma^2$) is called repeatability, and the total variation ($\sigma_L^2 + \sigma^2$) is called reproducibility. An example is given in Table 6.

Complications sometimes arise when a negative estimate of $\sigma_L^2$ occurs. This is not uncommon where $\sigma_L^2$ is small compared to $\sigma^2$. Statistical packages such as SAS (Statistical Analysis System) can obtain nonnegative estimates using more advanced procedures than those given here.

Another assay validation procedure involves the comparison of two methods where different batches are assayed by the two methods. In analytical chemistry this is called a crossover study and is related to the crossover studies in clinical trials in that two observations are made on the same blocking factor. A major difference between this crossover design and the crossover design in clinical trials is that here no carryover effect is even possible. The model is as follows:

| Sample | Assay 1 | Assay 2 |
|---|---|---|
| 1 | $x_1 = s_1 + a_1 + e_{11}$ | $y_1 = s_1 + a_2 + e_{12}$ |
| 2 | $x_2 = s_2 + a_1 + e_{21}$ | $y_2 = s_2 + a_2 + e_{22}$ |
| ⋮ | ⋮ ⋮ | ⋮ ⋮ |
| n | $x_n = s_n + a_1 + e_{n1}$ | $y_n = s_n + a_2 + e_{n2}$ |

$x_i$ = assay 1 value on i-th sample

$y_i$ = assay 2 value on i-th sample

$s_i$ = true value of i-th sample

$a_j$ = bias of assay j

$e_{ij}$ = assay j error on i-th sample

Table 6 Reproducibility Experiment[a]

| | Laboratory | | | |
|---|---|---|---|---|
| | 1 | | 2 | 3 |
| Replication | 98, 99 | | 96, 97 | 98, 100 |
| | Analysis of variance | | | |
| Source | SS | df | MS | EMS |
| Laboratory | 7.0 | 2 | 3.50 | $\sigma^2 + 2\sigma_L^2$ |
| Replication | 3.0 | 3 | 1.00 | $\sigma^2$ |

[a]Repeatability = $\sigma^2$ = 1.00; laboratory = $\sigma_L^2$ = 1.25; reproducibility = $\sigma_L^2 + \sigma^2$ = 2.25.

The random effects are $s_i$ and $e_{ij}$, which are assumed to have the normal distribution. Note that it is assumed that there is no method by sample interaction, but there is no assumption that the methods have equivalent variabilities. In fact, this is one of the hypotheses that will be tested. The assumption of no interaction is reasonable as long as the levels of $s_i$ do not vary greatly. This is usually the case when the samples are actual production batches. Let

$$\sigma_S^2 = \text{variation in s or product variance}$$

$$\sigma_1^2 = \text{assay 1 variance}$$

$$\sigma_2^2 = \text{assay 2 variance}$$

$$\sigma_x^2 = \Sigma(x_i - \bar{x})^2/n - 1$$

$$\sigma_y^2 = \Sigma(y_i - \bar{y})^2/n - 1$$

$$\text{cov}(x, y) = \Sigma(x_i - \bar{x})(y_i - \bar{y})/n - 1$$

Then it may be shown that:

$$\sigma_S^2 = \text{cov}(x, y)$$

$$\sigma_1^2 = \sigma_x^2 - \sigma_S^2$$

$$\sigma_2^2 = \sigma_y^2 - \sigma_S^2$$

Thompson (1963) suggests procedures for dealing with negative variance estimates.

The two main hypotheses of interest are whether the assays obtain equivalent expected results ($a_1 = a_2$) and whether the methods have equivalent variation ($\sigma_1^2 = \sigma_2^2$). It cannot be tested that the individual biases are equal to zero, $a_1 = a_z = 0$, but only that the biases are equal. This is acceptable because often a new method is being compared to a referee method.

The test for bias is the paired t-test. Define a new variable, d, as:

$$d_i = x_i - y_i, \qquad i = 1, \ldots, n$$

$$\bar{d} = \Sigma \frac{d_i}{n}$$

$$s_d^2 = \Sigma \frac{(d_i - \bar{d})^2}{n - 1}$$

The t statistic for the hypothesis test of equal bias is:

$$t = \sqrt{n}\,\frac{\bar{d}}{s_d} \quad \text{with } n - 1 \text{ d.f.}$$

The test for equal variances is derived from the Pitman-Morgan F test (Maloney and Rastogi, 1971) and the t statistic is given by

$$t = \frac{\{[\sigma_x^2/\sigma_y^2] - 1\}}{\sqrt{n-2}} / A \qquad \text{d.f.} = n - 2$$

$$A = \frac{\sqrt{4(1 - r^2)\sigma_x^2}}{\sigma_y}$$

r = correlation of x and y

This is two-sided test.

Table 7 gives an example.

## B. Process Validation

Process validation means demonstrating that a manufacturing process performs as designed. The two major methods in process validation are retrospective and prospective validation. Following the definitions of Chapman (1983):

**Table 7** Method Comparison

| Sample | Assay 1 | Assay 2 |
|---|---|---|
| 1 | 100 | 101 |
| 2 | 97 | 99 |
| 3 | 92 | 94 |
| 4 | 101 | 100 |
| 5 | 102 | 105 |

$\bar{x} = 98.4$, $\bar{y} = 99.8$; $s_x^2 = 16.3$, $s_y^2 = 15.7$;
cov(x, y) = 14.85, r = 0.9282. $\sigma_s^2 = 14.85$,
$\sigma_1^2 = 1.45$, $\sigma_2^2 = 0.85$.

Bias:

$$t(4 \text{ d.f.}) = \sqrt{5}(-1.4)/1.516 = -2.06$$
(not significant)

Precision:

$$t(3 \text{ d.f.}) = (\{[16.3/15.7] - 1\}/1.732)/.7581$$
$$= .03 \text{ (not significant)}$$

$$A^2 = 4(.1384)(16.3)/15.7 = 0.5747$$

Retrospective validation—establishing documented evidence that a system does what it purports to do based on review and analysis of historic information

Prospective validation—establishing documented evidence that a system does what it purports to do based on a preplanned protocol

Retrospective validation is used for products that have been manufactured for some time but were developed prior to the formalization of validation requirements, especially with respect to documentation. Prospective validation is used for newly developed products that have no prior manufacturing history.

In retrospective validation the definition of an acceptable or validated process is based on the ability of the process to produce material that consistently meets predefined quality control specifications. The requirements for an individual batch of a product are set forth in a quality control specification. These specifications are of the form "potency of the active ingredient is between 90 to 110% LC" or "meets the requirements of the USP XXI Dose Uniformity specification." Although these specifications define the quality of a product, companies often have internal specifications that are tighter than the quality definitions in order to have assurance that the batch truly meets the quality specifications. The assurance is necessary because of variation in both the sampling unit and the measurement. These tighter specifications are often called release specifications. In addition to individual batch acceptability, retrospective validation examines parameters of the process itself, such as trends in batch parameters, and the ability of the process to meet target values.

The strength of retrospective validation lies in its evaluation of a manufacturing process over an extended period of time. Its weakness is that for individual batches the information database is rather limited because it is based on the random samples taken from a batch to determine acceptability for release. On balance, however, many in the industry feel that retrospective validation may be the most conclusive proof that the process produces an acceptable product.

Table 8 lists the common criteria used in the retrosepctive validation of a product.

**Table 8** Retrospective Validation Criteria

| Criteria | Statistical test |
|---|---|
| 1. High percentage of batches meet specifications | Determine probability batch meets specifications:<br>a. Binomial confidence limits<br>b. Statistical tolerance limits<br>c. Individual test determination |
| 2. Process meets target value | Confidence limits on process average |
| 3. Lack of trends in key parameters | Trend tests such as mean square successive difference test (Nelson, 1980) |

The determination of the percentage of batches meeting specifications can take several forms, depending on the type of data. All specifications can be presented as pass/fail, i.e., whether the batch meets the specifications or not. The lower binomial confidence limit on the percent of batches actually meeting specifications provides assurance that the process produces acceptable batches. This is a robust procedure because no assumption about distributional form is necessary. Where observations exist as single determinations on a batch, such as average potency, statistical tolerance intervals may be used for assurance of acceptability if the distributional form is reasonable. Owen (1962) provides a table for normally distributed data. For tests such as dose uniformity and dissolution that have multiple criteria in their specification, more complicated procedures must be used. The within-batch variation is evaluated to determine whether large differences exist between batches. If there is no great difference between batches, the process average, between-batch variation, and within-batch variation are estimated. Depending on the form of the specification, these parameters are used to determine the probability that a batch will pass the specification. Sometimes this probability can be calculated exactly, but more often, because of the mixed variable/attribute nature of many specifications, the probability must be approximated using computer simulation. Pheatt (1980) presents an analysis for the dissolution specifications of Table 2.

If a process is targeted at a certain value, an acceptable process will achieve that target. While individual batches may vary slightly from the target, the overall process average should achieve the target. This criterian is evaluated using confidence limits on the process average. In potency evaluation some consideration must be given to possible assay inaccuracy, but any large deviation, say significantly more than 2%, must be evaluated and explained.

Significant trends in the process may indicate that process control is not adequate. Trends are evaluated on single determination per batch data, such as average potency, as well as summary statistics for tests such as dose uniformity and dissolution. For these tests, the average and within-batch standard deviation of the parameter (assay, dissolution) would be evaluated.

The major difference between retrospective and prospective validation is in the prospective case fewer batches are available for evaluation and the data are collected according to a planned experimental design. Usually for prospective validation at least three batches are examined, and data for evaluation are collected at each significant processing step. An example will be given of a tablet product that has only two significant processing steps: powder mixing and powder compression.

In the blending step, the active ingredient is mixed with the excipients to form a homogeneous mixture. For validation, samples are drawn from the blender to determine the homogeneity of the resulting blend. The homogeneity is demonstrated by a one-way analysis of variance resulting in no significant location effect. Locations within the blender are chosen according to the shape of the blender and ability to withdraw samples from the blender. In addition to homogeneity, the average potency is compared to the target by using confidence limits. Table 9 shows a typical design. An unresolved problem in blender studies is the determination of the amount of powder that should be sampled and assayed. Ideally this amount should be equivalent to the weight of a tablet, but experience has shown that the physical act of sampling both from the blender and in the laboratory adds

Table 9 Prospective Validation—Blending Study

*Blender Design*

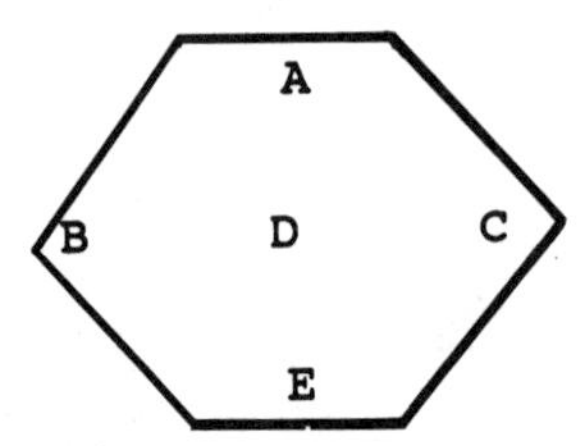


Duplicate samples withdrawn from locations A, . . ., E and assayed

*Analysis*

1. Homogeneity

   Analysis of variance

   | Source | d.f. |
   |---|---|
   | Location | 4 |
   | Error | 5 |

2. Target

   Confidence limit on assay average

---

a source of variation not present in a compressed tablet. On the other hand, too large a sample should not be taken, for nonhomogeneity could be masked.

In the compression step, the powder is compressed into tablets. This step often takes several hours. For validation, samples are taken during the compression run usually at intervals of start, 1/4, 1/2, 3/4, and end of the compression run. At each point during the compression run at least 10 tablets are sampled and assayed. The analysis consists of evaluating the consistency of potency and dissolution during the compression run for variance, using Cochran's or Bartlett's test (Winer, 1971), and average potency and dissolution using a one way analysis of variance. In addition to consistency during the production run other criteria are evaluated such as the probability of the batch meeting potency, dose uniformity, and dissolution requirements.

In prospective validation each individual batch is intensively studied as well as compared to the other batches in the validation study. The intercomparison of batches is made on the average parameters such as potency and the ability to meet the target as well as the within-batch variability. Although only potency, dose uniformity, and dissolution have been discussed, other parameters such as related compounds, moisture, and so on are also evaluated in the validation study, depending on the product.

## V. LONG-TERM STABILITY ANALYSIS

An important area in manufacturing and quality control is the confirmation of the stability of a pharmaceutical dosage form. Stability is defined as "the ability of a particular formulation, in a specific container, to remain within its physical, chemical, therapeutic, and toxicological specifications" (Lintner, 1973). The statistical applications in stability testing address the estimation of expiration dates, release limits, and amount of overage required in a product.

There are important differences between stability testing during the initial stages of drug development and stability testing after sales of the drug has begun. During the initial development stages sufficient time for evaluation of the product is not available. Also, the product being evaluated is usually from pilot or scale-up batches and not from normal production. Stability evaluation at later stages of drug development utilizes normal production and packaging materials and is focused toward confirming and refining expiration dates.

Stability evaluations at early drug development stages use accelerated storage conditions to obtain an early prediction of the product's stability. The accelerated conditions usually include increased temperature and humidity and the statistical analysis includes non-linear analysis of kinetic degradation models (see Chapter 4).

On the other hand, stability evaluation at later stages of drug development is centered on the product when stored under recommended conditions, and the product is under study for longer time periods. Usually, complicated kinetic models are not used, and the statistical analysis proceeds using linear regression techniques. This section will discuss linear regression concepts applied to refining the expiration date and determination of internal release limits.

The parameters examined during storage of a product include the physical, chemical, and biologic properties of the product. Each product dosage form has its own unique characteristics that are evaluated during the storage period. For all products the assay for active ingredient is a key measurement during the storage period. This is the parameter that will be discussed in the statistical analysis, but it should be remembered that other parameters also are evaluated during the study.

## A. Experimental Design

Samples of the product in their actual package are placed in storage at the recommended storage conditions. At least three production batches are placed on study for the confirmation of a tentative expiration date. The batch sample size of three has been chosen as a compromise between statistics and practicality. One objective in the analysis is to determine whether there is any batch-to-batch variability in stability profiles of batches from the production process. Clearly, from a statistical viewpoint, more than one or two batches are required. However, in order to make timely decisions it may not be wise to wait for a larger number of batches and the total number of assays required for the study may be unreasonable if more than three batches are required. Although only three batches may be studied, further control on the manufacturing process is obtained by requiring ongoing testing after the initial confirmation. Often, one batch a year is evaluated after the initial confirmation.

Samples from the production batches are assayed during the storage period. The usual study design assays samples at 0, 3, 6, 9, 12, 18, and 24 months and yearly thereafter. Often, additional assays are taken at the expiration date. The initial intensive sampling provides an early warning of any possible changes in the stability of a product. The intensive sampling at later time periods provides a better estimate of the true potency at the expiration date.

At each time point, multiple assays are made on multiple samples from the batch. There is disagreement as to how to evaluate these multiple observations. The multiple observations have several sources of variability including multiple sources from product variability and assay variability. Norwood (1986) discusses stability analysis using linear regression with nested error structures when there are two error sources. FDA Guidelines (Food and Drug Administration, 1984) discuss estimating these variability sources but do not comment on the regression analysis necessary. Current practice is to analyze only observations that are true replicates so that there are independent errors. Where observations are duplicates, as in multiple assays on the same sample, their average should be taken and used for statistical analysis. If the number of duplicates is the same across time points, this practice will produce acceptable estimates because the averages will be independent with constant error. For the remainder of this discussion it will be assumed that there is a single observation at each time point for each batch and that the observation errors are independent.

## B. Analysis

The basic procedures in stability analysis will be illustrated assuming only a single batch under study.

### 1. Single Batch

The observations on a single batch and the estimates obtained from linear regression are given in Table 10. The following equations are used to calculate the estimates.

Let $(x_i, y_i)$ $i = 1, n$ be the pairs of observations of time, potency. Then

$$\text{Slope} = b = \frac{\Sigma(x_i - \bar{x})(y_i - \bar{y})}{\Sigma(x_i - \bar{x})^2}$$

$$\text{Intercept} = a = \bar{y} - b\bar{x}$$

$$\text{Standard deviation} = s = \text{SQRT}\left[\frac{\Sigma(y_i - a - bx_i)^2}{n - 1}\right]$$

$$\text{Standard error—slope} = se(b) = \text{SQRT}\left[\frac{s^2}{n\Sigma(x_i - \bar{x})^2}\right]$$

$$\text{Standard error—intercept} = se(a) = \text{SQRT}\left[\frac{s^2\Sigma x_i^2}{n\Sigma(x_i - \bar{x})^2}\right]$$

Standard error—estimated value at given x = se(y | x)

$$= s\ \text{SQRT}\left[\frac{1}{n} + \frac{(x - \bar{x})^2}{\Sigma(x_i - \bar{x})^2}\right]$$

Table 10 Stability Analysis—Batch 1[a]

| Time (months) | Potency (% LC) |
|---|---|
| 0 | 101 |
| 3 | 99 |
| 6 | 97 |
| 9 | 100 |
| 12 | 98 |
| 18 | 95 |
| 24 | 97 |
| 36 | 94 |

[a] b = −0.1652, a = 99.85, s = 1.436,
se(b) = 0.045, se(a) = 0.794.
Expiration date (E):

```
(90 - 99.85 + 0.165 * E) ** 2
    = (3.775)(2.06)(1/8
    + (E - 13.50) ** 2/1008.0)

E = 42.9
```

Release limit (R) for E = 36:

```
R = 90 - 36(-0.165 - 1.943(0.045))
  = 99.1
```

The expiration date is defined as the time (x value) where the lower confidence limit reaches the lower potency specification. An individual confidence limit is used rather than simultaneous limits on the entire regression line because only the lower potency specification is of interest. A lower one-sided limit is used because only the worst possible degradation that could exist is of interest. Usually a 95% limit is used. The expiration date is then the solution for E to the quadrataic equation:

$$(P - a - bE)^2 = t^2 s^2 (1/n + (E - \bar{x}) ** 2/\Sigma(x_i - \bar{x})^2)$$

where

P = lower potency limit

E = expiration date

t = t statistic with n − 2 d.f.

The solution to this quadratic equation can pose some difficulty (O'Neill and Schuirmann, 1979). The following conditions must hold to estimate a valid expiration date:

1. There is not a statistically significant increase in potency,

$$\text{If } b > 0 \text{ then } \frac{b}{se(b)} < t$$

2. The estimated value at time 0 is significantly greater than P

$$\text{then } \frac{(a - P)}{se(y \mid x = 0)} > t$$

3. The following holds,

$$\frac{b^2}{se^2(b)} + \frac{(P - \bar{y})^2}{(s^2/n)} > t^2$$

These conditions are nearly always met in practice. When these conditions are met the following confidence intervals will be obtained:

1. Significant regression, negative slope,

   interval: $c_1$, $c_2$ where $0 < c_1 < c_2 < \infty$.

   The expiration date would be set to $c_1$.
2. Nonsignificant regression, positive or negative slope,

   interval: $-\infty$, $c_1$ and $c_2$, $\infty$ where $c_1 < 0$ and $c_2 > 0$.

   The expiration date would be set to $c_2$.

The release limit is defined as the potency at release, initial time 0, so that there is 95% confidence that the potency will meet the lower potency limit at the expiration date. This limit assists in decisions concerning release of individual batches where the initial time 0 potency is precisely known because of multiple assays. The release limit is defined as:

$$R = P - E[b - t\ se(b)]$$

The preceding discussion uses zero-order degradation kinetics for a potency model of

$$P = a + bT$$

A first-order model is sometimes applied to stability data.
This model relates potency as

$$\ln(P) = a + bT$$

In long term studies there is usually very little difference between the models because the degradation is so slight. The choice of a model is more important when the degradation is greater than 90% of the initial potency.
The analysis of the data in Table 10 is shown in Figure 2.

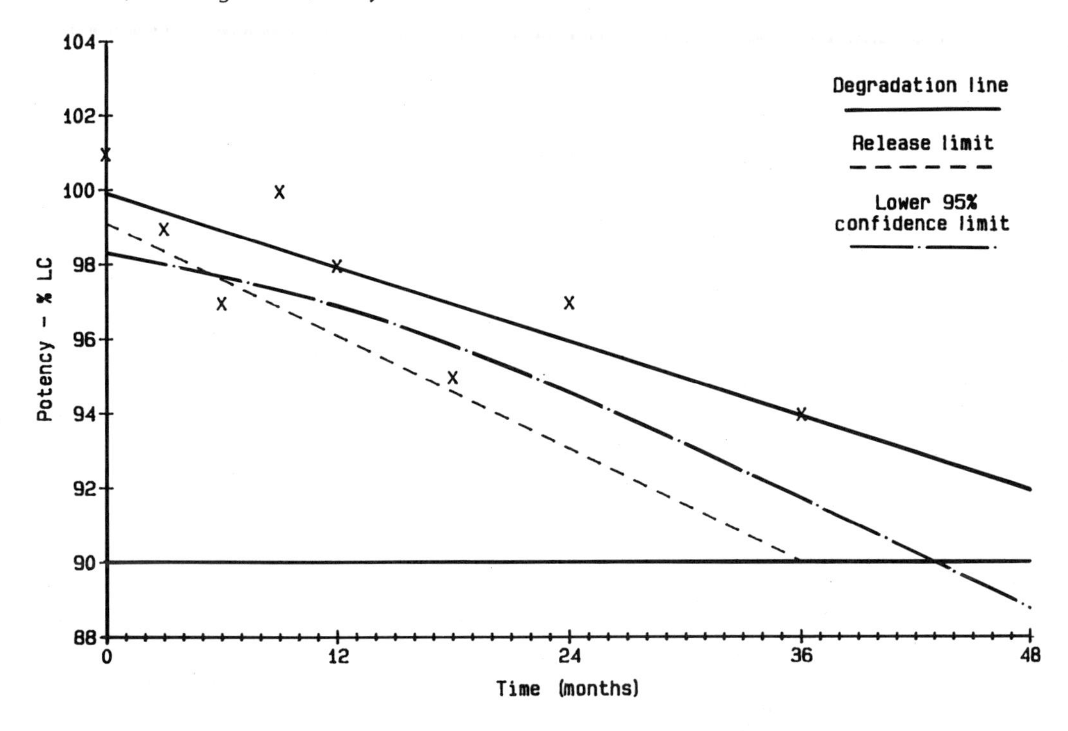


Fig. 2 Stability analysis, batch 1.

## 2. Multiple Batches

When there are multiple batches the analysis is more complex because the batches should be tested for equivalence in degradation profiles. The procedure given by Seber (1977) is recommended. This procedure uses three sequential tests for comparing straight lines: (1) test for parallelism, (2) test for equal intercepts given equal slopes, and (3) test for coincidence (equal slopes and equal intercepts). The tests use the methods of reduction in sum of squares and the exact formulas may be found in Seber (1977).

If there is no significant difference between the batches, the data may be pooled and the expiration date and release limit estimated using the procedures of simple linear regression. The analysis discussed in the single-batch situation may be applied.

If there is no significant difference between slopes, but a significant difference in intercepts then the more complicated model must be fit:

$$P = a_i + bT$$

where

$b$ = common slope

$a_i$ = intercept for batch i

In this situation, release limits could be calculated as previously discussed, but the expiration date depends on the initial potency of the batch, which has been demonstrated to vary significantly from one batch to the next. One solution to this problem is to choose an expiration date so that a high percentage of batches have an initial potency greater than the release limit for a specified expiration date. This situation is not unusual in long-term stability analysis because production batches rarely have exactly the same potency. The variation in batch potency, however, is small, but if a large number of batches are studied the power of the statistical test is sufficient to detect minor differences. Although there may be a statistical difference, it is usually not large enough to be practically important.

If batches are significantly different in slopes, a more difficult problem arises. Some individuals think that different degradation slopes between batches is indicative of a poor product formulation that should be corrected. Statistical procedures that model both the slope and the intercept as random variables have not been widely discussed as applied to stability analysis.

## VI. SUMMARY

This chapter has emphasized two areas in manufacturing and quality control that have a direct impact on the development of a pharmaceutical dosage form—long term stability analysis and validation. During initial product development, a formulation is tested for stability using a variety of accelerated conditions, but the true test of stability is found in the long-term evaluation of the product in the various packages in which it is marketed. Similarly, test methods and manufacturing processes are extensively studied to have the optimal method used by manufacturing and the chemical laboratories, but tests are usually done on a laboratory or pilot scale. Validation is a confirmation of the laboratory efforts that have preceded in the dosage form development process.

## REFERENCES

Abdel-Monem, M. M. and Henkel, J. G. (1978). *Essentials of Drug Product Quality*, Mosby, St. Louis.

Chapman, K. G. (1983). A Suggested Validation Lexicon, *Pharm. Technol.*, 7(8): 51–57.

Debesis, E., Boehlert, J. P., Givand, T. E., and Sheridan, J. C. (1982). Submitting HPLC Methods to the Compendia and the Regulatory Agencies, *Pharm. Technol.*, 6(9): 120–137.

Food and Drug Administration. (1984). *Draft Guidelines for Stability Studies for Human Drugs and Biologics—March 1984*, Center for Drugs and Biologics, Rockville, Md.

Lachman, L., Lieberman, H., and Kanig, J. L. (1976). *The Theory and Practice of Industrial Pharmacy*, 2nd ed., Lea & Febiger, Philadelphia.

Letterman, H. (1980). GMP Method Validation Procedures in an OTC Firm, *Pharm. Technol.*, 4(8): 44–49.

Lintner, C. (1973). Pharmaceutical Product Stability, *Quality Control in the Pharmaceutical Industry* (M. S. Cooper, ed.), Academic Press, New York.

Maloney, C. J. and Rastogi, S. C. (1971). Significance Tests for Grubbs' Estimators, *Biometrika*, 26: 671–676.

Nelson, L. S. (1980). The Mean Square Successive Difference Test, *J. Qual. Technol.*, 12: 174–175.

Norwood, T. E. (1986). Statistical Analysis of Pharmaceutical Stability Data, *Drug Dev. Ind. Pharm.*, 12: 553–560.

O'Neill, R. T. and Schuirmann, D. J. (1979). Statistical Considerations in the Expiration Dating of Ethical Drugs, unpublished paper presented at the Joint Statistical Meetings (Americal Statistical Association), Washington, D.C., August 13–16, 1979.

Owen, D. B. (1962). *Handbook of Statistical Tables*, Addison-Wesley, Reading, Mass.

Pheatt, C. B. (1980). Evaluation of U.S. Pharmacopeia Sampling Plans for Dissolution, *J. Qual. Technol.*, 12: 158–164.

Sampson, C. B. (1981). Quality Control of the Manufacturing Process, *Statistics in the Pharmaceutical Industry* (C. R. Buncher and J.-Y. Tsay, eds.), Marcel Dekker, New York.

Seber, G. A. F. (1977). *Linear Regression Analysis*, Wiley, New York.

Thompson, W. A., Jr. (1963). Precision of Simultaneous Measurement Procedures, *J. Am. Stat. Assoc.*, 58: 474–479.

USP XXI. (1984). United States Pharmacopeia, 21st rev. ed., Mack, Easton.

Winer, B. J. (1971). *Statistical Principles in Experimental Design*, 2nd ed., McGraw-Hill, New York.

# Index